Dmitry Yu. Murzin
Chemical Reaction Technology

Also of interest

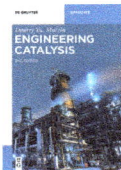

Engineering Catalysis
Murzin, 2020
ISBN 978-3-11-061442-8, e-ISBN 978-3-11-061443-5

Process Intensification.
Breakthrough in Design, Industrial Innovation Practices,
and Education
Harmsen, Verkerk, 2020
ISBN 978-3-11-065735-7, e-ISBN 978-3-11-046506-8

Chemical Product Technology
Murzin, 2018
ISBN 978-3-11-047531-9, e-ISBN 978-3-11-047552-4

Product and Process Design.
Driving Innovation
Harmsen, de Haan, Swinkels 2018
ISBN 978-3-11-046772-7, e-ISBN 978-3-11-046774-1

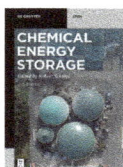

Chemical Energy Storage
Schlögl (Ed.), 2022
ISBN 978-3-11-060843-4, e-ISBN 978-3-11-060859-5

Dmitry Yu. Murzin

Chemical Reaction Technology

—

2nd Edition

DE GRUYTER

Author
Prof. Dmitry Yu. Murzin
Åbo Akademi University
Process Chemistry Centre
Biskopsgatan 8
20500 Turku/Åbo
Finland
dmurzin@abo.fi

ISBN 978-3-11-071252-0
e-ISBN (PDF) 978-3-11-071255-1
e-ISBN (EPUB) 978-3-11-071260-5

Library of Congress Control Number: 2021951603

Bibliographic information published by the Deutsche Nationalbibliothek
The Deutsche Nationalbibliothek lists this publication in the Deutsche Nationalbibliografie;
detailed bibliographic data are available on the Internet at http://dnb.dnb.de.

Cover image: zorazhuang/E+/Getty Images
Typesetting: Integra Software Services Pvt. Ltd.
Printing and binding: CPI books GmbH, Leck

www.degruyter.com

Preface to the first edition

There are quite a number of books available on the market dealing with industrial chemistry, oil refining, and production of petrochemicals and organic and inorganic chemicals. Many of them are of a very descriptive nature not involving any discussion of flow schemes.

There is a wealth of textbooks covering various aspects of unit operations, in particular chemical reactors. There are few handbooks, encyclopedia, and textbooks on chemical technology already available, including very recent textbooks of excellent quality by Moulijn, Makkee, and van Diepen entitled *Chemical Process Technology*, Jess and Wasserschied on chemical technology, and Bartolomew and Farrauto on industrial catalytic processes.

The aim of the textbook is not to replace these and other excellent literature sources focusing more on the chemistry of different reactions or chemical engineering textbooks addressing various issues of reactors and unit operations, but rather to provide a helicopter view on chemical reaction technology, omitting specific details already available in the specialized literature. Moreover, the author feels that there is a niche for such a textbook since the majority of the textbooks are dealing with oil refining and basic inorganic and, to a very limited extent, organic chemicals but not featuring the breadth of industrial organic transformations.

For a chemist and even for a chemical engineer who would like to be introduced to the field of chemical technology, it would be more natural and methodologically - stimulating to see how various types of chemical transformations are -implemented in the industry, rather than to read about apparently unconnected production technologies of different chemicals.

The textbook is based in part on a course on chemical reaction technology, which the author has been teaching to chemists and chemical engineers for almost 15 years, first covering the basics of chemical technology and also providing an overview of modern chemical and petrochemical industry. It then goes in depth into different chemical reactions, such as oxidation, hydrogenation, isomerization, esterification, etc., following the style of chemistry textbooks rather than product-oriented technical literature. Owing to a large number of products in the chemical industry, exceeding tens of thousands, such an approach with the focus on reactions, certainly not being a new one, will hopefully facilitate understanding of basic principles of chemical reaction technology and their implementation rather than force the students to memorize how certain chemicals are produced.

Variability of process technologies which can be applied industrially for the same reaction is another key feature that was specifically addressed in the textbook.

The author himself, while studying at Mendeleev University of Chemical Technology, took a course on chemical technology of basic organic chemicals based on a reaction-oriented approach and found it very stimulating and actually useful in the subsequent professional life.

https://doi.org/10.1515/9783110712551-202

Working as a trainee in a chemical plant, then as a researcher in a governmental research center and later in the industry, and currently in academia, the author has met in the last 30 years many brilliant chemists and chemical engineers who have developed new technologies that were implemented industrially and/or improved the existing ones. Some of their names appeared in the relevant patents, but a majority are seldom known outside of their respective companies. This book is dedicated to them.

Dmitry Murzin
May 2015, Turku/Åbo

Preface to the second edition

The first addition of the book apparently was able to find its readers, which prompted the publisher and then the author to consider a possibility of preparing the second edition.

The author is grateful to the editorial team at De Gruyter for efficient collaboration in making this edition possible.

The original text was revised and expanded, updating the processes already covered in the first edition and introducing some other reactions not touched initially. Moreover, the chapter on chemical processes and unit operations has been significantly enlarged. The author was keeping still the focus on chemical reaction technology, as it was not an intention to replace with the current work textbooks on chemical reaction engineering, chemical reactors, or design of chemical processes.

For the first edition, a substantial contribution to drawing of a large number of figures was done by MSc (Chem. Eng.) Elena Murzina, who passed away in 2019 after a long and difficult battle with an oncological decease.

This textbook is dedicated to her memory.

Dmitry Murzin
November 2021, Turku/Åbo

https://doi.org/10.1515/9783110712551-203

Contents

About the author

Dmitry Yu. Murzin studied chemistry and chemical engineering at the Mendeleev University of Chemical Technology in Moscow, Russia (1980–1986), and graduated with honors. He obtained his PhD (advisor Prof. M. I. Temkin) and DrSc degrees at Karpov Institute of Physical Chemistry, Moscow, in 1989 and 1999, respectively. He worked at Universite Louis Pasteur, Strasbourg, France, and Åbo Akademi University, Turku, Finland, as a post-doc (1992–1994). In 1995–2000, he was associated with BASF, being involved in research, technical marketing, and management. Since 2000, Prof. Murzin holds the Chair of Chemical Technology at Åbo Akademi University. He serves on the editorial boards of several journals in catalysis and chemical engineering field. He is an elected member of the European Academy of Sciences and the Finnish Academy of Science and Letters. Prof. Murzin is the co-author (with Prof. T. Salmi) of a monograph (*Catalytic Kinetics*, Elsevier, 2005, second edition 2016) and an author of textbooks (*Engineering Catalysis*, De Gruyter, 2013, second edition 2020, and *Chemical Product Technology*, De Gruyter, 2018). He holds several patents and is an author or co-author of nearly 850 journal articles and book chapters. In 2016, Prof. Murzin became Knight, First Class, Order of the White Rose of Finland.

https://doi.org/10.1515/9783110712551-205

Chapter 1
Chemical technology as science

1.1 Basic principles

Chemical technology can be defined as a science of converting natural resources or other raw materials into the desired products at the industrial scale using chemical transformations in a technically and economically feasible and socially acceptable way. Besides being based on sound economical considerations, chemical production should nowadays take into account ecological aspects, safety requirements, and labor conditions.

Chemical technology investigates chemical processes (whose structures had been given in Figure 1.1), which comprise feed purification, reactions *per se*, separation, and product purification.

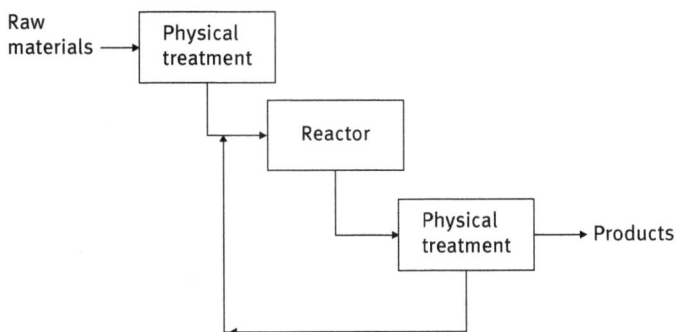

Figure 1.1: General structure of chemical processes.

Chemical technology is not limited only to chemical transformations *per se*, as there are other various physical, physico-chemical, and mechanical processes in the production of chemicals.

Criteria of a process quality are technological parameters (productivity, conversion, yield, product purity) as well as economical (costs, profitability, etc.) and ecological ones.

Success in implementation of a novel technology requires its robustness, reliability, safety, environmental compliance as well as significant gains over existing - processes.

Methods of chemical technology are used also in non-chemical industries, such as transport, metallurgy, building construction, electronic industry.

https://doi.org/10.1515/9783110712551-001

Chemical technology as a discipline is based on the following:
- Physical chemistry and chemical reaction engineering, covering stoichiometry, thermodynamics, mass transfer, kinetics, and various types of catalysis
- Unit operations, which include besides reactors also various separation-processes, such as absorption, adsorption, distillation, extraction
- General process considerations, viewing chemical production as a chemical technological system and applying principles of conceptual process design, process-intensification, and green chemical engineering

Let us consider as an example hydrogenation of benzene. For a physical chemist, the reaction will look like $C_6H_6 + 3H_2 = C_6H_{12} - \Delta H$, leading to a conclusion that the reaction is reversible and exothermal and that the parameters that could be used to alter equilibrium are temperature and concentrations (pressures) of reagents. When developing a process technology of benzene hydrogenation, other parameters aside from the issues mentioned above should be considered such as availability of the feedstock and energy, reactor type, other pieces of equipment needed, the phase in which the reaction should take place (gas or liquid), the optimal conditions from the viewpoint of economics, and minimization of the negative impact on the environment.

This simple example illustrates that chemical technology is different from organic (inorganic) and physical or other branches of chemistry.

1.1.1 Continuous or batch?

Chemical processes in oil refining and production of basic chemicals are mainly continuous, while in production of specialty and fine chemicals, they could be continuous and periodical. The latter mode of operation can be also used in the secondary processes (i.e., separation, catalyst regeneration) even for large-scale production in chemical process industries (sometimes abbreviated as CPIs).

Continuous processes typically require constant technological parameters (pressure, flows, temperature). Such processes are mainly aimed at production of a single product. In periodical (batch) processes, several products could be made under somewhat similar conditions.

Semi-periodical processes can be also applied in continuously operating units, with, however, a change of a product after a certain period of time.

The fine chemical industry, including the manufacture of active pharmaceutical ingredients (API) relies mainly on multipurpose batch or semibatch reactors. At the same time, the pharmaceutical industry is currently exploring the continuous manufacturing approach as a part of the move toward more sustainable chemical process technology. This approach has allowed a decrease in waste generation by minimizing solvent switching steps and/or product isolations, which both increase process complexity. The batch mentality still dominates in the pharmaceutical industry for a number of reasons, including segmented production of drugs in multi-purpose facilities and a need for quality control after every stage of the production process. Challenges in changing the manufacturing paradigm in industry are related to costs of implementing the changes, regulatory limitations, and, last but not least, lack of available technology for synthesis *per se* and for downstream processing. There are already a number of examples, however, showing that arranging chemical synthesis through continuous-flow processing helped to improve the manufacturing of API.

Continuous mode of operation allows constant quality of products, very efficient utilization of the equipment, high degree of process automation and control, and finally, much more efficient and safe processes.

In order to organize a continuous process, the following conditions should be fulfilled:
– separation of inlet and outlet in a reactor space
– continuous and substantially stationary flows of reactants and products (even if there are several successful examples of non-stationary operation)
– the products should be the same during the operation
– continuous flow of products inside reactors and other equipment and between them

These conditions can be well maintained when liquid and gaseous products are processed, while transport and handling of solids can be much more complicated.

The majority of process units can be utilized in a continuous mode. If, by some reason, a reactor dedicated for continuous operation cannot be used, reactors designed for discontinuous (batch) operations can be combined together in a cascade (Figure 1.2).

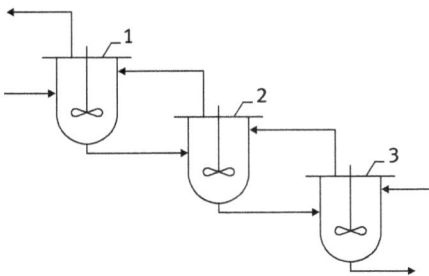

Figure 1.2: Reactors in a cascade.

Some disadvantages of continuous operation compared to the batch processes should also be mentioned, including production of potentially large quantities of off-specification products as well as less flexibility in technology as the equipment is designed and optimized for particular operating conditions. A need to operate throughout the year and on a large scale can conflict with seasonal variations in feeds or the product demand. Subsequently, large and expensive storage facilities are utilized. An option is to produce a range of similar chemicals (e.g. solvents or plasticizers) using essentially the same processes by several production campaigns which can last for few months.

There are operations that are inherently periodic, such as adsorption or absorption requiring regular regenerations of solid or liquid sorbents. To circumvent apparent difficulties in organizing such operations, two units can be coupled subsequently (for example, absorption or extraction columns followed by regeneration).

For adsorption, a cascade of adsorbers (Figure 1.3) having different operations at a particular moment in time (adsorption, drying and cooling of adsorbent, desorption) could be used. Alternatively, moving or fluidized beds of adsorbents can be arranged. In the first case, there is a need for a fast switch of large flows, while the drawback of the second case is attrition.

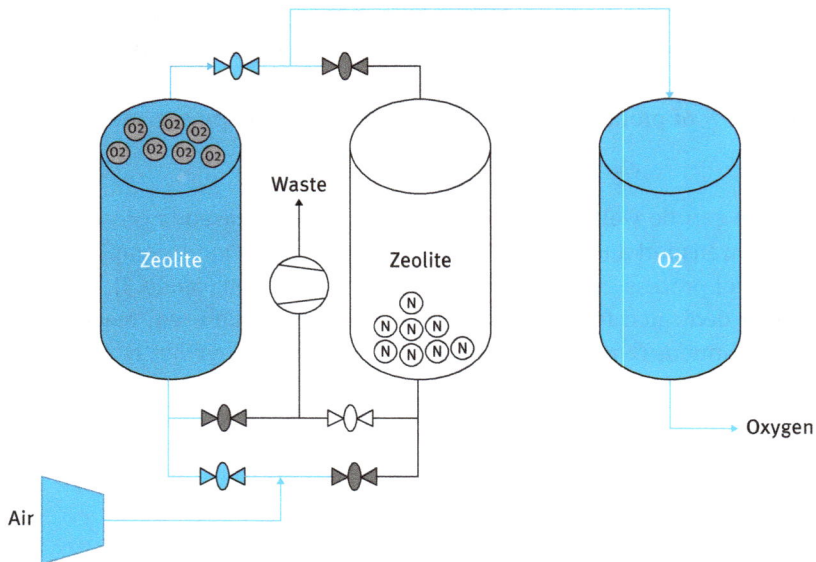

Figure 1.3: Vacuum pressure swing adsorption in two adsorbers. http://www.ranacaregroup.com/on-site-gas-systems/about-gas-generation.

Same difficulties could arise for continuous processes involving handling of solids, such as filtration or crystallization. Even in such cases, continuous processes are beneficial.

In particular, crystallization is a critical process manufacturing step in production of API as the crystallization conditions govern product purity and physical attributes, e.g., particle size and morphology. Sensitivity of crystallization to different parameters, such as temperature, mixing, and residence time, makes a transition from batch to continuous operation challenging, which is, however, necessary to ensure constant product quality by eliminating batch-to-batch variability, controlling the particle size, and improving the process scalability.

1.1.2 Multilevel chemical processing

The cornerstones of chemical technology will be considered in Chapters 1–3 of this textbook, but not in the level of detail available in the specialized literature. At the same time, the most essential features will be presented, targeting also chemists as potential readers, who might be less familiar with chemical engineering.

Processing of chemicals is very complex with several levels to be considered:
- Molecular level or level of a mechanism of chemical reactions. Such microlevel includes description of kinetics, molecular level catalysis, surface chemistry, processing of solids.
- Macrokinetics level, which addresses interactions and processes at a level of a catalyst granule, gas bubbles, etc., describing various heat and mass transfer processes.
- The level of a moving fluid (gas or liquid), which addresses the flow type (laminar or turbulent) and its characteristics (concentration and temperature gradients in axial and radial directions).
- Level of a reactor and other units, when reactor technology, unit operations, and scaling up are considered.
- Process technology level, which also includes process design and control.

A famous French chemical engineer, J. Villermaux, placed chemical technology in a broader context of space and time scale (Figure 1.4). The commercially attractive range of reactivity is rather limited, since reactions should occur within a reasonable time (or space time) in a reasonably sized reactor. The reactor dimensions might vary depending on the production capacity, in some cases having large volumes (100 m^3) or height (40–70 m). In the case of catalytic processes, higher activity and selectivity of a particular catalyst result in a decrease in equipment size and substantial savings in separation. This can also result in reduction of wastes. An example is the synthesis of acrylic acid done at BASF. In 25 years of process improvements, the amount of by-products was reduced by 75% due to development of better and more selective catalysts, which also resulted in less energy consumption for distillation and extraction, and even allowed catalyst regeneration.

Note that when the catalyst is too active, other processes, such as heat and mass transport, might become limiting; thus, in such cases, measures to improve a particular process should be aimed at overcoming transport limitations rather than developing a better catalyst. Such mass and heat transfer processes are extremely important at a reactor level; thus, scaling up from laboratory experiments done in intrinsic kinetics regime to the reactor level is very important and deserves special attention in process development.

Length (m)   Space and time scales

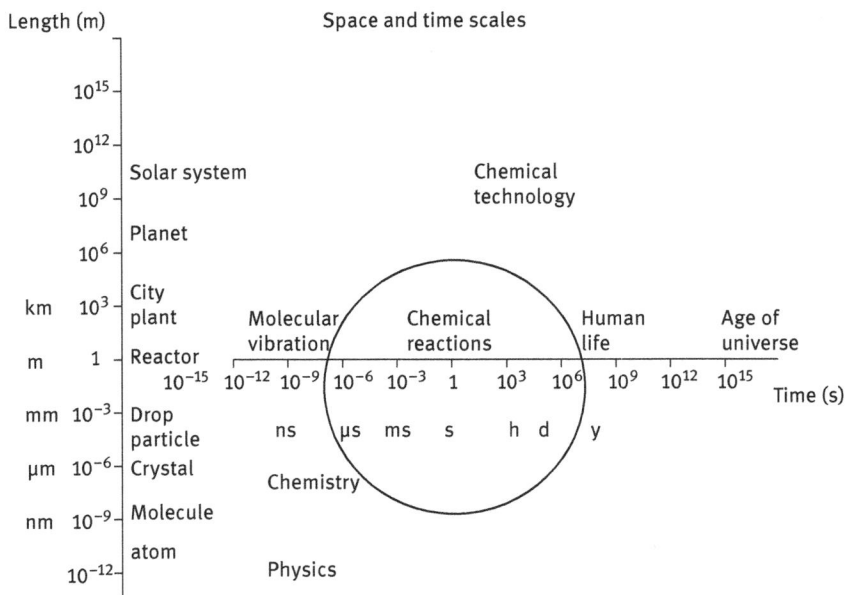

Figure 1.4: Chemical technology in the context of space and time.

Chemical reactions are the cornerstone of chemical technology. Chemical processes are often complex. By several consecutive and parallel reactions, not only the main products but also side products and waste are generated. Moreover, in real feedstock, impurities are present, either giving side products or influencing, for example, catalyst activity. Therefore, even if initial preliminary considerations of a particular process typically include only the main reactions and often a model feedstock, at some point, side reactions or other processes (such as catalyst deactivation) should be seriously considered if they have a significant impact on the product quality.

Analysis of potential alternatives for a chemical technological scheme with a subsequent process design should lead to a selection of a flow scheme and its further optimization. The latter task should give a process for production of the target product that is the most economically attractive (including impact on environment) as well as robust and safe. The number of alternatives is often limited by a number of restrictions related to physico-chemical, technological, and economical constrains.

Development of a chemical process consists of calculations of all material and energy flows, selection and design of equipment, calculations of all costs, consideration of various options for technological schemes, and the final selection of the scheme.

This is possible only after a study of all chemical transformations, physico-chemical properties of mixtures, and elucidation of all the boundaries that might appear at different stages.

Thus, the following issues should be considered in developing a chemical process: stoichiometry of the main and side reactions; temperature, pressure, concentration domains; thermodynamics including composition at chemical equilibrium; kinetics and the range of desired production rates; catalysts, their deactivation and regeneration; desired product purity at the available quality of the feedstock and influence of impurities; separation of reactants and products; such processing constrains as explosivity, corrosion, safety, and toxicity; phases in the reaction system.

The process design involves quite complicated flow charts and is not a straightforward application of disciplines on which chemical technology is based (chemistry, physical transport, unit operations, reactor design), but rather integration of this knowledge.

Complications also arise from a requirement of choosing from many possibilities, taking into account product markets, geographical location, social situation, legal regulations, etc. and that the final result must be economically attractive.

Combining such aims as minimization of energy and capital costs, generation of a product of the desired purity at the highest yield in robust and safe equipment with a minimal emission of wastes is very challenging, rarely possible. Thus, compromises are made still minimizing the costs to get a product of a desired purity with the lowest amount of wastes. The latter is important since hundreds of millions of tons of CO_2, hydrocarbons, SO_2, and NO_x are emitted globally. Moreover, modern technology for production of a particular chemical should also be aimed at utilizing not only the side products but also emitted heat through proper heat integration.

Chemical industry and chemical technology converting raw materials, such as coal, natural gas, crude oil, and biomass, into valuable chemicals can be subdivided into inorganic and organic chemical technology. The former deals with the production of basic inorganic chemicals (acids, bases, salts, fertilizers), fine inorganic chemicals (for example, materials for electronics), metals, silicates, glass, etc., while organic chemical technology comprises oil refining, synthesis of monomers and polymers, basic, and fine organic chemicals.

Description of various routes for production of such chemicals will be given in subsequent chapters of the book reflecting different reactions typical for oil refining and chemical industry.

Description of the production technologies will include not only the chemical transformations, but also the handling of the feedstock (i.e., purification) and the products (separation etc.). In chemical process industries, chemical transformations *per se* are combined with mass transfer processes, especially for reactions limited by equilibrium. This implies that, for example, extraction or distillation should be combined with chemical reactions.

In some cases, several chemical processes can be combined in one reactor, for example, exothermic and endothermal ones as in oxidative dehydrogenation of hydrocarbons.

Even if some features are the same for basic, specialty, and fine chemicals production, there are obvious differences. Thus, a large scale of manufacturing (millions tons per year) in the case of basic chemicals requires high capacity of production units that operate continuously. An increase in capacity results in a decrease of capital costs and costs of energy and water/steam.

It should be noted that continuous production typically leads to more constant product quality, avoiding batch-to-batch variations, which often happen in synthesis of fine chemicals, resulting in huge penalty if a product is off specifications (off spec).

1.1.3 Large or small chemical plants?

One of the megatrends in chemical technology along the years was the construction and operation of integrated production sites (the German word *Verbund*, which accounts for such a strategy and is strongly promoted by the largest chemical company BASF, was started to be used in English as well). Such strategy leads to integration of processes (a product of one plant can be a feedstock for another, emitted steam can be used for heating within the integrated site, etc.) and economically is more beneficial. However, such production sites cannot be within proximity to all customers, increasing thereby transportation costs of products (while diminishing them for reagents). In addition, superficially, it might look that, in integrated sites, there are more emissions, and as a consequence much worse environmental footprint. However, for such integrated sites, emissions could be even lower (per ton of product), as side products and waste of some units could be used as feedstock for the other. Energy integration is much better and even emission treatment could be more efficient (acid waste of one plant can be neutralized by a base waste from the other). Treatment of various types of wastes (cleaning strategy) will remain as the method of decreasing the ecological footprint of chemical technology at least for some time. This method requires large facilities, which occupy large space, consume energy and materials, lead to solid waste, etc.

A more promising way is to create a chemical technology that will be waste free (zero tolerance) or emit minimum waste (avoiding strategy).

When the process intensity is the same independent of the reactor volume, the reactor productivity is proportional to the volume. The latter can be approximately considered as $V = l^3$, where l is the reactor length. The costs of reactor materials (walls, internals) are roughly proportional to l^2. This implies that even if the overall capital costs increase, the capital costs per product unit decrease with an increased capacity, which is the basis of "economy of scale" concept. Variable costs (feedstock, energy, other materials) depend less on equipment size. At the same time, large capacity requires substantial capital investment, large energy consumption, and a need for a large territory with a proper infrastructure. Thus, the location of a plant is of vital importance. In addition to cheap energy and space requirements, proximity to the markets might also be an issue, if transportation costs are high and should be minimized. This is especially important for low-cost products.

Another trend in chemical technology opposite to the Verbund strategy is creation of specialized dedicated and decentralized plants with a limited product portfolio in close proximity to markets or feedstock.

This trend was clearly seen for decades in the wood-processing industry. Since transportation of solid feedstock, such as wood, for long distances (50–60 km) is not economical, pulp mills were built (at least in Nordic countries) in many locations. Lower consumption of paper in the recent time due to the evolvement of electronic media and strong competition from geographical areas where wood can grow faster than in Nordic countries resulted in difficult times for the pulping industry and they started looking for possible rejuvenation by making chemicals and/or fuels rather than paper. For example, one option for wood utilization could be decentralized wood pyrolysis, giving liquid bio-oil, while further processing of bio-oil can be organized in integrated sites with a large processing capacity.

Another disadvantage of a large plant is that when such a plant is shut down because of either regular maintenance or malfunction, units that are relying on the product from such plant might also experience a shutdown if there are no alternative supplies and the product is not in stock. The amount of waste would also be larger with a size increase, thus calling for creation of adequate waste treatment facilities.

The installation of large pieces of equipment and their maintenance could be also challenging, requiring, for example, special cranes and even transportation logistics (Figure 1.5).

Figure 1.5: Transportation of a chemical reactor. Adapted from http://www.chinaheavylift.com/ news/chinaheavylift-spmt-completed-the-663t-secondary-dehydration-tower-heavy-transportation-for-bp-chemical/.

The Verbund strategy, however, allows having certain integration of products. For example, in homogeneous catalytic hydroformylation of propylene, besides normal C4 aldehyde, an isoaldehyde is also formed. Some technologies of hydroformylation to be discussed in Chapter 13 rely on expensive Rh catalysts and quite sophisticated ligands used in excess. The targeted product is mainly used for the synthesis of 2-ethylhexanol, an intermediate in the production of plasticizers. An alternative is to utilize classical Co carbonyls, which are much less expensive even if they require higher pressures. Selectivity to normal butyraldehyde is, however, not that high, as in the case of Rh catalysts. Nevertheless, Co-based catalysts are still used in several locations, the reason being that isobutyraldehyde can be used for the synthesis of an important intermediate, neopentyl glycol (NPG), which can have a higher market price than 2-ethylhexanol. At some point, hydroformylation units using Rh with subsequent 2EH synthesis were running in "red" (the industrial jargon meaning that the plants were making losses), while an "outdated" Co carbonyl technology was profitable, since NPG enjoyed higher prices.

Another positive issue with large integrated sites could be that centralized supporting units might be more economical. The same holds for research and development (R&D) units, which might be even absent at small one-product-oriented plans, where only limited analytical facilities might be available.

1.2 Alternative production routes

An issue in the process design is the variability of the production of a particular product, which is reflected by the fact that the feedstock as well as the production routes can be different. For example, formic acid can be made in a dedicated process from CO and methanol, giving first methyl formate, or it can be generated as a by-product in the oxidation of naphtha (Figure 1.6).

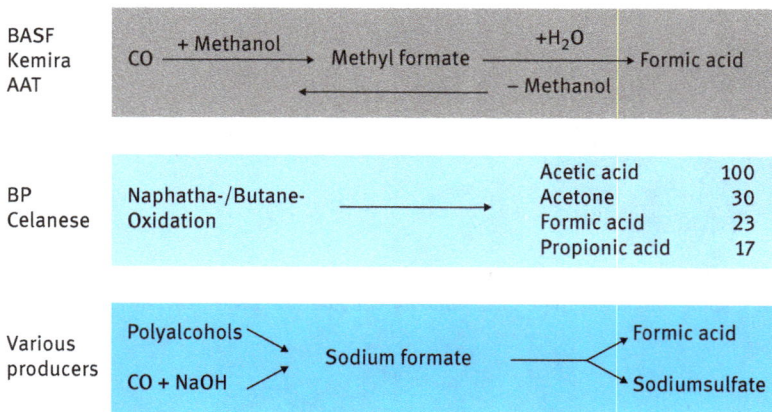

Figure 1.6: Several routes for formic acid synthesis.

There are also many commercial routes for production of such chemical as caprolactam as discussed in Chapters 13 and 15, emphasizing the fact that there is not a single option to produce this important monomer, and many variants for a particular technology are possible. One of the starting chemicals for caprolactam, phenol, can be made from oil, coal, and even biomass. Another example is the synthesis of styrene, which can be produced from ethylbenzene by dehydrogenation, oxidative dehydrogenation, or through a hydroxyperoxide.

Even a one particular reaction can have very many options. Hydrogenation of benzene can be made in the liquid phase or in the gas phase. The former technology can be implemented in a trickle-bed reactor with a concurrent downflow mode of operation or in a fixed-bed upflow reactor. A batch reactor with an impeller or a cascade of reactors can be used.

The same variability is seen in the separation units. In the synthesis of vinylacetate (VA), the light fraction can be separated first from the target product with subsequent separation of vinylacetate, or alternatively, it is distilled together with the light fraction first and separated thereafter. More that 30–40 variants of the VA synthesis could be proposed; thus, a detailed analysis comprising not only technological but also economic aspects is required when deciding on which technology to select.

Production of chemicals includes the production *per se*, storage facilities of the feedstock, products and intermediates, transportation facilities for the reactants, products and waste, additional buildings, as well as control, supply, and safety units.

The main focus in the textbooks on chemical reaction technology is obviously on chemical production *per se*. It should, however, be emphasized that, in addition to accidents in chemical industry originating from equipment failure or wrong design, storage can also be of crucial importance. A lot of explosions occurred in the chemical industry due to, for example, self-explosion of a particular fertilizer – ammonium nitrate. Thus, an explosion at BAFF Oppau site in 1921 resulted in the loss of 450 people, creating a crater measuring 80 m in diameter and 16 m in depth (Figure 1.7). As a consequence of a more recent explosion of the same chemical in Toulouse in 2001, there were 29 causalities.

Figure 1.7: Explosion at BASF Oppau site in 1921. http://www.landeshauptarchiv.de/filead min/blick/images/21.09.0.2.full.jpg.

Another example is accidents in oil storage facilities industry, when the technical and physical causes are related, among others, to mechanical failures, corrosion of materials, and leakage problems. In the so-called Petit-Couronne accident in 1990, a leak at the level of an elbow on a piping of fuel from the refinery resulted in more than 13,000 tons of hydrocarbons pumped into the groundwater. One house was destroyed, when the ignition of hydrocarbon vapors accumulated in a basement was triggered by turning on the hot water valve by the homeowner. In another accident in 2020, a fuel storage tank at Norilsk-Taimyr Energy's Thermal Power Plant failed, flooding local rivers with up to 17,500 tons of diesel.

1.3 Evaluation of chemical processes

Several technical, economical, and ecological metrics to evaluate performance of a certain production unit will be discussed below.

Productivity or capacity is related to the amount of product or processed feedstock per unit of time. Typically, the value is defined per hour or day. Often, reported numbers of annual production include regular turnarounds; thus, in order to relate the daily production with an annual one, it can be roughly assumed that a plant operates 8,000 h per year or 330 days.

Consumption coefficients illustrate the amount of feedstock or energy per unit (tonne or m^3 of product).

Product yields relate the real amount of a product to the theoretical one.

Relative capital costs are the costs of equipment calculated per unit of productivity. In order to organize a production of a certain chemical, obviously, there should be capital costs for equipment, reactors, pipelines, etc. The relative capital costs can be calculated either in tonnes of metal per ton of product per day or in monetary values.

Several metrics are used for environmental analysis and eco-efficiency. Financial metrics estimate environmental impacts or ecosystem services in terms of currency, thus giving a possibility for comparison with monetary transactions. Environmental (including health and safety) metrics estimate the potential for creating chemical changes or hazardous conditions in the environment. Safety metrics illustrate time between the accidents, while environmental metrics can simply measure emissions to the environment without consideration of pollutant degradation or formation of new pollutants. More complicated environmental metrics may include such factors as toxicity, reactivity, fate/transport of the pollutants.

Few basic indicators of process sustainability were proposed: (i) material intensity, (ii) energy intensity, (iii) water consumption, (iv) toxic emissions, (v) pollutant emissions, and (CO_2) emissions. Each metric is constructed as a ratio, with impact, either resource consumption or pollutant emissions, in the numerator and a representation of output, in physical or financial terms, in the denominator.

To calculate the metrics, all impact numerators and output denominators are normalized.

There are other indicators used by industrial companies. For example, environmental fingerprint of dyeing process of indigo dye was evaluated by BASF (Figure 1.8).

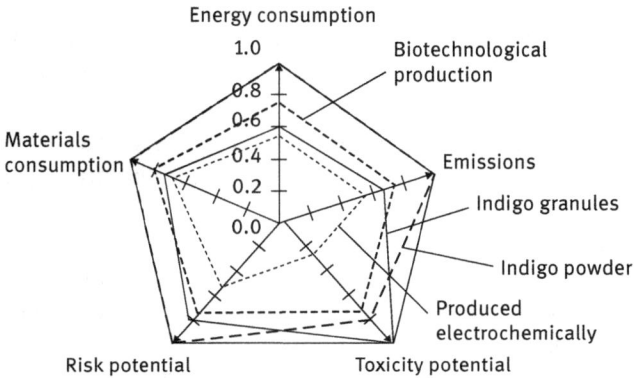

Figure 1.8: Environmental impact of indigo dye. http://www.corporate.basf.com/en/sustainability.

The scenario closer to the graph origin is the most advantageous. In this particular case, it was the electrochemical version of indigo production, which had the lowest environmental impact.

1.4 Chemical process design

1.4.1 Economic aspects

It is important to note that in chemical technology, the process should be viewed in its whole complexity, rather than as a combination of individual steps. For example, the performance of a reactor unit can depend on the performance not only on the units located upstream this reactor, but also downstream. Obviously, upstream units influence the inlet composition or the feed purity and thus have an impact on the reactor. The influence of downstream units is less obvious, but, for example, a loss of pressure downstream a reactor could lead to an increase in pressure in the reactor and might damage the catalyst support grid. Improvement of one unit (reactor) usually improves the overall performance. Thus, a slight improvement in catalyst selectivity would result in sometimes very large savings in the separation.

It should be also mentioned that the optimum conditions for a system element are not necessarily optimum for the system as a whole. Thus, optimization of a particular chemical reaction technology should include not only the reactor unit but other units as well.

There are several levels in the process design. Order-of-magnitude estimates with accuracy ca. 40% are based on similar previous cost data. A study estimate based on the knowledge of the major items of equipment gives an accuracy of ca. 25%. Even a higher accuracy of ca. 12% is achieved in preliminary budget estimates. A definitive estimate based on almost complete data provides accuracy ca. 6%. Finally, an accuracy of ca. 3% is reached during a detailed estimate based on complete engineering drawings and site surveys.

During process evaluation, the following questions should be addressed: is the production process technically feasible in principle; is it economically attractive; how big is the risk in economic and technological terms?

Technical risks are associated, for example, with exceeding the technically established limits such as too high dimensions of distillation columns, unfamiliarity of a company with a certain technology (utilization of high-pressure, continuous processes, fluidized beds, gases, etc.), use of units, difficult in scaling up, which is typical in processing of solids, or application of technically non-established equipment.

The probability of success of a new technology P is related to the number of innovations/uncertainties N and the level of confidence C: $P = C^N$. If, for example, there are five major innovations with 90% confidence, a project success probability is ca. 60%, which might be too low to justify investments in such process. Reducing the technical risk is related to an increased expenditure in R&D and development of various failure scenarios.

The main aim of any industrial company is making a profit. Sometimes, for economical reasons, a non-profit operation could be continued in a private company for the purpose of playing a game against financially weaker competitors, with an aim to eliminate the competitor from the market. Other reasons could be of social or political character when an otherwise unprofitable operation is subsidized by taxpayers.

Production of chemicals nowadays is governed by numerous regulations, thus imposing boundaries on emissions etc., influencing, for example, the capital and operation costs. The costs are typically divided into variable and fixed costs. Such division is useful when considering the costs of a single product or an individual production unit. The variable costs (raw material costs, energy input costs, royalty, and license payments) depend directly on a plant output. Fixed costs have to be paid independent on the production output even if it is temporarily shut down.

Raw material costs can be rather high in the total product costs and depend on the product type. In the case of basic chemicals, they could constitute 40–60% of the operating costs. The costs of catalysts and other materials (solvents, absorbents) should be also included. It is thus better to use an internally available feedstock than a purchased one. Energy input costs include steam, fuel oil, electricity, cooling water, lighting of plant structure, etc. Royalty or license fee, when made per ton of production, can be included in the variable costs. For example, a company producing catalysts can put such a royalty fee for the catalyst production inventors depending on the amount of catalysts produced. An alternative is to make an agreement on an annual basis; thus, such annual fees appear as a fixed cost in balance sheets.

The costs of packaging and transportation are largely variable costs. It should be noted that the products could be sold in different ways, thus not all and not always are the costs included in a price of a chemical. For example, if a certain chemical is sold "ex-works", it means that the customer is in charge of the product transport. Other delivery terms could be *ddu* (delivered duty unpaid) or *cpt* (includes packaging and delivery to a site, but not, for example, costs associated with custom clearance). The variable costs are essentially independent on capacity.

Fixed operational costs include labor and maintenance costs, laboratory staff, maintenance materials, depreciation, rates, and insurance as well as overheads. Depreciation reflects the diminishing capital value of a chemical plant during the years. In the chemical process industry, depreciation time is typically 10–15 years, although the real lifetime of equipment can be longer.

In fact, fixed capital costs represent the sum of all direct and indirect costs plus additional amounts for contractor's charges incurred in planning and building a plant ready for start-up. The costs can be specified as inside battery limits (onsite) – costs of installing the process plant equipment and materials within a specific geographical location (battery limits) and outside battery limits (offsite). The latter cover costs of facilities located outside the process plants battery limits: process building (control room, electric cabins), auxiliary buildings with services and furniture, site development (landscaping, site clearance, roads, fences, connections to roads), utilities production (steam, water, power, air, fuel, refrigeration, hot oil), distribution it to the plant battery limits, offsite facilities (waste disposal, incineration, flare, storage, loading, fire protection, non-process equipment (laboratory, workshop, maintenance, lifting, and handling equipment).

Fixed capital costs include equipment costs (cost of piping, steel structure, electrical equipment and materials, instrumentation and control equipment, insulation and painting), construction costs (civil works, mechanical erection, instruments and electrical erection, painting and insulation, vendors assistance), and contractor services (basic and detailed engineering, procurement activities, and site supervision).

Some of these costs are size-independent (for example, engineering), while the others (machinery, equipment) increase with plant capacity increase. As already discussed, the plant capital depends on the plant capacity and is proportional to the production sales to a certain fractional power, typically between 0.6 and 0.7.

It should be mentioned, that costs increase with an increase in distance from the major manufacturing centers in a slightly different way depending on the geographical location.

The initial investment in a chemical plant besides the total permanent investment mentioned above also includes working capital, which covers, for example, the costs associated with the initial catalyst load and inventories of the raw materials, etc.

Besides depreciation, which is needed to cover the investment, there are such costs as rates and insurance. This item is needed to cover local rates, which are location-specific. Finally, overhead charges not associated directly with production are needed to cover general management, administration, centralized facilities, such as legal services, patent office, supply, purchases, R&D, etc.

An important parameter in evaluating the process economics is income or net profit, which is the total income minus the operating costs minus depreciation minus tax. The calculation of income from operations is given in Figure 1.9.

Net sales	
– Distribution costs (provision; freight, insurance) – Costed interest – Cost of raw materials	Product variable costs
Contribution margin I	
– Fixed manufacturing costs (fixed personal, energy) – Shipping costs (storage, loading, etc) – Selling costs (sales, marketing, PR, services)	Fixed product costs
Contribution margin II	
– Difference in predetermined manufacturing costs – Cost of idle equipment (capacity untilization below 100%)	
Gross operating result	
+ Cost of idle equipment (capacity untilization below 100%) – R&D – Administration costs – Other costs	
Operating result	
+ Depreciation – Costed interest – Other costs	
Income from operation	

Figure 1.9: Calculation of income from operation.

By knowing income from operation and total capital costs, another important parameter, return on investment, ROI, can be calculated as ROI = income/total capital. This simple metrics is readily understood, but could be misleading. ROI provides a one-moment-in-time view, since it is difficult to predict future cash flows. Moreover, the generated cash can depend on a depreciation method used by a particular company. The charge of depreciation might change from year to year.

Note that the purchase/sales price sometimes could be less than the full product costs if by some strategic reasons there is a need to pay the price for entering the market or force a competitor to leave the market as already mentioned above.

Besides ROI, another important metrics is payback period, which is calculated as the total permanent investment divided by annual cash flow. During plant construction, cash flow is negative. After the start-up, the positive cash flow begins, which includes income from sales plus depreciation minus total direct and indirect production costs. A schematic view of payback time is given in Figure 1.10.

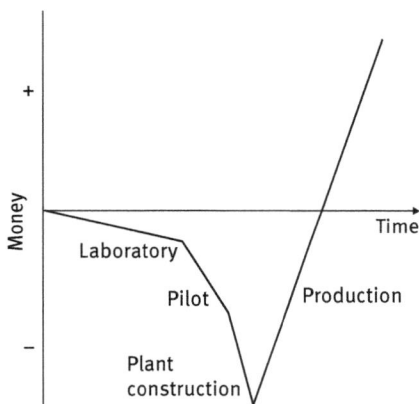

Figure 1.10: Illustration of money flow with time.

1.4.2 Flow schemes

Schematic structures (flow schemes or flowsheets) are typically used in presenting which types of chemical reactors and separation units are applied in a particular technology. Different symbols are used in flow diagrams, with some of them illustrated in Figure 1.11.

Before discussing conceptual process design, few words could be mentioned about flow diagrams in general. They should provide a clear and simple outline of the steps involved in the process, covering all the major steps in the process, including those before and after the main chemical processing.

Figure 1.12 demonstrates a flow scheme of cyclohexanol dehydrogenation in a multitubular reactor. An interesting feature is a heat exchanger upstream the catalytic reactor, which is used to preheat the reactants by the products. The scheme in Figure 1.12 also comprises a distillation column from which there is a stream of unreacted cyclohexanol back to the reactor indicating that conversion is not complete.

Usually, a block type diagram is sufficient in understanding chemical reaction technology evaluating the process flow. For these purposes, complex engineering drawings (Figure 1.13a) reflecting the industrial reality (Figure 1.13b) are not required.

P–1/C–101
Absorption

P–2/C–102
Stripping

P–6/MX–101
Mixing

P–7/CSP–101
Component
splitting

P–2/BX–101
Packaging

P–5/HX–102
Heat exchanging

P–4/HX–101
Heating

P–8/GR–101
Grinding

P–9/XD–101
Extrusion

P–3/V–101
Storage

P–6/BE–101
Bucket elevation

P–5/BC–101
Belt conveyling

P–3/PM–101
Fluid flow

P–4/G–101
Gas compression

P–6/INX–101
Ion exchange

P–8/BCF–101
Basket
centrifugation

P–9/GAC–101
GAC adsorption

P–4/FL–101
Flotation

P–7/C–101
Distillation

P–7/DO–101

P–1/TDR–101
Tray drying

P–6/V–103
Batch distillation

P–4/V–104
PF stolich fan

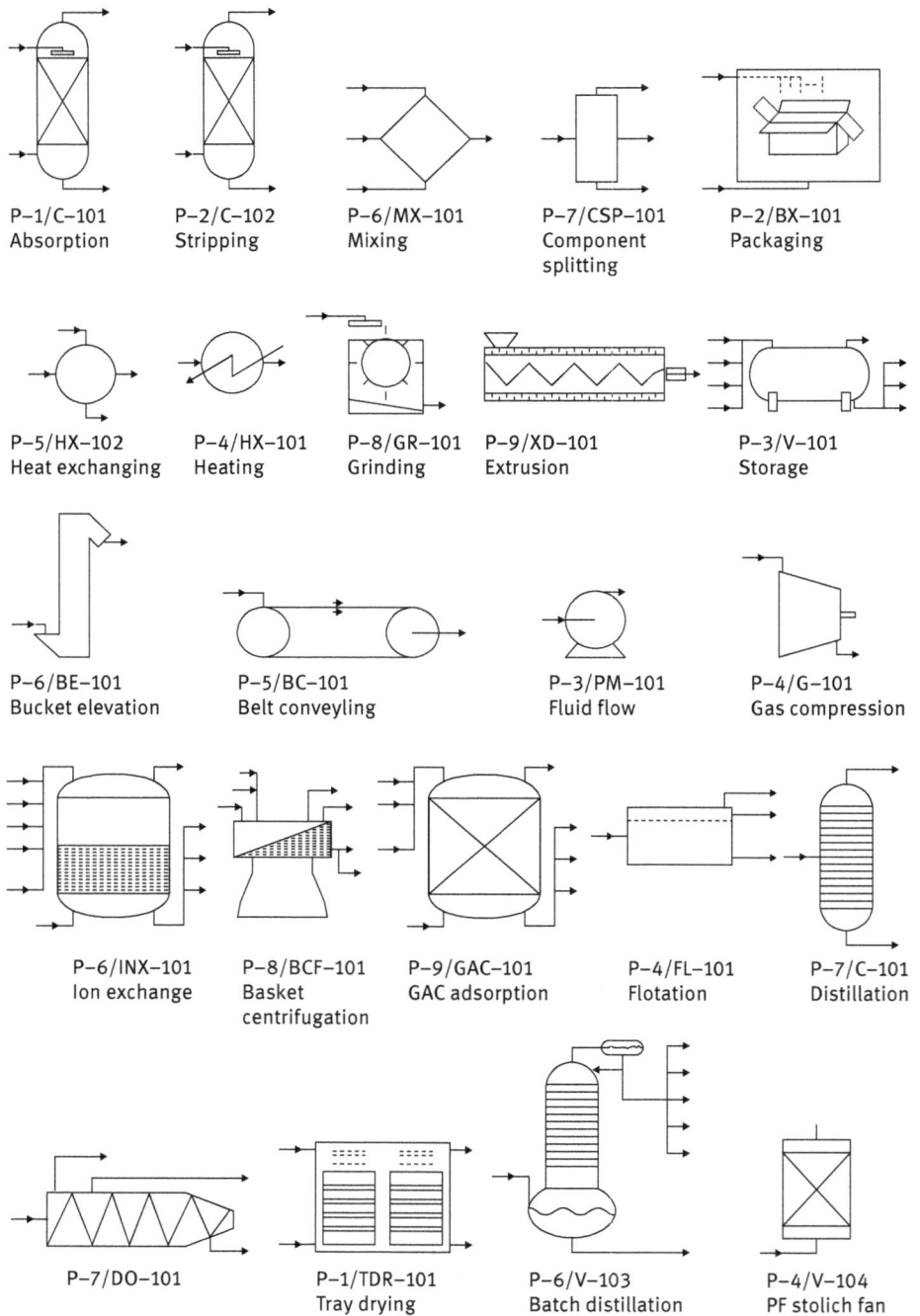

Figure 1.11: Symbols in flow charts.

P–9/MSX–101
Mixer-settler extraction

P–5/MF–101
Microfiltration

P–4/FL–101
Flotation

P–3/V–101
Decanting

P–1/V–101
Vessel procedure

P–2/FBDR–101
Fluid bed drying

P–3/CR–101
Crystallization

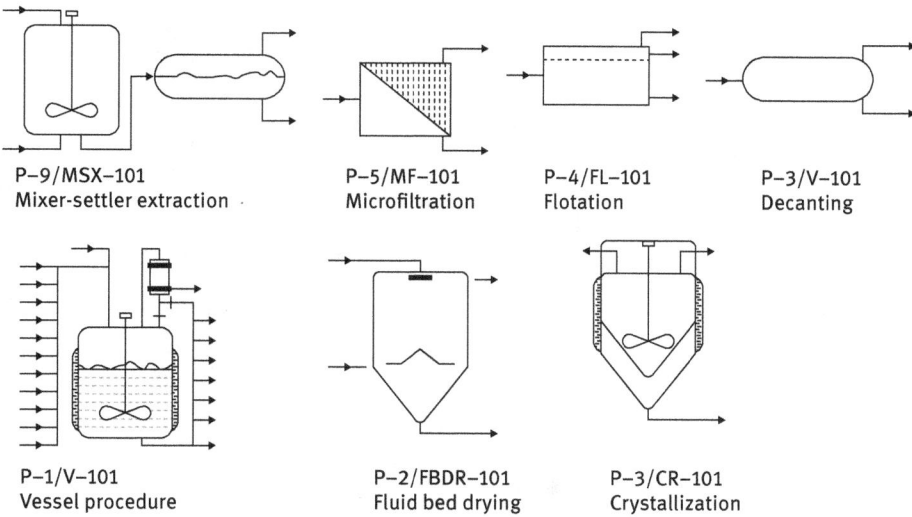

Figure 1.11 (continued)

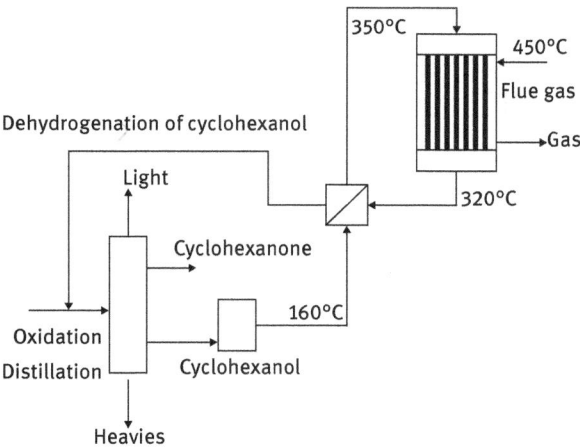

Dehydrogenation of cyclohexanol

350°C

450°C

Flue gas

Gas

320°C

Light

Cyclohexanone

160°C

Oxidation
Distillation

Cyclohexanol

Heavies

Figure 1.12: A flow diagram of cyclohexanol dehydrogenation.

Let us consider the synthesis of nitric acid following the treatment of V. S. Beskov and V. S. Safronov (*General Chemical Technology and the Fundamentals of Industrial Ecology*, Moscow, Khimia, 1999) as an example of conceptual process design. The process consists of several steps. Initially, ammonia is combusted to form nitric oxide:

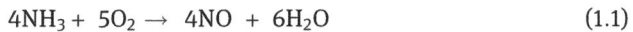

$$4NH_3 + 5O_2 \rightarrow 4NO + 6H_2O \tag{1.1}$$

(b)

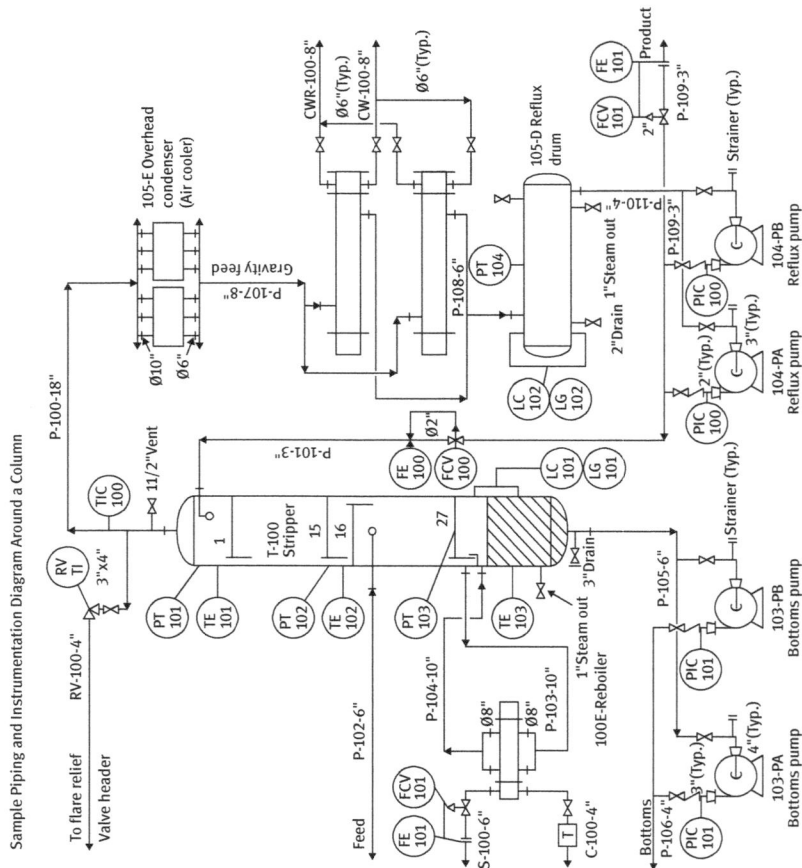

(a)

Figure 1.13: Example of (a) engineering drawings and (b) complex piping networks.

This reaction is highly exothermic and occurs at high temperatures of 850–900 °C over platinum catalysts. The residence time should be minimized (using high flow rates) to prevent side reactions, such as reduction of ammonia with NO:

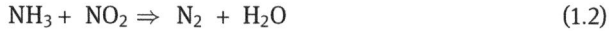

$$NH_3 + NO_2 \Rightarrow N_2 + H_2O \tag{1.2}$$

Minimization of residence time is done by applying rather untypical Pt gauzes. As the reaction rate is fast, external diffusion at such conditions can be prominent. Influence of reaction parameters is not straightforward. With temperature increase, the reaction rate increases at the expense of selectivity. An increase in pressure enhances the reaction rate but leads to higher metal losses. Under severe process conditions, the catalyst lifetime is limited usually to 1 year. Irreversible losses of platinum could be prevented at least partially by the metal recovery, which is done by placing a woven Pd-rich alloy gauze immediately below the oxidation gauze, affording 70% recovery.

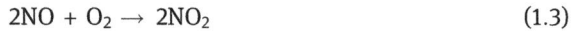

$$2NO + O_2 \rightarrow 2NO_2 \tag{1.3}$$

This is a non-catalytic reversible gas-phase reaction, occurring with a release of heat. Thereafter, nitrogen dioxide is absorbed in water to form nitric acid:

$$4NO_2 + O_2 \rightarrow 2H_2O \rightarrow 4HNO_3 \tag{1.4}$$

This heterogeneous gas-liquid process reaction is more complicated since besides NO_2, other components can react (NO, N_2O_3, N_2O_4, etc.). In fact, the main reactions happening in an absorption tower could be written in the following way:

$$N_2O_4 + H_2O \rightarrow HNO_2 + HNO_3 \tag{1.5}$$

$$3HNO_2 \rightarrow HNO_3 + H_2O + 2NO \tag{1.6}$$

The flow diagram must contain a cooler for the oxidation reaction to be as complete as possible. Reaction (1.4) is unusual in terms of its temperature dependence, being more active at low temperatures, and represents an example of trimolecular reactions.

Dosing of low boiling liquids is difficult; thus, ammonia evaporation should be also included. As mentioned above, oxidation of ammonia is an exothermic reaction. The reaction temperature is 850–900 °C and the adiabatic heat release is equivalent to 720 °C; thus, the reactor inlet temperature should be ca. 130–180 °C. This implies that a heat exchanger should be installed upstream the catalytic reactor.

Furthermore NO oxidation is also exothermic; thus, gases are heated during this reaction. As a result, absorption of NO_2 is worsened. In order to circumvent this, a cooler upstream absorber would be required.

An optimum ammonia/oxygen ratio in the reactor should be 1:1.8, but for the acid production 2 volumes of O_2 per 1 volume of NH_3 are needed, therefore extra air could be added to the absorber.

As catalyst is sensitive to impurities, air filtration should be installed. The gases coming from the reactor could be used to heat the reactants, and a heat exchanger upstream the reactor could be efficiently used. Since water is produced in ammonia oxidation, in the heat exchanger upstream absorber, low-concentration nitric acid could be produced due to water condensation. This in turn means that an extra stream could be introduced to the absorber.

Nitric acid will contain some NO_x, resulting in a low-quality product having a yellowish color, thus a bleacher (stripper) should be introduced to treat unwanted emissions of NO_x (Figure 1.14) through contacting the product acid with air.

Figure 1.14: NO_x emissions.

An example of the flow sheet for nitric acid production is given in Figure 1.15. An essential question that has to be addressed regarding the scheme in Figure 1.15 is at which pressure should the nitric acid process be operated, since absorption of NO_x should be preferentially performed at high pressure, while atmospheric pressure is beneficial for oxidation. In fact, several options exist, including carrying out the whole process at atmospheric pressure or combining oxidation at low pressure and absorption at high pressure. In the latter case, a compressor should be added between the ammonia conversion stage and the absorption stage. This will be discussed in more detail in Chapter 9.

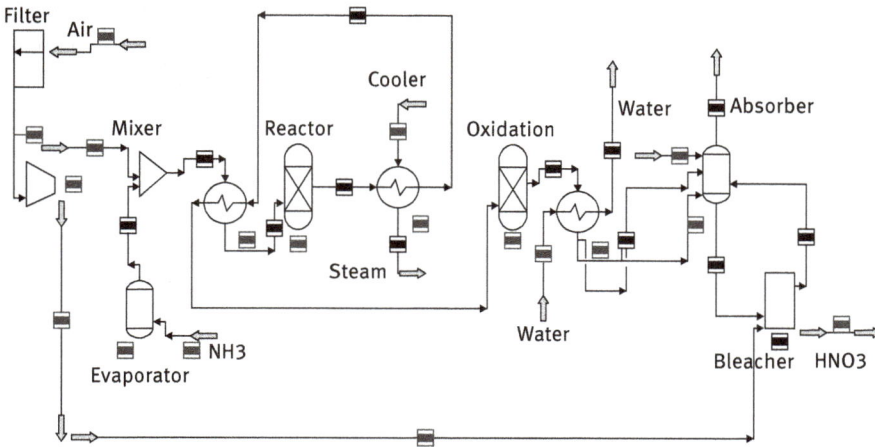

Figure 1.15: Flow diagram of nitric acid synthesis.

1.4.3 Sustainable and safe chemical technology: process intensification

As illustrated in the example above featuring nitric acid synthesis, a number of issues should be considered in conceptual process design. Some of them are related to chemistry, kinetics, thermodynamics, and catalysis.

Thus, conceptual process design should answer a number of questions. Some of them are presented below:
1. Is continuous or discontinuous processing to be preferred?
2. What are the optimal regions of process conditions?
3. Which process conditions are dangerous?
4. How is the reaction T reached?
5. Which type of reactor is preferred?
6. Is pretreatment of the reactor feed necessary?
7. How is the reaction mixture processed?
8. Are there any special measures in relation to co-products and waste required?

For instance, if conversion is not complete, there is an option of recycling (Figure 1.16), which should be carefully considered, as recycling at rather high conversions is not economical. Moreover, a simple recycling could lead to the buildup of impurities, which might be present in the feedstock, thus introduction of a purge stream is necessary.

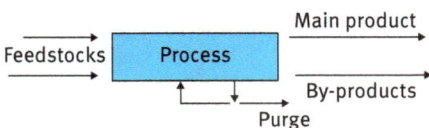

Figure 1.16: Input-output structure with purge.

Common sense rules are recommended to be used in the phase of conceptual design, such as those for separations: avoid unnecessary separations; do not separate fuel and waste stream products any further; do not separate and then remix.

If there are competing features for a system (productivity-product quality; productivity-feedstock consumption) one parameter should be set as an optimization basis, thus allowing a selection of the best process alternative.

Prior to optimization, the following tasks should be solved:

1. Selection of the optimization criterion (productivity, safety, reliability, production costs, capital costs, etc.). Production costs can be an optimization parameter especially in the case of optimization of a scheme for only one product. Important criteria are safety and reliability. Typically, reliable units are also safe, which do not mean, however, that they are optimized and the most efficient. Outdated simple technology with low productivity can be in fact the most reliable, but not the most desirable one.
2. Selection of variable independent parameters in optimization (temperature etc.).
3. Selection of boundaries for parameters (lower and upper limits on temperature or pressure etc.).
4. Selection of the optimization method.

Rather recently, the concepts of sustainable and green chemical technology started to be introduced in the chemical process design. Such principles include the following requirements:

1. The maximum amounts of reagents are converted into useful products according to the concept of atom economy.
2. Production of waste is minimized through reaction design.
3. Non-hazardous raw materials and products are used and produced wherever possible.
4. Processes are designed to be inherently safe.
5. Greater consideration is given to use of renewable feedstock.
6. Processes are designed to be energy efficient.

The following indicators were proposed for sustainability of a chemical process:

1. Waste minimization in terms of amount produced per ton of product for greenhouse gases, ozone-depleting gases, gaseous pollutants (NO_x, SO_x, VOC, HC), waste (solid, liquid, and gaseous, including catalysts and auxiliary), non-biodegradable material, cyto-, eco-, and phyto-toxic materials.
2. Process indexes:
   - Synthesis effectiveness: ratio between the desired product and the input materials (reactants, solvents, catalyst, auxiliary, etc.) flow rates (or weight for discontinuous reactor)
   - Process intensification: product to reactor volume or cumulative volumes for multistep reactions

- Process integration: number of steps, including separation, for the whole process
- Recycle: ratio between waste and by-products recycled and produced
- Energy efficiency: ratio between energy input (reactants, fuel, and other energy sources, including utilities) and output (as valuable products, including energy streams, which can be used, for example, steam)
- Intrinsic eco-efficiency: ratio between the product amount and end-of-pipe waste amount (gas, liquid, solid) to be treated before being externally discharged
- Safety control: number of process parameters under multiple automatic control with respect to parameters with single or human control, normalized to process degrees of freedom
- Operators' risk: number of operators exposed directly to hazardous chemicals with respect to those necessary for operations
- Intrinsic safety: ratio of intrinsic safe operations to those requiring human control
- Safety: time dedicated to training and safety operations (including maintenance) with respect to total working time
- Hazard storage index: amount of hazard chemicals stored (as reactant, intermediate, or end products) with respect to day production

3. Efficiency of the use of resources (amount per ton of product): freshwater used; solvent used and lost; equivalent oil barrels of energy input (all forms, from heat to electrical energy, to sustain the process, including utilities and services)
4. Eco-economics indexes: ratio between cleanup costs and product value; ratio between monetary compensation that must be paid due to toxic or pollutants release above legislation limits and total production value; ratio between monetary compensation that must be paid due to accidents and total production value; ratio between monetary compensation to local communities and total production value
5. Impact on local environment: change of biodiversity; degree of increase in persistent pollutants; degree of change of local use of land and water bodies for human activities

The risk of a particular process should be given a proper consideration during the process design. In the simplest form, the risk is expressed in the following form risk = hazard × exposure.

An inherently safer product and process design represents a fundamentally different approach to safety in the manufacture and use of chemicals. The designer is challenged to identify ways to eliminate or significantly reduce hazards, rather than to develop protective systems and procedures.

Design of inherently safer processes is based on the following principles (D. C. Hendershot, An overview of inherently safer design, *AIChE*, DOI 10.1002/prs.10121):

1. Minimize: Use small quantities of hazardous materials; reduce the size of equipment operating under hazardous conditions such as high temperature or pressure
2. Substitute: Use less hazardous materials, chemistry, and processes.
3. Moderate: Reduce hazards by dilution, refrigeration, and process alternatives that operate at less hazardous conditions.
4. Simplify: Eliminate unnecessary complexity; design "user-friendly" plants.

Hazards associated with typical chemical reactions and presented in Table 1.1.

Table 1.1: Hazards in different chemical reactions.

Reaction	Hazards
Oxidation	Highly exothermic; substrate and oxygen could be within explosion limits; risks for explosions
Hydrogenation	Highly exothermic; hydrogen is flammable; risks for explosions
Nitration	Highly exothermic; explosive products with several nitro groups
Chlorination	Highly exothermic; possibility for runaways; toxicity of chorine and products; problems with corrosion
Esterification	When reactants are flammable
Amination	Exothermic; toxicity of ammonia
Polymerization	Increase in viscosity during polymerization can lead to problems with heat removal.

Reducing the size of equipment obviously diminishes the quantity of a hazardous chemical. The former can be achieved by process intensification, which essentially means significantly smaller equipment. Some methods for process intensification are illustrated in Figures 1.17 and 1.18.

Process intensification can be achieved by the application of novel reactors (foam reactors, monoliths, microreactors, membrane reactors, rotating beds, spinning disk reactors, etc.), intense mixing devices, multifunctional equipment with several unit operations (reactive extraction, reactive absorption, reactive distillation, membrane adsorption), alternative ways of energy supply (microwaves, ultrasound, etc.). Safe smaller processes can be even cheaper than the conventional ones, contrary to an old wisdom that safety is about spending money.

Smaller reactors might offer much better heat transfer. In industrial settings, heat and mass transfer could in fact be the limiting factors; thus, a reaction that is

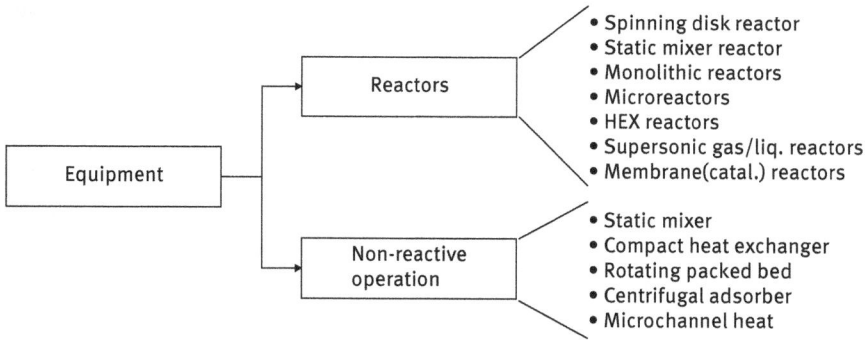

Figure 1.17: Process intensification equipment. Adapted from A. I. Stankiewicz, J. A. Moulijn, *Chemical Engineering Progress*, 2000, January, 22–34.

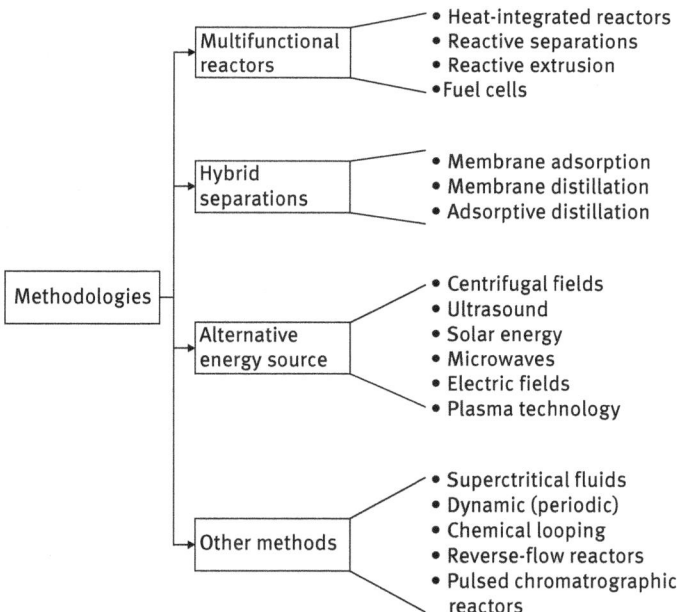

Figure 1.18: Process intensification methodologies. Adapted from A. I. Stankiewicz, J. A. Moulijn, *Chemical Engineering Progress*, 2000, January, 22–34.

slow in a batch reactor could be carried out in a continuous reactor with efficient gas-liquid or liquid-liquid mass transfer. The spinning disk reactor is an example of the latter case. The short residence time achieved in continuous reactors can be in some cases also beneficial from the viewpoint of selectivity for the intermediate product in consecutive reactions, as in a plug-flow reactor, there will be less back mixing.

The inventory of hazardous materials should be also reduced. Typically, chemical plants have an overstorage of raw materials to ensure smooth operation in the case of delays with supply due to transportation problems or other reasons. Thus, efforts should be devoted to ensure a reliable supply.

Two examples related to self-explosion of ammonium nitrate have already been mentioned in Section 1.2. In fact, there are even more examples related to the same chemical. During the Texas City explosion in 1947, several hundred people were killed, with as many as 4,000 as 2,500 t of ammonium nitrate exploded. In 2013, an ammonium nitrate explosion also occurred at the West Fertilizer Company storage and distribution facility in Texas, killing 15 people while more than 160 were injured, and more than 150 buildings were damaged or destroyed. A more recent example is the 2020 Beirut port explosion of approximately 2,750 metric tons of ammonium nitrate, which killed more than 200 people, left more than 7,000 people injured, 300,000 people homeless, and severely damaged critical health infrastructure and medical supplies.

Among other accidents due to large inventories, an accident in Ludwigshafen in 1948 is worth mentioning. A tank car filled with dimethyl ether exploded, leading to collapse of the surrounding buildings and death of at least hundreds of workers. The detonation causes the leak of other chemicals and a plume of smoke as high as 150 m. Almost 4,000 were injured suffering severe damages from the toxic gas, with many losing their eyesight.

A catastrophic explosion occurred in 2019 in Xiangshui County, resulting in the death of 78 people and the injury of 617. The fire was first observed in the solid waste warehouse, where hazardous chemicals generated during production were stored and were wrongly mixed. Most probably, the first explosion happened because of the accumulated heat and pressure generated by continuous chemical reactions in the warehouse. Furthermore, the fire generated by the first explosion ignited the adjacent natural gas station, leading to a large explosion.

Piping for hazardous materials could be also designed in a more reliable and safe way. For example, instead of pumping liquid chlorine to a plant site with subsequent evaporation, a vaporizer can be installed already in a storage area, reducing the inventory of chlorine in a pipe tenfold.

Dangerous chemicals can be even produced on site in smaller scale using, for example, microreactors, thus avoiding storage and transportation of them.

Previously, it was mentioned that large-scale plants are preferential from the view-point of capital costs. From the viewpoint of sustainable production, when possible, such scale economy can be replaced by small plants even with modular design.

Replacement of a hazardous chemical by another route is also recommended by the safe design approach.

The notoriously famous Bhopal disaster, when several dozens of tons of methylisocyanate were released to atmosphere, was due to the utilization of inherently unsafe route and storage of large quantitates of MIC (67 tonnes). This industrial disaster, the worst of the twentieth century, occurred at Union Carbide Corporation site on the midnight of December 2–3, 1984, in the city of Bhopal, India, which had about 1 million people. Over 40 tonnes of methyl isocyanate (MIC) as well as other lethal gasses including HCN leaked from the plant side to the city. There are different numbers available in the literature about the casualties. According to the Bhopal People's Health and Documentation Clinic, 8,000 people were killed in its immediate aftermath, and over 500, 000 people suffered from injuries.

On the night of the disaster, water that was used for washing the lines entered the tank containing MIC through leaking valves. The refrigeration unit designed to keep MIC close to 0 °C had been shut off in order to save on electricity bills. The entrance of water to the tank, which was full of MIC at ambient temperature, initiated an exothermic runaway process and subsequent release of the gases. The safety systems, which were not properly designed to handle such runaway situations, were non-functioning and under repair. Unfortunately, workers ignored early signs of disaster, since gauges measuring temperature and pressure in the various parts of the unit, including MIC storage tanks, were known to be unreliable.

It was supposed that MIC could be kept at low temperatures by the refrigeration unit, which, however, was shut off. In addition, the gas scrubber, meant to neutralize MIC if released, had been shut off for maintenance. In any case, the design was inappropriate, since the maximum designed pressure was only 25% of the actual pressure reached during the disaster.

Moreover, the flare tower, which was installed to burn off escaping MIC was not in operation, waiting for the replacement of a corroded piece of pipe. Even if it was operating it could only process a fraction of the gas released. There were some other reasons for the disaster such as too short water curtain, lack of effective warning systems, and failure of the alarm on the storage tank to signal temperature increase. Overfilling of the storage tank beyond recommended capacity and filling with MIC of a reserve tank, which was supposed to be empty, added to the overall picture.

In the case of the Bhopal disaster, there are many reasons for such unfortunate events, including poor maintenance, design, and inventory excess. After that disaster, the EU had allowed only a maximum of half a ton of MIC inventory.

The release of MIC can be prevented if the technology had been organized in another way. As illustrated in Figure 1.19, MIC was formed by the reaction of methylamine with phosgene with subsequent reaction with 1-naphtol. If the process had been designed in another way, e.g., phosgenation of naphthol, the release of MIC could have been avoided.

The alternative process still uses extremely toxic phosgene. Thus, other safer routes should be developed for production of this or similar type of carbamates applied on a large scale such as pesticides.

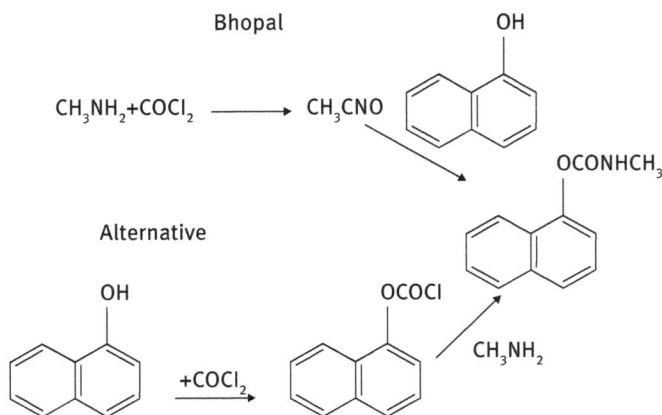

Figure 1.19: Bhopal, and alternative routes to *N*-methyl-2-naphthyl carbamate (carbaryl).

Overall, the decisions on the chemical routes are those that have the largest impact on the process success compared to the conceptual design and subsequent process optimization.

Thus, identification of alternative process chemistries should be done at the very beginning of any conceptual design influencing eventually the costs of the raw materials, the value of the byproducts, complexity and safety of the process, as well as the types of emissions and wastes and the overall environmental impact.

Another example of an alternative and inherently safer process design is the synthesis of phenol by oxidation of benzene with N_2O in the gas phase using a zeolitic catalyst containing iron (Figure 1.20). N_2O is generated as a side stream in the synthesis of adipic acid. The classical process is described in Chapter 12.

The advantage of this process compared to the classical one is clear, since in the cumene process (Chapter 12), benzene is first alkylated by propylene followed by oxidation to cumene hydroperoxide and decomposition to phenol and acetone. The latter is a low-value product. Moreover, synthesis of cumene hydroperoxide intermediate has inherent safety problems, which can be overcome in a direct synthesis method presented in Figure 1.20.

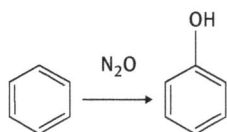

Figure 1.20: Synthesis of phenol by oxidation of benzene with N_2O.

Major process accidents often happen due to human errors. In 1974, the Flixborough site of Nypro company in the UK producing caprolactam was severely damaged by a large explosion. Twenty-eight workers were killed and a further 36 suffered injuries.

Approximately 2 months prior to the explosion, it was discovered that a vertical crack in one of the reactors for cyclohexane oxidation in a cascade of six reactors was leaking cyclohexane. Shutting down the plant after an initial investigation, a temporary 50 cm diameter piping was installed bypassing that reactor (Figure 1.21).

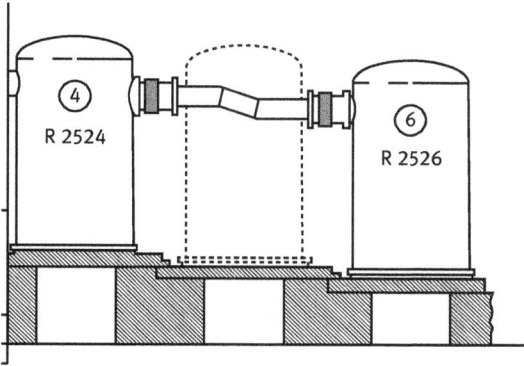

Figure 1.21: Diagram of the bypass between two reactors (https://www.aria.developpement-durable.gouv.fr/wp-content/files_mf/FD_5611_flixborough_1974_ang.pdf).

Unfortunately, no design was performed for a process operating at 150 °C and 10 bar and construction drawing was done with chalk on the floor. After the bypass system ruptured, a large quantity of hot cyclohexane (40 tons) was released in 30 s. A massive vapor cloud explosion caused extensive damage, started numerous fires on the site, and led to 28 fatalities (18 of the in the control room), 53 injuries, 1,800+ houses damaged, and the destroyed plant was never rebuilt. The number of fatalities would be higher if the explosion would not be on Saturday.

Another example of a human error is an accident at Seveso in Italy, which happened in 1976 at a small manufacturing facility producing 2,4,5-trichlorophenol (Figure 1.22).

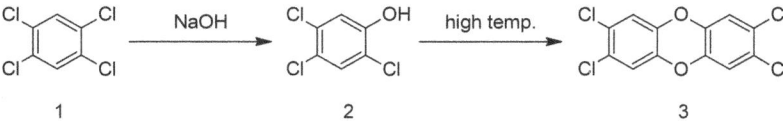

Figure 1.22: Production of 2,4,5-trichlorophenol (2) from 1,2,4,5-tetrachlorobenzene (1) and sodium hydroxide.

According to the national regulations, the plant operations had to be shut down over the weekend. When a batch process was interrupted by turning off the stirrer and isolating steam used for heating, the reactor operators were unaware of a much higher temperature of steam (300 °C vs 190 °C in a usual operation). The latter happened because of a dramatic drop in the load of the electricity-generating turbine as other parts of the site already started to close down. Without a running stirrer, the local temperature in the reactor approached the critical temperature for the exothermic side reactions starting a slow runaway. Eventual opening of the reactor relief valve caused the aerial release of 6 tons of chemicals including 1 kg of a highly toxic 2,3,7,8-tetrachlorodibenzo-p-dioxin (3 in Figure 1.22). Over 18 km^2 of the surrounding area became affected, 80,000 animals died or slaughtered, and the plant was shut down and destroyed.

Utilization of a broader feedstock base and its diversification in general is an important area that has lately been exploited, for example, in conjunction with the quest for biomass utilization as a base for chemicals and fuels. Shale gas could serve as another example.

One of the concepts used during sustainable design can be related to complete utilization of the raw materials.

Let us consider one example. Vinyl chloride can be obtained by chlorination of ethylene and subsequent pyrolysis of dichloroethane:

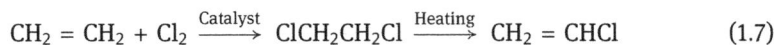

$$CH_2 = CH_2 + Cl_2 \xrightarrow{\text{Catalyst}} ClCH_2CH_2Cl \xrightarrow{\text{Heating}} CH_2 = CHCl \qquad (1.7)$$

The selectivity at each stage is 95%. The obtained HCl is considered as a waste; thus, the yield of vinyl chloride calculated per consumed chlorine is rather low (50%), while the yield per ethylene is ca. 90%. A change to the one-step process with a switch from one reactant to another (HCl),

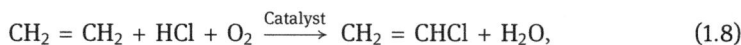

$$CH_2 = CH_2 + HCl + O_2 \xrightarrow{\text{Catalyst}} CH_2 = CHCl + H_2O, \qquad (1.8)$$

results in a process with 95% yield calculated per both reactants.

The excess of one reactant (typically a cheaper one) and the possibility to recycle it also result in a more complete utilization of the feedstock. For example, in steam reforming of natural gas, the stoichiometric ratio between methane and steam is 1:1, while in the industry, a much higher steam excess is used for several reasons, including a desire to shift equilibrium as well as to prevent formation of coke on the catalyst surface.

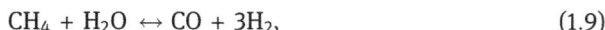

$$CH_4 + H_2O \leftrightarrow CO + 3H_2, \qquad (1.9)$$

Another option of a more complete utilization of the feedstock is to use countercurrent flows, affording higher driving forces for various separation processes. Absorption of CO_2 by water solutions of amines in the production of ammonia or absorption of NO_x by nitric acid could be mentioned as examples. In the latter case, the flow scheme (Figure 1.23) is organized in a way that at the top of absorber, the concentration of the nitric acid is the minimal one and the exhaust gases contain small amounts of NO_x, leading to almost complete absorption.

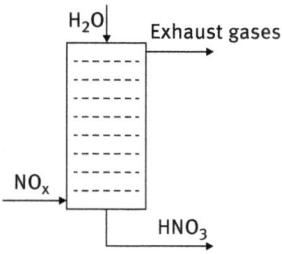

Figure 1.23: Absorption of NO_x.

Recycling of non-reacted feedstock is used when the conversion is far from complete. A typical example is the synthesis of ammonia, whose conversion can be 10–18% because of thermodynamic imitations; thus, after condensation (liquefaction) of ammonia and separation from nitrogen and hydrogen, the latter mixture is redirected to the ammonia synthesis converter. Production of ethylene oxide or methanol could be mentioned as other examples when the unreacted substrate is recycled.

Recycling with regeneration is used, for example, in already mentioned removal of CO_2 by amine solutions. After removal of unwanted CO_2 from the gas stream, the solvent containing CO_2 is regenerated in a desorber (stripper). One of the simplest versions of this technology is given in Figure 1.24, while other more energy efficient options will be discussed in Chapter 3.

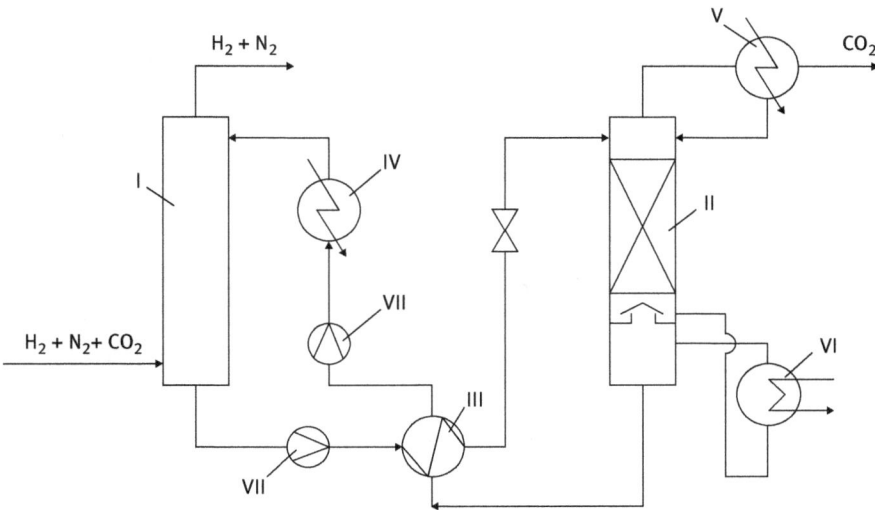

Figure 1.24: Removal of CO_2 with subsequent regeneration with monoethanolamine solutions. I, absorber; II, regenerator; III, heat exchanger; IV, cooler of the lean solution; V, cooler (condenser); VI, reboiler; VII, pumps.

One way to better utilize resources is to combine two or more processes when products from one process serve as a feedstock for another. For example, many ammonia plants also produce urea at the same site. In the synthesis of ammonia from natural gas after removal of sulphur-containing compounds from the natural gas, the latter undergoes steam reforming for generation of hydrogen (primary reformer). This is done in the excess of steam. Thereafter, during secondary reforming, air is introduced. As products, CO and CO_2 are formed in primary and secondary reforming. Removal of CO by absorption is very difficult due to its low solubility in aqueous media. Therefore, the water-gas shift reaction $CO + H_2O = CO_2 + H_2$ is conducted downstream reforming, giving extra hydrogen and forming CO_2. The solubility of the latter in potash or amine solutions is much higher than that of CO. Instead of emitting CO_2 to the atmosphere, it can be used for production of urea,

$$CO_2 + 2NH_3 = CO(NH_2)_2 + H_2O, \tag{1.10}$$

utilizing also ammonia as another substrate. These two production lines (urea and ammonia) can be linked not only by CO_2 and ammonia lines, but other links as well, making an integrated production (Figure 1.25).

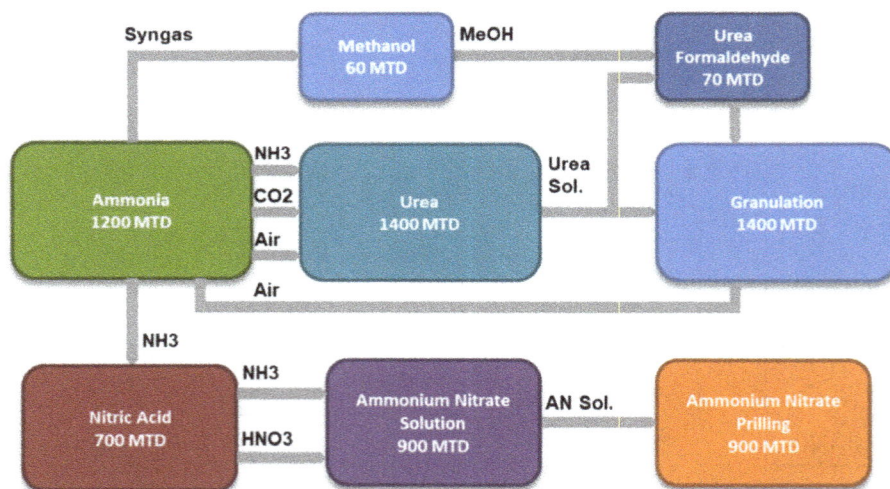

Figure 1.25: Integrated complex from natural gas. From F. Ferraria, Integrated Complexes for the Production of Ammonia, Urea, Nitric Acid and Solid Fertilizers SYMPHOS 2019 – 5th International Symposium on Innovation & Technology in the Phosphate Industry, Available at SSRN: http://dx.doi.org/10.2139/ssrn.3604237.

The concept of atom economy was mentioned above as one of the guidelines in the design of sustainable processes. This concept of atom economy or atom efficiency relates the molecular weight of the desired product by the sum of the molecular weights of all substances (Figure 1.26).

Atom efficiency: stoichiometric vs catalytic oxidation

Stoichiometric: the Jones reagent

$3PhCH(OH)CH_3 + 2CrO_3 + 3H_2SO_4 \rightarrow 3PhCOCH_3 + Cr_2(SO_4)_3 + 6H_2O$

Atom efficiency = 360/860 = 42% E_{theor} = ca.1.5

Catalytic:

$PhCH(OH)CH_3 + \frac{1}{2}O_2 \xrightarrow{\text{Catalyst}} PhCOCH_3 + H_2O$

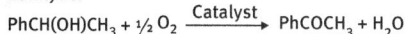

Atom efficiency = 120/138 = 87%

Byproduct: H_2O E_{theor} = ca.0.1(0)

(a)

Amount of waste/kg product:

	Product tonnage	E factor
– Bulk chemicals	10^4–10^6	< 1 – 5
– Fine chemical industry	10^2–10^4	5 – 50
– Pharmaceutical industry	10–10^3	> 50

(b)

Figure 1.26: Sustainability matrices: (a) concept of atom economy and (b) E-factor.

The atom economy concept is based on the reaction stoichiometry and does not consider solvents, other reagents, excess of some substrates, yield, etc.

Another concept, the so-called the E-factor proposed by R. Sheldon in the 1990s, is a measure of the amount of waste to the desired product. Contrary to atom economy, this factor also includes solvents. Smaller numbers indicate that less waste is produced per kilogram of product.

This concept became popular in the chemical and especially in the pharmaceutical industry. An example of process analysis using E-factor is the synthesis of benzotriol (Figure 1.27).

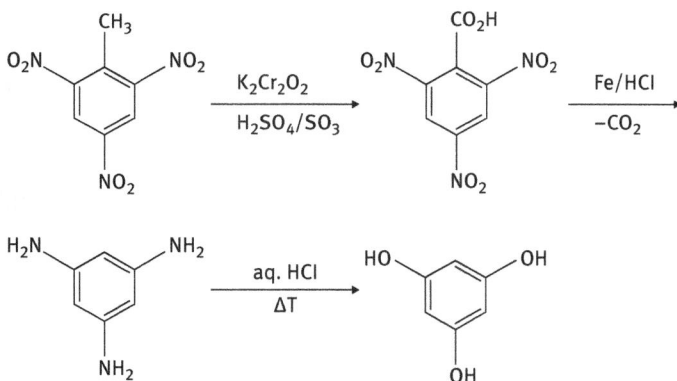

Figure 1.27: Production of benzotriol.

Production of this chemical has to be stopped since cleaning of wastes became more expensive than the product *per se.* In practice, 40 kg of solid waste $Cr_2(SO_4)_3$, NH_4Cl, $FeCl_2$, and $KHSO_4$ per 1 kg of product were generated, illustrating large values of E-factor in the production of specialty chemicals.

At the same time, it should be mentioned that such indicators as atom efficiency (atom economy) are better suited for organic synthesis and production of fine chemicals than for oil refining and petrochemistry.

Consider as an example selective oxidation processes when both air and oxygen can be used as oxidants. In general, it can be stated that application of oxygen allows the use of a much lower total pressure, which is advantageous from the viewpoint of energy consumption.

Oxidation of ethylene to ethylene oxide is performed in excess of ethylene to avoid explosive mixtures of the gases. Inlet composition of 20–40% ethene and 7% O_2 allows running the reaction above flammability limits. This, however, means that unreacted ethylene should be recycled. The oxygen-based process uses substantially pure oxygen, reduces the quantities of inert gases introduced into the cycle, and thereby results in almost complete recycling of the unreacted ethylene. The operation of the main reactor can be at much higher ethylene concentration than possible in air-based process. The high ethylene concentration improves catalyst selectivity because the per pass conversions are lower for a given ethylene oxide production.

At the same time, the drawbacks of oxygen-based processes associated with higher costs and lower process safety should not be undermined. Due to the absence of ballast (inert gases in air), knowledge of flammability limits, careful reactor design, presence of safety valves, etc. are needed to diminish explosion risks. Fluidized-bed reactors would be a much better option to control such highly exothermal reactions. This option is, for example, realized in oxychlorination of ethylene to 1, 2-dichloroethane, which is a part of vinyl chloride monomer synthesis. In the oxidation of ethylene, fixed-bed reactors are, however, still applied commercially.

The concept of atom economy being a part of "green chemistry" focuses rather on chemistry, than on technology, and thus does not consider the process from the viewpoint of sustainability or safety. Therefore, this concept, as some other green chemistry metrics (solvent recovery and reuse, use of benign solvents, etc.), being important for the pharmaceutical industry, is less relevant for oil refining or bulk chemicals production. Life cycle analysis is more valuable in the latter cases.

The reduction of energy intensity is one of the guidelines in sustainable technology design. Optimal utilization of energy can be achieved by various means including proper heat integration. Often, in order to have a certain reaction, it is important to heat the reactants, while after the reaction, the products should be cooled down, as it might be needed for better separation or other purposes. This can be done in a rational way by heating up the reactants using the exit stream from the reactor. This is done more efficiently in the case of exothermal reactions, such as hydrogenation or hydrotreating (Figure 1.28). The reactor in Figure 1.28 has several beds and can be thus considered as an example of combining several elements in one piece of equipment. In fact, such reactors can also have other elements (heat exchanges, flow distributors, etc.).

In the case above, basically different steps of the same reaction (or several reactions when reactants are injected in different places) could be combined in one

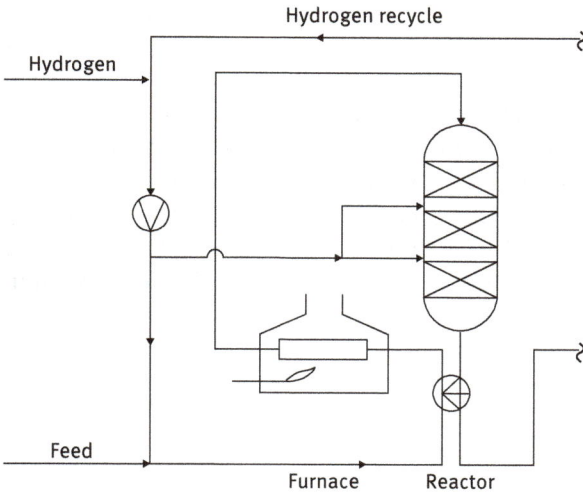

Figure 1.28: Heat integration in hydrogenation/hydrotreating.

reactor. Another option is to combine two processes in one when the second process has an impact on the first one. For example, a chemical (catalytic) reaction can be combined with separations, using distillation or absorption.

Such cases are typical when, for example, there is a need for shifting equilibrium by product removal. Let us consider an esterification reaction, when removal of water (Figure 1.29) through a membrane can drive the reaction to completion.

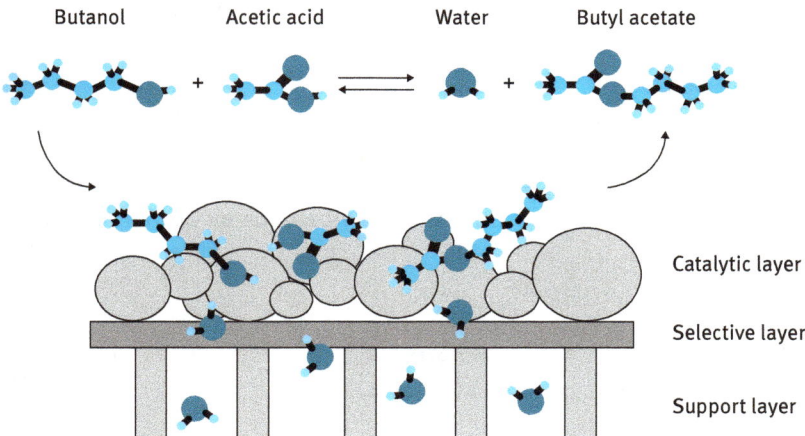

Figure 1.29: Esterification reaction with water removal. From T. A. Peters, J. van der Tuin, C. Houssin, M. A. G. Vorstman, N. E. Benes, Z. A. E. P. Vroon, A. Holmen, J. T. F. Keurentjes, *Catalysis Today*, 2005, 104, 288. Copyright Elsevier. Reproduced with permission.

Reactive distillation along with a membrane reactor could be also applied to separate products from the reaction mixture in the case of equilibrium-limited reactions such as the above-mentioned esterification. Conversion can be increased far beyond the equilibrium due to continuous removal of reaction products from the reactive zone.

Heterogeneous reactive distillations could be performed in distillation columns, illustrated in Figure 1.30. The reactor zone is the middle section containing a solid catalyst, while above and below the reaction zone, there are rectifying and stripping zones. A clear advantage of combined separation and reaction is that a single piece of equipment is used, making a considerable cost savings, as the need for additional fractionation, and reaction steps is eliminated, thus increasing conversion and the product quality.

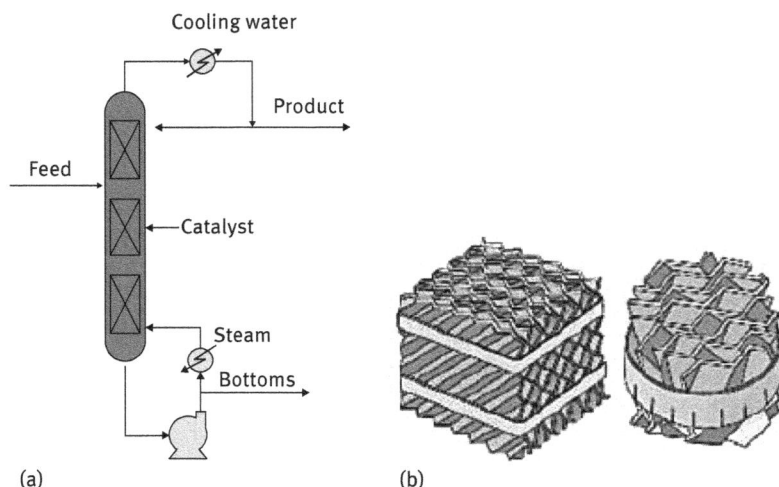

Figure 1.30: Reactive distillation: (a) general scheme (Koch Modular Process Systems, LLC. Pilot Plant Services Group, http://www.pilot-plant.com/reactions.htm) and (b) structured packing.

A similar strategy is applied in the synthesis of ethylbenzene by alkylation of benzene with ethylene:

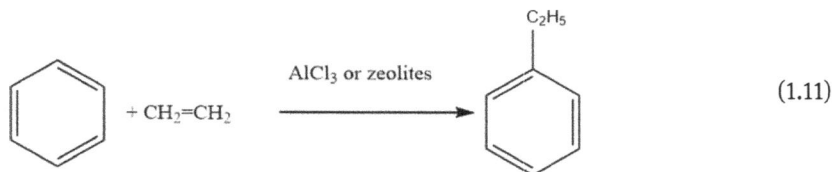

(1.11)

Benzene is fed to the top of the alkylation reactor (Figure 1.31), while ethylene is fed as a vapor below the catalytic distillation section, making a countercurrent flow of the alkylation reactants. In the catalytic distillation section, vapor-liquid equilibrium

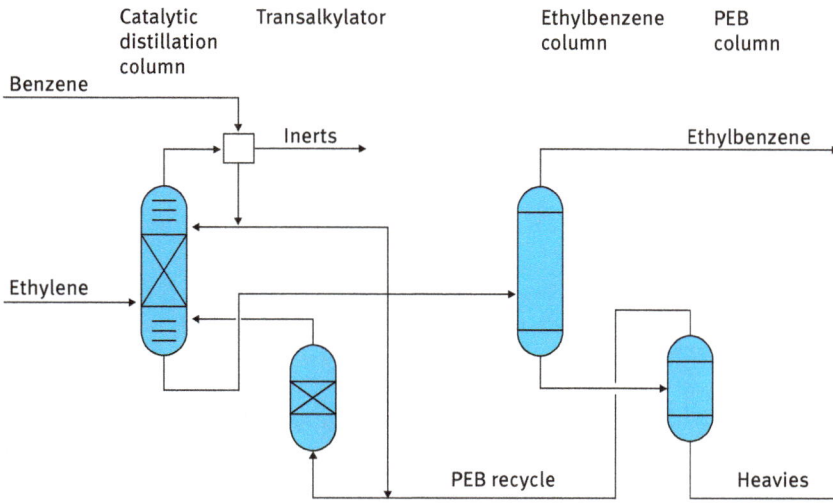

Figure 1.31: Flow scheme of CDTECH technology for ethylbenzene production. www.cdtech.com/
techProfilesPDF/CDTECHEB.pdf.

(VLE) is established with ethylene being mainly in the gas phase. The reaction heat
provides the necessary vaporization to influence distillation. The bottom of the reac-
tor/separator operates as conventional distillation columns.

The main advantages in catalytic distillation are decrease in equipment size
(lower capital costs), lower energy consumption, higher conversion and lower recycling costs, improved selectivity, breakage of azeotropes, isothermal operation, effective cooling, and efficient use of reaction heat.

The main interest in reducing energy consumption in distillation is because it is
the most energy-intensive unit operation.

Another approach for intensifying distillation is the concept of a dividing wall
column (Figure 1.32) when two columns are combined in one, which can decrease
installation and operation costs substantially and moreover improve process safety.

A vertical wall is introduced in the middle part of the column, creating a feed
and draw-off section in this part of the column. The dividing wall, which is designed to be gas- and liquid-sealed, permits the low-energy separation of the low
and high boiling fractions in the feed section. The medium boiling fraction is concentrated in the draw-off part of the dividing wall column.

The concept was introduced in ethylene oxide synthesis by ethylene oxidation.
This process often results in explosions due to that fact oxygen and ethylene can
form explosive mixtures. Introduction of the dividing wall column for ethylene
oxide processes leads to the reduction of substrate in the column, making the process inherently safer.

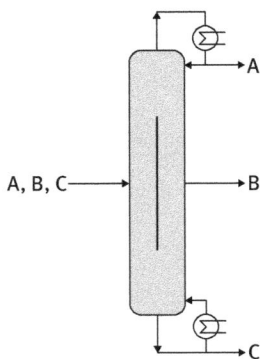

Figure 1.32: Dividing wall concept. From http://seperationtechnol ogy.com/wp-content/uploads/2012/05/44.png.

Similar to reactive distillation, when such barrier as azeotropes could be overcome, reactive crystallization processes could be applied in synthesis of pharmaceutical and agrochemical products, pigments, etc., since crystallization combined with a reaction can overcome the presence of eutectics in crystallization without any reaction.

1.4.4 Waste management

As clearly indicated above, the concept of "avoiding" pollution should be applied in the development of a process concept rather than focusing on cleaning the wastes. In this context, avoidance, reduction, and reclamation take priority over disposal by incineration and landfilling.

The product design and development can influence the environmental burdens by altering the chosen materials, technology, and manufacturing processes as well by taking the product-related actions during the product's life cycle.

In the context of the latter actions, some measures to improve waste reduction, such as good housekeeping and loss prevention, require little capital costs resulting in a high return on investment.

The careful transport and storage of the raw materials allow minimum spill during handling, which can be achieved with automatic loading and unloading facilities equipped with leakproof transport vehicles. Other measures include utilization of conveyor belts, or a proper design and utilization storage tanks (e.g. sealed) for the raw material.

Similar to avoiding spills and leaks during transport and storage, the wastes due to leaks from equipment during the process can also be minimized with appropriate practices, training of personnel, and regular inspection of equipment. Such measures should be obviously preferred over cleaning of chemical spills with adsorbents generating additional waste.

A proper segregation of hazardous from nonhazardous waste reduces the volume of the former diminishing the treatment or disposal cost. Keeping the waste streams of wastewater separated from contaminated water also contributes to the reduction of waste production. Recovery and reuse of cleaning solvents and wastewater not only decrease the environmental burden but also reduce the operational cost.

Technology changes usually require high capital cost and include addition of new unit operations or a complete replacement of outdated operations. Some examples of changes to mitigate air and water pollution are provided as follows:
1. Application of solvents with low volatility
2. Avoiding utilization of hazardous substrates and limiting the use of non-biodegradable chemicals
3. Installment of vents, scrubbing units, adsorbers, catalytic incinerators
4. Application of biological oxidation when possible

Nevertheless, a large number of different types of wastes is generated in chemical process industries (total direct and indirect greenhouse gas emissions, emissions of ozone-depleting substances; NO_x, SO_x, and other significant air emissions; water discharge, solid and liquid waste, and spills; etc.). Due to such large number and variety, it is not possible to give a general scheme of their utilization. Few technologies for waste handling will be briefly considered.

The unit operations used in handling wastes are basically the same as for the main processes, i.e., adsorption, sedimentation, filtration, distillation, extraction, crystallization, and thermal and chemical treatment.

Land filling and incineration were for a long time the main ways of handling wastes.

Incineration involves exposure of toxic and hazardous waste to a very high temperature destroying the hazardous compounds and converting them into gaseous and particulate matter. Even if, during incineration (combustion), the heat of this exothermic process can be recovered for generation of steam, incineration results in generation of char, tar due to incomplete combustion, and emission of toxic gases. The main advantage of incineration (Figure 1.33) is in its simplicity, while at the same time, during compete combustion of waste, some valuable chemicals are destroyed. Complications in combustion arise when sulphur-, chlorine-, phosphor-, or nitrogen-containing compounds are treated, since this leads to generation of HCl, sulphur, and nitrogen oxide, thus requiring gas cleaning before venting.

Widely used combustion systems include rotary kilns, fluidized bed furnaces, and multiple-hearth furnaces.

Rotary kiln (Figure 1.34) is a rotated cylindrical refractory-lined shell which is typically mounted at an angle from the horizontal level. The waste (solid or liquid) is fed from the top and moves inside the kiln. The post combustion chamber illustrated in Figure 1.34 can be added to the rotary kiln to guarantee the complete destruction of volatiles. Such incineration mode can handle numerous hazardous wastes, including, for example, chlorofluorocarbons, polyvinylchloride, or chlorinated coolant oils.

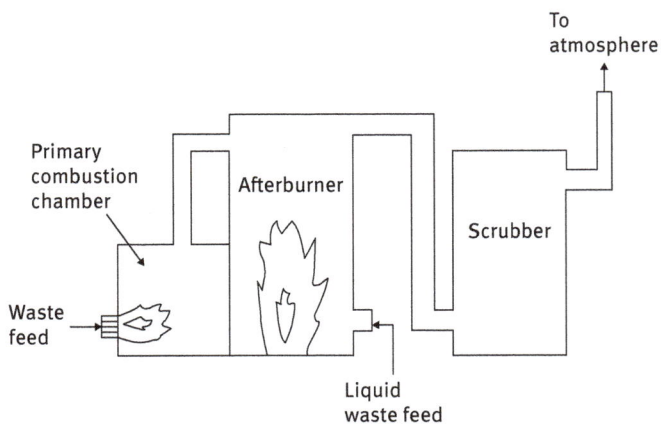

Figure 1.33: Configuration of an incinerator.

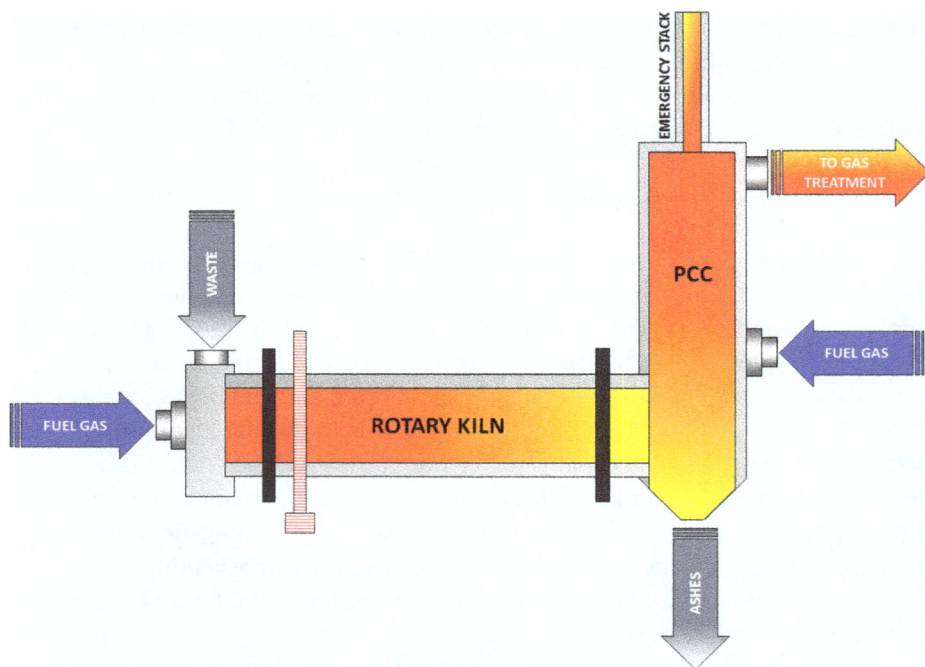

Figure 1.34: Rotary kiln incinerator with a post combustion chamber (PCC) (https://www.idreco. com/rotary-incinerators-for-industrial-wastes/).

The fluidized bed incinerators (Figure 1.35) are applied to burn finely divided solids, sludges, slurries, and liquid. There is a bed of a granular material (e.g. sand) which is suspended by air. The waste (e.g. sludge as illustrated in Figure 1.35) is conveyed into the fluidized bed and combusted into gases and ash. The posttreatment can include electrostatic precipitation and wet scrubbing before the gases are flared. The flue gas after energy recovery can undergo washing with, e.g., sodium hydroxide solution to remove and neutralize inorganic pollutants such as HCl or SO_2. Downstream selective catalytic reduction with ammonia as a reducing agent can be applied to remove NO_x. Water from the flue gas wash passes through a wastewater treatment plant in which heavy metals and dioxins are precipitated and filtered away. The remaining wastewater containing salts such as NaCl and Na_2SO_4 is treated into a biological treatment unit before the final release to the environment.

Fluidized bed incinerators have a simple design, low costs, and high combustion efficiency. Somewhat low-gas temperatures result in carbon buildup in the bed.

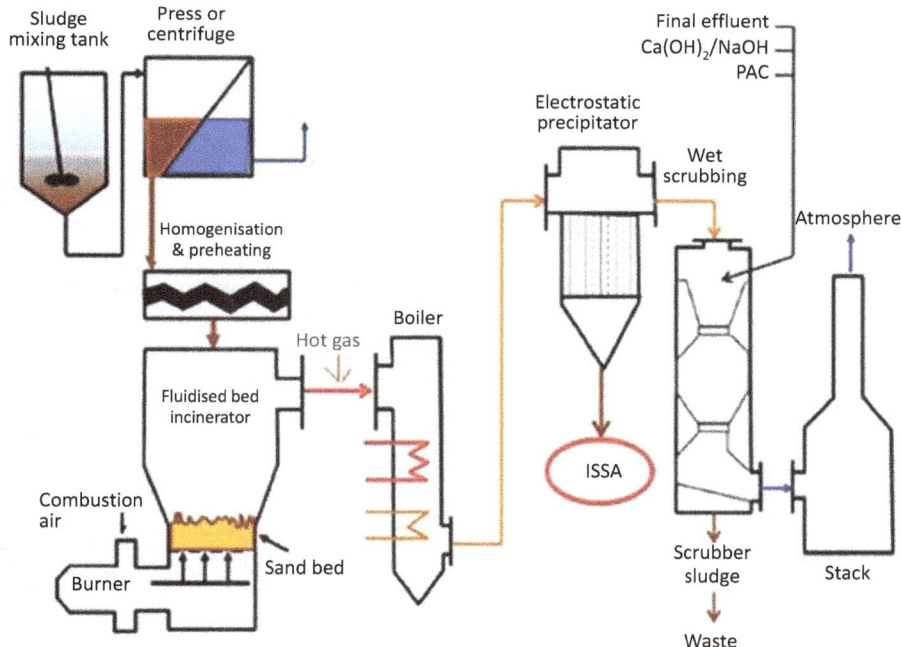

Figure 1.35: Fluid bed incinerator (from Vouk, D., Serdar, M., Nakić, D., Anić-Vučinić, A. (2016). Use of sludge generated at WWTP in the production of cement mortar and concrete, GRAĐEVINAR, 68 (3), 199–210, https://doi.org/10.14256/JCE.1374.2015).

Multiple-hearth incinerators (Figure 1.36) consist of a series of flat hearths laid in a series from bottom to top and lined with a refractory material. In the example in

Figure 1.36, the sludge waste is introduced to hearth 2, but in general the solid waste can be fed through the roof, while liquids and gases are introduced from burner nozzles. The waste falls from one hearth to another until discharged as ash at the bottom.

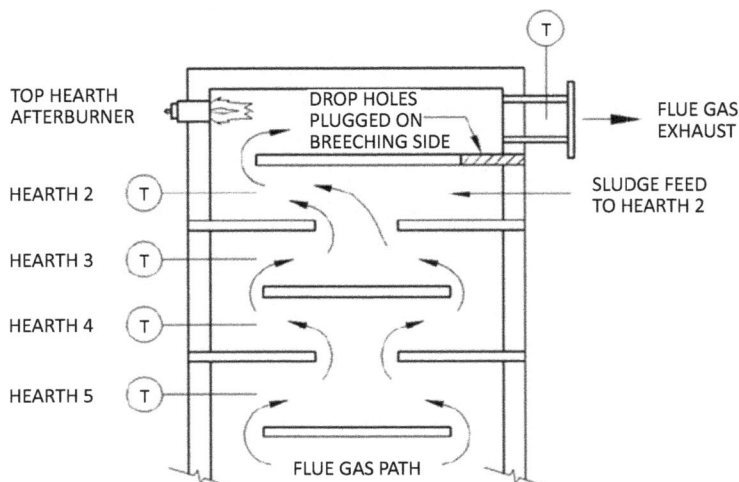

Figure 1.36: Multiple-hearth incinerator (https://www.researchgate.net/publication/228465188_Operating_Strategies_to_Reduce_Fuel_Usage_in_Multiple_Hearth_and_Fluid_Bed_Sludge_Incinerators#fullTextFileContent).

Hot flue gases are cooled, and their energy is recovered and used to raise steam before the cleaning stage.

When the gas phase contains only a small amount of impurities, catalytic oxidation (300–400 °C) is preferred over combustion/incineration (1,000 °C).

A special attention should be given to handling of solid microcrystalline or amorphous waste containing significant amounts (up to 80%) of water sludge. Such waste is produced after neutralization of liquid waste or during biochemical treatment of wastewaters. Various tars and heavy oil fractions could be also considered as sludge. Treatment of such waste includes filtration, drying, and finally combustion, giving secondary energy, which is utilized within a plant. Such utilization is important since combusting sludge (ca. 10–50% heavy oil waste) from oil refining (in some places as high as 7–10 kg per ton of oil, which translates for a refinery of 10 million t/a into 100, 000 t per year) requires extra energy.

The wastewaters from chemical process industries often contain significant amounts of oil and solids. Parallel plate separators (Figure 1.37) applied to separate the oil and suspended solids from their wastewater effluents include tilted parallel plate assemblies providing enough surface for suspended oil droplets to coalesce into larger globules.

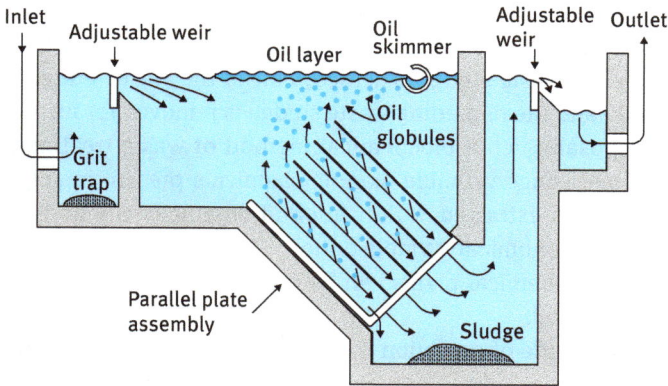

Figure 1.37: Wastewater treatment. http://en.wikipedia.org/wiki/Industrial_wastewater_treatment#mediaviewer/File:Parallel_Plate_Separator.png.

Separators presented in Figure 1.30 depend upon the specific gravity between the suspended oil and water. The suspended solids settle to the bottom of the separator as a sediment layer, the oil rises to top of the separator, and the cleansed wastewater is the middle layer between the oil layer and the solids. The oil layer is skimmed off and subsequently reprocessed or disposed, while the bottom sediment layer is removed by a scraper and a sludge pump. The water layer is further processed first for additional removal of any residual oil and then for removal of undesirable dissolved chemical compounds by biological treatment.

In an activated sludge process (Figure 1.38), which is a biochemical process, air (or oxygen) and microorganisms are used to biologically oxidize organic pollutants at 20–40 °C.

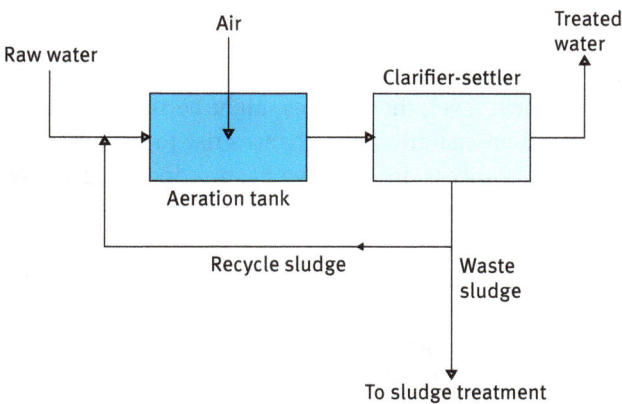

Figure 1.38: Activated sludge process. http://upload.wikimedia.org/wikipedia/commons/d/d5/Activated_Sludge_1.svg.

The flow scheme of the process consists of an aeration tank, to which air (or oxygen) is injected and thoroughly mixed into the wastewater, and a settling tank (a clarifier or settler), where the waste sludge is settled. Part of the waste sludge is recycled to the aeration tank, and the remaining waste sludge is removed for further treatment and ultimate disposal. Such a biochemical method of waste treatment might not be the most optimal, since valuable organic chemicals present in small amounts are oxidized rather than extracted. Valorization of organic compounds at the same time could be not an economically viable option. The schemes for biological treatment of wastewaters of chemical plants can be different depending on the type of the wastewater to be treated.

As an example of the magnitude of operation in wastewater treatment, the largest chemical site of BASF, in Ludwigshafen, Germany, should be mentioned, processing annually more than 90 million m^3 of industrial wastewater and an additional 20 million m^3 from the local communities, which in total correspond to a volume of wastewater for ca. 3 million people in private households.

1.4.5 Conceptual process design

After presenting process design aiming at sustainable and safe reaction technologies, it is worth to consider more conventional process design at a conceptual level. Plant design for specification products typically includes conceptual design and basic and detailed design. In conceptual design, the main steps are defined, the mass and heat balances are established, and the main process control is determined.

Such design relies on well-established procedures generated along many years and also on experience of oil, gas, chemical companies, and engineering contractors. Computer programs for design are available when gas/liquid flows and physical properties (boiling points, viscosity, etc.) could be defined through known thermodynamics.

At the initial level of R&D in the industry, a chance that a certain process will be realized might be 1–3%. At the next level, the chances might be 10–25%. In the case of a large pilot plant or even a demonstration, the chances rise to 40–60%.

A few basic rules for conceptual process design have been proposed. The raw materials and chemical reactions should be selected in such a way as to avoid, or reduce, the handling and storage of hazardous and toxic materials. Clearly, it is not always possible to follow this rule, and chemical process industries have to deal also with hazardous and toxic compounds.

An excess of one chemical reactant should be preferably used to consume completely a valuable, toxic, or hazardous chemical reactant.

A decrease in the number of operation steps (e.g. one-step dehydrogenation of *n*-butane to butadiene vs the stepwise dehydrogenation), if possible, will lead to substantial savings in capital and operational costs.

For production of nearly pure products, it is required to eliminate inert species upstream the reaction, as such separation would not need to handle, for example, a large reaction heat.

Some typical approaches to mixing, reaction, and separation are presented in Figure 1.39.

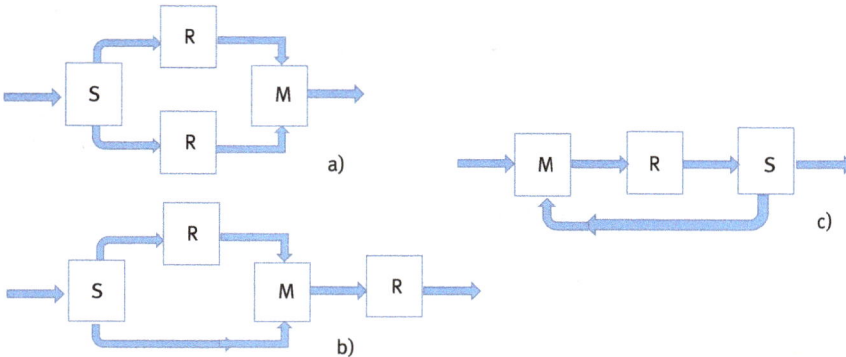

Figure 1.39: Examples of separation (S), reaction (R), and mixing (M) arrangements.

Systems with a split of the feed (Figure 1.39a) can have reactions operating at different conditions, while the bypass (Figure 1.39b) can be used for so-called quenching of the hot effluent from the reactor by the cold feed. Such arrangement increases the concentration of the reactants and at the same time decreases the inlet temperature for the subsequent catalyst bed. There are many cases in processes using adiabatic fixed-bed reactors for exothermal reactions (e.g. ammonia synthesis) when this approach is used. When the full conversion cannot be reached because of thermodynamic limitations (e.g. ammonia synthesis) or safety constraints requiring an excess of one reactant (e.g. oxidation of ethylene to avoid operation within explosion limits), recycling of the unreacted feedstock (Figure 1.39c) is done. This mode of operation also dilutes the incoming reactants obviously diminishing the average reaction rate compared to a technological system without a recycle. A decrease of the rates can be even desirable for fast reactions when there is a need to slow down the rates because of too excessive heat release or unfavorable selectivity.

Product recycling can also be arranged in cases when the product cannot easily be separated from recycled feed or formation of side unwanted products is retarded by the recycled product. Moreover, the product can act as a diluent to control the rate of reaction and/or to ensure operation outside of explosion limits.

Disadvantages related to the product recycling are larger size of the reactor and downstream equipment, larger recycle loop, and lower conversion and selectivity.

Upstream the reactor block, there could be different feed preparation blocks as, when the feed enters the process from storage, the concentration, temperature, and pressure are far from those which are required for the optimal reactor performance. An example of the heat integration by heating the feed with the product stream has been discussed above.

When there are minor species either introduced with the feed or generated during the reaction and such species in trace quantities are difficult to separate from the other chemicals, purge streams should be introduced to provide an exit. Lighter species can be removed in gas purge streams from gas recycling and heavier species can be removed in liquid purge streams. Valuable species or species that are toxic and hazardous even in small concentrations should not be purged; instead, separators to recover valuable species and reactors to eliminate toxic and hazardous species should be added. By-products generated in reversible reactions in minor quantities are typically not recovered in separators or purged, but instead recycled to extinction.

There are several rules for separations of reactions and products. Immediate separation from corrosive or hazardous components as well as reactive components or monomers should be done. Some example of such separations will be given later in the text.

Liquid mixtures should be separated using flash separation, distillation, stripping, enhanced distillation (extractive, azeotropic, and reactive), liquid-liquid extraction, crystallizers, and/or adsorption. Vapors are condensed with cooling water. An example of separation when the reactor exit is vapor is given in Figure 1.40.

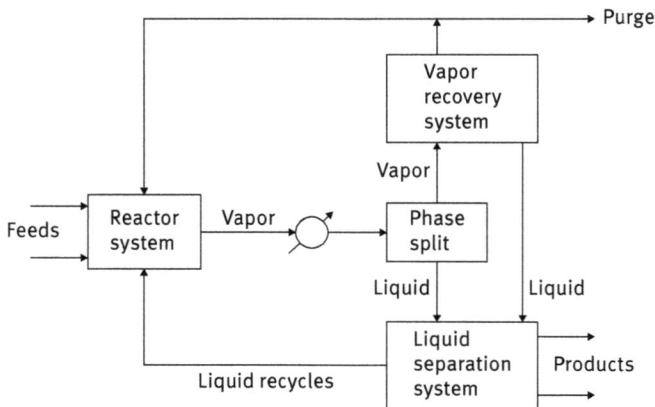

Figure 1.40: Separation of products with recycling when the reactor exit is vapor. Reproduced with permission from J. M. Douglas, *AIChE Journal*, 31, (1985) 353–362. Copyright © 1985 American Institute of Chemical Engineers. Reproduced with permission.

Distillation is usually considered as a first choice for separation of fluids when purity of both products is required. Typically, removal of the most plentiful or the lightest compounds is done first, while a high recovery or difficult separation of compounds with close boiling points is done last. Since the separation of such compounds in the latter case requires columns with many trays or excessive packing elements, it is better to do a preliminary removal of other compounds to get the minimal flow. An alternative to distillation might be required if the boiling points are very close, leading to unrealistically high distillation columns.

Some of these heuristic rules might be in contradiction with a particular process; thus, there could be some alterations not completely consistent with the approach above. For example, separation can be done based on the order of boiling points minimizing heat input.

For reversible reactions, when there is a need to drive the reaction to the right, separation can be done together with the reaction leading to a very different distribution of chemicals. Thus, reactive extraction or reactive distillation can be applied being beneficial not only from the reaction viewpoint but also for separation *per se* overcoming limitations set by phase equilibrium diagrams.

Gas absorption is applied to remove one trace component from a gas stream. Pressure swing adsorption to purify gas streams can be considered as an option, especially when one of the components has a cryogenic boiling point. Membranes can be used to separate gases of cryogenic boiling point and relatively low flow rates.

Extraction is considered as a choice to purify a liquid from another liquid, while crystallization is used to separate two solids or to purify a solid from a liquid solution. Concentration of a solution or a solid in a liquid can be done by evaporation, or in the latter case, by centrifugation. Removal of solids from a liquid is done by filtration. Separation of solids of different sizes or density can be done by screening or flotation, respectively. Solids from a solid mixture can be also removed through selective leaching. Reverse osmosis can be applied to purify a liquid from a solution of dissolved solids.

There are also several rules for efficient heat management. Removal of heat from a highly exothermic or endothermal reaction can be done using an excess of the reactant (typical in exothermal hydrogenations) or an inert diluent. Quenching by cold or hot shots is done for exothermal and endothermal reactions, respectively. For less exothermic or endothermic reactions, external heat exchangers (coolers or heaters, respectively), jacketed vessels or cooling (heating, respectively) coils can be applied. Another option includes heat exchangers (for cooling or heating) between adiabatic reaction stages when the total catalyst loading is separated in several fixed beds.

Heating as such can be done by primary or secondary energy sources. The second option does not diminish energy heat consumption in a particular unit but overall leads to more economical heat utilization in the whole plant. The outlet gases from a reactor can be are used not only for heating of the reactant but also in the boilers of distillation columns.

An important issue in process design is reactor selection. An economical option is to use fixed adiabatic beds when the temperature rise corresponds to the conversion.

In the case of adiabatic fixed-bed reactors, when too high temperatures should be avoided, several beds are applied with interbed cooling either using heat exchanges or quenching with cold reactants as already discussed above. Synthesis of ammonia or hydrotreating of various streams in oil refining are typical examples of this approach.

In very strongly exothermic or endothermic reactions, too many beds would be required in order to control temperature rise, thus hundreds (for benzene hydrogenation, or methane steam reforming) or thousands of tubes filled with solid catalysts (ethylene oxide, or phthalic anhydride synthesis) are arranged parallel to each other with cooling or heating in between the tubes (Figure 1.41). Such approach, due to better temperature control, might prevent excessive deactivation and/or is needed to improve selectivity.

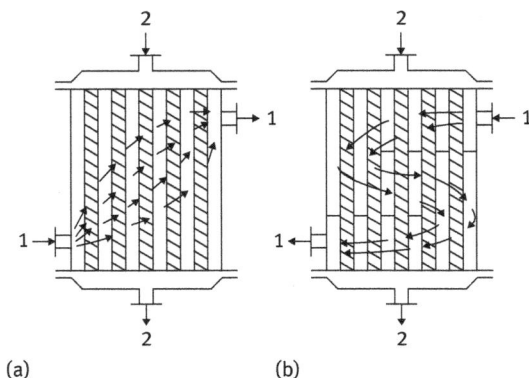

(a)   (b)

Figure 1.41: Different arrangements of heat media circulation in between the tubes: 1, heat medium; 2, reactants. (After A. S. Noskov, *Industrial Catalysis in Lectures*, Moscow, Kalvis, 2006).

Pressure can also be an important technological parameter and varies depending on a particular reaction. Large-scale production facilities typically operate at low and medium pressures (0.7–50 MPa), while few processes (e.g. ethylene polymerization) require much higher pressure, reaching even 300 MPa.

As a general rule, pressures between 0.1 and 1 MPa and temperatures between 40 and 260 °C do not cause severe processing difficulties. One of the reasons regarding the pressure conditions is the ability of most chemical processing equipment to withstand pressures up to 1 MPa. At larger pressures, additional capital investment is required for more expensive equipment with thicker walls. Operation under vacuum makes the equipment larger requiring special construction and elevating the costs. Moreover, potential leakage of air into the process is difficult to avoid.

Among the reasons of using elevated pressures, the following can be mentioned: increase of the reaction rates, maintaining the liquid phase, the shift in chemical equilibria or phase equilibria (e.g. ammonia synthesis or absorption, respectively), and a possibility to operate at lower temperature, which is important for reactions with low thermal stability of reactants or for processes when temperature elevation leads to excessive catalyst deactivation. One example of the latter case can be methanol synthesis over copper catalysts.

Catalyst deactivation in fact is an important issue because it often determines the type of reactors that are utilized in the industry. For example, for gas-solid catalytic reactions, if deactivation is not that profound (months to years), packed bed of catalysts can be used.

Poisons present in the feed can be removed if necessary by installing guard beds, which can be done either by using a separate adsorbent or oversizing the catalyst. In the latter case, this additional volume of catalyst is used to adsorb impurities.

Usually, if the catalyst lifetime is sufficiently long (several years), no regeneration is done and the catalyst is simply removed when it is considered uneconomical to continue with such catalysts. As an example, synthesis of ammonia could be mentioned, when the lifetime of catalysts is usually 14–15 years. In fact, since impurities do not influence the catalyst performance in this case and no carbon deposition occurs, the lifetime could be even longer. However, ammonia synthesis requires utilization of high pressure; therefore, according to local legislation, reactor vessels should be regularly inspected. For this reason, after 14–15 years, the catalyst charges are unloaded from the reactor. Other catalysts in the same ammonia train, such as catalysts for natural gas, primary and secondary reforming, high- and low temperature shift can operate for several (2–6) years without any regeneration. An interesting example in the same process is hydrodesulphurization (HDS), containing the so-called NiMo or Co-Mo catalysts. The lifetime of such catalysts depends heavily on the presence of sulphur in natural gas.

Note that the example of HDS demonstrates a case when in order to remove impurities in the feed (e.g., mercaptanes in natural gas), it is not sufficient to use just a bed with adsorbent or install more volume of the catalyst. In fact, a separate reactor is used upstream the main one, where steam reforming of natural gas on supported nickel catalysts is performed. Mercaptanes should be removed since they are poisons for nickel. In an HDS reactor, the following reaction occurs:

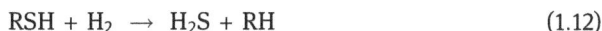

$$RSH + H_2 \rightarrow H_2S + RH \tag{1.12}$$

Additional hydrocarbons formed during this reaction are transformed along with methane to hydrogen, being exposed to steam over nickel catalysts. H_2S should be, however, removed upstream steam reforming, and this is done by putting it in contact with zinc oxide in a separate reactor, which reacts (non-catalytically) to zinc sulphide. The latter is discharged when all the zinc oxide is consumed.

After several years in operation, the activity of catalysts in adiabatic or isothermal fixed-bed reactors could decline. An engineering practice to compensate for activity losses during industrial operation is to increase temperature (Figure 1.42).

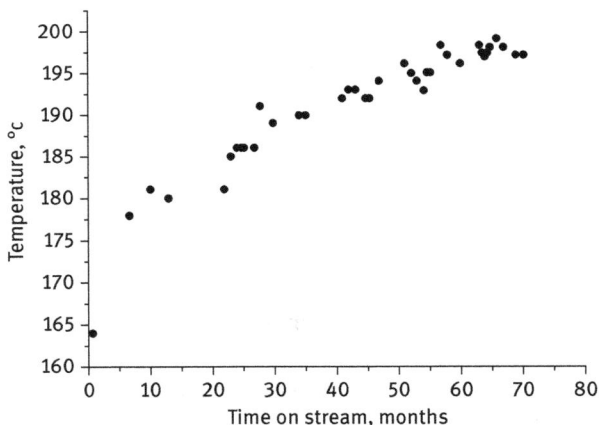

Figure 1.42: Compensation of activity losses by temperature increase.

Although the policy of temperature increase can allow the constant level of production output, the obvious drawback with this operation policy is that with temperature increase, side reactions are becoming more prominent, further deteriorating catalyst performance.

It is, however, possible that deactivation increases too strongly above a certain temperature. For example, copper catalysts are very sensitive to sintering, which prevents temperature increase during methanol synthesis over copper-containing catalysts. In such cases, the pressure could be gradually increased to compensate for the activity decline with operation time.

When deactivation is more profound because of coking and carbon deposition, it becomes necessary to regenerate the catalysts after several months. Fixed-bed reactors could be applied and regeneration (by coke burning) is done while the reactor is off-line. Since regeneration should be done very carefully in order to prevent, for example, catalyst sintering, this regeneration process could be rather time consuming. Consider as an example a hydrogenation process when the catalyst deactivates. As direct exposure of the catalyst to air (or oxygen) is very dangerous and can lead to explosions, because the catalyst can still contain some hydrogen, the reactor should be first purged with an inert gas; thereafter, the coke from the catalyst should be carefully oxidized, controlling the amount of oxygen in the feed. If it is not properly done, heat released during a highly exothermic coke oxidation can promote catalyst sintering and irreversible losses of catalyst activity. After the burning of coke and subsequent purging with an inert gas, the catalyst should be again activated.

These lengthy procedures can significantly influence production output; thus, typically, an additional reactor is installed in parallel. This allows the first reactor to be isolated and regenerated as the feedstock is rerouted to the second reactor, allowing the plant to operate continuously. After regeneration, the first reactor remains in a stand-by mode.

If the catalyst is active for several days or weeks, an option is to utilize a moving bed reactor with continuous catalyst regeneration. In a continuous process, a catalyst flows through the reactors in series, and this will be explained in more detail in the section devoted to catalytic reforming. The spent catalyst is continuously removed from the last reactor and transferred to the regeneration section, where it is regenerated in a controlled way and transferred back to the first reactor. This mode of operation with frequent regeneration allows to operate continuous catalytic reforming in more severe conditions than in a fixed-bed alternative.

Deactivation could be even more severe as in the case of fluid catalytic cracking when there is a continuous flow of the deactivated catalyst to a regenerator and a flow of the regenerated catalyst back to the riser reactor.

Examples of catalysts deactivation will be given in subsequent chapters covering a range of reactions with different deactivation time scale, from seconds to days, weeks, and even years.

If deactivation is not by coking but by irreversible poisoning, catalyst regeneration is not an option. In those cases, the technological scheme should include a purification section, as exemplified above for hydrodesulphurization of natural gas prior to steam reforming of methane.

Other issues that should be considered while selecting a reactor are injection and dispersion strategies. Reactants can be introduced in a one-shot mode, as in batch reactors, or in a step-function mode, as in continuous reactors. Staged injection is an intermediate case and is applied in semi-batch reactors. Another option is to apply pulsed feed, as in flow reversal type of reactors or semi-batch ones. Energy supply could be also envisaged in many ways. In adiabatic reactors, it can be done through quenching by the cold reactant or introducing intermediate heat exchangers. More details will be given in Chapter 3.

Application of fluidized-bed reactors can be an option for exothermal reactions, such as selective (or partial) oxidation of alkanes, i.e., n-butane to maleic anhydride (discussed in Chapter 9). In this particular case of butane oxidation, there is also a possibility to utilize a transport reactor. The advantage of such system, when a metal oxide oxidation (in this case V_2O_5) is changing during the reaction (from V^{+5} to V^{+4}), is that donation of oxygen from the catalyst lattice to the substrate with subsequent reduction of V^{+5} to V^{+4} is separated from oxidation of V^{+4} to V^{+5}. The latter process is conducted in a separate reactor, which prevents butane from being in contact with air, making the process much safer.

A number of oxidation reactions are conducted in a batch mode in slurry reactors. For such reactors, several options for energy removal could be imagined, including evaporative cooling, an external heat exchanger, or arranging heating through a double jacket or internal coils (Figure 1.43).

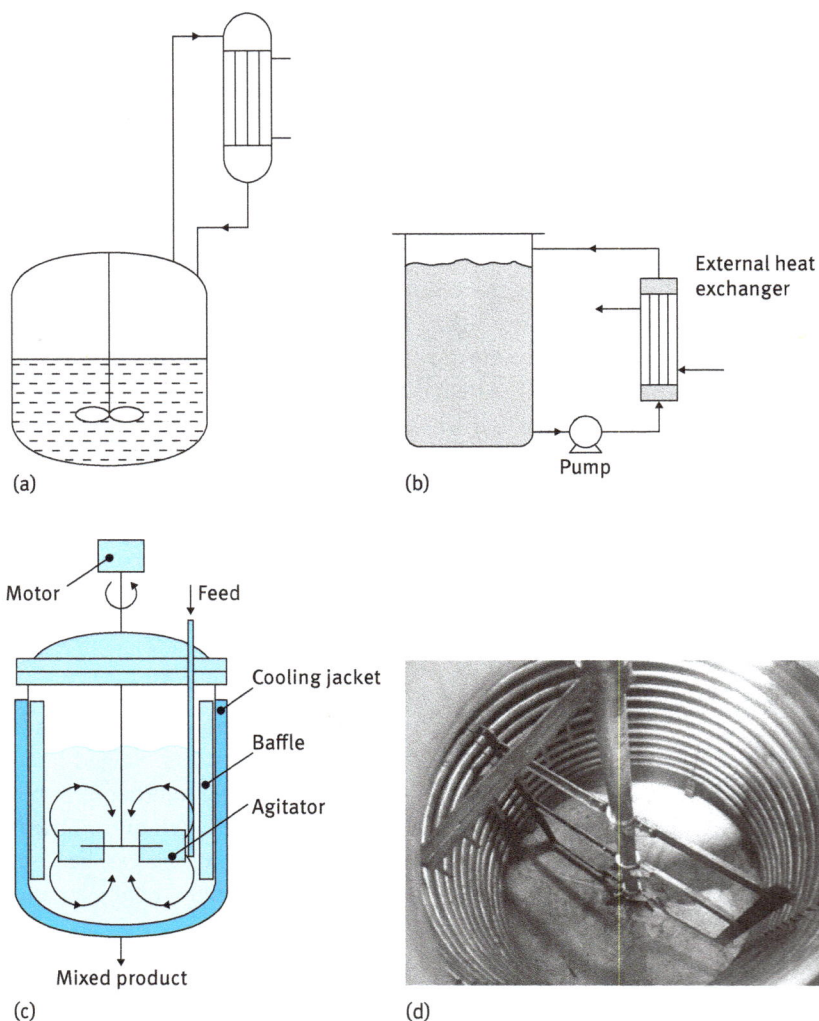

Figure 1.43: Slurry reactor with (a) evaporative cooling, (b) external heat exchanger, (c) double jacket (http://en.wikipedia.org/wiki/Continuous_stirred-tank_reactor) and (d) internal coil (http://pharmachemicalequipment.com/?page_id=2251).

Ways of energy input could be also different. In addition to the ones mentioned above, energy removal could be arranged through programmed temperature cooling.

From the viewpoint of concentration profiles, it could be more attractive to have a continuous plug flow reactor with no mixing of reactants. On the other hand, such type of arrangements for exothermal reactions leads to hot spots (i.e., spots with temperature much higher than in other places along the reactor length), which not only determine conversion and selectivity, but also catalyst lifetime, reactor materials, and safety of the whole process.

A specific way of energy removal could be to load catalysts with different activities along the tube. In such way, the amount of the active phase on a support can be profiled along the tube length. This counterbalances the excessive temperature increase and hot spots by deliberately minimizing the activity of the catalyst layer close to the reactor inlet. Another possibility is to keep the amount of metal or metal oxide on a support the same, but dilute the catalyst with the support at a different ratio changing along the reactor length (Figure 1.44).

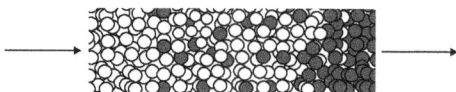

Figure 1.44: Catalyst profiling along the bed.

During reactor selection, a decision should also be made for a particular technology if *in situ* product removal will be beneficial for the process.

There are examples of reactions driven by equilibrium when, in order to shift equilibrium, the products have to be withdrawn. Typical examples are ammonia synthesis or production of sulphuric acid by oxidation of sulphur dioxide to sulphur trioxide over vanadium pentoxide catalysts.

In addition to the reactor selection principles discussed above, other requirements should be carefully considered, such as productivity, product and reactor costs, easiness of construction, delivery to a production site, and maintenance and operation including safety.

1.4.6 Process control (compiled together with Dr. Eugene Mourzine, University of Akron)

The goal of process control is to monitor and adjust various variables (e.g. flow, temperature, and pressure) during manufacturing, maintaining a relevant output variable within a desired range. To this end, a sensor is needed to measure a certain parameter (e.g. temperature) while the controller defines if the control element needs to be turned on or off.

Minimization of process variability and a control of the set point typically has economic benefits by reducing the operating costs. Moreover, a complex nature of chemical processing implies that changes in one process can cause variations downstream and even upstream. Moreover, manufacturing under a strict control of process variables can mitigate serious risks associated with production of chemicals, such as explosions, release of toxic chemicals, and fires.

Operating a complex chemical process thus requires that a large number of variables stays at their desired values. Process control involves setting the values of process variables to ensure the set point values. The process is monitored, and in case of deviations, the values of process variables are manipulated in such a way that the set point is restored.

The control loop thus contains the measurement sensors, transmitters of the measured values to some computerized devices for processing, a programmable logic controller to decide on the response to alterations of process conditions, and the magnitude of the process changes. The latter is needed to return the process to the desired (set point) values. The control output needs to be transmitted to an element that can change the value for the process variable.

From the viewpoint of process control, a reduction of process complexity is required as otherwise the complex nature of chemical processes with multiple steps can lead to complex solutions, which are difficult to design, implement, and maintain solutions. Modularization or decomposition has an obvious advantage because it is easier to maintain a smaller component than a large one and moreover already available process control solutions can be used for separately functioning modules.

In practice, in a chemical plant such process decomposition implies less interdependency of different units and mitigation of the effects that disturbances entering one module can bring to other modules. An example of a module boundary can be a storage tank between processing steps.

Typically, process control is needed to maintain the process at a desired operating condition, either at a constant value (e.g., pressure, temperature, flow rate, etc.) or at the desired change rate (e.g. heating or cooling). Another goal of the process control is to minimize the operating costs satisfying such process constraints as the product quality, production capacity, and the equipment operating ranges.

The process controlled variables and their range should be selected based on the understanding of the process, equipment constraints, and the source of possible disturbances. Typically, a few process variables, which should be reliably measured, are controlled. If direct measurements are not possible or are not unreliable, some other variables should be selected.

It is recommended to measure the flow rate of all streams, pressure, and temperature whenever there is energy exchange in an unit operation. Measurements of compositions are usually slow, expensive, being also of off-line character, and therefore should be used only when necessary, especially for maintaining the product quality. Some other physical quantities including viscosity, density, pH, conductivity, and stirring speed may be required for control or monitoring in some processes.

An adequate process knowledge is needed to develop the best control strategy and to determine which process variable(s) need to be adjusted to control the overall process.

Control valves should be placed, e.g., on the inlet feed streams, all utility streams entering the process, the purge and make-up streams, as well as on flows leaving vessels or reflux and distillate flows. Control of gas streams with very large flow rates or at extremely high temperatures and control solid streams should be avoided. Not surprisingly, handling hazardous process streams or critical unit operations requires additional control valves.

In a chemical process, a deviation can occur between the desired set point and the actual (measured) value of a process variable. Reasons could be disturbance inputs, process parameter variations, or imperfect modeling.

The controller structure and controller tuning parameters are designed to regulate the process by rejecting local disturbances and stabilizing the process and to ensure the production rate and the product quality. These local disturbances including fast changes in the process streams, pressure, inlet cooling water temperature, etc. are preferably rejected mitigating their negative influence on other process variables.

Temperature in general is controlled by adjusting heat input/removal with the steam or cooling water rate. In special cases, quenching with a coolant or manipulating the catalyst addition/withdrawal can also be used for the temperature control. Composition at the reactor outlet or of a distillation column is controlled by the operating conditions. In the latter cases, the distillate or the reflux rates can be adjusted. Typically, conditions in a chemical reactor are rather strictly controlled, as changes in the operation conditions can influence substantially the downstream processing. A case-by-case decision is made how fast a controller should react to changes in the variables. Too fast responses can lead to disturbances downstream, while if a response is too slow the product can be of poor quality.

A schematics of the feedback control with a closed loop shown in Figure 1.45 illustrates the measurement of controlled variable and the feedback to the controller through a loop without human intervention.

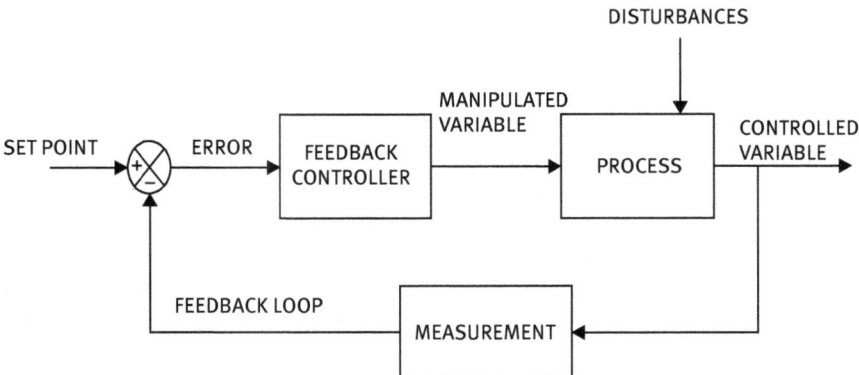

Figure 1.45: Feedback control.

The feedback loop verifies how well the process responded to the past manipulations. If the past adjustments do not produce desired outcome, the feedback control makes further adjustments until the controlled variable matches the set point. Such type of control takes corrective action only after there is a deviation between the controlled variable and the set point.

The most common and widely accepted feedback control algorithm is the proportional–integral–derivative control, which is a sum of proportional, integral, and derivative components. The proportional component depends on the difference between the set point and the measured variable, while the integral component is the sum of the error over time providing the accumulated offset that should have been previously corrected. The contribution from the integral term is proportional to both the magnitude of the error and the duration of the error. The final contribution comes from the derivative component, which is proportional to the rate of change of the process variable. Through differentiation the future behavior of the system can be anticipated.

An example of the alternative approach is the feedforward control of the open-loop type (Figure 1.46). It measures a disturbance, predicts the impact on the process, and manipulates the value of a process variable to eliminate the impact of the disturbance on the control variable. Such predictive control method does not validate the impact of variable manipulation or react to the error as the feedback control, but rather predicts the value of the output variable based on the values of measured input variables and thus should be based on a sound mathematical model of the process.

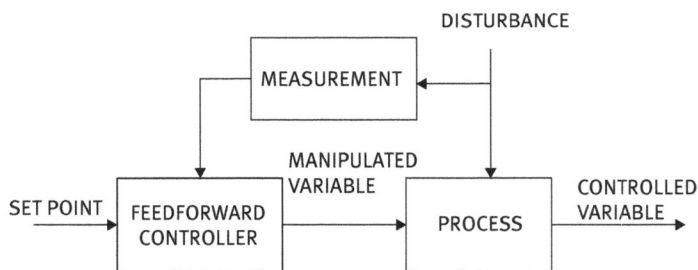

Figure 1.46: Feedforward control.

Among multiple limitations in the practical application of the feedforward control is the inability to handle unexpected or unmeasured disturbances. Moreover, such type of control requires accurate measurements and the perfect process model design.

Despite apparent limitations, feedback and feedforward control are widely used in industrial practice for system control.

More recently, predictive control has become the solution of choice for advanced multivariable control applications. It uses the current inputs and output measurements, the current state of the process and the process model and optimizes within the defined boundaries, and a finite decision limits a future control action. An analog is a walking in a darkroom when after sensing the surroundings, the decision about the best path into the direction of the goal is taken. After a single step, the surroundings are, however, reassessed and the next decision is taken. In essence, such control type includes the predictive model, optimization in a defined temporal window, and feedback correction. The feedback loop is essential because the measurements are used to update the optimization problem for the next time step.

1.4.7 Product design

Process design discussed in the previous section, or design of specification products, mainly focuses on optimization *versus* cost including many other factors such as safety, feedstock availability, handling of waste, liability, etc. The product purity specification is defined prior to design. Specification products (such as nitric acid, methylamine, sulphuric acid, and thousands of similar chemicals) are produced in different parts of the world by many companies with similar product purity specifications.

Process design for performance products focuses on the way they are produced (batch *versus* continuous); addresses various strategies for inputs and outputs; selects the type of reactors used, how recycles are organized, and how separation and heat integration are implemented.

Contrary to specification products, there are also performance products, such as cosmetics, detergents, surfactants, bitumen, adhesives, lubricants, textiles, inks, paints, paper, rubber, plastic composites, pharmaceuticals, drugs, foods, agrochemicals, and many more. Obviously, customers do not look for the cheapest alternative, but more on the performance. As a consequence, such product design is focuses on making a particular chemical of the desired performance rather than on making it in the most economical way. Specific performance is considered, which include many parameters such as color, taste, stability, etc. These parameters might not be even needed for the targeted performance but are required by the customers who might be eager to pay a premium price. In some instances (like pharmaceuticals), many customers with life-threatening diseases are certainly able to afford buying a more expensive product than it could be with a more rational production route. The differences in design for bulk (specification) and specialty (performance) chemicals are illustrated in Table 1.2.

Performance products business is thus much closer to the consumer market than synthesis of specification products. Product design needs to take into account physical chemistry and interfacial engineering and is often related to health, pharmaceutical, and medical sciences. Not only individual substances, but molecular systems are tailored to meet specific (end-use) properties. Typically, 4 to 50 components (i.e., molecules) can be found in a formulation or grade.

Table 1.2: Differences in process development for bulk and specialty chemicals.

Feature	Bulk chemicals	Specialty chemicals
Product life cycle	Long (>30 years)	Short (<10 years)
Focus of R&D	Driven by cost and environment Process improvement (modified or new technology)	Driven by yield Product improvement
Competing processes	One route usually the best	Competitors may use a variety of routes
Route selection		
Number of feasible routes	Few	Many, possibly hundreds
Number of feasible feedstock	Few	Many
Number of reaction steps	Usually one or two	Several (from 3 to 20)
Effort as a proportion of total R&D	Low	High
Impact of poor route selection	High	Moderate
Patent protection	On processes or technologies	On chemistry, often to block competitors' use of similar routes
Technology	Usually continuous	Batch and continuous
Scale-up	Pilot plants and simulation tools commonly used	Use of rules of thumb prevails

From J. A. Moulijn, M. Makkee, A E. van Diepen, *Chemical Process Technology*, 2013, 2nd Ed. Copyright © 2013, John Wiley and Sons. Reproduced with permission from Wiley.

Two examples will be given below. In a functional food product, Benecol developed in Finland in 1990s an active ingredient is sitostanol ester, which is obtained by chemical transformations of sitosterol (Figure 1.47), a by-product of pulping.

The structure of sitostanol resembles cholesterol; thus, the mode of action of Benecol is related to lowering the cholesterol level of human serum. The latter is considered as one of the main risk factors for heart and coronary diseases. Sitostanol obtained by hydrogenation of sitosterol has a larger effect than sitosterol has; however, poor solubility limits its direct utilization in vegetable fat products such as margarine or yogurt. Sitostanol esters in turn are fat-soluble and can be mixed

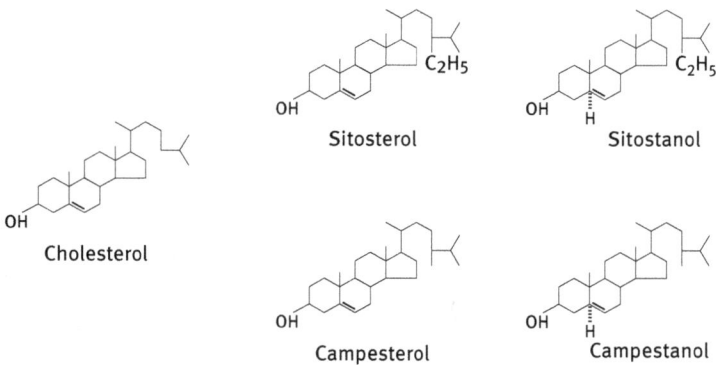

Figure 1.47: Structure of cholesterol and plant sterols.

easily with vegetable fats. Production of Benecol involves hydrogenation of a complex mixture of sterols available as a by-product of pulping followed by esterification of the product mixture with fatty acids (60% oleic acid). This technology gives a product which can easily be incorporated into a variety of foods. Obviously, prior to introducing Benecol to consumer market, extensive medical testing of the products had been conducted.

Another example of the performance products is laundry detergents (Table 1.3), which include not only the surfactant *per se* but also a number of additives, such as builders (agents that enhance the detergent action), anti-redeposition agents, brighteners and bleaches, cosurfactants, biocides, and sales appeal ingredients (e.g. fragrances).

To be an acceptable detergent, a surfactant, besides being a good wetting agent and economically competitive, should be able to displace soil into the washing fluid and act as an anti-redeposition agent. Other requirements include low sensitivity to water hardness, high solubility, neutral odor and color, low toxicity, and desired foaming.

Detergency builders are used to enhance performance of the detergent by raising pH and complexing Ca^{2+} and Mg^{2+} to avoid their interference with surfactants. Builders should satisfy a number of requirements including eye irritation, oral toxicity, compatibility with bleaching agents, and alkalinity. Zeolites, such as the most often used small-pore zeolite A, act as ion-exchange insoluble builders forming insoluble complexes with Ca^{2+} and Mg^{2+}. Moreover, they provide adsorption of water-soluble substances (e.g. dyes), heterocoagulation of pigments and solid fats, and even act as crystallization nuclei for sparingly soluble salts.

Bleachers in laundry detergents based mainly on various peroxides (e.g. inorganic peroxides and peroxohydrates) generate hydrogen peroxide anion in alkaline media, which oxidizes bleachable soils and stains. Excellent reactivity of hypochlorite bleaches allows, on the one hand, their application even at very low temperature,

Table 1.3: Detergents. Common ingredients are in italics (from D.Murzin, Chemial Product Technology, 2020, De Gruyter).

Purpose	Ingredients	Ingredients
Soil-capturing and soil-buffering agent	Monoethanolamine citrate	Tetrasodium ethylenediamine tetraacetic acid
Solvent	Diethylene glycol	
Soil-release or soil-capturing polymer	Polyethyleneimine propoxyethoxylate	
Emulsifying and dispersing agent	Sodium cumene sulfonate	
Optical brightener	Disodium diaminostilbene disulfonate	Disodium distyryl biphenyl disulfonate
Chelating agent	Diethylenetriaminepentaacetic acid	Trisodium ethylenediamine disuccinate
Builder and chelating agent		Sodium citrate
Breaking down polysaccharide-based soils and stains	Glucanase	
Breaking down fat-based soils and stains		Lipase
Removal of soil from cotton		Cellulase
Enzyme stabilizer	Calcium formate	
Foam regulator		Alcohol ethoxylate
Process aid	*Water*	*Water*
Defoaming agent	*Polydimethylsiloxane (dimethicone)*	*Polydimethylsiloxane (dimethicone)*
Traditional surfactant	*Sodium fatty acids*	*Sodium fatty acids*
Premium surfactant for hard water	*Alcohol ethoxysulfate*	*Alcohol ethoxysulfate*
Low-cost general-purpose surfactant	*Alkylbenzene sulfonate*	*Alkylbenzene sulfonate*
Soil-release or soil-capturing polymer	*Polyethylenimine ethoxylate*	*Polyethylenimine ethoxylate*

Table 1.3 (continued)

Purpose	Ingredients	Ingredients
Processing aid	Ethanol	Ethanol
Solvent and enzyme stabilizer	Propylene glycol	Propylene glycol
Soil-capturing agent and enzyme stabilizer	Sodium borate	Sodium borate
Breaking down protein-based soils and stains	Protease	Protease
Breaking down starch-based soils and stains	Amylase	Amylase
Breaking down guar gums	Mannanase	Mannanase
Balances electrolytes	Sodium formate	Sodium formate
Colorant	Blue dye	Blue dye

while on the other hand, textile dyes and most fluorescent whitening agents exhibit poor stability in the presence of chlorine.

Bleach activators are utilized to improve performance of less reactive sodium perborate and sodium percarbonate at temperatures below 60 °C. Acylating agents used as bleach activators at pH 9–12 preferentially react with hydrogen peroxide forming organic peroxy acids. These acids possess higher reactivity than hydrogen peroxide, improving thereby low-temperature bleaching properties.

Laundry detergents contain also antimicrobial agents (e.g. quaternary ammonium chlorides and alcohols), which either kill or inhibit the growth of microorganisms (bacteria, fungi, and viruses).

Sodium silicate is included in modern detergents as a corrosion inhibitor protecting metallic surfaces by deposition of a thin layer of colloidal silicate.

Fragrances are added in low quantities (<1%) to detergents providing a pleasant odor, as well as masking certain unpleasant odors.

Removal of various types of stain from proteins, fat, or carbohydrates (starch) can be done with the aid of enzymes (respectively proteases, lipases, and amylases), which also provide color and fabric care.

Optical brighteners enhance the light reflected from the fabric surface making fabrics, yellowing naturally with time, look brighter. Such brightening is done by converting UV into visible blue-violet light, which by interacting with the yellow

light emitted by the fabric produces white light. Optical brighteners for paper, textiles, and detergents are predominantly di- and tetra-sulfonated triazole–stilbenes
and a di-sulfonated stilbene–biphenyl derivatives.

Other ingredients of laundry detergents include preservatives to prevent spoil-
age during storage (e.g, glutaraldehyde), hydrotropes to keep the pouring characteristics, various processing aids, as well as foam regulators.

Product design and engineering deal with mostly complex media and particulate solids. Control of the end-use property and quality features, such as taste, feel,
smell, color, handling properties, ability to sinter, or biocompatibility, is important.
Complex media such as non-Newtonian liquids, gels, foams, polymers, colloids,
dispersions, emulsions, microemulsions, and suspensions for which rheology and
interfacial phenomena play a major role are often involved.

A general structure of product design is presented in Figure 1.48.

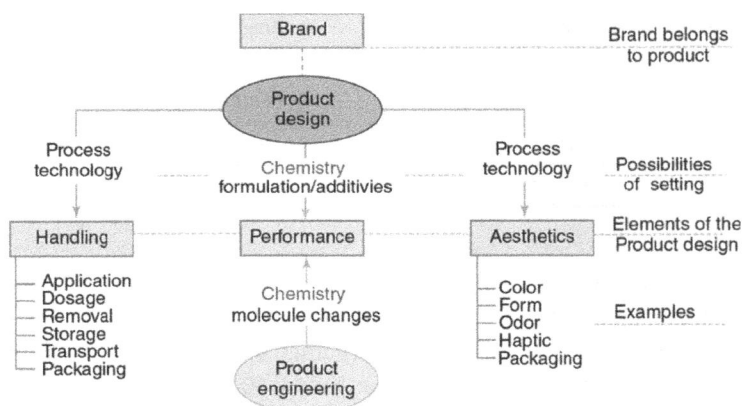

Figure 1.48: General structure of product design. Reproduced with permission from W. Rähse,
Chemical Product Design – A New Approach in Product and Process Development, in Industrial
Product Design of Solids and Liquids: A Practical Guide, 2014 Wiley-VCH. DOI: 10.1002/
9783527667598.ch1. Copyright Wiley. Reproduced with permission.

Figure 1.48 implies that in product design, performance in a particular application
and attractiveness to consumers are as important as the process technology *per se*.
The latter is often different from the technology considered in the current textbook
focusing mainly on reactions. For the product technology, the creation and the control of the particle size distribution in operations such as crystallization, precipitation, prilling, generation of aerosols and nanoparticles, as well as the control of the
particle morphology and the final shaping, and presentation in operations such as
agglomeration, calcination, compaction, and encapsulation are required.

A great number of operations on formulated products are performed on a batch
basis, in contrast to the continuous production processes most often encountered

with commodities, which are specification products. Typical operations such as granulation, extrusion, compression, spray drying, spray chilling, coating, emulsification, and gelation are carried out in place of classical unit operations (e.g., distillation, extraction, and absorption). Batch processing often results in irreproducible results and batch-to-batch variations, requiring also large inventories of (hazardous) chemicals. Scaling up of such processes as crystallization to full scale is risky, and pilot plants are typically needed. Crystallization can be challenging, since batch crystallization is often difficult to describe by rigorous mathematical models. One example worth mentioning is production of ritonavir,

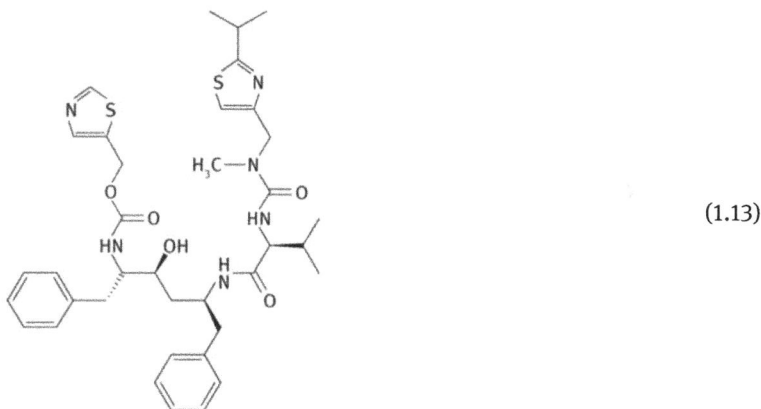

$$(1.13)$$

which proved effective in mitigating the effect of HIV infection. Two years after introduction of the product into the market, the control over quality was lost and a new polymorph with lower solubility in human blood appeared in the product diminishing the effect. The producer had to take the product from the market back to the development stage before the pharmaceutical was reintroduced into the market.

Not straightforward is also, for example, scaling up of catalyst preparation. This includes a selection of chemicals whose costs are rarely considered in lab scale, being an issue for industrial production. Other considerations include corrosiveness of the raw materials, potential gaseous emissions and their mitigation, disposal of effluents, and wastewater purification along with the process water recovery. Moreover, parameters such as stirring rate, heating and cooling rates, heat transfer, flow pattern, and drying and calcination conditions differ in lab and industrial scales, influencing, e.g., crystallization by hydrothermal synthesis. Large reactors have more prominent spatial inhomogeneity in temperature, pH, and concentration. Significant variations in the local temperature profiles can be present during drying and calcination. In summary, the scale-up of the catalyst preparation methods from a laboratory scale to pilot and industrial scales is challenging.

In the process design for performance products, there is also commercial pressure to introduce a product to market as soon as possible. Moreover, market situation can change quickly, thus there is a high risk of failure for new products, and many, especially in food consumer markets, disappear in few years.

Product development requires, therefore, besides skills related to technology, also marketing skills. It should be noted that information obtained from marketing surveys of a limited number of (sometimes poorly informed) consumers might be inaccurate and even contradictory.

Marketing skills are used to extend the product life by advertising to attract new customers, reducing the price, exploring new (geographical) areas and market segments, or even introducing new packaging of an established product. A more technology-oriented approach relies on introducing new features to a current product. These efforts are important because for structured and complex products the product life is much shorter than for specification chemicals, which are mainly discussed in this book.

In the product development, besides technical properties including performance and manufacturing also intellectual property issues, and safety and health aspects should be considered.

The patent literature, often neglected in the academic world, is also very important for the process development and will be briefly discussed below.

1.4.8 Patents

A very important aspect in chemical process industries is patenting strategy. Patentable inventions typically disclose a solution to a technical problem or a product, apparatus, process, and their use. Patents should function as described and be reproducible (not always the case). They should not describe just an idea but instead be working embodiments. Theories, computer programs, plant variations, to name a few, as well as unethical inventions are excluded from patenting.

Patentable inventions should be novel, not publicly disclosed before patent application in writing, orally, or even visually. In the language used for patenting, the so-called inventive steps should be substantially different from prior art and not obvious to a person skilled in the art. The following keywords are typically used in patents "new, different, improved, cheap, unexpected, surprising" to highlight the novelty.

Obtaining a patent is legally regulated. After invention, the company announces the patent in writing. Inventors either are paid a certain sum or have a right to royalties. The latter means that the inventors are paid for every ton of product made based on that patent.

Patents include background information about the prior art and describe the invention with examples. They consist of a number of claims that define the exact scope of the invention. The first claim is typically as broad as possible, while the following subclaims provide exact definitions. In Chapter 6, a key oil refining process – hydrocracking – will be considered. This is a catalytic process where the catalysts have a dual function. The cracking function is provided by the acidic support and the hydrogenation function by the metal. The acidic support can be amorphous or a combination of a crystalline phase (zeolite) and an amorphous phase (e.g. aluminosilicate). During the process, the catalyst is used in the form of extrudates. An example of a patent on a hydrocracking catalyst is given in Figure 1.49.

(12) **United States Patent**
  Domokos et al.

(10) **Patent No.:** US 9,199,228 B2
(45) **Date of Patent:** Dec. 1, 2015

(54) HYDROCRACKING CATALYST

(75) Inventors: László Domokos, Amsterdam (NL.); Cornelis Ouwehand, Amsterdam (NL.)

(73) Assignee: Shell Oil Company, Houston, TX (US)

(58) **Field of Classification Search**
  USPC 502/63, 64, 66, 67, 69, 74, 79, 85, 87
  See application file for complete search history.

(56)   **References Cited**

ABSTRACT

Process of preparing a hydrocracking catalyst carrier comprising amorphous binder and zeolite Y, which process comprises subjecting zeolite Y having a silica to alumina molar ratio of at least 10 to calcination at a temperature of from 700 to 1000° C., hydrocracking catalyst carrier comprising amorphous binder and zeolite Y having a silica to alumina molar ratio of at least 10, the infrared spectrum of which catalyst has a peak at 3690 cm^{-1}, substantially reduced peaks at 3630 cm^{-1} and 3565 cm^{-1} and no peak at 3600 cm^{-1}, hydrocracking catalyst carrier comprising an amorphous binder and zeolite Y having a silica to alumina molar ratio of at least 10, which catalyst has an acidity as measured by exchange with perdeuterated benzene of at most 20 micro-mole/gram, hydrocracking catalyst derived from such carrier and hydrocracking process with the help of such catalyst.

What is claimed is:

1. A process of preparing a hydrocracking catalyst carrier for use in the preparation of a hydrocracking catalyst having enhanced gas oil selectivity, wherein said hydrocracking catalyst carrier comprises an amorphous binder and zeolite Y, wherein said process comprises: calcining zeolite Y, having a silica to alumina molar ratio of at least 10, in the absence of added steam at a temperature in the range of from 700° C. to 1000° C. followed by mixing the obtained zeolite Y with said amorphous binder, comprising silica-alumina containing silica in an amount in the range of from 25 to 95% by weight as calculated on the carrier alone, and an acidic aqueous solution in amount so as to provide a mixture having a pH in the range of from 4.4 to 5.7 and an LOI in the range of from 50 to 65% such that said hydrocracking catalyst carrier has a monomodal pore size distribution, wherein at least 50% of the total pore volume is present in pores having a diameter in the range of from 4 to 50 nm; extruding said mixture to give an extrudate; and calcining said extrudate at a temperature of from 700 to 1000° C.

2. A process according to claim 1, in which process zeolite beta is also included in the mixing step and the mixture.

3. A process according to claim 2, in which process the hydrocracking catalyst carrier comprises of from 2 to 70% wt of zeolite and of from 98 to 30% wt of amorphous binder.

4. A process according to claim 3, in which process the calcining step is carried out during a time of from 20 minutes to 5 hours.

5. A process according to claim 4, wherein the pore volume present in said pores of said hydrocracking catalyst carrier is at least 0.4 ml/g, all as measured by mercury intrusion porosimetry.

6. A process according to claim 5, in which before calcination the zeolite Y has a bulk silica to alumina molar ratio above 12, a unit cell size in the range of from 24.10 to 24.40 Angstrom, and a surface area of at least 850 m^2/g.

7. A process according to claim 1, which process the hydrocracking catalyst carrier comprises of from 2 to 70% wt of zeolite and of from 90 to 30% wt of amorphous binder.

8. A process according to claim 7, in which process the calcination is carried out during a time of from 20 minutes to 5 hours.

9. A process according to claim 8, in which process the hydrocracking catalyst carrier has a monomodal pore size distribution, wherein at least 50% of the total pore volume is present in pores having a diameter in the range of from 4 to 50 nm, and wherein the pore volume present in said pores is at least 0.04 ml/g, all as measured by mercury intrusion porosimetry.

10. A process according to claim 9, in which process the hydrocracking catalyst carrier is prepared by shaping the mixture with added zeolite beta, wherein the mixture has a loss of ignition (LOT) in the range of from 55 to 65%.

11. A process according to claim 10, in which process, before the calcination the zeolite Y has a bulk silica to alumina molar ratio above 12, a unit cell size in the range of 24.10 to 24.40 Angstrom, and a surface area of at least 850 m^2/g.

Figure 1.49: An example of a patent.

Patents can contain drawings, and their preparation is often assisted by external patent agents or patent attorneys. They help in filing the application nationally or internationally. After patent examination, official actions and responses follow and a patent is granted. Certain fees are involved, which had to be paid every year for a duration of a patent (ca. 20 years).

The legal effects of a patent include a right to exclude others from using the invention for the patent duration and in patented countries/regions. Patenting also gives a right to compensation for unauthorized use. At the same time, it is patentee's own responsibility to monitor infringements.

Patents should be also used as a source of information (also for competitors), being public and disclosing new and old technical solutions. Patents serve as a business support protecting own business (technical developments/improvements) and possible future business areas as well as providing return on investment; promote own business by promoting the quality of products/processes and the company reputation and building a patent portfolio. Some companies view patenting as a weapon in competition, since it can exclude others from own (core) business, prevent others from patenting, and even mislead competitors, being, in this case, an expensive game. In the industry, it is thus a custom to monitor patenting done by competitors in order to get know-how for own technical development and knowledge of competitors' development as well as to avoid infringement.

An option practiced by several industrial companies is to keep an invention secret to protect a process from being publicly known as well as for inventions that are difficult to patent. An alternative could be a defensive publication about an invention in a local magazine, for example, hoping that such strategy in the long run can invalidate patents by competitors.

Chapter 2
Physico-chemical foundations of chemical processes

2.1 Stoichiometry

A chemical reaction can be written in a following way:

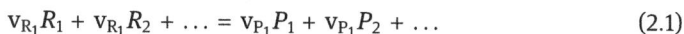

$$v_{R_1} R_1 + v_{R_1} R_2 + \ldots = v_{P_1} P_1 + v_{P_1} P_2 + \ldots \tag{2.1}$$

where R_i are the reactants, P_i are the products, v_{R1} are the stoichiometric coefficients of respective reactants, etc.

In the chemical industry, there are transformations that are described by just one chemical reaction (for example, ammonia synthesis, $N_2 + 3H_2 = 2NH_3$) or several reactions. In the oxidation of ethylene to ethylene oxide, besides the desired reaction ($C_2H_4 + 0.5O_2 = CH_2OCH_2$), unwanted total oxidation can also occur ($C_2H_4 + 3O_2 = 2CO_2 + 2H_2O$). Even if ammonia synthesis can be considered a simple transformation, this reaction is a complex one, since it requires the presence of a catalyst and consists of several elementary steps, which will be discussed later. Thus, the stoichiometry shows the ratio between reactants but does not necessarily reflect the real chemical transformations.

Stoichiometric equations could be written in a different way. In the example for ethylene oxide partial oxidation, its total oxidation can also be added ($C_2H_4O + 2.5O_2 = 2CO_2 + 2H_2O$). For mass balance calculations, such an uncertainty is not acceptable, and independent stoichiometric equations are used. For the sake of convenience, such equations are selected when the same compound (ethylene) is the substrate. Another example is steam reforming of methane:

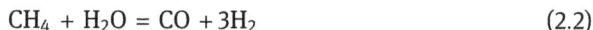

$$CH_4 + H_2O = CO + 3H_2 \tag{2.2}$$

Formed CO can be further transformed through the water-gas shift reaction:

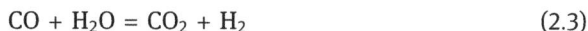

$$CO + H_2O = CO_2 + H_2 \tag{2.3}$$

For practical purposes of mass balance calculations, it is more convenient to use the system of two equations where methane is included in the left-hand side of both equations:

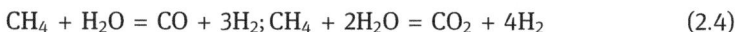

$$CH_4 + H_2O = CO + 3H_2; CH_4 + 2H_2O = CO_2 + 4H_2 \tag{2.4}$$

A parameter that relates the amount N and the concentration of the reactants is conversion. It is defined as the amount of the reacted substrate ($N_{Ro} - N_R$) to its initial amount N_{Ro}:

https://doi.org/10.1515/9783110712551-002

$$a = \frac{N_{R_o} - N_R}{N_{R_o}}.$$ (2.5)

If the initial reaction mixture does not have a stoichiometric ratio of reactants (for example, in steam reforming of natural gas an excess of steam is used), it is convenient to express the conversion through the substrate that is in deficit (the so-called key component).

For continuous processes, conversion is defined through the molar fluxes

$$a = \frac{\dot{n}_{R_{in}} - \dot{n}_{R_{out}}}{\dot{n}_{R_{in}}}$$ (2.6)

Besides conversion, another important reaction characteristic is selectivity, which reflects its ability to direct conversion of reactant(s) along one specific pathway to the desired product. For many reacting systems, various reaction paths are possible. In general, selectivity depends on pressure, temperature, reactant composition, conversion, and nature of the catalyst used. Therefore, selectivity should be referred to specific conditions.

Integral selectivity is defined as the ratio of the desired product per consumed reactant:

$$S = \frac{N_{R \to P}}{N_{R_o} - N_R},$$ (2.7)

while differential selectivity is the ratio of the rate of desired product formation to the rate of reactant consumption:

$$S = \frac{r_p}{-r_R}.$$ (2.8)

The product yield shows the fraction of the initial substrate transformed to a particular product. For a simple reaction, it is written as

$$y = \frac{N_{R \to P}}{N_{R_o}}$$ (2.9)

For a continuous process, the yield of a particular product is defined as

$$y = \frac{\dot{n}_{P_{out}} - \dot{n}_{P_{in}}}{\dot{n}_{R_{in}}} \frac{|v_P|}{v_R}$$ (2.10)

taking into account the stoichiometric coefficients in the case of a complex reaction. Consider as an example steam reforming of methane (eq. (2.2)), where selectivity for hydrogen and CO and yields are defined considering stoichiometric coefficients for batch:

$$S_{H_2} = \frac{N_{H_2} - N^0{}_{H_2}}{N^0{}_{methane} - N_{methane}} \frac{\nu_{methane}}{\nu_{H_2}} 100\%; \quad S_{CO} = \frac{N_{CO} - N^0{}_{CO}}{N^0{}_{methane} - N_{methane}} \frac{\nu_{methane}}{\nu_{CO}} 100\%$$

$$(2.11)$$

$$y_{H_2} = \frac{N_{H_2} - N^0{}_{H_2}}{N^0{}_{methane}} \frac{\nu_{methane}}{\nu_{H_2}} 100\%; \quad y_{CO} = \frac{N_{CO} - N^0{}_{CO}}{N^0{}_{methane}} \frac{\nu_{methane}}{\nu_{CO}} 100\% \qquad (2.12)$$

and continuous processes

$$S_{H_2} = \frac{\dot{n}_{H_2,out} - \dot{n}_{H_2,in}}{\dot{n}_{methane,in} - \dot{n}_{methane,out}} \frac{\nu_{methane}}{\nu_{H_2}} 100\%;$$

$$S_{CO} = \frac{\dot{n}_{CO,out} - \dot{n}_{CO,in}}{\dot{n}_{methane,in} - \dot{n}_{methane,out}} \frac{\nu_{methane}}{\nu_{CO}} 100\% \qquad (2.13)$$

$$y_{H_2} = \frac{\dot{n}_{H_2,out} - \dot{n}_{H_2,in}}{\dot{n}_{methane,in}} \frac{\nu_{methane}}{\nu_{H_2}} 100\%; \quad y_{CO} = \frac{\dot{n}_{CO,out} - \dot{n}_{CO,in}}{\dot{n}_{methane,in}} \frac{\nu_{methane}}{\nu_{CO}} 100\% \qquad (2.14)$$

While for batch reactors, the reaction time t is used; for flow reactors, the residence time τ is applied defined as the ratio of the reactor volume to the volumetric flow rate at reaction conditions:

$$\tau = \frac{V_{reactor}}{V(T, p)} \qquad (2.15)$$

Stoichiometric equations and conversion are used for calculating the amount of converted reactants and formed products in moles. In practice, concentrations are often used, which in fact could change during the reaction not only because of the reaction *per se*, but also due to changes in the volume of the reacting mixture. For liquid-phase reactions, changes in the volume are often negligible apart from the cases when the volume is changing, for example, by decomposition of the liquid compounds. In those cases of the gas- and liquid-phase reactions when the volume is changed, it can be defined through conversion in the following way:

$$V = (1 + \varepsilon\alpha)V_0 \qquad (2.16)$$

where V_0 is the initial volume, and ε is the expansion factor. For a continuous reactor, it subsequently holds

$$\dot{V} = (1 + \varepsilon\alpha)\, \dot{V}_0 \qquad (2.17)$$

In mass balance calculations, mass yields are also used, which reflect transformations on an industrial scale in a more natural way.

The space–time yield (STY), calculated as the amount of a certain product generated per unit time and unit volume, is often used to define reactor productivity. In industry, not surprisingly the mass of product is applied rather than moles. For a fair comparison, it should be stated if the volume of the catalyst bed or the volume of the catalyst without the void space is used in calculations.

To define the flows in industry, the weight hourly space velocity, liquid hourly space velocity, and gas hourly space velocity are also used reflecting, respectively, the mass flow or volumetric flow of the feed per reactor volume.

2.2 Thermodynamics

Evaluation of thermodynamics is the starting point in reaction engineering, since it addresses the overall feasibility of the reaction and defines the composition of reaction mixtures at equilibrium and thus maximum conversion and selectivity. In addition, the reaction rate depends on the approach to equilibrium or the distance of the reaction mixture composition to the equilibrium one, since at equilibrium, the rate of the forward and reverse reactions are equal to each other.

Thermodynamic conditions for the equilibrium system

$$aA + bB = cC + dD \tag{2.18}$$

can be expressed through the Gibbs energy, which must be stationary. It means that the derivative of G with respect to the extent of reaction, ξ, is zero, or the sum of chemical potentials μ (partial molar Gibbs energy) of the products and reactants are equal to each other:

$$a\mu_A + b\mu_B = c\mu_C + d\mu_D, \tag{2.19}$$

where chemical potentials of reagents depend on activity

$$\mu_i = \mu_i^o + RT \ln a. \tag{2.20}$$

In eq. (2.20), μ_i^o is the standard chemical potential. If a mixture is not at equilibrium, the driving force for the reaction to approach equilibrium, leading to subsequent changes in composition, is the minimization of the excess Gibbs energy (or Helmholtz energy at constant volume reactions). The activity-based equilibrium constant for eq. (2.18),

$$K_a = \frac{[a_A]^a [a_B]^b}{[a_C]^c [a_D]^d}, \tag{2.21}$$

is related to the standard Gibbs energy change for the reaction by the equation

$$\Delta G_T^0 = -RT \ln K_a^{eq}. \tag{2.22}$$

where R is the universal gas constant and T is the temperature.

For an ideal gas, the equilibrium constant for eq. (2.18), denoted as K_P, is given by

$$K_p = \frac{[P_A]^a [P_B]^b}{[P_C]^c [P_D]^d},$$ (2.23)

Or in a general form for transforming reactants (R) to products (P)

$$K_p = \prod \frac{[P_{P_i}]^{v_{P_i}}}{[P_{R_i}]^{v_{R_i}}},$$ (2.24)

For real gases, deviations from the ideal gas should be taken into account by replacing partial pressures with fugacities f_i

$$f_i = \varphi_i P_i$$ (2.25)

which are related to the partial pressures through fugacity coefficients φ_i.

The standard reaction Gibbs function, which can be calculated from the standard Gibbs functions of formation of all products and reactants, is thus directly related to the equilibrium constant of the reaction and thus the composition of the reaction mixture at equilibrium.

The Gibbs energy is defined through reaction enthalpy ΔH_T^0 and entropy ΔS_T^0:

$$\Delta G_T^0 = \Delta H_T^0 - T \Delta S_T^0.$$ (2.26)

The temperature dependence of the standard Gibbs energy change of the reaction is given by

$$\frac{\partial \Delta G^0}{\partial T} = \frac{\Delta G^0 - \Delta H^0}{T}$$ (2.27)

After rearrangement and division by RT, an expression relating reaction enthalpy and equilibrium constant is obtained:

$$\frac{\Delta H^0}{RT^2} = \left(\frac{1}{RT} \frac{\partial \Delta G^0}{\partial T} - \frac{\Delta G^0}{RT^2} \right) = -\frac{\partial \left(\frac{\Delta G^0}{RT} \right)}{\partial T} = \frac{\partial (\ln k)}{\partial T}.$$ (2.28)

The standard enthalpy changes are given by

$$\Delta HT_T^0 = \sum v_i \Delta H_{298K}^0 (\text{prod}) - \sum v_i \Delta H_{298K}^0 (\text{react}) + \int_{298K}^{T} \left(\sum_{i=1}^{N_i} v_i c_{p,i} \right) dT,$$ (2.29)

where $c_{p,i}$ is the temperature-dependent heat capacity, which for gases is given by

$$c_{p,i} = a'_i + b'_i T + c'_i T^2 + d'_i T^{-2}.$$ (2.30)

In eq. (2.30) a', b', c' and d' are coefficients. The rates of forward and reverse reactions are related in the following way. Expressing the rates of the forward and the reverse reactions, respectively,

$$r_+ = k_+ \prod_l C_l^{\alpha+,l}, r_- = k_- \prod_l C_l^{\alpha-,l}, \tag{2.31}$$

the ratio of rates is

$$\frac{r_+}{r_-} = \frac{k_+}{k_-} \prod_l C_l^{\alpha+,l-\alpha-,l} = K \prod_l C_l^{\alpha+,l-\alpha-,l}, \tag{2.32}$$

with the equilibrium constant defined as $K = k_+/k_-$. Equation (2.32) can be rearranged according to the De Donder equation:

$$\frac{r_+}{r_-} = e^{A/RT}, \tag{2.33}$$

where A is the reaction affinity

$$A = RT \ln K \prod_l C_l^{\alpha+,l-\alpha-,l}, \tag{2.34}$$

which is defined as the derivative of the Gibbs free energy with respect to the extent of the reaction

$$A = \left(-\frac{\partial G}{\partial \xi}\right)_{T,p}. \tag{2.35}$$

The relationship between the overall process rate and the rate in the forward reactions is also expressed through reaction affinity

$$r = r_+ - r_- = r_+\left(1 - e^{-A/RT}\right) = r_+(1 - \varphi(P)/K), \tag{2.36}$$

where r is the overall rate, r_+ is the rate of the forward reaction, $\varphi(P)$ is a function of only reactant concentrations, and K is the equilibrium constant. For a reversible reaction (2.18), assuming that stoichiometry corresponds to molecularity, the rate is given by

$$r = k_+ P_A^a P_B^b - k_- P_C^c P_D^d = k_+ P_A^a P_B^b \left(1 - \frac{k_- P_C^c P_D^d}{k_+ P_A^a P_B^b}\right) = r_+\left(1 - \frac{1}{K}\frac{P_C^c P_D^d}{P_A^a P_B^b}\right), \tag{2.37}$$

where P_A etc. are partial pressures (in case of gas-phase reactions) or concentrations (for liquid-phase reactions).

From the dependence of Gibbs energy on the equilibrium constant and an expression relating Gibbs energy and enthalpy and entropy, one gets

$$K_{eq} = e^{-\Delta G_T^0/RT} = e^{-\Delta H_T^0/RT} e^{\Delta S/R}. \tag{2.38}$$

For exothermal reactions, it holds that $\Delta H_T^0/RT < 0$ and thus the equilibrium constant decreases with temperature. The opposite is true for endothermic reactions (Figure 2.1).

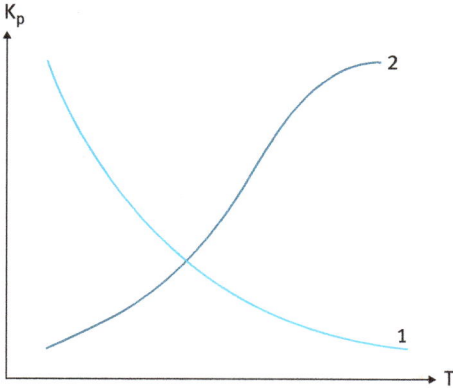

Figure 2.1: Changes of the equilibrium constant with T for exothermic (1) and endothermic (2) reactions.

Shifting equilibrium in various technological processes for equilibrium-limited reactions is a means to increase conversion and drive the reaction to more completion. The principle of Le Chatelier implies that by influencing a system at equilibrium, it will react in a way to minimize such influence.

Thus, if a temperature increase is related to heat supply, the system will react in an opposite way by consuming heat. In an endothermic reaction, with a temperature increase, there are more profound chemical transformations and thus more heat is consumed, which will drive the equilibrium to the product side and increase equilibrium conversion. In exothermal reactions, heat generation will decrease due to lower conversion levels with temperature elevation.

The heat generation rate calculated as the amount of heat released per unit of time per reactor volume,

$$q = \frac{Q}{tV}, \tag{2.39}$$

depends on the reaction rate and the heat of the reaction,

$$q = r(-\Delta H) = rQ_r \tag{2.40}$$

For a complex reaction, the heat generated is a sum of all reactions in the system,

$$q = \sum r_j(-\Delta H_j). \tag{2.41}$$

When a reaction is happening with a decrease in volume, elevation of pressure will drive the reaction toward the product side. An example is the synthesis of ammonia from nitrogen and hydrogen when industrial converters operate at 20–30 MPa to get technologically reasonable conversion levels. On the contrary, for reactions, when the volume is increased (dehydrogenation of ethylbenzene or propane), a decrease in pressure is needed to shift the equilibrium to the product side. Obviously, in the industry, operation under vacuum is seldom a feasible option; thus, a decrease in partial pressures is done through dilution with an inert substrate. In dehydrogenation of ethylbenzene or cracking of naphtha, dilution is typically done with steam, which in addition has another job of cleaning the catalyst surface or reactor walls from carbon deposits.

Another possibility to shift equilibrium is to use an excess of one of the components. In practice, the least valuable compound in used in excess to increase conversion of a more valuable one.

Finally, as a means of shifting equilibrium, it is possible to remove the product from the reaction mixture by combining reaction with separation *in situ*, for example, using reactive distillation or separation with a membrane as discussed in Chapter 1. Another example is reactive extraction of carboxylic acids from fermentation broths.

Let us consider from the viewpoint of thermodynamics dehydrogenation of butane to butene. This is an endothermic reaction with an increase in volume

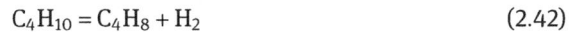

$$C_4H_{10} = C_4H_8 + H_2 \tag{2.42}$$

Based on the considerations above, the process should be organized at high temperature and low pressure or by dilution with an inert substrate. Application of pressures below atmospheric pressure has an inherent safety risk, since, owing to a potential leaking, outside air can penetrate inside the equipment and form an explosive mixture with the reactants. To avoid it, the feed is diluted with steam. Although temperature increase is beneficial from the viewpoint of thermodynamics, very high temperatures (above 600 °C) will lead to thermal destruction of hydrocarbons. Equilibrium conversion as a function of temperature is illustrated in Figure 2.2, where λ represents the ratio between steam and butane. If the target in an industrial process is to obtain a conversion of not less than 50%, the feed should be significantly diluted with steam ($\lambda > 8$).

2.3 Catalysis

Catalysis is a phenomenon related to acceleration of rates of chemical reactions. From the industrial viewpoint, higher rates lead to higher space–time yields and thus smaller reactors for the same production capacity meaning lower investment.

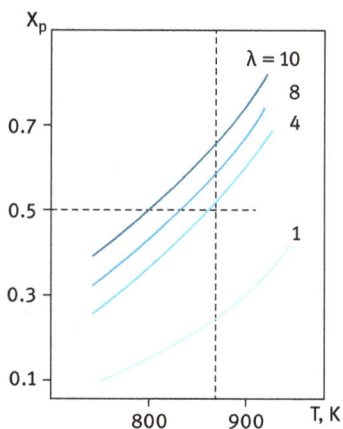

Figure 2.2: Equilibrium conversion for butane dehydrogenation.

Moreover, a catalyst of higher activity allows synthesis at lower temperature, which is beneficial from the viewpoint of thermodynamics for exothermal reversible reactions, such as the water-gas shift reaction or ammonia synthesis. In the latter case, a dream catalyst would operate at room temperature, giving much higher equilibrium conversion compared to the current industrial practice. For equilibrium-limited endothermal reactions, higher intrinsic catalytic activity implies the same productivity but at lower temperatures. This might be beneficial from the viewpoint of catalyst stability as less coke is formed or better selectivity avoiding side products. Moreover, in some cases, such as steam reforming of natural gas, efficient heat supply requires thinner walls of tubes containing the catalyst. This in turn puts lower requirements on the thermal properties of the tube material.

A catalyst was defined by Ostwald as a compound that increases the rate of a chemical reaction without being consumed by the reaction. This definition allows for a possibility that small amounts of the catalyst are lost in the reaction or that the catalytic activity slowly declines.

An important issue in catalysis is selectivity toward a particular reaction. For example, transformation of synthesis gas (mixture of CO and hydrogen) can lead either to methanol (on copper), methane (on nickel) or to high alkanes (on cobalt). For consecutive reactions, it could be desirable to obtain an intermediate product. In oxidation of ethylene, such target is ethylene oxide, but not CO_2 and water.

Thus, selectivity is generally more important than activity; otherwise, a valuable feedstock is just transformed into an undesired product.

For catalytic reactions, the change in the Gibbs free energy between the reactants and the products ΔG is the same, independent on the presence or absence of a catalyst, providing, however, an alternative reaction path. A lower value of activation energy implies higher reaction rates, which could be expressed through the dependence of rate constant k on temperature according to the transition state theory of Eyring and Polanyi:

$$k = AT^m e^{-Ea/RT}, \tag{2.43}$$

where A is the pre-exponential factor, E_a is the activation energy related to the potential energy barrier, and m is a constant.

If a catalyst is active in enhancing the rate of the forward reaction, it will do the same with a reverse reaction. Thus, in the case of a thermodynamically unfavorable process, there is no hope to find an active catalyst, which will beat thermodynamics.

The immense importance of catalysis in the chemical industry is manifested by the fact that roughly 85–90% of all chemical products have seen a catalyst during the course of production.

Besides heterogeneous catalysts (typically in the form of solids), when the reactants are either gases or liquids, there is also a great variety of homogeneous catalysts, such as metal complexes and ions, Brønsted and Lewis acids. Even if the industrial landscape is dominated by heterogeneous catalysis, due to several reasons including easier separation, catalysis by mineral acids and homogeneous transition metals is used in several industrial processes.

Among industrial applications of homogeneous catalysis, toluene and xylene oxidation to acids, oxidation of ethene to aldehyde, carbonylation of methanol and methyl acetate, polymerization over metallocenes, hydroformylation of alkenes, etc. could be mentioned.

Despite apparent difficulties in catalyst separation, involving for example co-solvent addition, or solvent distillation, homogeneous catalysts offer several advantages, especially in terms of selectivity. This is related to their well-defined molecular structure, which is somewhat difficult to achieve with heterogeneous catalysts.

Detailed description of the processes catalyzed by homogeneous catalysts will be illustrated in subsequent chapters.

Heterogeneous catalysis covers a wide length scale, from a molecular scale (nanometers) of an active site to a catalytic reactor (meters) scale. Catalysis occurs on the surfaces of solid materials, representing chemistry in two dimensions. Active sites of heterogeneous catalysts are related to a molecular- or atomic-level arrangement of atoms, responsible for catalytic properties. Chemical reactions proceed at this subnanometer level involving rupture and formation of chemical bonds.

Catalytically active particle have typically dimensions between 1 and 10 nm and are located inside pores of a support material. Chemical composition, texture of materials, and pore structure are very important issues related not only to catalysis *per se*, but also to transport of molecules through the pores to the active sites. Questions of interest are the size, shape, structure, and composition of the active particles.

Catalysts in the form of powders can be applied in industrial processes only in limited cases. Shaped catalysts, in the form of extrudates, pellets, and tablets on the millimeter length scale are introduced into industrial reactors. Application of such materials requires engineering porosity, mechanical strength, and attrition resistance in addition to activity, kinetics, and mass transfer. It should be noted that in industrial reality, the mass and heat transport through the catalyst bed can be as important as intrinsic kinetics.

The main requirements of a catalyst for an industrial process depend on the trinity of catalysis: activity, selectivity, and stability. The first property is related to an ability of conducting a process within a reasonable contact time, which influences the reactor dimensions and process capacity. Insufficient activity, in principle, could be compensated by higher catalyst amounts or some other means, such as higher temperature. Catalyst selectivity is probably the most important characteristics of a catalyst, which should also be sufficiently stable under operation conditions.

Small particles of the active catalyst alone cannot provide highly active thermostable catalysts due to their sintering at conditions of catalyst preparation and catalysis *per se*. Moreover, separation of reaction products from nanosized catalysts is far from being trivial, in many cases, even impossible. Therefore, an active phase (responsible for activity and selectivity) is usually deposited on a thermostable support, which also provides the required shape, mechanical strength, and pore structure.

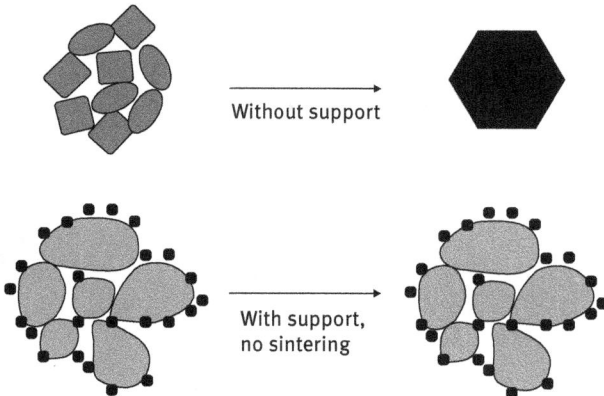

Figure 2.3: Supported active phase (J. W. Geus, A. J. van Dillen, Preparation of supported catalysts by deposition-precipitation, Handbook of Heterogeneous Catalysis, 2.4.1., Editors: H. Knözinger, F. Schueth, J. Weitkamp, Wiley-VCH, Weinheim. Copyright © 2008 Wiley-VCH Verlag GmbH & Co. KGaA. Reproduced with permission).

A catalyst can increase the rate of one reaction without increasing the rate of other reactions. In general, selectivity depends on pressure, temperature, reactant composition, conversion, and nature of the catalyst. Therefore, selectivity should be referred to specific conditions.

Although there is always a desire to have a stable catalyst for a particular reaction, strong catalyst deactivation does not necessarily mean that in such case, a particular catalyst cannot be applied at all. In fact, it is only in theory that catalysts remain unaltered during reactions. Actual practice is far from these ideal situations, as the progressive loss of activity could be associated with coke formation, attack of poisons, loss of volatile agents, changes of crystalline structure, which causes loss of mechanical strength. There are industrial examples such as catalytic reforming or fluid catalytic cracking showing successful industrial implementation of catalytic reactors in combination with continuous regeneration, when the catalyst life is merely few days or seconds, respectively. Summarizing this section, it can be concluded that the target priorities in catalyst development and applications typically are selectivity>stability > activity.

The main causes of deactivation in heterogeneous catalysis are poisoning, fouling, thermal degradation (sintering, evaporation) initiated by high temperature, mechanical damage, and corrosion/leaching by the reaction mixture (Figure 2.4).

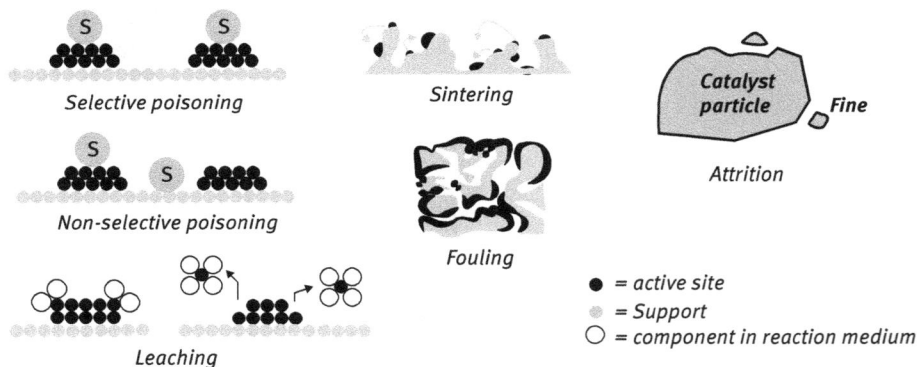

Figure 2.4: Deactivation mechanisms. From J.A. Moulijn, A.E. van Diepen, F. Kapteijn, *Applied Catalysis A: General*, 2001, 212, 3. Copyright Elsevier. Reproduced with permission.

Poisoning is caused by strong adsorption of impurities present in the feed and depends upon adsorption strength of such poisons relative to the other species competing for catalytic sites. Adsorbed poisons may not only block active sites but change the electronic or geometric structure of the surfaces as well.

Fouling is associated with covering of the catalyst surface by deposits, which are quite often hydrogen-deficient carbonaceous materials (i.e. coke), making the active sites inaccessible.

For liquid-phase reactions, catalysis leaching can be prominent with the leached metals being in fact responsible for high catalytic activity in some reactions. Such leaching is particularly problematic in oxidation catalysis because of the strong complexation and solvolytic properties of oxidants (H_2O_2, RO_2H) and/or the products (H_2O, ROH, RCO_2H, etc.).

Strong stresses of packed catalysts beds during start-ups, shutdowns, and catalyst regeneration could lead to mechanical deactivation. Finally, thermal degradation (because of sintering, chemical transformations, evaporation, etc.) could cause deactivation. In particular, prominent could be sintering, e.g., the loss of catalyst active surface due to crystallite growth of either the support material or the active phase.

Typical catalytic materials are shown in Figure 2.5. Most catalysts are multicomponent and have a complex composition. Components of the catalyst include the active agent itself and may also include a support, a promoter, and an inhibitor.

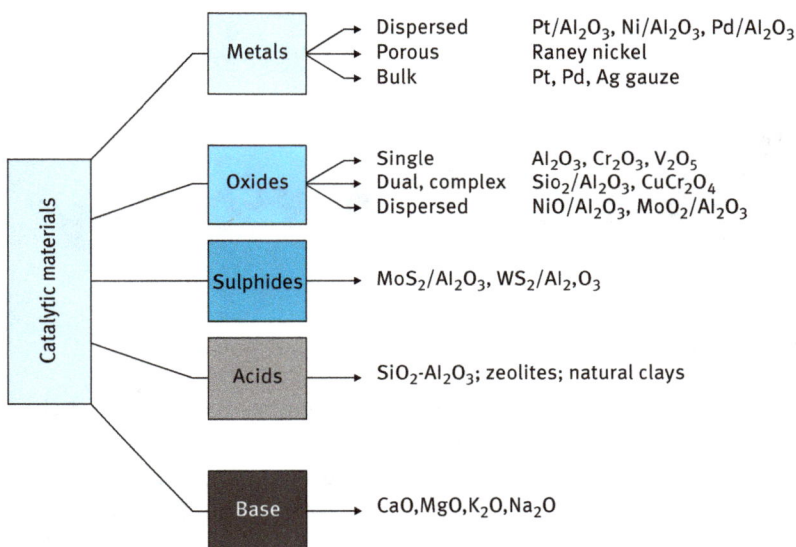

Figure 2.5: Catalytic materials.

Metal catalysts are used not only in their bulk form (e.g. Fe for ammonia synthesis) but are generally dispersed on a high surface area insulator (support) such as Al_2O_3 or SiO_2. Many heterogeneous catalysts need supports, which, from the economics viewpoints, are a means of spreading expensive materials and providing the necessary mechanical strength, heat sink/source, help in optimization of bulk density and in dilution of an overactive phase. There are also geometric (increase of surface area, and optimization of porosity, crystal, and particle size) and chemical functions (improvement of activity, minimization of sintering, and poisoning) provided by the supports.

Sintering of supports can start already at a temperature substantially lower than the melting point, thus supports with high melting points, such as alumina or silica, are applied. In particular, alumina is mainly preferred due to favorable bulk density, thermal stability, and price. It is applied in the form of γ-alumina in very many reactions such as hydrotreating, hydrocracking, hydrogenation, reforming, and oxidation. Another phase of alumina (α-alumina) being stable at very high temperatures is used, for example, in steam reforming of natural gas.

Silica has a lower bulk density, which means that for the same active phase loading on the support, more reactor volume is needed. Silica is more liable to sintering above 900 K than alumina and is volatile in the presence of steam and elevated pressures. There are still several applications (polymerization, oxidation, and hydrogenation) when silica is used. A very specific example of utilization of silica is oxidation of sulphur dioxide to trioxide (Figure 2.6).

Figure 2.6: Vanadium-based catalysts supported on silica for sulphur dioxide oxidation.

Although there were catalyst formulations using silica synthesized from water glass (sodium silicate), the catalyst manufacturers utilize predominantly heat-resistant diatomaceous earth (diatomite) kieselguhr. This is a form of silica composed of the siliceous shells of unicellular aquatic plants of microscopic size. The specific features of the diatomaceous earth are related to small amounts of alumina and iron as part of the skeletal structure and a broad range of pore sizes. The main components of the sulphur dioxide oxidation catalyst include SiO_2 (a support), vanadium, potassium and/or cesium, and various other additives. The reaction occurs within a molten salt consisting of potassium/cesium sulphates and vanadium sulphates, coated on the solid silica support. Vanadium is present as a complex sulphated salt mixture and not as vanadium pentoxide.

Interesting types of materials are alumosilicates, either amorphous or crystalline (zeolites), which can be used as supports and catalysts *per se* (as zeolites) due to their acidic properties. Another often applied support mainly for synthesis of fine chemicals in various hydrogenation reactions is active carbon. Besides these supports, also clays, ceramic (cordierite), titania (for selective oxidations and selective catalytic reduction of NO_x), as well as magnesium aluminate (steam reforming of natural gas) are used.

The shaping of catalysts and supports is a key step in the catalyst preparation procedure. The shape and size of the catalyst particles should promote catalytic activity, strengthen the particle resistance to crushing and abrasion, minimize the bed pressure drop, diminish fabrication costs, and distribute dust build-up uniformly. While small particle size increases the activity by minimizing the influence of internal and external mass transfer, bed pressure drop (Figure 2.7) increases. Thus, there is an apparent contradiction between the desire to have small catalyst particles (less diffusional length and higher activity) and to utilize large particles displaying lower pressure drop.

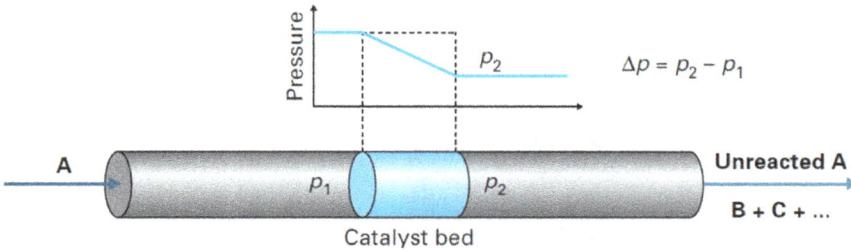

Figure 2.7: Pressure drop in fixed-bed reactors. Courtesy of Dr. E. Toukoniitty.

There are no precise guidelines about what should be the exact value of pressure drop. It is decided on a case-by-case basis depending on a particular process technology. As a rule of thumb, the size of catalyst particles in fixed beds is exceeding 1–2 mm to avoid high pressure drop, even if larger particles of 10–12 mm are also applied. In addition to the size, the shape is important affecting the bed porosity ε. Bed porosity for spheres, Raschig rings, and cylindrical particles varies between 0.35 and 0.4, 0.5 to 0.8 depending on the wall thickness, and 0.3 to 0.35, respectively.

Thus, the best operational catalysts have a shape and size that represents an optimum economic trade-off. The requirements of the shape (Figure 2.8) and size are mainly driven by the type of reactor. For reactors with fixed beds (see Chapter 3), relatively large particles are applied (several mm) to avoid pressure drop.

For calculation of the pressure drop in fixed-bed reactors with gaseous reactants often the Ergun equation is used:

$$\frac{\Delta P}{L} = \frac{150 \mu V_0 (1-\varepsilon)^2}{d^2{}_p \varepsilon^3} + \frac{7 \rho V^2{}_0 (1-\varepsilon)}{4 d_p \varepsilon^3} \tag{2.44}$$

ΔP is the pressure drop, L is the height of the bed, d_p is the particle diameter, ε is the porosity of the bed, ρ is the gas density, V_o is the superficial velocity defined as the volumetric flow rate per the cross section of the bed, and μ is the gas viscosity. Equation (2.44) containing two terms, the first one corresponding to the laminar and the second to the turbulent component, illustrates which parameters influence

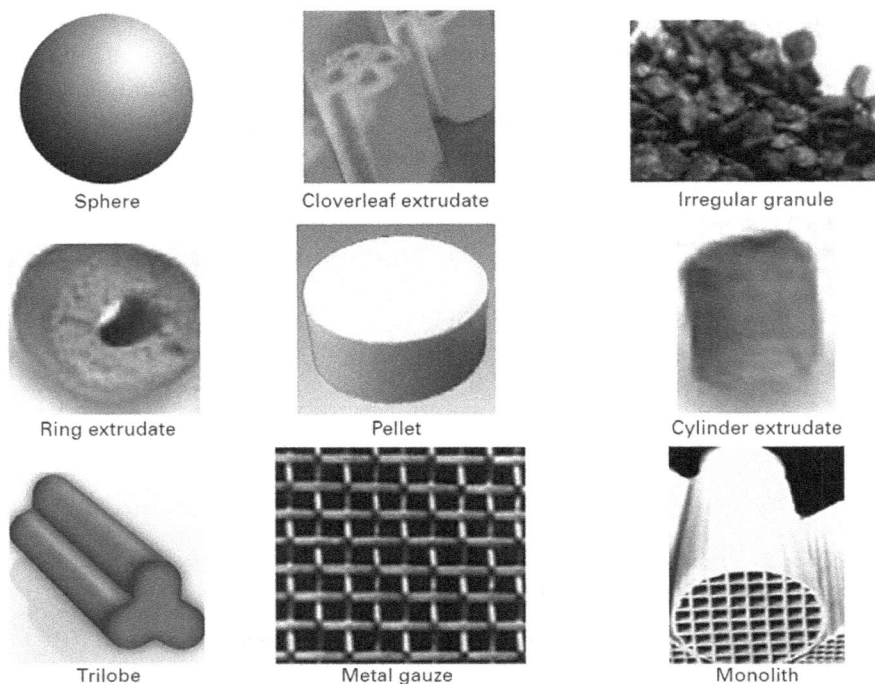

Figure 2.8: Various catalyst shapes.

the pressure drop. Apparently, lower sizes of catalyst particles and higher superficial velocity increase the pressure drop.

For moving-bed reactors (e.g. continuous catalytic reforming), spherical particles are preferred because they allow a smooth flow. Catalyst powders of various sizes are utilized in slurry three-phase reactors and in fluidized-bed reactors, where mechanical stress is found because of collisions between catalyst particles and with the reactor walls and formation of shear force due to cavitations at high velocities. In the case of slurry reactors, resistance to attrition is important, the size of the particles should allow easy filtration, while the bulk density is defined by settling requirements when easy settling is required. For fluidized-bed reactors, attrition resistance is important, as well as the particle size distribution.

Pressure drop can be regulated by making special types of extrudates, ranging from cylindrical to rings to cloverleaf extrudates (Figure 2.8). In the particular case of natural gas steam reforming, many types of catalysts with different shapes are available from catalyst manufacturers. Catalysts in the forms of rings, miniliths with several holes, wagon wheels, as well as some other complex geometrical shapes afford a low diffusional path in addition to low pressure drop. The mechanical stability typically deteriorates with an increase of complexity and decrease of the wall thickness. Extrudates with a typical aspect ratio of length L to diameter of ca. 3–6 have poorer strength compared to pellets (tablets), as the pellets possess good mechanical strength and a regular shape. This is a very common type of catalysts used in many hydrogenation, dehydrogenation and oxidation reactions.

Monoliths are mainly applied when high fluid flow rates (Figure 2.8) are required (off-gas cleaning for example), as they have a low pressure drop.

The manufacturing of monolithic catalysts is inherently more costly than of other shapes such as pellets or powders. The economic benefits of using monoliths should be thus clearly demonstrated by exceeding higher catalyst costs and investments in research and development. In particular, when the annual volume of each catalyst is small, it is difficult to justify the dedicated research. In addition, cordierite ($2MgO \times 2SiO_2 \times 5Al_2O_3$) has some limitations in terms of durability when in contact with alkali and alkaline earth above 700 °C.

Low residence time and low pressure drop are features of low surface area metal gauzes, which are utilized in very few specific cases, such as very exothermal oxidation of ammonia to NO, when longer residence times lead to excessive temperatures and volatilization of the active catalytic phase. Another example is oxidative dehydrogenation of methanol to formaldehyde on silver gauzes.

The choice of catalyst shapes is thus not straightforward and involves careful considerations of hydrodynamics, heat and mass transfer limitations, potential pressure drops, mechanical strength, thermal resistance to sintering and phase transition, efficient enough heat conductivity for strongly exo- and endothermic reactions, as well as manufacturing methods and associated costs.

Moreover, negative effects of a noncatalytic phase (carriers, binders, rheology improvers, lubricants, etc.) and solvents on catalytic behavior should be avoided.

2.4 Kinetics

Chemical kinetics is a discipline that concerns the rates of chemical reactions. It addresses how the reaction rates depend on concentrations, temperature, nature of a catalyst, pH, solvent, to mention a few reaction parameters.

A measure of activity is the reaction rate, which is defined through the extent of the reaction. The change in the extent of the reaction (number of chemical transformations divided by the Avogadro number) is given by $d\xi = dn_B/\nu_B$, where ν_B is the stoichiometric number of any reaction entity B (reactant or product) and n_B is the corresponding amount. This extensive property $d\xi/dt$ is measured in moles and cannot be considered as such as the reaction rate, since it is proportional to the reactor size.

For a homogeneous reaction, when the rate changes with time and is not uniform over a reactor volume v, the reaction rate is

$$r = \frac{\partial^2 \xi}{\partial t \partial v},$$ (2.45)

whereas for the constant reactor volume, it is defined as

$$r_i = \frac{1}{v_i} \frac{dC_i}{dt},$$ (2.46)

where i is the reactant or product with a corresponding stoichiometric coefficient v_i.

For a heterogeneous reaction occurring over a reaction space S (catalyst surface, volume weight, or number of active sites), the rate expression is given by

$$r = \frac{\partial^2 \xi}{\partial t \partial S},$$ (2.47)

leading to further simplifications when the rate is uniform across the surface

$$r = \frac{1}{S} \frac{\partial \xi}{\partial t}.$$ (2.48)

Rate laws express how the rate depends on concentration and rarely follow the overall stoichiometry. In fact, reaction molecularity (the number of species that must collide to produce the reaction) determines the form of a rate equation. Elementary reactions are those when the rate law can be written from its molecularity and which kinetics depends only on the number of reactant molecules in that step. For elementary reactions, the reaction orders have integral values typically equal to 1 and 2, or seldom 3 for trimolecular reactions.

Reaction orders m for a particular reaction can be fractional ($r_A = - kc_A^m$), indicating a complex reaction mechanism. Such formal kinetic equation with fractional orders can be useful to describe experimental data within a certain domain of parameters (concentrations), but a reliable prediction of a chemical process should be based on a mechanistically justified kinetics.

Some generalization of kinetic models is possible. For heterogeneous catalytic reactions $A + B = C + D$, the reaction rate takes the form

$$r = \frac{\text{kinetic factor} \ast \text{driving force}}{(\text{adsorption term})^n}$$ (2.49)

where the adsorption term includes adsorption coefficients of reactants multiplied by their concentrations (partial pressures); the power in the denominator corresponds to the number of species in the rate-determining steps, while the driving force is $(1 - P_C P_D / K_{eq} P_A P_B)$ or $(1 - c_C c_D / K_{eq} c_A c_B)$, with K_{eq} being the equilibrium constant.

For irreversible reactions, the kinetic expression is even simpler, not containing the term related to the driving force. As an example, we can consider the so-called Langmuir-Hinshelwood mechanism when two species A and B are adsorbed on catalyst active sites (*) in quasi-equilibrium steps, with subsequent surface reaction giving an adsorbed C. This surface reaction determines the reaction rate and is called rds or the rate-determining step.

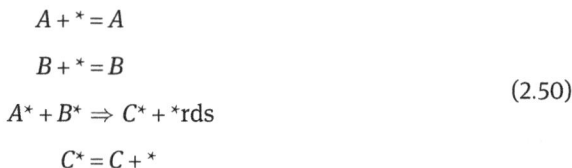

$$A + * = A$$
$$B + * = B$$
$$A^* + B^* \Rightarrow C^* + *\text{rds}$$
$$C^* = C + *$$

(2.50)

The rate in such case is often defined through partial pressures of reactants if the reaction occurs in the gas phase in the presence of a solid catalyst or through concentrations for liquid-phase reactions:

$$r_A = \frac{kK_AC_AK_BC_B}{(1 + K_AC_A + K_BC_B + K_CC_C)^2}$$

(2.51)

where k is the rate constant of rds. Strictly speaking, in the latter case of liquid phase reactions, activity should be applied instead of concentrations, and the equilibrium constant should also be determined through activity rather than concentrations.

One of the most important requirements for catalytic reactions as mentioned above is proper selectivity, which in a broad sense should be understood as chemoselectivity, regioselectivity, and enantioselectivity. Selectivity is the ability of a catalyst to selectively favor one among various competitive chemical reactions. Intrinsic selectivity is associated with the chemical composition and structure of surface (support), while shape selectivity is related with pore transport limitations (Figure 2.9). As can be seen in Figure 2.9, a branched alkane cannot penetrate inside the pores.

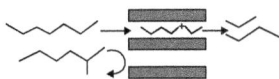

Figure 2.9: Reactant selectivity in catalysis by zeolites.

Chemoselectivity and regioselectivity describe the ability of a catalyst to discriminate among different and the same functional group, respectively, or several orientations. Diastereoselectivity defines the control of the spatial arrangement of the functional groups in the product, while enantioselectivity is related to the catalyst ability to discriminate between mirror-image isomers or enantiomers.

Since selectivity depends on conversion, it is extremely dangerous to compare selectivity for different catalysts at just one end-point or at a certain period of time. For parallel reactions, it still could be done, as selectivity for systems (1) $A \Rightarrow B$ and (2) $A \Rightarrow C$ with equal reaction orders is independent of the concentration of A and therefore of conversion.

2.5 Mass transfer

In any system, not only chemical reactions *per se* but mass and heat transfer effects should be considered as well.

First mass transfer and heat transfer effects in heterogeneous catalytic reactions will be discussed. These effects are present inside the porous catalyst particles and in the surrounding fluid films, resulting in concentration gradients across the phase boundaries and within the particle (Figure 2.10).

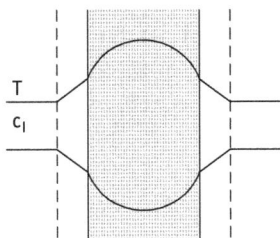

Figure 2.10: Concentration gradients and temperature profiles for an exothermal fluid-solid reaction with interphase (film or external) and intraparticle (internal) diffusion.

Due to heat and mass transfer, the observed rate in a catalytic reaction (macrokinetics) is different from the intrinsic rate of a catalytic transformation (microkinetics); thus, the modeling of a two-phase (fluid-solid) catalytic reactor includes simultaneous reaction and diffusion in the pores of the catalyst particle. In three-phase systems (gas-liquid-solid), the diffusion effects in the liquid films at the gas-liquid interphase (that is gas to liquid mass transfer) should also be considered.

The intraparticle and interphase mass transfer coefficients display lower temperature dependence than the intrinsic rate as visualized in Figure 2.11.

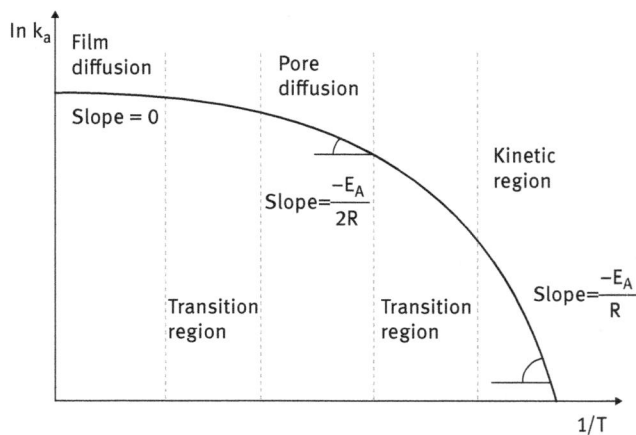

Figure 2.11: Temperature dependence of catalytic reactions (after D. Murzin, T. Salmi, *Catalytic Kinetics*, Elsevier, 2005).

For the determination of the mass transfer parameters from experimental data, the detailed reactor model containing kinetic and mass transfer could be used. The mass transfer parameters are estimated when the kinetic parameters are already available and are implemented as fixed parameters.

Elucidation of mass transfer can be done using dimensionless numbers. For example, in an isothermal case, when there is only transfer of mass from the bulk to the external surface of the catalyst and internal diffusion does not play a role, the external effectiveness factor η_{ext}, defined as the ratio of effective (observed) rate to the intrinsic chemical rate under bulk fluid conditions, takes a form

$$\eta_{ext} = 1/(1 + Da) \tag{2.52}$$

where Da is the Damköhler number $k_v/k_f\, a'$, i.e., the ratio of volumetric rate constant to the mass transfer coefficient times parameter a' (area divided by volume).

Large values of Da correspond to strong mass transfer limitations; therefore, the observed kinetics in the domain of mass transfer is of first order. In the case of strong external mass transfer limitations, increasing catalyst activity does not influence the rate. Catalyst poisoning, and deactivation might have an influence on the observed rate when the overall catalyst activity with operation time is decreased to such an extent that kinetics is becoming the limiting step.

It is clear that the effectiveness factor depends on the mass transfer coefficient, which in turn depends on the reactor, and hydrodynamic conditions, physical properties of the liquid, as well as the size of the catalyst grain. The mass transfer coefficient k_f depends on the velocity V and the diameter of catalyst particles d_p in the following way:

$$k_f \propto \left(\frac{V}{d_p}\right)^{0.5} \tag{2.53}$$

Thus, with increasing velocity and diminishing catalyst particle size, the impact of mass transfer on the intrinsic catalytic rate could be eliminated.

The mass transfer coefficient can be expressed through the diffusion coefficient $k_f \propto (D)^{2/3}$. The temperature dependence of the diffusion coefficient is defined for diffusion in the gas phase by the Chapman-Enskog equation:

$$D_{AB}\left(\text{cm}^2/\text{s}\right) = 1.8829 \times 10^{-3} \frac{(T)^{1.5}\left(\frac{1}{M_{rA}} + \frac{1}{M_{rB}}\right)^{0.5}}{p(\sigma_{AB})^2 \Omega_{AB}}. \tag{2.54}$$

Here, M_{rA} and M_{rB} are the relative molecular masses (dimensionless), p is the total pressure (kPa), σ_{AB} (nm) is the characteristic length (Lennard-Jones parameter) for a pair of molecules, Ω_{AB} is a collision integral and is a function of $k_B T/\varepsilon_{AB}$, where ε_{AB} (J) is another Lennard-Jones parameter and k_B (1.38×10^{-23} J/K) is the Boltzmann constant. Typical values of gas-phase diffusion coefficients are ca. $D_A \approx 10^{-5}$ m^2/s.

Equation (2.44) gives the following temperature dependence $D \propto T^{3/2}$, finally resulting in $k_f \propto (D)^{4/6} \propto (T^{3/2})^{2/3} \propto T$ and a very minor temperature dependence of the observed reaction rate with the apparent activation energy being below 5–10 kJ/mol. The diffusion coefficient according to the Chapman-Enskog equation is inversely proportional to pressure, and therefore, mass transfer is becoming more prominent with pressure increase.

For slurry reactors, the liquid-solid mass transfer coefficients κ_{LS} is

$$k_{LS} = \left(\frac{\varepsilon D^4 \rho}{\eta d_p^2} \right)^{\frac{1}{6}}, \tag{2.55}$$

where ε denotes the specific mixing power, D (or Do_{AB}) is the mutual diffusion coefficient of solute A in solvent B, ρ is the solvent density, η is the solvent viscosity, and d_p is the diameter of the catalyst particles.

A common test to verify if mass transport controls a catalytic reaction in three-phase slurry reactors is to vary the catalyst mass (for gas-liquid mass transfer) and the rate of agitation (for liquid-solid mass transfer). When the reactor productivity is independent on the catalyst mass, the observed rate is thus governed by the gas-liquid mass transfer.

For agitation, the situation is more complex, since energy dissipation can have a complex behavior depending on the selection of the solvent, stirring rate, liquid volume in comparison to the total volume, and design of the reactor internals, including impellers and baffles. In the majority of cases in three-phase reactors operating in the industry, agitation is not sufficient to overcome mass transfer limitations.

For the calculation of binary diffusion coefficients in the liquid phase, semi-empirical equations are often used, such as Wilke-Chang equation described in specialized literature, giving accurate results for diffusion coefficients of gases in liquids.

In a porous catalyst particle, the reacting molecules diffuse first through the fluid film surrounding the particle surface and then diffuse into the pores of the catalyst to the active sites. In a similar way, the reaction products are diffusing out of the catalyst grains. As an outcome of pore diffusion in the case of most common reaction kinetics, the reaction rates inside the pores have lower values than what would be expected with the concentration levels of the main bulk.

The effectiveness factor $\eta = r_{obs}/r_{kinetics}$, which relates the observed reaction rate with the intrinsic chemical rate, can be graphically (Figure 2.12) presented as a function of generalized Thiele modulus for an isothermal pellet:

$$\phi_p = \frac{V_p}{A_p} \sqrt{\frac{kc^{n-1}}{D_e}}. \tag{2.56}$$

This modulus is proportional to L- the ratio between the external pellet surface and volume, $L = V_p/A_p$, which is the characteristic length of diffusion. The effectiveness

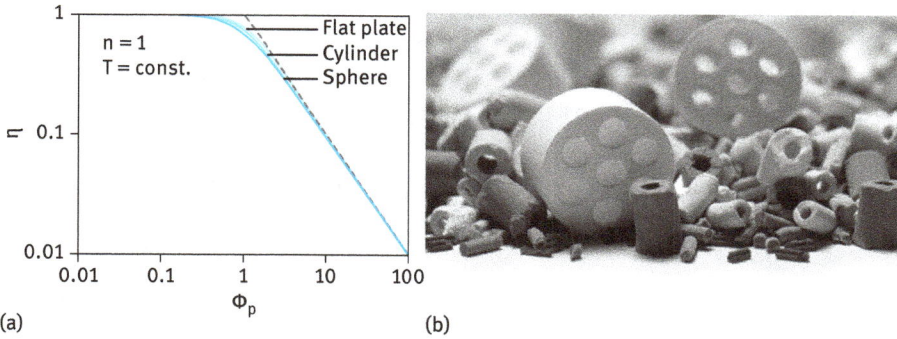

Figure 2.12: (a) Effectiveness factor η as a function of the generalized Thiele modulus φ_p for different pellet geometries (from R. Dittmeyer, G. Emig, Simultaneous heat and mass transfer and chemical reaction, in *Handbook of Heterogeneous Catalysis*, 2nd Ed., edited by G. Ertl, H. Knözinger, F. Schueth, J. Weitkamp, Copyright 2008, Wiley-VCH, Weinheim, ISBN: 978-3-527-31,241-2. Reproduced with permission) and (b) typical geometries of industrial catalysts (https://www.maxlab.lu.se/files/Topsoe_main-image(1).jpg).

factor for different geometries (flat plate, cylinder, sphere) in the case of isothermal, first-order irreversible reaction is shown in Figure 2.12, demonstrating that the particle geometry is of minor importance for the effectiveness factor.

In eq. (2.56), k denotes the rate constant and n is the reaction order. The effective diffusion coefficient is defined in the following way: $D_e = D(\varepsilon/\tau)$ being smaller than the diffusion coefficient *per se* since the diffusional cross section is smaller than the geometric cross section (thus, porosity ε is introduced) and the catalyst has irregular pore structure (expressed *via* tortuosity τ) as illustrated in Figure 2.13. Typically, ε/τ is in the range between 0.05 and 2.

Figure 2.13: Illustration of porosity and tortuosity.

For calculations of the diffusion coefficient D, molecular diffusion D_{AB} and Knudsen diffusion D_K are considered through the Bosanquet approximation,

$$\frac{1}{D} = \frac{1}{D_{AB}} + \frac{1}{D_{K_A}}, \tag{2.57}$$

where the first term stems from molecular diffusion and the second from Knudsen diffusion. The latter is important for materials with small pores when the mean free path becomes comparable to the size of the pore and molecules are colliding with the walls rather than with each other. Such diffusion is thus independent on pressure and is not observed in the liquid-solid catalytic reactions when the fluid density is much higher compared to the gas-solid catalysis.

The Knudsen diffusion coefficient is proportional to the pore radius r_e and the mean molecular velocity, giving then the proportionality of the Knudsen diffusion coefficient to the square root of temperature:

$$D_K = \frac{2}{3} r_e \sqrt{\frac{8\,RT}{\pi\,M}}, \tag{2.58}$$

where R is the gas constant, T is the absolute temperature, and M is the molecular mass.

Figure 2.12 clearly indicates that the effectiveness factor depends strongly on the size of catalyst grains. At small values of the Thiele modulus (i.e., small particle size), the effectiveness factor is approaching unity. The flat dependence of effectiveness factor on the Thiele modulus occurs at small catalyst particles, low catalyst activity (small k), large pore size, and high porosity (large D_e). When $\varphi \gg 3$, the following dependence is valid, $\eta \propto 1/\varphi$, and the effectiveness factor is inversely proportional to the Thiele modulus and thus to particle size. For large values of Thiele modulus, the overall rate is controlled by pore diffusion, and for very active catalysts or for catalysts with small pores, low porosity, and/or large diameter of catalyst particles, the reactant concentration approaches zero in the center of a particle.

Obviously, in laboratory-scale reactors, the size of catalyst particles can be rather small in order to diminish the impact of internal diffusion, while in fixed-bed industrial reactors, owing to increased pressure drop, the size of catalyst grains is unavoidably much higher, resulting in significant influence of internal diffusion. For slurry reactors, even at the pilot stage, the size of catalyst powder could be still in the range of 50–100 μm, which in most cases (i.e., when catalytic reactions are not very fast) is sufficient to eliminate internal diffusion. However, external diffusion limitations can still play a role.

For some heterogeneous catalytic reactions (oxidations, hydrogenations, dehydrogenations), substantial consumption or release of heat results in non-isothermal temperature profiles inside the catalyst particle and in the film surrounding the particle. For highly exothermic processes, the effectiveness factor can even exceed unity, due to temperature rise inside the particle and increased values of the rate constants, which are not overcompensated by the lower concentrations inside the pellet because of the diffusion. This effect is particularly visible at small values of the Thiele modulus.

Not only catalytic activity but also selectivity can be influenced by mass transfer phenomena. Differential selectivity in consecutive reactions $A \rightarrow B \rightarrow C$ depends on the values of the Thiele modulus φ and parameter λ. The value of the latter parameter is defined as $\gamma = \sqrt{k_B D_{e,A}/k_A D_{e,B}}$, with k_A and k_B being rate constants for reaction of A to B and B to C, respectively, while $D_{e,A}$ and $D_{e,B}$ correspond to their effective diffusion coefficients. The higher the value of parameter λ, the more pronounced is the influence of diffusion, resulting in lower selectivity toward intermediate B (Figure 2.14). This is an important conclusion pointing out that internal diffusion limitations, prominent in industrial conditions due to the large size of catalyst pellets, lead to diminished selectivity toward the intermediate product in comparison with the intrinsic kinetic conditions.

maximum shifts to lower conversion of A

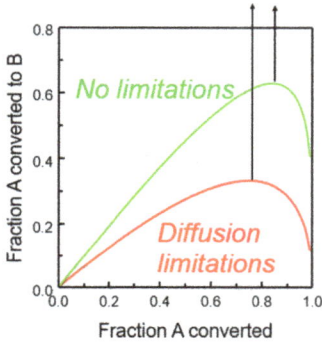

Figure 2.14: Fraction of reactant A is converted to the intermediate product B as a function of the fraction of A converted for a consecutive reaction A-> B-> C.

For parallel reactions $A \Rightarrow B_1$ and $A \Rightarrow B_2$, when the reactions are of the same order, the differential selectivity is independent on the presence of internal diffusion. If the desired reaction is of lower order then it is preferential to conduct the reaction in the diffusion region, since the highest penalty in the case of pore diffusion limitations is on the highest-order reaction.

The presence of two phases, namely gas and liquid, is characteristic of non-catalytic or homogeneously catalyzed reaction systems. Components in the gas phase diffuse to the gas-liquid interphase, dissolve in the liquid phase, and react with components in the bulk liquid phase. The liquid phase may also contain a homogeneous catalyst. Some of the product molecules desorb from the liquid phase to the gas phase and some product molecules remain in the liquid. The processes taking place in a gas-liquid reactor are displayed in Figure 2.15.

Two reactor types dominate in the synthesis of chemicals in the case of gas-liquid reactions: the tank reactor and the bubble column. Both types can be operated in continuous or semi-batch mode. In the semi-batch operation, the liquid phase is treated as a batch and the gas phase flows continuously through the liquid.

The following special cases can be distinguished according to the reaction kinetics: physical absorption, very slow reactions, slow reactions, normal reactions, fast reactions, and infinitely fast reactions. A more detailed description of the different reaction types is summarized in Table 2.1.

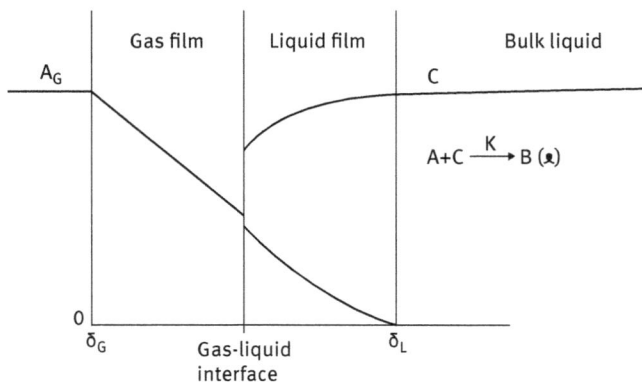

Figure 2.15: Mass transfer on gas-liquid interface.

Table 2.1: Summary of different reaction types.

Physical absorption	No chemical reaction in the liquid film and bulk
	Linear concentrations in the films
Very slow reaction	The same reaction velocity in the liquid film and in the liquid bulk
	No concentration gradients in the liquid film
Slow reaction	No reaction in the liquid film, chemical reaction in the liquid bulk
	Linear concentration gradients in the films
Finite speed reaction	Chemical reaction in the liquid film and in the liquid bulk
	Non-linear concentration profiles in the liquid film
Fast reaction	Chemical reaction in the liquid film
	No chemical reaction in the liquid bulk
	Non-linear concentration profiles in the liquid film
	The gas-phase component concentration is zero in the liquid phase
Infinitely fast reaction	Chemical reaction in the reaction zone in the liquid film
	The diffusion rates of the components determine the reaction velocity

Analytical expressions can be derived for the fluxes in the case of different reaction types and reaction kinetics, which are of importance for homogeneous catalysis, e.g., slow and finite speed reactions. The details are available in the specialized literature.

Chapter 3
Chemical processes and unit operations

3.1 Overview of unit operations

Unit operation is an important concept in chemical engineering, reflecting a basic step in the overall process. Each unit operation follows the same physical laws and may be used in all chemical industries. Thus, the approach of unit operations allows classifying different processes, such as separation, filtration, crystallization, independent on their chemical specificity and quantifying them based on the underlying physical laws. Obviously, many unit operations might be needed to obtain the desired product.

The following unit operations are typically present in chemical technology:
1. Mechanical and hydromechanical processes, which include transportation of solids and fluids, crushing, pulverization, screening, sieving, filtration.
2. Mass transfer processes, including absorption, adsorption, distillation, extraction, etc.
3. Heat transfer processes, including evaporation and condensation

In many textbooks on unit operations, chemical reactions are typically not discussed or considered separately. The unit operations in chemical technology can be also subdivided in three classes: combination (mixing), separation, and chemical transformations *per se*.

Industrial processes could be either continuous or discontinuous. In continuous processes, the materials (solids, liquids, and/or gases) are being processed continuously, undergoing chemical reactions, mechanical, or heat treatment. Continuous usually means operating 24 h per day, 7 days per week, with infrequent maintenance shutdowns, which could be done on semi-annual or annual basis. There are examples of chemical plants operating for more than 1 or 2 years without a shutdown. Continuous processes dominate in oil refining and synthesis of bulk chemicals, allowing advanced process control and constant product quality. Scaling up of the processes from the laboratory scale is done using the basic principles of chemical engineering and involves, for example, detailed modeling of separation processes and chemical reactors.

An alternative to continuous is batch production, which is more typical for smaller-scale manufacturing, such as synthesis of pharmaceutical ingredients, fine chemicals, inks, paints, adhesives, etc. Scaling up of batch processes is typically less complicated and a single production line can be used to produce several products that are typically manufactured on a campaign basis. As a consequence, equipment must be stopped and reactors must be cleaned if needed or even reconfigured. The output should be tested before the next batch can be produced. Even if only

https://doi.org/10.1515/9783110712551-003

one product is formed using a batch operation mode, there still is a need, for example, to load reactants, unload products, etc. The downtime (idle time between batches) may be rather long. In addition, such batch mode of operation can lead to variability in product quality and substantial losses if the obtained product is out of specification.

In the subsequent section, few of the most important unit operations will be considered in more detail.

3.2 Mechanical and hydromechanical processes

These processes in general are subdivided into
- separation of solids,
- separation of nonhomogeneous mixtures (sedimentation, filtration, flotation, defoaming, and cyclonic separations),
- dosing,
- mixing (of solids, pastes, mechanical and pneumatic mixing, preparation of dispersions, emulsions, foams, etc.),
- bubbling,
- forming (granulation, extrusion, tableting, etc.), and
- transportation of gases and liquids.

The separation principles for sedimentation and filtration are presented in Table 3.1.

Another view on separation is not considered the principles of separation, but rather the phases.

For homogeneous phases (i.e. gases and liquids), the classification is presented in Figure 3.1. They will be mainly considered in the following section, where mass and heat transfer-based separations will be discussed.

3.2.1 Sedimentation

Sedimentation relies on the difference in density between the liquid and the solid. Sedimentation is used either to increase the concentration of solids in the feed stream or remove small quantities of suspended particles giving a clear liquid.

Separation of solids from the liquid is achieved either because of gravity or centrifugal settling. The latter can be organized in gravity-settling tanks, which can have either vertical (Figure 3.2) or horizontal arrangements. In the former case, the solids fall in a countercurrent mode to the upward flow of the liquid (e.g. water as in Figure 3.2). The latter should be effectively distributed through spreaders and then collected in the upper part of the vessel. Flocculation agents may be added to enhance settling.

Table 3.1: Separation principles of mechanical separations (from A. de Haan and H. Bosch, Industrial Separation Processes. Fundamentals, 2013, Copyright de Gruyter. Reproduced with permission).

Method	Mechanical force	Technique	Applicable for particles (micron)	Principle
Sedimentation	Gravity	Settlers Classifiers	>100	Density difference
	Centrifugal	Centrifuges Cyclones	1–1,000	Density difference
	Electrostatic	Electrostatic precipitators	0.01–10	Charge on fine solid
	Magnetic	Liquid + Solid		Magnetism
Filtration	Gravity	Sieves filtration	>100	Particle size larger than the pore size of filter medium
	Pressure	Filtration Presses Sieves membranes	0.001 – 1,000	Particle size larger than the pore size of filter medium
	Centrifugal	Centrifuges	1–1,000	Particle size larger than pore size of filter medium
	Impingement	Filters Scrubbers Impact separators	0.1–1,000	Size difference

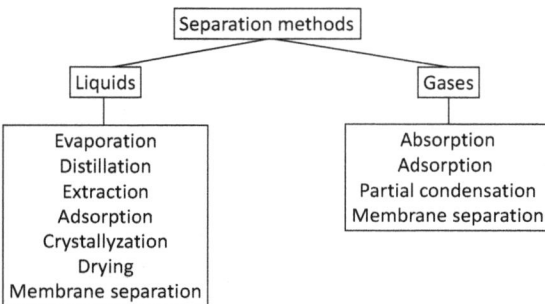

Figure 3.1: Main methods for separation of homogeneous mixtures.

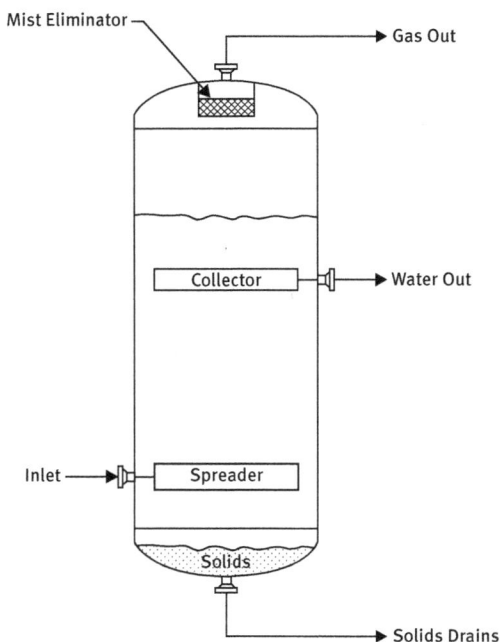

Figure 3.2: A scheme of a vertical settling tank. From https://ars.els-cdn.com/content/image/3-s2. 0-B9780750689700000049-gr6.jpg.

If separation by gravity is not sufficient, the force on particles can be increased using centrifugal sedimentation. This allows separation of finer particles than achieved just by gravity. Such separation is achieved either using sedimentating centrifuges or (hydro)cyclones.

In cyclonic separation, particulates are removed from gas or liquid streams through vortex separation using rotational effects and gravity. A high-speed rotating flow is in a cylindrical or conical cyclone, which has a helical pattern, beginning at the wide end (top) of the cyclone and ending at the narrow (bottom) end before exiting in a straight stream through the center of the cyclone. More dense particles having too much inertia strike the outside wall and fall to the bottom of the cyclone, thereafter being removed (Figure 3.3).

In a conical system, as the rotating flow moves toward the narrow end of the cyclone, the rotational radius of the stream is reduced, thus separating smaller and smaller particles. The size of the particles that are removed with 50% efficiency is the cyclone cut point and is determined by the cyclone geometry and the flow rates. Larger particles are more efficiently removed.

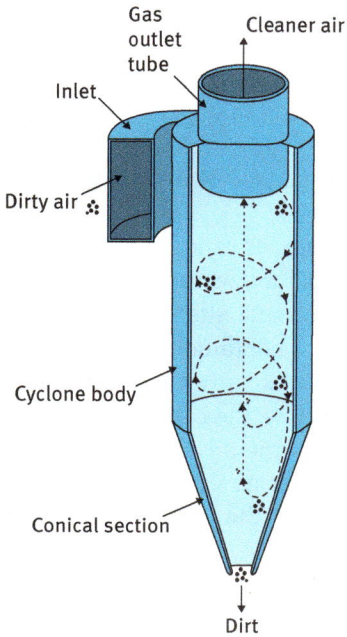

Figure 3.3: Separation with a cyclone. http://en.wikipedia.org/wiki/File:Cyclone_separator.svg.

Cyclones can be put in series to improve a poor sharpness or in parallel to improve retention efficiency. The latter arrangement (Figure 3.4) is needed, because high efficiency is achieved in devices of low diameters, which poses some limits on the throughput. Subsequently, several cyclones, as in the case of fluid catalytic cracking (Figure 3.4), operate in parallel.

Figure 3.4: Cyclones for a fluid catalytic cracking reactor.

3.2.2 Filtration

Filtration is the mechanical (with the help of filtration media) separation of suspensions into liquid and solid fractions depending on the size of solid particles.

In fact, separation of phases is not complete, and the separated solids (cake) contains some residual moisture, while the filtrate (separated liquid phase) often contains some solids, resulting in certain turbidity.

Filtration is done by application of vacuum, pressure, or centrifugal force (see Figure 3.5). Vacuum filtration requires generation of vacuum, while in the pressure filtration, the filter is placed within a pressure vessel. Pressure filters typically operate in a semi-continuous mode, being incorporated in a continuous process. This requires a surge tank upstream the filter and batch collection of cake downstream. In continuous pressure filters, it is more difficult to remove the cake; thus, the filters are mechanically complex and expensive. Besides the advantages of lower moisture content, there are several disadvantages, such as difficulties in cloth washing and cleaning of internals and inability to inspect the forming cake while the filter is in operation.

Centrifugal filtration is done in perforated centrifuge rotors.

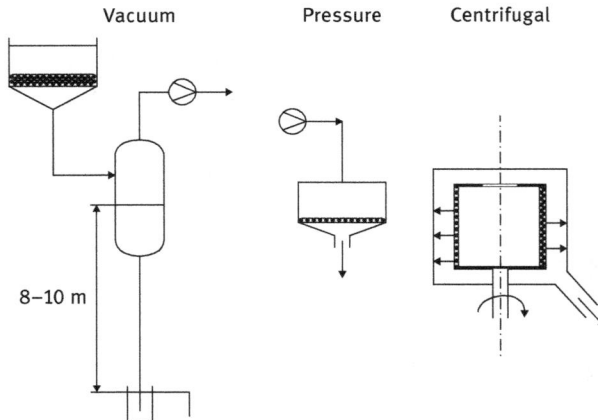

Figure 3.5: Different filtration types: (a) vacuum, (b) pressure, and (c) centrifugal force.

Cake handling is easy in vacuum filters and can be done automatically. At the same time, processing of hot liquids or solvents with high vapor pressure is troublesome. The pressure difference in vacuum filters is very limited. The residual moisture of the filter cake after vacuum filtration is thus higher than with pressure filters. In the latter case, handling of the filter cake is more complicated, resulting, however, in lower residual moisture content, which is important. Similar to pressure filtration centrifugal force yields solids with lower residual moisture.

Various models are used to describe filtration. In the often-used cake filtration model, it is supposed that filtration is done primarily not because of the filter material, but rather due to a homogeneous porous layer with a constant permeability that is formed during filtration. The pressure drop is thus linearly proportional to the amount of solid for an incompressible cake. After reaching a certain level of the filter cake, the latter should be removed, retaining only a small primary filtrate layer before restarting the filtration process starts again.

Filtering at constant pressure as implemented in vacuum filtration results in decline of filtration rates as the filter cake is growing in size. Too thick filter cakes lead to prolong filter cycles because of low filtering, dewatering, and washing rates.

The filter medium should be selected, taking into account the suspension properties, such as particle size and viscosity, and should be permeable with low pressure drop, chemically and mechanically stable, and moreover have a smooth surface for an easy cake removal.

Woven and non-woven fabrics of natural (e.g. wool or cotton, silk) or synthetic origin are often used in various types of filters, allowing to trap particles of the size exceeding ca. 10 µm.

As an example of filtration equipment, Nutsche filters (Figure 3.6) will be considered below. They are designed to operate under either vacuum or pressure (ca. 0.2–0.3 MPa). The latter version is often applied for batch-oriented industries, i.e., synthesis of fine chemicals, agrochemicals, etc.

Figure 3.6: Nutsche filter http://www.chimmash.com.ua/fv1.htm.

Nutsche filters can handle batches of 25 m^3 and a cake volume of 10 m^3 and are thus able to work with an entire charge of slurry. Sufficient holding volume is required for fast charging and emptying of the vessel. The difficulties of operation with such filters arise when cakes are slow to form and sticky and the product deteriorates during long downtime.

The operational sequence starts with filtration *per se* when the filter is charged with slurry and the pressure is applied. In the washing stage, the wash liquid is introduced over the cake, displacing the mother liquor. In the drying stage, air or gas purges the cake until the desired drying level. The final step is the cake discharge, and in some instances, washing the cloth or woven mesh screen with water to remove any cake residue.

In continuous large-scale vacuum filters, the suspension is introduced to the filter at atmospheric pressure. Vacuum applied on the filtrate side of the medium creates the driving force for filtration. The rotary vacuum drum filter is illustrated in Figure 3.7.

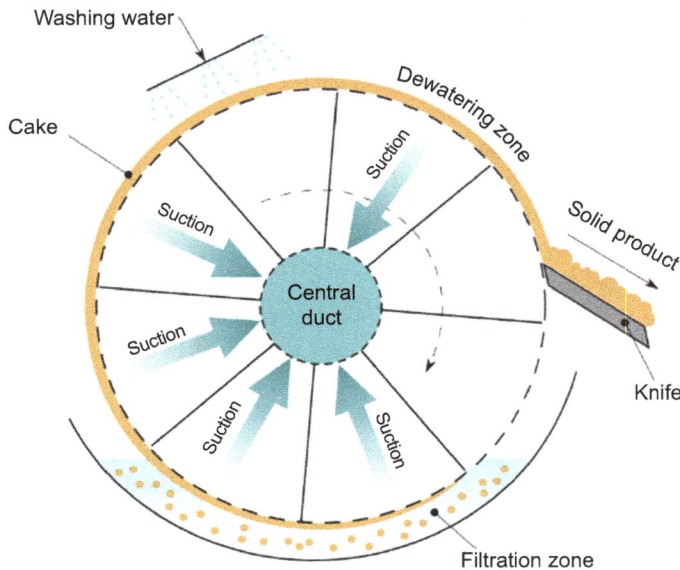

Figure 3.7: Schematic of a rotating vacuum filter. https://en.wikipedia.org/wiki/File:Rotary_vac uum-drum_filter.svg.

As the drum rotates, being partially submerged in the slurry, solids trapped on the drum surface are washed and dried. The cake discharge occurs at the end of the rotational cycle. The drum surface covered with a cloth filter medium can be pre-coated with a filter aid to improve filtration and increase cake permeability.

Horizontal filters, such as the horizontal belt filter presented in Figure 3.8, allow settling by gravity before the vacuum is applied. Horizontal-belt filters having a simple design and low maintenance costs have difficulties to handle very fast filtering materials on a large scale.

Membrane filtration allows to separate particles in submicron levels, which cannot be achieved using conventional filters. An important advantage of membrane separations is that, contrary to evaporation and distillation, they do not require additional heat.

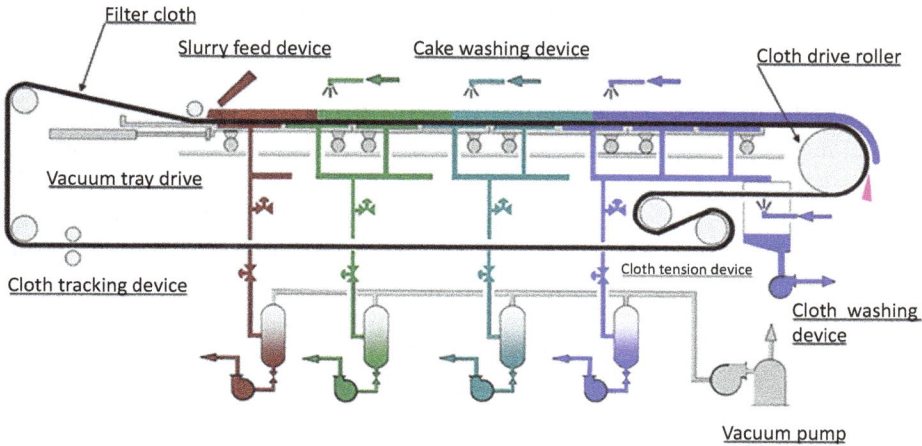

Figure 3.8: Schematic of a horizontal belt filter. From https://www.tsk-g.co.jp/wp/wp-content/up loads/2021/03/Horizontal-Belt-Filter001.png.

The size of the pores in a membrane should be smaller than the size of the smallest particles, otherwise the membrane cannot retain them.

The driving force in the most common applied methods of microfiltration and ultrafiltration is pressure (Figure 3.9).

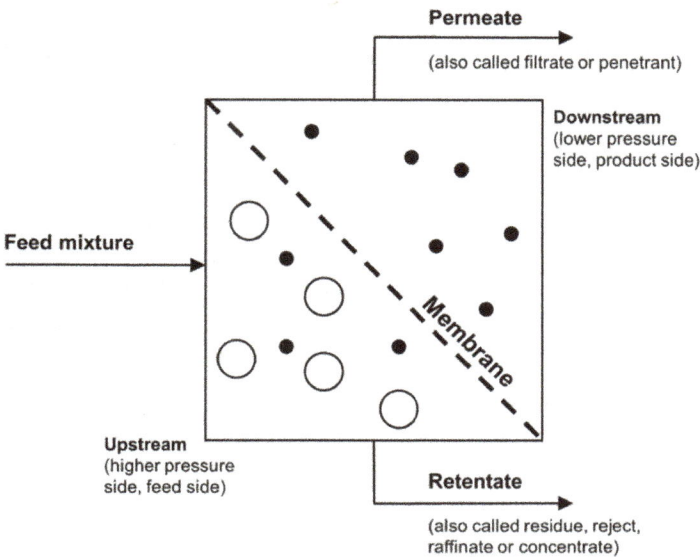

Figure 3.9: Principles of membrane filtration (from A. de Haan and H.Bosch, Industrial Separation Processes. Fundamentals, 2013, Copyright de Gruyter. Reproduced with permission).

In microfiltration membranes made of such polymers, as for example, poly-carbonate, polypropylene, and polyethylene, the pores are in the range of 0.05–3 µm. In ultrafiltration, cellulose acetate, polyvinylidene fluoride, and polysulfone are applied allowing removal of particles of the size from 0.005 to 0.1 µm. Other special membrane separation methods include, for example, electrodialysis and electrofiltration.

Application of membranes is limited by several factors including poor chemical resistance of polymers in some organic media, thermal stability of the membrane if elevated temperatures are needed in process technology, and fouling of a membrane or specific scaling up features. Contrary to other unit operations, which scale with 2/3 law in terms of costs, the capital costs for membrane technology scale linearly as such scaling is done by numbering up rather than installing larger modules. These reasons limit widespread utilization of membrane separation in process industries.

3.2.3 Mixing of emulsions

Emulsions are typically prepared by dissolving the emulsifying agent into the phase where it is most soluble. This is followed by adding the second phase and applying shear by efficient mixing. For o/w emulsions, such vigorous agitation can be crucial for making sufficiently small droplets. Thus, after an initial mixing, a second mixing with very high applied mechanical shear forces might be required. Several process parameters are important for proper homogenization. Energy density (energy input per volume) defines the minimum achievable droplet size. The latter typically decreases with an increase of energy density, unless mixing is inefficient. Energy efficiency influences heat losses and manufacturing costs, while production capacity depends on the volume flow rates. Some limitations on which type of materials can be homogenized are imposed by the product rheology.

The following devices can be used for preparation of industrial emulsions: vessels with high-speed stirrers, agitation or impact machines, centrifuges, colloid mills, metering pumps, vibrators, ultrasonic generators, and homogenizers. Some of the devices used for homogenization are presented in Figure 3.10.

3.2.4 Size reduction

Size reduction or comminution of solids is a unit operation to produce a desired particle size distribution needed for a final application. The ground product after milling is separated, and a coarser stream is returned to the mill feed. Such recycling helps to avoid overgrinding.

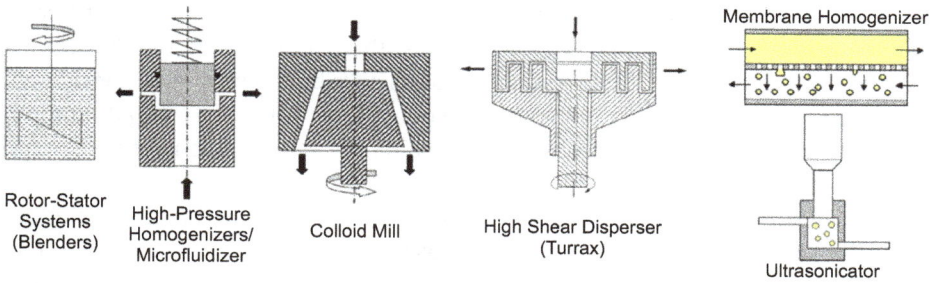

Figure 3.10: Devices for homogenization. From http://people.umass.edu/mcclemen/FoodEmul sions2008/Presentations(PDF)/(5)Emulsion_Formation.pdf.

Several options are used for size reduction. The necessary stress applied between two solid surfaces (crushing) either for single particles or a bed of particles is determined by the force applied to the solid surfaces. Size reduction can be also achieved by the impact of a particle against a solid surface or other particles. Two variants can be used: either a solid surface is accelerated to impact the particle or alternatively the particle is accelerated against a surface.

Other ways of size reduction include, for example, impact and cutting mills. In the former case, the stress is applied because of a machine–particle contact when the particles fly against an impact plate. Cutting mills imply cutting between rotating and static sharp edges with a narrow clearance and are used for such nonabrasive materials as polymers, rubber, or paper, which are too tough to be processed with other types of mills. In such cutting mills equipped with stationary knife bars and a rotor, the latter has several knife blades for cutting the product.

3.2.5 Size enlargement

Contrary to size reduction, size enlargement is needed to form a coarser product by agglomeration. Such size enlargement is employed for making a variety of products in manufacturing of fertilizers, pesticides, catalysts, or pharmaceutical products in different shapes (spheres, tablets, etc.).

Size enlargement can be done with or without external forces allowing forming larger agglomerates by, for example, pressure compaction and extrusion, tumbling, and other agitation methods, using chemical reactions or such physical methods as drying. In some methods, to achieve desired mechanical properties, binders are added to the feed. Catalysts made by extrusion can be one of the cases when binders are very often necessary as otherwise shaped materials will fall apart.

There are different agglomeration methods reflecting a variety of feedstock, which can be a dry solid or a wet paste or a fluid. In some cases, pumps can be used to transport the starting material, while in some other case this can be challenging. Some materials can be sprayed, while others not. Thus, the initial state as well as the final agglomerate shape, size, and its distribution should be considered. For some applications, the final materials should have sufficient strength and porosity, such as, for instance, in catalysis. Finally, the size enlargement or shaping methods should be selected based on the production capacity.

When size enlargement is done by growth agglomeration, capillary binding forces hold the particles together in the presence of aqueous media and binders, allowing to make typically spherical agglomerates of the size between 0.5 and 20 mm. Binders are often added in order to enhance particle-to-particle adhesion helping to keep the strength after drying. Inclined devices (drums, cones, pans, etc.) can be used for size enlargement (Figure 3.11).

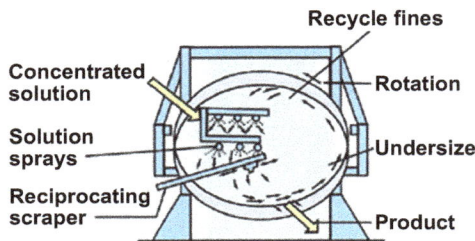

Figure 3.11: Continuous agglomerator. From http://cdn.chemengonline.com/wp-content/uploads/2017/12/12.jpg.

In an inclined agglomerator, the feed is from the top, while the product agglomerates of a rather uniform size are discharged over the rim. Large capacity, longer residence type, easy operation with dusty materials, and robustness are clear advantages of drum agglomerators.

Obviously, after a certain size no further agglomeration is possible in such agitated system as the destructive forces are becoming more prominent. Temporarily bonded conglomerates should undergo a curing step in tumble agglomeration methods to generate more permanent bonding. An example of such curing is drying, which in general is expensive. Nevertheless, tumble agglomeration can have a high throughput making fine particles of high surface area and porosity.

In spray drying (Figure 3.12) used for drying and agglomeration of pastes, suspensions, or solutions, the starting material is sprayed into the drying agent (e.g. hot air) by a suitable atomizer. This allows the formation of spray-dried particles of the size from 5 μm to 1 mm.

The following parameters influence the properties of the spray-dried particles: composition and nature of the ingredients, solid content, solvent type and viscosity, atomization pressure, feed rate, inlet air temperature, gas feed rate, and drying air humidity. Rapid solvent evaporation arranged by high inlet air temperature and selection

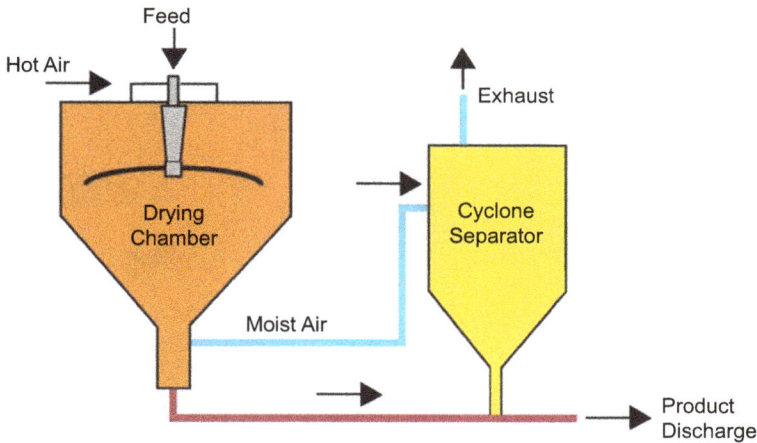

Figure 3.12: Illustration of spray drying. From https://www.eurotherm.com/spray-drying.

of a suitable solvent can give amorphous powders that are more moisture sensitive. Complex molecules exhibit lower chances to crystallize during spray drying. Drying chambers can be of different shapes determined by the spray pattern. Size of particles is important in defining the size of a drying tower, as larger particles dry slower.

3.2.5.1 Tableting

Particles with only low amounts of moisture can be agglomerated in tablets of few mm and larger briquettes of several centimeters in different types of presses. Advantages of this method include besides operation with dry solids, also the uniform shape of tablets, independence on the feed particle size, rather high throughput (up to 30 t/h), and a possibility to avoid cutting.

Often tableting is considered primary in connection with production of pharmaceuticals; however, in other areas, such as production of catalysts, tableting is also widely used.

Compression of the material loaded into a mold is done using two punches (upper and lower ones). In eccentric press, lower punch is usually stationary even if it can be adjusted. After completion of pressing, the tablet is removed by an expulsion stroke of the lower punch (Figure 3.13). Such tableting of relatively low capacity (3,000 tablets/h) can be used for production of small batches of catalysts.

Tableting problems are related to potentially poor flow properties of the substrate, too high moisture content, or too low lubrication. As a result of sticking during the tablet release, surface defects can appear. The interaction of the particle size distribution, the crystal shape, and other physical properties of the substrate are the main factors affecting the compressive strength, the disintegration time, and the release of the active ingredient.

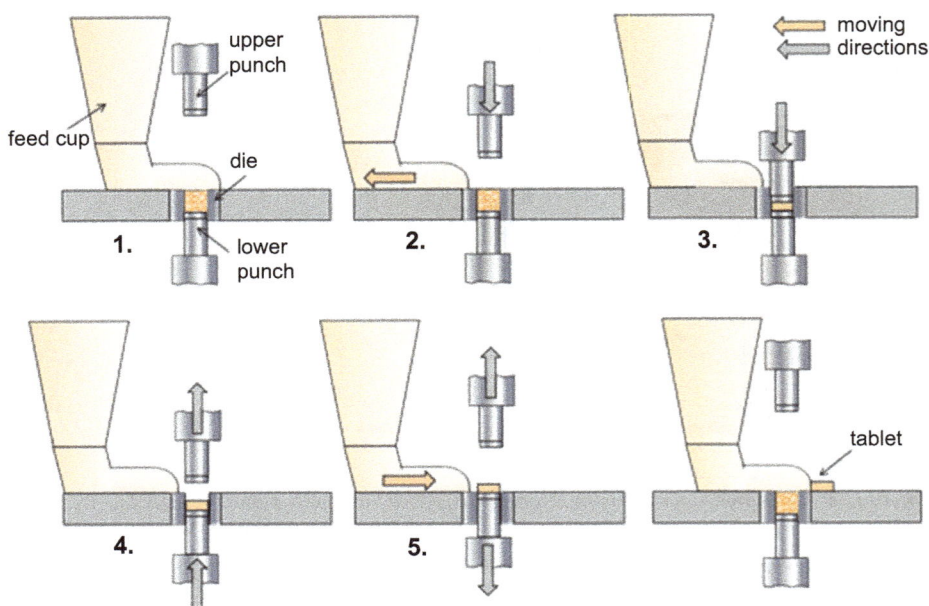

Figure 3.13: Schematic illustration of the compression process in an eccentric tablet press. From http://www.tankonyvtar.hu/hu/tartalom/tamop412A/2011-0016_01_the_theory_and_practise_of_pharmaceutical_technology/ch24.html.

The steps are: (1) the feed cup fills the die; (2) in order to fill the same volume, the feed cup removes the excess material from the surface of the die and moves away to ensure the free movement of the upper punch; (3) the upper punch compresses the particle aggregate inside the die; (4) the upper punch returns to its initial upper dead point position, while the lower punch ejects the tablet from the die – reaching its upper dead point position; (5) the feed cup rolls off the tablet; and (6) the lower punch returns to its lower dead point position and the feed cup fills the die.

Improvement of tableting can be done by modification of the filler properties, changes in a binder or a lubricant, pressure and speed, and more importantly humidity and compression rate.

As an alternative to eccentric tablet presses, rotating ones are used where compressive forces are generated between an upper and a lower pressure roll (Figure 3.14).

In a rotating press, compression is done not only from the top but also from beneath the tablet producing one tablet per revolution. Centrally rotating machines are capable of producing ca. 10–700,000 tablets per hour depending on the rotation speed and a number of punch pairs.

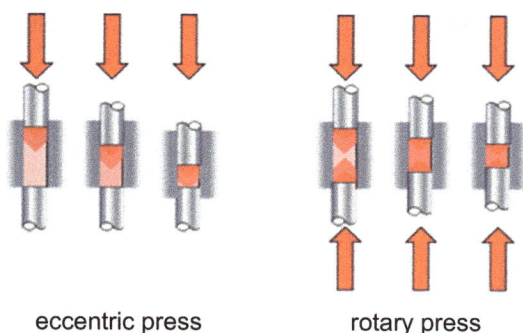

eccentric press   rotary press

Figure 3.14: Schematic illustration of the compression process in an eccentric and rotary press. From http://www.tankonyvtar.hu/hu/tartalom/tamop412A/2011-0016_01_the_theory_and_prac tise_of_pharmaceutical_technology/ch24.html.

3.2.5.2 Extrusion

Extrusion is the most widely used technique of processing several types of materials such as polymers and plastics, ceramics, food products (pasta, breakfast cereals, ready-to-eat snacks, etc.), catalysts, some drug carriers, or biomass briquettes. Behavior of materials during extrusion is mainly determined by rheology and can be tuned by modifying rheological properties. In what follows, two different type of materials will be considered, namely polymers, which are extruded as melts and catalysts, when extrusion is done with concentrated suspensions.

Extrusion of polymers is one of the methods for compounding polymers. In general, compounding is done when there is a need to alter properties of polymers or prevent degradation through introducing appropriate additives (antioxidants, UV and heat stabilizers, lubricants, pigments, dyes, and flame retardants). Polymers can only be processed in the rubber state or when molten. Therefore, polymer extrusion is done by pushing a polymer melt across a metal die which continuously shapes the melt into the desired form.

A typical extrusion apparatus shown in Figure 3.15 illustrates that a rotating (an extrusion) screw conveys the polymer fed from hopper to the die. The polymer pellets, powder, or flakes from the hopper fall through a feed throat (a hole) onto the extrusion screw placed inside the extrusion barrel. The screw pushes the polymer forward into a heated region of the barrel where the polymer melts because of external and frictional heating. The molten polymer moves forward until exiting through the die. The extrudate is immediately cooled and solidified typically in a water tank. Typically, feeding is done by gravity; therefore, for a sticky feed, forced feeding might be needed.

An extruder is a continuous pump without back mixing; therefore, consistent feeding rate from the hopper into the screw is needed to ensure constant composition and weight of extrudates. Physical and chemical characteristics of the feed (size and shape, and their distribution, solid density, and friction on the metal surface) as well as the hopper and the feed throat design determine the feeding rate.

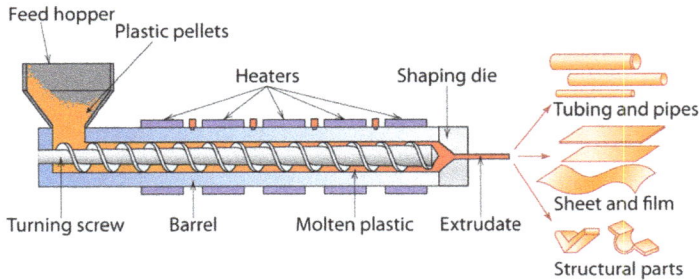

Figure 3.15: A typical extrusion apparatus for processing plastics. http://slideplayer.com/slide/4235773/.

Extrusion is also the most economic and commonly applied shaping technique for catalysts and supports. It is different from extrusion of polymers, which are melted. The polymer melt is essentially homogeneous, and the properties can be regulated by extruder temperature.

Pastes for catalyst extrusion are on the contrary highly concentrated dispersions, whose behavior is determined by rheological characteristics. The pore structure and mechanical stability of extrudates are determined by the properties of the paste and extrusion conditions. In case of catalysts and catalyst support, a certain pore structure should be developed allowing transport of reactants to the active sites. Typically, extrudates contain large transport pores of 300–600 nm in addition to mesopores (10–25 nm) being different from materials prepared by tableting as the latter have mainly monomodal distribution of mesopores.

A need to process concentrated dispersions leading after extrusion to a product with not only certain porosity but also catalytic activity means that only a restricted number of additives or binders can be used being not detrimental for the required catalytic properties.

The pressure, which is developed in the screw extruder as the paste moves toward the die, is affected by the screw geometry and the paste rheology. Unlike polymers, which are melted during extrusion, this is not happening with the catalyst pastes. Moreover, usually the catalyst powders obtained after the thermal treatments behave like sand, not possessing the required moldability and plasticity even after water addition. If the viscosity of the pastes is too low, it can result in unstable extrudates. On the contrary, in case of too viscous pastes, the extruder can be blocked.

In order to improve the flow and rheological properties, various additives are used in formulation of pastes, including clays and starch for better rheological behavior;

binders (e.g. alumina) to keep the active particles together; peptizing agents (diluted acids) for de-agglomeration; combustible porogens for porosity increase (carbon black, starch, etc.); plasticizers; lubricants, and water (typically 20 and 40 wt%).

Typically, inorganic binders such as alumina, silica sols, or clays are utilized because organic ones will be burned away at the calcination step. A special care should be taken on the surface properties of binders which can be themselves catalytically active. Application of binders can result in nonuniformity of the active component distribution. It implies that there could be zones with a high concentration of the active phase, and consequently, higher rate, maybe local overheating and appearance of zones which are controlled by mass transfer rather than kinetics.

The quality of the extrudates depends not only on extrusion *per se* but also on downstream drying and calcination. These steps require special attention in case of larger structures such as extruded monoliths (Figure 3.16). Such monoliths are, for instance, used in selective catalytic reduction of NO_x at power and waste incineration plants. Drying of the extruded monolith must be slow enough to prevent ruptures and cracks.

Figure 3.16: Extrusion of a monolithic structure.

Compared to other preparation methods, extrusion process affords high throughput at relatively low costs giving a variety of possible extrudate shapes. The downside of the method is a nonuniform shape of extrudates and lower abrasion resistance compared to pellets.

3.3 Mass transfer processes

Several most common separation methods are shown in Table 3.2, illustrating that the feed is often in a single phase. Otherwise, mechanical separations should preferably be installed upstream. The separation methods mentioned in Table 3.2 will be presented below.

Table 3.2: An overview of some separations methods (adapted from A. de Haan and H.Bosch, Industrial Separation Processes. Fundamentals, 2013, de Gruyter).

Type	Feed phase	Separation agent	Products	Principle
Rate control				
Gas adsorption	Vapor	Solid adsorbent	Gas + solid	Difference in adsorption strength
Liquid adsorption	Liquid	Solid adsorbent	Liquid + solid	Difference in adsorption strength
Leaching	Solid	Liquid adsorbent	Liquid + solid	Difference in solubility
Crystallization	Liquid	Heat transfer	Liquid + solid	Difference in solubility
Equilibrium based processes				
Distillation	Liquid and/or vapor	Heat transfer	Vapor + liquid	Difference in volatility
Absorption	Vapor	Liquid absorbent	Liquid + vapor	Difference in solubility
Extraction	Liquid	Liquid solvent	Liquid + liquid	Difference in solubility

3.3.1 Distillation

Distillation is an old (Figure 3.17) and the most common separation technique consuming enormous amounts of energy, both in terms of cooling and heating requirements (ca. 50–60% of capital/investment costs and 80–90% of energy costs in chemical industry).

In this process, a liquid or vapor mixture of two or more substances is separated into its component fractions of the desired purity. The cornerstone of distillation is richness of the boiling mixture vapor in the components with lower boiling points; therefore, after condensation, the condensate will contain more volatile components. This difference between liquid and vapor compositions is the basis for distillation operations.

Figure 3.17: Distillation in the medieval times. From J. French, The art of distillation, Richard Cotes, 1651.

Separation of components from a liquid mixture by distillation depends on the differences in boiling points of the individual components. The vapor pressure and the boiling point of a liquid mixture depends on the relative amounts of the components in the mixture.

Thermal stability of the components in the mixtures to be separated by distillations is of primary concern imposing a limit on operating temperatures. Another restriction is related to the medium used for the heat supply, which in industrial conditions is typically steam, whose pressure is directly related to its temperature. Low-, medium-, and high-pressure steam is typically available at oil refineries and chemical industry sites with the latter affording temperatures of 300 °C. Higher temperatures (up to 400 °C) can also be realized by heating with, for example, hot oil or using natural gas burners. The low-temperature limit is dictated by the type of coolant applied in overhead condensers, which is typically water. Subsequently, a minimum temperature in the distillation column is 40–50 °C.

Distillation columns can operate at overpressure (ca. 2.5 bar) as in so-called atmospheric distillation to increase the boiling point of low boiling point compounds. Alternatively, as in vacuum distillation of the bottom fraction from atmospheric distillation, vacuum (60–80 mbar) is applied to decrease the operating temperature to ca. 400 °C. Pressures lower than 2 mbar are seldom used resulting otherwise in high capital and operating costs.

Distillation columns are made up of several components, each of which is used either to transfer heat or mass. A typical distillation unit contains a column *per se*, column internals such as trays (plates) and/or packings needed to enhance separations, a reboiler providing vaporization for the distillation process, a condenser to cool and condense the vapor leaving the top of the column, and a reflux drum, which holds the condensed vapor from the column top of the column and allows liquid reflux back to the column (Figure 3.18).

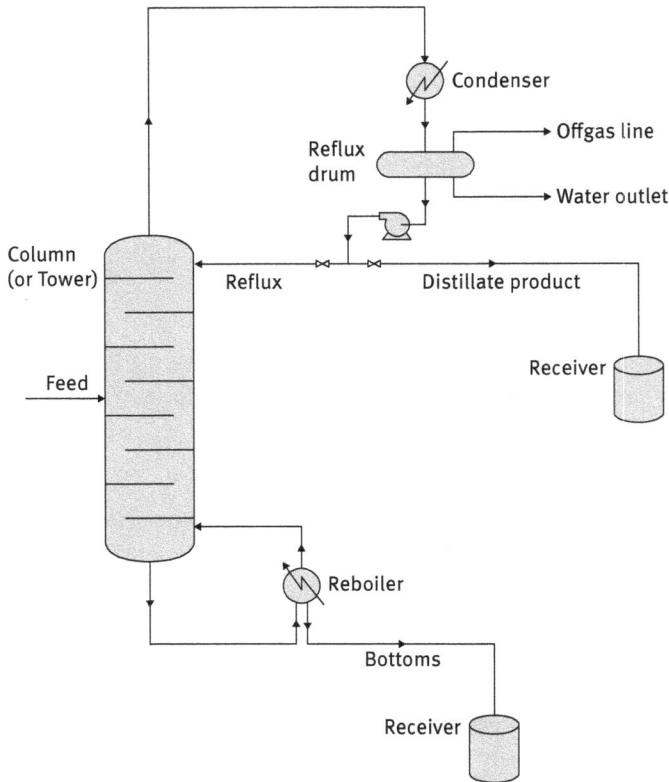

Figure 3.18: Distillation column. http://en.wikipedia.org/wiki/File:Distillation_Column.png#media viewer/File:Continuous_Binary_Fractional_Distillation_EN.svg.

As shown in Figure 3.18, a reboiler is used to supply heat and generate vapors, which move up the column, exiting at the top of the column. Recycling a part of the condensed liquid back to the column affords better separation. This is because of a contact between the vapor moving up and the liquid flowing down from the reflux, which results in partial condensation of higher boiling point substrates and partial evaporation of lower boiling point substrates.

The liquid mixture is introduced usually somewhere near the middle of the column to a feed tray, which divides the column into a top (enriching or rectification) section and a bottom (stripping) section. The feed flows down the column into the reboiler.

A special arrangement is made in crude oil distillation (to be discussed in Chapter 4) when different fractions are taken as sidestream drawoffs (Figure 3.19). Such sidestreams contain excessive volatile products, which are eliminated by stripping with steam leading to a reduction of partial pressure and thus partial revaporization. Parts of the side streams are returned as reflux. The total quantity of steam is between 1 and 3 wt% of the crude stream.

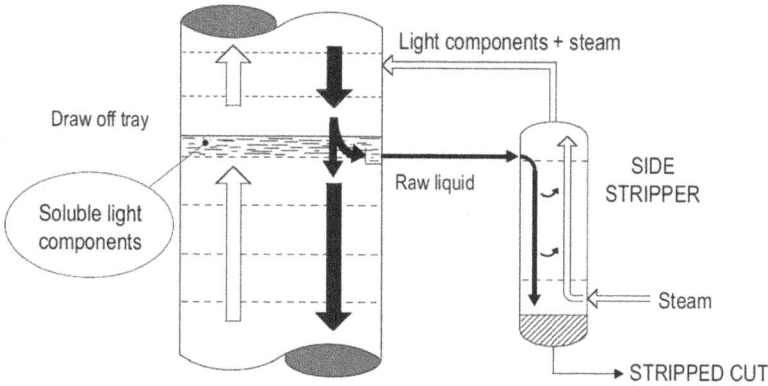

Figure 3.19: Stripping of sidestreams in crude oil distillation with steam.

More conventional arrangements for separation of multicomponent mixtures are shown in Figure 3.20. As illustrated in this figure, two columns are typically needed to separate three components which can be arranged in an alternative way. More complex mixtures give even more variability.

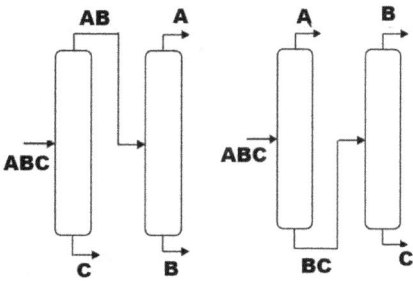

Figure 3.20: Possible distillation configurations for separating ternary mixtures.

To aid in designing the distillation column trains, some heuristic rules have been established, namely that corrosive and hazardous materials should be removed first, majority components should be eliminated first, and easier separations should be done prior to more complex ones (e.g. separation of azeotropic mixtures or separations with strict product specifications). Other rules include preference for removal of the components one by one in column overheads (i.e. the arrangement on the right in Figure 3.20) and favoring separations which a more even split of the feed between the distillate and bottoms. Vacuum distillation and refrigeration should be avoided, if possible.

Distillation columns are designed based on the boiling point properties of the components in the mixtures being separated. Thus, the sizes, particularly the height, of distillation columns are determined by the vapor-liquid equilibrium (VLE) data for the mixtures.

The performance of a distillation (the number of stages, separation efficiency) is determined by many factors, for example, feed conditions and composition, presence of trace elements, efficiency of internals. Some columns are designed to have multiple feed points if the feed can contain varying amounts of components.

An important parameter in distillation is reflux defined as the ratio between the reflux flow and the distillate flow. With an increase in the reflux ratio, more liquid rich in the more volatile components is recycled back into the column, allowing better separation. Minimum trays are required under total reflux conditions, i.e., when there is no withdrawal of distillate. An opposite of total reflux is the minimum reflux ratio, when an infinite number of trays is required for separation. Most columns are designed to operate between 1.2 and 1.5 times the minimum reflux ratio, corresponding to approximately the region of minimum operating costs (more reflux means higher reboiler duty).

After a theoretical number of trays is calculated for a distillation column, it is divided by the tray efficiency (typically 0.5–0.7 depending on the tray type, vapor, and liquid flow conditions), giving the actual number of trays. Foaming (expansion of liquid) providing even a high interfacial liquid-vapor contact leads to liquid buildup on trays if excessive. The foam can even mix with liquid on the tray above. Similar to foaming entrainment and flooding influence in a negative way flow characteristics and tray efficiency. Some foaming at the same time is needed to ensure the adequate interfacial area. A specific feature of tray columns is a requirement of a minimum gas flow velocity, as otherwise the liquid would not stay on a tray.

There are cases when there are extra feeds to assist separation either in the bottom product stream (extractive distillation) or at the top product stream (azeotropic distillation). Such streams are added when there is a minor difference in the boiling points. A typical example is isolation of aromatics from reformate and pyrolysis gas.

The drawback is that one extra step is required to remove the additional component that is recycled back to the azeotropic distillation column. An example of azeotrope distillation is separation of benzene and cyclohexane with close boiling points, to which acetone is added as an entrainer. This results in a new azeotrope between acetone and cyclohexane, which is taken from the top of the column, while benzene is at the bottom. Breaking of acetone-cyclohexane azeotrope is done by extracting acetone with water and subsequent separation by distillation (Figure 3.21).

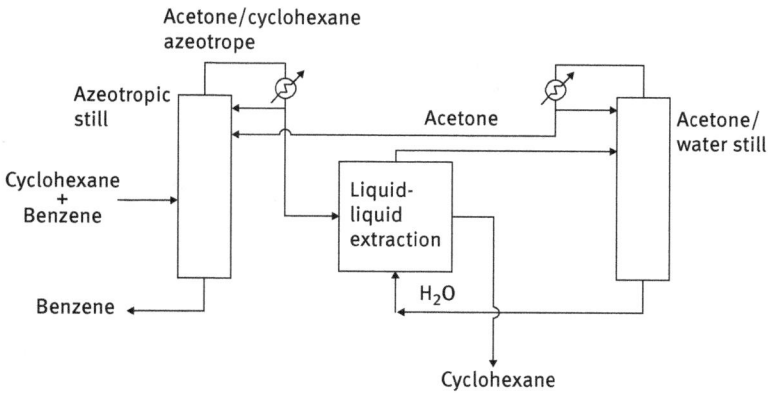

Figure 3.21: Distillation of cyclohexane and benzene.

Extractive distillation is a vapor-liquid process operation that uses a third component, or a solvent, in order to effect separation (Figure 3.22). The extractive agent and the less volatile component flow to the bottom of the distillation column; thereafter, the extracted component is recovered by distillation. The non-extracted species (raffinate) are distilled to the top of the extractive distillation tower.

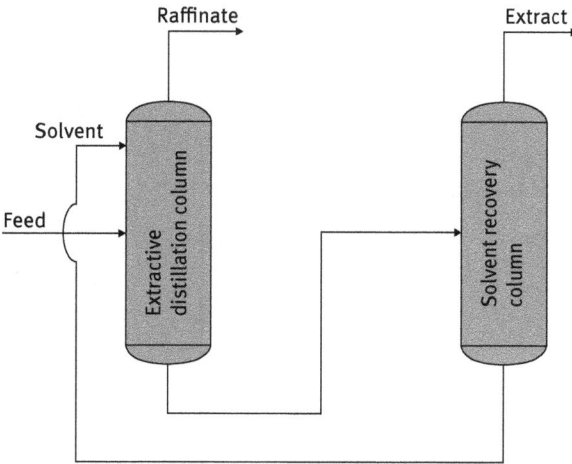

Figure 3.22: Extractive distillation.

An example when aromatics are separated from non-aromatic compounds is given in Figure 3.23 featuring the Uhde Morphylane process.

Both the terms "trays" and "plates" are used interchangeably to denote column internals. The most common trays are bubble cap, valve, and sieve trays. Trays typically have a distance of 0.3–1 m between them. Figure 3.24 illustrates the direction

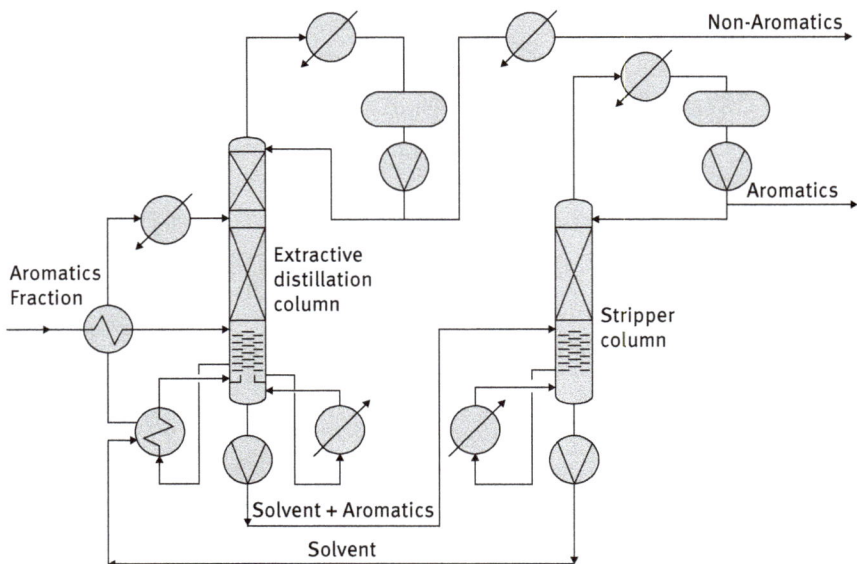

Figure 3.23: Separation of aromatics. http://www.thyssenkrupp-industrial-solutions.com/filead min/documents/brochures/uhde_brochures_pdf_en_16.pdf.

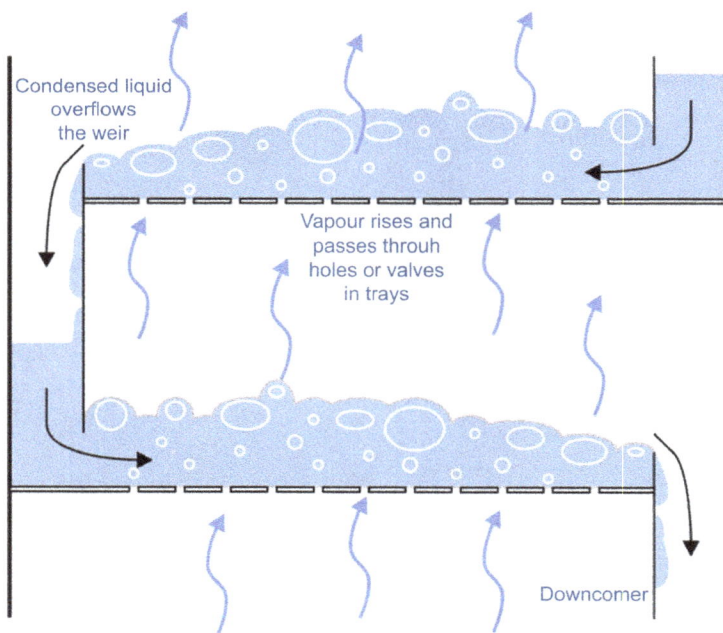

Figure 3.24: Liquid and vapor flows in a tray column. From http://www.wermac.org/equipment/dis tillation_part2.html.

of vapor and liquid flow across a tray and a column. A weir on the tray is designed to ensure a suitable height of the liquid on the tray covering the caps.

While Figure 3.24 corresponds to just one pass of the liquid from the upper tray to a lower one, other design options are also possible (Figure 3.25).

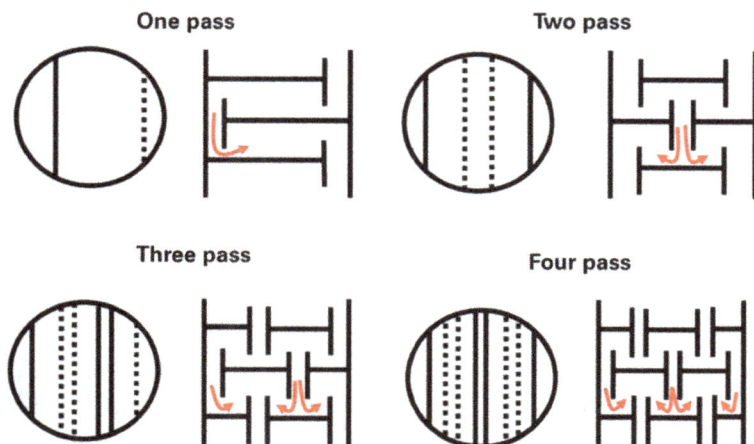

Figure 3.25: Different design options for liquid flows in a distillation column.

Bubble cap trays (Figure 3.26) have risers fitted over each hole, and a cap that covers the riser providing a space between riser and cap to allow the passage of vapor. In valve trays, perforations are covered by liftable caps, which are lifted by vapor, itself creating a flow area for the passage of vapor. The lifting cap directs the vapor to flow horizontally into the liquid, thus providing better mixing than is possible in sieve trays. Valve trays (Figure 3.27) are more likely to plug if solids are present and are more costly than sieve trays (Figure 3.28). In the latter version, the plates simply have large holes in them, which are easier to clean and thus sieve trays are relatively resistant to clogging. Plate spacing can be smaller than in bubble cap trays.

Trays are designed to maximize vapor-liquid contact. Such contacts can be improved by applications of packings (Figure 3.29), leading in general to shorter columns. These packing elements should enhance vapor-liquid contact without significant pressure drop and should be uniformly loaded to avoid channeling and bypassing close to the walls.

The Raschig rings are the simplest ones and cheapest from the manufacturing viewpoint. The Pall rings give lower pressure drop and are more efficient than the Raschig rings. The Lessing ring can have different structure to enhance the contact area between the gas and the liquid being at the same time more difficult to manufacture and therefore more expensive. They also have a lower free volume. Somewhat difficult to manufacture are also Berl rings.

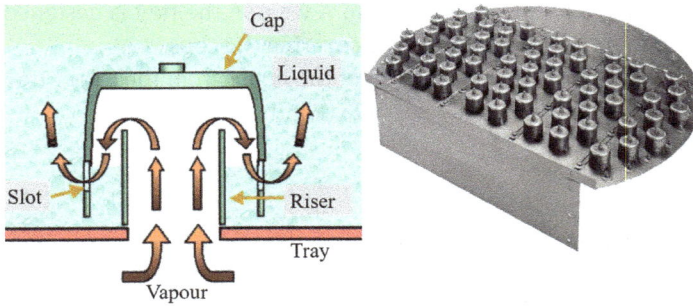

Figure 3.26: Bubble cap trays.

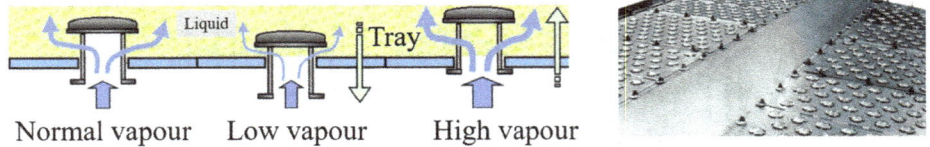

Figure 3.27: Valvecap trays.

Figure 3.28: Sieve trays.

Structured packings consisting of thin corrugated metal plates or gauzes affording a large interfacial area between different phases (higher effective exchange area per cubic meter of packing) and better liquid-distributing properties especially at very low liquid loads, are used, for example, in vacuum and atmospheric crude oil distillation or FCC fractionators.

Selection of packings and trays should ensure absence of clogging and flooding. The height of a segment for packings is chosen to allow a uniform distribution of the liquid. If the length of the segment is too high, the liquid will be poorly

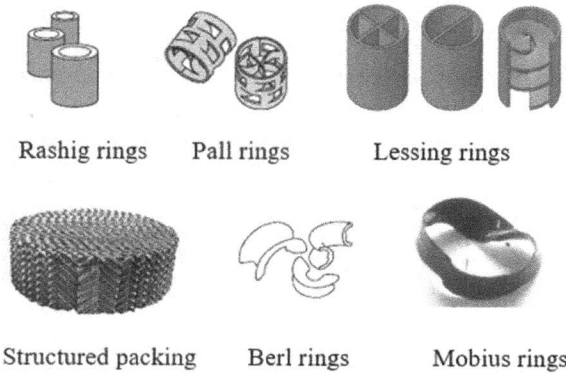

Rashig rings Pall rings Lessing rings

Structured packing Berl rings Mobius rings

Figure 3.29: Different packing elements including structural packing.

distributed and will flow closer to the walls. Therefore, the bed height is typically limited to 3–5 m followed by the distributors, which ideally provide uniform liquid distribution, have a lower pressure drop, are resistant to plugging and fouling, and ensure a minimal liquid residence time.

Moreover, if the packing is manufacturing from brittle materials (e.g. ceramics or polymers), the bottom of the packing can be destroyed because of an excessive weight also limiting the height of the packing bed.

Both tray and packing columns have their advantages and disadvantages also depending on the application. In this chapter, the distillation columns are discussed, however, in general the same packing elements are used in absorption or extraction.

In the case of distillation, it can be mentioned that trays are used in columns of a large diameter operating in a narrow range of gas flows and a much wider range of liquid flow rates. Valve trays allow more operational flexibility with regard to the gas load. Other features of tray columns include a relatively high liquid holdup and robustness against impurities in the feed. For smaller diameter columns, conventional packing is preferred, while structured packing can be used even for very large diameters. A small pressure drop which can be achieved with structured packings is their another advantage, allowing utilization of these internals in distillation columns operating under vacuum.

Finally, it is possible to combine chemical reactions with distillation in one unit through the so-called reactive distillation (Figure 1.30). For equilibrium-limited reactions such as esterification of acids with alcohols, chemical equilibrium can be shifted by continuous removal of reaction products from the reactive zone. This will lead to a reduction of capital and investment costs being an example of process intensification.

3.3.2 Extraction

Solvent extraction is widely applied in petrochemical industry allowing separation of heat-sensitive liquid mixtures according to their chemical type (e.g. aromatics vs non-aromatics, as discussed below) rather than by molecular weight or vapor pressure.

In the case of distillation, the second vapor phase is formed exclusively from the components of the initial (liquid) phase, while in extraction, a solvent is added to the liquid phase (Figure 3.30).

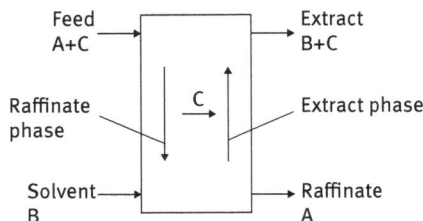

Figure 3.30: Principles of extraction.

Liquid-liquid extraction can be used for separation of compounds in a liquid mixture with close boiling points (e.g., separation of aromatics from aliphatic hydrocarbons), for compounds that are prone to decomposition or to undesired reactions at high temperatures (vitamins, acrylate), or for separation of azeotropic mixtures (extraction of acetic acid from aqueous media with such solvents as MTBE). Extraction is an isothermal process, normally carried out at ambient temperature and pressure.

A distribution ratio is often used as a parameter reflecting the efficiency of extraction. This ratio is equal to the concentration of a solute in the first phase (usually the extracting agent) divided by its concentration in the second phase, usually the raffinate. The key problem in extraction is a proper selection of the most suitable and commercially available solvent affording required selectivity and capacity.

The extracted component is usually separated from the solvent by distillation with the boiling point difference determining the reflux ratio in this distillation. A sufficiently large difference in densities between the two liquid phases should be ensured for the separation process. Several options of extraction apparatus with or without energy input are available.

After extraction, the solvent should be separated from the extract and recycled. Such separation is done by distillation column or another method requiring, in the former case, a distillation column. When the solvent is present also in the raffinate, a second separation process by, e.g., distillation is needed to recover the solvent as illustrated in Figure 3.31. Figure 3.31 shows a case of a low boiling point solvent recovered at the top of both distillation columns, while high boiling solvents are recovered as the bottom product.

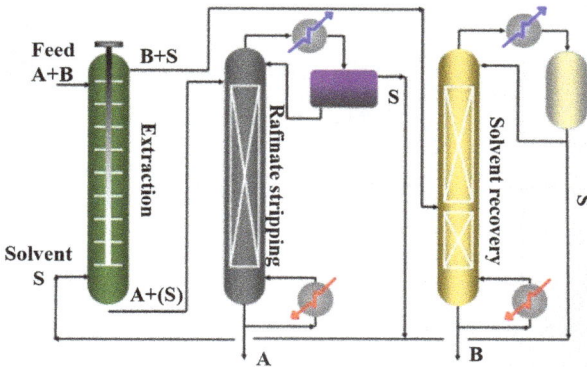

Figure 3.31: Separation of a mixture by extraction and solvent recovery with two distillation columns. Modified from https://kochmodular.com/our-work/articles-publications/white-paper/solving-separation-problems-using-liquid-liquid-extraction-lle/.

Separation of the solvent and the solute can be cumbersome if they have close boiling points resulting in distillation columns with many trays and a high reflux ratio. All this inevitably increase the process costs. Even if the distillation part of the extraction process is not that expensive, a necessity to use a solvent increases the overall separation complexity, and thus the costs. In some areas, such as, for example, removal of a high boiling point component present in small quantities, separation of heat-sensitive materials or compounds with close boiling points but different molecular structure (e.g. cyclohexane vs benzene) and different solubility or separation of mixtures forming azeotropes, extraction can be the preferred option compared to distillation.

The simplest extractors in terms of their construction are spray columns, which are used in operations such as washing and neutralization. Poor phase contacting and excessive backmixing in the continuous phase result in very low efficiency and thus limited applicability.

Packed columns (Figure 3.32) of better efficiency than spray columns have been adopted from distillation. The packing type is similar to the one applied in distillation including random and structured packings. Although mass transfer takes place mainly during the formation of a new interfacial area and is not that efficient, packed unagitated columns found their widespread utilization in industry because of simplicity and low cost.

Mass transfer can be substantially improved by application of additional mechanical energy in form of, for example, pulsation. The use of pulsed columns is, however, limited mainly to small and medium throughputs. In the pulsed sieve plate column, the trays are fixed and the entire liquid content of the column is vibrated; in the reciprocating-plate column (Figure 3.32), the plates are moving.

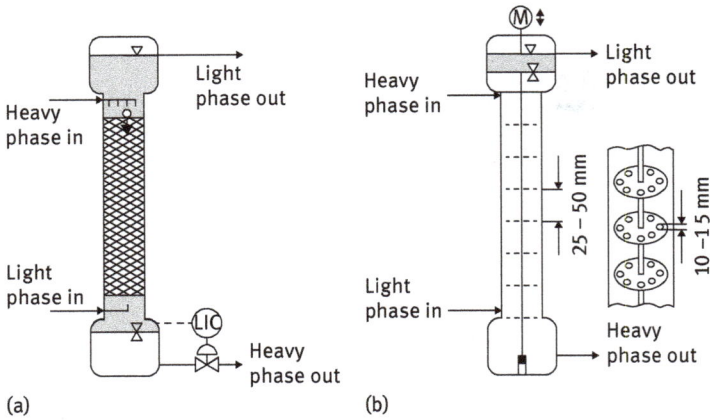

Figure 3.32: Packed and reciprocating-plate (pulse) columns for extraction.

The use of a countersolvent or water in some cases can be an option to improve separation. In the Carom process (Figure 3.33), for separation of aromatics from non-aromatics, there are three internal circulation loops: solvent, water, and hydrocarbon recycling. The feed enters the extractor somewhere between the middle and the bottom trays. The lean solvent, essentially free of hydrocarbon components, enters the extractor column at the top. The denser solvent phase travels down the column from tray to tray as the hydrocarbon phase travels upward. The solvent extracts aromatics from the feed and leaves the bottom of the extractor while substantially all the non-aromatic hydrocarbons leave extractor from the top as raffinate.

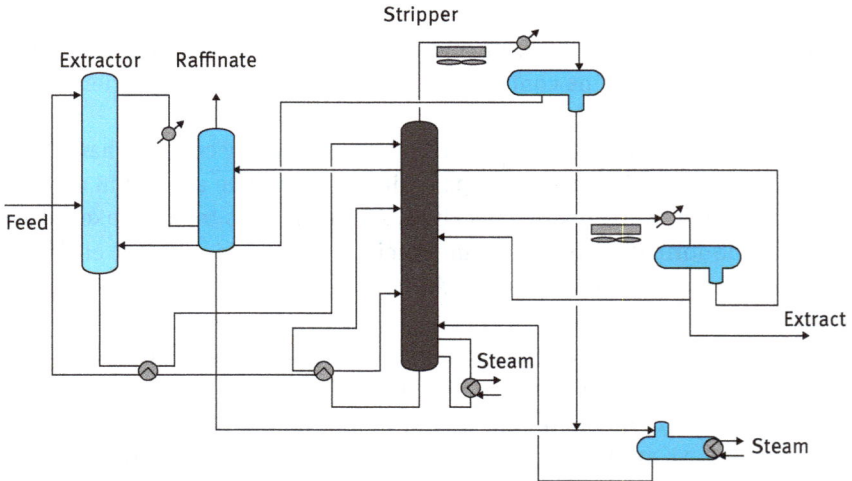

Figure 3.33: Carom process for separation aromatics from non-aromatics.

A recycle stream from the stripper overhead is introduced at the bottom of the extractor in order to back-extract the heavy non-aromatics from the solvent as it leaves the bottom of the extractor.

The rich solvent is sent to the top of the stripper. After the initial flash, the rich solvent is subjected to a combination of extractive distillation and steam stripping in the upper section of the stripper column. The non-aromatics are removed as stripper overhead vapors, along with a small amount of aromatics. The stripper overhead vapors and the vapor from the flash are combined in the overhead condenser. The hydrocarbon and water phases are separated in the stripper overhead receiver, and the hydrocarbon phase is recycled back to the bottom of the extractor. The water phase joins the spent wash water from the raffinate wash column to become the stripping water. By the time the rich solvent reaches the extract sidedraw on its way down the stripper column, only aromatics hydrocarbons remain in the solution. In the section below the sidedraw, the solvent is stripped of the aromatic by the combined action of the stripper reboiler and stripping steam injected near the bottom of the column. The extract vapors leaving through the sidedraw are essentially pure aromatic hydrocarbons. The extract is directed to the fractionation section where petrochemical-grade benzene, toluene, and C8 aromatics are recovered. At the bottom of the stripper column, hydrocarbon-free lean solvent is cooled to extraction temperature before it enters the top of the extractor column.

3.3.3 Adsorption

This process involves preferential partitioning of substances from the gaseous or liquid phase onto the surface of a solid substrate (adsorbent). Examples of industrial applications are presented in Table 3.3.

Table 3.3: Some industrial examples of separation processes based on adsorption (adapted from A. de Haan and H. Bosch, Industrial Separation Processes. Fundamentals, 2013, de Gruyter).

Separation	Application	Adsorbent
Separation and purification of gases	*n*-Paraffins, isoparaffins, aromatics	zeolite
	Nitrogen/oxygen	zeolite
	Sulphur compounds from organics	zeolite
	Hydrocarbons from vent streams	active carbon
Separation and purification of liquid	*n*-Paraffins, isoparaffins	zeolite
	Xylenes	zeolite
	Organics from aqueous streams	active carbon
	Water from organics	silica, alumina, zeolite

In chemical technology, adsorption is applied when there is a need to achieve high purity while removing low quantities of impurities. The adsorbed material is called adsorbate. This surface sorption process is different from absorption, which is a bulk process. Mechanism of adsorption (exothermal from thermodynamic point of view) includes physical adsorption with van der Waals forces and electrostatic forces between adsorbate molecules and the surfaces, as well as chemisorption (formation of chemical bonds). Adsorption capacity is related to the specific surface area of solid materials. An increase in this area comes along with the creation of small-sized pores, which impair the accessibility of incoming molecules when their cross section (kinetic diameter) is larger than the pore size.

Such adsorbents as alumina, silica gel, and various aluminosilicates (zeolites, clays, or silica-alumina) and carbonaceous or polymer adsorbents having different polarity are applied depending on the type of adsorbate (Table 3.3). Alumina, silica gel, and some zeolites are used for drying, while zeolites, clays, and active carbons are applied also for gas and liquid separations. Commercial adsorbents are generally produced in regular shapes (beads, pellets, extrudates, granules, etc.) to diminish pressure drop. Binders can be added in the amount of 10–20%, providing the so-called transport pores, which facilitate the transport of the adsorbate molecules from the bulk of the fluid phase to the adsorption sites.

Adsorbents may be energetically homogenous, containing adsorption sites of identical adsorption energy (heat of adsorption), or energetically heterogeneous, containing a distribution of sites of varying energies. Langmuir adsorption isotherm is often used to describe adsorption of equal-sized adsorbates on an energetically homogeneous (uniform) adsorbent without any lateral interactions:

$$\theta_i = \frac{K_i P_i}{1 + K_i P_i + \sum_{j=1, j\neq i}^{J} K_i P_i} \tag{3.1}$$

This equation implies that the surface coverage (θ_i) of a particular compound i approaches unity for high pressure P_i of this compound and absence of other compounds in the gas mixture. The equilibrium constant of adsorption K_i decreases with temperature increase according to the following expression

$$K_i = K_i^0 e^{\Delta H/RT} \tag{3.2}$$

From eqs. (3.1) and (3.2), it is clear that the amount of adsorbed substrate depends on temperature and pressure. Low temperature and higher pressure are beneficial for exothermal adsorption, while low pressure and high temperature promote endothermal desorption (Figure 3.34).

Adsorption technology requires integration of both adsorption and desorption steps, since the adsorbent should be regenerated and repeatedly used. There are,

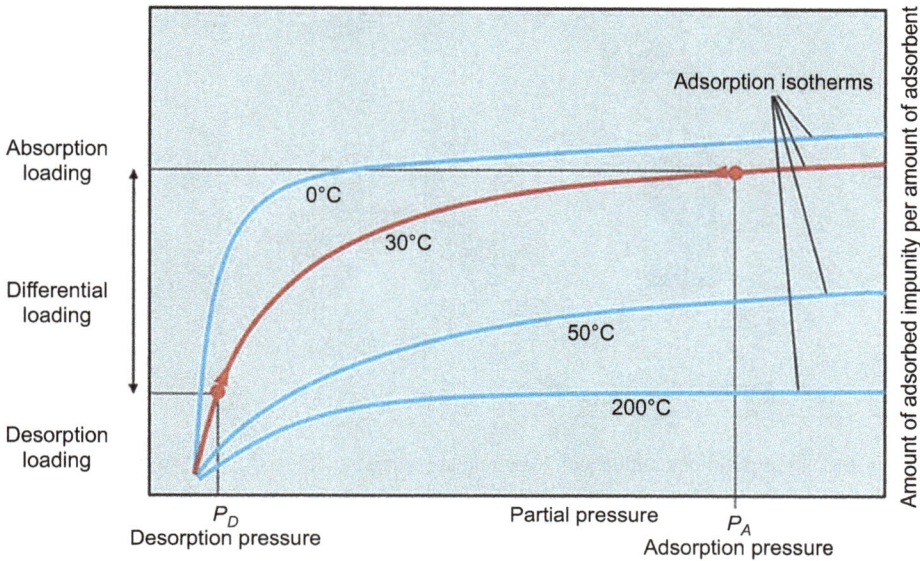

Figure 3.34: Illustration of adsorption isotherms (https://www.chemengonline.com/wp-content/up loads/2016/01/22.jpg).

however, exceptions to this rule, for example, ZnO applied for adsorption of H_2S after desulphurization is just disposed after complete saturation as it is transformed in ZnS.

Regenerative adsorption includes the so-called swinging of temperature, pressure, or concentration.

Adsorption typically has very fast kinetics, but it can be influenced by mass transfer (in gas-phase processes by molecular and Knudsen diffusion, as well as activated diffusion of adsorbed molecules inside the micropores).

For gas-phase adsorption, fixed-bed adsorbers are mainly used. An illustration of such adsorption unit is given in Figure 3.35. Initially, the mixture to be purified flows through the first adsorber with a fresh (regenerated) adsorbent, while the second adsorber undergoes regeneration.

Adsorption occurs in a layer-by-layer manner with the mass-transfer zone moving through the fixed bed with time (Figure 3.36). At some point, a breakthrough occurs and the impurity leaves the adsorber, indicating the end of the loading cycle.

Further operation of the bed will eventually lead to a situation when the concentration of the mixture leaving the adsorber is equal to the feed concentration.

At the end of the loading cycle, there should be a switch in the operation mode as discussed above by regenerating the first adsorber by desorption and starting to operate the second regenerated adsorber.

In temperature swing adsorption, desorption is effected by temperature, while lowering of the gas partial pressure or flowing a less selectively adsorbed liquid

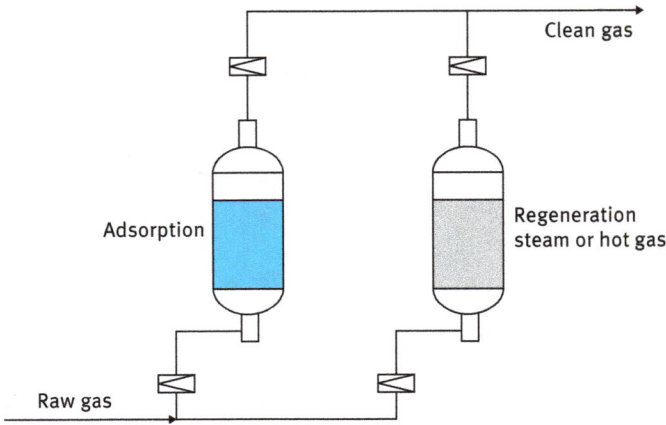

Figure 3.35: Adsorption-desorption with fixed-bed absorbers.

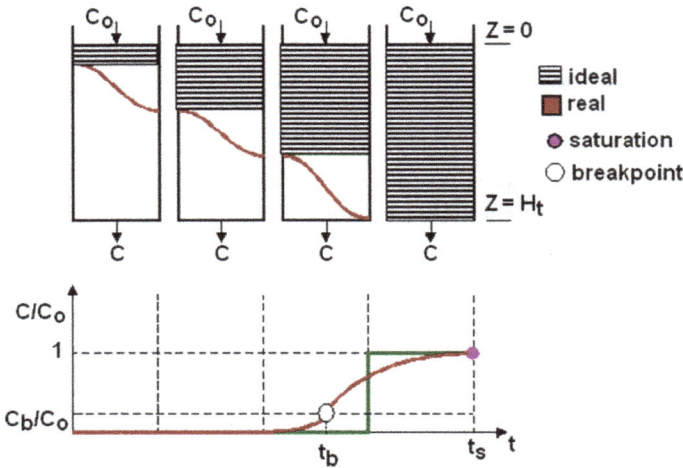

Figure 3.36: Layer-by-layer adsorption with a moving zone. From https://www.intechopen.com/media/chapter/44496/media/image2.png.

over adsorbent is done in pressure swing and concentration swing adsorption technologies, respectively.

One of the differences between these options is the time to switch from adsorption to desorption. Apparently, pressure and concentration can be changed much more rapidly than temperature, limiting the latter option to treating feeds with low adsorbate concentrations. Otherwise, regeneration time is too long, compared to the adsorption time, as regeneration also includes temperature increase and heating, desorbing and cooling of the bed are time consuming obviously increasing the separation costs. Examples of temperature swing adsorption include drying with zeolites or silica gel and removal of various pollutants with activated carbon.

In a pressure-swing adsorption (PSA) operating at a nearly constant temperature, there are several adsorbers alternating between the adsorption and desorption steps. Adsorption occurs at elevated pressures, while in countercurrent depressurization (Figure 3.37), strongly adsorbed species are desorbed. Additional heat supply/removal is not required as the exothermic heat of adsorption is used for desorption.

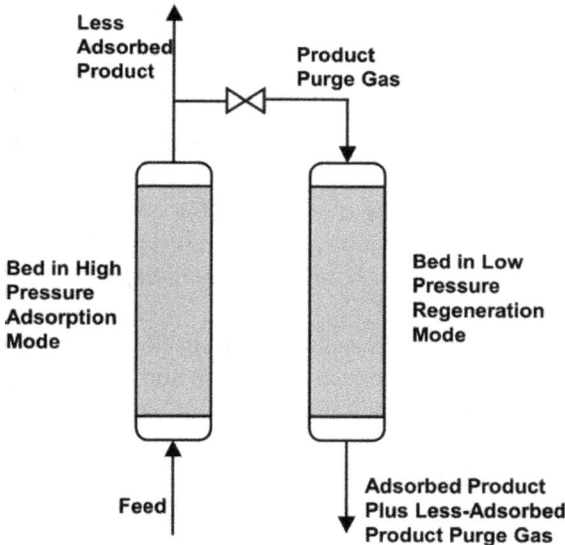

Figure 3.37: Pressure-swing adsorption.

Moving-bed and fluidized-bed adsorbers, which can be operated continuously, are much less frequently applied. In the moving-bed adsorber, the solid is moving downward through the column while the liquid is flowing upward. The adsorbent leaves the apparatus at the bottom and is returned to the top pneumatically.

3.3.4 Absorption

Absorption, being one of the main separation methods in the chemical industry, is a mass transfer operation when a soluble gaseous component is removed from a gas stream by dissolving it in a liquid with or without a chemical reaction. The absorbing liquid is then continuously regenerated and recycled. Thus, an absorption unit consists of at least two pieces of equipment – an absorber and a desorber (stripper) (Figure 1.24).

Typically, adsorbers are columns with trays and packing elements, similar to those discussed in the section on distillation. Their tasks is to improve gas-liquid mass transfer and provide better flow distribution, thus increasing pressure drop across the columns. In a packed column, there is a support plate for the packing at the base and a liquid distributor at the top of the column.

In general, for smaller installations, corrosive environments and liquids prone to foaming packed columns are more suited than tray columns. They offer a possibility to operate at very high liquid-to-gas ratios, otherwise leading to a high pressure drop. Moreover, such columns can also operate under vacuum where such low pressure drop is required. Obviously, packings can be more easily replaced than trays. At the same time, packing columns are less suited for large installations and low-to-medium liquid flow rates.

Tray columns, which, in general, have higher capital costs with trays of special design, and operate at a broader range of gas and liquid flows than a countercurrent packed column. In the latter, low liquid flow rate and, on the contrary, high gas flows will lead to flooding. Other advantages of tray columns are a possibility to install cooling coils and use tall columns without channeling of both vapor and liquid streams present in tall packed columns.

In an absorber, which is a device operating in a countercurrent flow mode, the fluid with the lower density (the gas stream) enters at the bottom and leaves from the top. The higher density fluid (solvent) enters at the top and exits from the bottom.

Either water (as such or with some organic or inorganic compounds) or low volatile organic liquids (methanol) can be used as absorbing liquids, affording saturation of the solvent with the gas leaving an absorber.

It is advantageous to have lower-viscosity solvents that are not corrosive, toxic, or flammable. Solvents could be either physical or chemical. In the former case, the main processes are physical in nature, with a typically linear relationship between the loading of the gas and its partial pressure (Figure 3.38) in the form of the Henry law:

$$p_i = x_i H_{i,L} (x_i \rightarrow 0) \tag{3.3}$$

where x is the molar content of the solute in the liquid and H is the Henry constant. In most cases, gas solubility decreases with temperature increase.

Chemical absorption also involves a chemical reaction with the absorbing medium. This provides higher loading but at the same time a certain saturation (Figure 3.38). Langmuir or more complicated isotherms such as logarithmic (Temkin) isotherm can be used to describe experimental absorption data depending on a system. Utilization of chemical solvents with high solubility reduces the amount of solvent to be circulated.

In industry, aqueous solutions of alkanolamines are widely used to remove hydrogen sulphide and carbon dioxide from natural gas or gaseous effluent by chemical absorption. One of the most commonly used chemical absorbents is *N*-methyldiethanolamine (MDEA) activated with piperazine (PZ)

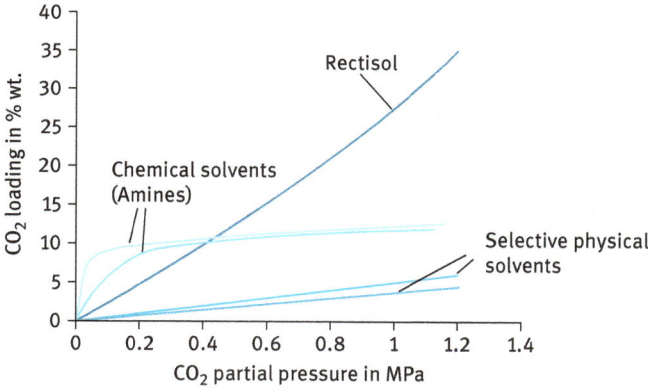

Figure 3.38: Physical and chemical adsorption for CO_2 loading as a function of its partial pressure.

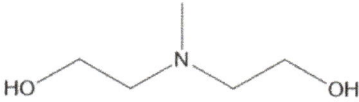

N-methyldiethanolamine   piperazine

The vapor-liquid equilibrium and chemical reactions in aqueous solution of PZ, MDEA, and CO_2 are shown in Figure 3.39 and include water dissociation, dissociation of bicarbonate (reactions 2 and 3), protonation of MDEA and PZ (reactions 4 to 6), as well as formation of PZ carbamate, PZ dicabamate, and protonated PZ carbamate (reactions from 7 to 9).

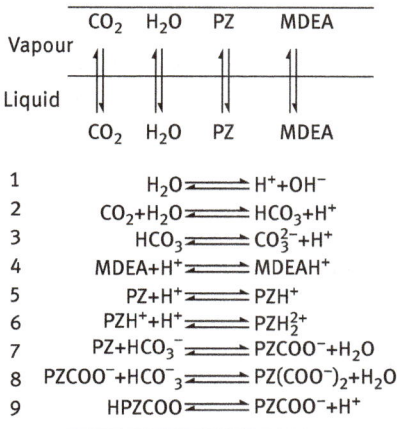

	CO_2	H_2O	PZ	MDEA
Vapour				
Liquid				
	CO_2	H_2O	PZ	MDEA

1 $H_2O \rightleftharpoons H^+ + OH^-$
2 $CO_2 + H_2O \rightleftharpoons HCO_3^- + H^+$
3 $HCO_3^- \rightleftharpoons CO_3^{2-} + H^+$
4 $MDEA + H^+ \rightleftharpoons MDEAH^+$
5 $PZ + H^+ \rightleftharpoons PZH^+$
6 $PZH^+ + H^+ \rightleftharpoons PZH_2^{2+}$
7 $PZ + HCO_3^- \rightleftharpoons PZCOO^- + H_2O$
8 $PZCOO^- + HCO_3^- \rightleftharpoons PZ(COO^-)_2 + H_2O$
9 $HPZCOO \rightleftharpoons PZCOO^- + H^+$

Figure 3.39: VLE and chemical reactions in the H_2O-PZ-MDEA-CO_2.

Typically, chemical solvents display more prominent temperature dependence than physical solvents. A higher purity of the treated gas is certainly an advantage of chemical solvents that comes at the expense of more energy-demanding desorption. On the contrary, desorption from physical solvents can be easily done by flashing (release of pressure). Thus, absorption with physical solvents is preferred for large concentrations of the impurities present in a gas at high pressure. Examples of absorption processes mentioned also in Chapter 1 include absorption of SO_3 and NO_x in water to make corresponding acids and removal of hydrogen sulphide and carbon dioxide by aqueous solutions of amines.

Such chemical solvents as water solutions of amines, after reaching their chemical capacity, start to behave as rather poor physical solvents, since the capacity of CO_2, for example, in water is very limited.

For physical solvents, the temperature in the absorber is typically rather low, while the regenerator operates at a high temperature. After stripping off the dissolved gas, the solvent is cooled and recirculated to the absorber. For chemical absorption, the reaction kinetics should also be considered, which increases with temperature; therefore, very low temperature cannot be used, resulting in very high absorbers. As illustrated in Figure 1.24, countercurrent operation is used in absorption equipment. The maximum gas flow is limited by the pressure drop and the liquid holdup that will build up and could lead to flooding.

In design of absorption units, the following should be considered: flow rate, pressure, composition, and temperature of the gas to be treated; the type of absorbent; and the desired purity level. All these influence the type of absorber, internals, pressure drop, geometry of the absorber and desorber, and presence of other pieces of equipment (e.g. vessels where a certain part of the gas is released just by flashing (decreasing pressure)).

The absorbent selection is based on several requirements. High solubility of the solute decreases the inventory of the absorbent, while low volatility to diminish the losses by evaporation. Other requirements include stability, noncorrosive nature, low viscosity, and high mass and heat transfer rates. In addition, is the sorbent does not lead to high foaming it will eliminate a need to use defoaming agents. Safer use in industrial settings requires that the absorbent is nontoxic and nonflammable. Moreover, availability of the absorbent and its costs also play a role in the selection of a suitable absorbent.

For absorption, high pressures and low temperatures are needed. At the same time, if the temperature is too low the absorption kinetics might become a limiting factor. For desorption (stripping), on the contrary, low pressures and high temperature are beneficial. Too high temperature will inevitably lead to side reactions and additional solvent losses. Operation under vacuum is expensive, thus desorption is typically done at pressures slightly above atmospheric pressure.

As an example of gas absorption, CO_2 removal from process gas in ammonia synthesis is illustrated in Figure 1.24 with one adsorber and one desorber (stripper) representing the so-called one-stage process with only one lean absorber where CO_2 is absorbed by a stripper-regenerated solution. The residual CO_2 loading in such solutions can be very low, < 50 ppm. The stripper always requires a reboiler, which generates the strip steam. The energy supply has to be high to keep the temperature at the bottom of the stripper at boiling conditions.

There are also other possibilities for adsorption process design. When the purity of the gas is not important, a one-stage absorber with two regeneration vessels, namely high-pressure (hp) flash (ca. 0.6–0.8 MPa) and lower-pressure (lp) flash (slightly above atmospheric pressure), can be applied (Figure 3.40).

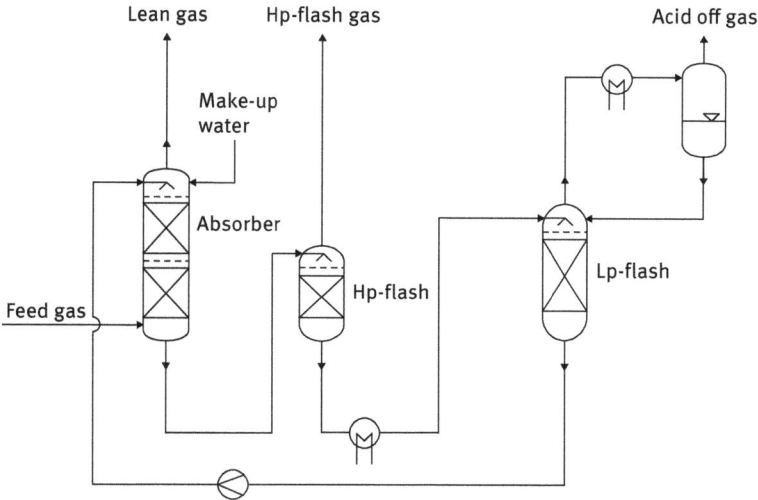

Figure 3.40: One-stage process (absorber + hp flash + lp flash).

This arrangement can also contain a stripper (Figure 3.41).

Some designs use, aside from a lean adsorber, a semi-lean absorber for lower energy consumption, where CO_2 is absorbed by flash-regenerated solution (Figure 3.42). This solution comes either from an lp flash or a vacuum flash. In practice, these two absorbers typically form one column, with the lean adsorber with the smaller diameter placed on top of a semi-lean absorber.

Moreover, lp flash is typically combined with the stripper (Figure 3.43)

The arrangement presented in Figure 3.43 is typical for hot potassium carbonate (hotpot) removal based on transformations of potassium carbonate to bicarbonate. Such systems experience severe problems with corrosion due to alkalinity and high temperature. In order to improve absorption corrosion, inhibitors should be added leading in turn to foaming. Moreover, sterically hindered amines (diethanolamine) added to the carbonate solutions to improve absorption result in degradation products and subsequent foaming. If the solution containing all additives is slipped to the

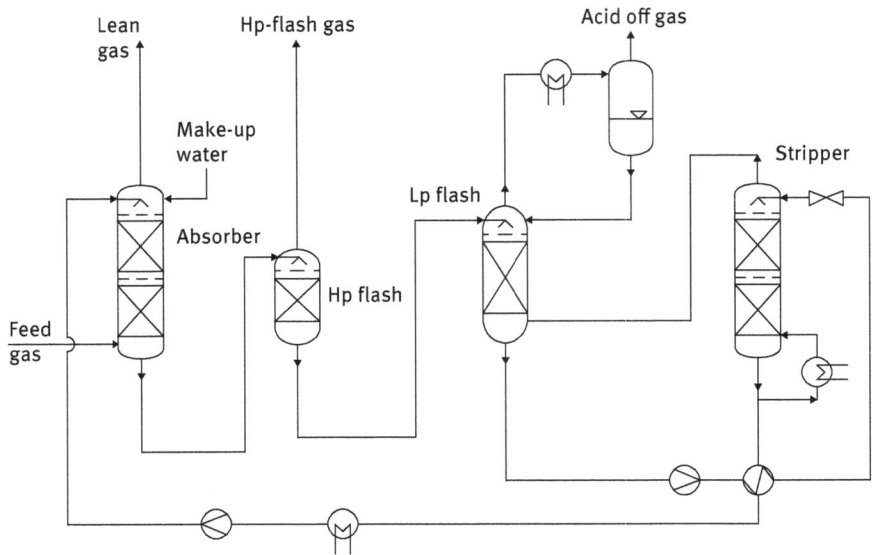

Figure 3.41: One-stage process (absorber + hp flash + lp flash + stripper).

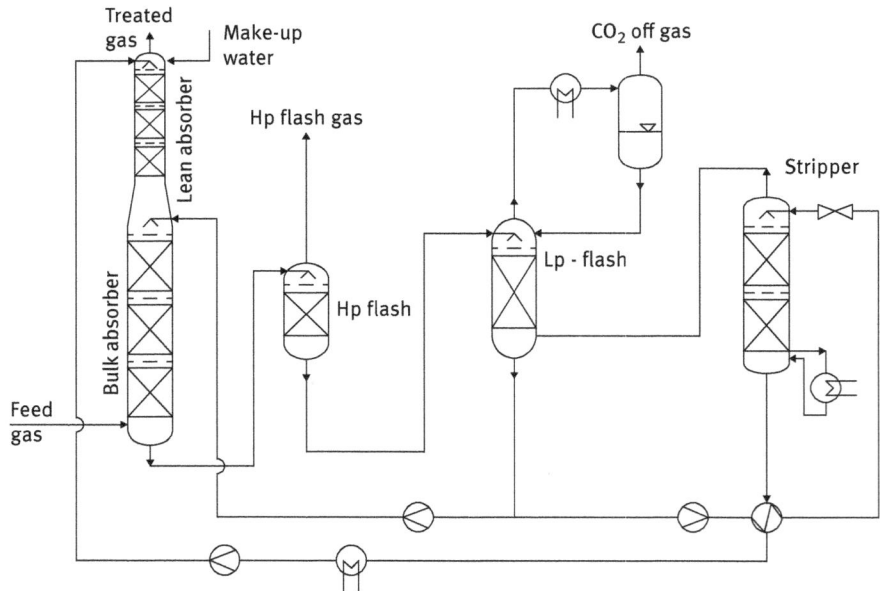

Figure 3.42: Two-stage process (absorber + hp flash + lp flash + stripper).

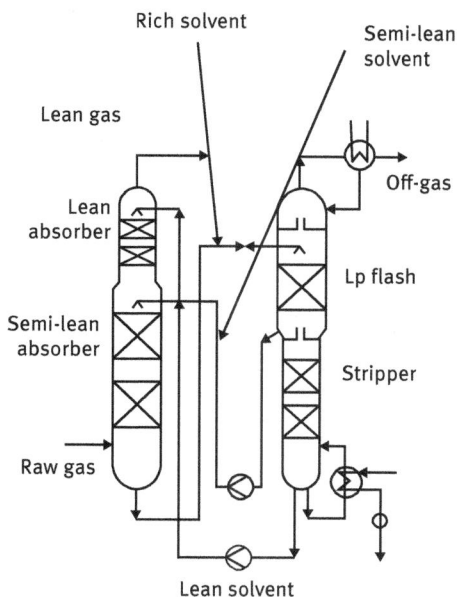

Figure 3.43: Two-stage process with lp flash on top of a stripper.

downstream units as in the case of ammonia synthesis, this will obviously influence in a negative way operation of a nickel catalyst in the methanator.

Such problems with potassium carbonate solutions were reasons for why, in industry, longtime solutions of monoethanolamine (MEA) were applied for a long time in removal of H_2S and CO_2. In addition to the schemes presented above (e.g. Figures 3.40–3.42), more complicated arrangements were employed in ammonia synthesis trains (Figures 3.44 and 3.45).

Apparent disadvantages using MEA are related to substantial solvent losses reaching for ammonia syngas 5–15% of the amine holdup per year. Moreover, monoethanol amine is rather corrosive (Figure 3.46).

Such disadvantages of MEA led to a revamp of units operating with MEA and utilization of activated methyldiethanolamine for CO_2 removal in ammonia synthesis plants. A scheme of two-step BASF aMDEA process comprising also of an lp flash and an hp flash in addition to the absorber and a stripper is shown in Figure 3.47. The high-pressure flash prevents the solution from extensive degassing, since the operating pressure is slightly higher than the CO_2 partial pressure in the feed gas.

The optimized two-stage technology of CO_2 removal in ammonia synthesis plants allows to achieve the concentration of CO_2 in the clean gas below 20 ppm with the energy consumption of 35 MJ/kmol CO_2, while the corresponding values for the one-stage without and with lp flash are 120 and 90 MJ/kmol CO_2, respectively.

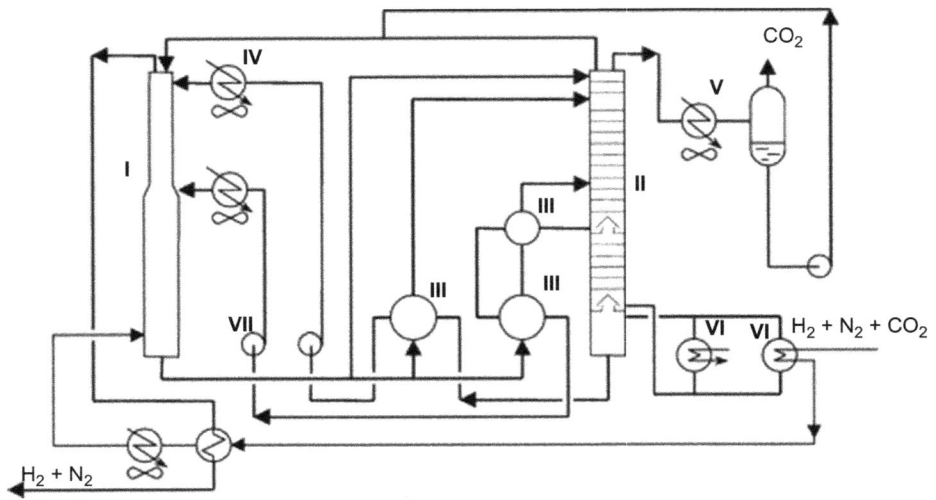

Figure 3.44: Two-stage process for purification of gas from CO_2 with MEA solution in 1,500 MTPD ammonia plant with three flows of saturated solution and two flows of regenerated solutions: *I – absorber, II – regenerator, III – heat exchanger, IV – cooler, V – cooler (condenser) for steam-gas mixture, VI – reboiler, and VII – pumps.*

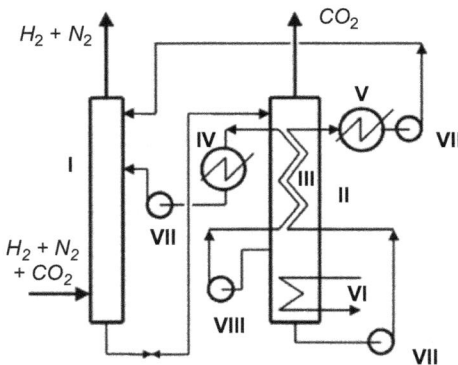

Figure 3.45: Process flow diagram for purification of gas from CO_2 with MEA solution in 1,500 MTPD ammonia plant with integration of solution regeneration and heat recuperation: *I – absorber, II – regenerator, III – heat exchanger, IV – cooler of solution, V – cooler (condenser) for steam-gas mixture, VI – reboiler, and VII – pumps.*

An interesting option, which recently was commercialized for H_2S/CO_2 removal, is the Rectisol process (Figure 3.48), typically operating below 0 °C with methanol as a solvent. Due to low temperatures, approximately 5% of the material in a Rectisol plant is stainless steel. The main disadvantages of this process are high complexity

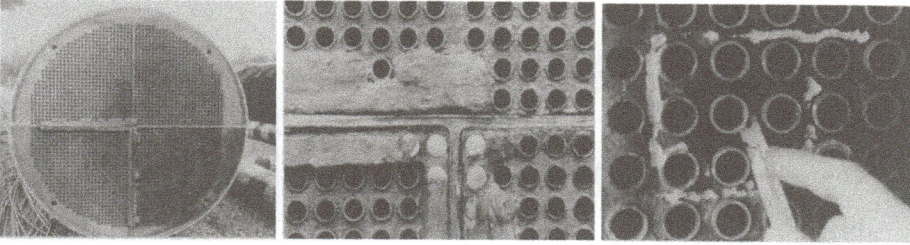

Figure 3.46: Amine–amine heat exchange after operation with MEA.

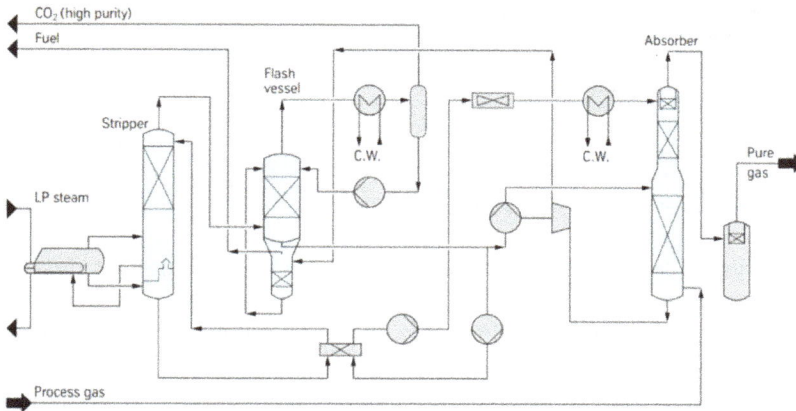

Figure 3.47: BASF aMDEA technology for CO_2 removal.

and subsequently high investment costs as well as high efforts for refrigerating. However, the heat requirements are very low, since desorption is relatively easy.

3.3.5 Crystallization and precipitation

The main applications of crystallization are related to generation of crystals from a solution of a dissolved solid, purification of a solid substrate through crystallization, formation of crystals with a special morphology, and crystal size or distribution.

Crystallization is important in the manufacturing of chemicals such as ammonium nitrate and phosphates and urea to name a few. Crystallization is governed by very complex variables being simultaneously heat- and mass-transfer process with a strong dependence on fluid and particle mechanics. Crystallization occurs in a multiphase, multicomponent system. The key processes in this operation are nucleation and crystal growth.

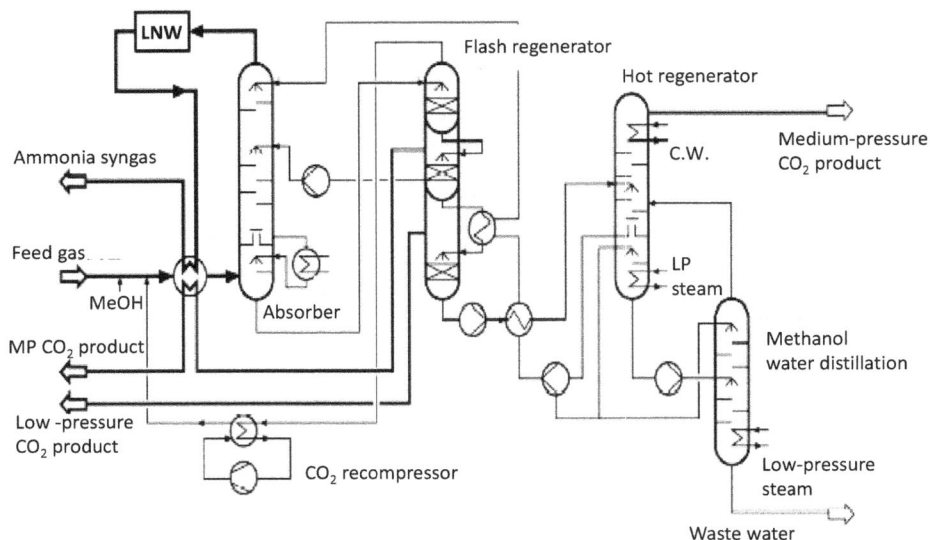

Figure 3.48: Rectisol technology for CO_2 removal.

Crystals can be grown from the liquid phase (solution or melt crystallization) or the vapor phase (desublimation), requiring in all cases supersaturation, i.e., the state when the liquid contains more dissolved solids than can be ordinarily dissolved at a particular temperature.

Crystallization can be done by evaporation resulting in elimination of the solvent and subsequent concentration of the dissolved compound above the saturation concentration. Another option to achieve supersaturation is cooling resulting in lower amount of solid needed for saturation. Vacuum can be also applied in crystallization with cooling and evaporation of a part of solvent due to pressure decrease, thus leading to supersaturation.

The sequence of process steps in crystallization is illustrated in Figure 3.49, demonstrating that the solids after crystallization are separated from the liquid, washed, and dried.

Crystallization is used to obtain a nearly pure solid product of a desired shape with a controlled size distribution. This method of separation contrary, for example, to distillation is less energy intensive and can also be used for temperature-sensitive feeds.

Design of crystallization requires knowledge on solubility, phases and their stability, nucleation and growth characteristics, as well as hydrodynamics of crystal suspensions.

Nucleation and growth depend not only on temperature and supersaturation, but also impurities. Supersaturation must first be achieved independent on from which phase crystals can be grown. In fact, such growth can be done either from the

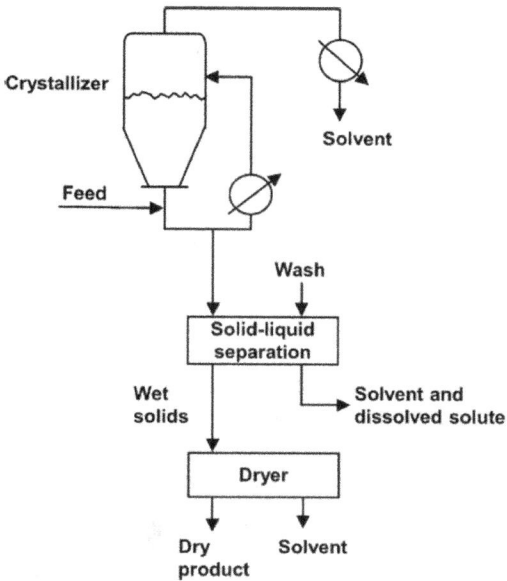

Figure 3.49: Crystallization technology (from A. de Haan and H.Bosch, Industrial Separation Processes. Fundamentals, 2013, Copyright de Gruyter. Reproduced with permission).

liquid (solution or melt) or vapor phases (desublimation). For the industrial crystallization of inorganic substances from solution, water is used as a solvent.

Solubility–supersolubility diagram (Figure 3.50) is often used to illustrate how concentration depends on temperature.

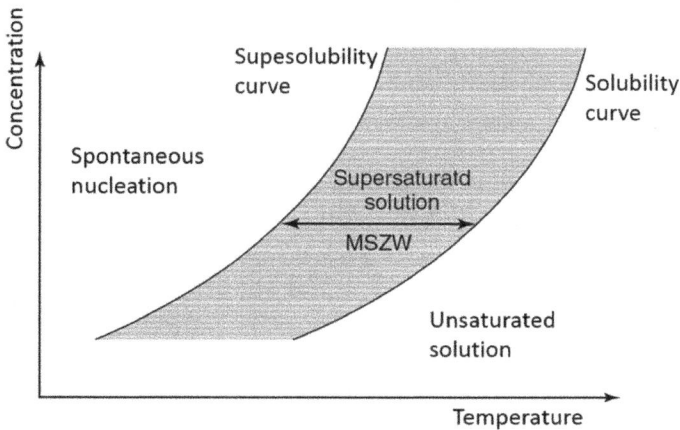

Figure 3.50: Solubility–supersolubility diagram with MSZW denoting the metastable zone width.

While position of the lower equilibrium solubility curve can be accurately determined, location of the upper supersolubility curve is less clear as it is influenced by many factors such as the rate of supersaturation, agitation intensity, and presence of crystals or impurities. The metastable zone width visible in Figure 3.50 is thus characteristic of a particular crystallization system.

In most crystal growth processes, not only diffusion but also surface reactions are important. The two most common surface processes are referred to as spiral (formation of screw dislocations on a crystal face) and polynuclear growth. The latter develops from monolayer nucleation on various parts of a crystal such as faces, corners, or edges.

As a result, the crystal growth kinetics often depends not only on the crystal size but also on the surface structure or perfection. Small crystals (<50 µm) typically have a slower growth rate than larger ones. It is important to predict the behavior of small crystals since it has a practical consequence in continuously operated industrial crystallizers. When new crystals with a size of 1–10 µm are constantly generated by secondary nucleation, their share in the overall crystal size distribution is decreasing significantly. Therefore, the ability to predict the growth rates of small crystals is useful in assessing the performance of crystallizers.

Another parameter influencing the growth rate is the presence of impurities, which because of their adsorption can block the crystal surface, thereby decreasing the growth rate. Moreover, impurities may change the crystal shape (or habit) altering growth at different crystal faces. Habit modifications can also be influenced in some cases by adjusting operating conditions such as the rate of cooling or evaporation, the degree of supersaturation, or crystallization temperature. This might be important from the process technology viewpoint, helping to improve rheological properties of the slurry, downstream unit operations (filtration or washing), and handling properties of the dried product and its storage stability. Other parameters that can be used to influence the crystal shape are proper solvent and pH selection. Moreover, some impurities can be added to the system influencing the crystal habit or alternatively can be removed from the crystallization milieu.

These precautions are done because of undesired caking (formation of hard materials) upon storage of crystalline materials. Such caking depends not only on storage *per se*, including its duration, humidity, pressure, and temperature as well as fluctuations in some parameters, but also on the crystal size, shape, and moisture content. Some measures to diminish caking include prevention of air contact with the crystals by careful drying, efficient packaging, and coating the crystals with a protective layer of an inert dust.

Despite these measures, the leading role in prevention of cake is the size of crystals, and more importantly, the size distribution and shape. Larger crystals are less prone to caking having a smaller number of contact points per unit mass. Caking can also be diminished by having a narrower size distribution and more granular shape. Operating conditions during crystallization and application of habit modifiers can influence, respectively, the crystal size distribution and shape.

Nucleation and growth are not separated in time as once few viable nuclei are formed, they start to grow.

For primary nucleation in industrial crystallizations, a simple empirical equation can be used:

$$J = K_n \Delta c^n, \tag{3.4}$$

which relates the primary nucleation rate J to the supersaturation Δc. In eq. (3.4), K_n and n are the rate constant and order of nucleation, respectively, the latter typically being higher than 2.

Crystallization is often explained by the empirical Avrami-Erofeev equation $1 - \varphi c = e^{-Kt^n}$, where φ_c is the crystalline volume fraction developed at time t and constant temperature, K, and n are suitable parameters; the former is temperature dependent. This equation corresponds to a characteristic S-shaped, or sigmoidal, profile where the transformation rates are low at the beginning and the end of the transformation but rapid in between. The initial slow rate is attributed to the time required for a significant number of nuclei of the new phase to be formed and to start growing. During the intermediate period, the growth is rapid and the nuclei continue to be formed. In the final state, there is little untransformed material for nuclei formation.

In the industry, large supersaturation is necessary to initiate primary nucleation. Once a stable nucleus has been created in a solution, it is capable of growing into a crystal, which is a two-step process involving mass transfer from the bulk solution to the crystal face and subsequent surface reaction when the growth units are integrated into the crystal lattice. In industrial reality, crystal nucleation, and growth interact with each other. The factors influencing crystallization are given in Figure 3.51.

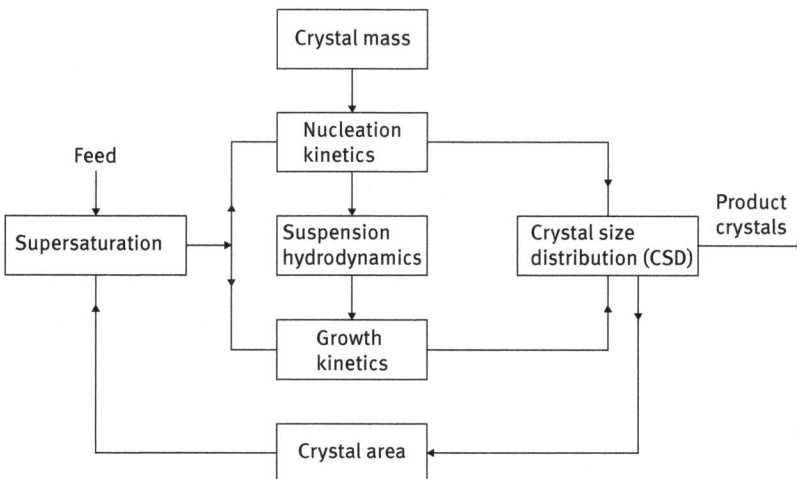

Figure 3.51: Interactions in continuous crystallization.

Three key variables control the rate of nucleation: temperature (T), interfacial energy (y), and the degree of supersaturation (S). The rate is increasing with increasing supersaturation, the latter giving, however, smaller crystals with limited filterability, thus requiring a delicate control of supersaturation.

Crystal size distribution (CSD) reflects how the size is distributed over the size range (Figure 3.52). The dominant crystal size is most often used as a representation of the product size.

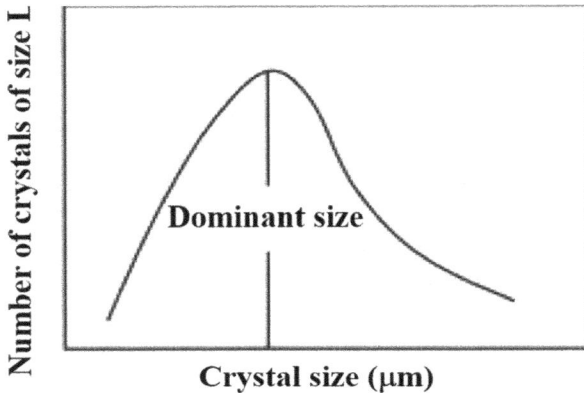

Figure 3.52: Crystal size distribution.

The form of the CSD in a continuous perfectly mixed crystallizer is determined by the residence time distribution and can be modified by selective removal of crystals altering such residence time distribution.

In crystallization processes, product requirements are a key issue in determining the ultimate success in fulfilling the function of the operation. Key parameters of product quality are the size distribution (including mean and spread), the morphology (including habit or shape and form), and purity. CSD determines several important processing and product properties. It is often important to control the CSD. Most favored is monodisperse, where all crystals are of the same size and dissolve at a known and reproducible rate. Critical phenomena that influence the CSD outside nucleation and growth are breakage and agglomeration. Breakage of crystals is almost always undesirable because it is detrimental to crystal appearance and it can lead to excessive fines and have a deleterious effect on crystal purity. Agglomeration is the formation of a larger particle through two or more smaller particles sticking together.

Several methods exist to initiate crystallization. One of the preferred approaches is to low temperature of the feed solution by direct or indirect cooling. Fast cooling and low molecular mass of the dissolved substrate give smaller crystals, which are more difficult to separate. However, there will be then fewer impurities. Another option, when solubility is less sensitive to temperature, is to remove solvent by alternatively adding heat. For the case of strong temperature dependence of solubility, vacuum cooling of the feed solution without external heating can be applied. Vacuum cooling can be combined with external heating when there is an intermediate dependence of solubility on temperature. Finally, a nonsolvent can be added for precipitating solute from a solution. Smaller crystals are obtained at higher temperature and supersaturation, as well as with more efficient stirring.

In general, continuous, steady-state operation is preferred in chemical process industries for different unit operations and chemical reactions *per se*. In crystallization processes, this is not always the case and batch operation can offer not only easier equipment and ability to clean the crystallizer after each batch, but also formation of the desired product in the required crystal form, size distribution, or purity. Higher operating costs and batch-to-batch variation are a downside of a batch system. At the same time, continuous crystallizers do not discharge the product under equilibrium conditions, thus a holdup tank is required to reach equilibrium; otherwise, there will be unwanted deposition in pipelines and effluent tanks. Moreover, there could be self-seed crystallization in continuous systems, leading to frequent shutdowns and washouts. Continuous crystallization is done typically for capacity exceeding 50 tons/day.

Continuous crystallizers may be operated in a series of well-mixed vessels with, for example, a temperature gradient in a cascade fashion with the crystals moved from one stage to another.

If crystallization is done without any stirring, the technology is very simple, since after introduction of a hot feedstock solution into an open vessel, cooling is done by natural convection often taking several days. Introduction of foreign metal particles promotes growth of crystals on their surfaces, thereby diminishing the amount of product which will end up at the bottom of the crystallizer. Slow cooling results in broad crystal size distribution with large interlocked crystals retaining some of the mother liquor and leading to impure crystals after drying.

Capital, operating, and maintenance costs are low in such type of crystallization, at the same time low productivity, high space requirements, and labor costs make this method economical mainly for small batches.

Agitation in open-tank crystallizers allows a decrease of batch time and formation of more uniform and pure crystals of smaller size. Better purity is possible since less mother liquor is retained and washing is more efficient. Apparently, introduction of an impeller increases costs, which is overcompensated by much higher productivity. In a large agitated cooling crystallizer, an upper conical section is needed to decrease the upward velocity of liquor, thereby preventing carryover of the crystals with the spent liquor. Cooling is provided by preferentially water jackets to avoid potential accumulation of crystals on cooling coils.

Evaporation of a part of solvent can be used to obtain the required degree of solution supersaturation if lowering temperature is not sufficient. In a vacuum crystallizer operating at reduced pressure, supersaturation is achieved by simultaneous evaporation and adiabatic cooling of the feedstock.

Large vessels and poor control over crystal form in batch crystallizers sparked an interest in fully continuous crystallization, resulting in development of different crystallization options such as, for example, forced-circulation Swenson crystallizers (Figure 3.53).

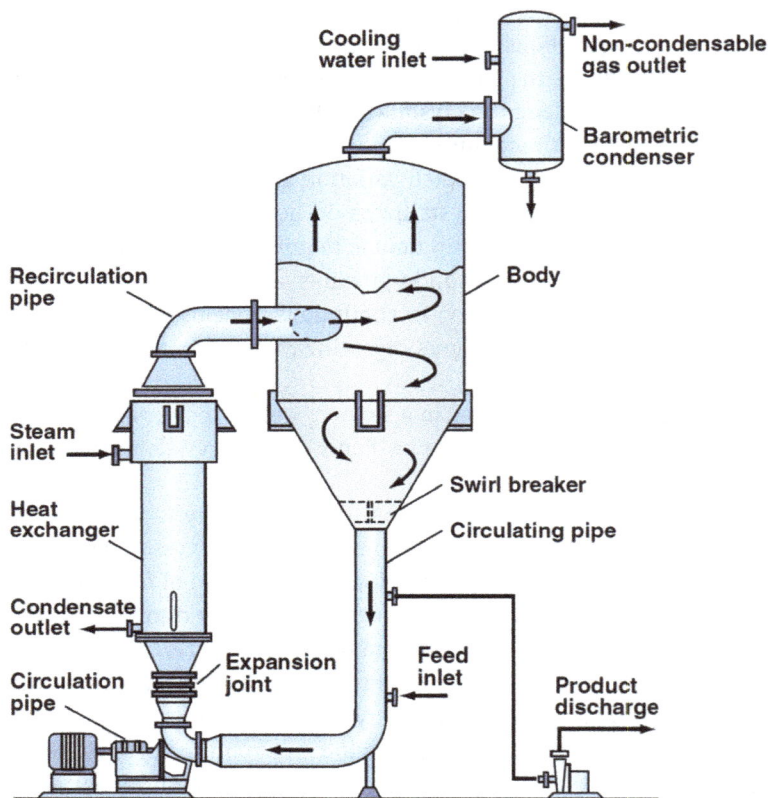

Figure 3.53: Forced-circulation Swenson crystallizer (https://www.chemengonline.com/wp-content/uploads/2011/07/fig1-768x840.png).

Swenson forced-circulation crystallizer (Figure 3.53) with an external heat exchanger operates under reduced pressure. High recirculation rate provides a good heat transfer minimizing encrustation. The crystal magma passing through the vertical tubular heat exchanger is reintroduced in the evaporator below the liquor level preventing sudden evaporation. Forced circulation crystallizers operate with a high circulation rate, limiting scaling.

Another option for continuous crystallization is to use draft-tube-agitated vac-uum crystallizers. One example (a so-called Swenson draft-tube-baffled vacuum unit) is illustrated in Figure 3.54. Location of a relatively slow-speed agitator is in a draft tube somewhat below the liquor level. At the base of the draft tube, there is an entrance of concentrated hot feedstock. Slow-moving impellers move slurry upward to the boiling surface, where supersaturation created by surface cooling and evapo-ration initiates crystal nucleation and growth. Crystal residence time is typically 4–6 h. The crystal size distribution is controlled by the vertical velocity of the mother liquor in the baffle areas and controlling the maximum crystal size that will be removed and dissolved.

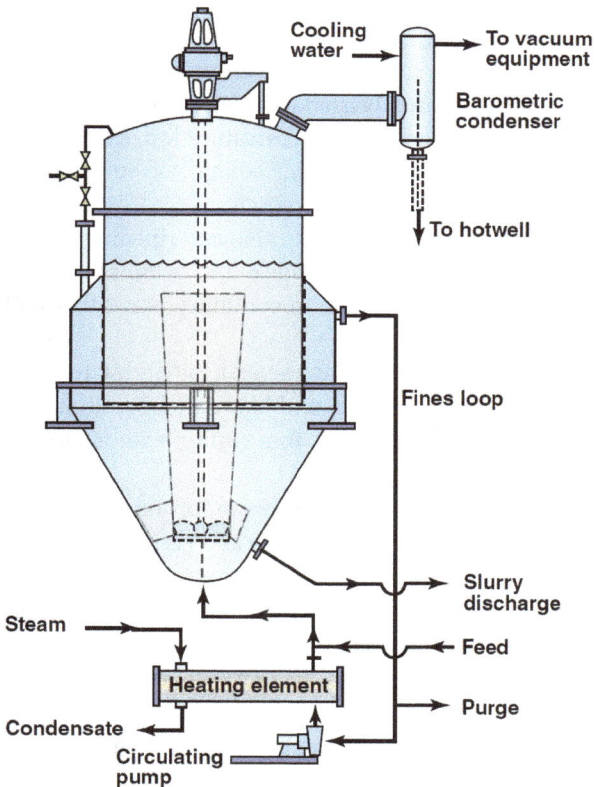

Figure 3.54: Swenson draft-tube-baffled (DTB) crystallizer (https://www.chemengonline.com/wp-content/uploads/2011/07/fig2.png).

Design of industrial crystallizers is far from being straightforward as it is difficult to characterize particle-fluid hydrodynamics and extrapolate data in lab-scale reactors on nucleation and crystal growth to large-scale units. Crystal suspension velocity is an important parameter for scaling up of fluidized-bed crystallizers, while for agitated vessels, the minimum rotational speed necessary to keep all crystals in suspension should be elucidated. Scaling up of agitated vessel crystallizers is often done based on constant power input per unit volume.

Precipitation is closely related to crystallization from solution being different to an extent that a precipitate is usually amorphous, not pure, and has a poorly defined size and shape, contrary to crystallization. In manufacturing of various chemical products including catalysts, paints, pigments, and pharmaceuticals, precipitation is often applied.

Precipitation can be carried out in one or in several steps gradually removing the desired components by such fractional precipitation, which, however, cannot be very sharp.

Precipitation can be considered as a first step in the overall process comprising precipitation and crystallization, where the latter is used for final purification.

Similar to crystallization, precipitation consists of supersaturation, nucleation, and growth of nuclei. Supersaturation can be done by, e.g., cooling, solvent evaporation, salting out, introduction of antisolvents, and pH alterations.

In saturated solutions, particle coarsening or so-called Ostwald ripening occurs by dissolution of smaller particles and deposition from the solute on the larger particles. The driving force for ripening is the difference in solubility between small and large particles.

Precipitation is a complex phenomenon which involves besides nucleation, also ripening, agglomeration, phase transformations, and habit modifications. Supersaturation, concentration of active impurities, and, to a certain extent, pH are the parameters influencing precipitation.

Several rules apply for precipitation. A reactant concentration increase gives first an increase of the median particle size of the precipitate at a certain time after mixing. This is followed by a subsequent decrease of the median crystal size. With an increase of the time interval, the maximum is shifted toward lower initial supersaturation and higher median particle sizes. For completed precipitation, the median size of the precipitate crystals decreases with an increase in the initially created supersaturation.

Highly supersaturated solutions can be made by mixing two reacting solutions in a very fast way, which might be challenging if reasonably uniform conditions should be maintained. Sequence of reactants mixing can be thus important. The induction period, which starts after mixing and continues until appearance of nuclei depends on a number of parameters, such as supersaturation, mixing characteristics, temperature, and impurities. This induction period is followed by rapid desupersaturation comprising primary and secondary nucleation and nuclei growth. Thereafter, particle coarsening might start because of ripening or agglomeration.

Similar to crystallization, a simple batch operation can be more advantageous than a continuous, steady-state operation mode. In industrial precipitation, good mixing should be ensured, which is essential to smooth-out supersaturation peaks in local regions. Both micromixing at molecular level being influenced by physical properties and macromixing, which is influenced by the impeller geometry and speed, as well as vessel geometry and other parameters, are important.

For instance, the feedstock entry point can influence the precipitate quality. In fact, two reactants can be introduced simultaneously near the surface or the impeller blades. Alternatively, the second reactant can be introduced when the first is already in the precipitation vessel. This can be done near the surface of the first reactant or near the impeller blades. In the latter case due to good local mixing, nucleation is less prominent and larger primary crystals are formed.

3.3.6 Leaching

Leaching is fluid–solid extraction when a certain substance is removed from a solid matrix using a suitable solvent, which can be an organic solvent, water, or supercritical fluids (e.g. carbon dioxide). Such method is often used in extractive metallurgy when the metals are leached from the ores using sulphuric acid. Another example is pulping when hemicelluloses and lignin are extracted from wood leaving the cellulose pulp behind. The solvent after penetration into the solid matrix remains partially in the solid residue; thus, total extraction is not possible.

The overall process is determined by penetration of the solvent in a porous structure, diffusion of the extract in the solid matrix, and dissolution into the solvent. Subsequently, the size of the solid matrix is important, necessitating grinding prior to leaching. High temperature and low viscosity are favorable from the process viewpoint improving solubility and decreasing the diffusion coefficient. At the same time, the solvent should be chemically and thermally stable enough at process temperature being also preferably nonflammable, nontoxic, and noncorrosive. Moreover, the solvent should selectively extract the desired component (in extractive metallurgy) or alternatively extract all other compounds with the desired one remaining in the solid form (e.g. pulping).

Extraction can be carried out in a batch mode or continuously. Apparently, drawbacks of batch extractors, which are less capital intensive than continuous ones, are their limited capacity, difficulties to maintain constant process conditions, and a need to periodically load and unload the equipment, which is time consuming. One example which will be discussed below is wood pulping, which can be carried out in a continuous or batch process. Considering disadvantages of batch digesters mentioned above, and moreover unfavorable heat economy of batch pulping systems, it is not surprising that modern pulp technology is dominated by continuous digesters (Figure 3.55).

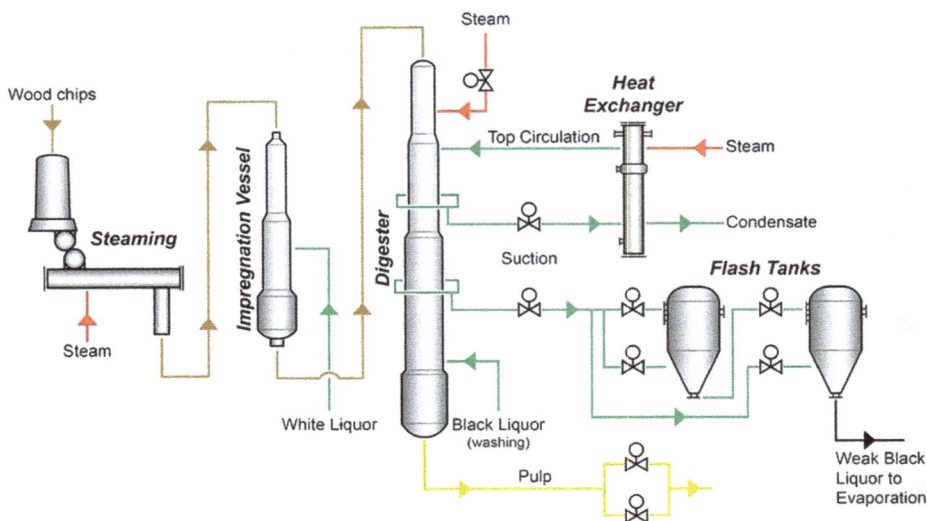

Figure 3.55: Continuous digester for Kraft pulping (https://naf.se/wp-content/uploads/2018/02/Continuous-Digester-diagram_v3.jpg).

Wood chips after steaming with low-pressure steam (150 kPa) are impregnated with the pulping liquor in an impregnation vessel. The impregnated chips taken from the bottom are conveyed to the digester along with the liquor and then heated to the maximum digestion temperature.

Washing liquor is added at the lower part moving upward countercurrently to wood chips. The cooked and partly washed pulp is taken out of the bottom of the digester.

The spent cooking liquor (black liquor) is separated by screens and is routed to the flash tanks, where the pressure is decreased to the below vapor pressure, and then undergoes evaporation. More details on the Kraft process will be given in Chapter 4.

3.4 Chemical reactors

The choice of reactors is a very essential part of the engineering process. Homogeneous and heterogeneous reactors will be briefly overviewed below, following mainly the textbook *Chemical Reaction Engineering and Reactor Technology* by T. O. Salmi, J.-P. Mikkola, J. P. Wärnå (CRC Press, 2018).

Prior to overviewing different reactor types few words should be mentioned about the reactor modeling. The complexity of the tasks, which a chemical engineer faces, justifies the necessity and the need of reliable kinetic and transport models, which are based on physicochemical understanding of catalytic processes. The reactor models cover several levels: an active site (mechanism and kinetics), a catalyst grain (mass and heat transfer and thus knowledge of catalyst effectiveness factor, interphase mass transfer, phase holdups) and a reactor itself (hydrodynamics, bed voidage profile, axial and radial dispersion, bubble properties, wall heat transfer, etc.). The degree of complexity of a reactor model is still a matter of debate. A very complex model with a very large number of adjustable parameters might be found unpractical by some industrial companies. On a more general level it can be stated that the reaction model should be clarified first before properly addressing reactor hydrodynamics. Although the actual hydrodynamics should be modeled adequately, the impact of inadequate reaction model is more serious than inadequate hydrodynamic model.

More details on the reactor modeling are available in textbooks on chemical reactors, including the one mentioned above.

3.4.1 Homogeneous processes

In a homogeneous reactor, just one phase (i.e. gas or liquid) is present. Three most common reactors used industrially are a batch reactor (BR), a tube reactor, and a tank reactor.

Batch operation is used in production of many different organic chemicals, typically belonging to fine and specialty chemicals, such as drug synthesis, and manufacture of paints or pesticides.

Vessel reactors can have different types of impellers for mixing viscous liquids. The reactant can be in the form of a gas and not only liquid illustrated in Figure 3.56.

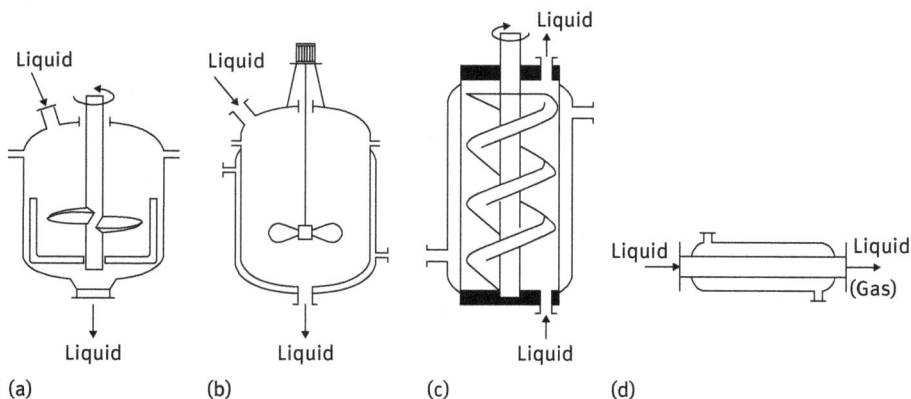

Figure 3.56: Reactors for homogeneous liquid-phase processes.

Such reactors are also used for homogeneous catalytic processes when the catalyst is in the same phase as reactants. Stirred tank reactors (STR) can be operated batchwise (batch reactor, BR; Figure 3.57a), semi-batch-wise (semi-batch reactor, SBR; Figure 3.57b) or continuously (continuous stirred tank reactor, CSTR; Figure 3.57c).

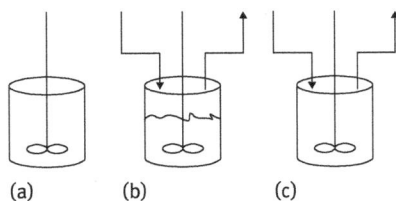

(a)   (b)   (c)

Figure 3.57: Stirred tank reactor (STR) operating (a) batch-wise, (b) semi-batch-wise, and (c) continuously.

A choice of the reactor type and the operation mode depends on the production capacity, operation conditions (e.g. temperature and pressure), mixing requirements, heat duty, and time needed for cleaning of the reactor between batches and construction material. Exothermic reactions obviously require efficient cooling, which might not be efficient just with a double jacket (Figure 3.56a). Introduction of cooling (heating) coils or a heat exchanger in the loop can be a solution for a proper heat management.

A stirred tank reactor operating batch-wise can be used for chemically different reactions, which is important for synthesis of fine chemicals of rather low production capacity. In essence, various reactions with different reaction times under different operation conditions can be organized in the same batch reactor.

Moreover, the batch operation allows changes of the operation conditions (e.g. temperature) during the course of the reaction, which might be beneficial for the reversible exothermic reactions. In such a case, initially a higher temperature is favorable from the kinetic viewpoint, while low temperature at the end of the reaction is favorable for thermodynamics.

The scale-up of batch operating reactors for lab to industrial scale is relatively straightforward.

For different types of reaction kinetics, higher yields are achieved in the batch mode compared to CSTR, and are similar to those obtained in a tube reactor. Obviously operating in a continuous mode in a tubular reactor does not have a need to spend additional time for reactor filling and emptying the reactor vessel between the batches.

A semibatch mode when one or several reactants are fed into the reactor during the course of the reaction can be used for strongly exothermic reactions, avoiding excessively high temperatures and fine-tuning the product distribution for some reaction types.

CSTR operates as a backmixed reactor meaning that the outlet composition is the same as the composition of the reaction mixture. If a high conversion is needed, then the reactor operates at low reactant concentrations and lower rates. A stirred tank thus results in lower yields compared with a tubular reactor and a lower level of intermediates than a tubular or a batch reactor operating at the same residence time. On the other hand, a continuous mode under stationary conditions guarantees the same product quality and efficient heat exchange. Moreover, in some specific cases CSTR can be more beneficial than a tubular reactor. Such cases include formation of the desired product in a reaction with the lowest reaction order among parallel reactions of different orders or autocatalytic reactions when the rate is enhanced in the presence of a product.

Stirred tank reactors can operate as a cascade with several CSTR in series approaching performance of a tubular reactor.

The latter is used industrially for both homogeneous and liquid-phase reactions. Tubular reactors installed in a furnace are often used for endothermic cracking and dehydrogenation reactions.

The tubes are bent, thus ensuring tubular flow conditions. An example of rapid gas-phase reactions occurring in a tubular reactor with bended tubes is shown in Figure 3.58.

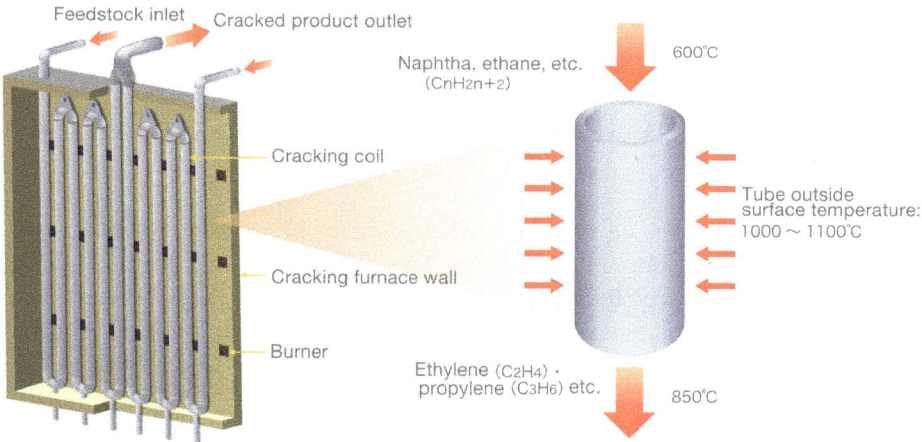

Figure 3.58: Thermal cracking furnace (https://www.kubota.com/products/materials/products/cracking_coil/img/index/img_03.jpg).

Tubular reactor affording the highest yield for the most common reaction kinetics still have their limitations, such as formation of hot spots.

3.4.2 Non-catalytic heterogeneous processes

The reactors for heterogeneous processes with a solid phase in a simple form consist of a solid phase through which the fluid phase is circulated (Figure 3.59a). Removal of H_2S by reacting it with ZnO is carried out in such reactor when the solid phase is gradually transferred from the oxide to sulphide. After the end of the reaction, the solid phases should be removed from the reactor, which can be done after

5–10 years depending on the amount of H_2S in the feed. When a continuous removal of the solid phase is required, the movement of the solid phase within the reactor could be arranged in several ways (Figure 3.59b–d). The solid is moved (Figure 3.59b) from one bed to another with a special devices (oxidation of FeS_2), a conveyer belt (Figure 3.59c), or in leaned reactors (Figure 3.59d) (ammonia neutralizer in monocalcium phosphate or superphosphate production).

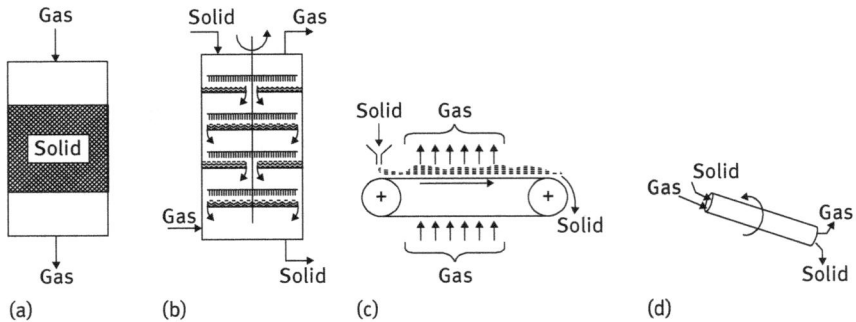

Figure 3.59: Reactors for gas-solid reactions: (a) fixed solid phase and (b–d) moving solid phase (explanations in the text).

Gas-solid processes can be enhanced when the size of the solids is decreased, which is difficult to achieve in the reactors mentioned above. The reason is that the smaller size of solid particles will lead to a higher probability of their agglomeration, and eventually, higher pressure drop. A way to avoid these problems is to use fluidized beds or reactors where the solid phase is transported (Figure 3.60).

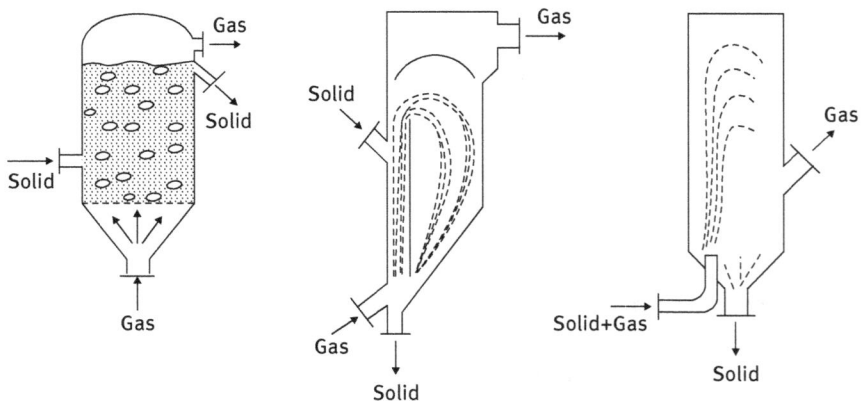

Figure 3.60: Gas-solid reactors with transport of solids.

Fluidized-bed reactors allow for efficient radial and axial mixing and thus give an even solid distribution, which eventually leads to the isothermicity of such reactors. Conceptually similar arrangements can be used for catalytic reactors, with the obvious difference that in the latter case, the solid phase (catalyst) is not consumed. Advantages and disadvantages of fluidized-bed reactors will be considered below when their utilization for catalytic reactions will be discussed.

In gas-liquid reactors, the contact between the gases and the liquids can be arranged in several ways (Figure 3.61).

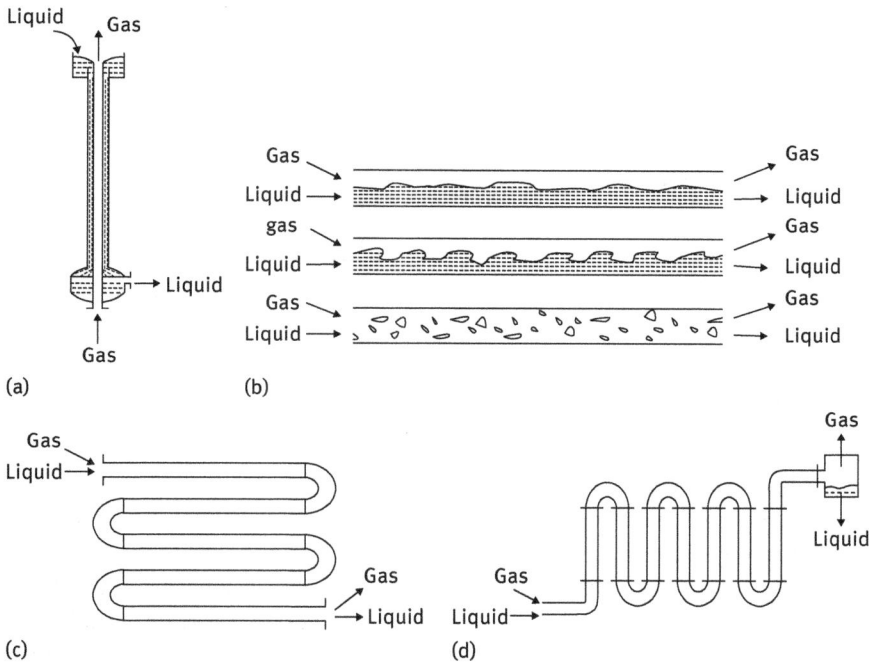

Figure 3.61: Reactors for gas-liquid reactors.

In the case of a continuous phases (Figure 3.61), the liquid is flowing along the reactor walls being in contact with the gas in either concurrent or countercurrent mode. This operation mode is applied when temperature regulation needs to be done in a precise way.

Alternatively, the liquid can be sprayed in the gas using packing elements (Figure 3.62a), improving the interfacial area substantially. The liquid is moving by gravity, which gives a possibility of saving on the energy input. This mode of operation, however, has a limit with respect to the gas flow rate, since high rates can lead to flooding. The spraying could be also done without any packing (Figure 3.62b).

The gas can be also bubbled through the liquid (Figure 3.63). The diameter of bubbles can be regulated in a narrow range, since the smaller bubbles would coalesce and

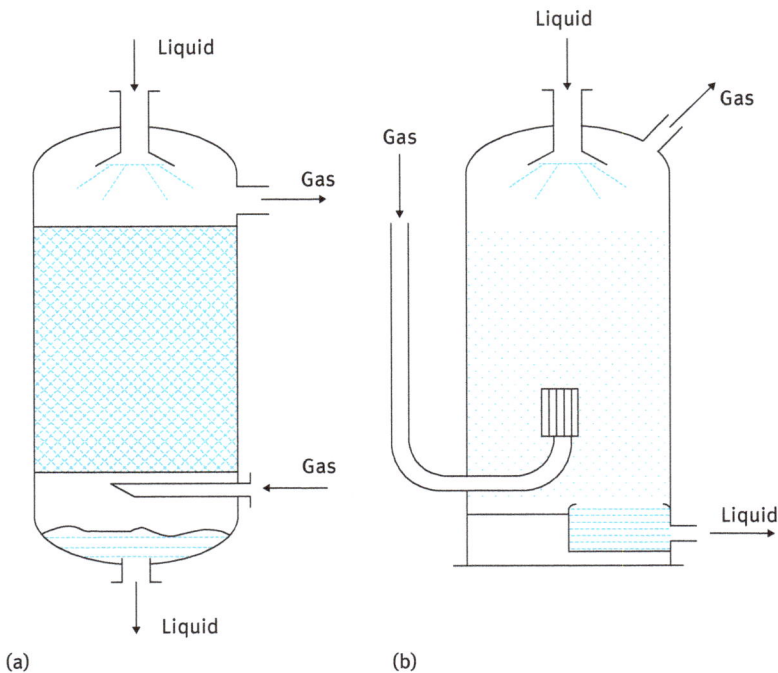

Figure 3.62: Gas-liquid reactors (a) with and (b) without spraying.

the larger ones are not stable and can be broken. The mass transfer coefficients in bubble columns are typically not high ($0.3\ s^{-1}$). The small size of bubbles can be somehow preserved when trays columns are applied (Figure 3.64).

An example of the solid–liquid reaction is the wet-acid process for manufacturing of phosphoric acid requiring relatively pure raw materials generates besides the desired products also ca. 5 tons of so-called phosphogypsum per ton of phosphate in phosphoric acid.

The technology can be viewed also as production of calcium sulphate by crystallization with a minimum amount of P_2O_5 and phosphoric acid as a by-product. Analysis of the process is done by considering the equilibrium diagram for the $CaSO_4$–P_2O_5–H_2O system (Figure 3.65).

Figure 3.65 illustrates several forms of calcium sulfate including dihydrate $CaSO_4 \bullet 2H_2O$; α-hemihydrate, α-$CaSO_4 \bullet \frac{1}{2}H_2O$, and anhydrite $CaSO_4$. A temperature increase to 75 °C elevates formation of α-hemihydrate at the expense of dehydrate, while the latter has better filtration efficacy. Moreover, at higher temperatures corrosion is more prominent; thus, the dihydrate processes operate in the DH region (Figure 3.65) just below the dihydrate–hemihydrate equilibrium line in the range of 70–80 °C at 28–32% P_2O_5 concentration. An apparent disadvantage of this approach is a low P_2O_5 concentration of only 28–30%. The market demand for product with 40–42% and

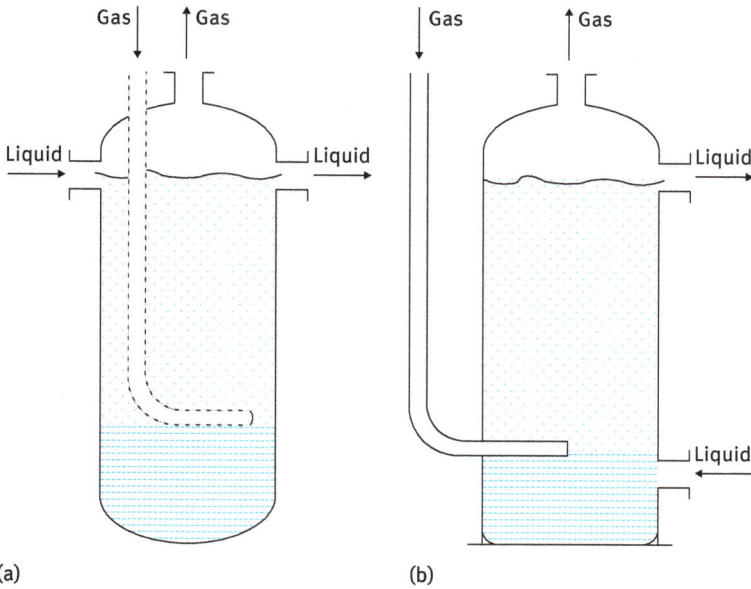

Figure 3.63: Gas-liquid reactors with bubbling of the gas.

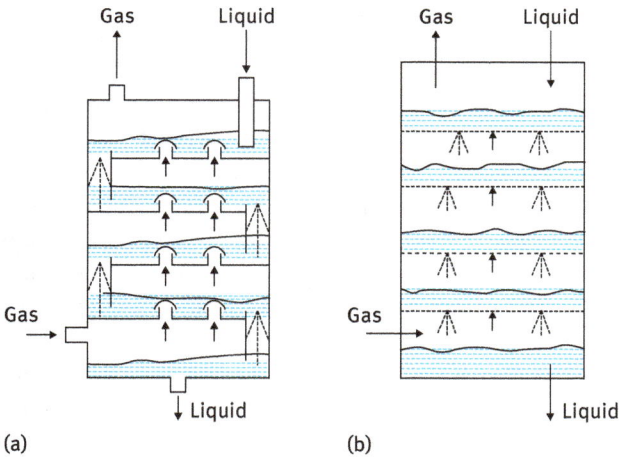

Figure 3.64: Gas-liquid reactions in tray columns.

52–54% of P_2O_5 requires adding a concentration section with either hot water or low-grade steam.

Several options exist to operate in the hemi-hydrate region of the $CaSO_4$–P_2O_5–H_2O system, shown as region A in Figure 3.65. In the dihemihydrate or the Central Prayon

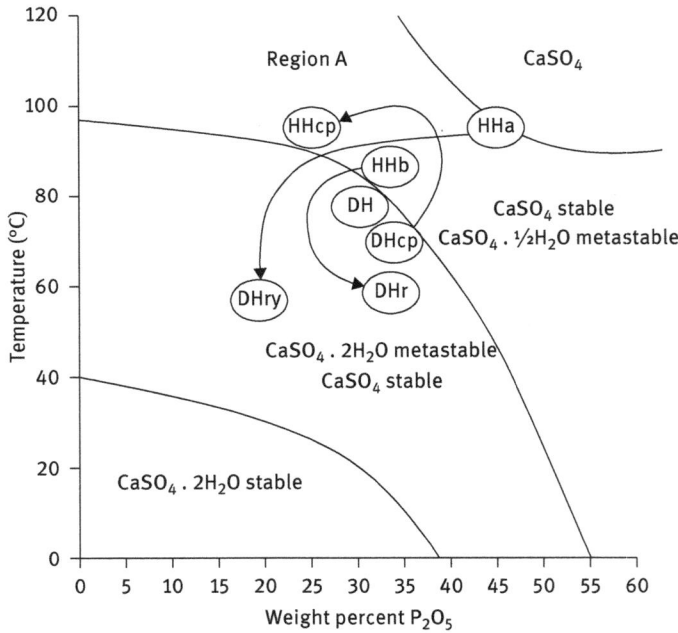

Figure 3.65: Equilibrium diagram for the $CaSO_4$–P_2O_5–H_2O system (adapted with permission from Dahlgren, S.-E., Fertilizer materials, calcium sulfate transitions in superphosphate, J. Agric. Food Chem., 8(5), 411–412. Copyright 1960 American Chemical Society. Reprinted with permission).

process, the reaction takes place in the dihydrate region at 32–35% P_2O_5 concentration (DHcp region). After the first filtration, the temperature is raised with steam and the dihydrate slurry is converted to high-quality hemihydrate, followed by second filtration, water washing, and recycling. Insufficient quality of the phosphate rock and control of crystallization often prevents gypsum formed in this process to compete at the commercial gypsum market because of impurities originating in the feedstock along with the presence of P_2O_5. Minimization of latter needs a good separation between crystals and the acid.

The production technology involves dissolution of the phosphate rock, its reaction with sulphuric acid, and gypsum crystal growth. In the Prayon Mark IV plants, multicompartment reactors are used with, e.g., six compartments of 245 m³ capacity each and a single agitator (Figure 3.66). Less reactive rocks might require more compartments. The phosphate rock is introduced to compartment 1 and mixed with the reaction slurry by the single agitator in compartment 1. Sulphuric and weak phosphoric acids are added to the second compartment 2. Moreover, sulphuric acid can also be added to compartments 3 and 4 to control supersaturation. One part of the slurry from compartment 4 passes to the subsequent compartment 5 where

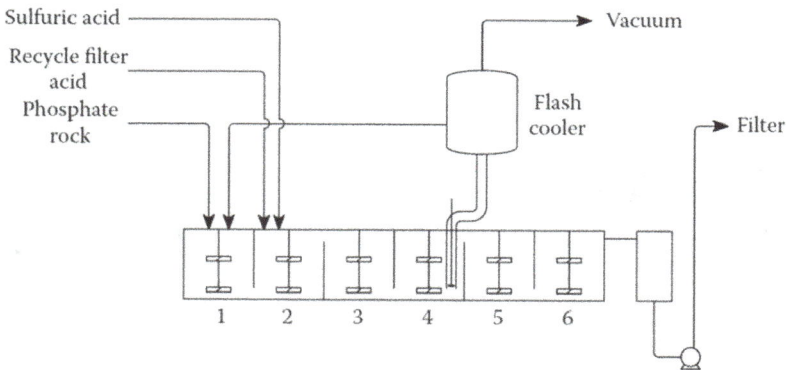

Figure 3.66: Prayon Mark IV reaction flow sheet. From R Gilmour. Phosphoric acid: Purification, uses, technology, and economics. Reproduced with permission from Taylor & Francis Group.

crystallization continues, while the other fraction after that is taken to the cooler where water is evaporated under vacuum. The cooled slurry is returned to compartment 1. The slurry from compartment 6 undergoes filtration.

There are several other processes using a single or two reactors.

Downstream the reactor section, there is filtration, which is done in all wet process plants to separate phosphoric acid from calcium sulfate crystal using suitable cloth filters. An example of a belt filter is illustrated in Figure 3.67.

Figure 3.67: A belt filter for the phosphoric acid production. From http://www.victorenvirotech. com/e_productshow/?50-Phosphoric-Acid-Filter-Belt-50.html.

For a dihydrate plant giving 28–30%, the filter acid is usually concentrated to 40–42% P_2O_5 needed for the fertilizer manufacture and then clarified to 52–56% P_2O_5 for the merchant market (Figure 3.68). In the clarification process, the solid-rich underflow from the phosphoric acid clarifier is either returned to reaction or filtered, usually on a small belt filter (Figure 3.68).

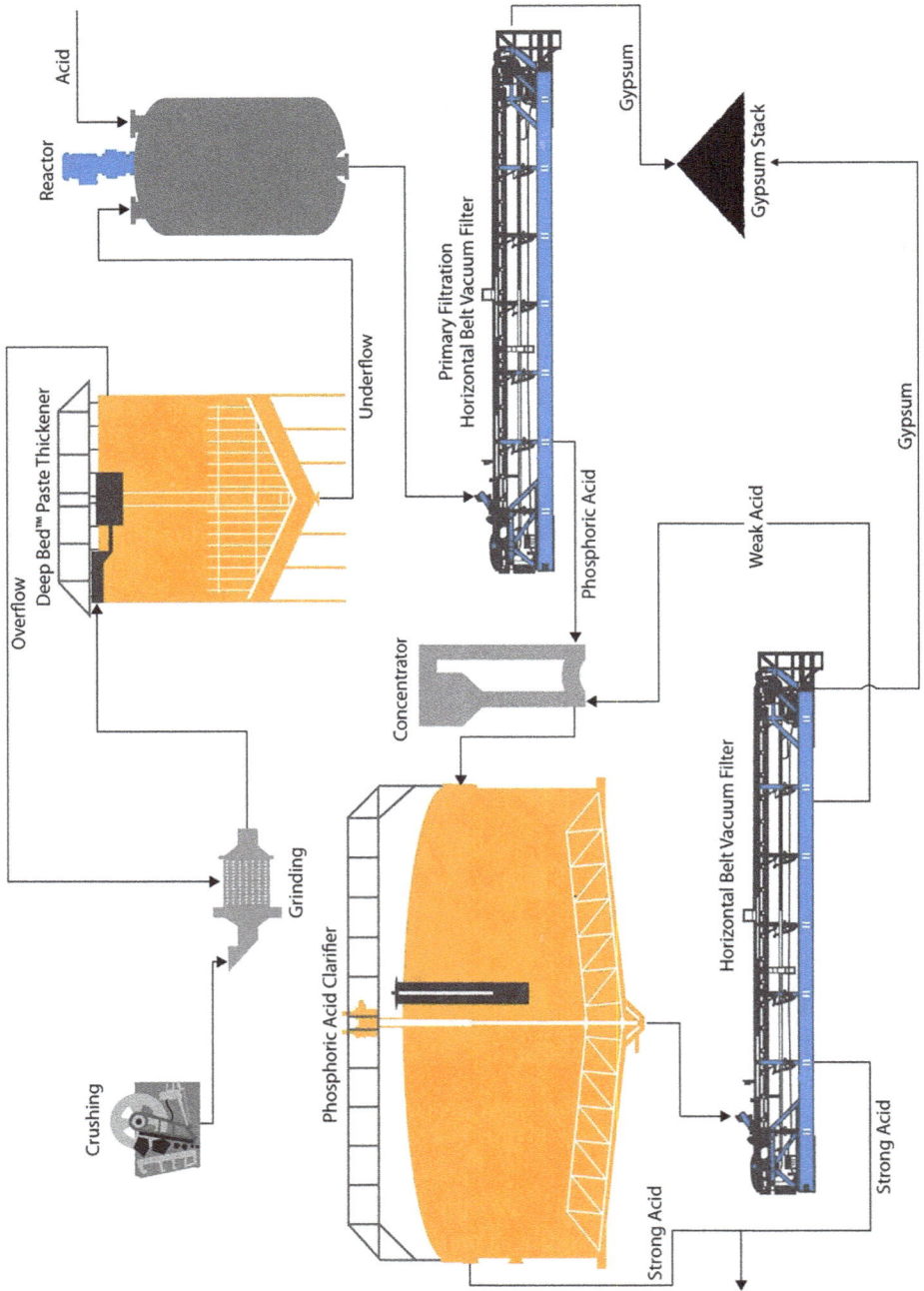

Figure 3.68: Phosphate processing simplified flow sheet (https://f.hubspotusercontent10.net/ hubfs/541513/image/blog/efficient-liquid-solids-separation-is-critical/phosphate-processing-flow-sheet-lg.jpg).

Presence of such impurities in the phosphate rock as iron, magnesium, aluminum, etc. makes wet processes challenging, limiting also the feedstock base as low-quality grades of 13–14% phosphate content cannot be efficiently processed. A recent development by JDCPhosphate allows production of high-quality phosphoric acid from lower grade phosphate rock, avoiding at the same time generation of phosphogypsum.

The Novaphos technology (Figure 3.69) uses a kiln-based process and can operate with the phosphate raw material containing ca. 14% of phosphate as P_2O_5 with high levels of impurities giving a product with 68% P_2O_5. The technology combines carbothermal reduction with petroleum coke and silica sand and then oxidation to P_4O_{10} with subsequent hydration to recover phosphate. A commercially useful aggregate for construction and road building is formed as a by-product.

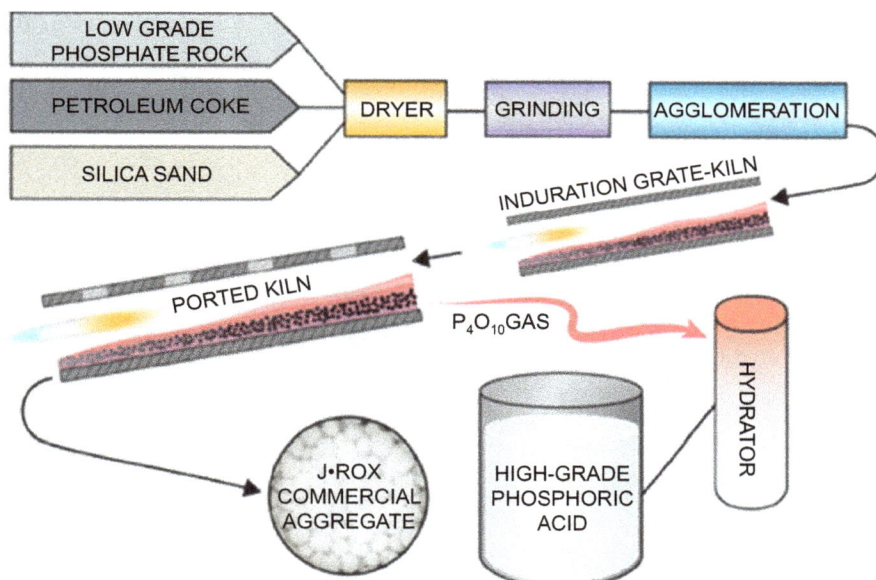

Figure 3.69: The Novaphos technology of phosphoric acid manufacturing (https://www.avenira.com/wp-content/uploads/2017/03/JDC-Phosphate-David-Blake.pdf).

3.4.3 Catalytic reactors

The discussion below will be done separately for two- (gas–solid or liquid–solid) and three-phase (gas–liquid–solid) reactors when the catalyst is in the form of a solid.

3.4.3.1 Two-phase reactors

Fixed-bed reactors with a packed catalyst bed are most commonly used in industrial practice (Figure 3.70) for the production of petrochemicals and various basic chemicals. The catalyst is loaded in a vessel, through which the fluid is directed. The catalyst particles are in the form of pellets (extrudates and tablets) of a size that does not lead to too high-pressure drop, while increasing the diffusional length at the same time. Figure 3.70 shows adiabatic fixed-bed reactors, when there is no temperature control in the bed and there is a monotonic temperature increase along the bed. For single-route reactions (such as, for example, water-gas shift), the conversion can be calculated simply from the adiabatic temperature increase or decrease. When heat should be removed to prevent damaging the catalyst particles or because equilibrium composition is unfavorable at a high temperature, multibed reactors (Figure 3.70B, C) can be used with heat removal achieved either using heat exchanges or cooling by cold reactants, respectively. Such procedures allow for adapting the temperature profile to the requirements of an optimal reaction pathway.

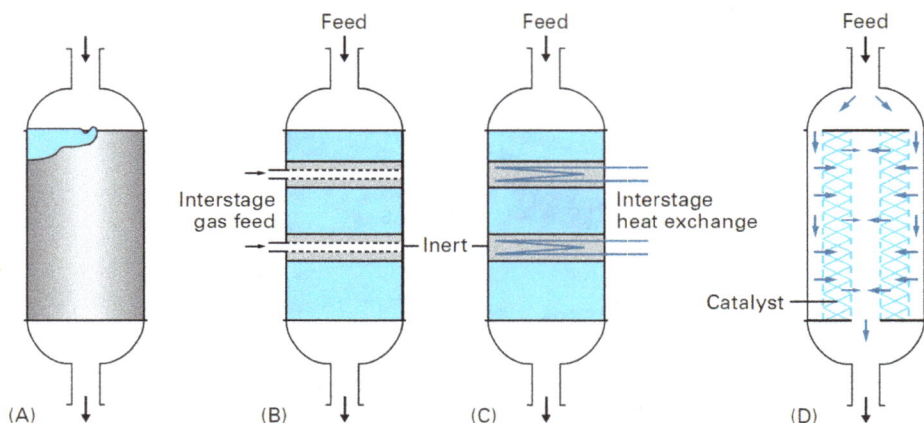

Figure 3.70: (A) Single-bed adiabatic packed-bed reactor. (B) Adiabatic reactor with quenching. (C) Multibed adiabatic fixed-bed reactor with interstage heat exchange (from Eigenberger, G., Ruppel, W. (2012) Catalytic Fixed-Bed Reactors, Ullmann's Encyclopedia of Industrial Chemistry. 10.1002/ 14356007.b04_199.pub2, copyright © 2012 by Wiley-VCH Verlag GmbH & Co. KGaA. Reproduced with permission). (D) Adiabatic reactor with radial flow.

Injection of hot or cold gas between the stages (or beds) as visualized in Figure 3.70B is a simple option. The disadvantages for constant tube diameter are cross-sectional loading, which increases from stage to stage, and energetically unfavorable mixing of hot and cold streams. As the composition is changing, it can have an either positive or negative effect on the reaction kinetics and thermodynamics. Adiabatic multistage reactors with interstage heat transfer are typically used in reactions giving a single product, but limited by equilibrium (sulphur trioxide, ammonia, methanol syn-theses), where intermediate cooling is used to displace the gas temperature in the di-rection of higher equilibrium conversion.

For strongly exothermic (partial oxidations and hydrogenations) or endothermic reactions (dehydrogenations), multitubular reactors with cooling or heating in between the tubes are applied. This allows for better temperature control (Figure 3.71), although in this case the temperature profile is characterized by a maximum (hot spot), which is related to enhanced heat release at the reactor inlet and a decline in heat release along the tube length caused by a diminished concentration of reagents. An important issue with a multitubular reactor is the uniform flow distribution among the tubes, which requires special measures when the catalyst is loaded in the reactor.

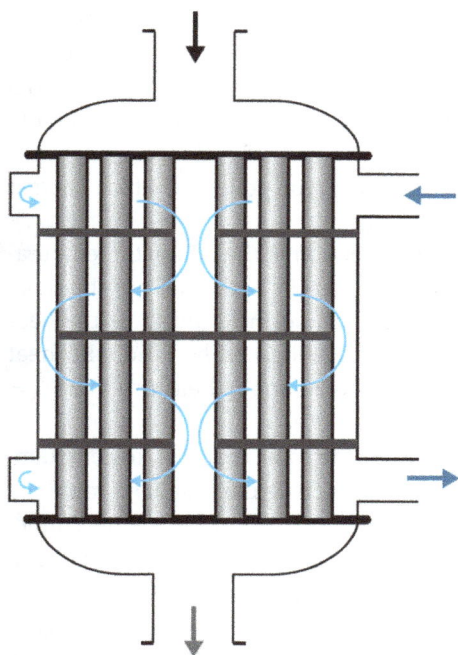

Figure 3.71: A multitubular reactor (from Eigenberger, G., Ruppel, W. (2012) Catalytic Fixed-Bed Reactors, Ullmann's Encyclopedia of Industrial Chemistry. 10.1002/14356007.b04_199.pub2). copyright © 2012 by Wiley-VCH Verlag GmbH & Co. KGaA. Reproduced with permission).

The flow conditions in packed-bed reactors are very close to the plug flow (i.e., minor back-mixing), which implies a higher conversion rate for the most common reaction kinetics and better selectivity toward the intermediate products for conse-cutive and parallel-consecutive reactions.

The construction of the packed-bed reactors is simple, not requiring any moving parts, and therefore this is generally a low-cost, low-maintenance reactor. Among other advantages it is possible to have large variations in operating conditions (including high pressure) and contact times, little catalyst attrition and subsequent catalyst losses, high catalyst to reactant ratio, and thus long residence time. In addition, principles of mathematical modeling of packed-bed reactors are well established. The most serious disadvantages are related to poor heat transfer in large fixed-bed reactors and thus the inability to control temperature properly. High-pressure drop is associated with limited productivity, while nonuniform flow patterns and wide-residence time distribution may lead to poor selectivity. As mentioned above, the temperature control problems could be overcome by introducing internal and external heat exchangers, interstage cooling or heating, and the application of diluents. The revamp of adiabatic axial flow operation to radial flow and the use of special geometrically shaped catalysts can improve pressure drop, simultaneously helping in removal of pore diffusional problems.

Regeneration and replacement of the catalyst are difficult and require shutdowns.

Uniform flow distribution could be achieved if the pressure drop is significantly high. It should be still ensured that the catalyst is packed uniformly in order to avoid bypassing. Such uniform flow distribution is more difficult to achieve if the pressure drop is low, i.e., in monolith reactors. The pressure drop in fixed-bed reactors depends very strongly on the void fraction, as follows from the Ergun equation, which is often applied for pressure drop calculations.

An important parameter during the reactor selection is the adiabatic temperature rise, which is defined as $\Delta T_{ad} = \Delta H_r\, C_o/c_p$, where ΔH_r is the heat of the reaction (kJ/mol), C_o is the initial reactant concentration (mol/m^3), and c_p is the heat capacity (kJ/m^3 K). When $\Delta T_{ad} < 300$ K, a multibed reactor can be used with heat exchanges or quenching in between the beds. If the adiabatic heat rise is $\Delta T_{ad} = 300$–700 K a multitubular reactor is preferable.

Sometimes two of these options are applied for the same reaction, as for example possible in synthesis of methanol. The low-cost radial axial adiabatic multibed reactor with quenching is used in plants when extra steam is required in the synthesis unit. In the isothermal reactor, the reaction heat is removed by partial evaporation of the boiler feed water generating 1 metric ton of medium-pressure steam per 1.4 metric ton of methanol. This isothermal reactor allows easier temperature control by regulating steam pressure, conditions close to isothermal, high heat recovery, and low by-product formation.

When the adiabatic temperature rise is above 700 K, fluidized bed reactors (Figure 3.72A) are usually used. When heat exchanges are located directly in the fluidized bed of catalyst, it is possible to maintain the temperature at, for example, 300–400 °C, and the order of magnitude is higher even in the adiabatic temperature rise. Because of intensive heat transfer, the temperature in almost all parts of the fluidized-bed reactor is the same, affording isothermal conditions.

When the catalyst undergoes strong deactivation, the application of packed-bed reactors is cumbersome, as continuous addition, removal, and regeneration of the catalyst are required. In the moving-bed reactor, the catalyst moves downward and the gas flows upward (Figure 3.72B, left). The removed catalyst is transported in a regenerator unit. For upflow operation, the catalyst, which is removed, is further deactivated because it is in contact with the feed most rich in the reactant, while less deactivated catalyst is in the reactor.

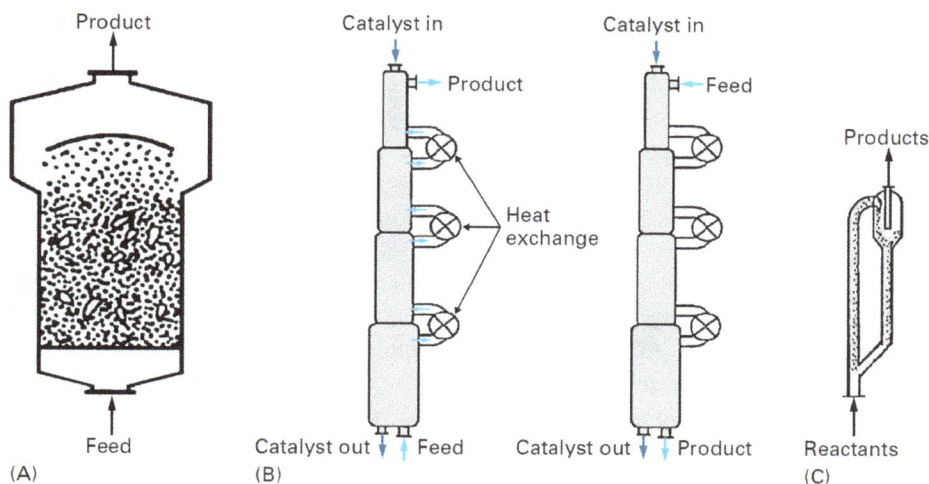

Figure 3.72: Reactors with moving catalysts. (A) Fluidized-bed reactor; (B) a moving-bed reactor (left, counter-current and right, concurrent flow of the feed and the catalyst); (C) entrained flow (riser) reactor.

For the concurrent operation mode (Figure 3.72B, right) applied for heavier feedstock, as for example in catalytic reforming, uniform catalyst activity along the reactor is maintained. Typically, in such reactors, the catalyst flows by gravity and the reaction mixture can be withdrawn from the reactor and cooled or heated. Heating is required in catalytic reforming, where one of the dominant reactions is dehydrogenation.

When deactivation is even more prominent and the catalyst life is just few seconds, an entrained flow or riser reactor is used when the catalyst powder is entrained by the catalyst in feed, thereafter being separated in a cyclone, regenerated (not shown in Figure 3.72C), and sent back to the riser.

The obvious advantages of the fluidized-bed reactors are associated with the possibility to regenerate the catalyst continuously, high thermal efficiency, and thus efficient temperature control. In some cases, for example, vapor-phase oxidations of organic compounds, fluidized-bed reactors are effective flame barriers that allow operation at high concentration of organic compound, possibly even within the explosion limits. Moreover, small catalyst particles (50–100 pm) are needed for fluidization, which leads to less prominent pore diffusional resistance and makes it possible to run a process free from mass transfer limitations.

Despite a small size of catalyst particles, the pressure drop in the fluidized-bed reactors is low due to high bed porosity.

High heat transfer coefficient is also an advantage limiting catalyst sintering.

The construction is, however, much more complicated than for simple packed-bed reactors, requiring extensive investments and higher maintenance costs. The flow conditions are complex and could vary from an ideal plug flow to complete back-mixing.

Other apparent disadvantages are catalyst attrition, formation of dust influencing efficiency of solid particles separation in cyclones, and backmixing having a negative impact on selectivity. For fluidized-bed reactors combined with catalyst regeneration, a part of still active catalyst is taken out from the reactor undergoing downstream regeneration. This leads to a regenerator block of a larger size and thus additional costs.

Below the minimum velocity required for fluidization, a fixed-bed regime is observed when gas velocity is below. In the ebullated bed, there is a smooth bed expansion with a well-defined bed surface. Higher velocity gives rise to bubbling fluidization when gas bubbles grow from the distributor, coalesce with other bubbles, and eventually burst at the surface of the bed. In small diameter reactors, a further size in fluidization velocity results in a slug flow regime with the bubble diameter approaching the reactor diameter. In a turbulent regime, the top surface of the bed cannot be distinguished. Transport of particles and their recycle back to the reactor is a feature of fast fluidization.

In three-phase fluidized beds, the solid volume fraction is much higher (10–50%) compared to slurry bubble columns (<5%).

Scaling-up of fluidized-bed reactors is challenging as the mass transfer characteristics of the fluidized bed are dependent on reactor dimensions. Moreover, some large gas bubbles are formed that could pass the reactor without being in contact with the catalyst, leading to lower conversion and requiring additional amounts of catalyst. The riser reactor (Figure 3.72C), however, approaches the plug flow regime.

An example of a riser reactor usage will be discussed in relation to an oil refining process – catalytic cracking. Some other examples of fluidized-bed reactors usage in synthesis of basic organic chemicals could be mentioned, such as oxychlorination of ethylene and synthesis of acrylonitrile.

Serious disadvantages of the fluidized-bed reactors are attrition, loss of catalyst, and small variations in the residence times, which are typically low. Attrition of already small particles produces fines, which are expensive to separate from the product gas. It is, however, possible in principle to combine the catalyst with an attrition-resistant binder or coat the catalyst with some attrition-resistant material.

The operational velocities should be between the velocity at which fluidization starts U_b and the velocity at which carry over starts, $U_{c'}$. For small catalyst particles of ~60 µm, the ratio between U_c and U_b is ~50, while for some large particles (100 µm), it can be around 10. Because for real industrial catalysts there is a certain particle size distribution, the values of velocities at which the fluidized bed should operate are also confirmed experimentally.

Formation of large gas bubbles (laminar fluidized bed) is undesirable because it decreases the interphase mass transfer. In order to achieve a desired turbulent regime, the gas-flow rate should be sufficiently high and the catalyst particle must have sufficiently low density; otherwise, a transition from the bubble fluidized-bed regime to the riser transport regime (carry over) will occur with an increase of the gas velocity.

Fluidized-bed reactors in the turbulent fluidization regime have such drawbacks as a high degree of gas solids back-mixing with wide residence time distributions. Therefore, already many decades ago (from 1960), there were attempts to introduce multi-stage fluidized beds, which achieve advantages such as: (i) excellent gas solids interaction in each stage; (ii) inhibition of gas and solids back-mixing between each stage; (iii) large gradients of mass and heat along the whole reactor; and (iv) the opportunity to establish different reaction atmospheres within a single reactor. The industrial experience with multistage fluidized-bed reactors is limited. There are, however, some recent examples (Figure 3.73) reporting apparently successful industrial application of such reactors in China for nitrobenzene hydrogenation to aniline, as well as for 10 kt/year industrial demonstration of methanol transformations to aromatics.

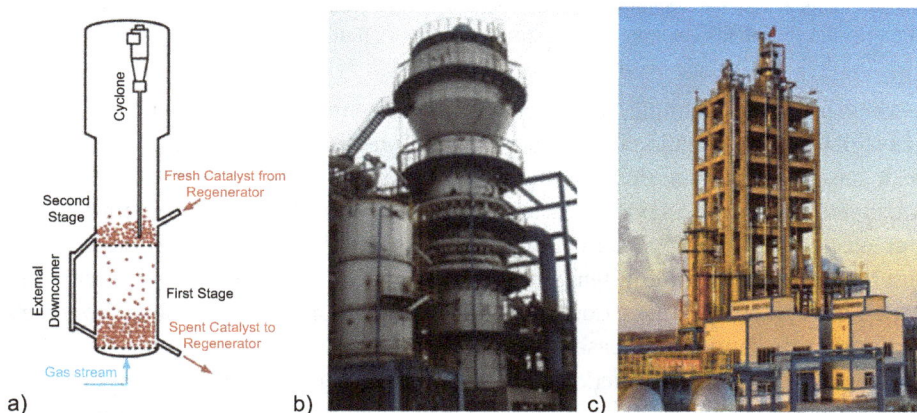

Figure 3.73: Multistage fluidized-bed reactors: (a) schematic (from C. Znag, W. Qian, Y. Wang, G. Luo, F. Wei, Heterogeneous catalysis in multi-stage fluidized bed reactors: From fundamental study to industrial application, *Canadian Journal of Chemical Engineering*, 2019, 97, 636–644, https://onlinelibrary.wiley.com/doi/full/10.1002/cjce.23425, Copyright © 2019, John Wiley and Sons, Reproduced with permission); (b) aniline production; and (c) methanol to aromatics demonstration plant.

3.4.3.2 Three-phase catalytic reactors

Catalytic three-phase (gas–liquid–solid) reactors are extensively used in industry in oil refining, synthesis of petrochemicals, and various basic, specialty, and fine chemicals; these processes require different types of reactors. In addition to reactors with fixed-catalyst beds using upflow and downflow of liquids and well as co- and counter-current flows of gases and liquid, slurry reactors (bubble columns and stirred tanks reactors, reactors with external loops) and fluidized-bed reactors are also applied (Figure 3.74).

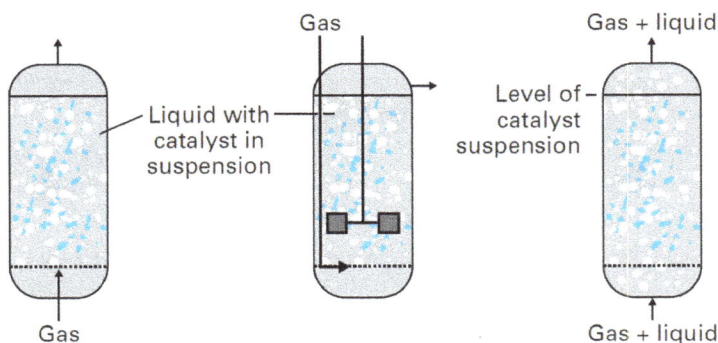

Figure 3.74: Typical three-phase reactors (adapted from Salmi, T.O., Mikkola, J.-P., Wärnå, J.P. (2019) Chemical Reaction Engineering and Reactor Technology. Second Edition, Boca Raton, FL: CRC Press). Left, bubble column; middle, stirred tank; right, fluidized bed.

In some cases (for example, hydrogenation of benzene or methanol synthesis), both gas–solid and gas–liquid–solid operations are possible. In general, in comparison with gas–solid reactors, three-phase reactors operate at low temperatures, which is translated into energy savings, also preventing potential damage to thermally unstable catalysts and support. In addition to better temperature control and more design options, low operation temperature usually also affords better selectivity and higher catalyst effectiveness. Conversely, the introduction of another phase boundary obviously results in an increase of mass transfer, while low temperature leads to a decrease in the chemical reaction rate. Moreover, batch slurry reactors are designed to operate in a limited pressure and (more importantly) a limited temperature range, which imposes some difficulties for catalyst reduction and *in situ* activation where they are needed. In addition, in industry, the power input for impellers is limited and the same holds for the minimum catalyst particle size, which is defined by available separation systems, and thus the mass transfer cannot be neglected. In general, with slurry bed reactors, the efficient separation of catalysts, which are either suspended in the slurry by stirring or bubbling gas up through the liquid, is difficult.

Packed-bed reactors can operate in different modes with the downflow and upflow of the liquid and gas, and the gas can flow either co- or counter-current. There are certain advantages for the liquid in operating either up- or downflow. For example, the downflow option in fixed-bed multiphase reactors results in lower energy input, lower pressure drop, and an absence of fluidization, while upflow operation leads to better liquid distribution, mixing, heat dissipation, surface wetting, easier heat exchange, higher gas/liquid mass transfer and effectiveness, as well as slow aging.

In the downflow operation mode, the liquid moves in a laminar flow downward, wetting the catalyst in a trickling flow, thus the name "trickle-bed reactors" (Figure 3.75). In these trickle-bed reactors, both the gas and liquid approach a plug flow condition.

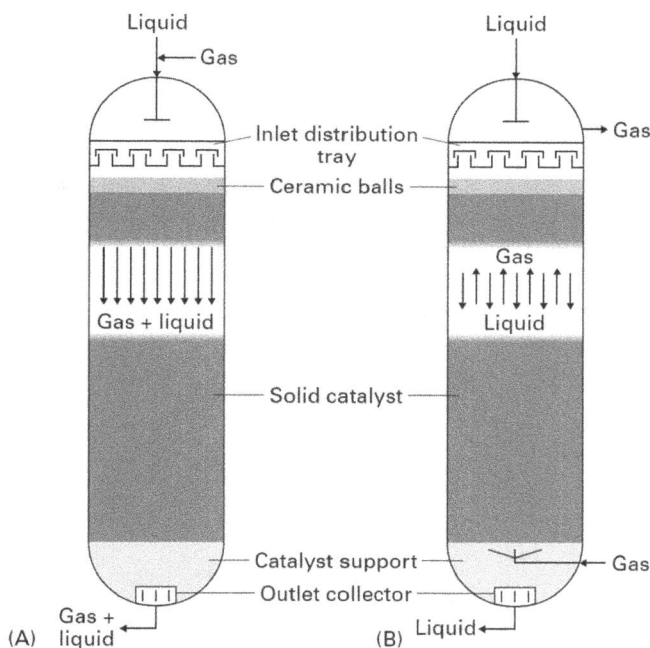

Figure 3.75: Trickle-bed reactors: (A) downward cocurrent, (B) counter-current gas flow. (From Mederos, F.S., Ancheyta, J. (2007) Mathematical modeling and simulation of hydrotreating reactors: Cocurrent versus countercurrent operations. Appl. Catal. A: Gen. 332: 8.Copyright Elsevier. Reproduced with permission).

Trickle-bed reactors usually operate in two regimes. At low gas and liquid flows, a trickle flow dominates, while at low gas and high liquid flows, the liquid phase is continuous and gas bubbles flow through in a pulsed flow regime.

In such a trickling flow, maldistribution and incomplete wetting and losses caused by attrition counterbalance the advantages, while upflow operation requires a larger energy input. In order to avoid stagnant regions and excessive catalyst load-ings, more complicated geometrical shapes are preferable for trickle-bed operations.

For upflow reactors, in addition to the bubble flow (low gas and high liquid flows) and spray flow (higher gas and low liquid flows), a slug flow with uneven dis-tribution of bubbles (high gas and low liquid flows) can also be developed, which gives in the latter case a plug flow for the gas phase, but partially back-mixed liquid.

In summary, trickle-bed reactors offer a simple structure, low investment costs, and closeness to plug flow. It should be mentioned that the ideal plug flow is the one that is usually desired.

In stirred tanks, cascades of stirred tanks and bubble columns, the back-mixing is significant, broadening the residence time distribution and making it more difficult to achieve high yields. There are, however, several advantages of the reactors with the moving catalyst beds, justifying their frequent application in industry.

Bubble columns (Figure 3.76) often operating in a semi-batch mode, i.e., continu-ously for gas and as batch for liquid, provide good mixing and heat transfer character-istics. Gas is distributed from the bottom of the reactor as bubbles, whose size depends depend on physical–chemical properties, sparger design, and superficial gas velocity. The distribution of gas bubbles in radial direction is nonuniform, assisting in internal circulation within the bubble column driven by buoyancy (an upward force exerted by a fluid). Overall, the back-mixing for the liquid and the solid catalyst particles is more intensive, thus the bubble columns operate as isothermal reactors.

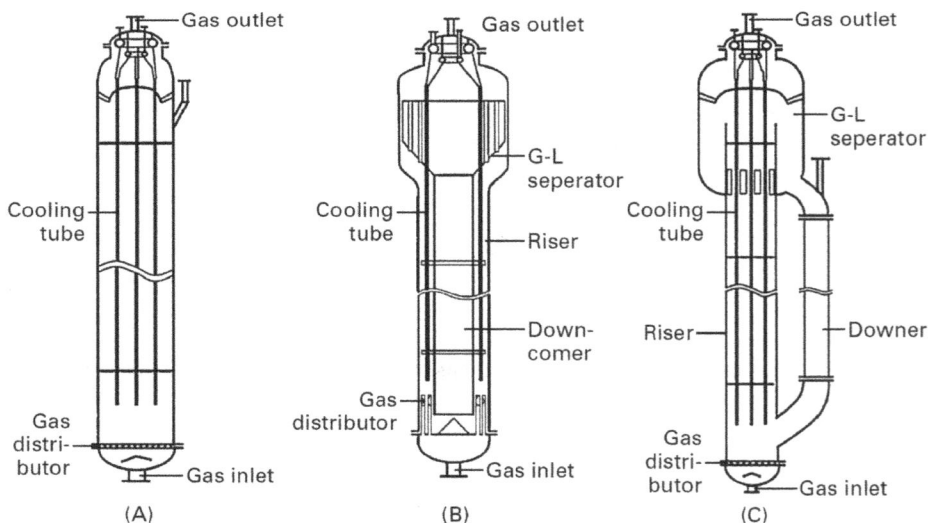

Figure 3.76: Moving-bed multiphase reactors: (A) bubble column, (B) internal loop airlift reactor, and (C) external loop airlift reactor (Reprinted, with permission, from Wang, T., Wang, J., Jin, Y. (2007) Slurry reactors for gas-to-liquid processes: a review. Ind. Eng. Chem. Res. 46:5824–5847).

Mixing can be enhanced by introducing gas (air) lift, ejectors, and circulation pumps. The airlift reactor consists of risers and downcomers that provide the flowing channel for global circulation of the liquid–solid slurry phase (Figure 3.76B, C).

Bubble columns exhibit very complex hydrodynamics due to spatiotemporal variations in interactions among gas, liquid, and solid phases.

Mechanically agitated stirred tanks (with complete back-mixing profiles) can be used in batch and semi-batch (continuous for the gas) modes, as well as CSTR with a possibility of catalyst recycling. In CSTR (Figure 3.77A), the mixing of the incoming reactant is instantaneous; thus, the concentration in the reactor is the same as the concentration at the outlet. Such a high degree of back-mixing is typically unfavorable for kinetics, resulting in a lower conversion of the reactants. Therefore, cascades of CSTR are used in industrial practice as discussed above for homogeneous reactors (Figure 3.77B).

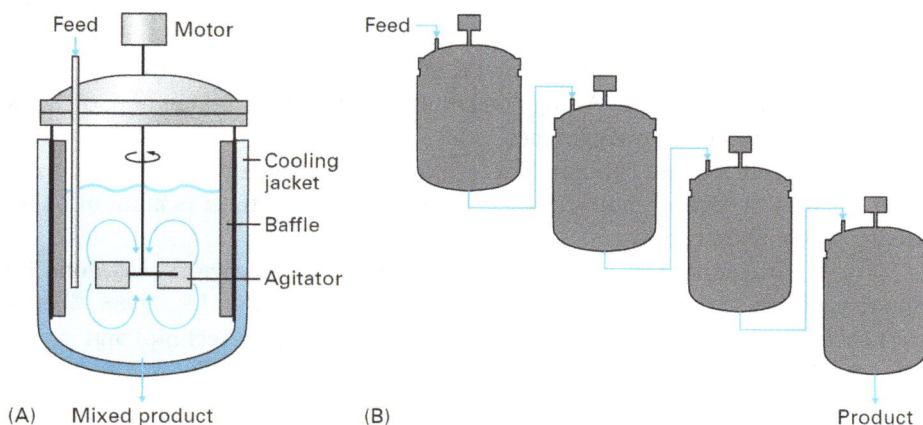

Figure 3.77: Stirred tank reactor. (A) CSTR and (B) cascade of stirred tanks.

The separation of fine catalyst particles can be challenging; therefore, somewhat larger particles are used in industrial practice with non-negligible diffusion resistance.

The fluid flow in stirred tanks is highly complex and requires computational fluid dynamics to get an exact solution of the partial differential equations associated with the Navier–Stokes momentum, energy, and continuity equations. Tangential, axial, and radial flows exist in stirred tanks depending on the impeller used.

Application of baffles (three or four flat plates attached to the walls), occupying up to 10% of the tank diameter, can stop the tangential flow. Such baffles cannot be recommended when the flow is laminar. Axial and radial flow impellers could be used to generate flow parallel or perpendicular to the axis (shaft), respectively. Pitched bladed turbines create a mixed flow. Axial flow impellers are recommended for turbulent mixing. Although variations in the size of catalyst particles mean that it is not possible to get suspensions that are completely uniform, the impellers are usually placed at one-quarter of the liquid level to achieve acceptable mixing. In order to increase the gas–liquid mass transfer, the gas bubbles should be dispersed by the impeller. It should also be remembered that because of the gas dispersed in the liquid, the volume of the liquid is increased (gas holdup). Heat transfer is typically unaffected by mixing and could be improved by increasing heat transfer area (for example, installing coils in addition to a heat transfer jacket).

Industrial reactors cannot operate at the same level of mixing efficiency as the small-scale reactors; thus, simple geometrical similarities are insufficient in scaling up, and location of the feed points should be carefully considered. Some engineering consultants even advise for process development to rather scale down and use laboratory reactors based on available large-scale reactors to mimic the industrial conditions.

Finally, fluidized-bed reactors should be highlighted as an option for conducting three-phase reactions. Such reactors typically operate in a concurrent mode with upflow of both gas and liquid. Conditions of high gas and low liquid flow (aggregative fluidization) resemble two-phase (gas–solid) fluidized beds, while at low gas flows, the flow pattern (bubble flow) is obviously similar to fluidized beds with only liquid and solid phases. A slug domain with uneven distribution of particles is formed in between these two extremes.

Compared to bubble columns, larger particles are usually used in fluidized beds because of higher liquid flow velocities. The flow pattern is more of back-mixing type than plug flow.

A general comparison between catalytic three-phase reactors was done by Trambouze et al. (Trambouze, P., van Landeghem, H., Wauquier, J.P., (1988) Chemical Reactors – Design/ Engineering/Operation. Paris: Editions Technip) and is presented in Table 3.4.

Table 3.4: Comparison between three-phase reactors.

Application criteria	Suspension catalysts	Three-phase fluidized beds	Fixed bed
Catalysts related			
Activity	Highly variable, but possible in many cases to avoid the diffusion limitations found in a fixed bed	Highly variable: intra- and extra-granular mass transfers may significantly reduce the activity, especially in a fixed bed. Back-mixing unfavorable	Plug flow favorable
Selectivity	Selectivity generally unaffected by transfers	As for activity, transfers may decrease selectivity. Back-mixing often unfavorable	Plug flow often favorable

Table 3.4 (continued)

Application criteria	Suspension catalysts	Three-phase fluidized beds	Fixed bed
Stability	Catalyst replacement between each batch operation helps to overcome problems of rapid poisoning in certain cases	Possibility of continuous catalyst renewal: the catalyst must nevertheless have good attrition resistance	This feature is essential for fixed-bed operation: a plug flow may sometimes be favorable because of the establishment of a poison adsorption front
Cost	Consumption usually depends on the impurities contained in the feed and acting as poisons		Necessarily low catalyst consumption
Technologies characteristics			
Heat exchange	Fairly easy to achieve heat exchange	Possibility of heat exchange in the reactor itself	Generally adiabatic operation
Design difficulties	Catalyst separation sometimes difficult: possible problems in pumps and exchangers because of the risks of deposit or erosion		Very simple technology for a downward concurrent adiabatic bed
Scaling up	No difficulty: generally limited to batch systems and relatively small sizes	System still poorly known, should be scaled up in steps	Large reactors can be built if liquid distribution is carefully arranged

Chapter 4
Chemical process industry

4.1 General overview

The chemical process industry converts raw materials (oil, natural gas, air, water, metals, and minerals) into more than 70,000 different products, such as fuels, various polymers and plastics, basic chemicals, consumer goods, and chemicals and products for agriculture, manufacturing, construction, pulp and paper production, life sciences, textile, and other industries. The overall turnover is close to $3 trillion.

The chemical reaction technology described in this textbook mainly covers oil refining, production of polymers, and manufacturing of some basic and specialty inorganic and organic chemicals.

The largest-volume polymer product, polyethylene (PE), is used mainly in packaging films. PE is mainly synthesized using catalysts, the most common being Ziegler-Natta catalysts or titanium (III) chloride. An alternative is the Phillips catalyst, prepared by depositing chromium oxide (VI) on silica. A more rigid product is produced with a higher melting (high-density PE) compared to low-density PE, which is made by a free radical process at 200–300 °C under high pressures up to 3,000 atm.

The second largest polymer products in volume are polypropylene and polyvinylchloride (PVC) mainly applied in construction markets. Commercially available are isotactic and syndiotactic polypropylene (Figure 4.1) made with Ziegler-Natta and metallocene catalysts.

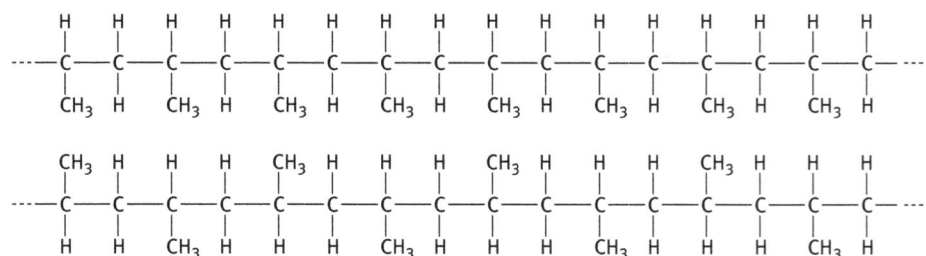

Figure 4.1: Segments of isotactic (top) and syndiotactic (bottom) polypropylene.

Commercial synthesis of syndiotactic polypropylene is carried out with the use of a special class of metallocene catalysts described in Chapter 15. They employ bridged bis-metallocene complexes of the type bridge-$(Cp_1)(Cp_2)ZrCl_2$, where the first Cp ligand is the cyclopentadienyl group, the second Cp ligand is the fluorenyl group, and the bridge between the two Cp ligands is $-CH_2-CH_2-$, $> SiMe_2$, or $> SiPh_2$. These

https://doi.org/10.1515/9783110712551-004

complexes are converted to polymerization catalysts by activating them with a special organoaluminum co-catalyst, methylaluminoxane.

About 80% of PVC production involves suspension polymerization, which will be discussed in Chapter 15. Suspension polymerizations affords particles with average diameters of 100–180 μm, whereas an alternative method of emulsion polymerization results in much smaller particles of ca. 0.2 μm.

Among other plastics, polystyrene and polyamide can be mentioned. The polymerization processes will be discussed in Chapter 15.

Basic inorganic chemicals are subdivided in metals, inorganic sulphur (sulphuric acid), nitrogen (nitric acid, ammonia, urea), and phosphorus (phosphoric acid) compounds as well as products of air separation (nitrogen, oxygen, noble gases) and some other important gases (hydrogen, CO). A mixture of the latter two, the so-called synthesis gas, is used in the production of various chemicals, as explained in Chapter 13, where production of urea is also discussed. Manufacturing of sulphuric and nitric acids is described in Chapter 9, which deals with oxidation reactions. Phosphoric acid is made in a wet process by adding sulphuric acid to apatite (tricalcium phosphate):

$$Ca_5(PO_4)_3X + 5H_2SO_4 + 10H_2O \rightarrow 3H_3PO_4 + 5CaSO_4 \cdot 2H_2O + HX, \tag{4.1}$$

where X may include OH, F, Cl, and Br. Commercial-grade phosphoric acid, which contains about 54% P_2O_5, is made by evaporating water from the initial solution.

Chemicals in the bulk petrochemicals and intermediates are primarily made from liquefied petroleum gas (LPG), natural gas, and crude oil. Typical large-volume organic products are ethylene, propylene, benzene, toluene, xylenes, methanol, vinyl chloride, styrene, butadiene, and ethylene oxide, and a broad range of their derivatives.

In large-scale production of those products, reactions such as various hydrogenations, dehydrogenations, oxidations, and acid-catalyzed reactions, such as alkylation, hydration, dehydration, condensation, should be mentioned. They will be addressed in relevant chapters of this textbook.

Typically, alkenes (ethylene, propylene, butane, etc.) are produced by steam cracking of naphtha (carbon number ranges from 4 to 12), as described in Chapter 6.

Further transformation of ethylene is given in Figure 4.2. Propene and benzene are converted to acetone and phenol *via* the so-called cumene process (Chapter 12). Propene is also used to produce, for example, isopropanol (propan-2-ol) by hydration or acrylonitrile by catalytic ammoxidation (Chapter 9). Other routes are given in Figure 4.3.

Butadiene and butane are mainly used as monomers or co-monomers in the formation of polymers, for example, synthetic rubber.

Many important chemical compounds such as phenol, nitrobenzene, chlorobenzene, and aniline are derived from benzene (Figure 4.4) by introducing a functional group instead of hydrogen. Chlorobenzene is synthesized by chlorination

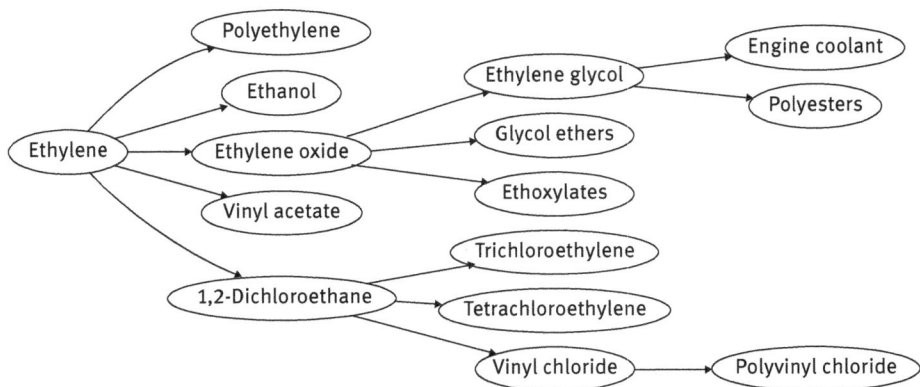

Figure 4.2: Ethylene as a platform chemical. http://commons.wikimedia.org/wiki/File:Ethylene_chemical_production_network.svg.

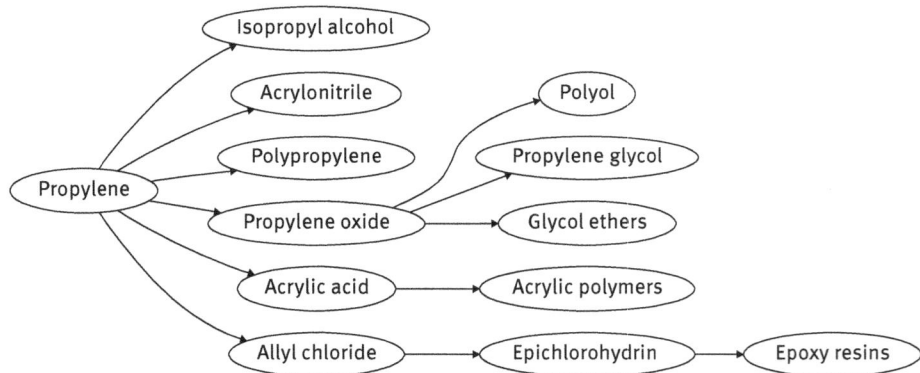

Figure 4.3: Propylene as a platform chemical. http://commons.wikimedia.org/wiki/File:Propylene_chemical_production_network.svg.

(Chapter 8). Phenol is produced first by alkylation of benzene with propylene (Chapter 12), leading to cumene (isopropylbenzene). Subsequent oxidation leads to hydroperoxide, which is further treated with sulphuric acid, resulting in an equimolar mixture of phenol and acetone. Ethylbenzene and some other alkylbenzenes could be synthesized through alkylation reactions (Chapter 12). Dehydrogenation of ethylbenzene leads to styrene, a monomer for polystyrene (Chapter 10).

Many higher alcohols are produced by hydroformylation of alkenes (Chapter 13) followed by hydrogenation. For terminal alkenes, linear alcohols are mainly formed:

$$RCH = CH_2 + H_2 + CO \rightarrow RCH_2CH_2CHO \tag{4.2}$$

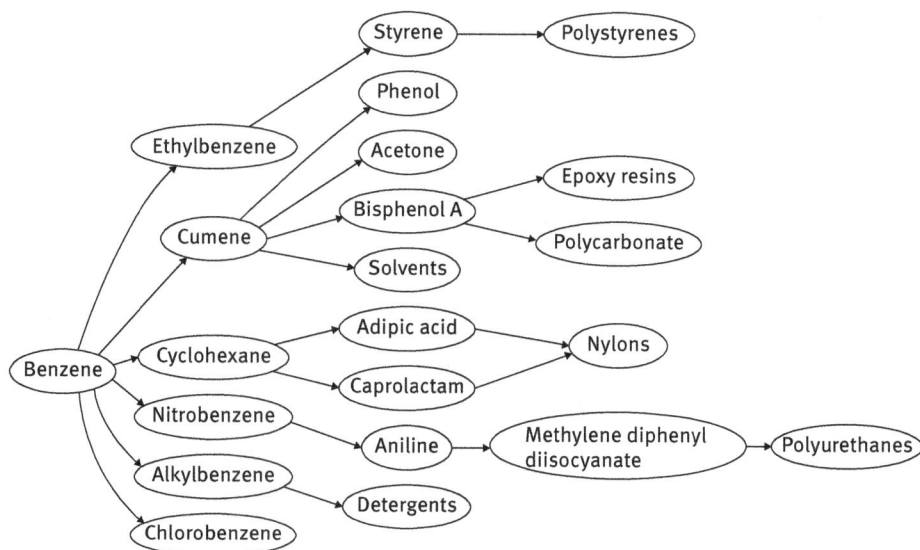

Figure 4.4: Benzene as a platform chemical. http://commons.wikimedia.org/wiki/File:Benzene_chemical_production_network.svg.

Low-molecular-weight alcohols of industrial importance (ethanol, isopropanol, 2-butanol, and *tert*-butanol) are produced by the addition of water to alkenes (Chapter 11). Hydration is also used industrially for the synthesis of ethylene glycol from ethylene oxide. Intermolecular dehydration of alcohols gives ethers (Chapter 11).

Aldehydes are synthesized by hydroformylation (butyraldehyde from propene), oxidation or oxidative dehydrogenation (formaldehyde from methanol), and oxidation of alkenes (ethylene to acetaldehyde, and propylene to acrolein). Oxidation processes are described in Chapter 9, which also covers the production of carboxylic acids, which can be done by, for example, oxidation of aldehydes with air using cobalt and manganese catalysts or oxidation of hydrocarbons (benzoic and terephthalic acids from toluene and *para*-xylene, acrylic acid from propene).

The most industrially significant amines (methylamine, dimethylamine, and trimethylamine) are made from ammonia by alkylation of alcohols (Chapter 12),

$$ROH + NH_3 \rightarrow RNH_2 + H_2O, \tag{4.3}$$

while aniline is manufactured by reduction of nitrobenzene.

Synthesis of halogen-containing compounds by various halogenations is described in Chapter 8. A special type of chlorination also addressed in Chapter 8 is oxychlorination of, e.g., ethylene using HCl in the presence of oxygen, giving dichloroethylene, which is subsequently converted to vinyl chloride by thermal elimination of hydrogen chloride.

Chemicals mentioned above correspond to the so-called commodities (Table 4.1). Besides commodities, specialty chemicals as well as fine chemicals are also produced in chemical process industries.

Table 4.1: Different types of chemicals manufactured in chemical process industries.

Commodities	Fine chemicals	Specialties
Single pure chemical substances	Single pure chemical substances	Mixtures
Produced in dedicated plants	Produced in multipurpose plants	Formulated
High volume/low price	Low volume (<1,000 mtpa)/high price (>$10/kg)	Undifferentiated
Many applications	Few applications	Undifferentiated
Sold on specifications	Sold on specifications, i.e., "what they are"	Sold on performance, i.e., "what they can do"

Specialty chemicals include a number of organic chemicals, such as adhesives, agrichemicals, cosmetic additives, construction materials, fragrances, lubricants, dyes and pigments, surfactants, and textile auxiliaries to name a few. Specialty chemicals are usually manufactured in batches using batch processing techniques. While commodity chemicals are mainly made on large-scale, single-product manufacturing units to ensure the economy of scale, specialty manufacturing units should be more flexible in responding to the customers' need.

Unit operations for specialty chemicals are also different from manufacturing commodity chemicals. The creation and the control of the particle size distribution requires crystallization, precipitation, prilling, agglomeration, calcination, compaction, and encapsulation. Therefore, typical operations are granulation, extrusion, compression, spray drying, spray chilling, coating, emulsification, and gelation, rather than classical unit operations (e.g., distillation, extraction, absorption). Product design is complicated by handling of solids, pastes, etc., which are difficult to calculate and whose physical properties are often not known. Utilization of the same equipment for different purposes leads to a situation that existing equipment might not be the optimal one for a particular product.

Fine chemicals are typically complex, single, or pure chemical substances produced in limited quantities in multipurpose plants by multistep batch chemical or biotechnological processes according to exacting specifications. Production of fine chemicals lacks the economy of scale; thus, manufacturing is far from being optimized since, due to shorter life cycles, there is an ongoing need for substitution of products. The focus in R&D is on product improvement rather than process improvement, contrary to production of petrochemicals and bulk chemicals.

Besides organic chemicals mentioned above, chemical process industries are involved in manufacturing of inorganic chemicals. The diversity of sources for inorganic chemicals reflects the fact that this group combines all the elements besides carbon. The important sources for inorganic chemicals are metallic ores and salts, while elements such as sulphur occur in an elemental form. Air, being a mixture of several gases, could be separated by liquefaction and subsequent distillation.

Main inorganic chemicals are salt, chlorine, caustic soda, soda ash, acids (nitric, phosphoric, and sulphuric acid), titanium dioxide, and hydrogen peroxide. Of particular interest are fertilizers, which include phosphates, ammonia, and potash chemicals.

The overall structure of chemical industry is presented in Figure 4.5.

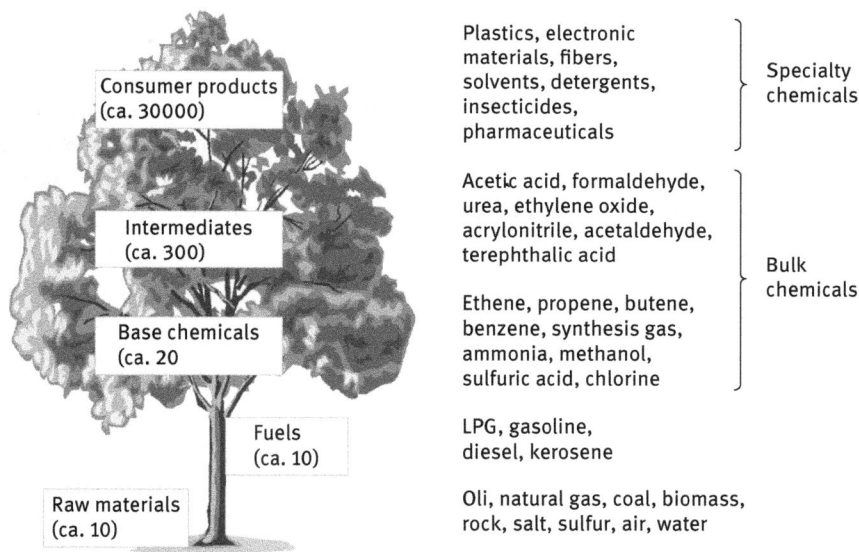

Consumer products (ca. 30000)

Plastics, electronic materials, fibers, solvents, detergents, insecticides, pharmaceuticals — Specialty chemicals

Intermediates (ca. 300)

Acetic acid, formaldehyde, urea, ethylene oxide, acrylonitrile, acetaldehyde, terephthalic acid

Base chemicals (ca. 20

Ethene, propene, butene, benzene, synthesis gas, ammonia, methanol, sulfuric acid, chlorine — Bulk chemicals

Fuels (ca. 10)

LPG, gasoline, diesel, kerosene

Raw materials (ca. 10)

Oli, natural gas, coal, biomass, rock, salt, sulfur, air, water

Figure 4.5: Chemical and petrochemical industry.

4.2 Feedstock for chemical process industries

Most organic chemicals are produced from crude oil and natural gas. In the past, the major source was coal, which is gaining more attention due to various predictions about the static range of various resources. Thus, the static rage of crude oil could be 40–50 years, for natural gas, 60–70 years, for coal, above 160 years. In the future, coal and biomass, which were along the years the origin of just 10% of chemicals might be much more important.

Biomass is generally considered as a renewable resource, and thus, a lot of attention is currently devoted to the so-called biorefinery concept, which would allow production of fuels and chemicals in the same way as it is done nowadays in classical refineries and large (petro)chemical complexes. At the same time, the amount of biomass available currently is limited to address all the demands with respect to fuels and chemicals. Thus, ca. 30% of the global arable land is needed to cover only 10% of the global fuel demand by 2030. It should be noted that, currently, the majority of oil is used for production of fuels, while only 5–8% of a crude oil barrel is used in manufacturing chemicals, while the turnover in monetary value is almost the same for fuels and chemicals. Since the price of fuels is much lower than for chemicals, in the future, such limited resource should be used mainly for chemicals, while the growing energy demand must be compensated by alternative energy sources (solar, hydropower, nuclear, etc.). Moreover, fossil fuels are not carbon dioxide-neutral, generating extensive emissions of carbon dioxide to the atmosphere and leading subsequently to the so-called greenhouse effect.

The composition of different feedstock in terms of hydrogen/carbon and oxygen/carbon ratios is given in Figure 4.6. Feedstock varies in terms of composition, for example, solid fuels such as coal, wood, or peat contain more than 50% carbon. Oxygen content in biomass as shown in Figure 4.6 is much higher than in coal or crude oil.

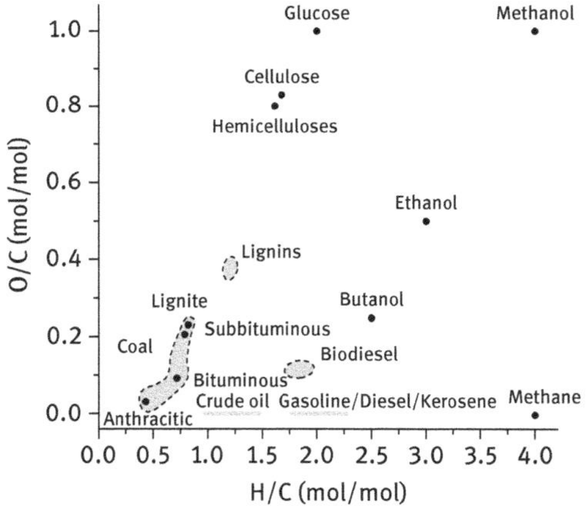

Figure 4.6: Hydrogen/carbon and oxygen/carbon ratio in various feedstock. From R. Rinaldi, F. Schueth, Energy and Environmental Science, 2009, 2, 610–626. Copyright RSC. Reproduced with permission.

The top three oil-producing countries are the USA, Russia, and Saudi Arabia. About 50% of the world's readily accessible reserves are located in the Middle East, according to the BP Statistical Review of World Energy 2019.

Petroleum includes not only crude oil but also lighter hydrocarbons. Because the pressure is lower at the surface than underground, these compounds come out from the crude oil as associated gas. This gas may contain heavier hydrocarbons such as pentane, hexane, and heptane, which condense at surface conditions, forming natural gas condensate, similar in composition to some volatile light crude oils. Light hydrocarbons in the petroleum mixture range from 50% in heavier oils to 97% by weight in lighter oils. The hydrocarbons in crude oil are mostly alkanes, cycloalkanes, and various aromatic hydrocarbons. There are compounds containing nitrogen, oxygen, and sulphur. Moreover, trace amounts of metals such as iron, nickel, copper, and vanadium are also present. Currently, only 30% of world oil reserves are conventional oils, while heavy and extra heavy oils constitute 15% and 25%, respectively, with the rest being oil sands and bitumen. The average composition of oil is 83–85% carbon, 10–14% hydrogen, 0.05–1.5% oxygen, 0.1–2% nitrogen, 0.05–6% sulphur, and < 0.1% of metals.

Four different types of hydrocarbon molecules appear in crude oil: alkanes, ca. 30% (range 15–60%); naphthenes, ca. 49% (range 30–60%); aromatics, 6% (range 3–30%); remainder, asphaltics. The alkanes from C5 to C8 are the basis of gasoline (petrol), C9 to C16 alkanes are refined to diesel, kerosene, and jet fuel, whereas heavier hydrocarbons serve as a feedstock for fuel and lube (lubricating oils). Paraffin wax has approximately 25 carbon atoms, while asphalt has a carbon number > 35. Due to limited application areas of such hydrocarbons, they are processed in modern oil refineries to more valuable products by cracking, as will be discussed in Chapter 6.

Petroleum gases with carbon number < 4 are flared off, liquefied to form LPG, or used inside refineries as a fuel. Flaring of the associated gas in many countries is prohibited, as it is negatively influencing the greenhouse effect and is a waste of energy.

Some of the molecules constituting oils are presented in Figure 4.7. Besides crude oil an interesting feedstock for generation of fuels is shale oil. It can be recovered from oil shale (organic-rich fine-grained sedimentary rock) containing 25–30% kerogen (a solid mixture of organic chemical compounds). The mineral content in oil shale is ca. 42–50%, the rest is moisture.

Natural gas is another important fossil fuel, which is a hydrocarbon gas mixture of mainly methane (70–90%), with various amounts of higher alkanes (0–20%) and also CO_2 (0–8%), oxygen (0–0.2%), nitrogen (0–5%), and hydrogen sulphide or mercaptanes (0–5%) and traces of rare gases.

Natural gas is used as a fuel for vehicles and as a feedstock for various organic chemicals. The key process of steam reforming of natural gas will be discussed in Chapter 5. Gas-to-liquid (GTL) technology for converting stranded natural gas into synthetic gasoline, diesel, or jet fuel through the Fischer-Tropsch process is described in Chapter 13.

Alkanes
  Normal CH_3-CH_2-R

  Branched CH_3-CH_2-CH-R
                             |
                            CH_3

Cycloalkanes (Naphthenes)

  Alkylcyclopentanes

  Alkylcyclohexanes

  Bicycloalkanes

Phenanthrene

Pyrene

Chrysene

Aromatics

  Alkylbenzenes

  Aromatic-cycloalkanes

  Fluorenes

  Binuclear aromatics

1,2-Benzanthracene

3,4-Benzopyrene

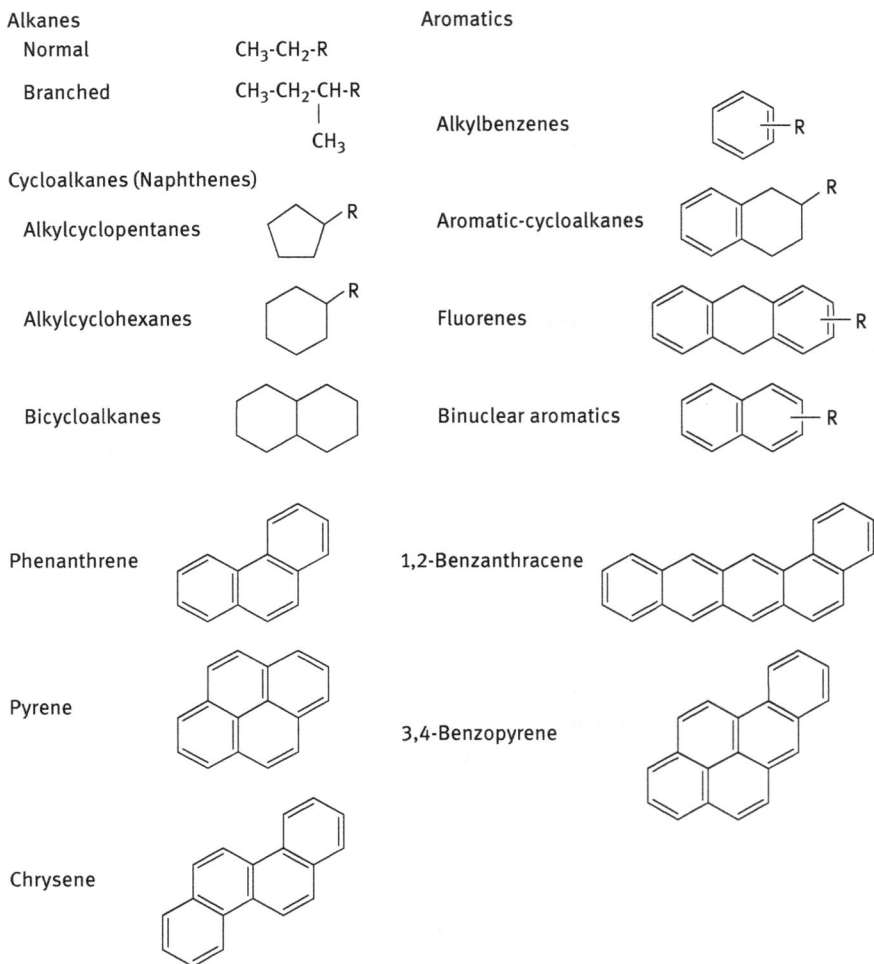

Figure 4.7: Representative molecules in crude oil.

The world's largest reserves of natural gas are in Russia, Iran, Qatar, USA, Saudi Arabia, and Turkmenistan. There are "unconventional" gas resources available such as shale gas (Figure 4.8), which is a natural gas found trapped within shale formations. The role of shale gas is becoming increasingly important, greatly expanding energy supply and reshaping manufacturing landscape in a number of countries.

One of the concerns about shale gas is the potential risk associated with fracking, namely groundwater contamination, since for the gas to be released, shale gas wells

Hydrogen sulfide	H_2S	Thiophenes

Mercaptans		Thiophene
Aliphatic	R-SH	
Aromatics	SH	Benzothiophene

Sulfides		
Aliphatic	R-S-R	Dibenzothiophene
Cyclic	S / CH$_2$-CH$_2$-R	Substituted Dibenzothiophenes

Disulfides		
Aliphatic	R-S-S-R	
Aromatic	S-S-R	

Basic nitrogen compounds   Non-basic nitrogen compounds

Pyridines		Pyrroles	
Quinolines		Indoles	
Acridines		Carbazoles	

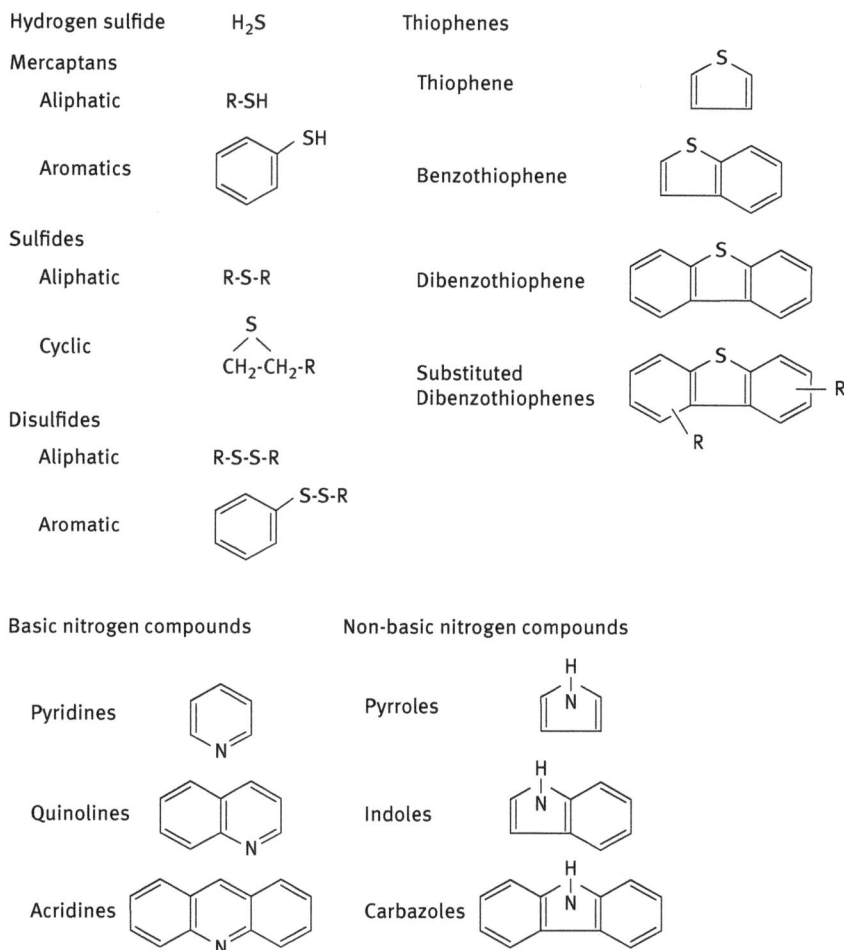

Figure 4.7 (continued)

need fractures. They are artificially created by hydraulic fracturing, which requires that a fracturing fluid containing sand, water, and chemicals be pumped into a well.

Coal, which is a combustible black or brownish-black sedimentary rock composed primarily of carbon with variable quantities of other elements, such as hydrogen, sulphur, oxygen, and nitrogen, along with oil and natural gas, is not considered as a renewable feedstock. The H/C molar ratio in coal 0.85:1 is much lower than in oil. Contrary to oil, heteroatoms are present in substantial amounts and macromolecules in coal can have molecular weight of up to 1,000 (Figure 4.9).

Acidic

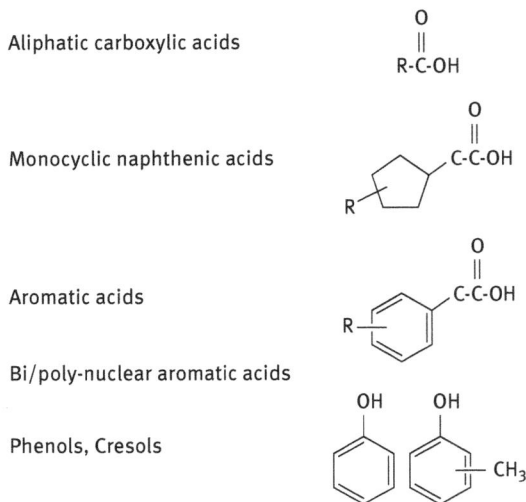

Aliphatic carboxylic acids

Monocyclic naphthenic acids

Aromatic acids

Bi/poly-nuclear aromatic acids

Phenols, Cresols

Non-acidic

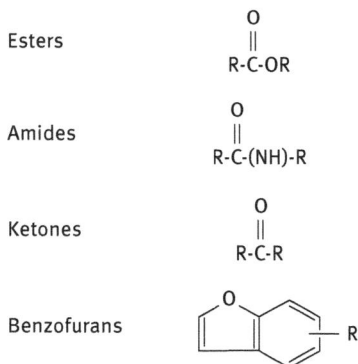

Esters

Amides

Ketones

Benzofurans

Figure 4.7 (continued)

Formation of coal is related first to the conversion of dead plant matter peat, followed by further transformations into lignite, then sub-bituminous coal, bituminous coal, and finally, anthracite. Coal is the largest source of energy for electricity generation and one of the largest anthropogenic sources of CO_2 release. Coal is extracted from the ground by coal mining. The world's top coal producer is China, with ca. 3,700 million tons of coal from ca. 8,000 million tons produced worldwide in 2019. The elemental composition of coal is given in Table 4.2.

Schematic geology of natural gas resources

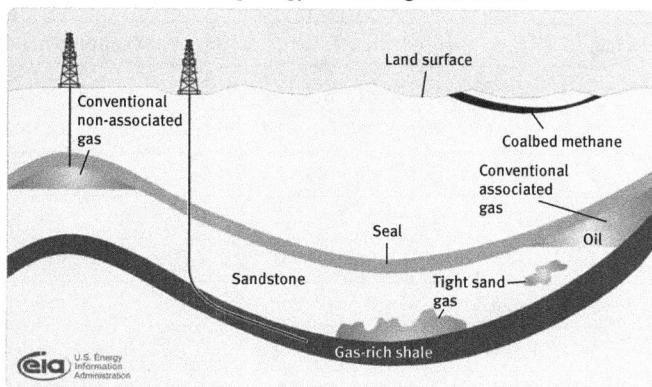

Figure 4.8: Schematic geology of natural gas. http://en.wikipedia.org/wiki/File:GasDepositDia gram.jpg.

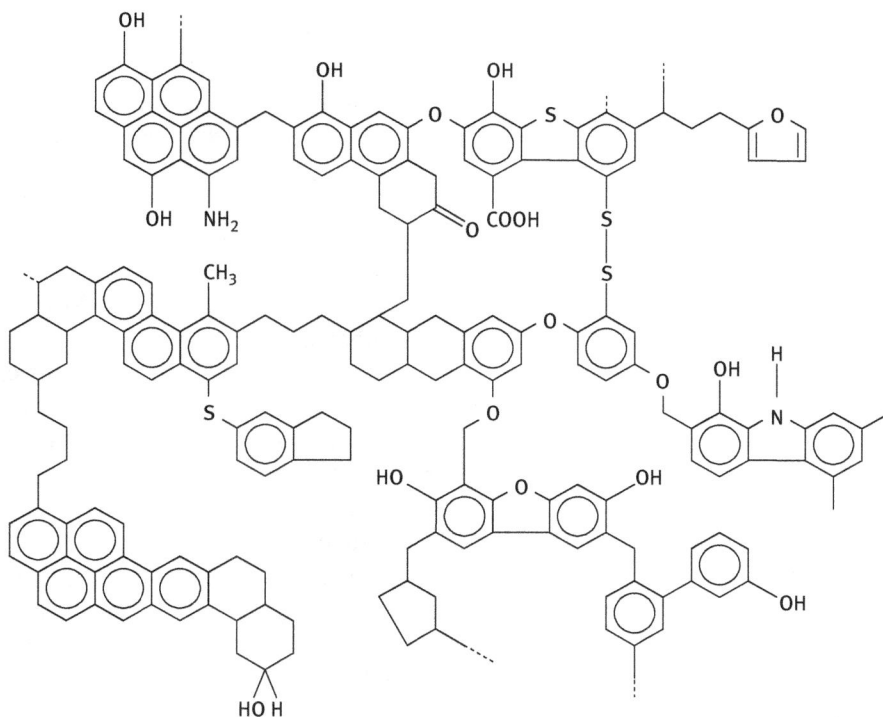

Figure 4.9: Macromolecules in coal. http://en.wikipedia.org/wiki/File:Struktura_chemiczna_w%C4 %99gla_kamiennego.svg.

Table 4.2: Elemental composition of different types of coal.

Type	Volatiles (%)	C (%)	H (%)	O (%)	S (%)	Heat content (kJ/kg)
Lignite (brown coal)	45–65	60–75	6.0–5.8	34–17	0.5–3	<28,470
Flame coal	40–45	75–82	6.0–5.8	>9.8	~1	<32,870
Gas flame coal	35–40	82–85	5.8–5.6	9.8–7.3	~1	<33,910
Gas coal	28–35	85–87.5	5.6–5.0	7.3–4.5	~1	<34,960
Fat coal	19–28	87.5–89.5	5.0–4.5	4.5–3.2	~1	<35,380
Forge coal	14–19	89.5–90.5	4.5–4.0	3.2–2.8	~1	<35,380
Non-baking coal	10–14	90.5–91.5	4.0–3.75	2.8–3.5	~1	35, 380
Anthracite	7–12	>91.5	<3.75	<2.5	~1	<35,300

Peat has industrial importance as a fuel in some regions, for example, Finland (where 90% of extracted peat is used as fuel), or Ireland. Brown coal or lignite is the lowest rank of coal and used almost exclusively as fuel for electric power generation. Sub-bituminous coal is used primarily as fuel for steam-electric power generation and as a source of light aromatic hydrocarbons. Bituminous coal is applied for steam-electric power generation as a fuel, for heat and power applications in manufacturing, and in making the so-called coke by temperature treatment in the absence of oxygen at 1,000 °C. Metallurgical coke is applied, for example, as a reducing agent in smelting iron ore in blast furnaces in substantial amounts (0.5 t per tonne of pig iron); thus, ca. 20% of the global coal production is used in blast furnaces.

Tightening regulations for mobile emissions and for fuel quality and other trends regarding the static range of feedstock substantially expand the interest toward biomass as a source of chemicals and fuels. In particular, lignocellulosic biomass, not competing with the food chain, has attracted a lot of attention. Lignocellulose (Figure 4.10) is the fibrous material that makes up the cell walls of plants, containing cellulose ($\approx$40 wt%), hemicellulose ($\approx$25–30 wt%), lignin (ca. 20%), extractives (fatty acids, terpenes, stilbenes, and lignans), and inorganic components, including metals. With 700 billion metric tons available on Earth, cellulose is the most abundant organic material. Cellulose is a linear polysaccharide built from glycoside units. The number of glucose monomers can be as high as 15, 000; the average molecular mass is $(3–5) \times 10^5$ g/mol. The units are linked by β-1, 4-glycoside bonds, and this linkage results in the formation of a crystal structure of cellulose with intramolecular and intermolecular hydrogen bonds.

Due to the limited space around the glycosidic bridging oxygen in cellulose, it is apparent that the access of homogeneous or heterogeneous catalysts to this bond is limited. One of the challenges in valorization of cellulose is thus related to a possibility of performing reactions in suitable solvents. However, cellulose is soluble in

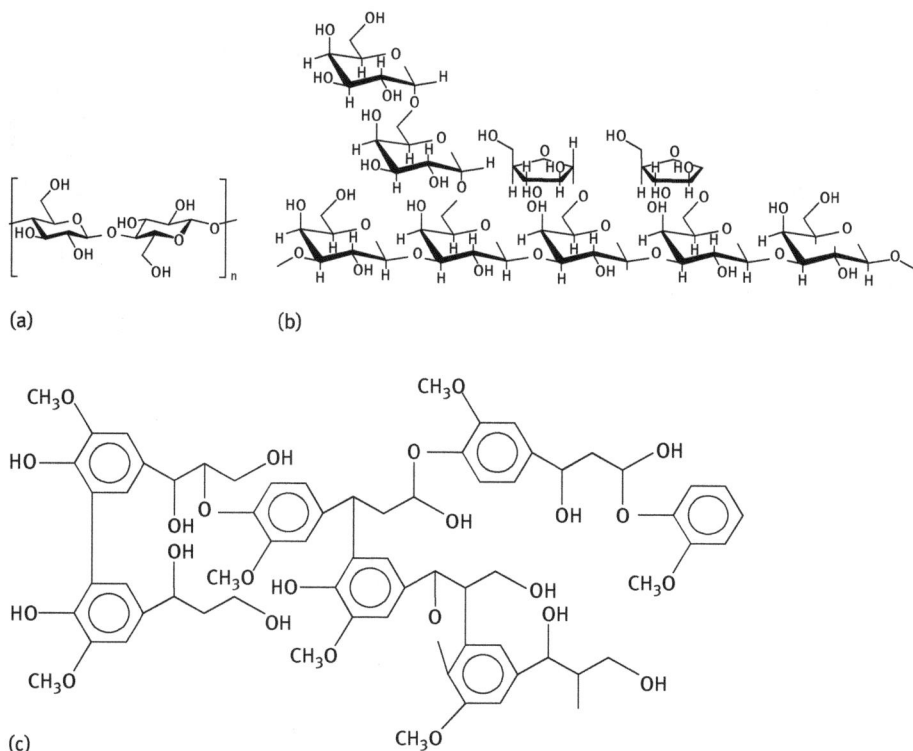

Figure 4.10: Structure of the main constituents of lignocellulosic biomass: cellulose (ca. 40%), hemicellulose (arabinogalactan as an example, ca. 25–30%), and lignin (ca. 20%) (http://www.inte chopen.com/source/html/43233/media/image12_w.jpg).

some rather unusual solvents, e.g., concentrated $ZnCl_2$ solutions and ionic liquids, and therefore, development of economically competitive technologies would allow more widespread use of cellulose.

Hemicelluloses are composed of several monomers, for example, arabinogalactan of larch (Figure 4.10) has bonded (1, 3) β-galactopyranose as its backbone and d-galactopyranose, L-arabinofuranose, and D-glucuronic acid units in the side chain. The average ratio of galactose, arabinose, and glucuronic acid is close to 5:1:0.08; the molecular mass is $(2-10) \times 10^4$ g/mol. The C6 carbohydrates in hemicellulose are composed of mannose and galactose in addition to glucose, and C5 carbohydrates contain xylose in addition to arabinose. Steric hindrance, due to the presence of side chains and axial hydroxyl groups in different sugars of hemicelluloses, prevents formation of a crystal structure; thus, hemicelluloses are amorphous, have much better solubility, and are more reactive than cellulose.

Lignocellulose contains not only polysaccharides, but also lignin (see Figure 4.10), which is a three-dimensional polymer composed of propylphenol units. The latter polymer acts not only as a binder but protects lignocellulose from various microorganisms.

Like cellulose, another polysaccharide-starch is a biopolymer, which consists of glucose units connected with glucosidic bonds. Typically, the composition of starch is 20–25% amylose and 75–80% amylopectin (Figure 4.11). Because of its tightly packed structure, amylose is more resistant to digestion than other starch molecules. The number of repeated glucose subunits is usually in the range of 300 to 3,000. Amylose contains very few $\alpha(1{\to}6)$ bonds.

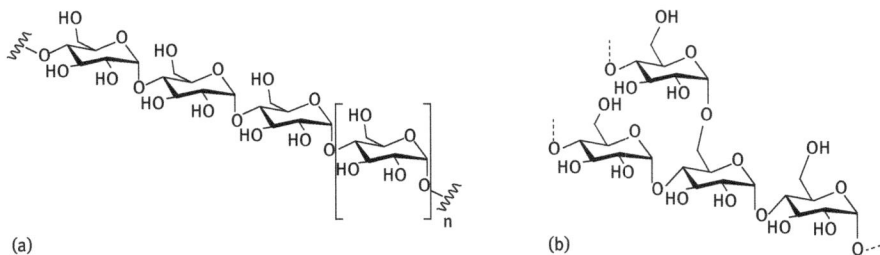

(a)

(b)

Figure 4.11: (a) Amylose (http://commons.wikimedia.org/wiki/File:Amylose_3Dprojection.cor rected.png) and (b) amylopectin (http://en.wikipedia.org/wiki/File:Amylopektin_Sessel.svg).

Amylopectin is a polymer of glucose that is soluble due to its branched nature. Glucose units are linked in a linear way with $\alpha(1{\to}4)$ glycosidic bonds. Branching takes place with $\alpha(1{\to}6)$ bonds occurring every 24 to 30 glucose units.

4.3 Oil refining

Processing schemes of refineries depend on the market requirements. First, the crude oil is distilled in oil refineries, giving different fractions. The separation of chemically similar components by distillation is based on the differences in boiling points, which is mainly related to the number of carbon atoms in the molecule.

The products of refining can be grouped into a number of main classes (Figure 4.12): (a) industrial and domestic fuels (gas oils, fuel oils, or LPG), (B) motor fuels (gasoline, diesel, kerosene, LPG), (C) feedstock for the chemical and petrochemical industry (virgin naphtha, LPG, olefins), and (D) other products (lubricating oils, bitumens, paraffins, solvents, sulphur).

A refinery is a complex plant consisting of various components. The most important and characteristic part consists of the plants, or process units, used to refine crude oil. The various processes schemes include as a first step crude oil distillation. All refineries have at least one atmospheric distillation (primary distillation or topping)

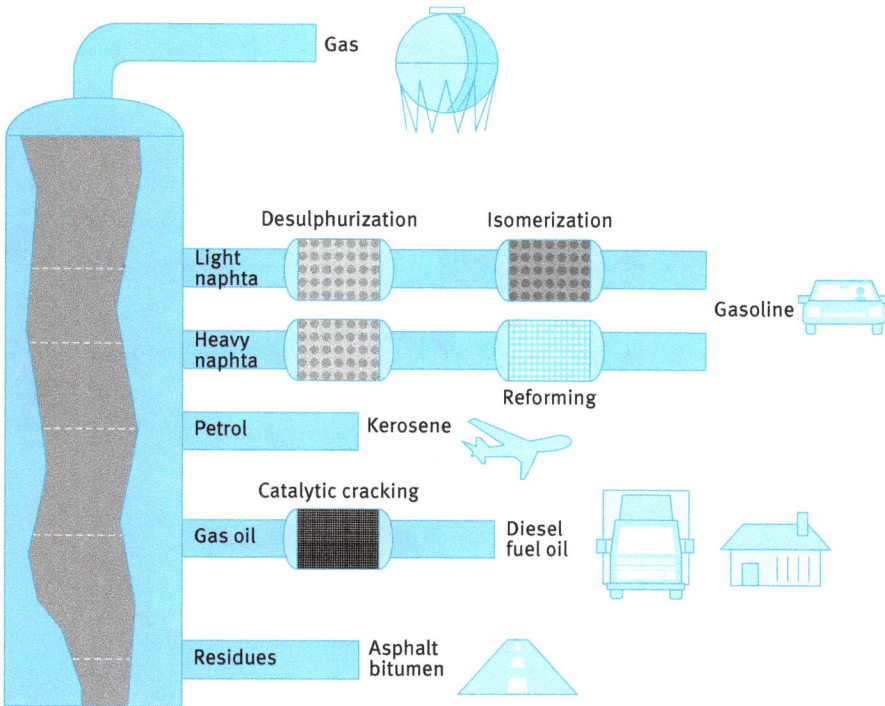

Figure 4.12: Basics of oil refining. From *Future Perspectives in Catalysis*, 2009, ISBN: 9789081408615, p. 30, http://www.vermeer.net/pub/communication/downloads/future-perspectives -in-cata.pdf.

unit for separating the crude oil into various fractions with different boiling ranges. Desalting of crude oil is done prior to distillation.

The operating pressure of the atmospheric distillation column is related to condensation conditions, namely temperature, in the reflux drum at the top of the column. Temperature of ca. 40 °C implies that the corresponding pressure should be slightly higher than the atmospheric pressure (1.5–3 bar). The crude inlet temperature is between 340 and 385 °C, being limited by thermal stability of hydrocarbons.

Vacuum distillation is made downstream of atmospheric distillation to recover additional distillates from the residue without, however, increasing the temperatures and breaking the heavier hydrocarbons. A diagram of crude oil distillation unit is given in Figure 4.13.

Presence of water (ca. 0.1–0.6 vol%), mineral salts (from 20 to 300 g per ton of crude oil), and sediments requires crude oil desalting prior to distillation. The salts originating either by contamination from seawater during transportation or from the wells consist of Na, Mg, and Ca chlorides. Similar to sediments, the salts can cause poor flow behavior or plugging when they cake out inside, for example, crude

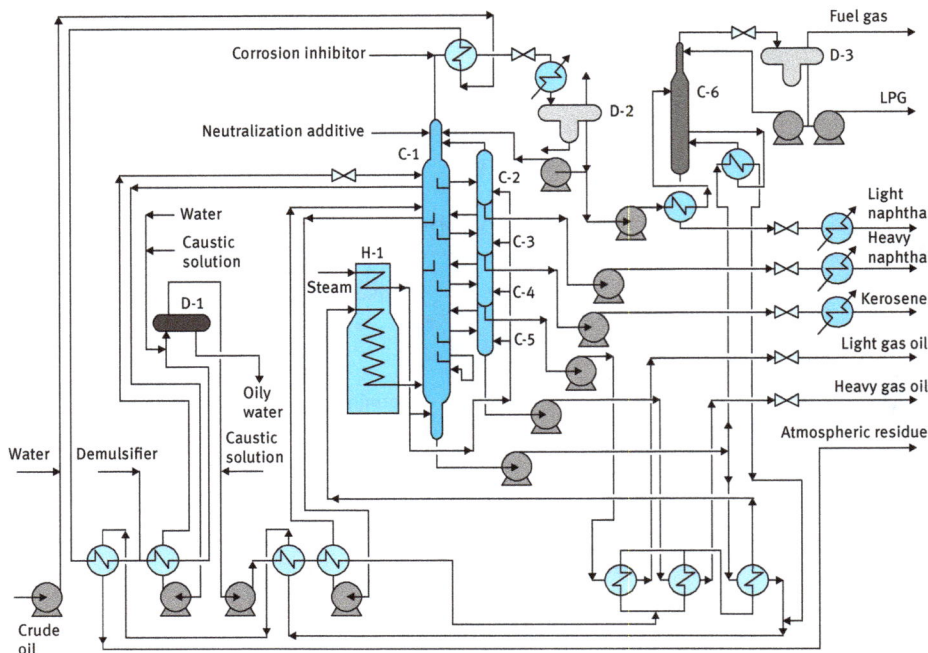

Figure 4.13: Simplified layout of a crude oil atmospheric distillation unit. D-1, desalter; C-1, principal column; C-2–C-5, stripping columns; C-6, stabilization unit; D-2 and D-3, reflux accumulators; H-1, heater. Modified after http://www.treccani.it/export/sites/default/Portale/ sito/altre_aree/Tecnologia_e_Scienze_applicate/enciclopedia/inglese/inglese_vol_2/089-104_ ING3.pdf.

oil preheating heat exchanges or on fractionation trays. Moreover, such depositions influence the heat transfer efficiency in a negative way.

Flash vaporization of crude oil can lead to formation of hydrochloric acid vapors by hydrolysis of calcium and magnesium chlorides. Condensed HCl is highly corrosive affecting the sections where steam is likely to condense, namely, the top of the crude atmospheric distillation column and the condenser. While sodium chloride is the least harmful from the viewpoint of HCl generation, and therefore sodium hydroxide is injected into the crude oil desalting system. On the other hand, sodium can influence catalysts in downstream processing in the negative way.

Desalting requires temperature of 110–150 °C, especially for heavier and thus viscous crude oil facilitating decantation of water from the crude. Large sizes of desalters (Figure 4.14) are determined by the residence time of ca. 20–30 min.

Besides water (3–8 vol% of the crude), deemulsifiers are also added to crude oil to prevent formation of stable emulsions due to asphaltenes present in crude oil.

Desalting can be performed in either one or two stages (Figure 4.15). Dehydration efficiency reaches 95% in a single stage and can be up to 99% in two stages.

Figure 4.14: Crude oil desalting unit. From https://howebaker.com/wp-content/uploads/2017/07/Desalter-1-768x495.jpg.

Figure 4.15 illustrates that chemicals (e.g. corrosion inhibitors, biocides, and oxygen scavengers) are added during the dehydration/desalting process. The latter is needed if there is no further treatment of produced water, which is disposed directly to, e.g., a nearby river.

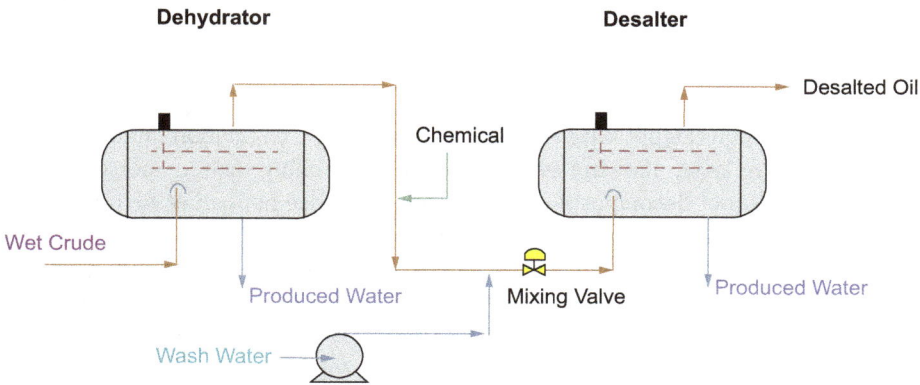

Figure 4.15: Dehydration/desalting unit. From https://i2.wp.com/www.arab-oil-naturalgas.com/wp-content/uploads/wet-crude-treatment-train.jpg?w=655&ssl=1.

Some mineral salts still remain in the desalted oil, thus caustic soda is injected to it prior to the heat exchange train to convert a part of the residual salts into sodium chloride. Another option to prevent corrosion is injection of ammonia or another basic neutralizing agent to the top of the atmospheric column and the condenser system. In such sections, water is more likely to condense.

The desalted and dehydrated crude oil feed is directed to a preheat train for heat recovery from various streams. After passing through a fired heater, the feed enters the main column where the cut points are controlled by adjusting the reflux and temperature profile. As in any distillation, the lower boiling and higher boiling hydrocarbons accumulate in the upper and lower parts of the column, respectively. As shown in Figure 4.13, fractions with higher boiling points are taken as side stream draw-offs and stripped with steam to reduce the partial pressure or by heating and reboiling. Parts of the side streams are returned as reflux.

Table 4.3 list various oil fractions with their boiling range and application areas. The low quality of straight-run gasoline, in particular with low octane number, requires its upgrading to the desired quality.

Table 4.3: Various oil fractions with their boiling range.

Fraction	Boiling range (°C)	Use
Gases	<20	Methane (65–90%), ethane, propane, butane
Naphtha	70–170	Base for gasoline; used for chemicals. Number of carbon atoms: 6–10
Kerosene	170–250	Number of carbon atoms: 10–14
Gas oil	250–340	Jet, diesel, and heating fuel. Number of carbon atoms: 14–19
Heavy fraction	350–500	Lubrication, boiler fuel. Number of carbon atoms: 19–35
Bitumen	>500	Number of carbon atoms: > 35

Vacuum distillation (Figure 4.16) is similar to atmospheric distillation but operates at a lower pressure level, allowing evaporation of high boiling hydrocarbons at lower temperature without decomposition, which will otherwise happen at high temperature. The inlet temperature of the residue is generally between 390 and 420 °C.

The residue from atmospheric distillation is first preheated using heat from distillates prior to be heated in the fired heater. The high specific volume of vapor at low pressure demands large column diameters, particularly in the upper part (Figure 4.16, right). The lower part of the column (below the feed inlet) is used as a stripping section operating with steam to reduce the partial pressure of hydrocarbons in the vapor phase. The vacuum within the column of ca. 60 mbar at the column top is generated by the evacuation unit comprising a series of steam ejector stages.

Besides crude oil distillation units (Figure 4.17), oil refineries have a number of processing units, including: (a) separation units, (b) conversion (cracking) units to turn heavy fractions into lighter fractions, (c) units to improve the quality of some fractions by, for example, isomerization changing the octane number, (D) units for the removal of unwanted components (such as sulphur), and (e) units in a limited number of refineries to produce lubricating oils.

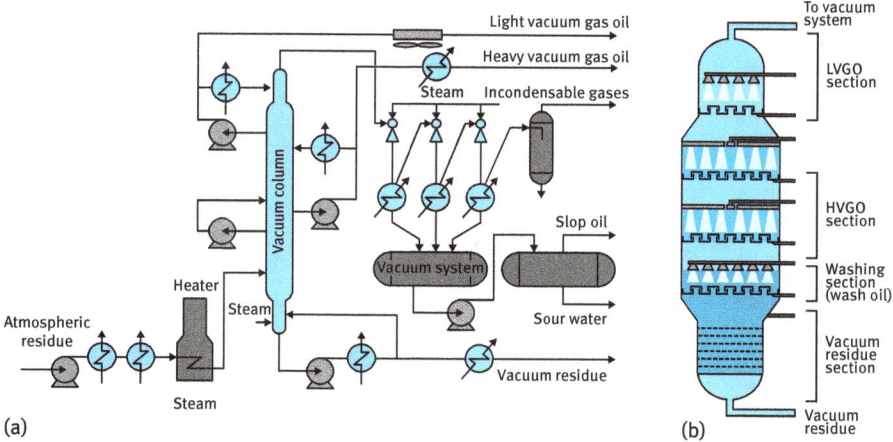

Figure 4.16: Vacuum distillation. Modified after http://www.treccani.it/portale/opencms/han
dle404? exporturi = /export/sites/default/Portale/sito/altre_aree/Tecnologia_e_Scienze_
applicate/enciclopedia/inglese/inglese_vol_2/105-112_ING3.pdf.

Figure 4.17: Crude oil distillation with a vacuum column on the left and atmospheric crude
distillation on the right. From https://www.emerson.com/resource/image/1469794/landscape_ra
tio4x3/414/310/2abb06010f23415dccc80a6b1990a541/ve/h003-crude-vacuum-unit.jpg.

Two identical refining schemes are unlikely to exist, thus a scheme in Figure 4.18
gives a general overview of oil refining including various reactions and generated
products.

It should be mentioned that commercial products are made in oil refining by
blending products from different oil refining units and introducing also some additives to the final commercial product. As an example, Figure 4.19 shows how blending of six bases enables production of gasoline.

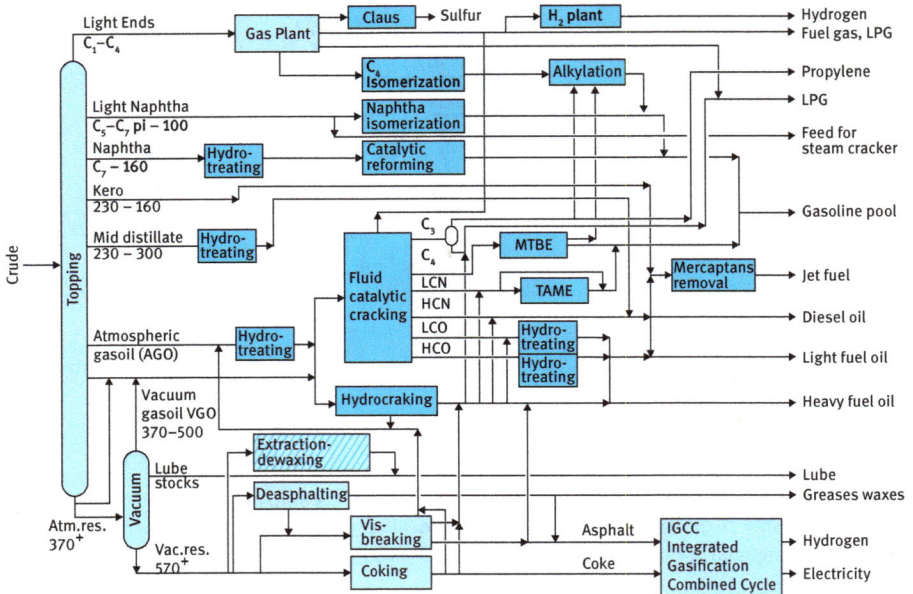

Figure 4.18: Processes in a generic oil refinery. LCN, light-cut naphtha; HCN, heavy-cut naphtha; LCO, life cycle oil; HCO, heavy cycle oil; LPG, liquefied petroleum gas. From http://www.pa.ismn. cnr.it/scuolagic2010/presentazioni_docenti/Sanfilippo.pdf.

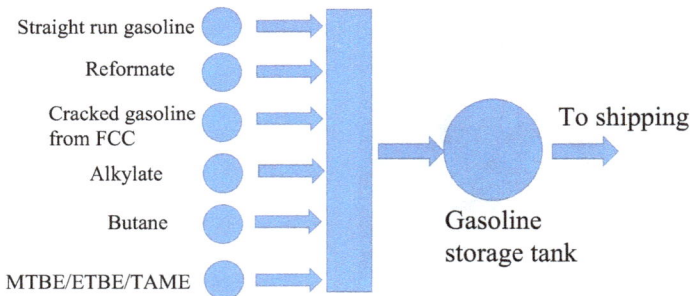

Figure 4.19: Illustration of blending for production of gasoline.

While, as shown in Figure 4.18, oil refineries are mainly focused on production of fuels, there are also some non-fuel applications. LPG and light naphtha fractions are used as feedstock for the petrochemical industry in steam cracking (Chapter 6). Special boiling-range naphthas are used as solvents, while paraffin waxes are applied in a range of industries. Heavy fractions are utilized for manufacturing of lubricants and bitumens.

Refinery schemes have the following typical cycles:

1. Simple-cycle (hydroskimming) refineries with crude oil distillation plants, desulphurization units, and reforming units to increase the octane number of gasoline.

2. Thermal conversion cycle containing scheme with visbreaking (and thermal cracking) and with coking. Such conversion schemes involve distillation residues (visbreaking and coking units) or distillates (thermal cracking). The vacuum residue is thermally cracked due to the presence of metals such as nickel or vanadium; otherwise, deactivating zeolite catalysts in the case of catalytic cracking. Visbreaking and coking can treat the residues of atmospheric and vacuum distillation in a relatively simple and economical way, which, however, do not meet the current environmental regulations.

3. Catalytic conversion cycle: schemes with cracking (fluid catalytic cracking, FCC) and schemes with hydrocracking (HDC). Catalytic crackers were traditionally used to treat the heavier distillate fractions obtained from vacuum operation (and in part also the residues), and HDC units.

4. Deep conversion cycle: a scheme including the hydroconversion of residues and a scheme with deasphalting and gasification. HDC developed to treat heavy paraffinic distillates is used nowadays for processing residues.

As already mentioned, some refineries have units for production of lubricants. Typical yields in a conversion refinery are given in Figure 4.20.

The catalytic and thermal processes in oil refining will be discussed later in detail in relevant chapters. Isomerization of paraffins into branched ones and catalytic reforming described in Chapter 7 are used to improve the octane number of light gasoline. These reactions are accompanied with aromatization in the case of heavier gasoline fraction. Due to high sulphur content in oil, nearly all refinery streams after distillation undergo hydrodesulphurization (HDS), which is addressed in Chapter 10. The higher boiling point fractions are converted to lighter ones by various thermal and catalytic cracking processes presented in Chapter 6. HDC requires the presence of plants for the production of hydrogen by steam reforming (Chapter 5). The alkylation units aimed at treating the olefinic fractions C3 and C4 from crackers and producing branched C7 and C8 components are described in Chapter 12.

The capacity of oil refineries (ca. 700 globally) depends on a number of factors, such as geographical location, the demand in a particular region, availability of energy and feedstock, transportation costs. A current trend in oil refining is to have larger production of distillates with higher quality diesel. The economy of scale is also applied to oil refineries, with large plants being more profitable, at the same time requiring longer distances for fuel delivery to customers from centralized locations.

In some countries, such parameter as refinery yield (%) is used as a measure of refinery efficiencies. This parameter represents the percentage of finished product produced from input of crude oil and net input of unfinished oils. Another important

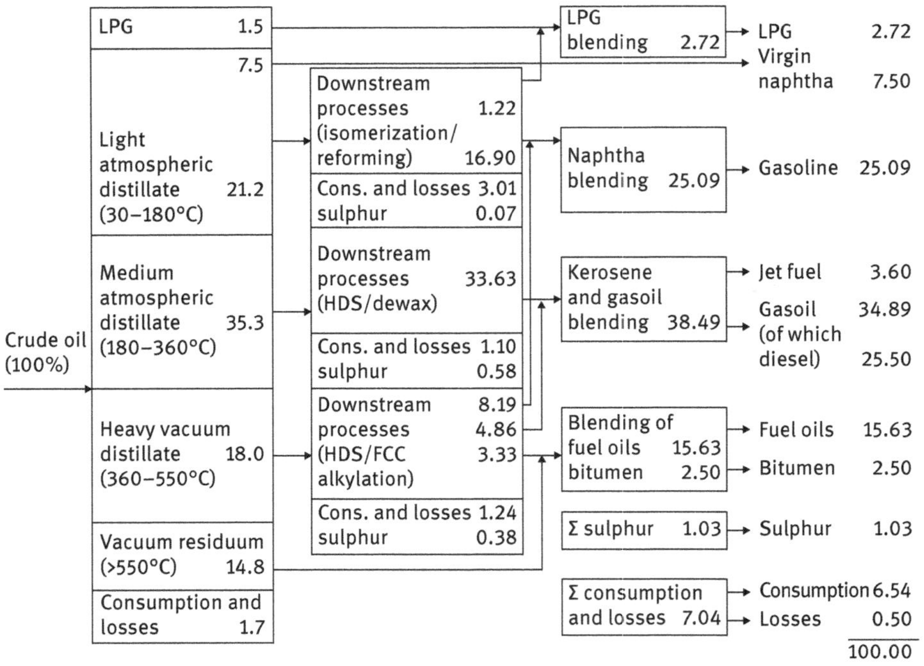

LPG 1.5

7.5

Light atmospheric distillate 21.2 (30–180°C)

Medium atmospheric distillate 35.3 (180–360°C)

Crude oil (100%)

Heavy vacuum distillate 18.0 (360–550°C)

Vacuum residuum (>550°C) 14.8

Consumption and losses 1.7

Downstream processes (isomerization/reforming) 1.22 16.90
Cons. and losses 3.01 sulphur 0.07

Downstream processes (HDS/dewax) 33.63
Cons. and losses 1.10 sulphur 0.58

Downstream processes (HDS/FCC alkylation) 8.19 4.86 3.33
Cons. and losses 1.24 sulphur 0.38

LPG blending 2.72

Naphtha blending 25.09

Kerosene and gasoil blending 38.49

Blending of fuel oils 15.63 bitumen 2.50

Σ sulphur 1.03

Σ consumption and losses 7.04

LPG 2.72
Virgin naphtha 7.50

Gasoline 25.09

Jet fuel 3.60
Gasoil 34.89 (of which diesel) 25.50

Fuel oils 15.63

Bitumen 2.50

Sulphur 1.03

Consumption 6.54
Losses 0.50

100.00

Figure 4.20: Typical yields in oil refinery.

characteristics of a refinery is the ratio between gasoline and diesel. The plants focusing more on gasoline have catalytic cracking and alkylation units, while production of diesel requires HDC. In a high-conversion refinery combining FCC and HDC, the feed to the latter is vacuum gas oil.

In a refinery scheme with HDC-coking, the production of middle distillates is maximized. As a coker feed vacuum residue is used and a combination of vacuum gas oils and heavy coker distillates is applied as HDC feed, to which extra hydrogen is supplied from steam reforming.

It is also worth elaborating about the lubricating oil plants. The total annual production capacity of lubricating oils, which are classified in five groups (Table 4.4), is above 40 Mt/year.

The residue from the vacuum distillation process (Figure 4.21) contains heavier base oils such as the bright stocks. Removal of aromatics is done by solvent extraction and/or HDC and selective catalytic hydrogenation. Other impurities, besides aromatics, including sulphur and nitrogen compounds, are removed by extraction with solvents such as furfural. Aromatics deteriorate the quality of base oil due to their potential oxidation.

After aromatics removal by solvent extraction or HDC, the subsequent step is dewaxing. This can be done using solvents such as methyl-ethyl-ketone or furfural,

Table 4.4: Classification of lubricating oils.

Group	Sulphur (wt%)	Saturates (wt%)	Viscosity index	Main refining technology
I	>0.03	<90	80–120	Solvent extraction
II	≤0.03	≥90	80–120	HDC and hydroisomerization or HDC and solvent extraction
III	≤0.03	≥90	≥120	HDC and hydroisomerization
IV	Poly-α-olefins			Synthetic
V	All base oils included in groups I, II, III, and IV			Solvent extraction or hydrogenation or synthetic

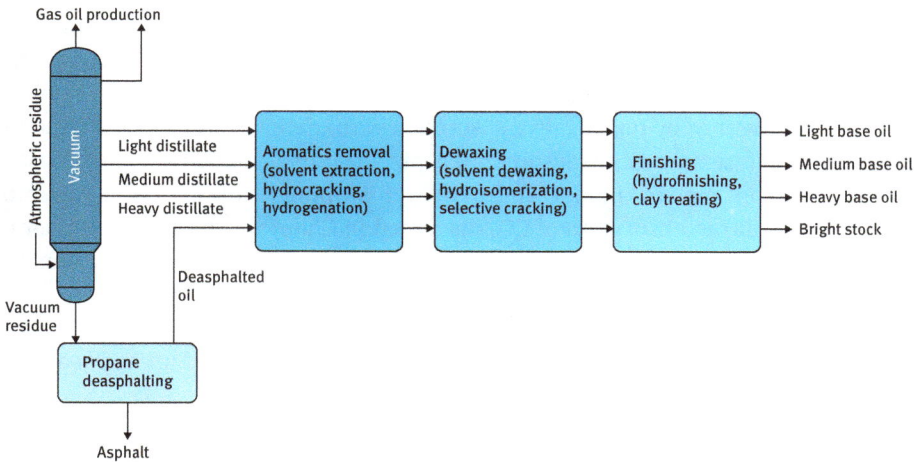

Figure 4.21: Base oil refining. Modified after http://www.treccani.it/export/sites/default/Portale/sito/altre_aree/Tecnologia_e_Scienze_applicate/enciclopedia/inglese/inglese_vol_2/339-350_ING3.pdf.

which are mixed with the waxy oil and cooled to allow wax crystallization. The wax crystals are then removed by filtration. Alternatives to such solvent dewaxing are catalytic dewaxing and wax hydroisomerization. The former applies catalytic cracking of long *n*-paraffins and waxy side chains, while in wax hydroisomerization, desirable branched molecules are obtained without cracking. In the finishing step, traces of contaminants are removed from the base oils by hydrofinishing or by clay treatment. Lubricating oils are produced in blending plants, where a variety of additives are added to base oil.

The current trend of lower demands for gasoline and other fuels because of carbon emission mandates more fuel-efficient cars and increasing popularity of electric vehicles, and forces oil companies to reconsider their long-term strategy. In the future, refining complexes will be much more oriented toward production of chemicals, which can influence existing petrochemical markets as refineries have a larger scale compared with a chemical plant.

Some new refineries are targeting significant petrochemical production instead of transportation fuels as in a conventional refinery with the aim of converting 40–60% per barrel of oil to chemicals, for example, *p*-xylene, benzene, ethylene, and other downstream petrochemicals. An increase of *p*-xylene capacity will certainly influence downstream polyethylene terephthalate production.

4.4 Natural gas processing

Processing of raw natural gas starts from collecting it from a group of adjacent wells and then removal of free liquid water and natural gas condensate. The condensate is then usually transported to an oil refinery and the water is disposed as wastewater. The raw gas after removal of water by, for example, cooling and subsequent separation of liquid water or by regenerative absorption with triethylene glycol or regenerative adsorption (pressure swing adsorption) is pipelined to gas-processing plants. Gas purification from carbon dioxide and hydrogen sulphide can be done by amine treatment (similar to absorption of CO_2 discussed in Chapter 3) or using polymeric membranes. Various units in a typical natural gas processing plant are presented in Figure 4.22.

To the sales gas pipelined to the domestic market, small amounts of sulphur-containing compounds are added, since natural gas is odorless and any leakage would otherwise go unnoticed.

Natural gas is mainly used for heating and synthesis of chemicals (ca. 40%), power plants (30%), and domestic purposes (30%). The main chemical applications of methane are depicted in Figure 4.23.

4.5 Processing of coal

The main processes of coal utilization are gasification, pyrolysis, combustion, and liquefaction under high hydrogen pressure. Coal gasification with oxygen and steam discussed in Chapter 5.2 is used to produce synthesis gas, a mixture of CO, and hydrogen:

$$C(as\,coal) + O_2 + H_2O \rightarrow H_2 + CO \tag{4.4}$$

Further processing of syngas is done, for example, in an integrated plant to produce fuels and chemicals from coal based on Fischer-Tropsch technology (Chapter 13) in a facility operating at Sasol in South Africa in the late 1950s (Figure 4.24).

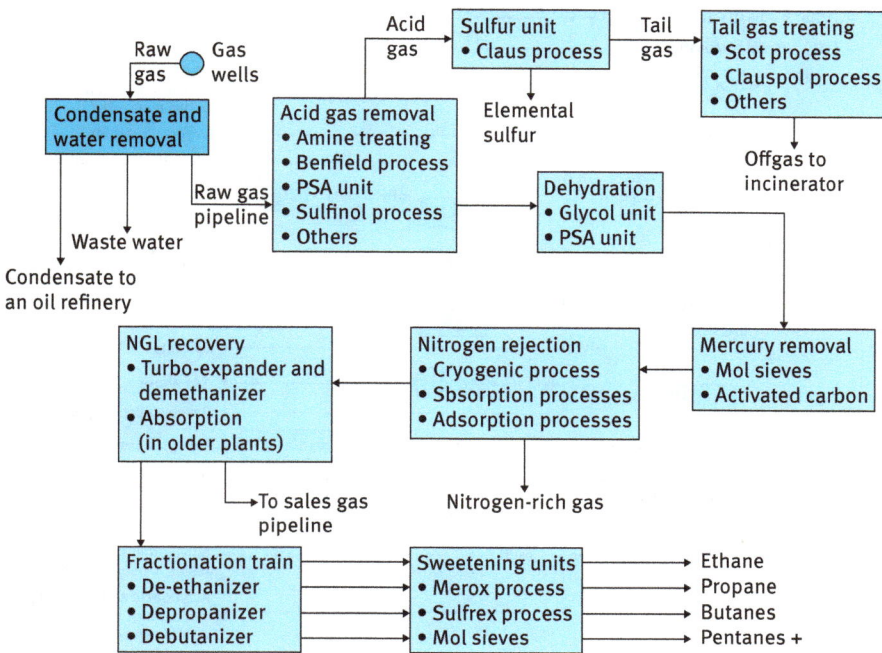

Figure 4.22: Units in a typical gas processing plant. http://en.wikipedia.org/wiki/File:NatGasPro cessing.svg.

Conversion of coal to synthetic fuels can be done by direct liquefaction processes. In low-temperature carbonization between 360 °C and 750 °C coal tars richer in lighter hydrocarbons are produced and further processed into fuels. In the Lurgi-Ruhrgas process (see Figure 4.25), crushed coal is rapidly heated in a mixer to 450–600 °C by contact with recirculating char particles, heated by partial oxidation in an entrained-flow reactor. Separation from particulates and condensation results into liquid products (ca. 18%). The rest of the products are char (ca. 50%) and gas.

Treatment at a higher temperature (so-called semi-coking) is done at 500–580 °C; giving semi-coke or a solid with appreciable volatile matter. High-temperature processing (coking) done at 900–1,200 °C leads to coke with almost no volatile matter; thus, no further volume changes occur after heating. Such high-temperature carbonization/ pyrolysis conducted at ca. 900 °C is shown in Figure 4.26. The coal is conveyed to the coke oven bunker. The ovens are loaded from a mobile larry car located on top of sealed ovens where coking takes place. Typically, coke ovens are 14–16 m in length and 4–7 m in height. The number of adjacent ovens in a battery, where ovens share heating fuels, could be 68–78 or even higher. During coking, a number of processes occur, including, to name a few, destruction of coal, cracking of alkanes, dehydrogenation of naphthenics, and condensation of aromatics.

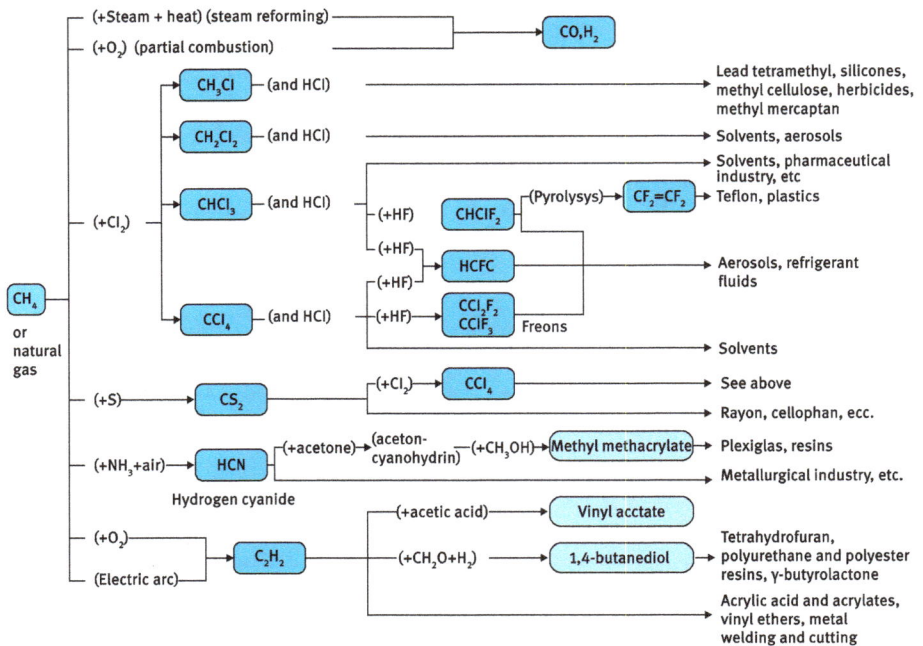

Figure 4.23: Main chemical applications of methane. Modified after http://www.treccani.it/export/ sites/default/Portale/sito/altre_aree/Tecnologia_e_Scienze_applicate/enciclopedia/inglese/in glese_vol_2/405-454_ING3.pdf.

After the carbonization process, the coke oven doors are opened and the red-hot coke is transported to the quenching unit. After cooling and drying in a wharf, the dry coke is transferred to the screening and loading unit, while the by-product vapors and gases are collected and sent to the recovery and processing units. Coke oven gas contains a significant amount of water vapor (47%). On dry basis, besides hydrogen (55%), methane (25%), nitrogen (10%), CO (6%), and carbon dioxide (3%), some light hydrocarbons (ethane, propane, etc.) are also present.

Another by-product namely coal tar (3.5–4% of coke produced) can be fractionally distilled (Figure 4.27) to give light oil (boiling point, < 200 °C), containing aromatics such as benzene, toluene, xylenes, and styrene; middle oil (boiling point < 370 °C) with tar acids (phenols), tar bases (pyridine, anilines, and quinolines), and neutral oils (naphthalene); and heavy oil (boiling point < 550 °C) with fused rings, such as anthracene and phenanthrene. A solid residue after distillation, coal tar pitch, contains polycondensed aromatic compounds.

Direct conversion of coal to liquids in the past was done by the hydrogenation (Bergius) process developed in 1913; however, plants operating with this technology ceased operation long ago. In the Bergius process, dry coal was mixed with heavy oil recycled from the process and hydrogenated in the presence of a number of catalysts

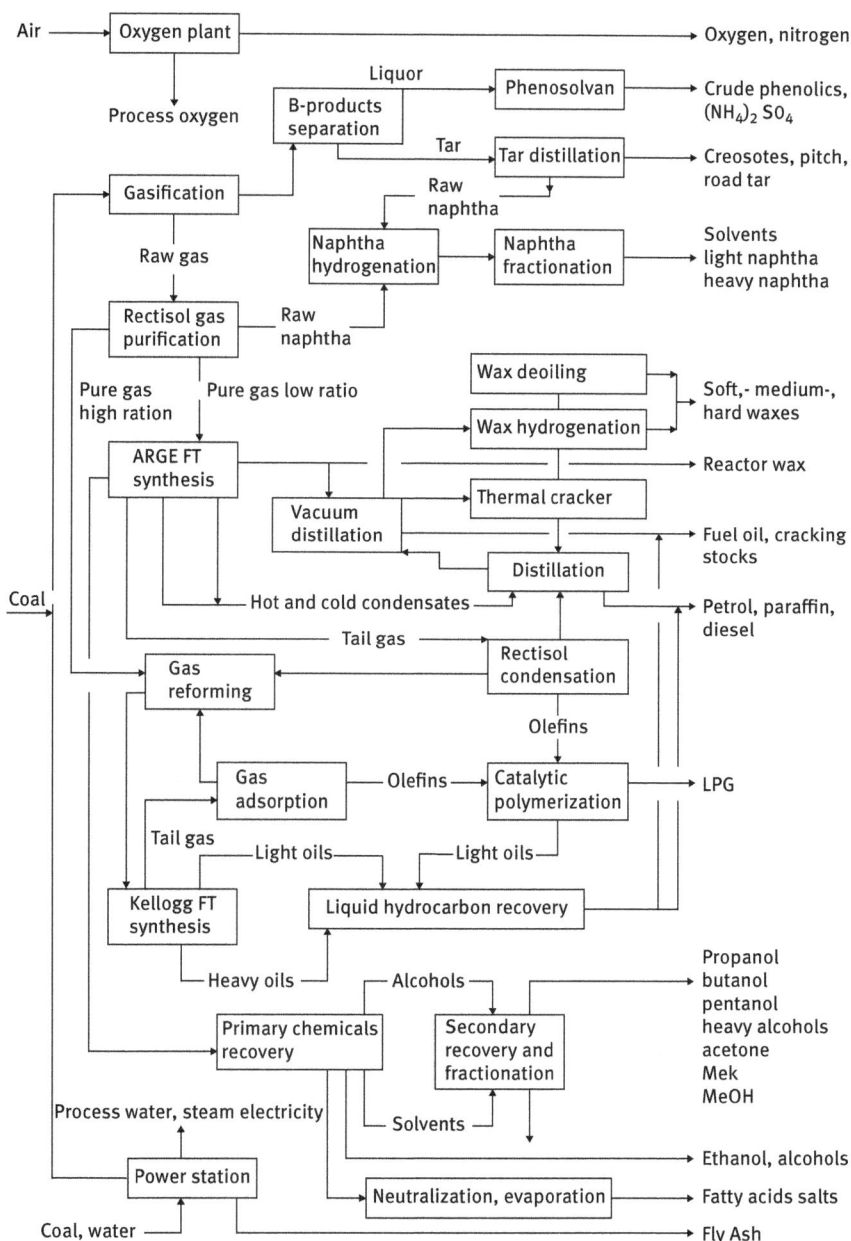

Figure 4.24: Sasol 1 integrated HTFT-LTFT plant in Sasolburg, South Africa (late 1950s). Reprinted with permission from D. Leckel, Diesel production from Fischer-Tropsch: The past, the present, the new concepts. *Energy and Fuels,* 2009, 23, 2342. Copyright (2009) American Chemical Society. Reproduced with permission.

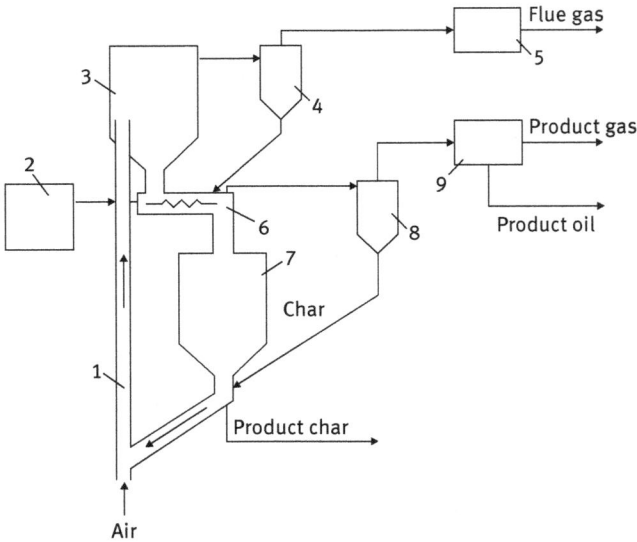

Figure 4.25: Lurgi-Ruhrgas process: 1, transport reactor and lift line; 2, coal preparation; 3, collecting bin; 4, cyclone; 5, heat recovery; 6, mixer-carbonizer; 7, surge hopper; 8, cyclone; 9, condenser.

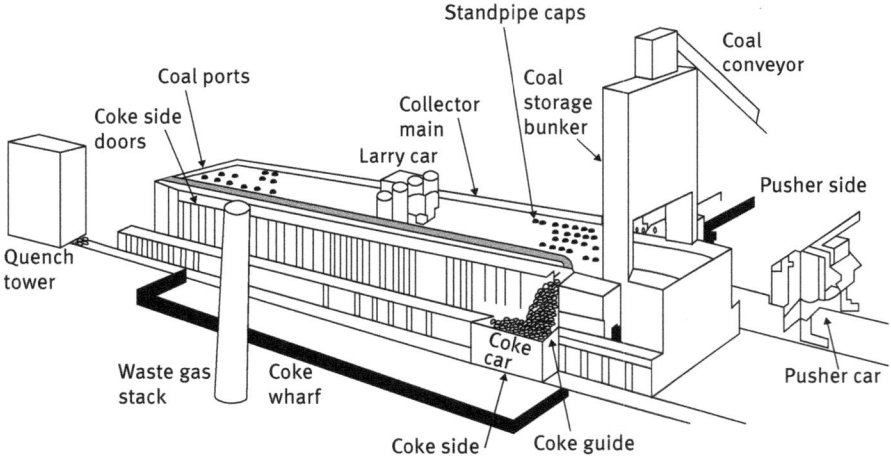

Figure 4.26: High-temperature coking of coal.

such as alumina supported tungsten or molybdenum sulphides at 400–500 °C under 20 to 70 MPa hydrogen pressure to produce heavy oils, middle oils, gasoline, and gases. High pressure prevented polymerization and polycondensation reactions and thus catalyst coking. The amount of gaseous and liquid products decreases

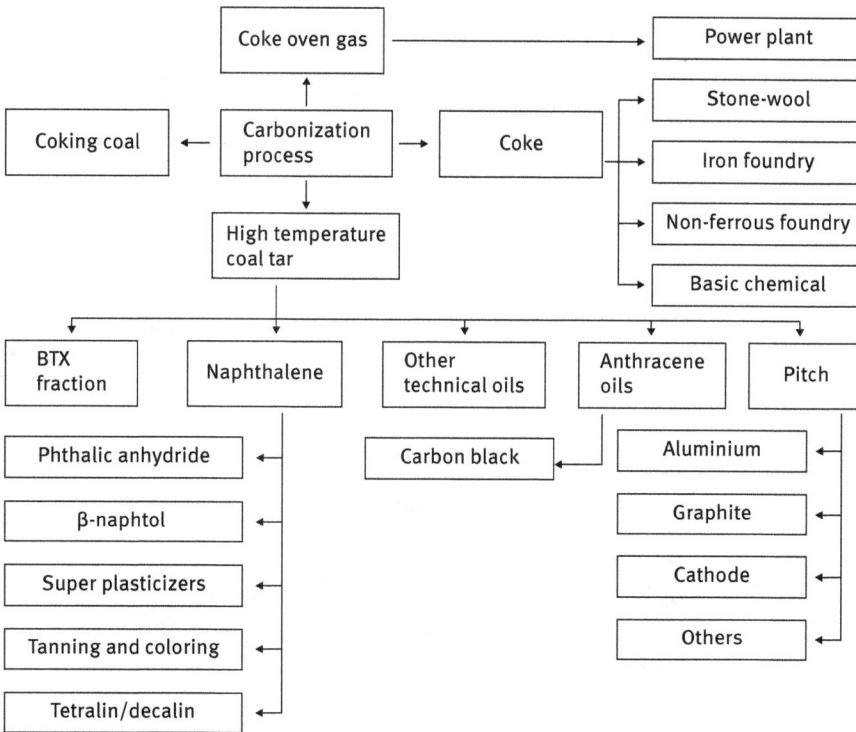

Figure 4.27: Products of coking coal.

with an increase in carbon content, disallowing utilization in the Bergius process of coals with large carbon content such as anthracite.

4.6 Biomass processing

There are several options to process biomass depending on the type of product. For long time, generation of ethanol from lignocellulosic biomass used hydrolysis of wood chips. Details are provided in Chapter 11. Nowadays, if the main target is cellulose, and eventually paper, Kraft processing (Kraft pulping or sulphate process) of wood chips is mainly used (Figure 4.28). The product of Kraft processing is wood pulp consisting of cellulose fibers. Kraft pulping removes most of the lignin present originally in the wood, in contrast with, e.g., mechanical pulping processes, wherein most of lignin is left in the fibers. The basis of the Kraft pulping process is treatment of wood chips with a mixture of sodium hydroxide and sodium sulphide (white liquor). In this process, hemicelluloses and lignin are separated from cellulose.

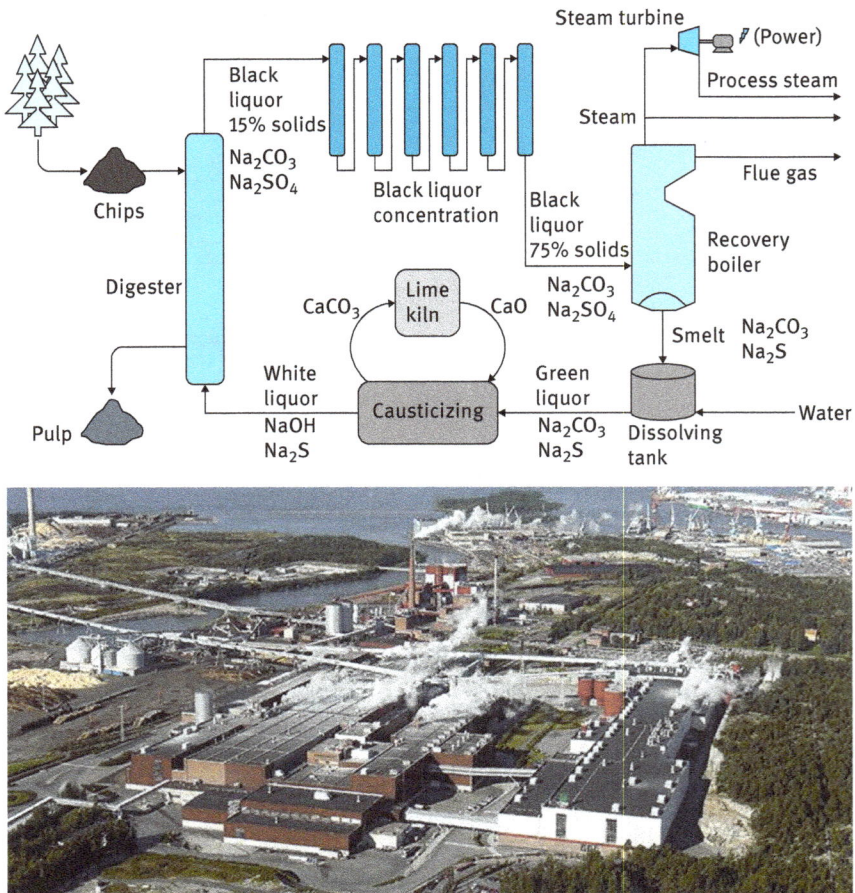

Figure 4.28: Kraft pulping process: (a) flow scheme and (b) a pulp mill (https://www.upm.com/news-and-stories/articles/2016/07/the-giant-of-rauma/).

Wood chips 12–25 mm in length and 2–10 mm in thickness are first presteamed. The impregnation by cooking liquors at temperature below 100 °C is done before or after the chips enter the pressurized vessels (digesters), assuring homogeneous cooking and consuming ca. 40–60% of all alkali consumption.

Digesters can operate in batch or continuous modes. In the latter option, the feeding rate should correspond to the completion of pulping. Degradation of hemicelluloses and lignin requires several hours at 170–176 °C. After the delignification process, the solid pulp (ca. 50 wt% based on the dry wood chips) has a brown colour. The Kraft pulp is darker than other wood pulps, but it can be bleached, giving very white pulp. Such bleaching process can include a sequence of operations, for example, treating with oxygen and then ozone, washing with sodium hydroxide, and then treating in sequence with alkaline peroxide and sodium dithionite. Fully bleached Kraft pulp is used to make high-quality paper where strength, whiteness, and resistance to yellowing are important.

The combined liquids, known as black liquor with products from hemicellulose and lignin breakdown along with sodium sulphate and carbonate, are concentrated. The excess black liquor is at about 15% solids and is first concentrated into 20–30% solids with removal of rosin soap. The latter is further processed to tall oil. Subsequent evaporation gives heavy black liquor with 65–80% solids, which is burned in a recovery boiler to recover heat and inorganic chemicals for reuse in the pulping process. In fact, the recovery boiler is a source of energy for a paper mill and neighboring industries and communities. During recovery, sodium sulphate is reduced to sodium sulphide by the organic carbon in the mixture.

The smelt (molten salts) from the recovery boiler gives a solution of sodium carbonate and sodium sulphide (green liquor). The white liquor used for the pulping *per se* is regenerated by reacting the green liquor with calcium oxide, which is transformed in the solution to calcium hydroxide. The latter reacts with components of the green liquor

$$Na_2S + Na_2CO_3 + Ca(OH)2 \leftrightarrow Na_2S + 2NaOH + CaCO_3 \tag{4.5}$$

forming calcium carbonate precipitate, which is separated and calcined in a lime kiln

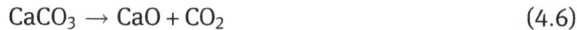

$$CaCO_3 \rightarrow CaO + CO_2 \tag{4.6}$$

to recover calcium oxide (lime).

The main by-products of Kraft pulping are crude sulphate turpentine and tall oil soap.

An example of the product distribution at Metsä Fibre bioproduct mill at Äänekoski, Finland which started operation in 2017, is shown in Figure 4.29.

Apart from pulping, which is aimed at making cellulose and eventually paper, there are other approaches to transform biomass illustrated in Figure 4.30.

Thermal conversion of biomass comprises several routes as illustrated in Figure 4.31.

Gasification (above 650 °C) or pyrolysis of biomass (above 450 °C) at rather high temperatures giving synthesis gas or pyrolysis oil on the one hand and hydrolysis to sugars (180–200 °C) on the other hand are the two main technological lines for the biorefinery concept. The synthesis gas and sugars are further processed to value-added products *via* several reactions. Various extractives could be transformed through catalytic routes to primarily specialty and fine chemicals.

Due to the structure of biomass, which is different from petroleum, technologies of biomass processing are different from those widely used in oil refining, e.g., in the processing of hydrocarbons. In particular, the oxygen/carbon ratio in lignocellulosic biomass is close to unity, which is an undesirable ratio in the production of transportation fuels. Since a slight amount of oxygen in the fuel is in fact beneficial for

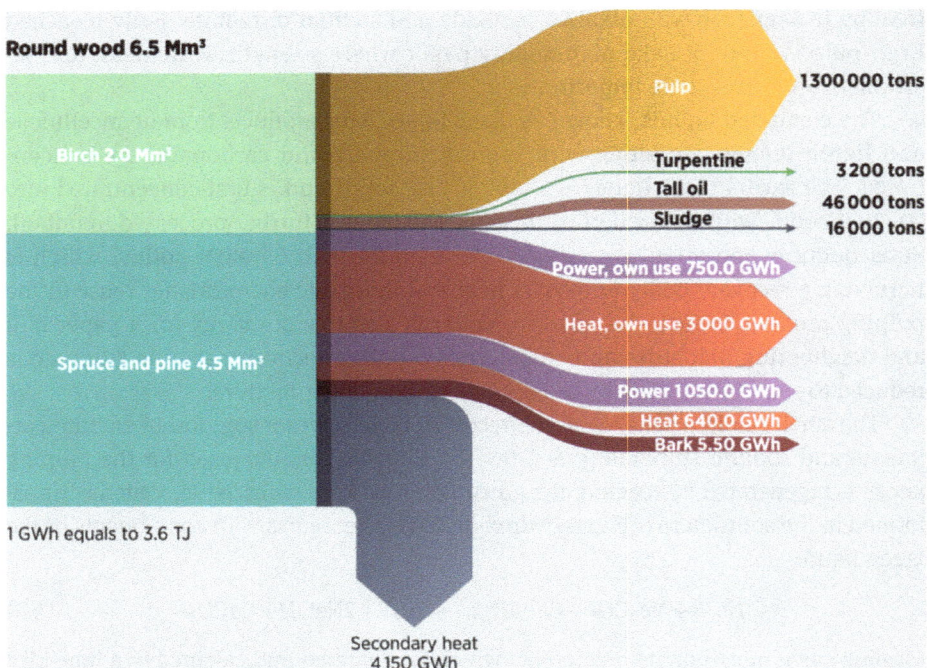

Figure 4.29: Product distribution of Äänekoski mill. https://www.irena.org/-/media/Files/IRENA/ Agency/Publication/2018/Mar/IRENA_Bioenergy_from_Finnish_forests_2018.pdf.

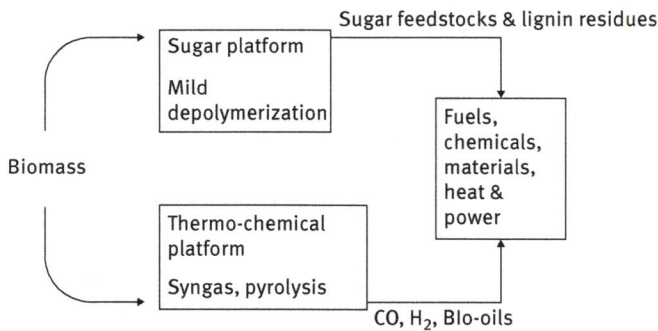

Figure 4.30: Integrated biorefinery concept, comprising two routes to chemicals, fuels and heat and energy through sugar and thermochemical platforms.

gasoline engines, the strategy for using biomass for motor fuels should therefore include substantial, but not total, removal of oxygen. On the other hand, a large number of intermediates in the chemical industry contain oxygen. This implies that such substances as methanol, acetic acid, and ethylene glycol with the carbon/oxygen ratio equal to unity do not require a very high degree of oxygen removal.

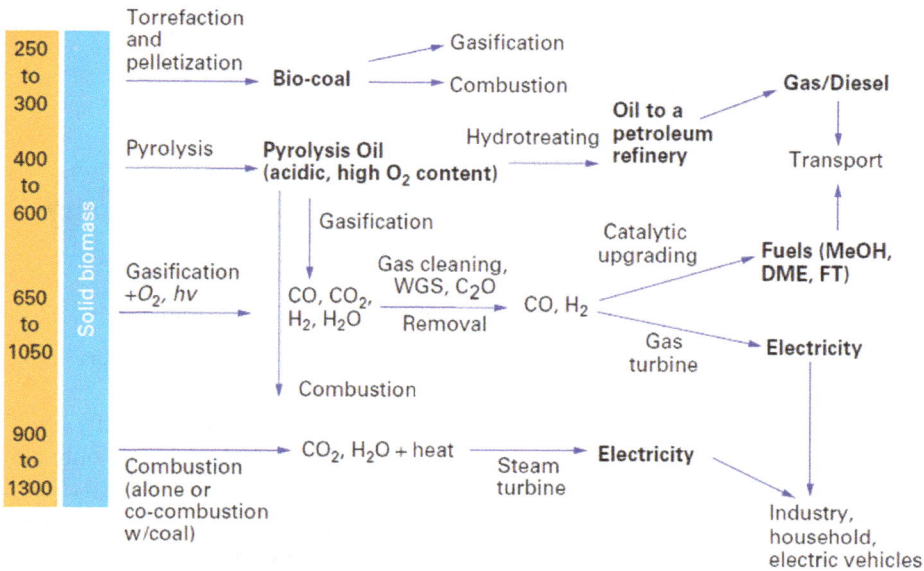

Figure 4.31: Possible routes for thermal conversion of biomass. Temperature is given in degree Celsius.

Gasifier technologies include fluidized beds (bubbling and circulating), moving beds (concurrent and countercurrent), grate (both moving and stationary), and entrained flow. Fluidized beds need to be operated below the ash melting temperature to avoid bed agglomeration and the resulting defluidization of the bed. Moving beds can be operated above or below the ash melting temperature, while entrained flow gasifiers generally operate above the melting temperature of the salts.

Synthesis gas obtained after gasification is characterized by an unfavorable CO/H ratio if the subsequent step of biomass processing is Fischer-Tropsch synthesis, which requires H_2/CO ratio of 2. Thus, water-gas shift reaction is conducted prior to Fischer-Tropsch synthesis. Another option is to make syngas and then methanol together with dimethylether (DME). The latter mixture can be transformed to gasoline, as in the combined process making green gasoline from wood with Carbona Gasification and Topsøe TIGAS technologies (Figure 4.32). It is fair to state that various biomass gasification processes are nowadays mainly at the pilot and demonstration scale.

An interesting option is co-combustion of biomass with coal or natural gas. For partial substitution of coal by biomass, the biomass can be fed with the coal through the same burner/feeder, or it can be fed to a separate burner. Alternatively, the biomass can be pyrolyzed/gasified first, and then the products can be burned in the boiler. Biomass gasification followed by combustion in a boiler has been commercially applied and is illustrated in Figure 4.33. The potential advantage of co-combustion for

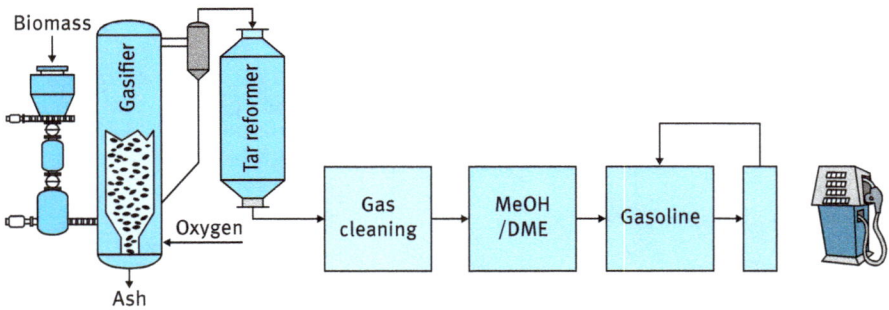

Figure 4.32: Carbona Gasification and Topsøe TIGAS technologies for making gasoline from wood.

energy generation is that higher steam temperatures, and thus higher electricity generating efficiencies, can potentially be reached compared to a boiler burning only biomass. Furthermore, in co-combustion pyrolysis, oils can also be used.

Pyrolysis, the thermal degradation of an organic material in the absence of oxygen, is conducted at lower temperatures than gasification, giving a solid residue called char, a liquid called bio-oil, and uncondensed gases. The process can be optimized for maximal bio-oil production (ca. 75%). Important requirements in this optimization that can be achieved in fluidized-bed reactors of different types (Figure 4.34) are moderate temperature, optimal temperature around 500 °C, rapid heating of the biomass

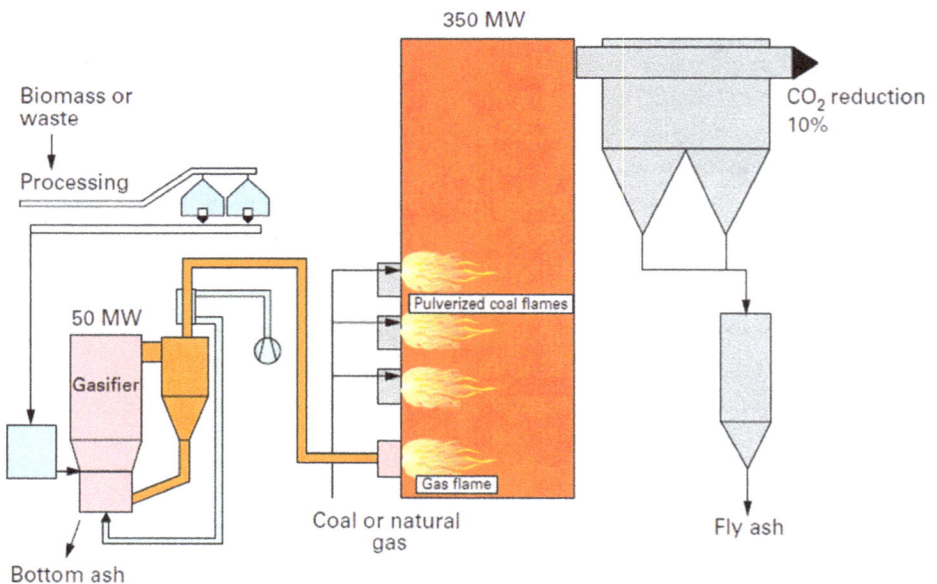

Figure 4.33: Illustration of Foster Wheeler gasifier in Lahti, Finland, used for gasification of waste or biomass. The gas from the gasifier is then burned in a boiler, which can process coal or natural gas.

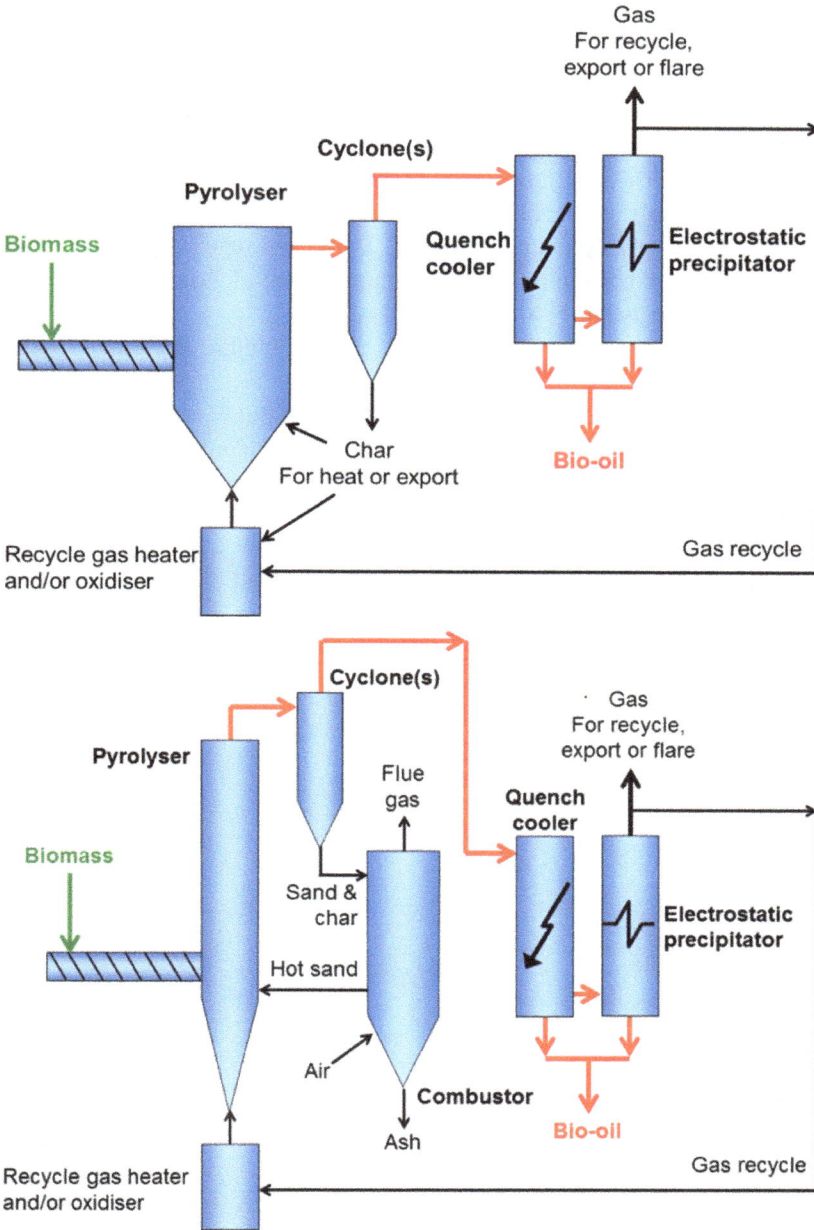

Figure 4.34: Biomass pyrolysis: (a) bubbling fluid bed reactor with electrostatic precipitator and (b) circulating fluidized bed. From https://task34.ieabioenergy.com/pyrolysis-reactors/.

particles, short residence time (typically under 2 s) of the pyrolysis vapors, and fast quenching of the pyrolysis vapors to condense the bio-oil.

Similar to gasification, there are several demonstration and pilot-size units in operation (Figure 4.35). An example of the pyrolysis plant integrated with the combined heat and power production plant delivering 50,000 tons of bio-oil per year is illustrated in Figure 4.35.

The heat needed for the pyrolysis is provided by the fluidized-bed boiler by feeding hot sand to the pyrolysis unit at 800 °C. Cold sand together with the char (at ca. 500 °C) is recycled back to the boiler. The uncondensed gases are also fed to the boiler and then combusted.

The integrated concept is easy and smooth to operate and has lower investment costs when integrating into existing boilers.

Bio-oil is a water-containing, rather unstable, brown liquid with several hundred different highly oxygenated organic molecules with acidic pH (ca. 2.5) due to presence of organic acids. The heating value is mediocre, being 16–19 MJ/kg.

Depolymerization of biomass does not require such harsh conditions (high temperatures) as pyrolysis or gasification. The industrial pulping of wood is performed at 150 °C to 180 °C as discussed above. The acid hydrolysis of biomass for production of sugars is described in Chapter 10. An alternative to hydrolysis with acids or

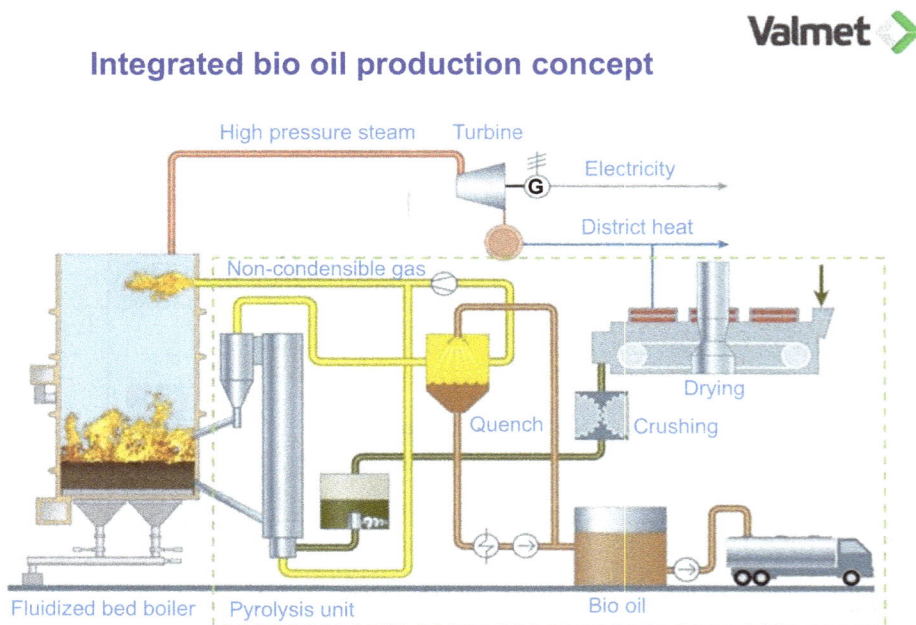

Figure 4.35: Integrated bio-oil production concept. From http://pyrowiki.pyroknown.eu/images/f/f1/Valmet2.jpg.

bases is enzymatic hydrolysis, which requires utilization of highly specific cellulases. Enzymatic hydrolysis of wood is a heterogeneous process depending on such properties as feedstock crystallinity and the degree of polymerization. The rates are typically low, and hydrolysis is accompanied often with by-product degradation.

Production of high-value products from biomass relies on efficient transformations of sugars. Hydroxyacids, sugars, and polyols obtained from biomass (and their numerous derivatives) can serve as platform chemicals. These compounds include levulinic, succinic, fumaric, lactic, and 3-hydroxypropionic acids, glycerol, sorbitol, and xylitol. A few commercial processes produce chemicals such as lactic, levulinic, and succinic acids from sugars through chemical or fermentation routes. The Biofine process for manufacturing levulinic acid from cellulose relies on the chemical route and consists of two distinct acid-catalyzed stages (Figure 4.36).

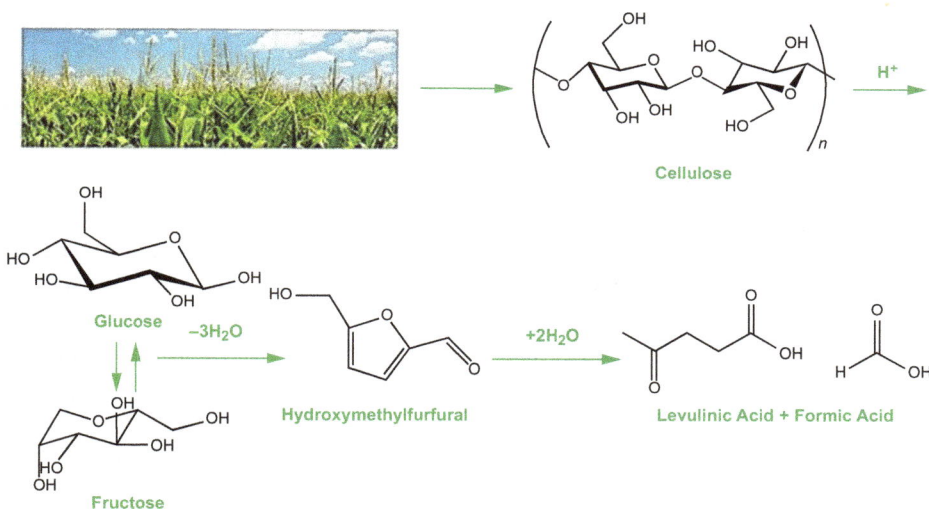

Figure 4.36: The Biofine process. From F. D. Pileidis, M.-M. Titirici, levulinic acid biorefineries: New challenges for efficient utilization of biomass, ChemSusChem, 2016, 9, 562–582, Copyright © 2016 WILEY-VCH Verlag GmbH & Co. KGaA, Weinheim. Reproduced with permission.

The first stage is performed in a plug flow reactor at 210–220 °C, 25 bar, and a short residence time of ca. 12 s. Carbohydrate polysaccharides are hydrolyzed in the presence of an acid catalyst to soluble intermediates (e.g. HMF). The second reactor also operates in the presence of an acid catalyst, and levulinic and formic acids are mainly formed at 190–200 °C, 14 bar, and a residence time of ca. 20 min. A commercial plant processing 50 dry tons of feedstock (paper sludge, agricultural residue, and waste paper) per day has been constructed in Caserta, Italy.

Potential transformation of levulinic acid to various chemicals is listed in Figure 4.37.

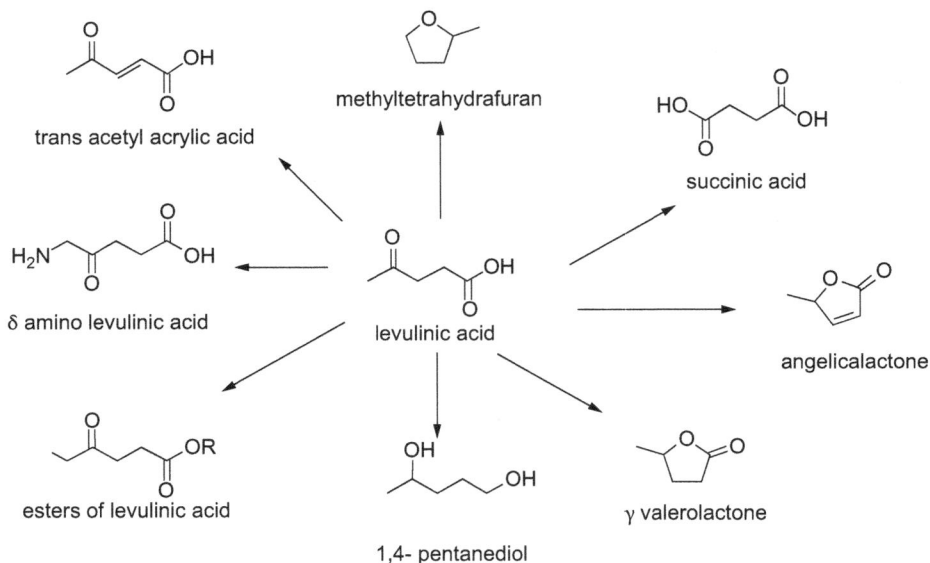

Figure 4.37: Transformations of levulinic acid.

Some potential transformation of other acids which can be obtained from cellulosic biomass, namely succinic and lactic acid are shown in Figures 4.38 and 4.39.

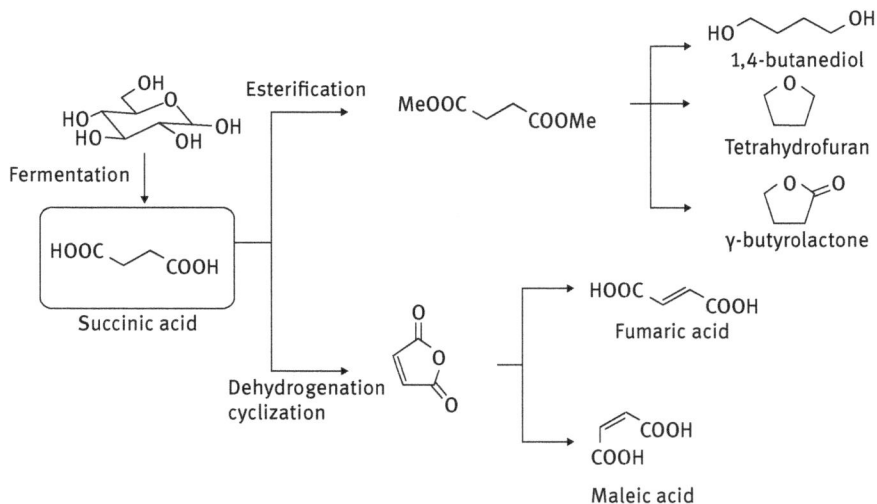

Figure 4.38: Catalytic transformations of succinic acid. http://bioweb.sungrant.org/TechnicalBio products/Bioproducts + from + Carbohydrates/Organic + Acids/Organic + Acids.htm.

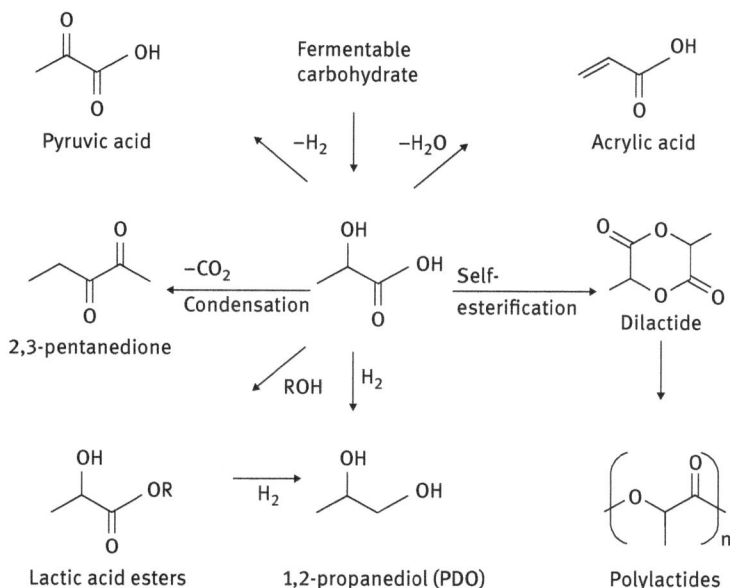

Figure 4.39: Lactic acid and derivatives.

Among the key compounds obtained from biomass, lactic (2-hydroxypropionic) acid should be emphasized, which has a great potential for use in the production of chemicals, including monomers. Esterification of lactic acid results in corresponding esters, which can be used as so-called "green" solvents. Dehydration of lactic acid to acrylic acid opens many possibilities to synthesize polymers. Unfortunately, the yields in the dehydration step are low; moreover, acidic catalysts are prone to deactivation. Hydrogenation of lactic acid leads to propylene glycol, which is widely used as a solvent and a valuable intermediate in, e.g., the synthesis of propylene oxide.

A multi-step ring opening polymerization of lactide (Figure 4.40) with various metal catalysts (typically tin octoate) results in a biodegradable polymer – polylactid acid.

The lactic acid is first condensed to form low-molecular-weight prepolymer. The next step is formation of lactide (the cyclic dimer) by controlled depolymerization of the prepolymer. Distillation is done to purify the lactide, which then undergoes ring-opening and controlled catalytic polymerization, giving a product with a high molecular weight.

The properties of the polylactic acid resin obtained through this process can vary depending on the amount of *meso*-lactide (Figure 4.41) formed in the lactide reactor. The highest molecular weight product is obtained from L-lactide.

(a)

(b)

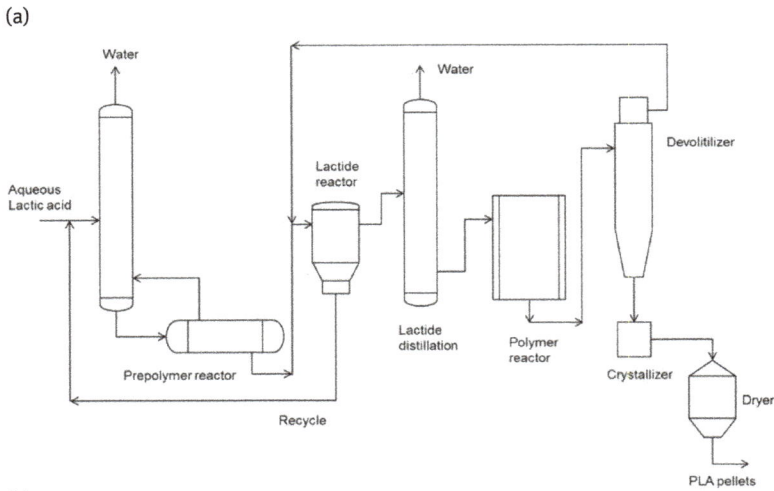

Figure 4.40: Production of high-molecular-weight polylactic acid (a) flow scheme, from https://ars. els-cdn.com/content/image/1-s2.0-S0169409X16300965-gr4_lrg.jpg, and (b) NatureWorks Ingeo plant, from https://www.biobasedpress.eu/wp-content/uploads/2013/07/NatureWorks-Ingeo-Plant.jpg.

Fats and vegetable oils are other feedstock for production of fuels and chemicals, such as surfactants, cleaning agents, and cosmetic products. In vegetable oils, the dominant fatty acids are C18 with different degrees of saturation. Another

D-lactide   L-lactide   *meso*-lactide

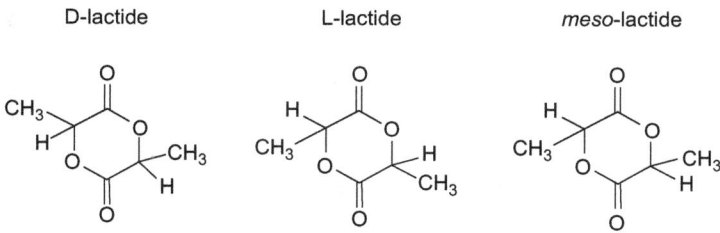

Figure 4.41: Stereoisomers of lactide.

potential feedstock is animal fat, which is produced globally in quantities fivefold lower than vegetable oil. Composition of various vegetable oils is given in Table 4.5.

Table 4.5: Fatty-acid contents of different vegetable oils.

Fatty acid	Jatropha oil	Pongamia (karanja oil)	Sunflower oil	Soybean oil	Palm oil	Rapeseed oil
Lauric acid (C12:0)*	–	–	0.5	–	–	–
Myristic acid (C14:0)	–	–	0.2	0.1	–	0.1
Palmitic acid (C16:0)	14.2	9.8	4.8	11.0	40.3	4.8
Palmitoleic acid (C16:1)	1.4	–	0.8	0.1	0	0.3
Stearic acid (C18:0)	6.9	6.2	5.7	4.0	3.1	1.9
Oleic acid (C18:1)	43.1	72.2	20.6	23.4	43.4	61.9
Linoleic acid (C18:2)	34.4	11.8	66.2	53.2	13.2	19.8
Linolenic acid (C18:3)	–	–	0.8	7.8	–	9.2
Arachidic acid (C20:0)	–	–	0.4	0.3	–	0.6
Gadoleic acid (C20:1)	–	–	–	–	–	1.4
Behenic acid (C22:0)	–	–	–	0.1	–	–
Saturated	21.1	16.0	11.6	15.5	43.4	7.4
Unsaturated	78.9	84.0	88.4	84.5	56.6	92.6

*– first number – carbon number, second – number of double bonds.

The use of biomass-derived raw materials for the preparation of biofuel (particularly biodiesel) has in fact a long history. The first engine designed by Rudolf Diesel in 1900 operated on vegetable oils, including pure peanut oil. The drawbacks intrinsic to vegetable oils (e.g., high viscosity, low stability) led to widespread utilization of petroleum fuel.

Direct utilization of triglycerides as fuels is limited because of their high density and viscosity. Two major processes are applied to transform triglycerides to fuels – transesterification and (hydro)deoxygenation (Figure 4.42). The first approach leads to the so-called biodiesel (or fatty acids methyl ester), while the second results in the formation of hydrocarbons, referred sometimes to as "green" diesel. Deoxygenation in the absence of hydrogen can result in unsaturated alkenes; thus, hydrogenation of double bonds in triglycerides and fatty acids can also be an important step in the conversion of triglycerides into fuel-range hydrocarbons. Oxygen is removed from classic biodiesel because biodiesel is only partially compatible with conventional diesel motors. An additional disadvantage is the formation of glycerol as a by-product in biodiesel synthesis, whose utilization is currently limited.

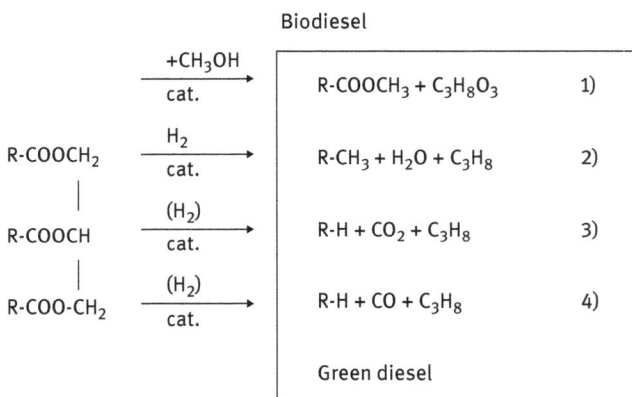

Figure 4.42: Scheme for transformation of triglycerides to biodiesel and green diesel.

Purification (degumming) of triglyceride oils (Figure 4.43) is performed in order to decrease the amount of phospholipids, since phosphorus is known to be a catalyst poison influencing further transformations of triglycerides.

Degumming is conventionally performed with water, and this process removes both phospholipids and mineral impurities. Lecithin is formed as a by-product from the degumming process and it is separated *via* centrifuging. The temperature during the centrifuging should be below 75 °C in order to achieve an efficient removal of phospholipids. A very low temperature decreases oil yield. Alternatively, enzymatic and acid-aided processes can be used. An acid, such as citric acid, acts as a chelating agent for phospholipids in the corresponding degumming process.

Biodiesel (methyl esters of fatty acids) is synthesized *via* transesterification of triglycerides with methanol with mainly homogeneous catalysts. Commercially available biodiesel is produced mostly by base-catalyzed transesterification, when either sodium or potassium hydroxides are utilized, while homogeneous acid catalysts,

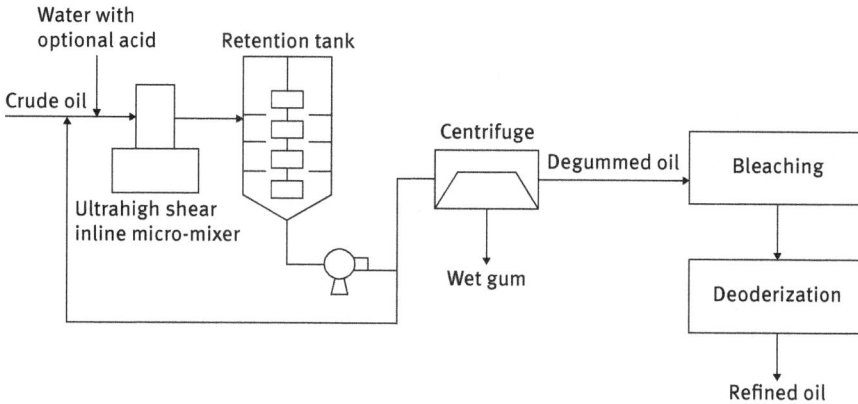

Figure 4.43: A simplified scheme for degumming of triglycerides.

such as sulphuric, phosphoric, and hydrochloric acids are less popular due to significantly slower reaction rates.

Green diesel, contrary to biodiesel, does not contain oxygen and is prepared by deoxygenation, which, broadly speaking, covers hydrodeoxygenation, decarboxylation, and decarbonylation. Hydrodeoxygenation process (Figure 4.44) is already applied on the industrial scale by Neste Oil (NExBTL process). There are several units (one of them is illustrated in Figure 4.45), operating with a combined nominal capacity of around 2 million tons per year.

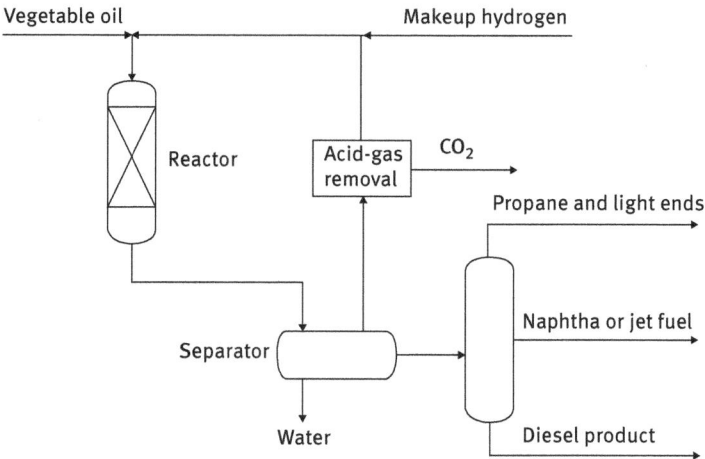

Figure 4.44: A simplified flow scheme for production of green diesel via hydrodeoxygenation.

Figure 4.45: Neste Singapore refinery. https://scandasia.com/wp-content/uploads/2018/12/nes teoil-singapore.jpg.

Besides vegetable oils (palm, soybean, rapeseed, etc.), waste and residues (waste animal fats, used cooking oils, etc.) are also currently used in this hydro-treating process with supposedly sulphided NiMo or CoMo catalysts.

As mentioned above, tall oil is a side product of Kraft pulping process. Crude tall oil components are separated *via* fractional distillation to two main fractions, tall oil rosin and distilled tall oil. The content of resin and fatty acids in crude tall oil is 85%. Fatty acids being important compounds used for production of various chemicals, such as conjugated fatty acids, can also be used for the manufacturing of renewable diesel by hydrodeoxygenation (Figure 4.46).

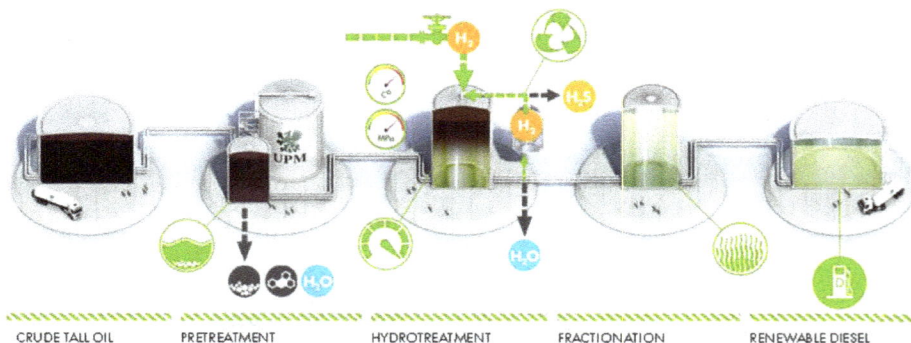

CRUDE TALL OIL   PRETREATMENT   HYDROTREATMENT   FRACTIONATION   RENEWABLE DIESEL

Figure 4.46: A flowchart of renewable diesel manufacturing from tall oil at UPM Lappeenranta plant.

In addition to fatty acids, crude tall oil contains about 10–15% neutral, unsaponifiable compounds, especially sterols such as siosterol. In the industrial production of sitosta-nol fatty acid esters, which are applied as functional food that lowers cholesterol values, hydrogenation of β-sitosterol is carried out over supported palladium catalysts followed by esterification of β-stanol mixture with a fatty acid.

Another important source of biobased fine chemicals is turpentine, with the main compounds being α- and β-pinene, limonene, and camphene. For example, the reaction between β-pinene and formaldehyde gives an unsaturated alcohol, Nopol (Figure 4.47), as a product, which is used as a fragrance and a pesticide.

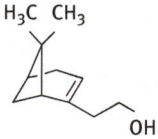

Figure 4.47: Structure of Nopol.

The list of chemicals and fuel compounds mentioned above is far from being exhaustive, and currently, a lot of research globally is devoted to finding ways for efficient transformations of biomass to valuable products.

For example, many very complex natural products of limited structural repertoire are difficult and expensive to synthesize on an industrial scale. Instead, naturally occurring compounds can be used as templates for semisynthetic modifications. Therefore, it is not surprising that natural products and related drugs are used to treat 87% of all categorized human diseases, acting as antibacterial, anticancer, anticoagulant, antiparasitic, or immunosuppressant agents.

For example, betulin, extracted from birch bark, can lead to the synthesis of β-alaninamide of betulonic acid, a compound with a potential to be used in chemotherapy (Figure 4.48).

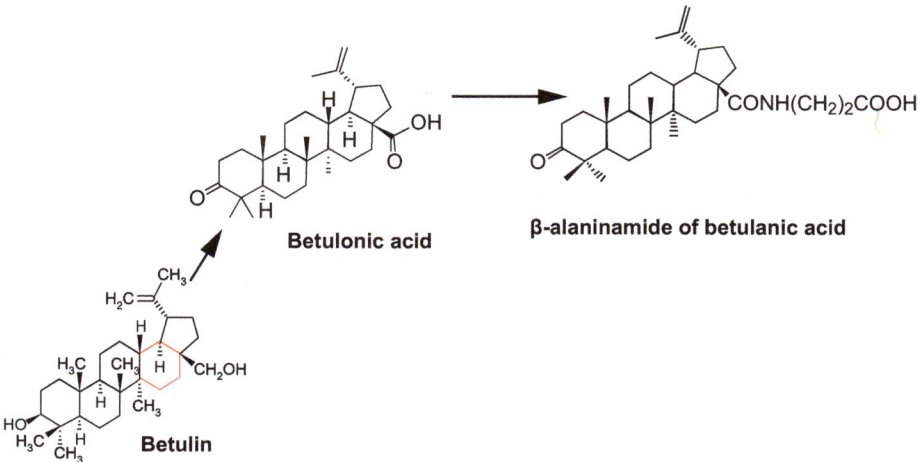

Betulonic acid

β-alaninamide of betulanic acid

Betulin

Figure 4.48: A schematic route from betulin to β-alaninamide of betulonic acid.

Chapter 5
Hydrogen and syngas generation

5.1 Steam reforming of natural gas

Steam reforming of natural gas or methane is the most common method of producing commercial hydrogen as well as hydrogen used in the industrial synthesis of methanol and ammonia (Figure 5.1). Besides natural gas, other hydrocarbons containing streams such as associated gas, liquid petroleum gas, and naphtha boiling up to 220 °C can be fed, for example, to ammonia plants requiring hydrogen, while application of higher hydrocarbons can lead to excessive coke formation on the catalysts.

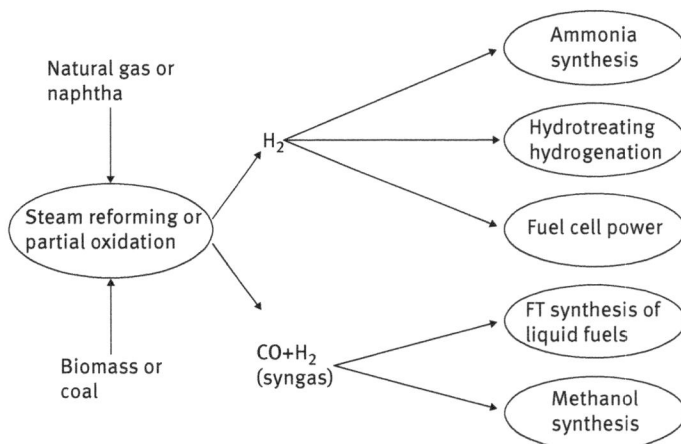

Figure 5.1: Routes to hydrogen and syngas.

Reversible endothermic reaction of steam reforming is typically carried out at high temperatures (700–1,100 °C) over nickel-based catalyst

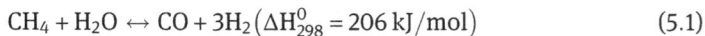

$$CH_4 + H_2O \leftrightarrow CO + 3H_2 \left(\Delta H^0_{298} = 206\,kJ/mol\right) \tag{5.1}$$

yielding CO and hydrogen (syngas).

Although natural gas consists of some other compounds, this reaction is the main one occurring during steam reforming and is accompanied by extensive coke formation, which leads to significant catalyst deactivation by methane decomposition

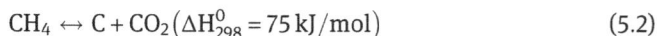

$$CH_4 \leftrightarrow C + CO_2 \left(\Delta H^0_{298} = 75\,kJ/mol\right) \tag{5.2}$$

https://doi.org/10.1515/9783110712551-005

or by Boudouard reaction

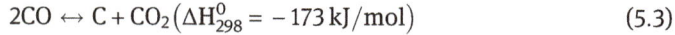

$$2CO \leftrightarrow C + CO_2 \left(\Delta H^0_{298} = -173 \, kJ/mol\right) \tag{5.3}$$

The equilibrium gas composition is given in Figure 5.2. Obviously, hydrogen and carbon monoxide content increases with temperature increase. From the thermodynamic viewpoint, steam reforming does not require high pressures. Generation of syngas is done, however, at high pressures, since downstream applications of syngas in methanol and ammonia synthesis require high pressures. Although application of high pressure is unfavorable for equilibrium and thus requires high temperature and excess of steam, expensive compression of syngas is avoided. The reformer size is also smaller.

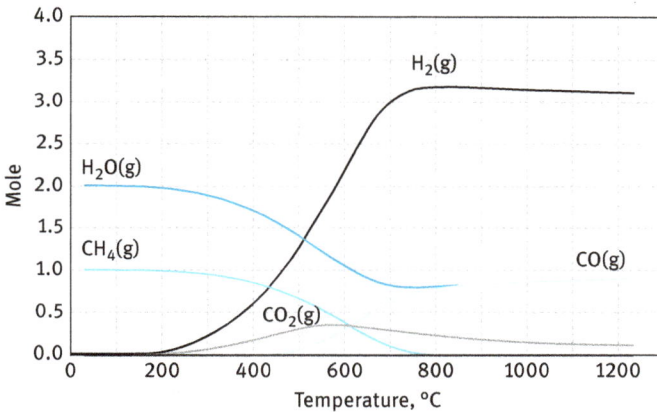

Figure 5.2: Equilibrium gas composition at 0.1 MPa as a function of temperature at steam/methane molar ratio 2:1.

Regarding kinetics for most practically relevant conditions such as industrially applied temperature, steam/carbon ratios, and pressures, first-order kinetics can be used.

At higher temperatures, dry reforming of methane is also becoming important; thus, the concentration of CO_2 passes through a maximum (Figure 5.2):

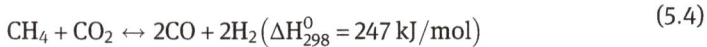

$$CH_4 + CO_2 \leftrightarrow 2CO + 2H_2 \left(\Delta H^0_{298} = 247 \, kJ/mol\right) \tag{5.4}$$

The conversion of methane to CO and hydrogen is favored at high temperature, low pressure, and high steam/methane ratio (or S/C, steam/carbon).

In practice, increase in temperature is limited by the thermal and mechanical stability of tubes, since the process requires utilization of multitubular reactors. In particular, thick walls would worsen heat transfer, while tubes with thin walls might be subjected to severe damage at high temperature, including bending, formation of cracks, etc.

The operating steam/hydrocarbon ratio must be higher than the stoichiometric level to avoid carbon formation on the catalyst by cracking reactions and to provide enough steam to operate the water-gas shift reaction

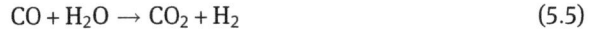

$$CO + H_2O \rightarrow CO_2 + H_2 \tag{5.5}$$

later in the process. Thus, in steam reforming of natural gas, the steam/carbon ratio is typically 3–3.8, more often, between 3.3 and 3.6.

The presence of excess steam in the process gas to the reformer results in the formation of carbon dioxide by the water-gas shift reaction. Thus, the gas leaving the steam reformer also contains between 7% and 15% carbon dioxide.

In ammonia plants, the methane-reforming reaction from the tubular (primary) reformer is continued in the secondary reformer *via* the introduction of air to the reactor (Figure 5.3). The combustion of the air produces temperatures around 1,250 °C, resulting in further reforming of methane.

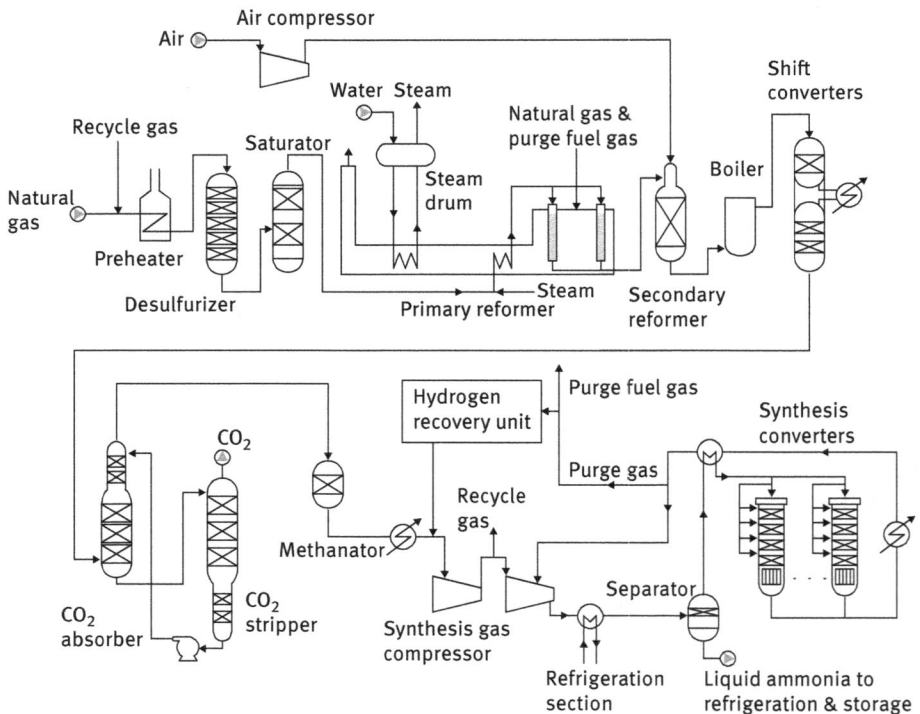

Figure 5.3: Single train ammonia synthesis. http://csd.newcastle.edu.au/chapters/Fig1_2.png.

As mentioned above in ammonia or hydrogen production units, additional hydrogen is obtained by a mildly exothermic stand-alone water-gas shift reaction, which occurs at lower temperature than steam reforming. In the industry, usually two-process steps are applied with iron- and copper-based catalysts called the high- and low-temperature shift reactions, respectively. Thereafter, carbon dioxide is removed by absorption using either KOH or amine (monoethanolamine or activated methyldiethanol amine) scrubbing systems, and traces of residual carbon monoxide are converted to methane over a nickel-based methanation catalyst.

The final step is ammonia synthesis *per se*, which will be discussed in Chapter 10.

Since the hydrocarbon feedstock for the production of hydrogen or synthesis gas contains catalytic poisons, which are detrimental for nickel catalysts, feed purification is performed in order to remove such poisons, namely sulphur- and chlorine-containing compounds. They are removed first by hydrogenation of organic sulphur, nitrogen, or chlorine compounds (e.g., mercaptanes) to hydrogen sulphide using, for example, cobalt/molybdenum hydrodesulphurization catalysts. These reactions are exothermic and require hydrogen, which is generated in steam reforming. The cobalt/molybdate component is sulphided during commissioning and operates thereafter in a similar same way as in refinery hydrotreating. The latter reaction will be described in Chapter 10.

Olefins and aromatics lead to carbon deposits and catalyst deactivation even in regions that are carbon-free according to equilibrium conditions. Therefore, if the presence of these compounds cannot be avoided, a recycling of product hydrogen is often used to hydrogenate the unsaturated hydrocarbons. The hydrogen recycling ratio depends on the amount of unsaturated hydrocarbons expected and will often be ca. 3–5 mol% of the carbon content of the feedstock. Hydrogenation is usually performed simultaneously with the hydrogenation of sulphur-containing compounds, and the same type of catalyst (a sulphided cobalt or molybdenum catalyst) can be employed.

The spent cobalt/molybdate catalyst at the end of the lifetime in a typical ammonia plant can contain 0.5–3.0 wt% sulphur and 2–10 wt% carbon depending on operation severity. As adsorbed hydrogen and carbon deposits could be present in hydrodesulphurization catalysts, making them pyrophoric, after use, catalyst discharge is organized in a way that, prior to it, the reactor must be flushed with an inert gas until the catalyst temperature has decreased to ca. 25–30 °C.

Due to very low levels of sulphur compounds in the feedstock, a temperature rise in the fixed-bed reactors used for hydrodesulphurization is seldom observed. This step is followed by scrubbing hydrogen sulphide with ZnO:

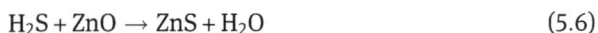

$$H_2S + ZnO \rightarrow ZnS + H_2O \qquad (5.6)$$

This reaction is not a catalytic one, and the front of produced ZnS is slowly moving along the reactor length. An example of a hydrodesulphurization/steam reforming unit in a plant producing 1,360 t of ammonia per day is given in Tables 5.1 and 5.2.

Table 5.1: Conditions in hydrodesulphurization/steam reforming in an ammonia plant producing 1,360 t of ammonia per day (GIAP AM76 design).

Parameter	Hydrodesulphurization	ZnO	Primary reformer
Catalyst volume (m^3)	24	60	22.4
Bed height (m^3)	5	6.8	10.8
Diameter (m)	3.2	3.2	0.072 (internal) 0.114 (external)
Pressure (MPa)	3.5	3.4	3.1
T inlet (°C)	350	350	540
T outlet (°C)			780
CH_4, out (vol%)			8.6
Number of tubes			504
Lifetime (years)	5–25		4–8

Table 5.2: Conditions in hydrodesulphurization/steam reforming in an ammonia plant producing 1,360 t of ammonia per day (GIAP AM76 design).

Flow	Amount	T (°C)
Hydrocarbons (in vol%): C1, 98.7; C2, 0.3; C3, 0.03; C4, 0.04; N$_2$, 1; CO$_2$, 0.02; H$_2$S, < 10 ppm	41, 000 m^3/h	471
H$_2$ recycling (in vol%): H$_2$, 61.5; N$_2$, 20.3; CH$_4$, 0.2; Ar, 0.2	1,500 m^3/h	45
Steam to reformer	6,800 kg/h	367

The novel development in the area of hydrodesulphurization was done in 2010 by Johnson Matthey in collaboration with Prof. Avelino Corma (Valencia). The new catalyst provides three-in-one functionality combining full HDS conversion, H$_2$S absorption, and ultrapurification. For a new plant design with an HDS vessel upstream of a ZnO-based H$_2$S removal system, the HDS vessel can be completely removed from the design.

In steam reforming of natural gas for ammonia production, the conditions for most of the plants are designed to afford certain methane content in the outlet of the primary reformer to be adequate for the downstream secondary reforming where the ratio between hydrogen and nitrogen of 3 mol/mol should be achieved.

Some examples of operation conditions for steam reforming in ammonia synthesis plant are given in Table 5.3.

Table 5.3: Plant data for steam reforming in ammonia synthesis of 1,000 metric tons per day with 19 m^3 of steam-reforming catalyst.

Lifetime months	9	15	24
Feed (N m^3/h)	29, 850	33, 198	29, 846
S/C (mol/mol)	3.45	3.43	3.41
Tube inlet (°C)	489	470	478
Tube outlet (°C)	798	799	796
Pressure inlet (MPa)	3.25	3.27	3.29
Pressure drop (MPa)	0.24	0.23	0.28
CH_4 leakage (vol%)	9.6	9.3	9.1

Inlet methane, 85–88%; ethane, 5–6%.

The amounts of CO and CO_2 for such conditions, as mentioned in Table 5.3, would be almost in the same range as methane (9–11%), with somewhat higher concentration of CO_2 (11% CO_2 *versus* 9% CO).

Increased heat duty (outlet temperature) in a primary reformer leads to lower methane leakage; however, it increases the risk of excessive temperature rise in the combustion zone of a secondary reformer.

Proper choice of the steam/carbon (S/C) ratio in steam reforming is important, and conventional reformers for ammonia synthesis production were designed with this ratio between 3.3 and 3.6 mol/mol. Economic evaluations indicate that feedstock consumption is slightly less at low S/C ratios and investment costs also decline with smaller gas flows; therefore, the optimum S/C ratio should be low. At the same time, carbon formation is avoided if high S/C ratios are chosen because the equilibrium of the feed gas is shifted from the region of carbon formation. When the aim is not hydrogen, but synthesis gas, the S/C ratio during steam reforming also influences the ratio between hydrogen and CO in syngas. Otherwise, a high S/C ratio favors hydrogen production.

Therefore, the optimum conditions for hydrogen production including steam/carbon ratio could be different from ammonia synthesis plants. They are dictated by, for example, such considerations as the feedstock. Hydrogen plants can operate based on natural gas or light hydrocarbons through naphtha. As in refineries, hydrogen production is integrated with other refinery units, and feedstock to steam reformer could contain a mixture of off-gases from catalytic reformer, catalytic cracker, and other units.

In a steam reformer (Figure 5.4), the process gas passes up through vertical reactor tubes with supported nickel catalyst, typically 10 m long and 10 cm in diameter. The long catalyst tubes are aligned vertically in rows within the furnace. Heat is supplied to the tubes by burners. For effective operation, heat must be transferred rapidly from the furnace itself to the surface of the catalyst, particularly at the top of the tubes.

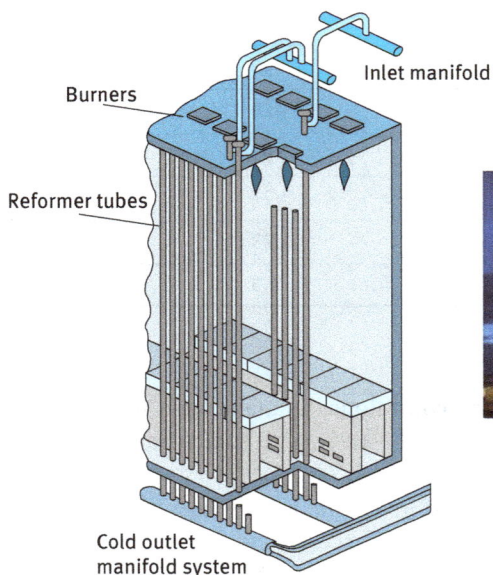

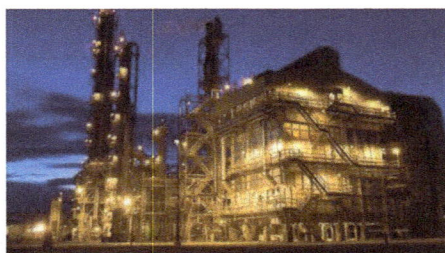

Figure 5.4: View of a steam reformer: (a) a schematic view from http://www.thyssenkrupp-industrial-solutions.com/products-solutions/chemical-industry/hydrogen/steam-reforming/process/steam-reformer-box.html; (b) a photo from http://3.bp.blogspot.com/_jwzEb3tcs7U/TFKw_9xfJ8I/AAAAAAAAAGY/C1UMo5yhAoE/s1600/pr.jpg.

Heat distribution in conventional reformers is influenced by the geometry of burner arrangements; the pitch of the tubes; the type and length of the flame; the radiation of the flames, flue gases, and refractory walls; size, shape, and thickness of the tubes; materials; and the way the tube surfaces are processed.

In the steam reformer, the hydrocarbon should react rapidly with the steam, helping to avoid cracking and subsequent catalyst deactivation. For inactive catalyst, overheating of the reactor tubes can occur. One of the means to avoid overheating, besides introducing an active catalyst, is also shaping the catalyst in a way that allows rapid gas mixing, efficient heat transfer from the wall of the tube to the catalyst, and uniform temperature gradient throughout the tube. Careful loading of the catalyst is also very important to achieve uniform gas flow and pressure drop through each tube. When these conditions are not reached, flow maldistribution occurs, leading to variable tube temperatures and hot spots.

A penalty of operating a reformer with hot tubes, for any reason, is that a temperature only 10 °C above the design level can reduce the tube life by up to 50%. The cost of tube failures resulting from the use of low-activity catalysts or decreasing the steam ratio is very high.

Commercial tubular reformers operate at space velocities of 2,000–8,000 h^{-1} (standard temperature and pressure), which are dictated by heat transfer and to lesser extent mass transfer. High flow rates and a need to minimize pressure drop require application of rather large catalyst pellets. Therefore, a variety of shapes (rings or spheres with multiple holes, wagon wheels, clover leafs, etc.) have been proposed to minimize pressure drop on the one hand and improve catalyst effectiveness factor on the other, which is rather low (0.03–0.1).

Catalyst performance in primary reforming reactors is evaluated by several criteria, such as methane leakage (approach to equilibrium), pressure drop, tube wall temperature (hot spots), resistance against failures, and catalyst lifetime.

The tube wall temperature is influenced by the catalyst *per se* (Ni content, surface area, shape, resistance to poisoning by S, Cl, As, degree of reduction), and process conditions (steam/carbon ratio, average heat flux, content of higher hydrocarbons, hydrogen content, pressure, inlet/outlet *T*, space velocity, and operating failures).

A central problem in steam reforming is to balance the heat input through the tube with the heat consumption by the endothermic reforming reaction, while at the same time limit the stress on the tubes by minimizing the maximum tube wall temperature and the maximum tube wall temperature difference.

While the process gas is fed downward inside the tubes, the flue gas can go downward in top-fired reformer furnaces or upward in side wall-, terrace wall-, or bottom-fired furnaces. There are four types of burner configurations used in tubular reforming as shown in Figure 5.5.

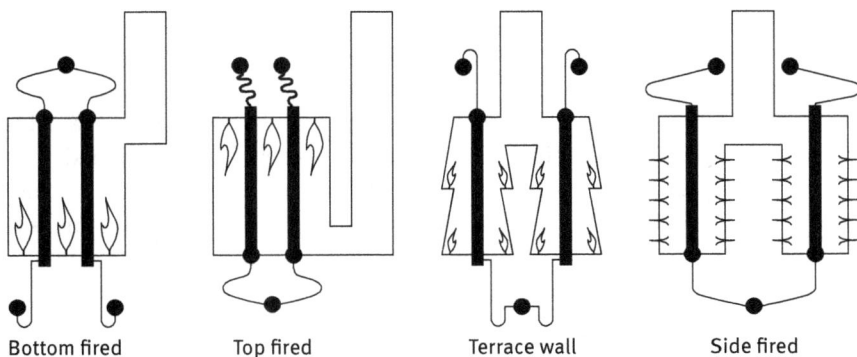

| Bottom fired | Top fired | Terrace wall | Side fired |

Figure 5.5: Typical configurations of reformer furnaces.

The tube wall temperature is the highest on bottom-fired furnaces compared to other types of furnaces for the same process gas outlet temperature.

Top-fired reformers can have several parallel rows of tubes, while wall- or side-fired reformers only can have one row. The homogeneity of the heat transfer to the tubes is determined by burner geometry, flame length and diameter, tube-to-tube and row-to-row spacing, fired tube length, and distance from the flame to the reformer wall. The temperature of the flame is approximately 1,800 °C at the hottest place and 1,100 °C at the coldest place. Heat transfer in the wall-fired reformers is mainly by the radiant side wall, while in top-fired reformers, the heat is transferred through radiation from the flame and hot flue gases.

The top-fired reformer is characterized by a temperature peak in the top, and it has the highest heat flux where the metal temperature is at its maximum. The side-fired reformer allows a better temperature control, and the maximum temperature is at the outlet of the tube. The highest heat flux is at a rather low temperature. The side-fired reformer has a higher average heat flux than the top fired. Moreover, the short residence time in the flames in the side-fired reformer ensures very low emissions of NO_x in the flue gases.

The catalyst in steam reforming of hydrocarbons generally operates at severe conditions; thus, high thermal stability is important. In addition, since deactivation due to formation of carbon deposits and coke could be detrimental for the catalyst performance, the catalytic phase, promotes or supports should be carefully selected.

Carbon formation is favored at low steam/carbon ratio and acidic catalyst support, while presence of hydrogen decreases generation of carbon.

There are peculiar features in carbon or coke formation during steam reforming of methane, making it different from coke in many other reactions related to hydrocarbon transformations. In particular, carbon forms needles, growing as filaments with the active catalytic metal (nickel) at the top. This can lead to structural damages of the catalysts and should be avoided, as it leads to catalyst replacement.

Although noble metals display higher intrinsic activity than nickel and are more resistant to coke, in industrial formulations, nickel is applied due to availability and costs. The choice of support is also crucial, since the proper carrier material should ensure long-lasting activity and mechanical stability. The surface area of supports is rather low (5–80 m^2/g). Among thermally stable supports with low acidity, α-alumina, magnesium aluminate, MgO, and CaO should be mentioned. The last one, although possessing the highest thermal stability, has relatively low mechanical strength; therefore, it is used as calcium aluminate. Calcium aluminate is very hard and has high initial crushing strength. In the case of high pressure of carbon oxides, calcium can, however, react with them, somewhat diminishing the crush strength. The silica content should be kept minimal (<0.2 wt%), as it is volatile under normal operation conditions.

MgO should be used with a precaution, since it can hydrate at below 425 °C, giving magnesium hydroxide, dramatically weakening the catalyst. Reformer startups and shutdowns with catalysts including MgO in the formulations should be thus done in a dry atmosphere (without steam) when temperature is below the critical hydration temperatures.

Potassium compounds are sometimes used as promoters, as they are effective in lowering the acidity and allowing gasification of carbon with steam. The drawback of utilizing potassium is its volatility at high temperature. For example, one commercial supplier offers for primary reforming of naphtha the catalyst containing 25% NiO and 8.5% K_2O on calcium aluminate, while for steam reforming of natural gas, the recommended catalyst is 10-hole ring (19 × 16 mm) containing 14% NiO on $CaAl_{12}O_{19}$ without potassium. For a mixed feed (natural gas/LPG), the catalyst formulation from this supplier is 18% NiO and 1.6% K_2O on $CaK_2Al_{22}O_{34}$ in the form of 10-hole ring (19 × 12 mm). Other companies manufacture catalysts containing nickel oxide on α-alumina, calcium aluminate, magnesium aluminate.

In the old formulations for methane steam reforming, the amount of nickel could be high, reaching 25% with rather homogenous nickel distribution along catalyst grains. Nowadays, nickel profiling is applied. For an average nickel concentration ca. 15%, it is higher at the exterior of the catalyst and is diminishing along the grain depth.

Catalyst shape is extremely important for steam-reforming catalysts, since it influences activity and the heat transfer coefficient. The shape should be designed to maximize the heat transfer rate, and subsequently the reaction rate, and minimize the tube wall temperature. An optimum shape also decreases pressure drop by increasing the voidage and the size. In addition, with increased geometric surface area, the activity could also be increased by decreasing the tube wall temperature.

Although utilization of Raschig ring-shaped catalysts with different sizes was a common practice during several decades, modern steam reformers with increased heat transfer duty require more complex shapes (some of them are presented in Figure 5.6). Such complexity is created using multiple holes or adding flutes to the outside structure.

(a) (b)

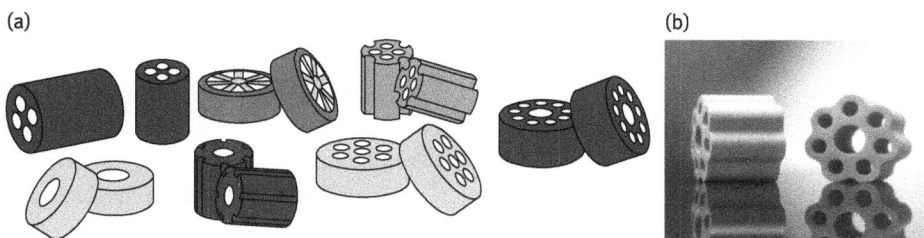

Figure 5.6: (a) Complex shapes for steam-reforming catalysts (from http://chemengservices.com/sitebuilder/images/Proc_Tech-4-02_P1_graphic2-736x249.png. (b) A recent development of a steam reforming catalyst of a floral shape with eight holes (from https://www.clariant.com/en/Corporate/News/2020/04/Clariants-new-Reformax-330-LDP-Plus-catalyst-increases-energy-and-production-efficiency).

It should be noted, however, that each hole diminishes mechanical strength, which decreases with an increase in hole sizes. At the same time, geometric surface area is improved, pressure drop is lowered affording also better heat transfer from the tube wall to the reacting gases and thus lower tube wall temperature (Figure 5.7).

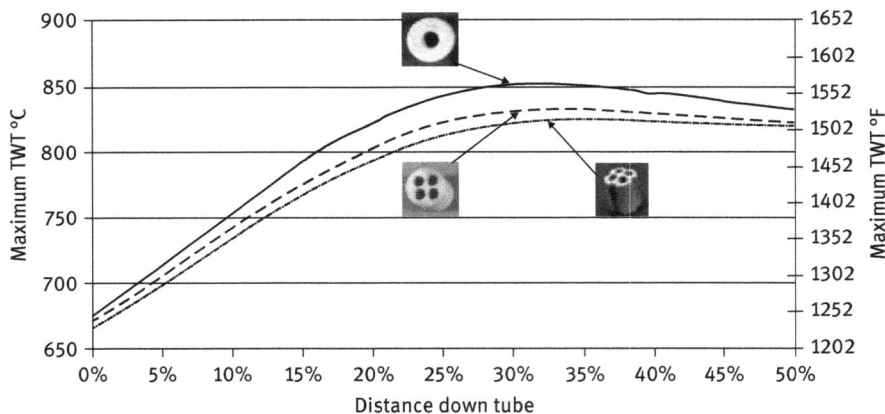

Figure 5.7: Maximum wall temperature for different shapes. From F. Beyer, J. Brightling, P. Farnell, C. Foster, Steam reforming – 50 years of development and challenges for the next 50 years, *AIChE 50th Annual Safety in Ammonia Plants and Related Facilities Symposium*, Toronto, 2005, Copyright © Uhde GmbH 2005.

Careful loading is important in every steam reformer tube in order to have even distribution and avoid side reactions, such as carbon formation. There are several ways of loading the catalyst including utilization of socks filled with the catalyst (Figure 5.8a) which are then moved upward, special equipment for dense loading with brushes attached to a rope (Figure 5.8b) or the SpiraLoad™ loading equipment (Figure 5.8c). The latter consists of a number of equally sized tube sections with spiral-shaped guide elements placed along the inner walls. The tube sections can be assembled and inserted in the reformer tubes to form one long loading tube.

For sock loading after filling of each sock, vibration of the tube could be done to ensure uniform packing and minimization of the shrinkage during operation. Thereafter, pressure drop is measured for all tubes. Such pressure drop measurements could be done after half filling and complete filling of the tube. Tubes displaying a pressure drop outside ±5% from the average should be refilled or additionally vibrated if the pressure drop is too low.

Recent developments in steam reforming include introduction of the structured reforming catalysts in the form of thin metal foils shaped into modules (Figure 5.9).

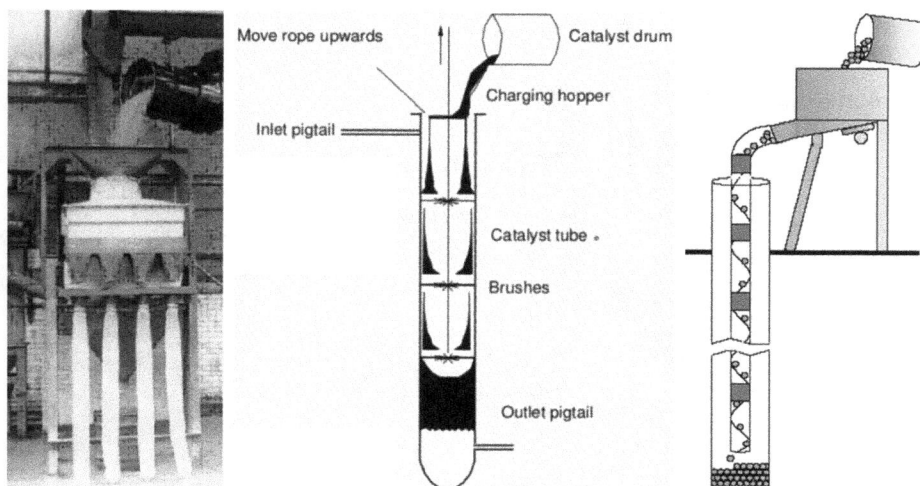

Figure 5.8: Equipment for loading tubes for steam reforming using (a) socks filled with the catalyst, (b) a rope with brushes (from https://www.slideshare.net/GerardBHawkins/steam-reforming-catalyst-loading), and (c) SpiralLoad™.

Figure 5.9: Structured reforming catalysts (from https://matthey.com/-/media/files/products/chemical-processes/catacel-ssr-tailored-catalyst-technology-web-9-26-2019-updated.pdf).

The fans coated with a promoted nickel-based steam reforming catalyst are stacked one upon the other and can be pushed into tubes being rather flexible. The outer edges of the fans are located close to the tube's internal surface without touching it. Such structures ensure a controlled flow pattern, reproducibility of the behavior in different tubes, as well as higher voidage and lower pressure drop. The fan height and different density of folds can be adjusted depending on the specificity of a reformer. Moreover, such catalyst allows a decrease in tube wall thickness or the number of tubes.

Primary reforming is followed by secondary reforming (Figure 5.10), where steam reforming is combined with catalytic partial oxidation of part of the feedstock.

This reaction is needed to obtain high equilibrium temperatures (950–1,100 °C) minimizing methane concentration, which is not feasible in fired steam reformer tubes. Moreover, in the case of ammonia synthesis, this is a smart way to introduce

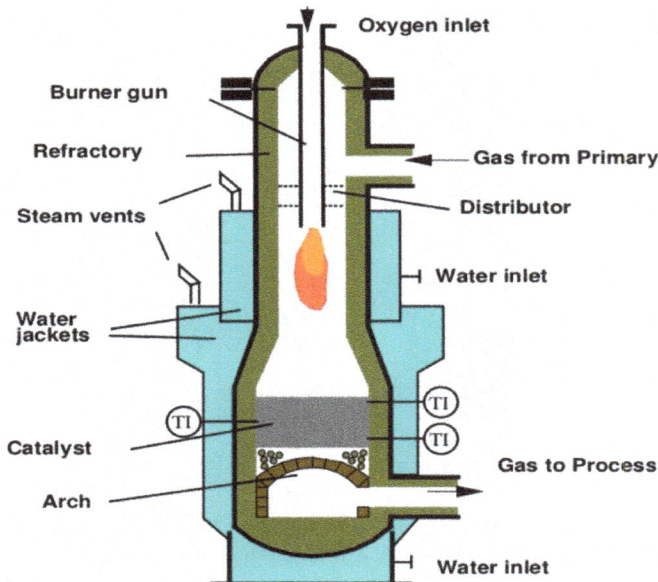

Figure 5.10: Secondary reformer (from https://www.slideshare.net/GerardBHawkins/secondary-reforming-burners).

nitrogen, which reacts downstream with hydrogen. Thus, air or turbine off-gases are supplied to secondary reformer in the ammonia synthesis plants. The reactions in autothermal reforming can be described by the following equations with steam

$$4CH_4 + O_2 + 2H_2O \rightarrow 10H_2 + 4CO \tag{5.7}$$

and CO_2

$$2CH_4 + O_2 + CO_2 \rightarrow 3H_2 + 3CO + H_2O \tag{5.8}$$

A highly exothermic reaction of hydrogen and oxygen also occurs giving water.

A critical parameter for satisfactory secondary reformer performance is efficient mixing of the process gas and air or oxygen.

Uneven mixing can result in large temperature variations above and into the catalyst bed, causing variations in the degree of methane reforming achieved and often yielding a poor overall approach to reforming equilibrium, even with a highly active secondary reforming catalyst. The efficiency of gas mixing is primarily a function of the burner design. In addition to causing inefficient gas mixing, a poorly designed burner can damage the vessel walls, refractory or even the burner itself (Figure 5.11), when there is impingement of hot gas and/or flame in these areas.

Figure 5.11: Furner after operation (https://www.slideshare.net/GerardBHawkins/secondary-reforming-burners).

Nickel catalysts with larger dimensions than in primary reformers are used in the catalyst bed, which is often divided into two parts. The upper part (ca. 15–20% of the bed length) contains nickel catalyst with lower nickel loading (ca. 7%) and larger diameter (in the case of rings, the diameter, height, and hole are 22, 22, and 12 mm, respectively). In the lower part, ca. 20 wt% Ni on a thermostable support is introduced, with diameter, height, and hole dimensions being 16, 16, and 6 mm, respectively. An alternative to this is the application of a nickel catalyst on a magnesia alumina spinel carrier of a ceramic type with a fusion point above 2,000 °C. Such temperature is well above the highest temperatures typically observed in secondary reformers; therefore, there is no need for a special heat shield material on top of the catalyst bed. Such catalyst can be loaded in the entire reactor, topped off by target bricks or alumina lumps. The latter are serving as a hold-down material to prevent agitation of the catalyst by the incoming process gas streams (Figure 5.12).

Figure 5.12: A protective bed (from https://www.slideshare.net/GerardBHawkins/secondary-reforming-burners).

More recently, not only catalysts in the form of Raschig rings but also other shapes such as rings with seven holes or more sophisticated shapes with four or five holes (Figure 5.13) started to be used commercially.

Figure 5.13: Shaped catalysts for secondary reforming.

The residual methane content after secondary reforming is 0.2–0.3% on the dry gas base. Removal of heat is done using a waste heat boiler (Figure 5.3), and the gas is cooled to ca. 330–380 °C before entering shift converters.

Autothermal reforming (Figure 5.14) combining endothermic steam reforming with exothermic partial oxidation can be applied for ammonia, methanol, and hydrogen plants as well as for gas-to-liquid plants as will be discussed in Chapter 13. Such plants can operate at an S/C ratio as low as 0.6 which along with high reforming temperatures enable larger single-train capacities. The generated syngas has low H_2/CO and high CO/CO_2 ratios. For methanol plants, higher CO/CO_2 ratios lead to lower methanol synthesis reactor and thus lower capital costs. In ammonia plants, substantially lower steam throughput diminishes the equipment and piping size.

With an autothermal reformer, tubular reforming in general is not necessary as the pre-reformed natural gas is sent directly to the autothermal reformer along with oxygen.

The reactor for autothermal reforming is presented in Figure 5.15.

In the case of ammonia synthesis, conventionally nitrogen is introduced to the secondary reformer, while with autothermal reforming nitrogen is directly introduced to the ammonia synthesis reactor (Figure 5.16).

The technology illustrated in Figure 5.16 includes besides a high-temperature shift reactor another one operating at medium to high temperature contrary to high and lower temperature water-gas shift reactors in a conventional technology (Figure 5.3). Moreover, instead of methanation, a nitrogen wash to remove CO is used. Another feature of the technology shown in Figure 5.16 is a purge-free synthesis loop.

Less radical way of technology modification is incorporating dry reforming (i.e. reaction of methane with carbon dioxide) to methane steam reforming. Such dry

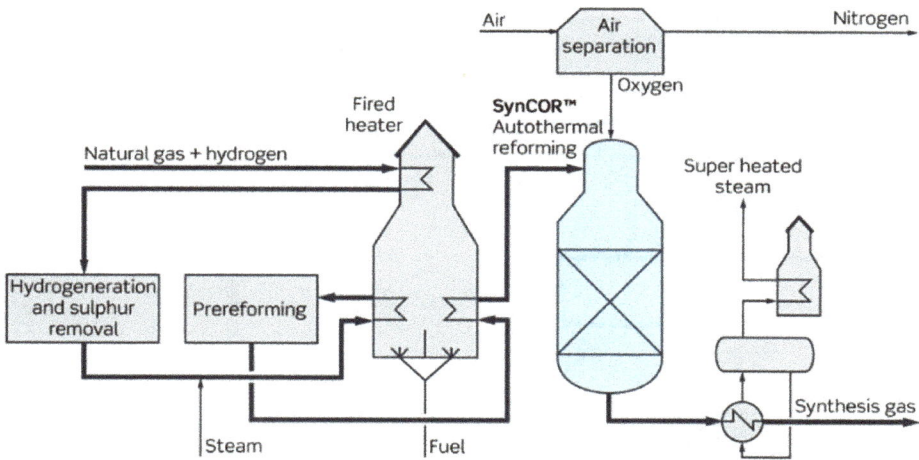

Figure 5.14: Autothermal reforming (from https://www.topsoe.com/products/equipment/syncortm-autothermal-reformer-atr).

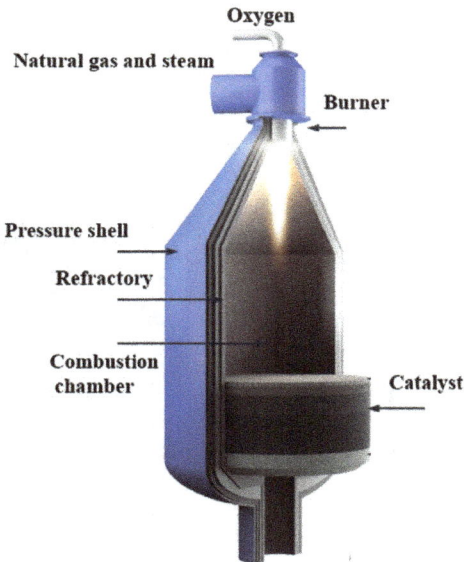

Figure 5.15: The reactor for autothermal reforming.

reforming will not lead to a substantial decrease of greenhouse gas emissions but can be rather used to process hydrocarbon feeds with high CO_2 content or inexpensive CO_2 waste streams.

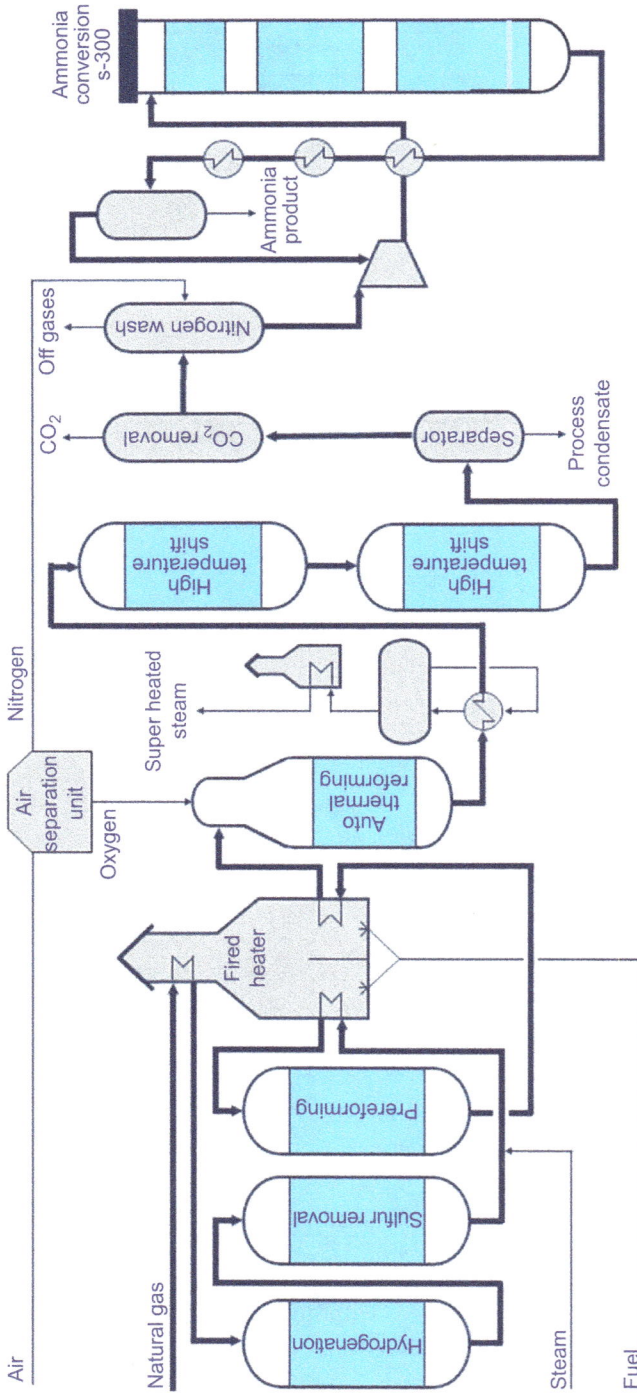

Figure 5.16: Simplified process sheet of Topsøe's SynCOR Ammonia™ plant (from https://www.cewindia.com/merethe_features.html).

Similar to steam reforming, dry reforming is an endothermal reaction requiring high temperatures to achieve high conversions. Dry reforming results in a relatively low hydrogen to CO ratio close to unity or below, which is suitable for production of higher alcohols or acetic acid.

From the viewpoint of thermodynamics, low pressures are preferred; however, because compression of synthesis gas downstream the reformer is more expensive than compression of the feed, the reformers using CO_2-rich natural gas operate at elevated pressures (Table 5.4).

Table 5.4: Industrial plant data for reforming of CO_2-rich natural gas with Ni supported on magnesium aluminate.

Feed	Pressure (barg)	T_{exit} (°C)	H_2O/CH_4 inlet	CH_4, inlet (dry, mol%)	CO_2, inlet (dry, mol%)	H_2, inlet (dry, mol%)	H_2O/CO outlet
NG	20	960	1.3	70	12	17	3.0
NG	12	960	2.1	74	23	2	2.7
NG	25	950	1.5	63	20	16	2.7
NG	15	815	2.0	28	71	1	1.2
NP	25	900	1.7	47	38	15	2.1
NP	22	920	1.7	40	41	17	1.9
NP	20	900	2.0	32	56	12	1.7

NG, natural gas; NP, naphtha.
Adapted from http://dx.doi.org/10.1016/j.apcata.2015.02.022.

All plants were in operation several years without problems related to carbon formation. As a measure of catalyst deactivation, an approach to equilibrium is used. The latter is defined as the difference between the temperature for which equilibrium is reached for the actual gas composition and the actual temperature at the reactor outlet. A value of ca. 20 °C is a good indicator of catalyst stability, while larger values indicate lower activity and thus catalyst deactivation. The plants utilizing naphtha-based reforming (Table 5.4) have a significant slip of CO_2 and thus operate with a CO_2 recycling system (Figure 5.17) comprising, for example, pressure swing adsorption or chemical absorption.

One of the big challenges for dry reforming *per se* is carbon formation under the conditions when the stoichiometric ratio between methane and carbon dioxide is used. Subsequently, high-severity CO_2 reforming has been commercialized rather than the pure dry methane reforming. The catalysts for such reforming are similar to those for steam reforming. Besides nickel on magnesium aluminate mentioned in Table 5.4, also other nickel-based catalysts, which can operate under a low steam to carbon ratio (<1.8) with low carbon formation, are commercially available.

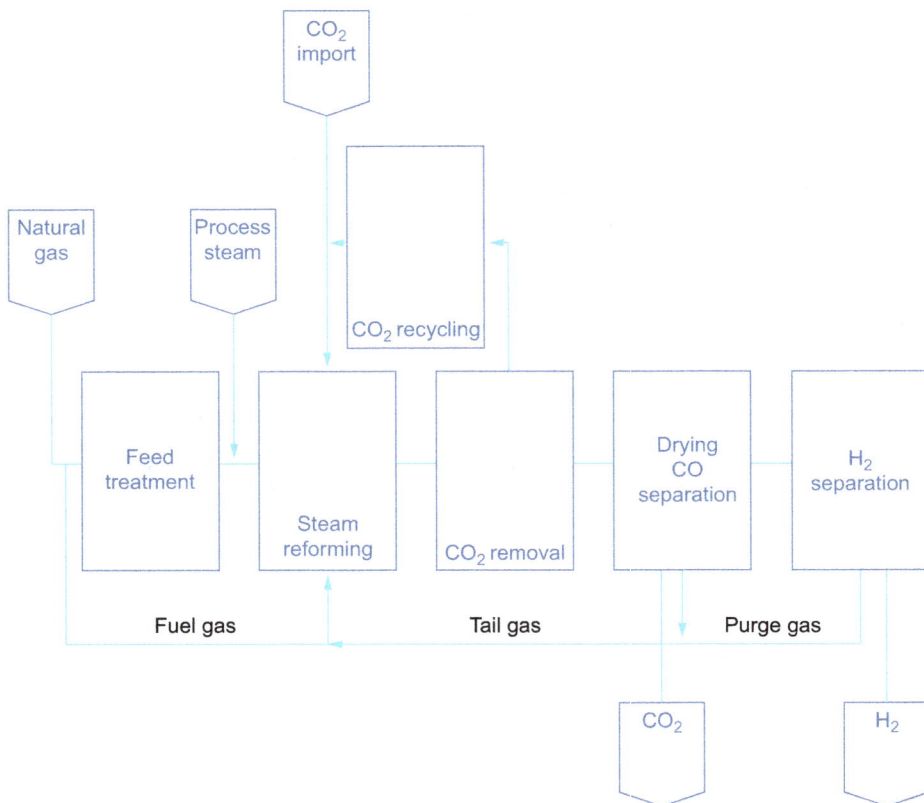

Figure 5.17: A scheme for CO_2-rich methane reforming. Modified from https://www.engineering. linde.com/dryref.

5.2 Gasification

Generation of hydrogen can be also done using not only natural gas but solid feed-stock such as coal or biomass. Partial oxidation of coal generates a syngas consisting of carbon monoxide and hydrogen in various ratios, with their sum up to ca. 85% of the total volume and minor quantities of carbon dioxide and methane. The overall gasification scheme is presented in Figure 5.18.

In the combustion chamber of gasifiers, there are zones where the formation of carbon is possible; therefore, in syngas, there is always a certain quantity of soot, e.g., carbonaceous particulate. The gas produced through gasification is purified by the removal of soot and other impurities including hetero-elements (sulphur and nitrogen). Gasification is an attractive option, since this technology permits practically the total removal of pollutants from the syngas. Sulphur is mainly recovered

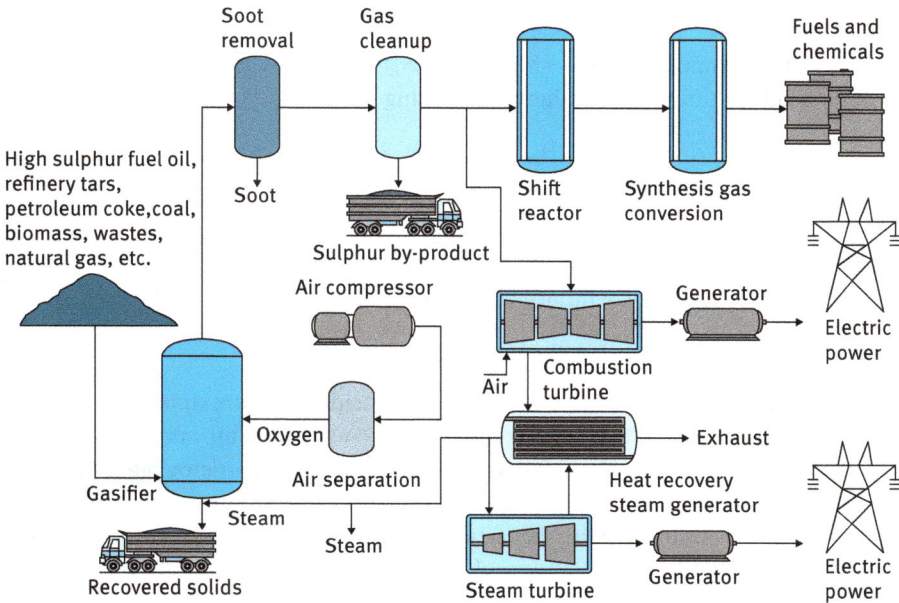

Figure 5.18: Overall gasification scheme. Modified after http://www.treccani.it/export/sites/de fault/Portale/sito/altre_aree/Tecnologia_e_Scienze_applicate/enciclopedia/inglese/inglese_vol_ 2/325-338_ING3.pdf.

in an elemental form, while the cleaned gas can be sent for chemical production purposes and/or be used to produce electricity. Besides the main products, traces of numerous other products can be formed. Such product as formic acid makes the process condensates corrosive. When the feedstock is petroleum residues and pet-coke, significant quantities of metals such as nickel and vanadium are present. Significant quantities of ash (inorganic combustion residues) can be present in coal.

The composition of the hydrocarbon feedstock is very heterogeneous, leading to complicated gasification chemistry. Numerous reactions occurring in the system include

– strongly endothermal reforming

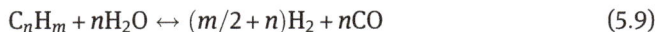

$$C_nH_m + nH_2O \leftrightarrow (m/2+n)H_2 + nCO \tag{5.9}$$

$$C_nH_m + nCO_2 \leftrightarrow (m/2)H_2 + 2nCO \tag{5.10}$$

– strongly exothermal combustion

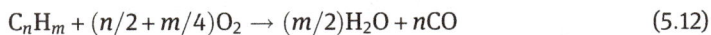

$$C_nH_m + (n+m/4)O_2 \rightarrow (m/2)H_2O + nCO_2 \tag{5.11}$$

$$C_nH_m + (n/2+m/4)O_2 \rightarrow (m/2)H_2O + nCO \tag{5.12}$$

$$C + 0.5O_2 \rightarrow CO \tag{5.13}$$

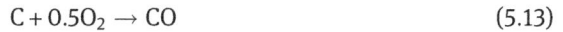

- mildly exothermic water-gas shift (eq. (5.5)) as well as formation of carbonaceous residues through endothermal cracking

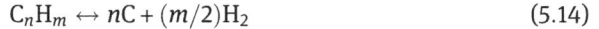

$$C_nH_m \leftrightarrow nC + (m/2)H_2 \tag{5.14}$$

- exothermal Boudouard reaction

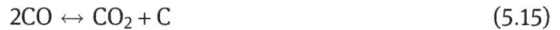

$$2CO \leftrightarrow CO_2 + C \tag{5.15}$$

- and its consumption by exothermal gasification

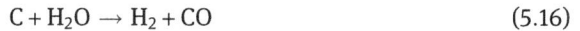

$$C + H_2O \rightarrow H_2 + CO \tag{5.16}$$

Due to the composition of coal, m and n can be considered approximately the same.

Gasification is done either with oxygen (95–99%) or with air and/or steam at a high temperature and variable pressure and under stoichiometric oxygen. Gasification, or in other words, partial oxidation, is a non-catalytic process occurring at 850–1,500 °C depending on the gasification type and feed. Application of lower temperatures minimizes consumption of oxygen at the expense of slower kinetics. On the contrary, in the temperature region 900–1,000 °C, thermodynamic equilibrium can be practically achieved and all reactions are very rapid.

Gasifiers are grouped according to the relative flow of the feedstock and oxidant in three main categories: moving-bed, fluid-bed, and entrained-flow gasifiers. These gasifiers operate at different temperatures, resulting in different composition of the products (Figure 5.19). The cold gas efficiency, which relates the energy of syngas (in terms of heating values) compared to the feed, is typically lower in entrained flow reactors than in moving or fluid beds, but can still reach 80–83% for this gasifier, somewhat approaching efficiency of the other types of the gasifiers.

Moving-bed gasifiers operating at around 2.5–3.0 MPa consist of a bed in which the coal (size up to 50 mm) is moving downflow due to gravity. Gasification occurs when the oxidant (oxygen and steam) distributed through a rotating grate passes counter-currently. Several temperature zones can be identified in this configuration corresponding to drying, volatilization, gasification, and combustion. The outlet temperature of the syngas is ca. 600 °C, being much lower than in the center of the bed where the ash melting temperature can be reached (2,000 °C).

In this type of gasifier, the reaction temperature is relatively low to protect the reactor internals, which leads to a high excess of steam. As a consequence of lower temperature in comparison with other gasifiers, the hydrocarbon conversion is far from complete and larger amounts of by-products are generated, requiring more elaborate gas cleaning than in alternative processes.

In the dry ash gasifier of this type, the ash is discharged in the solid form at about 1,100 °C (Figure 5.20a), while discharging the ash in the liquid form (slagging) is also possible at a temperature that is higher than the melting point (1,200 °C).

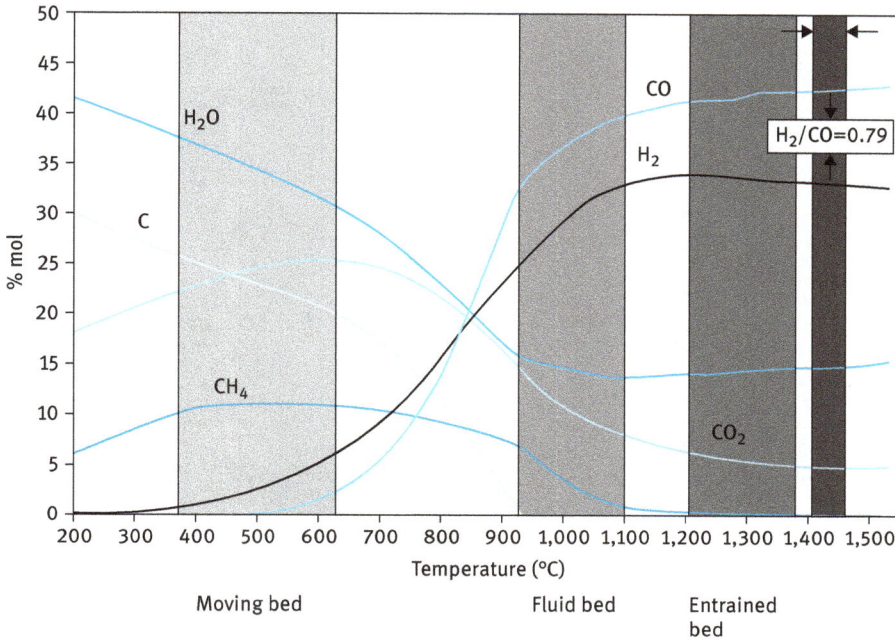

Figure 5.19: Equilibrium composition corresponding to different gasification conditions.

In the fluid-bed gasifier (Figure 5.20b), the size of ground coal is smaller (maximum size 10 mm). Oxygen (or air) and steam act not only as oxidants, but also as a fluidizing agent. Oxidant flow is directed to the base of the gasifier for fluidization and above the bed to gasify the entrained coal particles. High-temperature Winkler gasifier acts as a back-mixed reactor providing intensive mixing and excellent heat and mass transfer. The feedstock is fed into this reactor under pressure using a pressurized screw feed system. The reactor operates at atmospheric pressure and uniform temperature. Cyclone (Figure 5.20b) is required for removal of ash. The temperature in the process should be below the ash-softening point (950–1,100 °C for coal and 800–950 °C for biomass). Higher temperatures would lead to aggregation of the ash and dropping to the gasifier bottom, which is difficult to remove. The reactor design, the back-mixed character, and the moderate temperature result in removal of carbon along with the ash and the product gas, lowering the conversion to 97%. In order to compensate for it, highly reactive coals of lower grade (brown coal or lignite, peat) and biomass are preferred for fluid-bed gasifiers.

In the entrained-flow gasifiers (Figure 5.21), which can operate in the range of 2–8 MPa with practically any finely ground solid feed (<100 μm), pumpable liquid, or gaseous feed, due to the flow arrangement and low residence time (2–5 s), high temperatures (1,200–1,500 °C) are required for appreciable conversion.

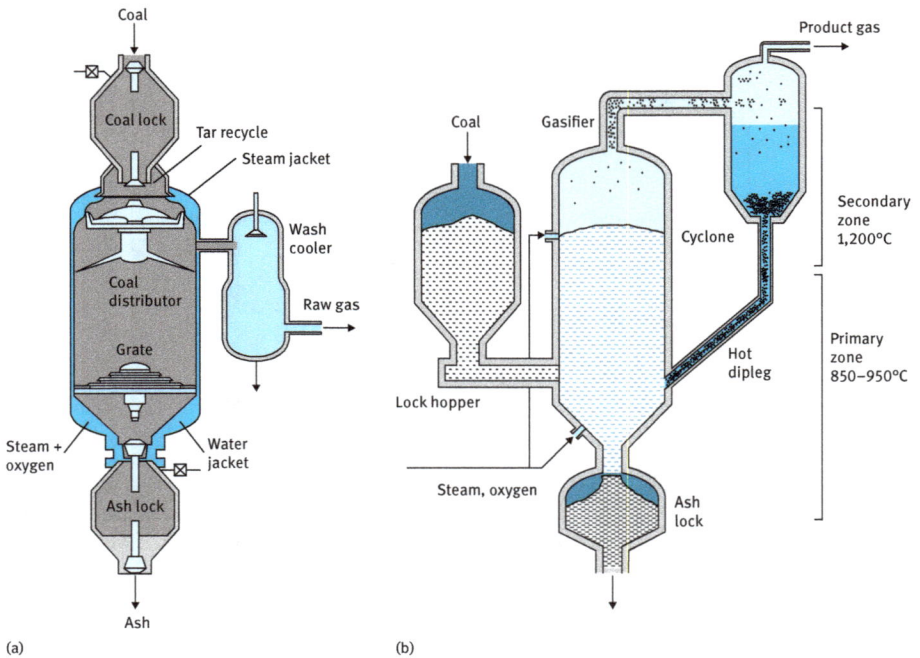

Figure 5.20: Gasifiers. (a) Dry ash moving-bed gasifier. (b) Fluid bed gasifier. Modified after http://www.treccani.it/export/sites/default/Portale/sito/altre_aree/Tecnologia_e_Scienze_applicate/enciclopedia/inglese/inglese_vol_2/325-338_ING3.pdf.

Such a high temperature leads to high coal conversion (up to 99%) and larger oxygen consumption than for other gasifiers and very small methane content in syngas. Due to larger oxygen consumption, lower quantity of steam, and higher H/O ratio, the product gas has a higher CO/H ratio compared to a moving-bed gasifier. Entrained flow gasifiers operate with upflow or downflow of the feed, oxygen, or air, and gas cooling done either by gas or water quenching and/or heat exchange.

One example of utilization of an entrained flow gasifier is the Shell process for solid feedstock operating with dry solids and oxygen in a one-stage upflow mode with a membrane as a reactor wall. Cooling is arranged through quenching and installation of a syngas cooler. The gasifier operates at the outlet temperature of 1,300–1,600 °C and pressure of 3–4 MPa.

There are usually four burners in this process, positioned around the circumference of the lower part of the reactor. Compared with an alternative water-coal slurry feed, dry feeding permits savings of 20–25% of oxygen, but is less economical.

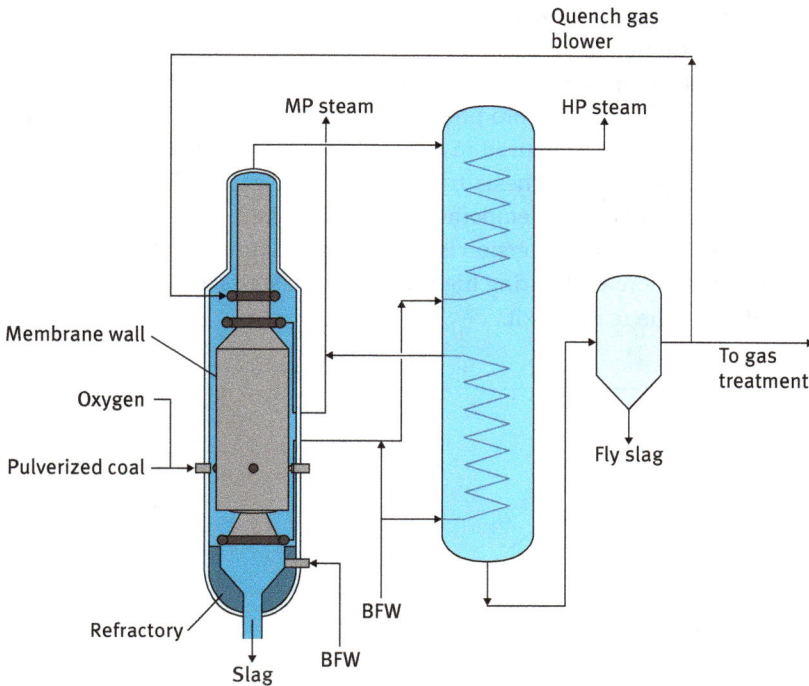

Figure 5.21: Entrained-flow gasifiers. Modified after http://www.treccani.it/export/sites/default/Portale/sito/altre_aree/Tecnologia_e_Scienze_applicate/enciclopedia/inglese/inglese_vol_2/325-338_ING3.pdf. BFW – boiler feed water, MP – medium pressure, HP – high pressure.

5.3 Water-gas shift reaction

Exothermic water-gas shift conversion

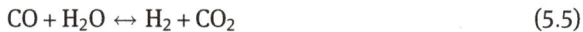

$$CO + H_2O \leftrightarrow H_2 + CO_2 \qquad (5.5)$$

typically follows steam reforming or gasification. Therefore, the water-gas shift equilibrium among carbon monoxide, carbon dioxide, hydrogen, and steam depends on temperature but is almost independent of pressure in the industrial range (elevated pressures up to 7.0 MPa), while pressure influences activity and selectivity. From the viewpoint of equilibrium, it is better to use a low temperature; however, less expensive iron catalysts operate in the range of 350–450 °C and are therefore used for water-gas shift reaction.

The gas after, e.g., secondary reformer (Figure 5.3) is cooled down to 350 °C and still contains a fair amount of carbon monoxide, which can be converted into hydrogen by high-temperature water-gas shift (or high temperature shift, HTS) reaction over an iron-chromium oxide-based catalysts with trace additions of compounds containing sodium and potassium.

The typical composition of the catalyst is Fe_2O_3 86% and Cr_2O_3 8%. In addition, graphite (4%) and promoters (<2%) are present. Catalyst lifetime is 5–8 years. One-stage adiabatic reactors are mainly used in HTS even if there are examples of radial flow reactors in the ammonia synthesis trains. The water-gas shift reaction is an exothermic one, with an adiabatic temperature rise in the bed. The outlet temperature is ca. 400–450 °C and the conversion is incomplete. Figure 5.22 illustrates how the bed temperature is increased during high-temperature shifts and also shows the concentration of CO at the reactor exit.

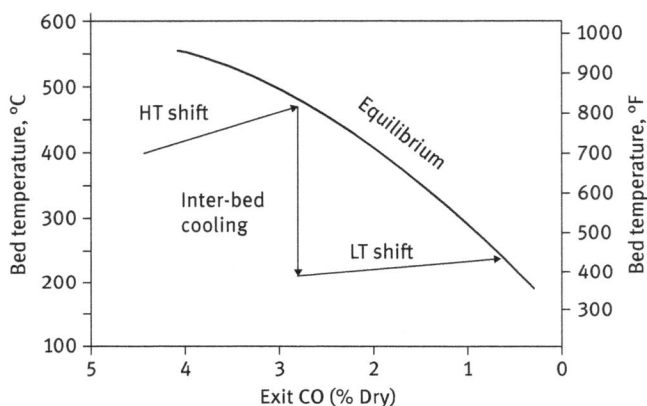

Figure 5.22: Typical CO variation in HTS and LTS catalyst beds.

After the HT shift, for example, in a classical process of hydrogen production or in ammonia synthesis trains, the gas is cooled (an interbed cooler shown in Figure 5.22) and fed into a low-temperature water-gas shift reactor (low temperature shift, LTS) operating at 180–270 °C.

In the LTS step, the carbon monoxide concentration is decreased to ca. 0.3 vol% on dry basis. Adiabatic reactors are applied, and thus, the outlet temperature increases to ca. 230 °C. A typical composition of copper-based catalysts for LTS: CuO, 40%; ZnO, 40%, with Al_2O_3 being the balance. Such catalysts with traces of potassium as a promoter should be highly active at low T and possess high thermal and mechanical stability and high selectivity, generating low amounts of by-products (alcohols). Amounts of by-products increase at higher T and pressure and longer residence time. High catalyst activity also leads to overall more by-products; however, less methanol is produced with less active catalysts.

In low-temperature shift with copper catalysts, there is an upper temperature limit due to accelerated recrystallization of copper above 270 °C. Very low temperatures cannot be used, impairing catalyst activity. Moreover, the gas inlet temperature should be enough above the dew point of the gas, as otherwise, steam can condense, damaging the catalyst. The catalysts can be also poisoned by sulphur or halogen compounds. Recent catalyst development includes incorporation of cesium and metal oxides with the latter serving as barriers between the nano-sized copper particles present in the catalyst, thereby stabilizing them from sintering.

The LTS step is followed by removal of CO_2 using either chemical or physical solvents. Detailed description of scrubbing (absorption) was presented in Chapter 3. CO_2 content after CO_2 scrubbing depends on which type of system is employed, chemical or physical and can be in the range ca. 50–300 ppm, respectively. When CO_2 is almost removed, there is still a certain amount of CO present in the gas. In the case of ammonia synthesis, the carbon monoxide concentration in the gas must be decreased, since it is a poison for ammonia synthesis catalysts.

This is done using a methanation reaction (reverse reaction of natural gas steam reforming) resulting in methane, which is not a poison for ammonia synthesis catalyst

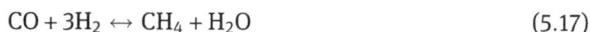

$$CO + 3H_2 \leftrightarrow CH_4 + H_2O \qquad (5.17)$$

Under typical methanation conditions (high hydrogen partial pressure, low methane and water content), the equilibrium is shifted to the right side. Catalysts similar to those applied in steam reforming of natural gas are utilized, but with much higher metal (30–60%) content using alumina or diatomaceous earth as the catalyst support. The concentration of carbon monoxide and carbon dioxide at the outlet is typically below 5 ppm. In the case of ammonia synthesis, the resulting gas contains also methane (ca. 1%) and argon (0.3%), the rest being a stoichiometric (3:1) mixture of hydrogen and nitrogen. For hydrogen-generating plants, hydrogen purity of 97–99% can be obtained.

An alternative route to hydrogen production with LTS and CO_2 removal by scrubbing is to use pressure-swing adsorption (PSA) described in Chapter 3. For production of pure hydrogen, this version is a standard one, allowing hydrogen purity of usually > 99.9%. Lower temperatures are required for the exothermic adsorption process; therefore, after high-temperature shift conversion, cooling the gas to ambient temperature, condensing, and removing water, the gas is fed to a system of several (up to 12) parallel adsorbers. Such adsorbers are run in a predetermined cyclic operation with the cycle time of 3 to 10 min. The components to be removed are adsorbed at high system pressure and high partial pressure and are desorbed at low system pressure.

Chapter 6
Cracking

6.1 General

Cracking is a process of breaking carbon-carbon bonds in complex organic molecules into simpler ones, which are more valuable. The cracking rate and selectivity depend on the presence of a catalyst and temperature. This chapter describes thermal cracking (visbreaking and steam cracking) and several catalytic processes such as fluid catalytic cracking (FCC) and hydrocracking.

Thermal cracking is currently applied for transformation of either very heavy fractions at milder temperature (ca. 500 °C) or to produce light fractions such as ethylene, propylene, and other feedstock for petrochemical industry by pyrolysis in the presence of steam (steam cracking) at higher temperatures (ca. 750 °C to 900 °C or higher). FCC produces a high yield of gasoline and liquefied petroleum gas (LPG), while hydrocracking is a major source of diesel, jet fuel, LPG, and naphtha.

These processes (visbreaking, hydrocracking, FCC, and steam cracking) will be considered below following an increase in the process severity.

6.2 Visbreaking

The quantity of heavy residues is expected to increase in the future in view of the progressively increasing heavier nature of the crudes. The most promising technologies to process such feedstock involve the conversion of vacuum residue and extra heavy crude oil into light and middle distillate products.

Visbreaking (viscosity reduction, viscosity breaking) is a relatively mild thermal non-catalytic cracking process mainly used to reduce vacuum tower bottom viscosities and pour points and to reduce the amount of cutting stock required for residue dilution to meet fuel oil specifications. Heavy fuel oil production can be reduced from 20% to 35% and cutter stock for dilution by 20% to 30% by visbreaking (Figure 6.1). This increases the yield of more valuable distillates directly converted from visbreaking or used as catalytic cracker feedstock.

Visbreaker feeds comprise saturates, aromatics, resins, and asphaltenes. Saturates are found to have an average carbon number in the range C38–50 with relatively low heteroatom content. They consist of long alkyl chains with few or negligible naphthenic and aromatic rings. The aromatics have a carbon number in the range C41–53. Asphaltenes are the heaviest and the most polar class of components in the crude oil, containing larger amounts of heteroatoms than the rest of the components in the crude oil. They comprise molecules with 4–10 condensed aromatic

https://doi.org/10.1515/9783110712551-006

Figure 6.1: A visbraking unit. From https://www.metesta.com/turnaround-2007-visbreaking-unit-at-refinery-mazeikiai/.

rings, 4–6 alkyl carbon chains, and 20–24 aliphatic carbons, with an average H/C atomic ratio of about 1.2.

Cracking, dehydrogenation, H-abstraction, cyclization, aromatization, and condensation are the major reactions that occur within the molecules during thermal processing of the feed.

In particular, there is a reduction in the paraffin chain length from C50 to C30 in saturates, dealkylation of aromatics and resins resulting in shortening of alkyl chains, saturation of naphthenic rings, an increase in highly polar asphaltene content, and a change in the internal polarity distribution within the asphaltenes themselves.

Thermal cracking of hydrocarbons is always accompanied by fouling and coke formation that deposits on the furnace walls and results in increased pressure drop, reduced heat transfer rates, hot spot formation due to uneven flow distribution, and corrosion by carbonization.

The residue comprises a large number of complex organic components, and a detailed kinetic model involving the cracking behaviour of each of the components is impractical.

Visbreaking conditions range from 455 °C to 510 °C at a short residence time, while in the delayed coking process (Figure 6.2), residence times are much longer and thermal reactions are allowed to proceed to completion. Liquid-phase cracking takes place under these low-severity conditions to produce some naphtha as well as products in the kerosene and gas oil boiling range. The gas oil may be used as additional feed for catalytic cracking units or as heating oil.

Figure 6.2: Delayed coking unit. From https://johnbrear-plantintegrity.com/?page_id=159.

A 5%–10% conversion of residuum to naphtha is usually sufficient to afford at least an approximate fivefold reduction in viscosity. Reduction in viscosity is also accompanied by a decrease in the pour point. An alternative option is to use lower furnace temperatures and longer times, achieved by installing a soaking drum between the furnace and the fractionator.

Thus, there are two types of visbreaker operations: coil furnace cracking and soaker cracking. Coil cracking uses higher furnace outlet temperatures (470–500 °C) and few minutes of reaction times.

The feed (Figure 6.3) is introduced into the furnace with a dedicated soaking coil and heated to the desired temperature and quenched as it exits the furnace with gas oil or tower bottoms to stop cracking. Conversion is controlled by regulating the feedstock flow rate (residence time) through the furnace tubes. The quenched oil is sent to a fractionator, where the products of the cracking (gas, LPG, gasoline, gas oil, and tar) are separated. Typical yields of the visbreaking process in wt% are gas (C4) 2–4, naphtha 5–7, gas oil 10–15 and tar 75–85. Steam ($\sim$1 wt%) is injected along with the feed to (i) generate the turbulence and, hence, increase the heat transfer coefficient at the wall, (ii) control the liquid-phase residence time, and (iii) prevent the coking along the tube walls.

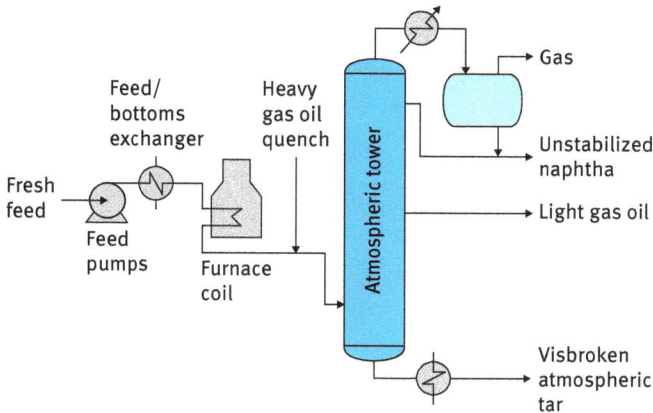

Figure 6.3: Coil visbreaking. Modified after http://www.treccani.it/portale/opencms/handle404?ex porturi=/export/sites/default/Portale/sito/altre_aree/Tecnologia_e_Scienze_applicate/enciclope dia/inglese/inglese_vol_2/229-238_ING.pdf.

The main advantage of the coil-type design is the two-zone fired heater that provides better control of the material being heated. In such type of design of a coil visbreaker, decoking of the heater tubes is accomplished more easily using steam-air decoking.

The higher heater outlet temperature specified for a coil visbreaker is an important advantage of coil visbreaking. Such higher temperature is used to recover significantly higher quantities of heavy visbroken gas oil. This capability cannot be achieved with a soaker visbreaker without the addition of a vacuum flasher.

An alternative to coil visbreaker is soaker cracking, when the feed after the furnace passes through a soaker drum for additional reaction time before it is quenched (Figure 6.4).

Conversion within the heater is thus much lower, and the main transformations occur in a soaker, which is a vessel operating at an elevated temperature. Soaker cracking uses lower furnace outlet temperatures (430–450 °C) and longer reaction times (15–25 min). Soaker visbreaking has the advantages of lower energy consumption and longer run times before having to shut down to remove the coke from the furnace tubes. Due to lower temperature, only cracking of higher molecules occurs without further cracking. The run time of soaker-type visbreaking can be up to 18 months, which is significantly longer than for the coil type (3–6 months).

A disadvantage of using soaking visbreaking is the need to remove coke from the drum, which is difficult and time consuming. This operation mode requires more equipment for coke removal and handling. The customary practice of removing coke

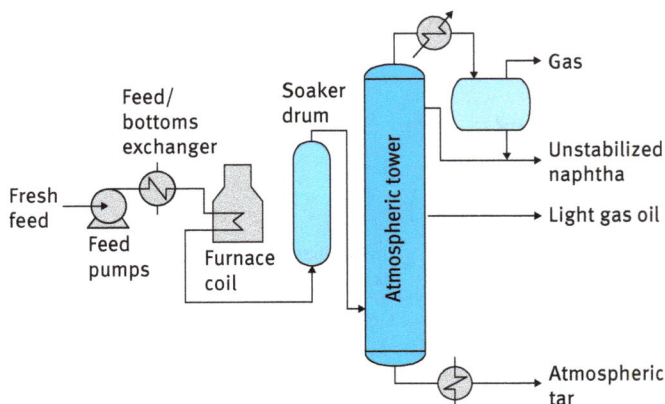

Figure 6.4: Soaking visbreaking. Modified after http://www.treccani.it/portale/opencms/han
dle404?exporturi=/export/sites/default/Portale/sito/altre_aree/Tecnologia_e_Scienze_applicate/
enciclopedia/inglese/inglese_vol_2/229-238_ING.pdf.

from a drum is to cut it out with high-pressure water, thereby producing a significant amount of coke-laden water that needs to be handled, filtered, and then recycled for use again.

In terms of product yield, there is little difference between the soaker visbreaker and the coil visbreaker.

Pressure is an important design and operating parameter, usually from 0.5 to 0.9 MPa for liquid-phase visbreaking and 0.1–0.2 MPa for partial vaporization at furnace outlet.

The soaker drum can be considered similar to a bubble-column reactor in which the gas phase consists of product gas, vaporized low-boiling components, and the added turbulizing steam.

Overall, the main limitation of the visbreaking process is formation of unstable products due to the presence of unsaturated compounds.

6.3 Hydrocracking

Hydrocracking (HC) process (Figure 6.5) was developed to produce high yields of distillates with better qualities than can be obtained by fluidized-bed catalytic cracking (FCC) which is described in the previous section. FCC is performed under more severe conditions than HC and is mainly aimed for gasoline production. Milder conditions of HC allow also producing middle distillates thus making the process more flexible.

Figure 6.5: Hydrocracking unit.

An example of how hydrocracking can be incorporated in an oil refinery with some other catalytic processes is presented in Figure 6.6.

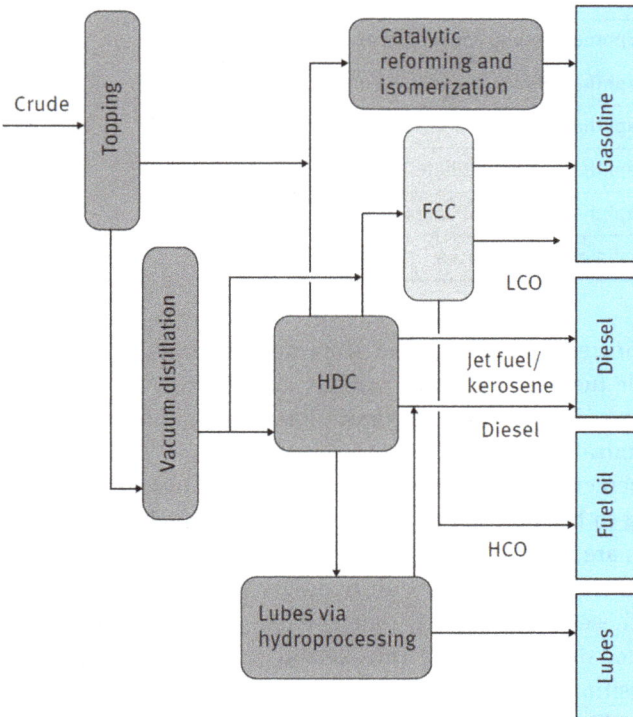

Figure 6.6: An example of utilization of hydrocracking in oil refining for fuel production. From http:// www.treccani.it/portale/opencms/handle404?exporturi=/export/sites/default/Portale/sito/altre_ aree/Tecnologia_e_Scienze_applicate/enciclopedia/italiano_vol_2/273-298ITA3.pdf&%5D.

Importance of hydrocracking especially in Europe and the USA is illustrated by the fact that currently hydrocracking of vacuum gas oil represents ca. 7–8% of processes in oil refining in terms of volume. Capacity of hydrocrackers can be up to 3–4 million tons per year. In the USA, hydrocrackers were traditionally used to produce naphtha from low-value aromatic streams including light cycle oil (LCO) product from fluid catalytic cracking, however, more units are shifting to production of middle distillates.

Hydrocracking feedstock is typically flashed distillates from vacuum distillation, from catalytic and thermal cracking, or deasphalted oils (Table 6.1). Operating conditions in the reactor section of hydrocrackers are usually about 400 °C and 8–15 MPa.

Table 6.1: Feedstock and products in hydrocracking.

Feedstock	Products
Kerosene	Naphtha
Straight run diesel	Naphtha and/or jet fuel
Atmospheric gas oil	Naphtha, jet fuel, and/or diesel
Vacuum gas oil	Naphtha, jet fuel, diesel, lube oil
FCC light cycle oil (LCO)	Naphtha
FCC heavy cycle oil (HCO)	Naphtha and/or distillates
Coker LCO or HCO	Naphtha and/or distillates
Deasphalted oil	Olefin plant feedstocks

Major advantages of the process include the full-scale production of high-quality products, i.e., no low-grade fuel oil remains as residue, and the high flexibility toward product yields, i.e., possible gasoline maximization, or gasoline – kerosene, or kerosene – gas oil maximization.

The reactions in hydrocrackers take place on metal sulfide catalysts in the presence of hydrogen resulting in hydrotreating, in which impurities such as nitrogen, sulfur, oxygen, and metals are removed from the feedstock, and hydrocracking per se when carbon–carbon bonds are cleaved with hydrogen addition over bifunctional catalysts. Besides these reactions, there are several others such as hydrogenation of olefins generated during cracking and skeletal isomerization of alkanes giving equilibrium composition of normal and isoalkanes. From the viewpoint of catalyst stability, hydrocracking catalysts operating under milder conditions than in FCC and in the presence of hydrogen are less prone to catalyst deactivation.

Both hydrocracking and FCC give minimal C_1 and C_2 products maximizing C_{3+} compounds with significant product branching. Contrary to hydrocracking with minimal olefin and coke formation, FCC gives significant aromatics and coke formation along with a moderate olefin formation. The latter can be boosted either by introducing additive to the conventional FCC catalyst or by developing a technology, which is targeted for e.g., propylene formation.

The hydrocracking process is carried out in fixed catalytic reactors at 280 °C and 475 °C and hydrogen pressure between 3.5 and 22 MPa depending on the feedstock properties and the products desired. Heavier feedstock requires higher temperatures and pressures and longer residence time. Lower pressures (3.5–8 MPa) correspond to mild hydrocracking.

Typically, cracking conversion is between 30% and 70% per pass to achieve required selectivity to naphtha or middle distillates, and can be increased by recycling of the unconverted feed close to 100%.

A schematic view of hydrocracking reactions is presented in Figure 6.7.

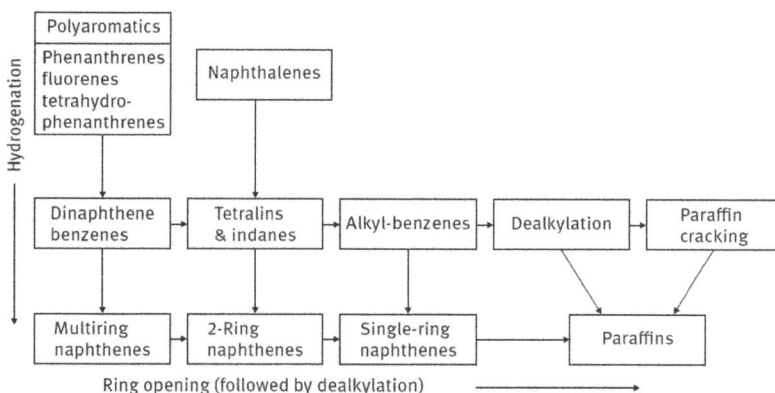

Figure 6.7: Schematic hydrocracking chemistry.

The first step in the mechanism of hydrocracking is dehydrogenation of normal alkanes to alkene intermediates at very low near-equilibrium concentrations, which react on Brønsted-acid sites giving alkoxides. Subsequent acid-catalyzed isomerization and beta-scission cracking is followed by desorption from the acid sites and shuttling to metal sites, where isomerized and cracked species are hydrogenated. In parallel to this route, alkenes can form coke. Keeping low concentrations of alkanes allows maintaining the activity of hydrocracking catalysts for several years, while FCC catalysts as presented below deactivate in seconds.

One of the design parameters for hydrocracking units is hydrogen pressure, which elevation diminishes coking and extends time between regeneration. Obviously, in bifunctional catalysis with consecutive reactions, the rates of dehydrogenation/hydrogenation on metal sites and reactions on acidic sites should be harmonized. The same holds for location of these functions, which in an ideal hydrocracking catalyst is for a long time and supposed to be in a relatively close proximity. Such proximity ensures that the equilibrium composition of alkenes is maintained at the acid sites.

An example of the product distribution of vacuum gas oil at different temperature is given in Table 6.2, while compositions of the product depending on the main target of hydrocracking is illustrated in Table 6.3.

Table 6.2: Yield of vacuum gas oil hydrocracking products at different reaction temperature.

Product	360–400 °C	400–420 °C
H_2S	2.3	2.3
C1–C2	5.7	6.5
C3–C4	4.3	10.6
C5–C6	2.6	17.6
Gasoline (b.p. up to 180 °C)	12.7	33.4
Diesel (180–350 °C)	66.9	8.3
Recycle (above 350 °C)	7.9	8.3
Hydrogen consumption, vol%	2.4	4.1

Table 6.3: Yield of vacuum gas oil hydrocracking depending on the desired main target.

Yield (vol%)	Gasoline	Jet	Diesel
Butane	16.0	6.3	3.8
Light naphtha	33.0	12.9	7.9
Heavy naphtha	75.0	11.0	9.4
Jet fuel	–	89.0	–
Diesel	–	–	94.1
Hydrogen consumption, m^3/m^3 of feed	361	312	260

Occurrence of two main types of reactions, hydrotreating and hydrocracking, allowed to split processes in modern hydrocrackers into units with two reaction stages – a hydrogenation step or desulfurization, denitrogenation, and deoxygenation using cobalt – molybdenum catalysts, and a hydrocracking step using nickel – tungsten catalysts. Hydrotreating catalysts are used upstream of conventional hydrocracking resulting in higher yields of gasoline. There is also a possibility that using milder conditions for hydrocracking will lower catalytic activity and higher yields of middle distillates.

Hydrocracking catalysts (more than 200 available) combine acid and hydrogenation components, therefore, during hydrocracking acid-catalyzed isomerization and cracking reactions as well as metal-catalyzed hydrogenation reactions occur resulting in products with lower aromatic content, more naphthenes and highly branched paraffins. As mentioned above, olefins are completely hydrogenated. An example of a portfolio of catalysts from one manufacturer is illustrated in Figure 6.8.

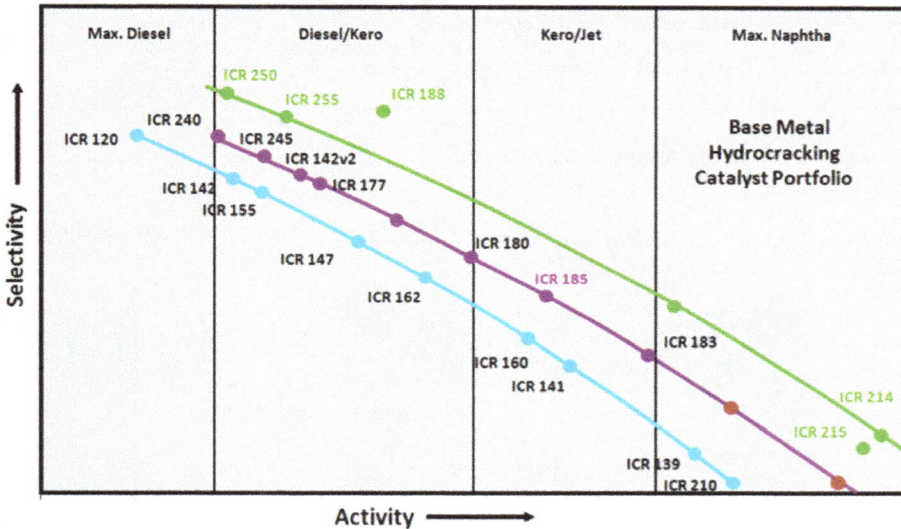

Figure 6.8: Dependence of selectivity vs. activity for some hydrocracking catalysts.
https://www.chevrontechnologymarketing.com/Documents/HE_NOV10_39-47.pdf.

Hydrogenation function is provided by noble metals and combinations of certain base metals. Either platinum and palladium or sulfide forms of molybdenum and tungsten promoted nickel or cobalt are applied. The noble metal hydrocracking catalysts are commonly used in very-low-sulfur or sulfur-free environments with the metal loading typically below 1 wt%. The base metal content is substantially higher, with 2–8 wt% cobalt or nickel, and 10–25 wt% molybdenum or tungsten. The cracking function is provided by one or a combination (Figure 6.9) of zeolites (for example, zeolite Y) and amorphous silica-alumina selected to suit the desired operating and product objectives.

Acidity of amorphous silica-alumina hydrocracking catalysts is moderate requiring higher temperature and longer residence time. At the same time, such catalyst can give a higher yield of diesel fraction and prevent extensive hydrocracking. As illustrated in Figure 6.10, there is also a possibility to have a complex catalyst loading.

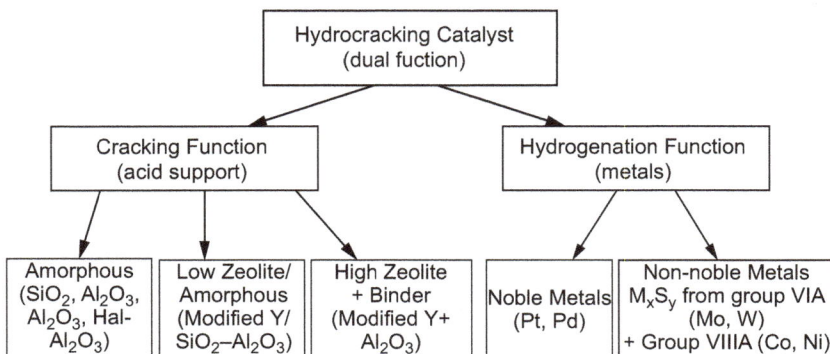

Figure 6.9: An overview of hydrocracking catalysts.

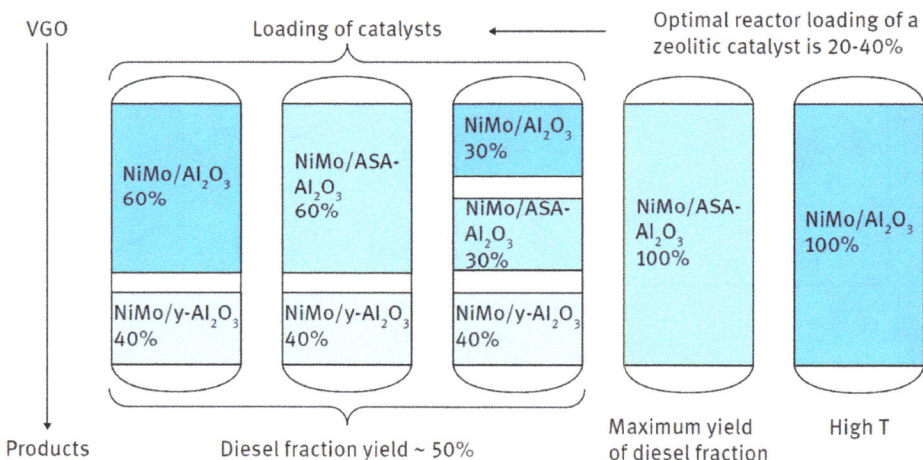

Figure 6.10: Options for using complex catalyst loading with different acidic catalysts (ASA stands for aluminosilicates).

Unlike amorphous-based catalysts, zeolite-containing materials are usually more acidic and thus selective to lighter products. Microporous character of zeolites with narrow channels restricts accessibility of larger organic molecules to the active sites. Y-type zeolites are thus modified by removing aluminum through hydrothermal treatment (steaming at temperatures above 540 °C) or first acid washing with subsequent steam calcination. An alternative approach is to improve acidity of amorphous aluminosilicates.

A hydrocracking catalyst support can therefore be tailor-made depending on how the catalyst is to be operated and the product type needed. To obtain maximum conversion to gasoline at the lowest operating temperature, up to 80% Y zeolite and about 20% of a peptized alumina binder are used as support. High middle-distillate conversion is obtained at higher operating temperatures with a support containing about 10% dealuminated Y-zeolite, 70% alumina or silica/alumina in varying proportions, and 20% of the binder. The appropriate Y zeolite and the proportion to be used in the catalyst can be selected for the products, as required.

For a long time, the concept of ideal hydrocracking was used as a guideline for development of bifunctional hydrocracking catalysts, as well as skeletal isomerization of hydrocarbons. It implies careful optimization of the cracking activity of the support and the hydrogenation activity of the metal component allowing maximum conversion with different operating conditions and feeds.

An ideal hydrocracking catalyst for a long time was supposed to have a close proximity between the metal and acid sites, thus an optimal location of the metal function in bifunctional catalysts was thought to be in the micropores of the zeolite. More recently, it was shown that the metal (sulfide) particles can be located on the binder or on the surface/in the mesopores of the zeolite.

Hydrocracking catalysts during operation gradually loses activity which is compensated by temperature increase. Hydrocracking catalysts typically operate for cycles of several years between regenerations depending on process conditions. Catalyst regeneration involves combusting of coke in an oxygen environment (dilute air) either in the reactor or externally (Figure 6.11). Until the mid-1970s, hydroprocessing catalysts were regenerated in situ in the unit reactors, lately ex-situ regeneration started to be mainly used for many reasons including corrosion, safety, time, and better activity recovery.

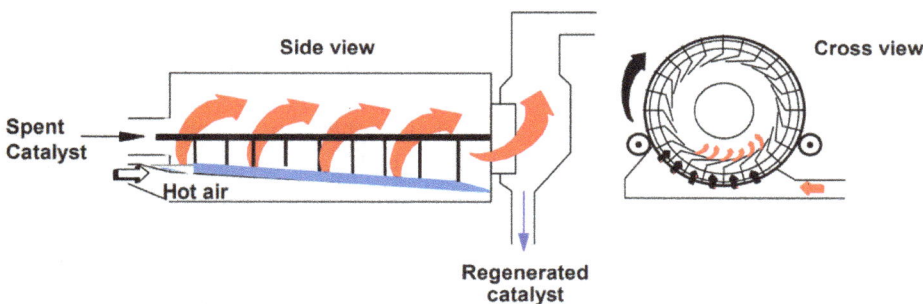

Figure 6.11: Ex-situ regeneration of spent hydrocracking catalyst by combustion in air.

Such regeneration procedure used for nickel–molybdate or nickel–tungstate catalysts is followed by resulfidation. The situation with deactivated palladium supported on zeolite is more complicated because of metal sintering. Treatment with an excess of ammonium hydroxide solution dissolves agglomerated palladium giving a tetramine complex, which is decomposed after calcination in air. In this way, the original palladium distribution and activity are restored.

Activation of nickel catalysts by sulfidation is done either ex situ or in situ using the same type of organic sulfur compounds and procedures as implemented for hydrotreating catalysts. Palladium catalysts should be carefully reduced in hydrogen at ca. 350 °C, avoiding overheating and thus sintering.

Development of commercial hydrocracking catalysts is aimed not only on optimizing the chemical composition, but also on the catalyst shape. Some of the shapes are presented in Figure 6.12. In Criterion's Advance Trilobe Technology, a replacement of a cylindrical catalyst to a trilobe can elevate the diesel yield by 1.5%, which for a hydrocracker capacity of 50,000 bbd capacity gives daily additionally 750 bbl.

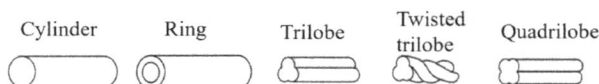

Cylinder   Ring   Trilobe   Twisted trilobe   Quadrilobe

Figure 6.12: Some shapes of commercial hydrocracking catalysts.

Sulfided Ni–Mo catalysts, supported on a suitable γ-alumina with the nickel molybdate content higher than in typical hydrotreating catalysts, are used for first-stage hydrocracking reactions. Sulfided Ni–W or palladium, supported on zeolite or silica/alumina, are applied in the second hydrocracking stage. Palladium catalysts are active if the feed contains residual sulfur compounds but are not active for the hydrogenation of benzene rings.

The following factors affect product quality and quantity and the overall economics: catalyst type; process configuration (e.g., one or two stages) and operating condition (e.g., conversion, hydrogen pressure, liquid hourly space velocity, feed/hydrogen recycle ratio).

Besides the role of catalysts discussed above, other parameters will be analyzed here. One of the parameters in hydrocracking, which prolongs catalyst lifetime by diminishing deactivation, is hydrogen pressure. At the same time, an increase of pressure enhances hydrogenation of aromatic compounds diminishing the octane number.

Another important operation parameter is catalyst average temperature (CAT), which is the volume average temperature of catalyst bed temperatures. Higher CAT allows lower residence time and thus higher throughput for the product quality or improved quality at the same feed rate. Moreover, more difficult feedstock with, for example, higher sulfur content can be processed. The downside is increased deactivation and thus shorter cycle time.

Cracking of large multiaromatic compounds requires first saturation of rings. Since hydrogenation is an exothermic reaction, from the viewpoint of equilibrium, it should be conducted at lower temperature. Thus, higher temperature makes it more difficult to saturate and subsequently cracks such aromatic compounds. An increase of T also enhances production of naphtha and light gases.

Gas-to-oil (hydrogen-to-oil) ratio in most modern hydrocrackers is designed to be four to five times larger than hydrogen consumption in various reactions. Increased hydrogen recycle improves catalyst stability, minimizes over-cracking, and acts as a heat sink limiting the temperature rise in the catalyst beds. At the same time, very high recycle rate is not economical.

Several flow schemes of hydrocracking are possible. The once through arrangement is the most cost-effective option and processes very heavy, high boiling feed. There is also a possibility to have a separate hydrotreating unit, which is installed when there are special feed considerations, such as high content of product in the feed or high nitrogen content.

Otherwise, either single or two state arrangements are commonly applied. The first option is cost-effective for moderate capacity and gives a moderate product quality, while the two-stage configuration is more effective for large capacity units affording high product quality and can be applied for difficult feedstock.

Flow schemes for hydrocracking unit comprise such sections as a high-pressure reaction loop; a low-pressure vapor–liquid separation section; a product fractionation section; and a make-up hydrogen compression section. As mentioned above, the gas-to-oil ratio is higher than hydrogen consumption, thus the excess of hydrogen is recovered. A high-pressure loop can account for 70% to 85% of the installed cost of the hydrocracker. The reactors are typically trickle-bed downflow reactors with multiple beds.

Such multibed arrangement is needed as exothermic cracking and saturation reactions result in a large heat release, therefore, a cold recycle gas is introduced between the beds to quench the reacting fluids and to control the extent of temperature rise and the reaction rate. An alternative for heat management is to use heat exchangers between the beds. Depending on the conversion level and the type of feed processed, the number of beds can vary from two to eight.

Important for trickle bed reactors is uniform reactant distribution across the catalyst, otherwise maldistribution leads to hot spots which deteriorate catalyst performance and life.

A typical example of a two-stage hydrocracker is shown in Figure 6.13. In the two-stage flow scheme, feedstock is treated and partially converted in a first reactor section. Products from this section are then separated by fractionation. The bottoms from the fractionation step are sent to a second reactor stage for complete conversion. This flow scheme is most widely used for large units as mentioned above.

The partially cooled reactor effluent is often flashed in a hot high-pressure separator (HHPS). This allows withdrawal of the heavy unconverted oil at a high temperature minimizing heat input to the product recovery section. The vapor from HHPS is first cooled in a heat exchanger. Addition of water is needed to prevent formation of solid ammonium hydrogen sulfide (NH_4HS) from ammonia and H_2S, which can otherwise deposit on the air-cooler tubes, reduce heat transfer, and eventually plug the tubes.

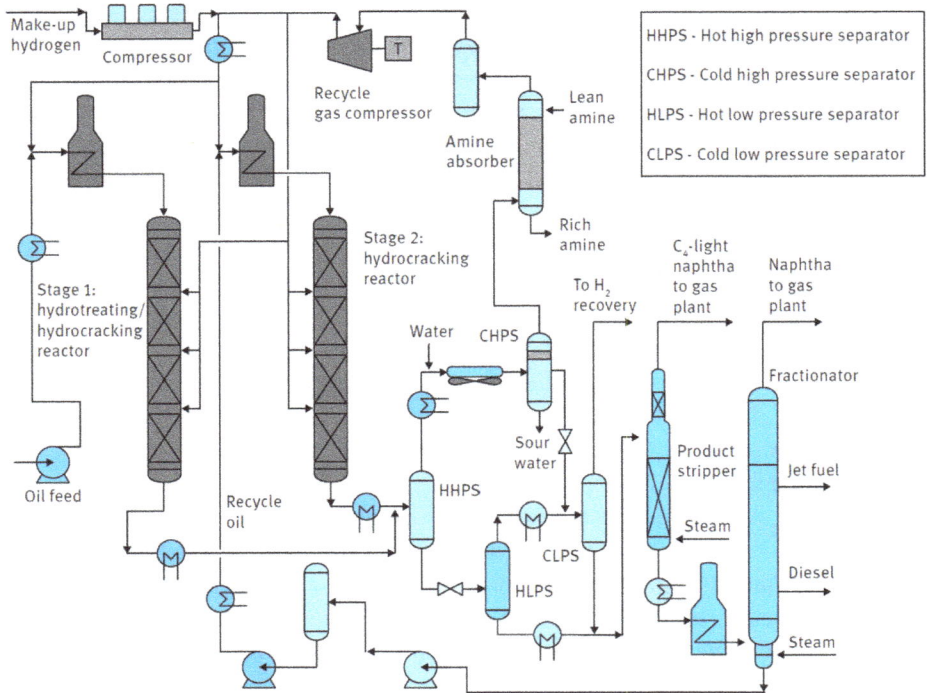

Figure 6.13: Two-stage hydrocracking. From http://www.treccani.it/export/sites/default/Portale/sito/altre_aree/Tecnologia_e_Scienze_applicate/enciclopedia/inglese/inglese_vol_2/273-298_ING3.pdf.

After cooling, the HHPS effluent vapor in an air cooler in the downstream cold high-pressure separator (CHPS) hydrogen-rich vapor is separated from water and hydrocarbon liquid phases. The latter is sent to the cold low-pressure separator (CLPS), while hydrogen-rich gas is either directly recycled to the reactor as feed or quench gas with gas compressor or is first purified from H_2S using absorption with methyl diethanol amine or diethanol amine.

After expansion (pressure reduction), the liquid from the HHPS is sent to the hot low-pressure separator (HLPS). Hydrocarbon liquid from the CHPS is combined with the vapor from the HLPS and cooled in an additional air cooler before entering the CLPS. Hydrogen still present in the vapor from CLPS is recovered, for example, by pressure swing adsorption.

The fractionation section consists of a product stripper and an atmospheric column separating reaction products into light ends, heavy naphtha, kerosene, light diesel, heavy diesel, and unconverted oil.

When a refiner wishes to convert all the feedstock to lighter products, the fractionator bottoms stream can be recycled back to the reactor and co-processed with fresh feed.

The single-stage flow scheme with recycle is illustrated in Figure 6.14. Typically, there are two reactors with the first one for hydrotreating and the second for hydrocracking catalyst. In most modern units, recycle oil is blended in with fresh feed before processing in the first reactor.

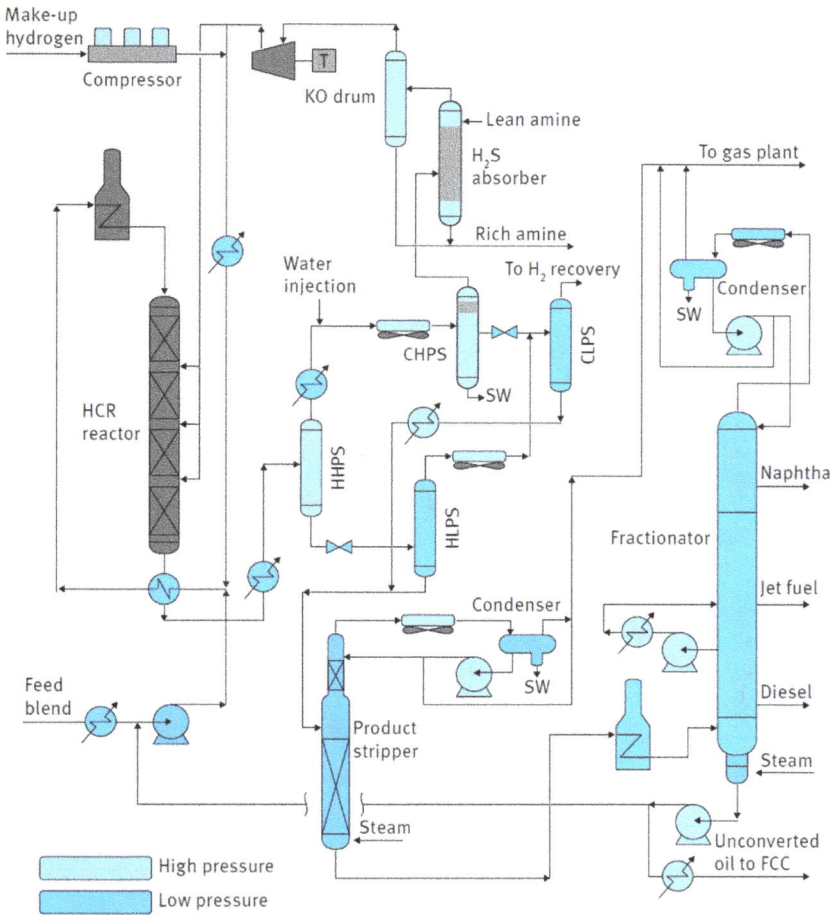

Figure 6.14: One-stage hydrocracking. From http://www.treccani.it/export/sites/default/Portale/ sito/altre_aree/Tecnologia_e_Scienze_applicate/enciclopedia/inglese/inglese_vol_2/273-298_ ING3.pdf.

Hydroconversion of heavy oil feeds (with up 4% sulfur and more than 400 ppm of metals) – gas oils, petroleum atmospheric and vacuum residues, coal liquids, asphalt, bitumen, and shale oil – is extremely important for several reasons. In petroleum refining, a substantial amount of crude oil ends up as vacuum residue (60 wt%), which should be processed to fuels and chemicals.

The role of unconventional/heavy oils containing large amounts of high-molecular-weight polynuclear aromatic (Figure 6.15, Figure 6.16), coke-forming and organometallic compounds, especially with nickel and vanadium (Table 6.4).

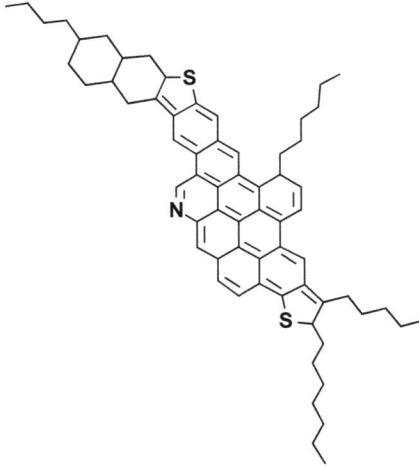

Figure 6.15: Figure 6.15. An example of a possible structure of asphaltenes: from https://upload. wikimedia.org/wikipedia/commons/thumb/e/eb/Possible_asphaltene_molecule.svg/440px-Possible_asphaltene_molecule.svg.png.

Table 6.4: Composition of some crude oils in wt% of total (after R. Sahu, B.J. Song, J. S. Im, Y-P. Jeon, C.W. Lee, A review of recent advances in catalytic hydrocracking of heavy residues, Journal of Industrial and Engineering Chemistry, 2015, 27, 12–24. Copyright Elsevier. Reproduced with permission).

Component	Hydrocarbons	S	N	V	Ni
Maya crude oil					
Saturates	20.7	0.9	3.3		
Aromatics	26.5	24.6	8.2	0.4	3.3
Resins	29.9	39.0	39.6	17.9	17.7
Asphaltenes	20.6	36.3	48.9	81.7	79
Arabian heavy crude oil					
Saturates	20.1	<1	6.7		
Aromatics	31.0	29.6	8.4	3.4	10.4
Resins	31.2	46.3	43.8	25.2	28.2
Asphaltenes	23.9	23.9	41.1	71.4	61.8

Figure 6.16: A lab sample of asphaltenes, derived from crude oil. Fromhttps://https://www.slb.com/-/media/images/insights/oilfield-review/pdf-03185-01-hi-res.ashx?h=710&w=1574&la=en&hash=B9E10488312756FE22BBA7E6855AF562.

Development of vacuum residue (Figure 6.17) or heavy oils upgrading is related to improvement in catalysts, namely the active metal (different from conventional Ni, Mo, Co, and W) or another phase (e.g., carbides or nitrides instead of sulfides), improvement in supports (texture, acidity, and chemical composition) including introduction of promotors, such as phosphorus, fluorine, and boron.

Figure 6.17: Visual appearance of vacuum gas oil (VGO) in comparison with the fuel products.

Moreover, the process technology is also improving, which is reflected by industrial application of different reactor types, such as fixed, moving, and ebullated bed reactors.

Residuum hydrocracking requires temperatures between 385 °C and 450 °C, hydrogen partial pressures from 7.5 to 15 MPa and space velocities from 0.1 to 0.8 vol h^{-1}. During hydrocracking, besides cracking of large molecules, heteroatoms (S, N, O) and metals (vanadium, nickel) are removed. Removal of nitrogen is more difficult than sulfur removal, thus some nitrogen compounds are merely reduced only to lower boiling range of nitrogen compounds rather than to ammonia.

Metal sulfides formed during hydrocracking are absorbed on the catalyst, plugging the pores and resulting in catalyst deactivation. Minimization of coking can be done by elevation of hydrogen pressure, introduction of aromatic diluents, and continuous physical removal of coke precursors from the reactor loop.

As catalysts in residuum fixed-bed hydroprocessing extrudates (8 mm to 16 mm) are used with the active metals (Co, Ni, Mo, W, etc.) impregnated on alumina. The pores should be large enough to allow transport of large asphaltene molecules.

Different catalysts can be used in different reactors, for example, a demetallization catalyst in the first reactor, and highly active Ni–Mo desulfurization catalysts in the subsequent two reactors.

Catalyst addition to optimize catalyst usage is done in a countercurrent fashion with installation of the fresh catalyst into the third reactor. After a certain time, it is withdrawn and reused in the second reactor. Finally, the catalyst from the second reactor is reused in the first reactor. Accumulation of carbon can be as high as 40% in the case of vacuum bottoms.

Few types of hydroprocessing reactors rely on moving, or ebullating catalyst beds with catalyst grain size less than 1 mm in size to facilitate suspension by the liquid phase in the reactor. Ebullated bed reactors operating also with NiMo or CoMo catalysts with the catalyst bed expansion of about 30–50% have up to 30 m height and 5 m diameter (Figure 6.18). The feed and hydrogen are added from the bottom. In LC-fining technology, several reactors are arranged in a series with the final product going to a separator. The vapor stream containing hydrogen is treated with amine in an absorber, purified with pressure swing adsorption, recompressed, and recycled back to LC-Fining reactors. A decrease of pressure before heat exchange, removal of condensates, and purification gives considerable savings in investment compared to a conventional high pressure recycle gas purification system. The liquid stream after a separator is sent to the hydrotreated distillate fractionator.

In the ebullated bed reactor, a part of the liquid product is recycled through a large pan at the reactor top and the central downcomer by means of a pump mounted in the bottom head of the reactor. Such flow is needed to keep the bed in an expanded (ebullated) state and assure near isothermal reactor temperature. Catalyst is added and withdrawn from the reactor to maintain a stable activity without a need for unit shutdown.

Fixed-bed hydrotreater/hydrocracker can be also put downstream of the ebullated bed reactors operating at the same pressure level. The feed for such fixed bed reactor is the vapor stream from the ebullated bed reactors and the distillate recovered from the

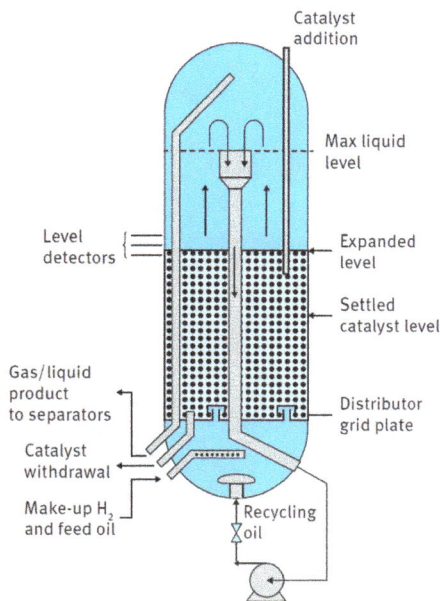

Figure 6.18: Ebullated bed reactor for LC fining. From 31.http://www.treccani.it/portale/opencms/handle404?exporturi=/export/sites/default/Portale/sito/altre_aree/Tecnologia_e_Scienze_appli cate/enciclopedia/inglese/inglese_vol_2/309-324_ING3.pdf&%5D.

heavy oil stripper overhead and the straight run atmospheric and vacuum gas oils. This design makes use of excess hydrogen in the effluent vapors to hydrotreat the distillates fractions. Additional hydrogen, equivalent only to the chemical hydrogen consumed in the fixed bed reactor, is introduced as quench to the second and third catalyst beds.

In LC-fining technology with ebullated bed reactors hydrocracking is performed at ca. 410–440 °C, pressure 110–180 bar, giving HDS efficiency of 60–85% and the following product distribution (% w/w): C4 = 2.35, C5–177 °C – 12.6; 177 °C–371 °C – 30.6; 371 °C–550 °C – 21.5; above 550 °C – 32.9.

An ebullated-bed process shows more selectivity, high conversion of feeds, high liquid yield, and relatively low hydrogen consumption. However, the back mixing of the reactants, high operating costs, high investment, and low reactor efficiency are the main hurdles.

Eni Slurry Technology (Figure 6.19), with the capacity of two million tons annually, is operating at Eni's Sannazzaro refinery at 15–16 MPa and 410–450 °C with molybdenite (MoS_2) nanosized catalyst, generated in situ from oil-soluble precursors. An attractive feature of such technology is absence of aging and plugging of pores, avoiding, therefore, catalyst substitution with subsequent shutdown of hydrocracking plants typical for other catalytic hydrotreating processes.

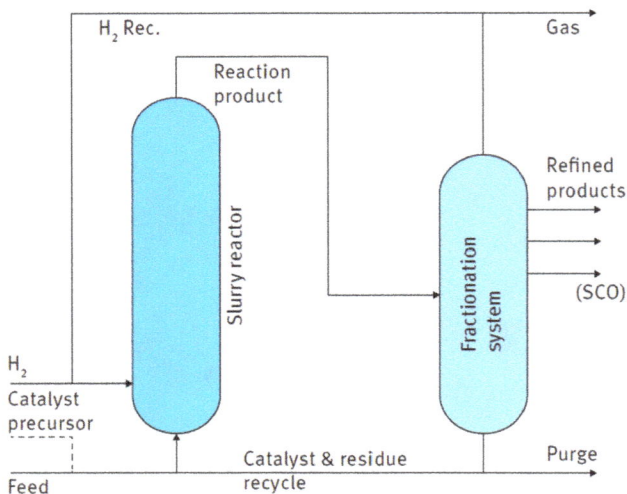

Figure 6.19: Eni Slurry moving bed hydrocracking technology: (a) scheme, (b) reactor (from https://www.eni.com/en_IT/attachments/innovazione-tecnologia/technological-answers/scheda-est-eng.pdf).

Small size of catalysts leads to the absence of mass transfer limitations increasing the overall catalyst activity and allowing to keep a very low (few thousand ppm) catalyst concentration.

High degree of backmixing in a bubble column reactor (height 58 m, diameter 5.4 m, weight 2,000 t) operating in slurry phase ensured almost flat axial and radial temperature profiles making the reactor intrinsically safe against temperature runaway.

The separation system in this technology is similar to other hydrocracking units with the main difference that the unconverted bottom material is recycled back to the reactor with the dispersed catalyst. A small purge (<3%) needed to limit the buildup of metals (Ni and V) fed when the heavy feed is processed.

6.4 Fluid catalytic cracking

The thermal cracking process was patented in 1891 and substantially modified later. Thermal cracking is still used to upgrade very heavy fractions or to produce light fractions or distillates, burner fuel, and/or petroleum coke.

Catalytic cracking, as introduced by E. Houdry in 1928, is a flexible process to reduce the molecular weight of hydrocarbons. The first full-scale commercial fixed-bed catalytic cracking unit began production in 1937. In the first plants cyclical fixed-bed operations were used which were replaced by a more efficient operation with continuous catalyst regeneration in 1941.

The reasons for such modifications were associated with the fact that during the catalytic cracking process, the catalysts deactivated after a short time because of coke deposition. Although coke can be removed and regenerated by burning, the regeneration time is relatively long compared to the reaction time. Application of a fixed-bed reactor with very frequent shutdowns for regeneration is not an economically viable option, while use of two reactors, one for hydrocarbon cracking and another for catalyst regeneration along with moving of catalyst between these two, is much more efficient. The first continuous circulating catalyst process used a bucket elevator.

Although this technology of the moving bed solved the problem of moving the catalyst between efficient contact zones, the catalyst beads used still were too large. Large catalyst particles result in large temperature gradients within catalyst particles, which restrict regenerator temperatures and result in a large regenerator and catalyst holdup.

The next step in the development of catalytic cracking was introduction of the FCC. This process uses fine-powdered catalysts that can be fluidized. The first commercial circulating fluid bed process was put on stream in 1942 and by the 1970s FCC units replaced most of the fixed and moving-bed units.

Initial process implementations were based on synthetic low activity silica–alumina catalysts and a reactor where the catalyst particles were suspended in a rising flow of feed hydrocarbons in a fluidized bed.

A dramatic increase of catalyst activity and gasoline selectivity was achieved with the introduction of zeolite cracking catalysts modified with rare earths in 1962 by Mobil. Thus, conversion and gasoline yield could be increased from 56% and 40%, respectively, for silica–alumina gels to 68–75% for conversion and 52–58% for gasoline yield depending on the catalyst.

A significant development in the 1980s was the introduction of ZSM-5 zeolite into the catalyst matrix as an octane enhancer.

Even today, FCC remains the dominant conversion process in petroleum refineries, producing a high yield of gasoline and LPG. Approximately 1,500 tons per day are consumed worldwide in more than 720 refineries.

In the 1950s, kerosene gas oil fractions were mainly applied as the feedstock, while later vacuum gas oil and even heavier feedstock started to be used. The boiling range of the feedstock could be 540–550 °C with a molecular mass 1.5 times larger than for a lighter feedstock. It also means that vaporization of the feedstock is not complete.

A switch to heavier feedstock also required efforts aimed at a decrease of the coke yield, which for vacuum gas oil can reach 5–6%. The trend of using heavier metal-containing feedstock, such as atmospheric residues and synthetic crude, is continuing and will only accelerate. The atmospheric residue can contain Ni and V in the amounts of up to 10 ppm and even higher, while the content of these metals is below 1 ppm in vacuum gas oil. Undesirable effects of nickel and vanadium are associated with increased gas and coke selectivity for Ni and loss of activity for V. In addition, this heavier feedstock can contain sulfur. One of the options of handling it on a catalyst level is to incorporate an additive into the catalyst, which will capture SO_x formed during regeneration. Other possibilities might include a hydrotreating unit upstream FCC.

Typical FCC unit feeds and operating conditions are presented in Table 6.5.

Table 6.5: Typical FCC unit feeds and operating conditions.

Feedstock	Operation	
	Vacuum gas oil	**Atmospheric residue**
API gravity	25.5	22.4
Sulfur (wt%)	0.7	0.8
Nickel (ppm)	0.4	3
Vanadium (ppm)	0.6	3.5
Conradson carbon (wt%)	0.2	4.0
Conversion (vol%)	86	76
Fuel gas (wt%)	4.4	4.6
Total C_3 (vol%)	13.9	10.4
Total C_4 (vol%)	18.6	13.2
C_5 + gasoline (vol%)	62.6	57.4
LCO (vol%)	6.8	11.6
Slurry (vol%)	6.9	11.9
Coke (wt%)	5.6	7.5
Reactor temperature (°C)	535	535
Regenerator temperature (°C)	720	720

[From L. Lloyd, (2011) Handbook of Industrial Catalysts. Springer.]

The amount and the quality of FCC products depend on the feedstock characteristics as well as the process parameters leading to fuel gas, gasoline, and LCO. LCO, when used as a blending component in heating oil, can have a greater value than that of gasoline. An example of the product distribution is given in Table 6.6. Under such circumstances, many refineries adjust their FCC unit operation to increase LCO yield at the expense of gasoline.

LCO compared to diesel fractions has lower cetane number and higher sulfur content. The cetane depends on the feedstock and operation temperature. Obviously, an increase of LCO yield can be achieved by diminishing FCC unit cracking severity, eventually leading to higher yields of heavy products (LCO, heavy cycle oil, and clarified oil) and lower yields of light products (gasoline, LPG, and gas) as well as coke. Reducing catalyst activity, lowering reactor temperature, and reducing the catalyst/oil (C/O) ratio can lead to increase in LCO yield.

Table 6.6: Product distribution of FCC unit (from https://inside.mines. edu/~jjechura/Refining/07_Catalytic_Cracking.pdf. Copyright © 2017 John Jechura).

Product	Yield, vol%	Yield, wt%
Light gases (C2–)	–	4.93
Propane	2.56	1.43
Propylene	5.80	3.34
Isobutane	5.59	3.48
n-butane	1.96	1.27
Butenes	7.61	5.06
Gasoline (C5+)	57.05	47.15
Light cycle oil (LCO)	21.20	20.60
Heavy cycle oil (HCO)	6.80	7.84
Coke	–	4.90
Total	108.57	100
Cycle oils	28.00	28.44
Total LPG	23.52	14.59

Capacity 25,000 bbl/d or 65.61 m^3/h. Conversion of gas oil = 72 vol.%.

Many reactions happen during catalytic cracking, such as isomerization, cracking, proton transfer, alkylation, polymerization, cyclization, dehydrogenation, condensation, and coking. Some of them are primary reactions, while the majority are secondary.

A somewhat simplified overview of the reactions is given in Figure 6.20.

Long chain alkanes

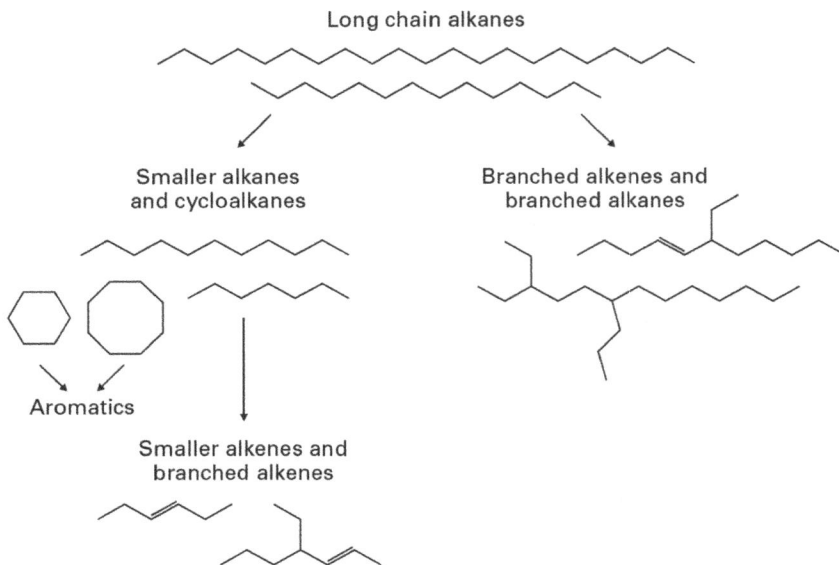

Smaller alkanes and cycloalkanes

Branched alkenes and branched alkanes

Aromatics

Smaller alkenes and branched alkenes

Figure 6.20: Diagrammatic example of the catalytic cracking of petroleum hydrocarbons (from http://en.wikipedia.org/wiki/File:FCC_Chemistry.png.).

During the cracking of normal paraffins, the cracking reactions are the dominant ones. The products contain mainly paraffins of lower molecular weight and olefins:

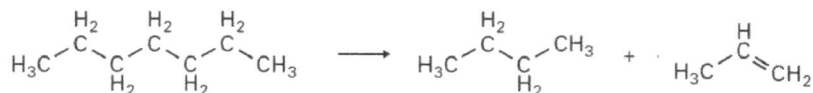

The yield of the latter increases with the increase of the feedstock molecular mass. Heavier fractions are less stable and can be cracked more easily than light fractions. Isoparaffins crack easier than paraffins.

Naphthenes are cracked at the side chain. Alkylnapthenes or aromatics with shorted side chains are formed first as they are stable, especially with methyl and ethyl substituent. Cracking of olefins, which is a secondary reaction, leads to smaller olefins:

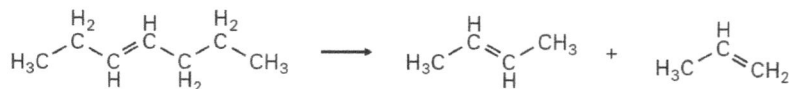

Initiation of the cracking process of olefins occurs (as for many other reactions of olefins) as a result of formation of carbocations, catalyzed by Brønsted and Lewis acids. A carbenium ion is a positively charged tricoordinated carbon atom, while a carbonium ion is a positively charged pentacoordinated carbon atom. If a carbenium ion is large enough (C6+), it can crack, forming alkene and another carbenium ion which isomerizes if possible into a secondary or tertiary ion. When the carbenium ion is not large (C3–C5), it will be terminated into an olefin through proton abstraction to the catalyst or another olefin, or through termination into a paraffin via addition of a hydride.

The mechanistic scheme for primary catalytic cracking is presented in Figure 6.21.

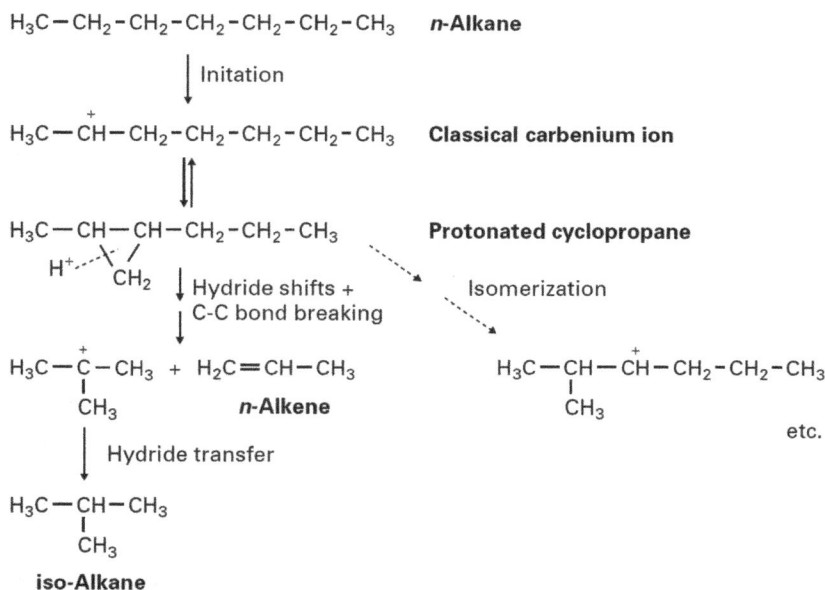

$H_3C-CH_2-CH_2-CH_2-CH_2-CH_2-CH_3$ **n-Alkane**

| Initation

$H_3C-\overset{+}{C}H-CH_2-CH_2-CH_2-CH_2-CH_3$ **Classical carbenium ion**

↕

$H_3C-CH-CH-CH_2-CH_2-CH_3$ **Protonated cyclopropane**

$H^{+}\cdots$ \ /
    CH_2 | Hydride shifts +   Isomerization
             | C-C bond breaking

$H_3C-\overset{+}{\underset{CH_3}{C}}-CH_3 \ + \ H_2C{=}CH-CH_3$ $H_3C-CH-\overset{+}{C}H-CH_2-CH_2-CH_3$
                                                          |
      n-Alkene CH_3
                                                                   etc.

| Hydride transfer

$H_3C-\underset{CH_3}{CH}-CH_3$

   iso-Alkane

Figure 6.21: Mechanistic scheme for catalytic cracking. From J. A. Moulijn, M. Makkee, A E. van Diepen, *Chemical Process Technology*, 2013, 2nd Ed. Copyright © 2013, John Wiley and Sons. Reproduced with permission from Wiley.

Branched polycyclic aromatics are often dealkylated

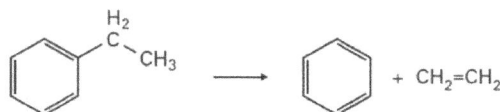

or can also be cracked in the side chain:

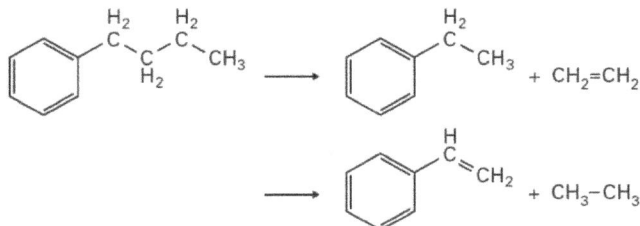

Carbon–carbon bond rapture occurs at the ring and is not very profound when an alkyl chain contains less than three carbon atoms. Polycyclic naphthenes are transformed to monocyclic naphthenes and alkenes. Naphthenes can react with alkenes through hydrogen transfer, leading to aromatics and alkanes:

A large amount of aromatic compound is obviously also produced from paraffins, whose structures allow cyclization.

Straight-chain alkenes can undergo acid-catalyzed skeletal isomerization reactions, leading to branched alkenes:

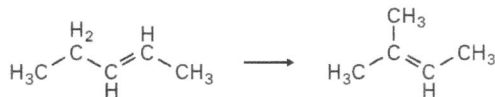

This is an important reaction because it increased the octane rating of gasoline range hydrocarbons. In addition to skeletal isomerization, double-bond migration can also occur. In addition, dehydrogenation can take also place, which together with cyclization gives polyaromatics, which are adsorbed on the catalyst surface, causing catalyst deactivation through coke formation.

The secondary reactions also produce unwanted light gases in addition to the formation of coke, which could be very substantial. In a conventional operation, around 2% to 5% coke is typically deposited on the catalyst.

Among secondary reactions, ring opening of cycloalkanes could be highlighted, as well as alkyl group transfer:

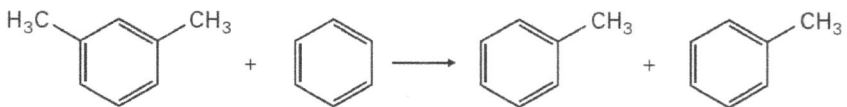

and dehydrocyclization:

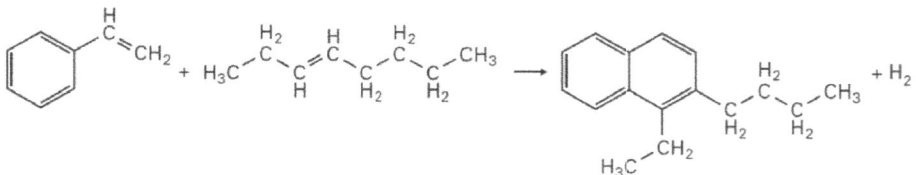

An overview of cracking reactions and the products is provided in Table 6.7.

Table 6.7: Catalytic cracking reactions.

Hydrocarbon	Initial products	Further products
Paraffins	Branched paraffins and olefins mainly in the C_3–C_{10} range.	Olefins crack and isomerize and are also saturated by hydrogen transfer to give paraffins. Olefins also cyclize to naphthenes.
Naphthenes	Crack to olefins. Dehydrogenate to cyclic olefins. Isomerize to smaller rings.	Further dehydrogenation to aromatics by hydrogen transfer.
Aromatics	Alkyl groups crack at the ring to form olefins.	Further dehydrogenation and condensation forms coke.
	Dehydrogenation and condensation to polyaromatics.	
Typical products (approximate)	Light gas (3%) LPG (17%) Naphtha (52%) LCO (16%) HCO (5%) Coke (5%)	H_2, CH_4, C_2H_6, C_2H_4, C_3H_6, C_3H_8 C_4H_8, C_4H_{10} Light, 40–110 °C, Heavy 110–220 °C Jet fuel 220–340 °C Kerosene, diesel, heating oil recycle. Higher than 340 °C

(From L. Lloyd, (2011) Handbook of Industrial Catalysts. Springer.)

Dehydrogenation reactions do not usually play a significant role during the cracking of high-molecular-mass paraffins, while for low-molecular-weight paraffins these reactions could be important as they result in the formation of valuable components – olefins and from low-value ones – paraffins.

Coke, which usually has an atomic ratio between H and C from 0.3 to 1 and spectral characteristics typical for polycyclic aromatic compounds, can be classified into several groups:
- catalytic coke formed on acid sites
- dehydrogenated coke, which results from dehydrogenation reactions over metals coming with the feedstock and deposited in the catalyst
- chemisorbed coke, which is a result of irreversible chemisorption of high boiling point polycyclic arenes
- reversible coke, resulting from incomplete desorption after exposure to steam

Catalytic coke is directly related to cyclization of olefins, condensation, alkylation, and hydrogen transfer. The highest yield of dehydrogenated coke occurs over metals such as Co, Ni, Cu and to a lesser extent V, Mo, Cr, and Fe.

One of the criteria for determining the significance of the secondary reactions is the ratio between gasoline and coke. If the ratio is high, it shows the dominance of the desired reactions, provided that the octane number of gasoline is high enough. Isomerization, hydrogenation, cyclization, and mild aromatization of olefins could be considered as desired reactions. They lead to high yields of paraffins, including branched ones, as well as aromatics with a boiling range close to that of gasoline and to a favorable ratio between iso- and normal paraffins. Excessive cracking, dehydrogenation, and polymerization of olefins, as well as condensation of aromatic hydrocarbons results in high yields of hydrogen and coke, low yields of olefins, giving a heavier product with lower yields of gasoline and a lower octane number. The amount of coke is dependent both on properties of the catalyst and the feedstock, and on the kinetic parameters of the process.

Because of the high complexity of cracking reactions and hundreds of them occurring at the same time, it is not realistic to develop a detailed model for cracking, therefore for modelling purposes empirical and semi empirical simplified models are usually applied when several compounds are lumped together. Although a detailed kinetic analysis is challenging, it is however worth considering various kinetic parameters and their influence on the catalytic cracking. Catalytic cracking is conducted in the absence of any heat supply and is thus considered adiabatic. The main process parameters are temperature, pressure, the ratio between the feedstock and the catalyst, as well as the circulation ratio.

In catalytic cracking, the catalyst is circulated continuously between the two reactors and acts as a vehicle to transport heat from the regenerator to reactor. The spent catalysts, which are partially covered with oil, are sent to a steam-stripping unit to remove the adsorbed oil before entering the regenerator. The stripped oil vapors are sent to a fractionation tower for separation into streams with desired boiling range.

Catalytic cracking is usually conducted at 450–550 °C decreasing from bottom to top in a riser as the cracking reactions are endothermic. Typical temperature profiles for catalysts with different amounts of rare-earth metals are illustrated in Figure 6.22.

After stripping, the catalyst, which still contains typically 0.8–1.3 wt% coke, is fed to the regenerator where the coke is burnt off. The temperature in the regenerator is 540–680 °C and is being carefully controlled as high temperatures can deactivate catalysts. The residence time is in the order of minutes. Pressure is 1.6–2.4 bar in the reactor and 1.3–3.1 bar in the regenerator. Because steam generated by combustion has an undesired effect of dealuminating the framework of the zeolite and degrading zeolite crystallinity, the fresh catalyst is fed to the unit continuously to partially replace a spent catalyst. Thus, in FCC, the catalyst (coined as "equilibrium catalyst") is a physical mixture of the fresh catalyst and regenerated catalyst with a broad age distribution circulating, within the FCC column.

Temperature in the riser has a following influence on the reactor performance. With temperature increasing to 470–480 °C, the gasoline yield is increasing, passing through a maximum at 490 °C as secondary cracking of formed hydrocarbons is becoming more prominent and the yield of gaseous products as well as coke is

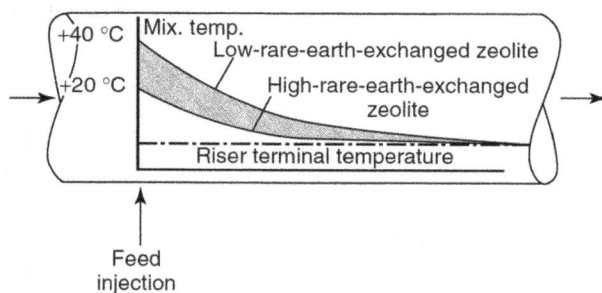

Figure 6.22: Typical riser reactor temperature profile. From Zeolites and Catalysis: Synthesis, Reactions and Applications, J. Cejka, A. Corma, S. Zones, (Ed.). Copyright Wiley, 2010. Reproduced with permission.

increasing. Moreover, with a temperature increase, the octane number is increasing, along with the C1 to C3 in the gases, while C4 is decreasing. With the pressure increase there is a decrease in the gasoline and C1–C3 yields, as well as the total amount of olefins and aromatics. The coke yield is almost independent of pressure. More catalyst than feed is introduced to the riser; the ratio of the catalyst to oil is 5–30, typically ca. 10. Increase of this ratio increases conversion and diminishes the coke content on the catalyst surface, while the yield of the desired products (gasoline) is increasing.

Another important parameter is the flow rate (16–20/h in a riser) or the contact time (few seconds). The process usually operates at 75–80% conversion, which also determines the contact time. Longer contact times will result in secondary cracking and undesired low–molecular-weight gases and coke.

In summary, the yield of the main products can be influenced by the following parameters. Dry gas (an undesirable product) can be diminished by decrease in temperature and residence time, metal content, and aromatic content of the feed. The yield of the desired product is increased by increasing the C/O ratio, operation at the maximum possible temperature (without over cracking), and increasing the catalyst activity. Production of LCO can be maximized by decreasing reaction temperature, decreasing C/O ratio and a decrease in catalyst activity.

Cracking catalysts are solid acids, therefore amorphous silica-aluminas were initially used for this process. Acidic strength is usually higher for crystalline zeolites compared to amorphous ones, also displaying higher activity, which is essential for the yield of the desired product. Because of such features, compact riser reactors with zeolites replaced rather large fluidized bed reactors.

The highest acid strength was found in zeolites containing the lowest concentrations of AlO_4 tetrahedra, for example, ZSM-5 and Y. Some early zeolite catalysts contained X-zeolite, probably because it was cheaper to manufacture, however, later on it was demonstrated that Y zeolite was more stable under typical operating conditions. One of the reasons could be that zeolite X is hydrophilic (contrary to Y zeolite, which is hydrophobic).

It should be also noted that if a catalyst is too active this might lead to higher yields of gas and coke, and during regeneration substantial amounts of heat would be released leading to the collapse of the zeolite framework. Undiluted zeolites are too active for use in existing units and would be immediately deactivated as coke is deposited on the surface.

Thermal stability increases with increasing Si/Al ratio, thus ultrastable zeolite Y (prepared by ion exchange with rare earth ions, La^{3+} and Ce^{3+} or by dealumination through treatment with steam), found widespread application in fluid catalytic cracking. Note that the regeneration of zeolites includes treatment with steam, thus hydrothermal stability is essential.

In addition to the modification of acidity and stability as described above in order to achieve better heat removal (~10%), the catalyst also contains a matrix in addition to zeolite. Initially, the matrix was simply a diluent and binder to form porous particles strong enough to resist attrition while circulating continuously between the reactor and the regenerator. As this matrix is also used to transport reactants, large pore materials are applied. Nowadays, alumosilicate is applied as an amorphous matrix, also providing the required mechanical stability. Typical commercial cracking catalysts are thus a mixture of Y zeolite and SiO_2–Al_2O_3. Alumina solutions made by dissolving pseudoboehmite in a monobasic acid, such as formic acid, had previously been used to bind silica/alumina catalysts. Other binders now include silica sols and aluminum chlorhydroxide. A fresh FCC catalyst consists of spray-dried spherical, fluidizable, attrition-resistant particles, typically containing 20–40% ultrastable Y (USY) zeolite, a binder, a catalytically active acidic matrix, and various additives.

The use of cracking catalysts containing zeolite increased rapidly following their introduction in 1962, and by 1972, they were being used in at least 90% of the FCC units in the United States. The most stable and successful catalysts for producing high yields of gasoline were made from rare-earth zeolites. Various manufactures prepared FCC catalysts with different amounts of rare earths depending on operational severity and product objectives.

The most important achievement of the application of zeolites compared to alumosilicates was not the activity increase, but rather an increase of selectivity toward some products. For example, the amount of olefins is much less over zeolites, which is associated with a higher rate of hydrogen transfer on zeolites than on alumosilicates, and stronger adsorption of olefins compared to paraffins. Because olefins have a longer residence time on the active sites, they are able to be saturated with hydrogen, which is generated because of naphthene dehydrogenation to aromatics. Yields of gasoline are higher on zeolites, while the yields of coke and gas (<C4) are not less than on alumosilicates.

During catalyst preparation, the initial exchange with rare earth and ammonium chloride removes sodium ions from the supercages. When calcined, the rare-earth oxides and hydroxides decompose and migrate into the sodalite cages, where they can exchange for more sodium ions. In the calcination process, some of the rare-earth ions are converted to cationic polynuclear hydroxy complexes that provide additional acid sites.

Because the demand for gasoline grew with time, the level of rare earths in the catalyst formulation refiners tended to increase, being on average ~3%. A recent spike in the prices of rare earth metals has put an additional pressure on catalyst manufacturers as well as refiners to diminish the role of rare earths for FCC.

Although better conversion and an increased yield of gasoline could be achieved with rare-earths Y zeolite catalysts, it was found that octane levels were lower because fewer olefins were produced. As gasoline quality could be improved by either increasing cracking severity or using antiknock lead that contained compounds such as tetraethyl lead, this was not a problem before new constraints were enforced.

In the mid-1970s, the application of small pore ZSM-5 (0.55×0.5 nm channel) as a co-catalyst with rare-earths Y zeolite increased the octane number of gasoline. Straight-chain C6–C10 olefins produced by normal cracking were further processed in a shape-selective cracking over ZSM-5, resulting in an increase in the gasoline octane number with the same dry gas, heavy oil, or coke production at the expense of decreasing gasoline yield by up to 2%. This loss could be compensated because propylene and n-butylene are used in an alkylation unit. ZSM-5 in an inert matrix is added along with Y zeolite on a daily basis to FCC units. ZSM-5 framework is extremely stable, affording less deactivation than typical Y zeolites.

Low hydrothermal stability of ZSM-5 leads to relatively fast dealumination and the relative activity loss if no special measures are taken. Modification with phosphorus is used to limit this relative activity loss.

The feedstock contains small amounts of metals even after hydrotreating (in particular, Ni and V), which deactivate zeolites, acting as dehydrogenation catalysts and giving more coke, thus passivators such as Sb, P, Sn, and B compounds are added because they react with the impurities. Nickel, which is a concern when its concentration on the equilibrium catalysts exceeds 800 ppm, may be passivated by addition of low levels of antimony or bismuth to the feed. A negative side of this method is an increase of NO_x emissions and potential bottoms fouling.

A more permanent solution is to use a material within the catalyst which is able to deposit Ni. In the current resid FCC catalysts, a special alumina is integrated in the catalyst throughout the catalyst microsphere in order to trap nickel-forming nickel aluminate. More efficient use of alumina could be achieved if it is located only at the outer layer of the catalyst. Such catalysts were introduced into the market.

Further development of more stable catalysts was related to introduction of boron, which migrates within the catalyst by solid-state diffusion to passivate Ni.

This innovation helps to increase the liquid products yields with heavy resid feeds by better metals passivation and lower hydrogen and coke production (Figure 6.23).

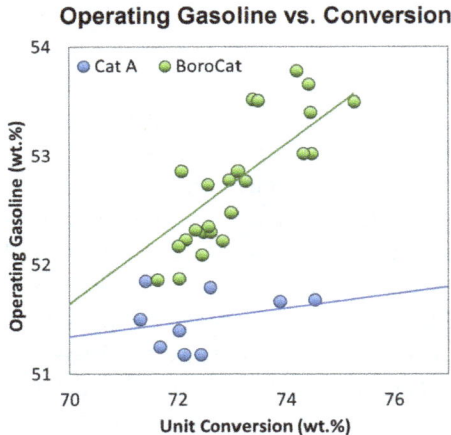

Figure 6.23: Comparison of boron-containing catalyst with a reference commercial catalyst (from https://refiningcommunity.com/wp-content/uploads/2017/11/Boron-Based-Technology-an-Innovative-Solution-for-Resid-FCC-Unit-Performance-Improvement-Clark-BASF-FCCU-Galveston-2018.pdf.).

A decrease in the amount of carbon on the regenerated catalyst and an improvement in burning CO in the regenerator can be achieved by adding a CO combustion promoter, such as Pt, Pd, Ir, and Ru, in small amounts (0.01–50 ppm) to the catalyst of the regenerator.

Another key property of FCC catalysts is attrition resistance, as it determines the catalyst addition rate (typically less than 1% per day) and the catalyst fluidity. This property increases with increasing bulk density and decreasing pore volume. Conversely, catalyst activity and resistance to deactivation by the plugging of pores increases with a decrease in bulk density and increased pore volume. The application of microspherical catalysts in alumosilicate amorphous matrix (10–150 μm) improved abrasion resistance and diminished the formation of fines.

As discussed in Chapter 4, there is a growing demand in propylene driven mainly by the polypropylene market. Steam cracking (section 6.5) is the main supplier of propylene, while a significant amount (ca. 1/3) is produced by FCC. Industrial trials (e.g., at Slovnaft Bratislava refinery) demonstrated that standard FCC units with vacuum gas oil as the feed can produce up to 10–12 wt% propylene by adding an additive to the catalyst feed, such as ZSM-5. Addition of large amounts of ZSM-5 (3–5 wt% of the catalyst inventory) diluting the main FCC catalyst did not have an adverse effect on FCC conversion.

In its simplest form, the FCC unit (Figure 6.24) consists of three sections: reactor section, fractionation, and gas concentration.

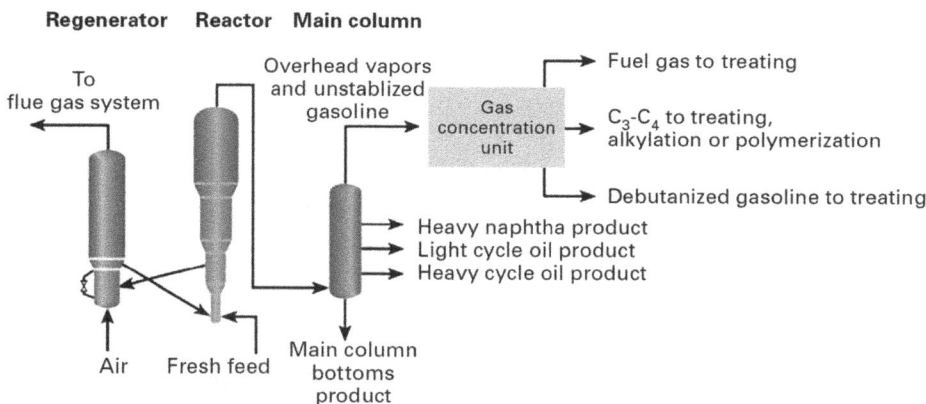

Figure 6.24: Flow scheme of a FCC unit (http://www.uop.com/fcc.).

In the riser reactor design introduced by UOP in 1971, the preheated raw oil feed, together with hot catalyst from the regenerator, enters at the lower part of the reactor (Figure 6.25). In FCC processes, the equilibrium catalyst is a physical mixture of varying proportions of fresh catalyst and regenerated catalyst or aged catalyst, circulating within the FCC column.

Vaporization of oil (vaporizing temperature depends on the feed, usually between 350 °C and 450 °C) occurs because of heat provided by the catalyst. In fact, the catalytic cracking unit (Figure 6.26) also has a furnace (Figure 6.26b), which is required during the start-up or insufficient heat supply from the regenerator.

The cracked hydrocarbons, separated from the catalyst in cyclones (Figure 6.26c), leave the reactor overhead and go to the fractionation column for separation into fuel gas, LPG, gasoline, naphtha, LCOs used in diesel and jet fuel, and heavy fuel oil. A large part of cracking occurs within a few seconds (2–5 s).

Minimum gas velocity should be at least 10 times higher than the minimum fluidization velocity of the catalyst particles. Because of the increase in volume inside the reactor, the gas velocity increases axially and the highest gas velocity at the outlet of the riser can be as high as 25 m s^{-1}.

The spent catalyst first falls down into the stripping section within the reactor vessel. Steam (3–5 kg of steam per 1,000 kg of the circulating catalyst) removes most of the hydrocarbon vapor and the catalyst then flows down a standpipe to the regenerator. If these hydrocarbons are not removed, temperature in the regenerator will be increased beyond that required for smooth operation limits. The residence time of the catalysts in the stripper is 0.5–1 min.

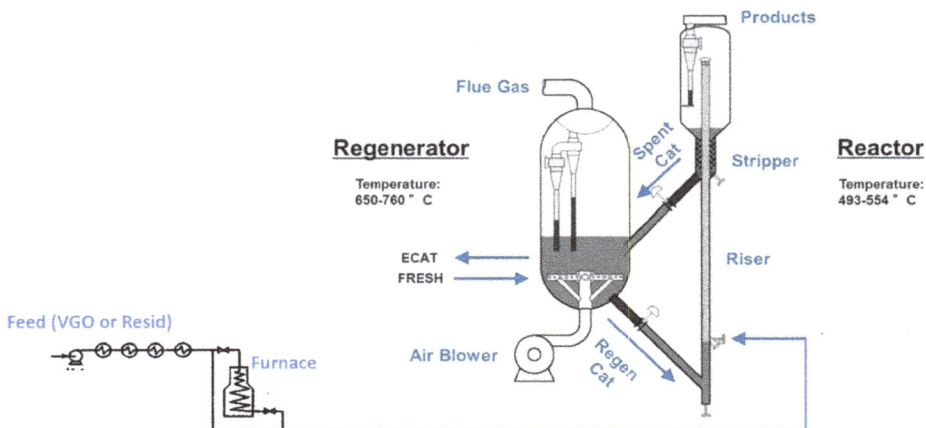

Figure 6.25: Fluid catalytic cracking with reactor and regenerator (modified from E.T.C. Vogt, B.M. Weckhuysen, Fluid catalytic cracking: recent developments on the grand old lady of zeolite catalysis, Chem. Soc. Rev. 2015, 44: 7342-7370 with permission from the Royal Society of Chemistry, CC BY-NC 2.0).

The spent catalyst mixes with air and clean catalyst at the base of the regenerator, where the coke deposited during cracking is burned off, thus regenerating the catalyst and additionally providing heat for the endothermic cracking reactions. The temperature in the regenerator can rise from 500 °C to 650 °C because of the exothermicity of the coke burning. Gases leaving the regenerator contain substantial amounts of CO and have significant heating value, which is used to generate steam, thus the flue gas is treated in a CO boiler where CO is transformed to CO_2.

Catalysts are partially deactivated in each cycle. Moreover, because of catalyst transport in the reactor and the regenerator and catalyst attrition, losses of catalyst occur. To compensate for this, make-up catalyst is added to the process continuously, with a replacement rate of about 1–3% of the total inventory every day. Catalyst attrition in the unit is directly related to the catalyst circulation rate. Increasing superficial gas velocity (with increase of feed rate, reactor temperature, and quench steam or decrease of operation pressure) increases catalyst entrainment to the cyclones and results in increased catalyst losses.

More than 2,000 tons of catalysts are replaced every year in a FCC unit with an inventory of 200 tons of catalyst and a replacement rate of 3% a day.

Because of the constant addition of catalysts, FCC units operate continuously and are very seldom closed to replace the catalyst inventory completely. Such shutdowns would obviously lead to significant decreases in production capacity. One major contributor to unscheduled FCC unit shutdowns is unexpected cyclone failure due to high-temperature erosion (Figure 6.27).

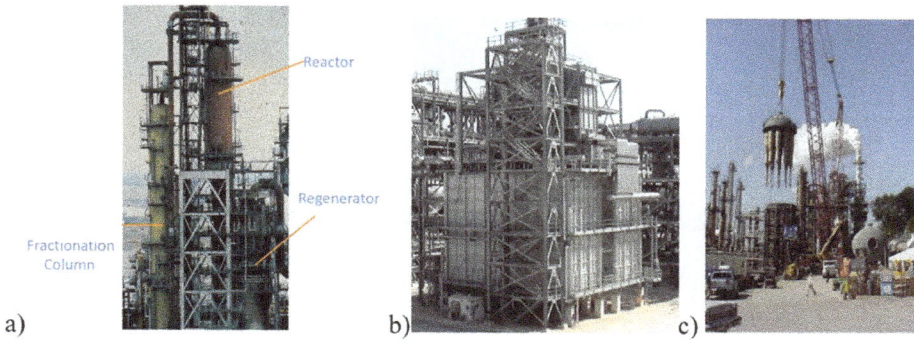

Figure 6.26: Photo of (a) the reactor section with regenerator, (b) furnace for heating the feed, and (c) cyclones.

Interestingly, highly loaded first-stage cyclones normally experience little to no cone erosion, whereas the lightly loaded second-stage cyclones can have severe cone erosion. For high-solids loading and low gas velocity typical for an FCC primary cyclone (Figure 6.28), the vortex length is much shorter because the high-solids loading dampens formation of a robust vortex. In the FCC, second-stage cyclone with solids loading of just a tiny fraction (10^{-3} to 10^{-4}) compared to the first-stage cyclone and a high gas velocity, the relatively long rapidly rotating vortex accelerates the solids stream causing intensive rotation and significant erosion.

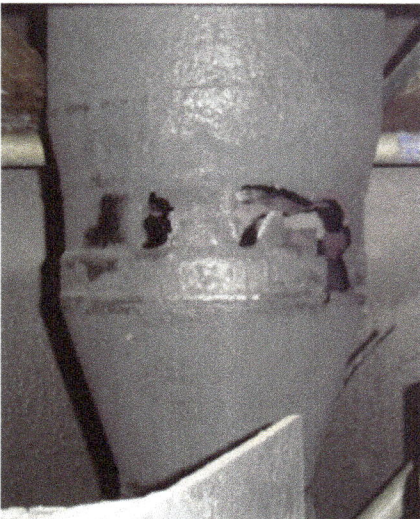

Figure 6.27: FCC cyclone damage by erosion.

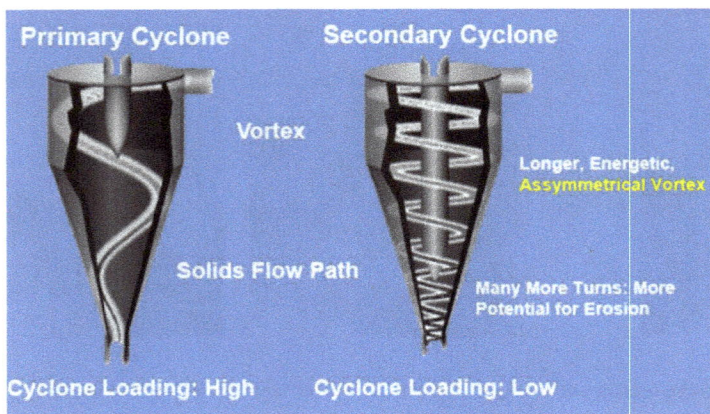

Figure 6.28: Vortices in FCC primary and secondary cyclones (from https://dc.engconfintl.org/cgi/viewcontent.cgi?article=1073&context=cfb10).

As mentioned above, the demand for propylene and ethylene is growing rapidly. To produce more propylene from heavy oil catalytic cracking process, there are two pathways. One is the introduction of additives (ZSM-5) in the conventional FCC reaction-regeneration system, while another option is to develop special FCC processes especially with residues as the feedstock. The main challenges are to minimize coking, excessive catalyst losses, and high-temperature excursions. Formation of coke, for example, inside the reactor cyclones can result in significant catalyst losses requiring often immediate unit shutdown. Utilization of the residue feedstock can lead to more frequent catalyst loading and unloading as well as larger catalyst addition rate increasing catalyst losses from the reactor and regenerator cyclones. An increase in the absorber off-gas yield compared to gas oil cracking has a negative effect on recovery of C3/C4 fractions. Larger levels of nitrogen and sulfur in the residue feed increase emissions of NO_x and SO_x and thus costs associated with mitigation of such emissions.

The R2R process (Figure 6.29) was initially developed by Total to process feedstock with a high residue content and is now licensed by Axens/IFP and Stone and Webster. In addition to a riser, a stripper, a disengager, and two standpipes, there is a regeneration system in this process composed of two regenerators linked by a lift. The first regenerator acts as a mild precombustion zone at a temperature not higher than 700 °C to achieve 40–70% of the coke combustion. The partially regenerated catalyst with less than 0.5 wt% coke is transported to the elevated second regenerator where complete regeneration is achieved at almost 900 °C with slight air excess and under a low steam partial pressure. Such conditions allow almost complete removal of coke.

Several processes developed to process heavy feedstock maximize gaseous olefins and gasoline production. SINOPEC deep catalytic cracking (DCC) and a

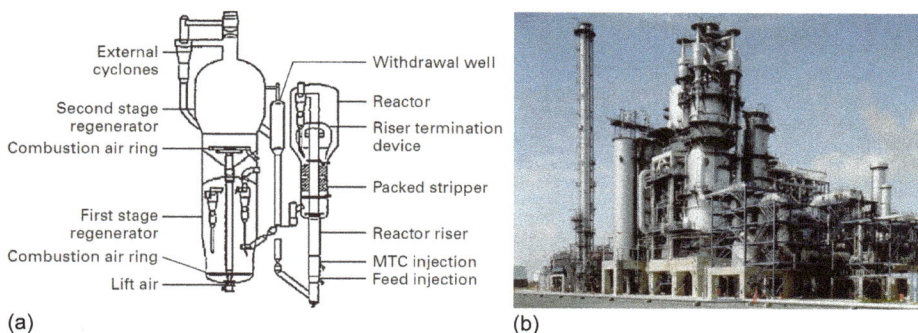

Figure 6.29: R2R FCC process (a) schematics from J. L. Fernandes, J. J. Verstraete, C. I. C. Pinheiro, N. M. C. Oliveira, F. R. Ribeiro, Dynamic modelling of an industrial R2R FCC unit, *Chemical Engineering Science*, 2007, 62, 1184. Copyright Elsevier. Reproduced with permission, b) photo of a unit.

catalytic cracking process for the production of clean gasoline (MIP-CGP) as well as PetroFCC, PetroRiser, and MILOS processes by UOP, Axens, and Shell respectively allow to get the yield of propylene up to 10–20 wt% which is much higher than in conventional FCC. These processes have multi-reaction zones and/or high operation severity. For the DCC and MIP-CGP processes (Figure 6.30), there is a diameter-enlarged stage in the middle of the riser. As shown in Figure 6.30, the temperature as well as the weight hour space velocity in the second reaction zone are lower. The reaction time in that zone is also different being longer.

The DCC process was developed in the 1990s and is very close to a standard FCC operating, however, at more severe conditions to optimize light olefins production. There are several changes however, introduced to meet the process objectives. Namely, the reaction time T is 560 °C and 530 °C in the first and second variants of the technology, respectively; the residence time is longer to enable cracking to light olefins; higher catalyst circulation and a higher C/O ratio (ca. 1.5–2 fold compared to classical FCC) provide the reaction heat; lower hydrocarbon partial pressure (and thus more steam) gives higher propylene yields and less coke. The conditions are less severe as for steam cracking discussed below requiring also less steam. The paraffinic feeds are preferred improving the olefins yields. A riser followed by a fluid bed is used to obtain longer residence time giving more dry gas yield and coke. Yields of up to 20 wt% propylene and 15 wt% butylenes were claimed for this configuration using 530–575 °C, C/O ratio of 8–15, steam to feedstock ratio of 20–30%.

The Axens Petroriser technology involves a second riser (Figure 6.31) as recycling of light naphtha to a separate riser at a temperature higher than the main riser allows cracking of the C_5 and C_6 olefins and also paraffins to attain more LPG and less C_5–70 °C naphtha. Moreover, such technology leads to high aromaticity of the naphtha produced.

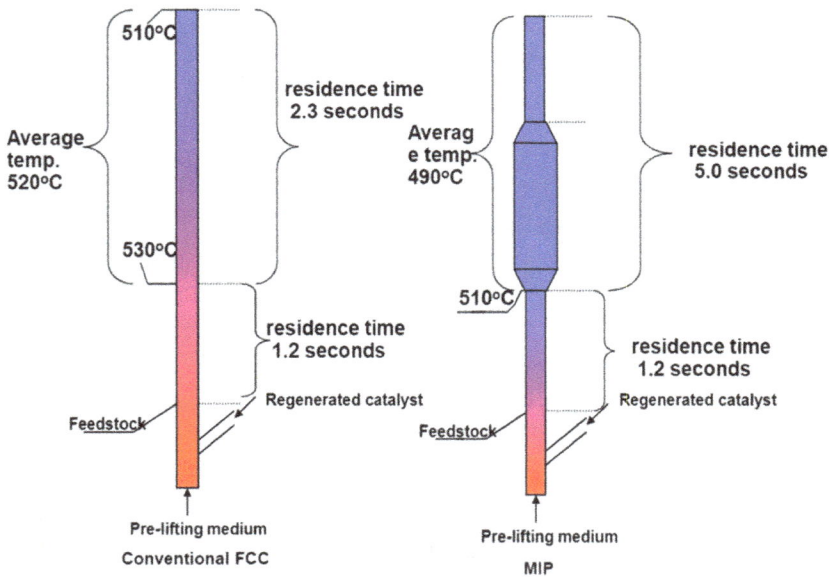

Figure 6.30: The two reaction zone risers in MIP process in comparison with a conventional FCC. From https://docplayer.net/60515871-Dcc-deep-catalytic-cracking-mip-maximizing-isoparaffin-and-cgp-clean-gasoline-propylene-from-Sinopec.html.

In Petro FCC process of UOP using severe cracking operation conditions in the riser and the RxCat technology (Figure 6.32), the spent catalyst from the stripper is recycled directly to the reactor and mixes with the regenerated catalyst, which has a higher temperature than the catalyst mixture. Such operation allows lower thermal cracking and the dry gas yield. The reactor has two different contact time zones in the same riser: the first one with higher temperature and very short contact time, followed by a second one with a lower temperature and a longer contact time.

An increase in the riser C/O ratio by blending carbonized catalyst with the regenerated one improves conversion at a constant reactor temperature and catalyst activity or decrease the reactor temperature and thermal cracking still maintaining or exceeding the original conversion level. This process gives high yields of propylene, light olefins, and aromatics from feedstock which can include conventional FCC feeds as well as higher boiling or residual feeds. The catalyst used in PetroFCC technology comprises a large pore zeolite (e.g., Y zeolite with a limited content of rare-earth elements) and a second component with a medium pore zeolite (e.g., ZSM-5). Presence of the latter improves selectivity to light olefins by further cracking the lighter naphtha range molecules.

The ratio of the coked to the regenerated catalyst is in a range of 0.3 to 3.0 weight bases. The following conditions are typically used in PetroFCC process to maximize the olefins yield and selectivity: a reactor temperature of 510–620 °C, a C/O ratio of at

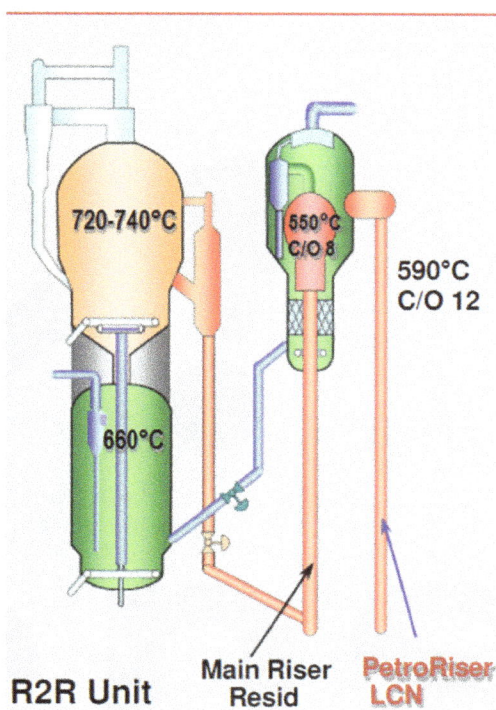

Figure 6.31: The Axens Petroriser technology (from https://cupdf.com/document/ax-ens.html).

least 10:1 and preferably between 15 and 25 wt/wt. The steam diluent flow rate is between 10 and 55 wt% of the feed flow rate to minimize the hydrocarbon partial pressure. The contact time is lower than 2 s.

The propylene yield of ca. 20 wt% with atmospheric residue as the feedstock was also proven in several commercial installations in China operating the two-stage riser catalytic cracking of heavy oil (TMP process of China University of Petroleum). The content of dry gas and coke is ca. 14.3 wt%. Figure 6.33 shows the schematic diagram of the TMP process (the reaction-regeneration system).

For the first riser, the feed comprises butenes (C_4) and the atmospheric residue, while the light gasoline (mainly C_5 and C_6 olefins) and the recycling oil are fed into the second riser. The outlet conditions of both risers are shown in Figure 6.33; for the second riser, the outlet temperature is 520–550 °C, being only slightly higher than the conventional FCC process.

An interesting technology giving up to 25% of propylene by converting heavy hydrocarbon feedstock under severe FCC conditions was developed by several companies including Saudi Aramco and JX Nippon Oil and Energy Corporation. The main features of this HS-FCC technology implemented at the Mizushima refinery

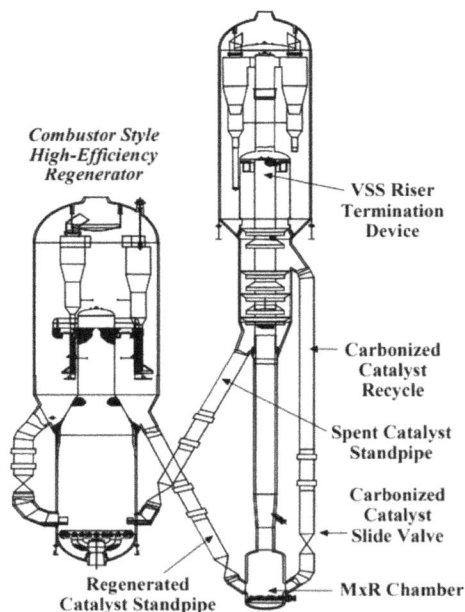

Figure 6.32: RxCat reactor of UOP (from https://honeywell-uop.cn/wp-content/uploads/2011/02/UOP-FCC-Innovation-tech-paper1.pdf).

(Figure 6.34a) which was in operation from 2011 to 2014 are a down-flow reactor (Figure 6.34b), high reaction temperature, short contact time and high C/O ratio.

In general, high temperature (550 to 650 °C) and high C/O ratio result in both thermal cracking and catalytic cracking. The former leads to dry gas while catalytic cracking enhances propylene production. Application of contact times lower than 0.5 s at the same time minimizes thermal cracking. In order to compensate for lower conversion at shorter contact times a high C/O is used, enhancing catalytic over thermal cracking.

A down-flow reactor, allowing a higher C/O ratio because lifting of the catalyst by vaporized feed is not required, has been added to HS-FCC technology. The catalyst and the feed are flowing downward by gravity through the reactor, minimizing back mixing and ensuring a narrower residence time distribution (Figure 6.35). The latter is favorable for production of gasoline and light olefins.

6.5 Steam cracking

In steam cracking, saturated hydrocarbons such as naphtha (boiling range 30–200 °C, mainly in Europe and Japan), LPG (mainly in the USA), ethane, propane, and butane are thermally cracked in the absence of oxygen and in the presence of steam in a bank of pyrolysis furnaces to produce mainly alkenes, such ethylene and propylene. The

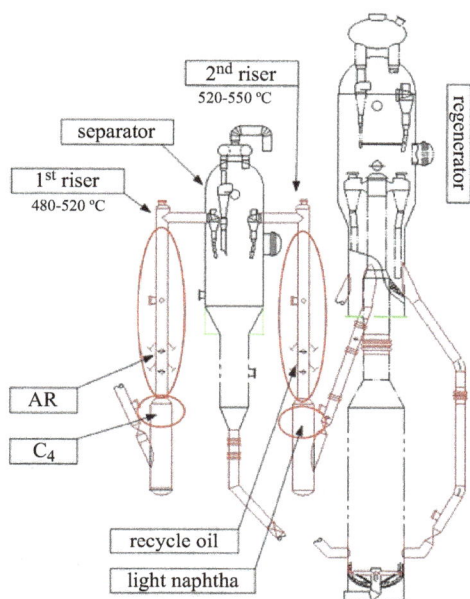

Figure 6.33: Schematic diagram of the TMP process. From C. Yang, X. Chen, J. Zhang, C. Li, H. Shan, Advances of two-stage riser catalytic cracking of heavy oil for maximizing propylene yield (TMP) process. *Appl Petrochem Res.*, 2014, 4, 435–439, https://doi.org/10.1007/s13203-014-0086-6. Open Access.

product composition depends on the feeds, steam/hydrocarbon (steam/carbon) ratio, cracking temperature, and residence time. Polyethylene is the main outlet (ca. 50%)

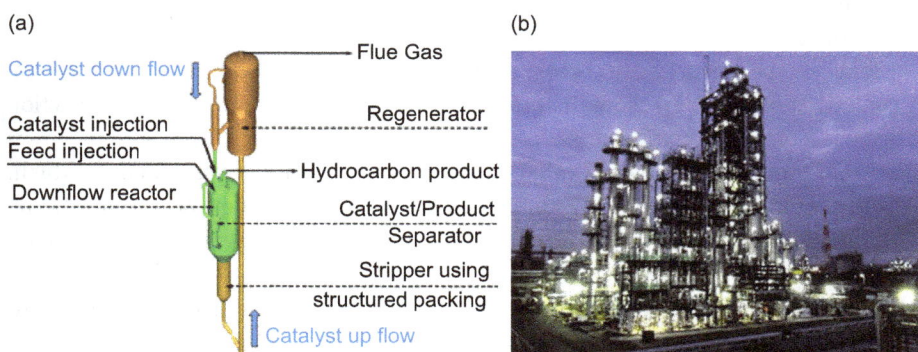

Figure 6.34: HS-FCC process (a) photo of a semi-commercial plant at Mizushima refinery of JX, (b) flow scheme (from R. Partasarathi, S. S. Alabduljabbar, HS-FCC High-severity fluidized catalytic cracking: a newcomer to the FCC family, *Appl Petrochem Res.*, 2014, 4, 435–439, https://link.springer.com/article/10.1007/s13203-014-0087-5/figures/3. Open Access).

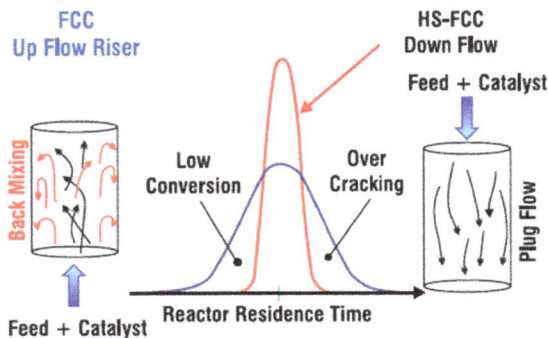

Figure 6.35: Up flow vs. down flow residence time profiles.

for ethylene produced in steam cracking. Besides valuable olefins, hydrogen, and aromatics, significant amounts of CO_2 are also produced, and NO_x emissions are generated. Light hydrocarbon feeds including light naphtha give product streams rich in the lighter alkenes (ethylene, propylene, and butadiene), while heavier hydrocarbons (full range and heavy naphtha), besides light alkenes, also generate aromatic hydrocarbons in the so-called pyrolysis gasoline C5+ fraction.

The stream (Figure 6.36a) enters a fired tubular reactor (radiant tube or radiant coil) and is heated from 500–650 °C to 750–875 °C. The residence time is ca. 0.1–0.5 s, allowing cracking of hydrocarbons in the feed into smaller molecules. Such a short residence time is required for high selectivity toward the main desired products (ethylene and propylene), while secondary reactions will be much more prominent at longer residence time. Endothermicity of cracking requires high energy input. The products are immediately cooled within 0.02–0.1 s to 550–650 °C in order to prevent degradation of the highly reactive products by secondary reactions, such as, polymerization of olefins or condensation reactions to produce cyclodiolefins and aromatics.

Radical-type reactions are commonly assumed to take place during steam cracking. A somewhat simplified scheme is presented in Figure 6.37, while the real reaction scheme is much more complex.

The presence of olefins and aromatics in the feed is unfavorable. Olefins, which, in fact, should be produced in steam cracking, give rise to condensation and coking, while aromatics undergo C–H rather than C–C cracking, resulting in dehydrogenation and condensation.

The fundamentals of furnace design and the main influence of the different parameters can be understood taking into account thermodynamics and kinetics of pyrolysis. A high temperature is needed for achieving high conversion because unfavorable thermodynamics is also dependent on the substrate structure with smaller alkanes requiring higher temperature. With the increase in molecule size, cracking occurs close to the end of the molecule, leaving a light alkane and a heavier product, which can be cracked further. A decrease in conversion naturally follows

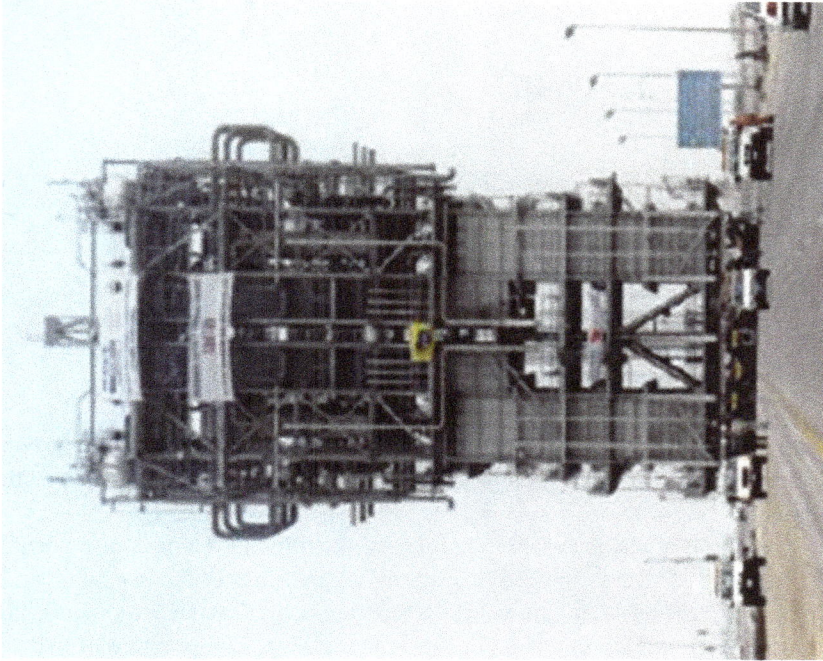

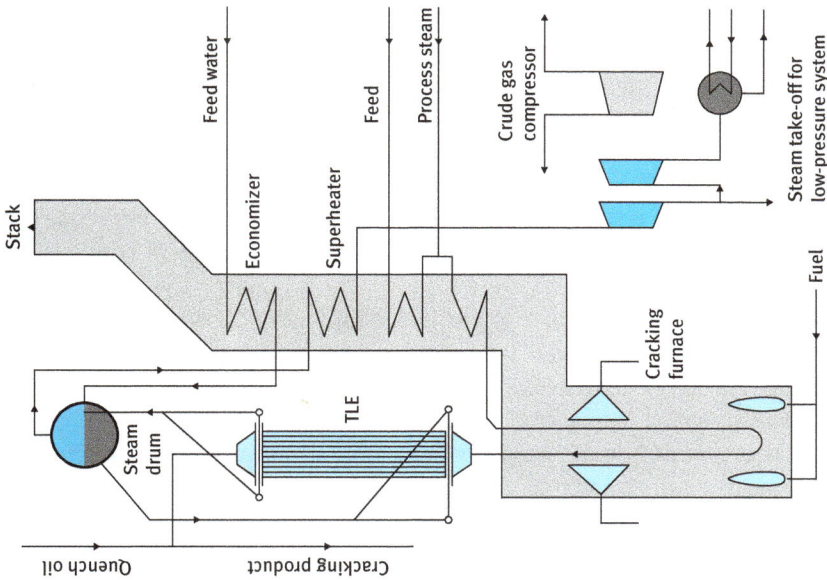

Figure 6.36: Steam cracking reactor: (a) detailed scheme with transfer line exchanger (TLE). http://www.shg-schack.com/uploads/media/Transfer_Line_Exchangers.pdf. b) Photo.

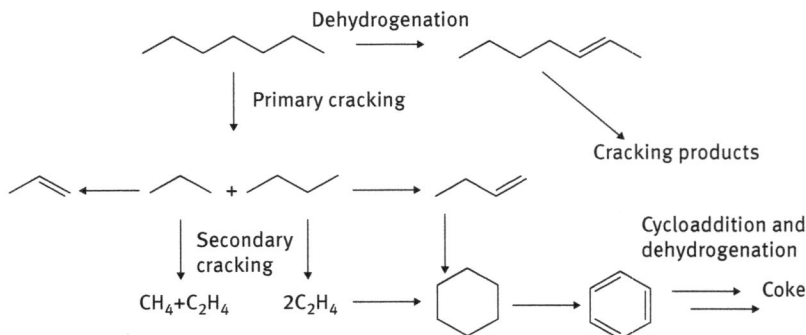

Figure 6.37: Thermal cracking reactions.

from the thermodynamic viewpoint with pressure increase, since in cracking, two molecules are formed from one substrate. Instead of using vacuum, dilution with steam is applied.

The reaction rate of steam cracking is increasing with chain length and partial pressure of reactants. At the same time, higher partial pressures of the reactants, and therefore of the products, lead to unfavorable secondary reactions, such as condensation reactions and generation of coke. For this reason, the partial pressure of hydrocarbons must be kept low by diluting with steam, favoring formation of primary products and increasing olefin yield. Presence of steam also reduces the partial pressure of high boiling, high-molecular-mass aromatics, and heavy-tarry materials, preventing their deposition on tubes, forming coke.

The quantity of steam, expressed as steam ratio (kilograms of steam per kilogram of hydrocarbon), depends on the feedstock, cracking severity, and design of the cracking coil, being in the range of 0.4–0.5 for steam cracking of naphtha. The downside of steam utilization is generation of significant amounts of condensate, which contains hydrocarbons as impurities.

The oligomerization reactions involved in the formation of secondary products are favored by lower temperatures; therefore, special temperature profiles are applied along the cracking coil to avoid long residence times at low temperatures and ensure rapid temperature increase in the inlet section.

The sulfur content of the feed is also important, as sulfur passivates active Ni sites of the cracking coil material by forming nickel sulfides. The latter, contrary to nickel per se and nickel oxides, do not catalyze coke gasification. Such gasification leading to high content of CO (up to 1% in the cracked gas) in the case of sulfur-free feedstock can be prevented when ca. 20 ppm of sulfur in the form of, for example, dimethyl sulfide is added to sulfur-free feedstock.

In summary, maximum ethylene production requires a highly saturated feedstock, high coil outlet temperature, low hydrocarbon partial pressure, short residence time in the radiant coil, and rapid quenching of the cracked gases.

The yields of the products in cracking various feedstock are given in Table 6.8.

Modern cracking furnaces typically have one or two rectangular fireboxes with vertical radiant coils located centrally between two radiant refractory walls. Firebox heights of up to 15 m and firebox widths of 2–3 m are standard design practice in the industry.

Firebox length is determined by the total ethylene production rate desired from each furnace and the residence time of the cracking operation. Short-residence-time coils, which are in fact fired tubular chemical reactors, can be as short as 10–16 m per coil. Long-residence-time coils can have lengths of 60–100 m per coil. The number of coils required for a given ethylene capacity is determined by the radiant coil surface, which is in the range of 10–15 m^2 per ton of feedstock for liquid feedstock. The production rate for each coil is determined by its length, diameter, and charge rate, which translates into a certain heat flux on the radiant coil. Coil alloys are normally protected by an oxide layer on the inner surface. The number of coils with an internal diameter of 25–180 mm can range from just a few to 200, allowing annual capacity per furnace based on ethylene up to 250×10^3 t.

Tube wall temperature rises with time because of coke deposition inside the coil. Coke acts as a thermal insulator inside the coil, requiring increased tube wall temperature for a given furnace loading. A steam-cracking furnace is thus run only for a few months at a time between decokings. The latter process requires the furnace first to be isolated from the process followed by exposure to a flow of a steam/

Table 6.8: Yields of products in steam cracking.

Parameters	Gas feed			Liquid feed		
	C_2H_6	C_3H_8	n-C_4H_{10}	Naphtha	Gas oil	Vacuum gas oil
Yield (mass %)						
Ethylene	48	36.7	31.6	31.3	26	23
Propylene	2.1	14.0	17.8	12.1	9.0	13.7
Butenes	1.1	3.1	2.4	2.8	2.0	4.9
Butadiene	–	–	1.7	4.2	4.2	6.3
Dry gas	8.4	33.2	32.2	18.2	16.2	11.2
Arenes						
C6–C8	1.7	5.0	10.3	13.0	12.6	16.9
Gasoline (without arenes)	–	–	–	9.0	8.0	–
Heavy tars	–	–	–	6.0	19.0	21.0
Unreacted feed (mass %)	38.7	8.0	4.0	–	–	–
Conversion (mass %)	61.3	92	96	Complete	–	–

air mixture at ca. 800 °C, generating CO and CO_2 from coke. Overheating of the radiant coil should be avoided; thus, the concentration of air should be increased carefully. An alternative is steam-only decoking, which does not require disconnection of the furnace and can be used at higher temperatures (950–1,000 °C) if the downstream processing equipment can handle the required amount of steam and CO.

The furnace should never be shut down or cooled without radiant-coil decoking, as it will lead to rupture of coils because of differences between the thermal expansion coefficients of coke and the metal tube wall.

An example of the catastrophic collapse of the coils is shown in Figure 6.38 being a result of the naphtha feed pumps going off service, unsuccessful immediate intervention of the steam emergency automatic injection, and blocking of the fuel oil to the burners. All coils have fractures with a high thickness coke layers inside.

Furnace capital costs represent ca. 20% of the total cost of an ethylene plant, with approximately one third for radiant coils, which have service lives of 5–7 years depending on the position in the furnace.

Novel developments in steam cracking technology are related to design of electrically heated steam cracker furnaces, which will significantly reduce CO_2

Figure 6.38: Collapse of the furnace coils (from https://media-exp1.licdn.com/dms/image/
C5122AQE7sciD9B54-Q/feedshare-shrink_2048_1536/0/1577899217620?e=1623888000&v=
beta&t=KmdrM_cx_Em3NjRLfIcXr3xQlUWZNTj-W2u0sfDvjDI).

emissions from the burning of fossil fuels. The biggest challenges in integration of an e-furnace plant into the steam cracker are related to the materials choice for low-voltage higher amperage electric furnaces and electricity storage, and ensuring a continuous power supply to the electric cracker furnace.

After leaving the cracking furnace, the gas is quickly quenched to stop the reaction. This is done preferably in a transfer line heat exchanger, TLE (Figure 6.36a) or inside a quenching header using a more rapid quench oil injection.

Indirect quenching in the TLEs permits heat recovery at a temperature high enough to generate valuable high-pressure steam. The temperature of the cracked gas leaving the primary TLE depends on the feedstock type varying from 300 °C for light feedstock, to 420 °C for naphtha, to 600 °C for atmospheric gas oil and hydrocracker residues.

The main downstream processing steps (Figure 6.39) are the removal of the heat contained in the cracked gas, condensation of water and heavy hydrocarbons, compression, washing, drying, separation, and hydrogenation of certain multiply unsaturated components.

The principle ethylene co-products are processed in the following way. Hydrogen is purified and used in different hydrogenations; methane can be used as a fuel, while ethane and propane are also applied as a fuel or recycled. Compounds with the triple bond such as ethyne (acetylene) and propyne are hydrogenated to monoenes. C4 and C5 fractions are refined for obtaining olefins. The main products such as ethylene and propylene are used in polymerization or for producing various chemicals. Pyrolysis gasoline is either sold as a motor gasoline component after hydrotreatment or is used as feedstock for the production of aromatics. Other minor products (fuel oil, naphthalene, tar, etc.) are also recovered.

Processing of cracked gas resulting from liquid feedstock is complex, as high amounts of heavy hydrocarbons are condensed and must be removed in this section. Tar and oily compounds are separated in a primary fractionator from the cracked gas with circulating oil. In subsequent water quench, the cracked gas still containing all steam is cooled to near-ambient temperature by contacting it with a large stream of circulating quench water. This results in the condensation of the major part of steam and a heavy gasoline fraction. Condensed gasoline is sent to the top of the primary fractionator. Steam is also recycled. As the separation of tarry material from water is not straightforward, online or offline washing with high-boiling, aromatic-rich solvents can be used to prolong the on-stream time.

Further processing of cracked gas requires compression to ca. 3,200–3,800 kPa. Interstage cooling during compression with water prevents polymerization by keeping temperature below 100 °C. Gasoline fraction containing the C6 to C8 aromatics formed during naphtha cracking is condensed in the interstage coolers and after separation from water of the compressor undergoes fractionation to remove C4 and lighter components.

After cooling of the cracked gas and removal of the main part of gasoline fraction, carbon dioxide and hydrogen sulphide are removed from the cracked gas by

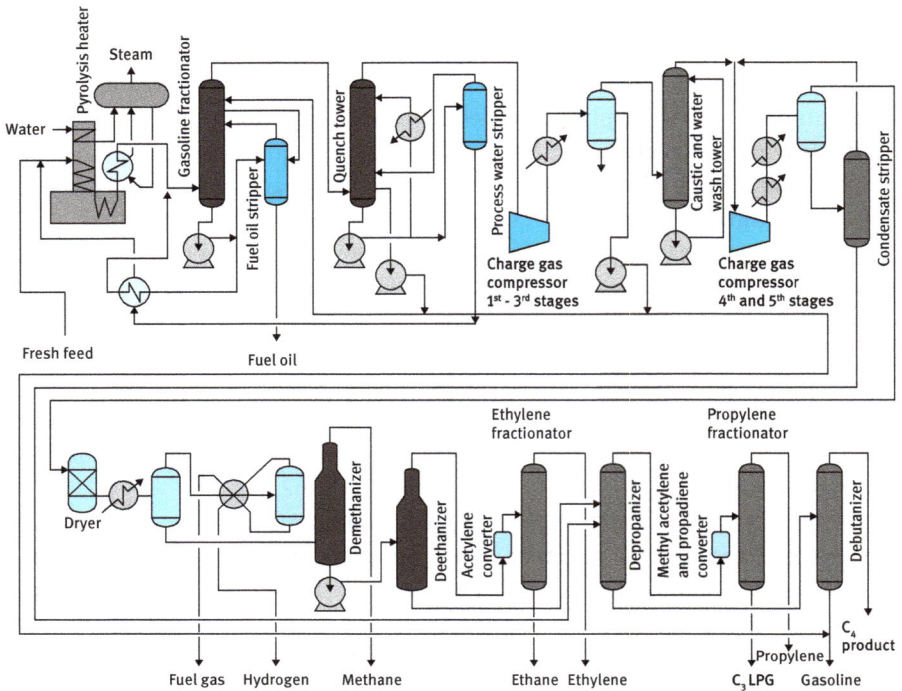

Figure 6.39: Processing of products from steam cracking. Modified after http://www.treccani.it/ex port/sites/default/Portale/sito/altre_aree/Tecnologia_e_Scienze_applicate/enciclopedia/inglese/ inglese_vol_2/551-590_INGL3.pdf.

once-through caustic wash and regenerative solvent scrubbing with amines. Such acid gas removal prior to drying is required because CO_2 can freeze at low T in heat exchange and fractionation equipment, while H_2S is a catalyst poison and corrosive.

Drying is done prior to fractionation by applying multiple adsorption beds. Subsequent fractionation of hydrogen and hydrocarbons is predominantly done using cryogenic separation.

Several process options are possible, including demethanizer first (such as in Figure 6.39) with tail-end hydrogenation, deethanizer first with front-end hydrogenation, and depropanizer first with front-end hydrogenation. The most energy efficient is front-end deethanizer (Figure 6.40).

In this option, after the first fractionation step, an overhead stream contains ethane and lighter components, while a bottom stream has C3 and heavier materials. Selective hydrogenation of acetylene present in the lighter stream is needed, as acetylene acts as a poison for polymerization catalysts. During the chilling and demethanizing, the overhead is separated into methane and not-very-high-purity hydrogen (80–95 vol%) containing up to 0.1% of CO. Because carbon monoxide is a

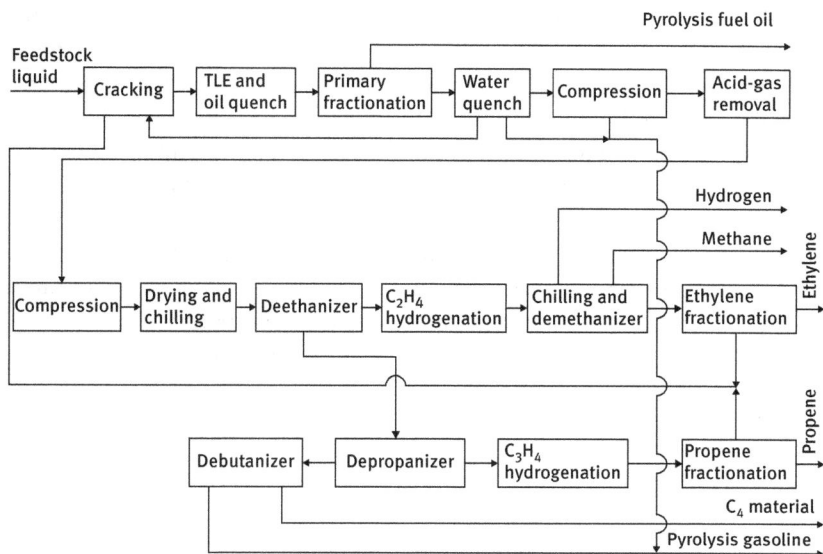

Figure 6.40: Simplified process flow diagram for the production of ethylene by liquid cracking with a front-end deethanizer.

poison for palladium catalysts used in multiple hydrogenation steps, hydrogen is purified by methanation of CO with subsequent pressure swing adsorption.

The bottom fraction, after demethanizing, contains ethylene and ethane and is separated with ethane stream by fractionation to yield ethylene and ethane. The latter is recycled back to cracking furnaces. High-pressure fractionation requires a high reflux ratio (ca. 4) and 90–125 separation stages giving ethylene purity above > 99.9 wt%. Heating and cooling of the fractionation columns is integrated into the refrigeration cycle, with a low-pressure refrigerant being vaporized in the top condenser, providing chilling duty. When this vaporized refrigerant is compressed, the vapours are condensed in the reboiler providing heat duty.

C3 + fraction is separated into C3 and C4 + fractions. The former undergoes selective hydrogenation to remove triple bond in methylacetylene and double bond in propadiene, which both account for 2–6% of the C3 fraction. Hydrogenation of C2 and C3 fraction is performed with palladium-based catalysts in adiabatic bed reactors (with one or two beds) operating at 2 to 3.5 MPa and 25–100 °C in the presence of excess hydrogen as well a minute amounts of CO in the parts-per-million range to improve selectivity. The operation cycle is typically several months, during which the catalyst is gradually deactivating. Such deactivation is compensated by temperature increase and regenerations at the end of a cycle.

The C4 cut from the top of the debutanizer contains besides 1, 3-butadiene, also normal and isobutenes as well as small amounts of vinylacetylene, ethylacetylene, and butanes. Different hydrogenation options for this stream are possible, depending on a particular situation in a steam cracker and downstream processing technology. For example, if there is no use for a C4 fraction, then full hydrogenation can be performed with the formed butanes recycled to cracking furnaces. Other options can include selective hydrogenation of ethylacetylene and vinylacetylene to butadiene if the latter is the desired product or selective hydrogenation of butadiene to butenes.

C5 + fraction or pyrolysis gasoline contains mainly aromatics, such as benzene, toluene, styrene, and xylenes, as well as some non-aromatics such as unsaturated hydrocarbons with triple and conjugated double bonds. The processing of this fraction depends on the application, for example, valuable aromatics used in various chemical transformations are recovered by extraction.

Chapter 7
Isomerization

Different types of isomerization reactions are practiced industrially. In this chapter, just few examples will be considered, limited to heterogeneous catalytic skeletal isomerization of alkanes, including skeletal transformations combined with dehydrogenation (catalytic reforming), as well as epimerization of aldoses to ketoses, catalyzed by enzymes.

7.1 Skeletal isomerization

In the subsequent section, catalytic reforming of straight-run naphtha to gasoline will be considered. Reforming is done to improve the octane number (octane rating) of gasoline, which is used in high-performance petrol engines requiring higher compression ratios. Fuels with a higher octane number can withstand more compression before detonating (igniting). A list of some compounds with their octane numbers is given in Table 7.1. As can be seen from Table 7.1 skeletal isomerization of straight chain alkanes, cyclization and aromatization significantly improve the octane number.

Table 7.1: Octane numbers and boiling points for some organic compounds.

	Octane numbers	Boiling points (K)
n-Pentane	62	309
2-Methylbutane	90	301
Cyclopentane	85	322
n-Hexane	26	342
2, 2-Dimethylbutane	93	323
Benzene	>100	353
Cyclohexane	77	354
n-Octane	0	399
2, 2, 3-Trimethylpentane	100	372
Methyl-tertiary-butyl-ether	118	328
Straight-run gasoline	68	

https://doi.org/10.1515/9783110712551-007

Table 7.1 (continued)

	Octane numbers	Boiling points (K)
Fluid catalytic cracking light gasoline	93	
Alkylate	95	
Reformate (CCR)	99	

In gasoline engine, the premixed air fuel is ignited by a spark and combustion proceeds in a progressing flame front. Typical for non-branched alkanes present, for example, in straight-run gasoline is undesired uncontrolled self-ignition. Contrary to gasoline engines in diesel engines, the fuel is injected in hot compressed air present in the cylinder relying thus on efficient self-ignition.

The octane number, determined by burning gasoline in an engine under controlled conditions before a standard level of knock appears, thus represents not burning efficiency but resistance to premature detonation.

For production of high-octane gasoline, a C7–C9 cut is the preferred choice. When the target is BTX (benzene, toluene, xylene) C6–C8 cut rich in C6, is mainly employed.

Catalytic reforming, which also comprises dehydrogenation, is limited to feeds in the boiling range of 100–180 °C. Typical operating conditions of catalytic reforming do not allow sufficient conversion in skeletal of lower boiling normal C5–C6 paraffins, ensuring a meaningful increase in the octane number. The skeletal isomerization reaction is limited by equilibrium being preferred at low temperatures, which imposes kinetic restrictions and requires an efficient catalyst. Operation at the lowest feasible temperature allows a higher fraction of disubstituted rather than monosubstituted isomers, which is beneficial for the octane number increase.

Historically, aluminum chloride effective at 115–120 °C was used, leading, however, to acid corrosion of the equipment and formation of sludges. Further catalyst development was associated with introduction of platinum catalysts supported on alumina (chlorided), sulphated zirconia, and zeolites.

Similar to C5–C6 fraction, skeletal isomerization of normal butane to isobutane can also be done over platinum on chlorided alumina catalysts. The aim of such reaction is to generate a feedstock for subsequent alkylation or production of methyl tert-butyl ether (Chapter 12).

The presence of the metal is because of the bifunctional nature of the process comprising first dehydrogenation of alkanes to olefins on the metal sites, skeletal isomerization of olefins on acid sites, and finally hydrogenation on the metal sites. From the chemical reaction technology viewpoint, an important step was related to separation of low octane paraffins from the product with molecular sieves and recycling of the unconverted feed driving the overall conversion to almost 100%.

Platinum (0.3–0.5 wt%) deposited on highly chlorinated alumina (8–15 wt% Cl_2) operates at low temperature (ca. 130 °C) to improve the equilibrium yield and hydrogen pressure (ca. 15 bar). Standard hydrotreating (Chapter 10) upstream skeletal isomerization is required to remove impurities (e.g. sulphur), which are otherwise detrimental for the catalyst. Presence of hydrogen as an astoichiometric component (ratio 2:1) helps to mitigate deactivation due to coking, but at the same time results in elution of chloride as hydrogen chloride. Subsequently, for catalyst activation, CCl_4 is injected into the feed. An alternative to chlorided alumina not requiring such additive is the proton-form mordenite zeolite as a catalyst support, which is more stable and water tolerant. Lower activity of a more stable zeolite-supported catalyst requiring operation at 250–270 °C and 30 bar can be compensated by recycling normal paraffins. The research octane numbers for chlorided alumina and zeolite-supported catalysts in the schemes with recycling can be, respectively, 90–91 (or even 93) and 89–90. The lifetime of mordenite-supported catalysts operating at 60+ conversion and ca. 97% selectivity can be as long as 7–8 years before replacement.

The Pemex process developed by UOP for upgrading C5/C6 light straight-run naphtha feeds uses chloride-promoted Pt on alumina catalysts. Without recycling, the research octane number is 83–86. The process flow diagram is shown in Figure 7.1. The scheme with recycling and fractionation recycles back to the reactor methylpentanes and *n*-hexane.

A process of Sinopec for light naphtha isomerization using a bifunctional catalyst with a zeolitic support is displayed in Figure 7.2.

The Pt catalyst on a zeolitic support operates at 220–280 °C and 1.5–3 MPa, the hydrogen to hydrocarbon molar ratio of 1 to 3 and is applied in the form of extrudates with the diameter of 1.6–2.2 mm and length of 3–10 mm. The catalyst operates for 2–3 years prior regeneration and can last up to 8 years. Several plants operate in China using this technology with the largest one having a capacity of 340 kt/a.

A Pt/sulphated-zirconia catalyst is applied in ISOMALK-2™ process of GTC Technology Company (Figure 7.3).

As can be seen, the scheme is essentially the same as presented in Figure 7.2 comprising two reactors and a heater. Such catalysts forming the basis of the new generation of isomerization catalyst have, however, considerably higher activity than the zeolitic ones, allowing to lower the reaction temperature by ca. 85 °C and thus to improve the octane number by 3 points.

Par-Isom process (Figure 7.4) is different from the scheme above as only one reactor is used with a possibility for catalyst regeneration *in situ* by burning carbon and subsequent catalyst reduction under hydrogen.

Introduction of sulphated zirconia-based catalyst into industrial practice can lead to revamps of existing zeolitic isomerization units, thereby increasing their capacity. Sulphated zirconia is not permanently deactivated by water or oxygenates in the feed, thus such catalyst can be used in units where due to contaminants in the feed even more active chlorided alumina catalysts cannot be used. Operation

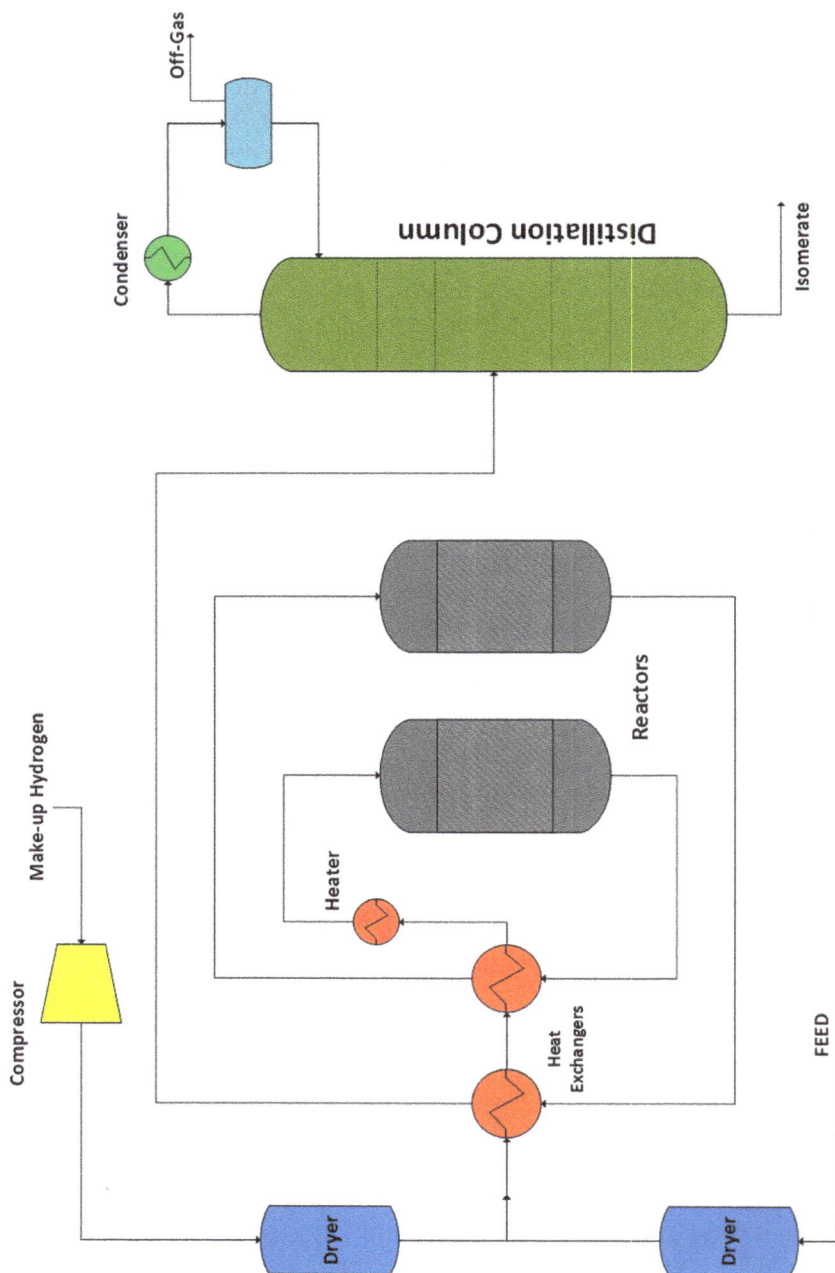

Figure 7.1: Pemex process for isomerization of C5/C6 light straight-run naphtha feeds. From https://www.linkedin.com/pulse/processing-routes-cleaner-high-performance-gasoline-da-silva-mba/?trk=read_related_article-card_title.

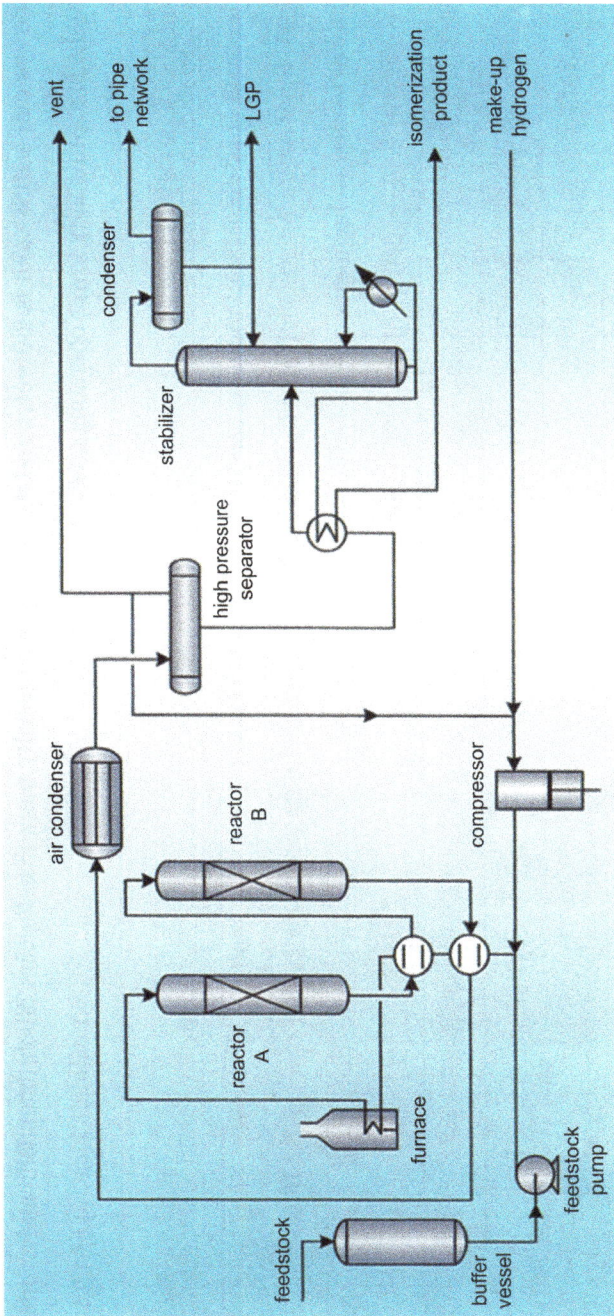

Figure 7.2: Skeletal isomerization of light naphtha, Sinopec RISO process.

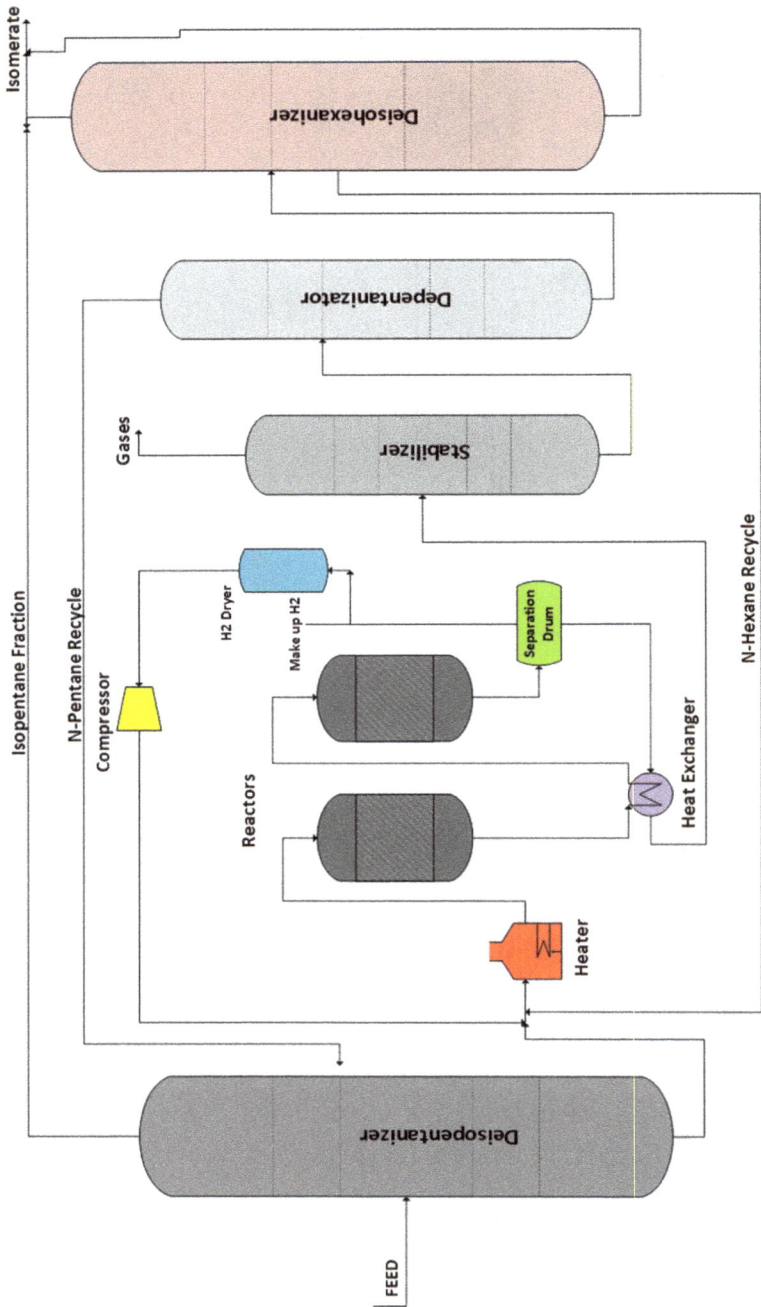

Figure 7.3: Flow scheme for ISOMALK-2 process (GTC Technology Company).

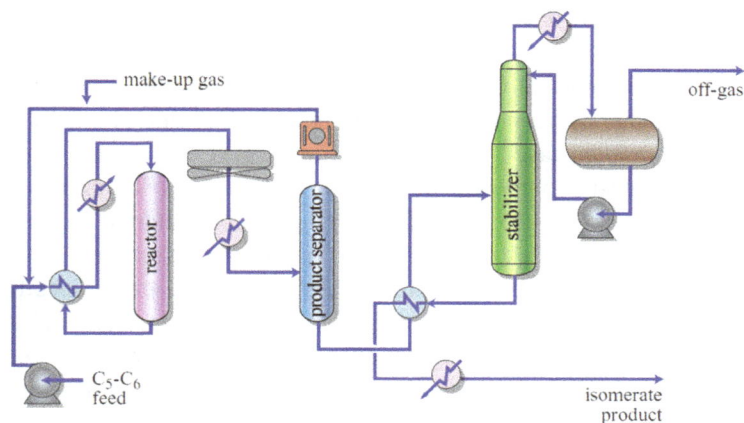

Figure 7.4: Par-Isom process scheme operating with a Pt/sulphated zirconia catalyst.

under higher hydrogen ratios than in a Pemex process requires a high-pressure separator and a recycle compressor. The same pieces of equipment are also visible in Figures 7.2 and 7.3.

A fired heater is not used in the Par-Isom process as the reaction temperature is lower than for Pt on a zeolite. In the product separator, hydrogen is separated, compressed, and recycled to the reactor. In the stabilizer column, the light ends and dissolved hydrogen are separated from the stabilized isomerate which can be used for gasoline blending. Alternatively, the isomerate undergoes distillation giving n-hexane and low-octane methylpentanes as a side-cut stream. The latter is used as a feed for the isomerization reactor similar to the Pemex process. The lights comprising isopentane, 2,2-dimethylbutane, and 2,3-dimethylbutane are recovered for gasoline blending.

7.2 Combining skeletal isomerization and dehydrogenation: catalytic reforming of gasoline fractions

In the reforming processes, gasoline fractions (complex mixtures like hydrotreated straight-run naphtha typically containing C6–11 paraffins, naphthenes, and aromatics with more than 300 compounds) with a low octane number are converted into a high-octane reformate, which is a major blending product for motor gasoline. Non-straight-run naphtha such as fluid catalytic cracking naphtha or visbreaker/coker naphtha can be also processed after substantial hydrotreatment with removal of olefins and dienes to prevent severe coking.

Currently, it is estimated that, globally, there are 1,500 catalytic reforming units with the overall annual capacity of 500 million tons.

Catalytic reforming is mainly based on the catalytic conversion of normal paraffins and cycloparaffins into aromatics and isoparaffins, e.g., combining isomerization and dehydrogenation. The most valuable by-product is hydrogen, as the catalytic reformer is the only indigenous hydrogen source of the refinery. The other important by-product is liquefied petroleum gas.

In terms of reactivity, most naphthenes react rapidly and efficiently to form aromatics, while reactions of paraffins are more difficult. As a result, a preferred fraction for catalytic reforming is naphtha rich in naphthenes with low paraffin content.

Efficiency of aromatization can be apparently problematic, as the aromatic content in reformate could reach 50%, while the regulations for class 5 gasoline require no more than 35% of aromatics. Prefractionation of the feed to minimize benzene, cyclohexane, and methylcyclopentane or post-fractionation of the reformate and further processing of the light reformate are the options to meet current regulations.

The catalysts (Figure 7.5) applied are either single metal type (platinum-on-alumina) or bimetallic ones, where platinum (0.25–0.8 wt%) is used in combination with a second metal, such as catalytically active Ir, Rh, or Re or inactive Sn or Ge.

Figure 7.5: Naphtha reforming catalysts.

Bimetallic catalysts have a better operational stability but are more sensitive toward poisoning by sulphur. Various commercial catalyst types also contain halogens, usually chlorine.

Typically, PtRe catalysts are used in fixed-bed reactors and PtSn in reactors with continuous regeneration. Coke formation is the main reason for catalyst deactivation in naphtha reforming, thus some of additives such as tin are utilized to prevent coke deposition, also enhancing selectivity to aromatics. Aromatization is also improved by addition of In. When addition of a promoter (Ir) substantially enhances hydrogenolytic activity in order to prevent significant heat release in C–C cleavage, sulfiding pretreatment is done prior to regular operation.

Innovation in reforming catalysts continues with the aim to improve the liquid and hydrogen yields, diminish sensitivity to poisoning, and decrease metal loading.

More active catalysts are typically more expensive affording at the same time better performance in terms of the yield and/or run length.

Several reactions happen during catalytic reforming (Figure 7.6).

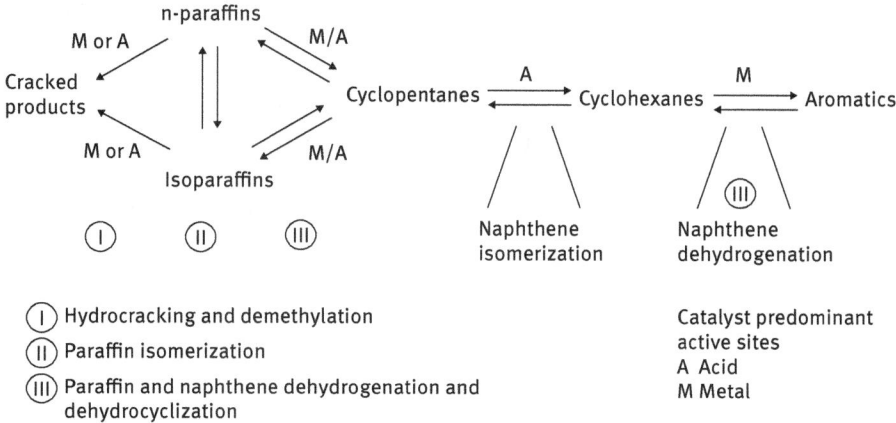

Figure 7.6: Reactions in catalytic reforming.

These include dehydrogenation of naphthenes to aromatics, dehydroisomerization of alkyl naphthenes to aromatics first through skeletal isomerization to cyclohexane with subsequent dehydrogenation, and dehydrocyclization of paraffins and their skeletal isomerization. These reactions as follows from Table 7.1 improve octane number. Moreover, during reforming, hydrocracking of paraffins and naphthenes to smaller molecules can also happen. Hydrogen is consumed in hydrocracking, and even if paraffins disappear from the gasoline boiling range after cracking, thereby improving the octane number, the amount of the overall liquid product is diminished, making hydrocracking unfavorable. Apparently, longer residence time or low space velocity is beneficial for formation of aromatics simultaneously promoting cracking.

Catalytic reformers should generally be operated at low pressure to achieve high liquid yields. However, the hydrogen partial pressure must be high enough to avoid the formation of unsaturated compounds, which may polymerize and cause increased coke deposition on the catalyst and hence its deactivation.

The dehydrogenation reactions are endothermic and require appropriate heat input (ca. 35 kJ/mol). Dehydrocyclization reaction is slower than dehydrogenation *per se*, which requires higher temperatures, being the most difficult reaction in catalytic reforming. Skeletal isomerization reactions are very mildly exothermic. Hydrocracking is exothermal, being favored at high temperatures and high hydrogen pressures. In dehydrogenation pressure increase decreases conversion. Dehydrogenation is also preferred at high temperature.

Hydrogenation and dehydrogenation reactions require metal function, while acidic function in the catalyst is needed for isomerization (Figure 7.6). Due to thermodynamics, the isomers that are more highly branched are favored, improving the product octane number.

Catalytic reformers are usually available in three different versions, namely the semi-regenerative, the fully regenerative, and the continuously regenerative reformer.

The first option is the most common (ca. 60% of all units). This process is characterized by continuous operation over long periods, with decreasing catalyst activity due to coke deposition. By decreasing the activity of the catalyst, the yield of aromatics and the purity of the by-product hydrogen decrease. In order to maintain the conversion nearly constant, the reactor temperature is raised as the catalyst activity declines. When the reactors reach end-of-cycle levels, the reformer is shut down to regenerate the catalyst *in situ*. The shutdown of this unit occurs approximately once every 6–24 months. Burning of the coke (up to 20 wt% of the catalyst) is done by first purging with nitrogen, then oxidation with 0.5–1% of oxygen in an inert gas followed by purging with nitrogen, and only after that reduction in hydrogen. Such a tedious procedure is required not only to avoid explosions when hydrogen is put in contact with oxygen over a highly active metal catalyst but also to prevent metal sintering. Start-up of the reactor is done, feeding several ppm of thiophenes along with the feed and switching several days thereafter to a regular feed.

Semi-regenerative reformers (Figure 7.7) are generally built with three to four catalyst beds in series (Figure 7.8)

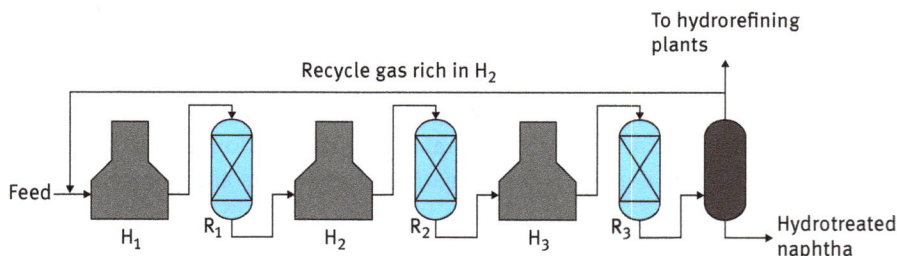

Figure 7.7: Semi-regenerative catalytic reforming. Modified after http://www.treccani.it/portale/opencms/handle404?exporturi=/export/sites/default/Portale/sito/altre_aree/Tecnologia_e_Scienze_applicate/enciclopedia/inglese/inglese_vol_2/161-170_ING3.pdf.

The liquid feed is pumped up to the reaction pressure (5–45 atm) and is joined by a stream of hydrogen-rich recycled gas. The resulting liquid-gas mixture is preheated by flowing through a heat exchange. The preheated feed mixture is then totally vaporized and heated to the reaction temperature (495–520 °C) before the vaporized reactants enter the first reactor. As the vaporized reactants flow through the fixed

Figure 7.8: Reactors in semi-regenerative reforming with increasing reactor size.

bed of catalyst in the reactor, the major reaction is the dehydrogenation of naphthenes to aromatics (which is highly endothermic and results in a large temperature decrease between the inlet and the outlet of the reactor). Temperature selection is a compromise between high catalyst activity (higher octane number) at elevated temperatures and at the same time lower yields because of cracking and lower stability.

To maintain the required reaction temperature and the rate of reaction, the vaporized stream is reheated in the second fired heater before it flows through the second reactor operating at the same temperature as the first one. The temperature again decreases across the second reactor and the vaporized stream must again be reheated in the third fired heater before it flows through the third reactor. As the vaporized stream proceeds through the three reactors, the reaction rates decrease and the reactors therefore become larger. At the same time, the amount of reheat required between the reactors becomes smaller. As an example, let us consider two units operating in the same oil refinery with different catalysts (I.V. Yakupova et al., Procedia Engineering, 2015, 113, 51–56, https://doi.org/10.1016/j.proeng.2015.07.288).

In the first case, the pressure is 2.6–2.7 MPa and the catalyst operates at the inlet temperatures of 489–502 °C. The unit capacity is 600 kt/a. In the three reactors, the reactor temperature drop is 72/30/5 °C while the catalyst split between the reactors is 1:2:4, respectively. The hydrogen yield in this case is 2.1–2.2 wt%. An alternative catalyst operating in four reactors at 1.6–1.7 MPa and the inlet temperatures of 470–490 °C results in the reactor temperature drop of 67/33/15/9 °C and the hydrogen yield of 1.8–1.9 wt%. Both units afford reformate octane number of 94–95.

Higher reactor temperature may be required to compensate for declining activity or feedstock quality reducing the yield and the run length while improving the octane rating. Higher pressure suppressing the formation of coke simultaneously promotes hydrocracking.

The hot reaction products from the third reactor are partially cooled by flowing through the heat exchanger where the feed to the first reactor is preheated and then flow through a water-cooled heat exchanger before flowing through the pressure controller into the gas separator. Most of the hydrogen-rich gas from the gas separator vessel returns to the suction of the recycled hydrogen gas compressor and the net production of hydrogen-rich gas from the reforming reactions is exported for use in other refinery processes.

Lower rate of hydrogen recycle results in higher amounts of coke while excessive H_2 recycle gives less aromatics. The minimum amount of hydrogen needed for $1 m^3$ feed is $1,000 nm^3$.

The liquid from the gas separator vessel is routed into a stripper. The overhead off-gas product from the stabilizer contains the by-products methane, ethane, propane, and butane and some small amounts of H_2, while the bottom fraction is high-octane liquid reformate. The off-gas is routed to the refinery's central gas processing plant for removal and recovery of propane and butane. The residual gas after such processing becomes part of the refinery's fuel gas system.

In the cyclic catalytic reformer units (ca. 12% of all units), an extra spare or swing reactor exists, which, together with other reactors, can be individually isolated. Thus, each reactor can undergo *in situ* regeneration while the other reactors are in operation. In this way, only one reactor at a time has to be taken out of operation for regeneration, while the reforming process continues in operation. Catalyst development along the years resulted in a possibility to diminish the reactor pressure, increasing C5 + and hydrogen yields. The pressure drop at lower pressures can become an issue leading to less efficient flow distribution; therefore, modern designs of reforming units utilized radial flow reactors.

The third option is continuous catalyst regeneration (CCR) version in the form of a sophisticated catalyst moving-bed system, which was realized in the UOP Platforming process with the reactors arranged in a "stacked" construction (Figure 7.9). The catalyst trickles through the system from the top of the first reactor to the bottom of the lowest reactor, where it is collected and lifted in portions to the regenerator section. The fired heaters are installed between the reactors.

Reaction and regeneration are done separately; thus, specific optimal conditions for each operation can be applied. Moreover, it is even possible to shut down the regeneration section temporarily for maintenance. Because of continuous regeneration, catalyst selectivity rather than resistance to deactivation is more important in CCR, while in semi-regenerative units, resistance to deactivation by coke is very important. During regeneration, addition of Cl containing compounds

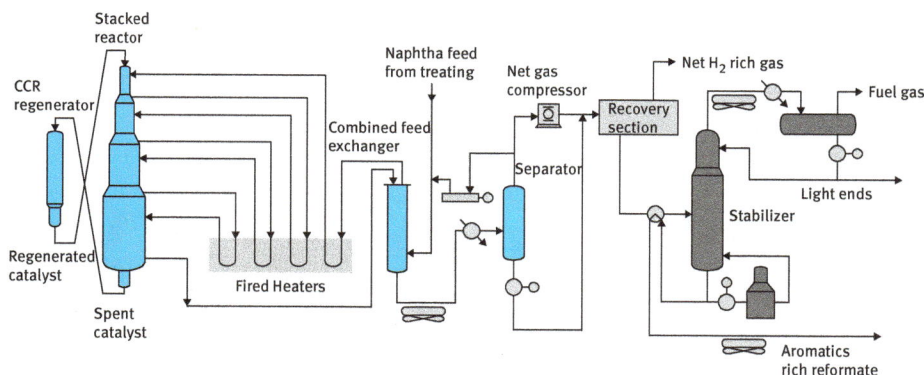

Figure 7.9: Schematic process diagram of CCR reformer. http://www.uop.com/reforming-ccr-platforming/.

is done which is needed because of a low pressure in the reaction zone. Table 7.2 illustrates operation parameters of two units at the same oil refinery.

Table 7.2: Comparison of semi-regenerative and continuous catalytic regeneration reformers.

	SR	CCR
RON	95	100
Pressure, bar	15	8
Capacity, Mt/a	1	1
Reformate, %	84.9	83.5
Reflux (C3−C4), %	1	3.2
Off gas	6.5	1.6
Hydrogen gas	7.1	12.5
Losses	0.5	–

Continuous catalytic regeneration reformers offer high utilization of the feed due to low pressure, feedstock flexibility, higher octane number of the product, economical design of stacked reactors, long unit operation cycle, and optimized heat and compression integration among other advantages. An apparent disadvantage is a more complicated process and a higher investment burden.

An alternative to the stacking system (Figure 7.10) is a side-by-side design of Axens in their CCR process (Figure 7.11), which has several advantages such as easier access for construction, inspection, and reactor modifications.

Figure 7.10: Catalytic reformer for processing 1 million tons of the feed annually. The unit at Atyrau refinery in Kazakhstan gives the product with RON of 103.

The flow can be arranged in a radial fashion and the reactor design does not have height or thermal expansion constraints.

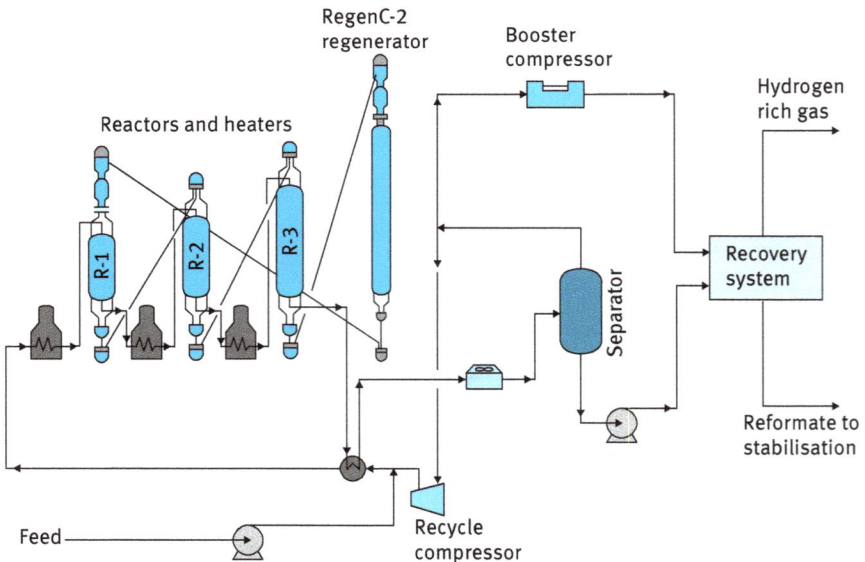

Figure 7.11: Side-by-side design in CCR (Axens design). http://www.axens.net/document/703/octanizing-reformer-options/english.html.

Revamping of a conventional semi-regenerative reformer can be done by adding a new moving-bed reactor and regeneration system to produce higher-octane reformate. Such revamp increases hydrogen production with affordable capital cost. Several such revamps or Dualformers (Figure 7.12) are currently in operation.

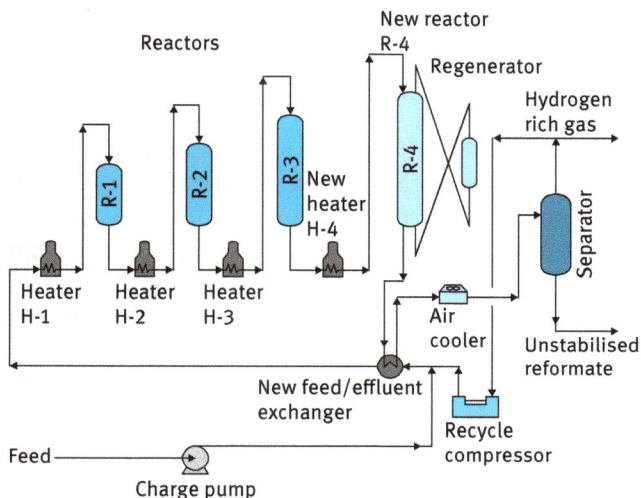

Figure 7.12: Dualforming. http://www.axens.net/document/703/octanizing-reformer-options/english.html.

Addition of a fourth reactor allows to have more severe operation, which, in essence, means lower hydrogen pressure, thus improving the yield of the reformate and especially hydrogen. Such a low-pressure operation cannot be easily achieved in fixed-bed reactors because of coking. The combined unit operates at a pressure (1.7 MPa) lower than the original unit. As a consequence of utilizing low pressure, more selective Pt-Sn catalyst can be used in the last reactor.

Several pieces of equipment are installed during revamp such as high-efficiency feed/effluent heat exchanger, an additional inter-heater, a booster compressor to export hydrogen at the original design pressure, and the regeneration system.

Even with addition of a continuous regenerator, there is still a need to periodically shut down the fixed-bed SR section of the reforming unit for regeneration.

7.3 Epimerization

In this section, epimerization of glucose to fructose syrups with a high sweetness using immobilized glucose isomerase will be considered. Such reaction with the global annual capacity of several million tons

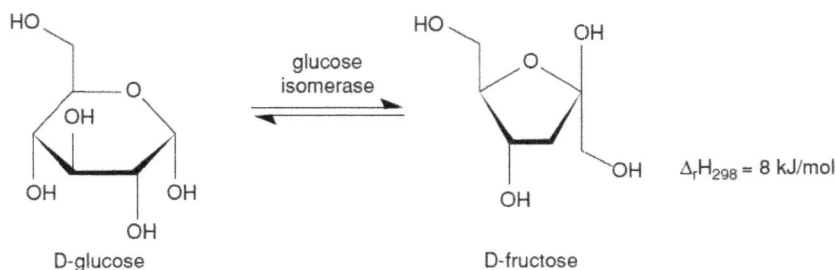

$\Delta_r H_{298} = 8$ kJ/mol

D-glucose   D-fructose

is limited by thermodynamics giving ca. 48–52% fructose between 300 and 350 K.

The epimerization reaction can also be done in the presence of an alkaline catalyst, however, resulting in large amounts of by-products.

A simplified flow scheme of the continuous production of high-fructose corn syrup (HFCS) starting from glucose is illustrated in Figure 7.13. Starch hydrolysis with α-amylase is typically done for preparation of the glucose syrup. This syrup is treated by mechanical filtration to remove insoluble impurities and then by carbon and with ion exchangers of both anionic and cationic types. The concentration of the syrup used as feed for isomerization is ca. 40–45%.

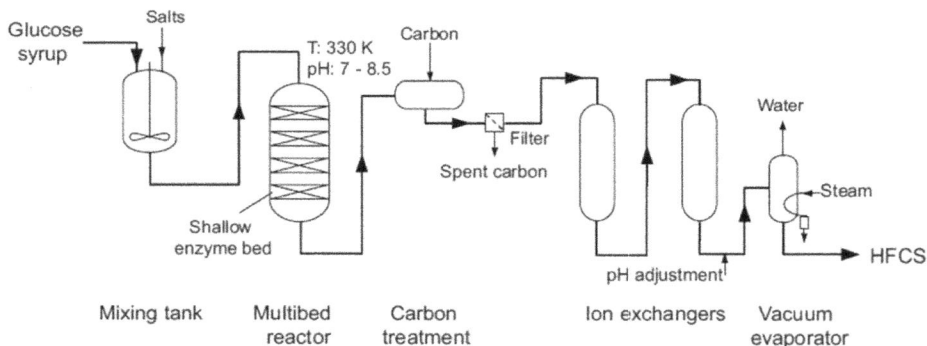

Figure 7.13: Continuous production of high-fructose corn syrup (HFCS) with immobilized glucose isomerize as a catalyst. From J. A. Moulijn, M. Makkee, A E. van Diepen, *Chemical Process Technology*, 2013, 2nd Ed. Copyright © 2013, John Wiley and Sons. Reproduced with permission from Wiley.

The enzyme (glucose isomerize) is immobilized on, for example, a composite support comprising 30% DEAE-cellulose, 30% titania, and the rest – polystyrene. The optimal conditions for isomerization are 55–57 °C and pH = 7.5–7.8. Cofactors (i.e. metal ions) are needed for the stable enzyme operation. Subsequently, salts (e.g. magnesium or cobalt sulphates and sodium hydrosulphite) are introduced to the mixing tank by also adjusting the pH. The multibed reactors (Figure 7.14) have shallow beds to avoid enzyme compression and diminish the pressure drop.

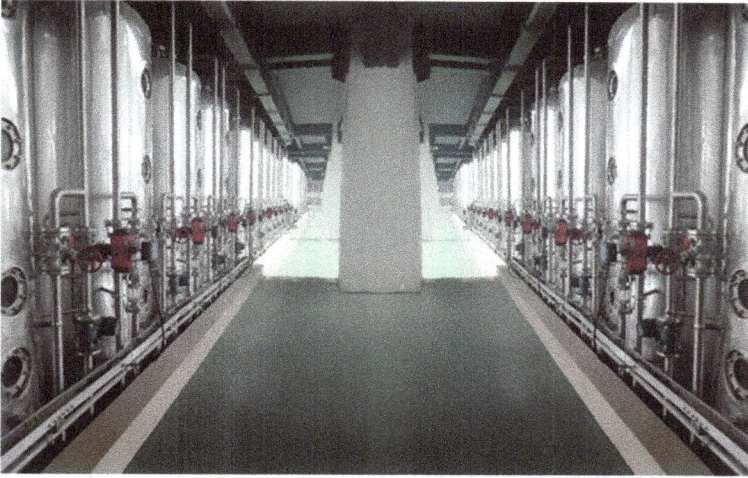

Figure 7.14: Equipment for isomerization of glucose. From https://www.syrupmachine.com/uploads/1806/2-1P615130943957.jpg.

Downstream the reactor, the product is treated with cation and anion ion exchangers and carbon. Such treatment is needed for the removal of added salts and impurities as otherwise the product will have undesired coloring. The final process step is concentration of the syrup by vacuum evaporation giving the product with ca. 24% water and the rest fructose and glucose. The fructose content in the HFCS is different in different countries. In the European Union, HFCS has 20–30% fructose content compared to 42% (HFCS 42) and 55% (HFCS 55) in the United States, where it produced exclusively with corn, while manufacturers in the EU utilize both corn and wheat.

Chapter 8
Halogenation

Halogenation is one of the important processes in organic synthesis required for production of chlorinated intermediates (for example, 1, 2-dichloroethane), chloro- and fluoro-containing monomers (vinyl chloride), chloro organic solvents (dichloromethane), and halogen-containing pesticides. Typically, halogenation is either radical or catalytic.

8.1 Radical chlorination

During chlorination, unbranched chain reactions take place, starting from the dissociation of Cl_2 by the influence of temperature (thermal chlorination), light (photochlorination), or initiators (chemical initiation with, for example, peroxides $C_6H_5C(O)OOC(O) C_6H_5$)

$$Cl_2 \Rightarrow 2Cl \qquad (8.1)$$

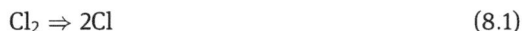

The chain propagation is characterized by repetition of reactions

$$Cl^{\cdot} + RH \Rightarrow R^{\cdot} + HCl; R^{\cdot} + Cl_2 \Rightarrow RCl + Cl^{\cdot} \qquad (8.2)$$

Finally, termination occurs when radicals recombine with the involvement of a wall or third particles present in the system, since recombination of radicals in the bulk is impossible, as the energy excess should be taken up by a third body. Therefore, the termination rate depends on the surface/volume ratio of a reaction vessel. Presence of such inhibitors as phenols, sulphur-containing compounds, or oxygen negatively influences the chlorination rate. As a consequence, it is beneficial to use as a reactant chlorine, obtained by the evaporation of liquefied chlorine because it is free from oxygen.

Activation energy of chlorination depends on the way chlorine is activated and is the highest for thermal chlorination (125–170 kJ/mol) and the lowest for photochlorination (20–40 kJ/mol).

When a chlorine atom is already present in a molecule, reactivity diminishes, with the strongest influence being for the neighboring carbon atom. Therefore, chlorination of ethylchloride gives predominantly 1, 1-dichloroethane rather than 1, 2-dichloroethane.

https://doi.org/10.1515/9783110712551-008

$$CH_3-CH_2Cl \xrightarrow[-HCl]{+Cl_2} \begin{cases} CH_3-CHCl_2 \quad 80\% \\ \\ ClCH_2-CH_2Cl \quad 20\% \end{cases} \tag{8.3}$$

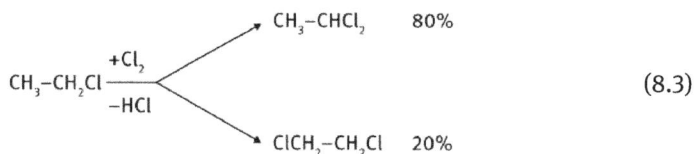

The next neighboring atom of carbon is less influenced; as a consequence, chlorination of 1-chloropropane gives a mixture of 1, 1 and 1, 3 dichloropropane:

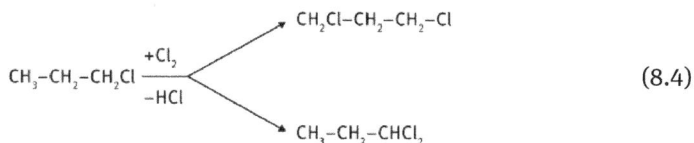

$$CH_3-CH_2-CH_2Cl \xrightarrow[-HCl]{+Cl_2} \begin{cases} CH_2Cl-CH_2-CH_2-Cl \\ \\ CH_3-CH_2-CHCl_2 \end{cases} \tag{8.4}$$

8.1.1 Liquid-phase chlorination

Liquid-phase chlorination is carried out at relatively low temperatures, which requires utilization of initiators or irradiation and is thus less economically attractive. However, when application of high temperature gives side reactions through elimination of HCl or when several chlorine atoms are incorporated in the reactant, thus giving rise to large heat release, it would be beneficial to use liquid-phase chlorination.

Technically, chlorination is made in bubble columns when chlorine gas is dissolved in the liquid reactant. Such liquid-phase chlorination can be done using either photochemical initiation or chemical initiators. In the latter case, the reactor is less sophisticated; on the other hand, utilization of initiator represents an additional cost. Photochemical initiation leads to higher capital and energy costs and needs more complex design. At the same time, a chemical initiator is not required and the reaction mixture is not contaminated by its decomposition products.

During photochemical initiation, typically, low temperature is sufficient since the activation energy is rather low, thus, for example, chlorination of benzene to hexachlorocyclohexane is done at 40–60 °C.

A chemical initiator such as benzoyl peroxide requires 100–120 °C for its decomposition. Considering the overall process costs, the uptake or consumption of the initiator should be minimized. This happens with temperature and initiator concentration decrease, leading at the same time to a decrease in chlorination rate. Consumption of the initiator could be diminished when it is introduced at several positions along the reactor length (in continuous processes) or periodically for batch processes.

Several types of chlorination reactors are given in Figure 8.1.

The first reactor operating in a continuous mode is a bubble column with an external heat exchanger, when the circulation is done either using a pump or, due

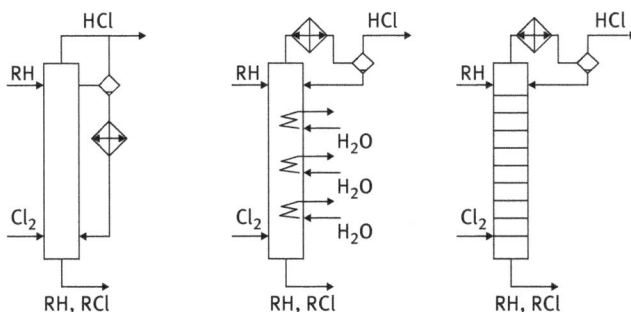

Figure 8.1: Reactors for radical chlorination.

to differences in densities, between the relatively hot liquid inside the reactor containing dissolved gas and a cold gas-free liquid in the circulation loop. Polychloroparaffins are produced in such reactors.

An option with an internal heat exchange is shown in Figure 8.1 (center). This reactor also contains a condenser. The flow of the liquid and the gas in such a reactor, which can be arranged in a cascade mode with other reactors of the same type, is countercurrent.

For chlorination of low boiling compounds, the heat can be recovered by evaporating them in a stream of HCl (Figure 8.1 right). This operation mode does not need a heat exchanger.

Chlorination plants also contain units for chlorine evaporation and heating it to the reaction temperature. Drying of gaseous chlorine can be done with sulphuric acid.

An important issue in chlorination technologies is cleaning of the outlet gases. The most effective way is to first remove the vapors of the reactant by absorption. Cooling of the outlet gas with water might be a feasible option in the case of nonvolatile compounds. This is followed by removal of HCl. When as a chlorination product gaseous HCl is formed, it is absorbed in water, giving 20–30% HCl.

Separation of HCl from the liquid products requires its removal by contact with air (Fig. 8.2a) when non-volatile products are formed (benzyl chloride or chloroparaffins).

Otherwise, extraction of HCl with water and NaOH (Fig. 8.2b) removes HCl, giving, however, a lot of wastewater, which should be properly treated. Alternatively, distillation of the reactant could be also done with its subsequent separation from HCl (Fig. 8.2 c).

An example of the liquid-phase radical chlorination of 1, 1-dichloroethane is presented in Figure 8.3.

The solution containing the porophor in 1, 1-dichloroethane is chlorinated in reactor 3 to which Cl_2 is added as a gas. Due to formation of a higher boiling point product, there is a temperature gradient in the reactor. The recuperation of heat is

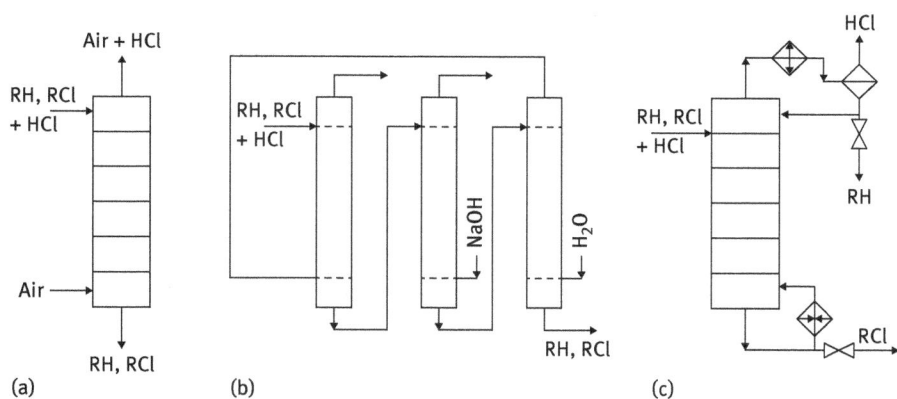

Figure 8.2: (a) Dry cleaning, (b) extraction with water and base, and (c) distillation.

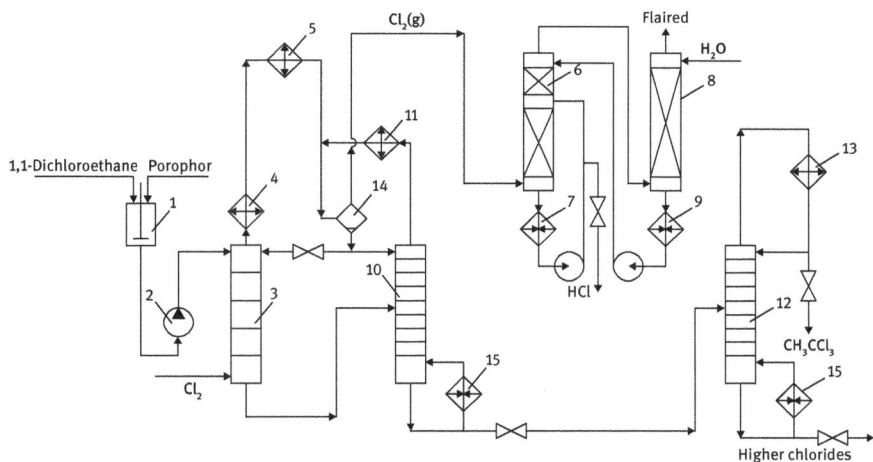

Figure 8.3: Liquid-phase chlorination of 1, 1-dichloroethane: 1, collecting vessel; 2, pump; 3, reactor; 4 and 5, heat exchangers/condensers; 6 and 8, scrubbers; 7 and 9, heat exchangers; 10 and 12, distillation columns; 11 and 13, condensers; 14, separator; 15, boiler. After N. N. Lebedev, *Chemistry and Technology of Basic Organic and Petrochemical Synthesis*, Chimia, 1988.

done by evaporation of the reactant in HCl under 0.2–0.3 MPa. The vapors of the unreacted 1, 1-dichloroethane after condensation (pos. 5) are sent back to the reactor. HCl free from organics is routed to an absorption column (pos. 6) where diluted hydrochloric acid flows from the top. Finally, another absorption column (pos. 8) operates with water. The liquid product (bottom of reactor 3) is sent to distillation column 10, where HCl and dichloroethane (column top) are separated from the products. Final distillation is done in column 12, separating the desired product 1,

1, 1-trichloroethane from 1, 1, 2 trichloroethane and higher polychlorinated products (i.e., tertachloroethane).

8.1.2 Gas-phase chlorination

A limited range of products, such as chloromethane or allylchloride, are synthesized by gas-phase chlorination. Such gas-phase chlorination is done at atmospheric pressure with efficient mixing of the reactants in the reactors presented in Figure 8.4.

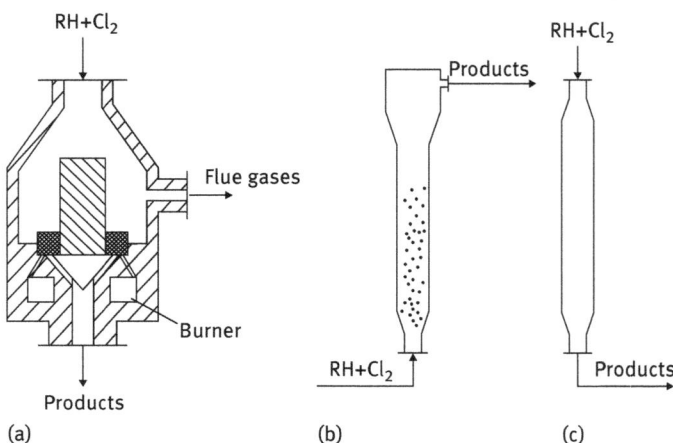

Figure 8.4: Gas-phase chlorination (a) reactors with a burner, (b) fluidized-bed reactor, and (c) reactor with a preheated reaction mixture.

The heat release can be quite high; thus, cold reactants should be introduced into the reactor and heated already inside it (Figure 8.4a) using a heat accumulator. An alternative is to use a fluidized bed of a catalyst or a heat carrier (Figure 8.4b). For highly exothermal reactions (synthesis of polychlorinated methane), cold CCl_4 can be injected in the reactor.

When the heat of the reaction (allylchloride synthesis) is moderate, the reactants could be preheated (Figure 8.4c). In such case, the reaction can start already at the mixing stage and the reactor is a hollow column with high length/diameter ratio.

As an example of the gas-phase chlorination, production of allylchloride by chlorination of propene will be considered (Figure 8.5)

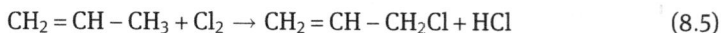

$$CH_2 = CH - CH_3 + Cl_2 \rightarrow CH_2 = CH - CH_2Cl + HCl \qquad (8.5)$$

with enthalpy equal to 113 kJ/mol at 298 K.

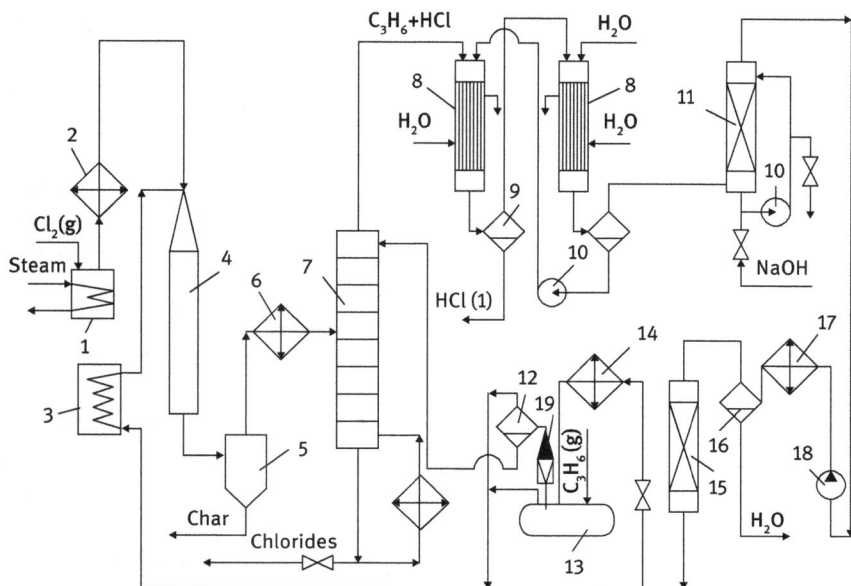

Figure 8.5: Production of allylchloride: 1, evaporator; 2, heater; 3, furnace; 4, reactor; 5, cyclone; 6 and 17, heat exchanger; 7, prefractionator; 8, film absorber; 9, 12, and 16, separators; 10, pump; 11, scrubber; 13, vessel for liquid propene; 14, condenser; 15, adsorber; 18, compressor; 19, valve. After N. N. Lebedev, *Chemistry and Technology of Basic Organic and Petrochemical Synthesis*, Chimia, 1988.

At 300–600 °C, chlorination proceeds by a free-radical chain mechanism, whereby the hydrogen atom in the allyl position is substituted preferentially by the chlorine. The most important variables in the industrial propene chlorination to allyl chloride are the temperature and the ratio of propene to chlorine, whereas pressure and residence time have only a slight effect on the allyl chloride yield. Because the dominant reaction below 200 °C is addition to form 1, 2-dichloropropane, the mixing temperature of propene and chlorine must be kept above 250–300 °C. If the reactor temperature is increased further, spontaneous pyrolysis occurs, with the formation of soot and high-boiling tars. In general, only propene is preheated. If chlorine is also preheated, expensive construction materials must be used to avoid the danger of a "chlorine fire."

Formation of by-products decreases with increasing propene excess, at the expense of the higher processing costs. Influence of pressure is minimal on the reaction *per se* and is selected based on pressure drop considerations.

Industrial-scale reactors mainly operate adiabatically, even though higher yields would be expected when operating isothermally. The technically simplest and oldest reactor type is the tube reactor, but all reactors are designed to achieve the mixing of the two reactants as rapidly and as thoroughly as possible in order to reduce the secondary reaction to form 1, 3-dichloropropene. The process scheme is given in Figure 8.5.

Liquid propene is vaporized (pos. 13), then preheated to 350–400 °C (pos. 3) and fed, together with gaseous evaporated chlorine, into the reactor (pos. 4). The chlorine reacts completely, thereby increasing the temperature to 500–510 °C. Small amounts of carbon are always formed, catalyzing chlorination. A protective film of vitreous carbon deposits on the reactor walls. This material, which also contains highly chlorinated substrates and tar, is separated in cyclone (pos. 5). The gas stream leaving the chlorination reactor is precooled (pos. 6) and led to a prefractionator (pos. 7), the overhead temperature of which is maintained at ca. – 40 °C by feeding liquid propene. This effectively separates all chlorinated hydrocarbons; the bottom product is free of propene and hydrogen chloride.

The gaseous mixture drawn off overhead is separated by absorption with water (pos. 8) into aqueous hydrogen chloride of commercial quality and propene. Propene is then washed with caustic soda in a scrubber (pos. 11), leading to removal of hydrogen chloride traces. After compression to 1.2 MPa, propene is liquefied in a condenser (pos. 17). Water is separated (pos. 16) and liquid propene is dried by adsorption (pos. 15) and returned to the storage tank (pos. 13). The bottom product of the prefractionator (pos. 7) contains 80% allyl chloride, 16% dichlorides (1, 2-dichloropropane and *cis*- and *trans*-1, 3-dichloro-1-propene), as well as some low and heavy boiling compounds that are separated by distillation.

8.2 Catalytic chlorination

Several chlorination reactions can be classified as catalytic ones, including, for example, chlorination of olefins, hydrochlorination, substitution reactions in the aromatic ring.

Thus, chlorination of olefins

$$RCH = CH_2 + Cl_2 \rightarrow RCHCl - CH_2Cl \tag{8.6}$$

is accelerated when $FeCl_3$ is used as a catalyst. The mechanism of this reaction is electrophilic addition. The stability of the intermediate cation determines the reactivity of the olefins increasing in the order $RCH=CH_2>CH_2=CH_2>CH_2=CHCl$.

Among the side reactions, substitution of hydrogen by halogen can be mentioned, which is a radical-type reaction. In order to suppress formation of by-products, either low temperature or inhibitors should be applied. One of such inhibitors is oxygen, which is already present in chlorine.

Since catalytic chlorination is rather selective, the feedstock should not be very pure, although drying of gases is mandatory.

Chlorination can be done through bubbling of gaseous reactants through the liquid product at 70–100 °C in the presence of catalyst under a slight stoichiometric access of olefin to ensure complete consumption of chlorine. Typical reactor config-urations are given in Figure 8.6 comprising, for example, a case with an external heat exchanger (left). In the case of lighter products (dichloroethane), heat integra-tion can be done through evaporation of the product (Figure 8.6, center). Moreover, reactive distillation approach can be also applied (Figure 8.6 right), when, for ex-ample, dichloroethane is separated from heavier trichloroethane, which is concen-trated at the bottom of the reactive distillation column.

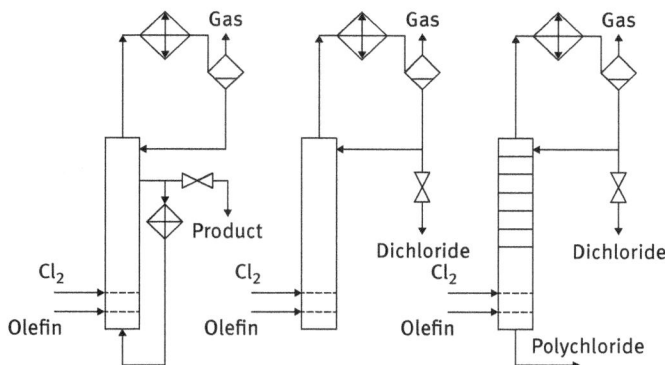

Figure 8.6: Reactors for catalytic chlorination: (left) external heat exchanger, (center) cooling by evaporation; (right) reactive distillation.

Another example of catalytic chlorination is synthesis of phosgene, notoriously known as a warfare agent in World War I. It has also been extensively used in pro-duction of isocyanate intermediates in manufacturing of pesticides and diisocyanates in synthesis of polyurethane plastics. High level of toxicity requires exceptional safety efforts in processing of phosgene. Alternative commercial methods for synthe-sis of isocyanates without phosgene have been only partially successful.

Phosgene is produced commercially by a strongly exothermal reaction of car-bon monoxide and chlorine over activated carbon. Carbon monoxide is added in a slight excess compared to stoichiometry with the purpose of avoiding formation of undesirable products from Cl_2 during further processing. The flow scheme is shown in Figure 8.7.

The reaction is carried out in a multitubular reactor of carbon or stainless steel. The tubes of the diameter 50–70 mm are filled with the granulated carbon catalyst of the size 3–5 mm. The lifetime of a catalyst can be several years. The reaction is carried out under normal or slight excess pressure at 40–150 °C. Due to the exo-thermicity of the reaction, the hot spot can reach 400 °C. The downstream treat-ment includes condensation with water or a coolant and the absorption in a solvent which is used in the subsequent process steps. The non-condensed gaseous com-pounds including CO and phosgene are fed to the waste-gas treatment.

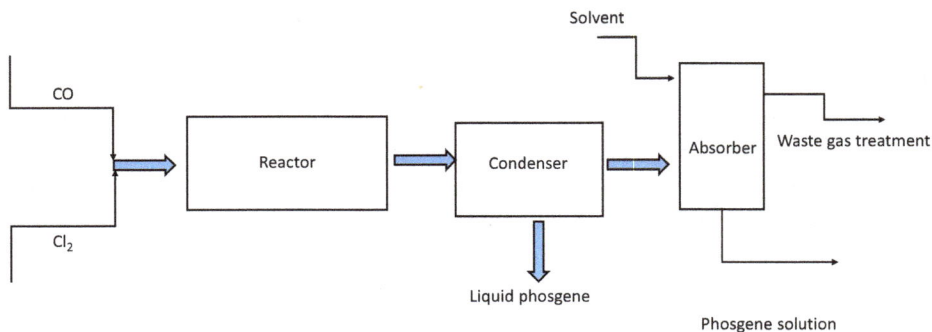

Figure 8.7: Simplified flow sheet for the production of phosgene.

Phosgene is mainly used industrially for the production of isocyanates by first reacting it with primary amines resulting in carbamoyl chlorides, from which isocyanates are formed by elimination of HCl at elevated temperature:

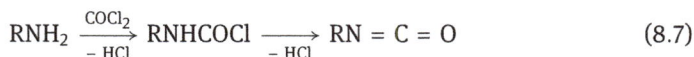

$$RNH_2 \xrightarrow[-HCl]{COCl_2} RNHCOCl \xrightarrow{-HCl} RN=C=O \qquad (8.7)$$

An important implementation of this reaction is synthesis of toluene diisocyanate (TDI), which is generally manufactured in a continuous process involving three steps: nitration of toluene to dinitrotoluene, hydrogenation of dinitrotoluene to toluenediamine, and finally phosgenation of the latter. The estimated global annual production capacity exceeds 1.5 million tons. Phosgenation is typically done by reacting diamine with phosgene at 0–50 °C in a solvent such as toluene, chlorobenzene, and o-dichlorobenzene to give a mixture of carbamyl chlorides and amine hydrochlorides.

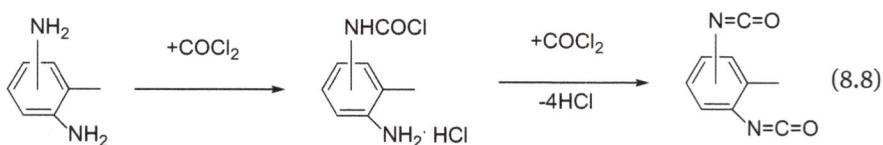

The subsequent processing with phosgene at 170–185 °C eliminates HCl and forms diisocyanates. Phosgene applied in excess is recycled after separation from HCl. After recovery and recycle of excess phosgene and the solvent, the crude TDI is distilled to remove heavies. The selectivity to TDI is 97% (based on diamine).

The by-product HCl is purified to a quality sufficient for its subsequent oxidation to chlorine (Chapter 9) which is used in the synthesis of phosgene. Purified gaseous HCl can be used as a feedstock for ethylene oxychlorination (discussed below) to produce vinyl chloride or can be absorbed in water giving hydrochloric acid.

An alternative method is phosgenation of the toluenediamine hydrochlorides, which are produced by dissolving toluenediamines in o-dichlorobenzene and injecting dry HCl. One-step phosgenation at elevated temperatures under strong agitation gives diisocyanates and HCl, which are removed with an inert gas stream.

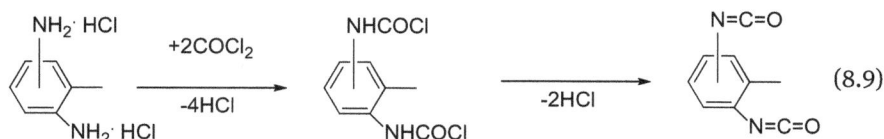

$$(8.9)$$

In this process, variant purification is also done by distillation.

Yet another option developed recently by Bayer is the gas-phase phosgenation at temperatures above 300 °C in a tubular reactor, which improves safety as the risk of phosgene leakages is significantly diminished, when not operating under pressure.

8.3 Hydrohalogenation

This reaction is the electrophilic addition of HCl or HBr to alkenes, forming haloalkanes

$$RCH = CH_2 + HX \leftrightarrow RCHX - CH_3 \qquad (8.10)$$

For example, ethylchloride can be obtained by addition of HCl to ethylene using $AlCl_3$ as a catalyst. In order to shift equilibrium and diminish the extent of polymerization, the process is carried out at low T (-10 °C to 30–40 °C) under a minor excess of HCl. In some cases, when the low boiling products are formed (chloroethane), high pressure can be applied to keep the product in the liquid phase. Bubble columns similar to Figure 8.6 (left) can be applied. Excess of HCl is removed by adsorption with water.

The same type of reaction can be used for hydrochlorination of acetylene to vinyl chloride (Figure 8.8)

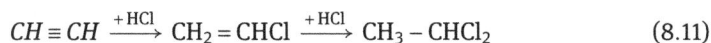

$$CH \equiv CH \xrightarrow{+HCl} CH_2 = CHCl \xrightarrow{+HCl} CH_3 - CHCl_2 \qquad (8.11)$$

In the gas-phase process, the gaseous reactants are brought into contact with the catalyst at slightly increased pressure (0.1–0.3 MPa) and 100–250 °C (contact time, 0.1–1 s) and then quenched and partially liquefied. The reaction products are separated, recycled, or sent to final purification. The molar feed ratios, varying from almost equimolar to a tenfold excess of HCl, depend heavily on the catalyst performance. Acetylene conversions of 95–100% with almost quantitative yields are achieved. The hydrogen chloride must be free of chlorine to avoid explosion and should not contain chlorinated hydrocarbons, which could also act as catalyst poisons. Water must be entirely

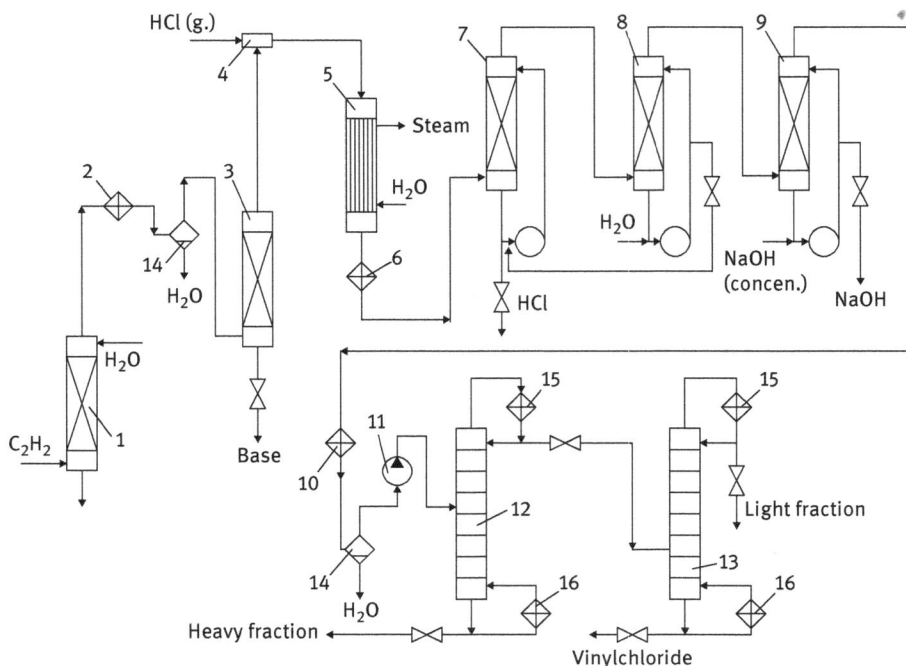

Figure 8.8: Hydrochlorination of acetylene: 1, fire protection; 2, 6, and 10, coolers; 3, drying column; 4, mixer; 5, reactor; 7–9, scrubbers; 11, compressor; 12 and 13, distillation column; 14, separator; 15, condenser; 16, boiler. After N. N. Lebedev, *Chemistry and Technology of Basic Organic and Petrochemical Synthesis*, Chimia, 1988.

excluded to avoid corrosion. Moreover, mercury(II) chloride on activated carbon, used primarily as a catalyst, promotes hydration of acetylene.

Acetylene is first dried by condensation of water in a cooler (pos. 3) and then with solid sodium hydroxide (pos. 3). Thereafter, it is mixed with HCl (pos. 4) and routed to the reactor (pos. 5). Conversion is ca. 97–98%, the product contains 93% vinyl chloride, 5% HCl, 0.5–1.0% C_2H_2 and 0.3% of acetaldehyde, and 1, 1-dichloroethane each. The catalytically active phase is carried out from the reactor due to its volatility; thus, the gases are cooled in a heat exchanger (pos. 6), separated from $HgCl_2$ and HCl in scrubbers (pos. 7–9) with hydrochloric acid (20%), water, and sodium hydroxide.

Volatility of Hg catalysts is a serious disadvantage of hydrochlorination as mercury chloride is easily sublimed especially at the high reaction temperature. Moreover, hot spots are formed, making losses of mercury even more severe. Such losses not only lead to catalyst deactivation, but cause serious problems as substantial amounts of mercury are released directly into environment.

After drying in a heat exchanger 10, the gas is compressed (pos. 11) up to 0.7–0.8 MPa. Separation is done in distillation columns 12 and 13, removing first the heavy boiling compounds (1, 1-dichloroethane) and then the low boiling compounds (acetylene, small amounts of acetaldehyde). Hydrochlorination of acetylene was mainly used in the past, since it was based on the coal-derived feedstock acetylene. Availability of ethylene at competitive prices encouraged development of ethylene-based processes, such as oxychlorination.

Still in some places, like China, a substantial amount of vinyl chloride (ca. 13 million tons annually) is produced from acetylene with a total installed capacity of carbon-supported $HgCl_2$ of ca. 15,000 tons. In a single reactor, the catalyst loading is 6–7 tons. The annual losses in ca. 90 plants across the country are estimated to be 1,000 tons. A viable alternative demonstrated at a full scale is to use as a catalyst 0.1 wt% Au/C extrudates.

8.4 Oxychlorination

Oxychlorination of ethylene to dichloroethylene over $CuCl_2$ catalyst ($\Delta H = -295$ kJ/mol) followed by thermal cracking of the latter

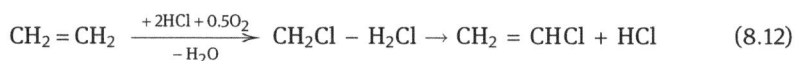

$$CH_2 = CH_2 \xrightarrow[-H_2O]{+2HCl + 0.5O_2} CH_2Cl - H_2Cl \rightarrow CH_2 = CHCl + HCl \qquad (8.12)$$

is a key step in the synthesis of vinyl chloride, a monomer for the second largest plastic – polyvinyl chloride. Oxychlorination reaction is strongly exothermic, and a careful temperature control is essential to both ensure high selectivity and prevent rapid catalyst deactivation.

In the so-called balanced process (Figure 8.9), ethylene oxychlorination and dichloroethylene pyrolysis are coupled along with direct chlorination of ethylene in a single process to increase the throughput of vinyl chloride (often abbreviated as VCM). In this way, there is no net consumption or production of HCl. Vinyl chloride is made only by thermal cracking of 1,2-dichloroethane, formed by either direct chlorination or oxychlorination of ethylene.

In industry, the reaction is carried out in the vapor phase in either fluidized-bed reactors (Figure 8.10a) or fixed-bed reactors (Figure 8.10b) with oxygen being supplied as pure gas (oxygen-based process) or as conventional air (air-based process). Fixed-bed oxychlorination generally operates at 230–300 °C and 150–1,400 kPa, while lower temperatures can be used in fluidized-bed reactors (220–245 °C) operating at 150–500 kPa. More efficient heat removal can be achieved in fluidized-bed reactors.

As catalysts, copper chloride as an active phase with promoters on porous supports such as alumina (e.g. 8–16 wt% $CuCl_2/\gamma$-Al_2O_3) is applied. The mechanism of oxychlorination involves a three-step redox process: reduction of $CuCl_2$ to CuCl with simultaneous chlorination of ethylene

Schematic diagram of a VCM plant

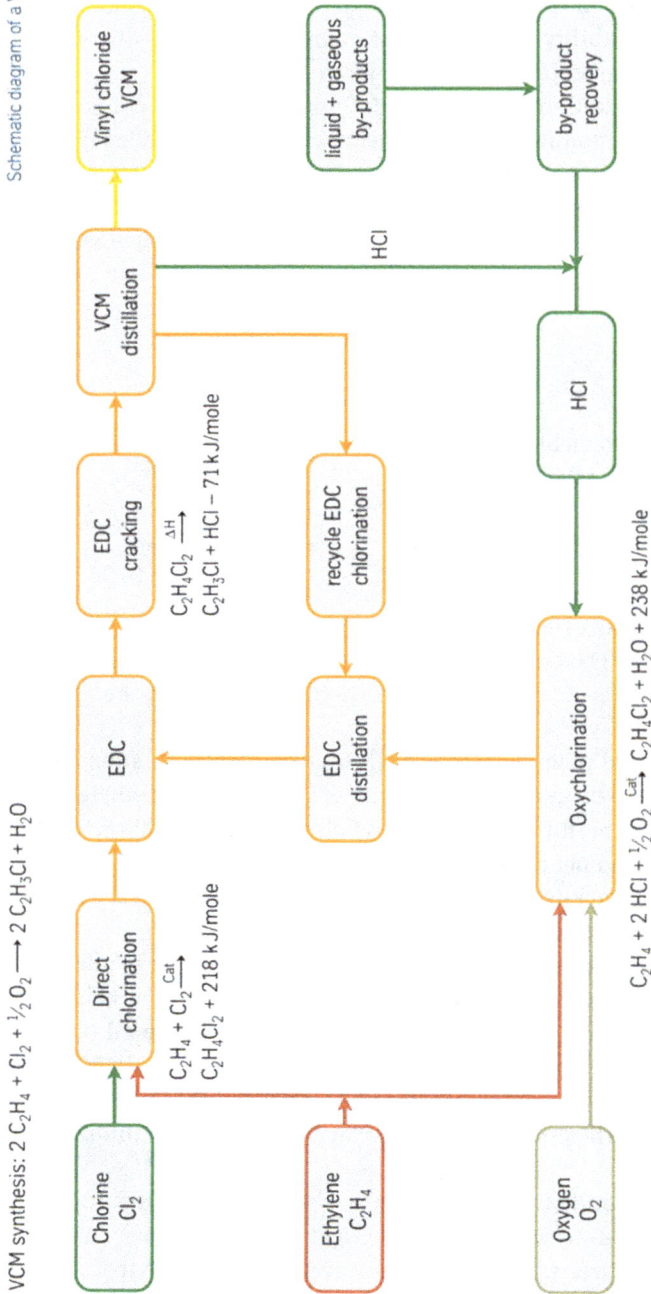

VCM synthesis: $2 C_2H_4 + Cl_2 + \frac{1}{2} O_2 \longrightarrow 2 C_2H_3Cl + H_2O$

Direct chlorination: $C_2H_4 + Cl_2 \xrightarrow{Cat} C_2H_4Cl_2 + 218 \text{ kJ/mole}$

EDC cracking: $C_2H_4Cl_2 \xrightarrow{\Delta H} C_2H_3Cl + HCl - 71 \text{ kJ/mole}$

Oxychlorination: $C_2H_4 + 2 HCl + \frac{1}{2} O_2 \xrightarrow{Cat} C_2H_4Cl_2 + H_2O + 238 \text{ kJ/mole}$

Figure 8.9: Balanced process of VCM manufacturing. From https://ucpcdn.thyssenkrupp.com/_legacy/UCPthyssenkruppBAIS/assets.files/products___services/chemical_plants___processes/polymer_plants/uhde_brochures_pdf_en_3.pdf.

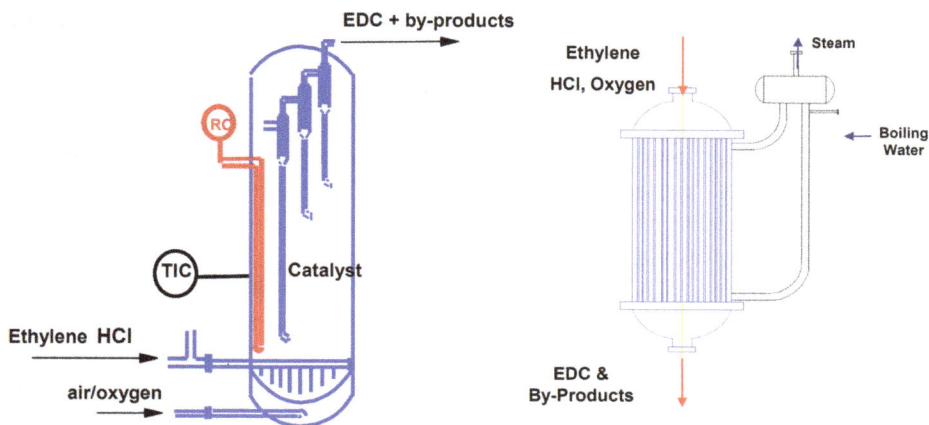

Figure 8.10: Reactors for ethylene oxychlorination.

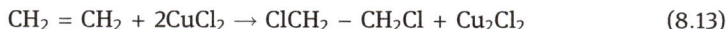

$$CH_2 = CH_2 + 2CuCl_2 \rightarrow ClCH_2 - CH_2Cl + Cu_2Cl_2 \tag{8.13}$$

(ii) oxidation of CuCl to give an oxychloride, and (iii) closure of the catalytic circle by rechlorination with HCl, restoring the original $CuCl_2$. The last two steps give the overall reaction:

$$Cu_2Cl_2 + 2HCl + 0.5O_2 \rightarrow 2CuCl_2 + H_2O \tag{8.14}$$

The active site probably involves an isolated Cu_xCl_y complex which is anchored to the high-surface-area γ-Al_2O_3 support.

A particular challenge with copper chloride is its volatility at reaction temperature, which can be decreased by adding potassium chloride. The latter results, however, in lower activity, improving at the same time selectivity by decreasing formation of ethyl chloride. The control over copper chloride content is needed as an excess would lead to catalyst caking. $KCuCl_3$ and K_2CuCl_4 have low melting points, and the eutectic mixture with copper chloride melts at 150 °C. Other alkali metals and rare earth chlorides (e.g. $MgCl_2$ or $LaCl_3$) can also be added to the catalyst formulation as promoters or to inhibit by-product formation. Their role is also in maintaining a high Cu^{2+} concentration, as it is critical for high activity, selectivity, and stability.

Catalysts are usually prepared on a large scale by impregnating a suitable alumina support with aqueous solutions of cupric and potassium chlorides. In fluidized-bed reactors, microspheroidal catalyst powder with a particle size of 40–60 μm is used, while in fixed-bed reactors loaded with catalyst pellets a much larger size of catalyst particles in the mm range should be utilized to avoid too high pressure drop.

During operation because of attrition, the average size is decreasing which is compensated by adding the fresh catalyst. As an example, a vinyl chloride plant with a capacity of 300,000 tons per year with 50 tons of catalyst hold-up requires 15 tons per catalyst to compensate for the losses. For adequate fluidization, the fraction of the particles below 40 μm should be 25–35%. The copper content does not practically change.

One of the main process problems with oxychlorination is a possible tendency for the catalyst particles to cake or to stick to the cooling bundles within the reactor. This results in poor temperature control, poor fluidization of agglomerated particles, local overheating, and a loss in selectivity. Erosion in the industrial reactor walls leads to a release of iron, whose penetration into the catalyst worsens selectivity by increasing the yield of the total oxidation products.

Temperature control for a very exothermic reaction is easier to organize in fluidized beds, compared to fixed-bed tubular reactors. Higher temperature besides promoting dehydrochlorination of ethylene dichloride to vinyl chloride and formation of other chlorinated products also leads to increased oxidation of ethylene, elevated formation of coke, and more losses of copper loss by sublimation.

In fluidized-bed reactors (Figure 8.11), the heat is controlled by adding cold reactants and having an internal heat exchanger generating steam. Due to high gas flow rates, industrial fluidized-bed reactors often operate in the turbulent fluidization regime. These reactors despite periodical makeup of the catalyst to compensate for losses in cyclones and mechanical degradation allow effective reaction heat removal, near isothermal operations, negligible impact of external mass- and heat-transfer limitations, as well as minimal internal mass transport limitations. The latter is important because operation in the internal mass transfer regime negatively influences selectivity to the intermediate product.

The reaction conditions for oxychlorination, which are optimum from the viewpoint of the ratio between reactants, require operation within the explosive limits. An appropriate design of fluidized-bed reactors with certain measures against radical chain propagation is thus needed. Special devices for mixing reactants to exclude formation of large volumes of explosive mixtures outside the fluidized bed allow safe operation, low by-product selectivity, and strict feed control less critical. Besides adequate mixing of reactants, conditions for fluidization should be properly ensured, which impose limitations not only on the flow rate and the catalyst particles size but also on the reactor diameter and thus production capacity.

Clear advantages of fluidized-bed reactors are their isothermal operation and a possibility to compensate the catalyst losses. This allows to maintain constant catalyst activity, and therefore a stable reactor performance, product quality, and feed consumption. The drawbacks include corrosion of the reactor internals needed for heat removal and lower conversion compared to the tubular reactor.

Tubular reactors allow higher conversion, especially HCl exceeding 99%. Heat is removed by evaporation of steam condensate, thus a uniform temperature profile is difficult to achieve. Formation of hot spots with too high temperature can be mitigated by diluting the catalyst with inert diluents of different type and material and/or by varying the copper chloride content in the catalyst.

Figure 8.11: Oxychlorination reactor. From ttps://ucpcdn.thyssenkrupp.com/_legacy/
UCPthyssenkruppBAIS/assets.files/products___services/chemical_plants___processes/
polymer_plants/uhde_brochures_pdf_en_3.pdf.

Several options of catalyst loading in tubular reactors can be used. For instance, in a single reactor up to four different layers of a catalyst containing the
same amount of $CuCl_2$ (8.5 wt%) can be loaded with different degree of dilution.
The first layer contains 7% of the catalyst, followed by 15%, 40%, and finally 100%
of the catalyst without any dilution.

Mixing of the catalyst and the inert material should be properly done to avoid
any dust and local overheating, minimize nonuniform pressure drop along the reactor, and ensure good performance in terms of catalyst's life and selectivity. This
was in particular important when silica particles with irregular shape were used as
dilutants. Mixing can be done by a catalyst manufacturer, which delivers catalyst
mixtures for direct loading in the reactor.

A better control with a diluted catalyst can be obtained in a series of several separate reactors, which could be further improved by having different catalyst compositions with varying amounts of $CuCl_2$. Spherical catalysts supplied previously by

Stauffer Chemical Company did not require dilution with an inert. Dilution of catalysts containing 6%, 10%, and 18% $CuCl_2$ and promoted with 3%, 3%, and 2% KCl with inerts was also reported in the first two beds, while the last bed containing only the more concentrated catalyst was used.

Conditions in different reactors are different. Obviously, they are more severe in the first reactor; thus, the catalyst life is shorter (1 year) and can be increased to 18 months and 3 years, respectively, in the second and third reactors. When several types of catalysts are applied, hot spots are present close to the position in tubes where the catalyst type changes.

In tubular reactors, which can operate with air or oxygen, control of the oxygen content is required to ensure operation outside of explosive conditions. Another option is to split air between several beds. Half of the overall amount of the air can be added to the first reactor, while the rest is split between the second and third reactors to minimize hot spots. Replacement of air with oxygen reduces the amount of vented gas.

Oxychlorination can be done either in air or oxygen atmosphere. The flow scheme of an air- based process is given in Figure 8.12. In this process ethylene and air are fed in slight excess of stoichiometry enabling high conversion of HCl. As the vent system is large because of the presence of nitrogen with the highest molar flow rate such operation is aimed at minimizing the loss of excess ethylene in the vent stream. In the oxygen-based process only a small portion of the vent gas is purged, being significantly lower than in the air-based process. Typical feedstock conversion is ca. 94–99% based on ethylene and 98.0–99.5% for HCl. Selectivity to dichloroethane is 94–97%.

Direct chlorination is done in the reactor 1 where there is a constant level of liquid containing the catalyst ($FeCl_3$). More sophisticated catalyst systems additionally include, for example, an alkali metal halide, usually sodium chloride. Such catalytic systems are better suited for prevention of subsequent chlorination of EDC into 1,1,2-trichloroethane even at comparatively high temperatures.

Heat of the reaction is removed by evaporation of 1,2-dichloroethane, which is partly condensed and partly goes to distillation. Oxychlorination is done for example in a fixed bed reactor 5 operating under 0.5 MPa at 260–280 °C. Ethylene, recycle gas and HCl are preliminary mixed, thereafter oxygen (as air) is added to the reactor. The mixing sequence and composition are selected to avoid expositions. The heat of reaction is removed by generation of steam. The outlet gases (unconverted ethylene, oxygen and HCl, as well 1,2-dichloroethane) are cooled in a heat exchanger 6 with a mixture of water and 1,2-dichloroethane. Partially cooled mixture is treated in a scrubber 9 with alkali removing HCl and CO_2. After cooling in a heat exchanger 10, the condensate is separated (pos.11), while the recycled gases (ethylene, oxygen and inerts) after compression are returned to oxychlorination. Part of this stream is purged. Condensate from 11 flows to separator 12, where heavier 1,2-dichloroethane is separated from water. It still contains some water, thus drying is done in column 14.

Two streams of dichloroethane (one from direct chlorination and another – unreacted dichloroethane after pyrolysis) are then combined and distilled (pos. 16). Pyrolysis of 1,2-dichloroethane into vinylchloride and HCl is done in a furnace 19 at 1.5–2.0 MPa and 500 °C. The gases are cooled in a heat exchanger 20 by a recycle of 1,2-dichloroethane and thereafter cooled with water. Distillation in column 21 allows to separate HCl, which is then send to oxychlorination, and the bottom part containing vinyl chloride and unreacted 1,2-dichloroethane. The latter mixture is separated in column 22 operating under pressure. The bottom part – unreacted 1,2-dichloroethane is sent back to distillation (pos. 16).

Utilization of oxygen allows a significant reduction in the volume of the vent system and operation at lower temperatures resulting in improved operating efficiency and product yield. Larger than stoichiometric excess for ethylene is used in oxygen-based process (Figure 8.13) compared to air-based process. Downstream processing includes cooling of the reaction effluent, its purification from the traces of unconverted HCl, separation of dichloroethane and water also by condensation as in the air-based process, recompression, reheating, and recycling.

Higher selectivity to dichloroethane is achieved by lowering per pass conversion of ethylene. A purge stream of ca. 2–5% is needed to avoid accumulation of such impurities as for example carbon oxides, argon or light hydrocarbons, either introduced with the feed or formed during the reaction.

In fluidized bed reactors the preheated gas mixture enters the base of the vessel and temperature is controlled either by cooling coils or a bundle of cooling tubes in the fluid bed. The oxychlorination reaction takes place at a relatively low temperature, around 220–225 °C, and at a low pressure of about 2 bar. At higher temperature more by-products are formed as both ethylene oxidation and dichloroethane cracking become more prominent. The latter gives by-products with higher levels of chlorine substitution. Moreover, elevation of temperature enhances sublimation of $CuCl_2$ and thus catalyst deactivation.

By-products in oxychlorination are chloral CCl_3CHO, 1,1,2-trichloroethane, chloroform, 1,2-dichloroethylene, and ethyl chloride. Removal of chloral is especially important as it can polymerize in the presence of strong acids. Caustic washing is done for this purpose preventing thereby polymerization and formation of solids, which can otherwise clog processing equipment downstream.

8.5 Fluorination

Fluorination was a difficult process to realize industrially due to its explosive character, thus requiring efficient heat removal, which is done by performing the reaction at low temperature, improving mass transfer by intensive mixing, and dilution of F_2 with inert gases (N_2 or CO_2) and organic substrates with inert solvents.

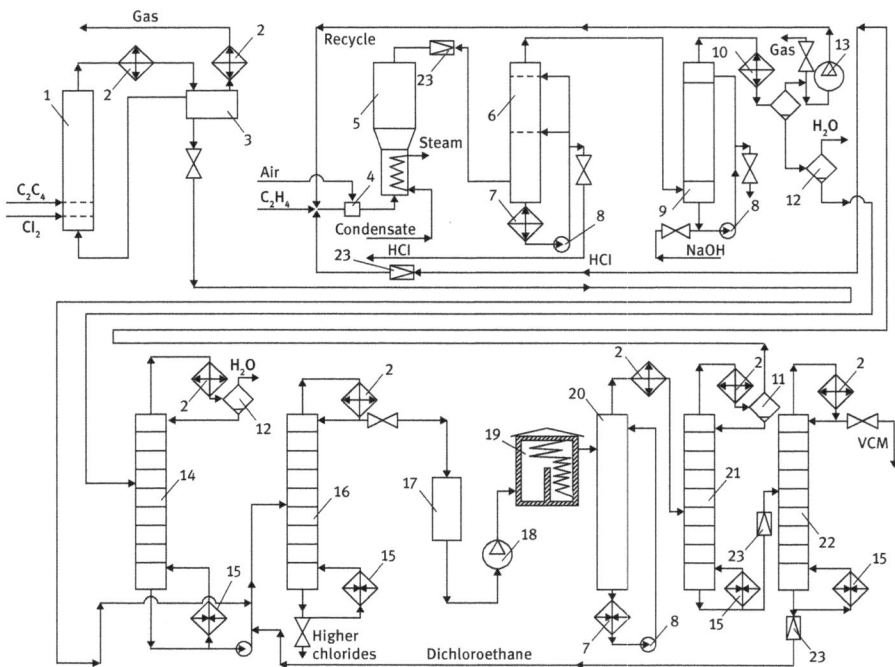

Figure 8.12: Synthesis of vinyl chloride: 1 – chlorator; 2 – condenser; 3 – collecting vessel; 4 – mixer; 5 – reactor; 6, 20 – direct heat exchanger; 7, 10 – heat exchangers; 8 – circulation pump; 9 – scrubber; 11, 12 – separators; 13 – compressor; 14 – drying column; 15 – boiler; 16, 21, 22 – distillation columns; 17 – vessel; 18 – pump; 19 – furnace; 23 – expansion valve. After N.N. Lebedev, Chemistry and Technology of Basic Organic and Petrochemical Synthesis, Chimia, 1988.

The mechanism of fluorination is different from chlorination reactions, which is connected with very weak electrophilic abilities of F_2, allowing only chain reactions. For such reactions, initiators are not even needed, since initiation occurs through a reaction of F_2 with hydrocarbons. Another feature of fluorination is that high reactivity results in low selectivity; thus, addition and substitution reactions occur simultaneously. In the case of fluorine addition, such high reactivity results in substitution of hydrogen in the formed fluoro hydrocarbons. Fluorination is thus conducted in the industry to produce perfluorocarbons. As an example, direct fluorination in the gas phase in a tubular reactor packed with silver- or gold-plated copper turnings is described below (Figure 8.14).

Nitrogen after being bubbled through evaporator 1, containing liquid hydrocarbons, enters reactor 2, to which fluorine diluted with nitrogen (volume ratio $N_2/F_2 = 2:1$) is added. The reaction is performed at 200–325 °C in an autothermal mode. The reactor is filled with copper tunings plated with gold or silver. Before the reaction, fluorine is passed through the tunings to transform silver into AgF_2. The packing has a high heat

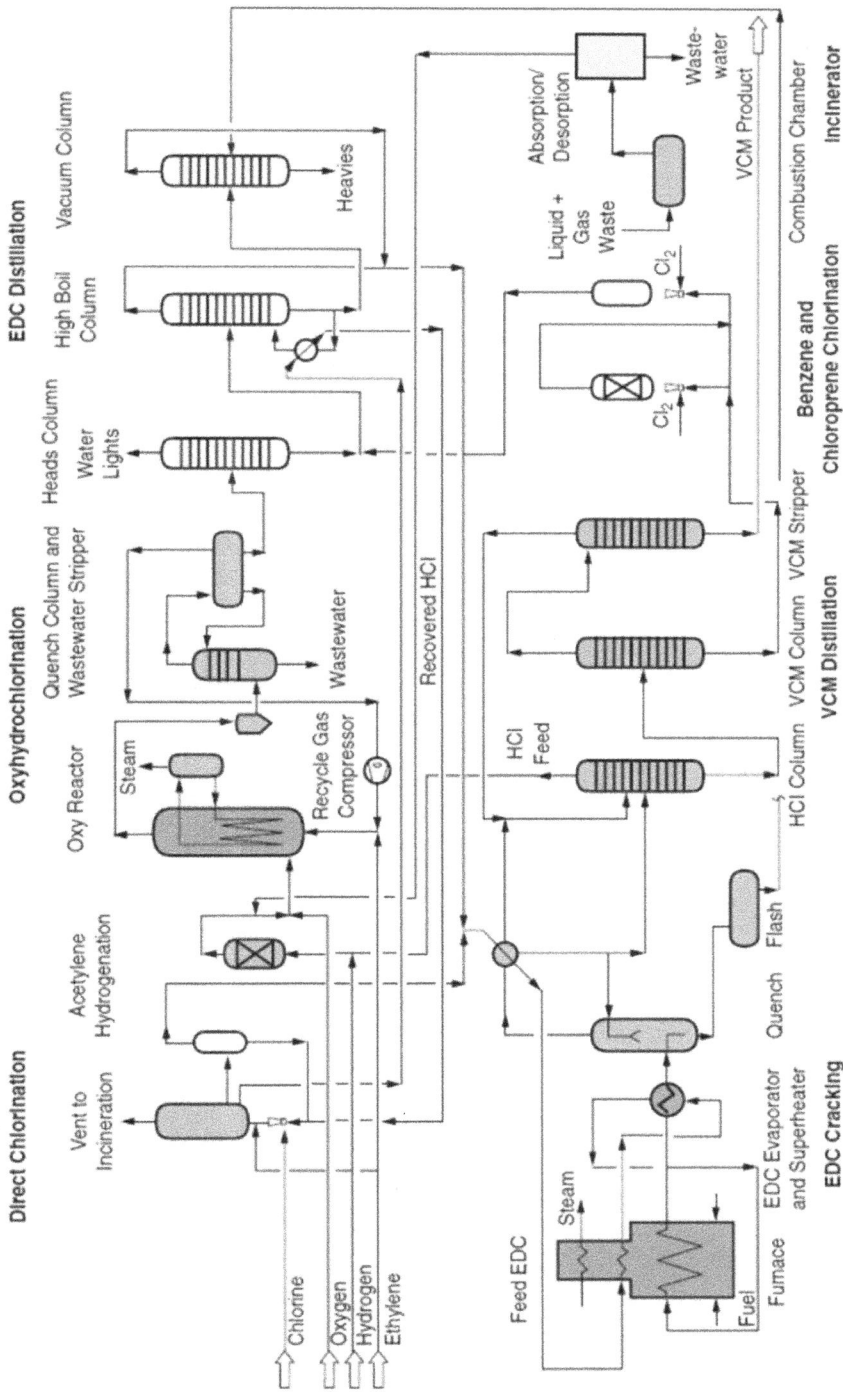

Figure 8.13: Synthesis of vinyl chloride using oxygen-based oxychlorination. https://www.accessengineeringlibrary.com/binary/mheaeworks/
75587e25c3abe6a1/5e021566ff8b3277ea6b32d3e2308a0b503534b1c62b90688e1c0978592b2c1b/18_1x01.png.

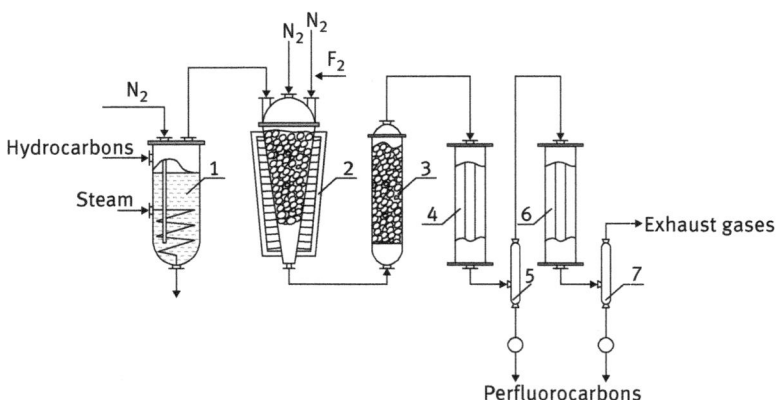

Figure 8.14: Gas-phase fluorination of hydrocarbons: 1, evaporator; 2, reactor; 3, adsorber; 4 and 6, condensers; 5 and 7, separators.

capacity, thus allowing efficient heat removal. A slight excess of fluorine is used to diminish formation of incompletely substituted fluorocarbons. The reaction products at the outlet of the reactor 2 are passed through adsorber 3 filled with spherical balls of NaF, which completely adsorb HF. After condensation and separation in condensers 4 and 6 and separators 5 and 7, respectively, the products are separated further in distillation columns (not shown).

Chapter 9
Oxidation

9.1 Oxidation of inorganic compounds

Two examples of industrial processes in the oxidation of inorganic compounds will be considered here: oxidation of ammonia in the synthesis of nitric acid and oxidation of sulphur dioxide to sulphur trioxide in the production of sulphuric acid.

9.1.1 Nitric acid

Catalytic highly exothermic oxidation of ammonia at temperatures of 850–900 °C over platinum catalysts to form nitric oxide is the first step in the process:

$$4NH_3 + 5O_2 \rightarrow 4NO + 6H_2O \qquad (9.1)$$

The residence time in this reaction should be minimized to prevent side reactions, giving dinitrogen monoxide N_2O and nitrogen. Such short residence time is achieved using precious-metal catalysts typically in the form of closely stacked, fine-mesh gauzes that have smooth surfaces at the beginning of the service. The upstream gauzes have a wire diameter of 0.076 mm and the downstream ones 0.060 mm. After a short time, due to microstructural changes and volatilizing of the catalyst constituents, the catalyst surface area increases substantially, which increases the yield of NO. Platinum is usually alloyed with 5–10% rhodium, improving the strength and reducing catalyst losses.

Downstream of the catalyst gauzes, recovery palladium gauzes have been developed containing a small amount of nickel to improve strength. Platinum, which is captured, forms an alloy with palladium. The level of platinum recovery is more than 80%, while up to 30% of rhodium can be captured.

Irreversible losses of platinum could be also prevented using mechanical filters instead of recovery gauzes. Mechanical filters made of glass, mineral wool, ceramic, or asbestos fibers are installed at mono-high-pressure plants, which have the advantage of low capitassl costs and disadvantage of large pressure drop and subsequently high operating costs. Mechanical filters can recover 10–20% platinum.

Temperature increase enhances the reaction rate but deteriorates selectivity. Reaction temperature to 950 °C could be realized at the expense of, however, catalyst losses mainly due to vaporization. Despite catalyst deactivation and metal losses, the temperature is raised toward the end of a campaign in order to maintain the conversion. The catalyst lifetime is limited typically to 1 year. Elevation of pressure being also beneficial from kinetic viewpoint leads to higher metal losses similar to temperature increase.

https://doi.org/10.1515/9783110712551-009

A pack installed in the reactor (Figure 9.1) can contain up to 40 platinum-rhodium gauzes with the catalysis occurring in just few gauzes, while the rest helps to mitigate deactivation and improve heat and flow distribution.

Monolith non-platinum catalysts for oxidation of ammonia have been introduced in commercial nitric acid plants (Figure 9.2) for partial substitution of Pt Rh gauzes to decrease platinum losses.

Figure 9.1: An assembled pack of platinum-rhodium gauzes.

Figure 9.2: Monolith non-platinum catalysts for oxidation of ammonia in the production of nitric acid.

Explosivity of ammonia-air mixtures and decrease of conversion with too high ammonia concentration lead to application of lower than stoichiometric (14.38%) ammonia/air ratios being up to 11% of NH_3 in high-pressure and up to 13.5% in low-pressure systems.

The subsequent steps after ammonia oxidation are non-catalytic reversible exothermic third-order homogeneous gas-phase oxidation of nitric oxide to nitrogen dioxide

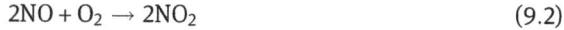

$$2NO + O_2 \rightarrow 2NO_2 \tag{9.2}$$

and absorption of nitrogen dioxide in water to form nitric acid in a heterogeneous gas-liquid process

$$4NO_2 + O_2 + 2H_2O \rightarrow 4HNO_3 \tag{9.3}$$

A cooler downstream the combustion reactor and upstream of nitric oxide oxidation to nitrogen dioxide is needed to achieve low temperatures in the second oxidation step. This is because reaction (9.2) has an unusual temperature dependence and is more active at low temperature.

In mono-pressure (single-pressure) processes, ammonia combustion and NO_x absorption take place at the same working pressure. These include medium-pressure (230–600 kPa) and high-pressure (700–1,100 kPa) processes. In dual-pressure (split-pressure) processes, the absorption pressure is higher than the combustion pressure. Modern dual-pressure plants feature combustion at 400–600 kPa and absorption at 900–1,400 kPa.

Due to cooling after the oxidation reactor, and if necessary, compression, part of nitrogen monoxide is oxidized to nitrogen dioxide or dinitrogen tetroxide, which is converted to nitric acid by absorption in water in a plate column.

The main reactions happening in an absorption tower could be written in the following way:

$$N_2O_4 + H_2O \rightarrow HNO_2 + HNO_3 \tag{9.4}$$

$$3HNO_2 \rightarrow HNO_3 + H_2O + 2NO \tag{9.5}$$

Extra air could be added to the absorber because for the acid production 2 vol of O_2 per 1 vol of NH_3 are needed, however, substoichiometric ratios are applied in the reactor.

A flow scheme of a medium-pressure process (ca. 550 kPa) is given in Figure 9.3.

Ammonia evaporator (pos. 1) is included in the scheme because dosing of low-boiling liquids such as ammonia is difficult. Ammonia stripper (pos. 2) is applied to remove water from liquid ammonia. After evaporation, the vapors are heated in a preheater (pos. 3) and filtered (pos. 4). Air filter (pos. 6) is needed to remove detrimental for the catalyst impurities from air. The latter is, then sucked by the air compressor (pos. 7) and mixed with superheated ammonia in the ammonia-air mixer (pos. 5). The minor amounts (the so-called secondary air) is sent through a tail-gas preheater (pos. 8) to the bleacher (pos. 24) for stripping raw acid. The exit stream from the bleacher contains NO_x and is added to the nitrous gas before entering the absorption tower (pos. 16). In the reactor (pos. 9), an exothermic reaction of ammonia with oxygen occurs over Pt-Rh gauze catalyst. The exit gases are cooled in the waste-

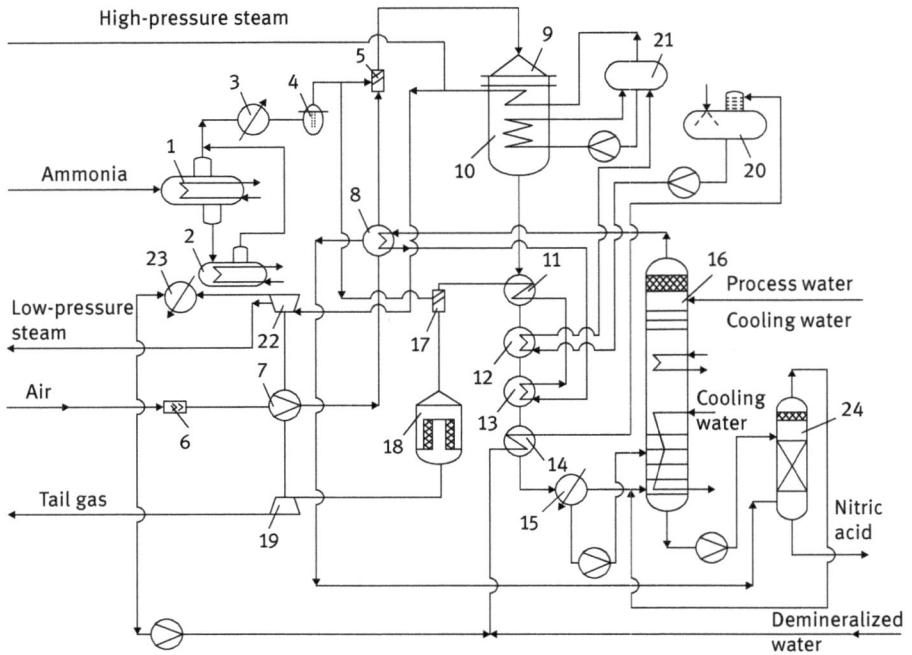

Figure 9.3: Simplified flow sheet of a medium-pressure process: 1, ammonia evaporator; 2, ammonia stripper; 3, ammonia preheater; 4, ammonia gas filter; 5, ammonia-air mixer; 6, air filter; 7, air compressor; 8, tail-gas preheater III; 9, reactor; 10, waste-heat boiler; 11, tail-gas preheater I; 12, economizer; 13, tail-gas preheater II; 14, feedwater preheater; 15, cooler-condenser; 16, absorption tower; 17, ammonia-tail-gas mixer; 18, catalytic tail-gas reactor; 19, tail-gas expansion turbine; 20, feedwater tank with deaerator; 21, steam drum; 22, steam turbine; 23, steam turbine condenser; 24, bleacher.

heat boiler (pos. 10), generating steam. Homogeneous reaction of NO oxidation is happening in the downstream piping and equipment. Due to exothermicity of this reaction, the nitrous gas stream is heated. Tail gas preheaters (pos. 11 and 13) and the economizer (pos. 12) are used for heat recovery. Condensate is formed not only in the cooler-condenser (pos. 15), but also in the feedwater preheater (pos. 14), located upstream. Therefore, this condensate is sent to a tray with the appropriate concentration in the absorption tower (pos. 16), while the nitrous gas stream, after being mixed with the secondary air containing NO_x, is routed to the bottom of the absorber. Nitric acid is taken from the bottom of the adsorption tower. The acid is pumped to the bleacher (pos. 24), where stripping of dissolved NO_x is done through contacting the product acid with air. This is necessary since otherwise, nitric acid containing some NO_x results in a low-quality product having a yellowish color. Finally, the acid is stored in tanks. Catalytic NO_x reduction in tail gases containing typically 500–1,000 ppm NO_x is done with ammonia over a predominantly supported vanadium pentoxide or other catalysts (for example, zeolites) in a $deNO_x$ reactor (pos. 18).

Absorption of NO_x is preferred at high pressure, while oxidation of ammonia is more beneficial at atmospheric pressure. Several options therefore exist in nitric acid production plants including carrying out this process at atmospheric pressure or conducting oxidation at low pressure with pressure elevation for absorption. Such dual-pressure design is preferred for larger plants.

In dual-pressure systems (Figure 9.4), a compressor should be added between the ammonia conversion stage and the absorption stage, which operates at 1.1 MPa.

Industrially produced nitric acid contains 50–70 wt% HNO_3, which is sufficient for production of fertilizers, but too low for nitration processes described in Chapter 14. Increasing the concentration of nitric acid to the desired specification (98–100%) by simple distillation cannot be done because of azeotrope formation of nitric acid with water (68.4% of acid at atmospheric pressure). Instead, as in the indirect processes, extractive distillation and rectification with sulphuric acid or magnesium nitrate is done. Direct processes rely on removal of water by rapid cooling. Nitrous gases are then reacted with the azeotropic acid. Subsequent distillation separates concentrated and azeotropic acid, with the latter either being recycled or used for production of weak acid. An alternative option is to separate completely oxidized NO_x in the liquid form by absorption in concentrated nitric followed by its reaction with oxygen and water (or weak acid) under pressure yielding concentrated nitric acid.

9.1.2 Sulphuric acid

Sulphuric acid is used in the manufacturing of fertilizers (70%), mining, chemical, and petrochemical industry for alkylation, synthesis of inorganic chemicals and pigments, production of organic chemicals, rubber, and plastics as well as water treatment and some other purposes. The key reaction in the synthesis of sulphuric acid is catalytic oxidation of sulphur dioxide to sulphur trioxide done with vanadium-containing catalysts, achieving in modern plants 99.8% conversion:

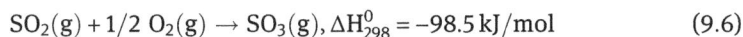

$$SO_2(g) + 1/2\ O_2(g) \rightarrow SO_3(g), \Delta H^0_{298} = -98.5\,kJ/mol \qquad (9.6)$$

This is an exothermic equilibrium-limited reaction, which is one of the few oxidation reactions that is conducted under adiabatic conditions. In order to achieve high final SO_2 conversion, proper removal of considerable reaction heat should be organized; therefore, the total catalyst mass is divided up into several catalyst beds with heat elimination through quenching by cool gas or heat exchangers between the beds. Figure 9.5 illustrates a development of conversion with the catalyst bed

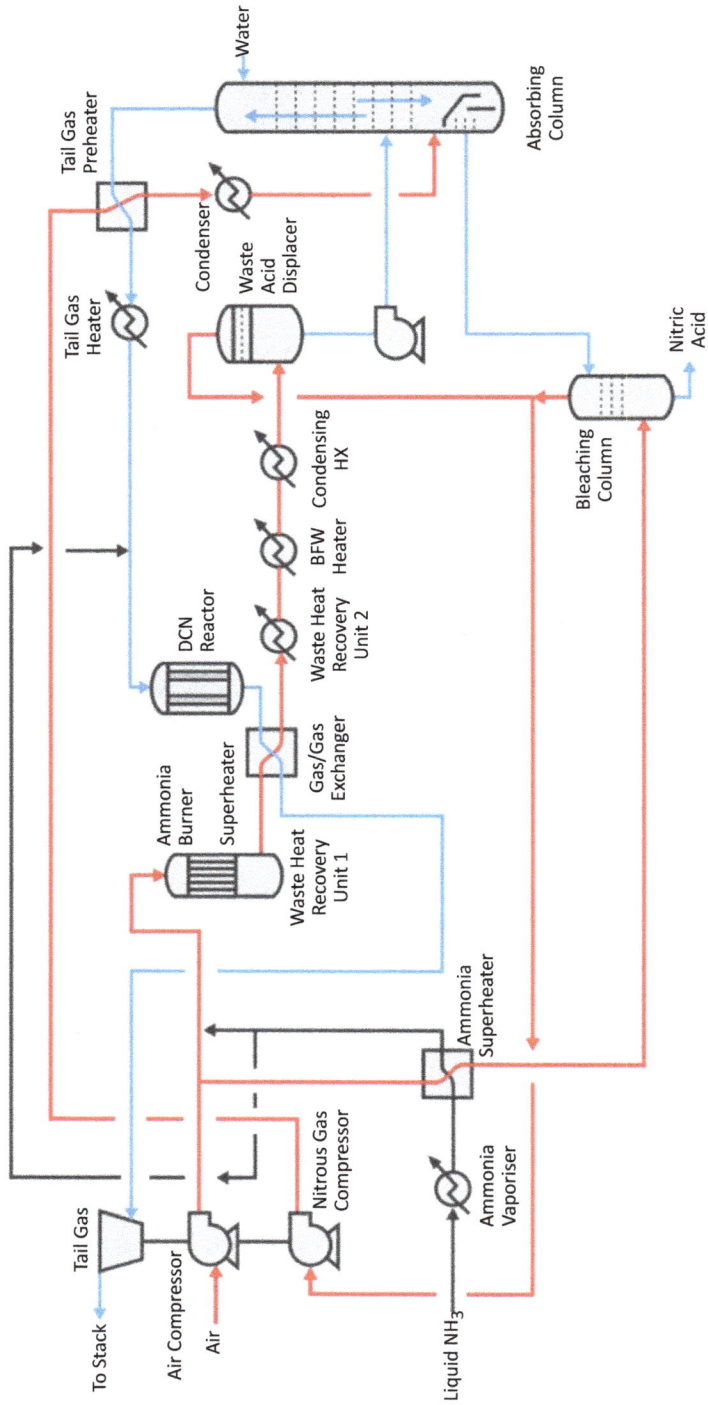

Figure 9.4: A process flow diagram of a dual- pressure system, https://cleantechnologies.dupont.com/technologies/mecs/technologiestechnologies-mecsdupont-clean-technologies-mecs-processes/nitric-acid/.

temperature for a four-bed conventional contact process with absorption of SO_3 after the final catalyst bed (the so-called single absorption), which is the final process step in sulphuric acid production:

$$SO_3(g) + H_2O(g) \rightarrow H_2SO_4(l), \Delta H^0_{298} = -130.4 \, kJ/mol \qquad (9.7)$$

The equilibrium line for such a case (line b in Figure 9.5) shows that there is a limit (ca. 98%) in the final equilibrium conversion of SO_2. A practical way for improving the overall conversion of sulphur dioxide to sulphur trioxide and to have less SO_2 emissions in the atmosphere is to have an intermediate adsorption (removal) of SO_3. This operation is beneficial not only from the thermodynamic but also from the kinetic viewpoint. In practice, it is done by taking reaction gases at the outlet of the second and third catalytic beds, cooling them, and sending to the intermediate absorption. Thereafter, the gases free from sulphur trioxide are reheated in indirect gas-gas heat exchangers and further react in one or two remaining catalytic beds. The equilibrium line for the $2 + 2$ case with an intermediate absorption after the second bed and the final absorption after the fourth bed (Figure 9.5, line a) illustrates that the final conversion is further boosted (>99.5%), which also translates to substantially lower SO_2 emissions.

Equilibrium in the oxidation reaction depends on temperature, total pressure, and partial pressures of the reactants. For an exothermic reaction, low temperatures should be beneficial from the thermodynamic viewpoint; however, temperatures around ca. 400 °C are necessary for a meaningful catalytic activity. More precisely, the inlet temperature is ca. 410–430 °C for a conventional catalyst described below and ca. 380–390 °C for cesium-doped catalysts. Since oxidation reaction is typically done in multibed adiabatic reactors, temperature rise along the bed is an important parameter. The upper operating temperature limit is determined by the thermal stability of the catalyst, which is ca. 600–650 °C. Such conditions are achieved in reality only in the bottom of the first catalyst bed.

According to stoichiometry, sulphur dioxide oxidation requires a stoichiometric O_2/SO_2 ratio of 0.5:1; however, in the industry, a much higher (close to equimolar) ratio is used. Higher than stoichiometric oxygen partial pressure is beneficial from the viewpoint of SO_2 equilibrium conversion, reaction kinetics, and for maintaining the vanadium catalyst activity. O_2/SO_2 composition is also related to the way this mixture in produced.

The first step in the process of sulphuric acid manufacturing is burning of sulphur (Figure 9.6)

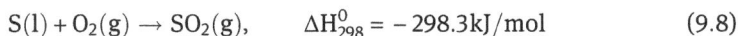

$$S(l) + O_2(g) \rightarrow SO_2(g), \qquad \Delta H^0_{298} = -298.3 kJ/mol \qquad (9.8)$$

or in some locations, pyrite with air coming to the burner through an air filter and a drying tower with 98% circulating acid. In the burner, sulphur is burned to sulphur dioxide.

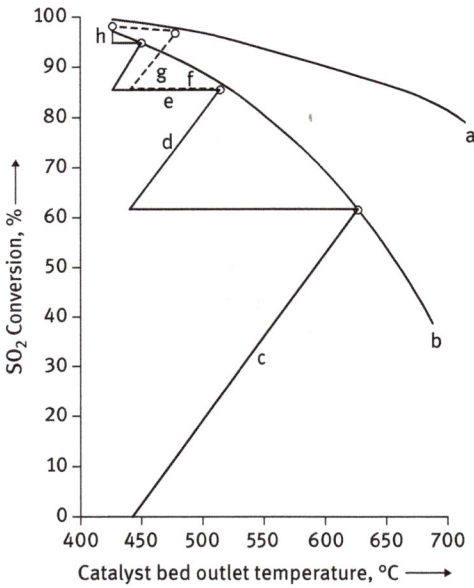

Figure 9.5: Comparison of reaction profiles and SO_2 conversion for four-bed normal contact (single absorption) and (2 + 2) double-absorption processes (feed gas: 8.5 vol% SO_2): (a) double-absorption process equilibrium curve after intermediate absorption; (b) equilibrium curve for normal contact process; (c) adiabatic reaction in bed 1; (d) adiabatic reaction in bed 2; (e) cooling and intermediate absorption; (f) cooling; (g) bed 3; (h) bed 4.

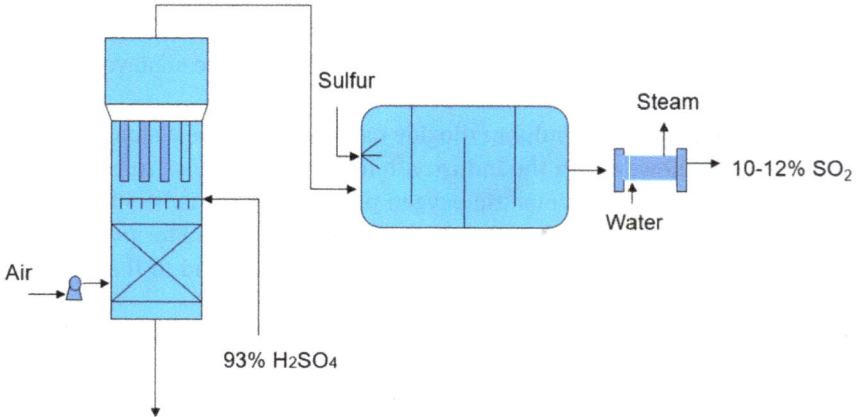

Figure 9.6: Generation of sulphur dioxide.

The SO_2 gas stream after a horizontal spray sulphur burner is 11.5% SO_2, 9.5% O_2 with nitrogen being the balance. There is a limit of adding extra air to the burner, as higher air content will give lower SO_2 concentration and more nitrogen, which at some point will be economically undesirable. It should be noted that O_2/SO_2 ratio due to the initial oxygen excess is changing along the catalyst length.

The service of the catalyst depends on the bed position and could range from 1 year at the top of the first bed to ca. 15 years in the last bed operating under very mild conditions.

Another important issue in practical operation is the presence of water in the feed gas. Water as such does not influence the catalyst; however, condensation of sulphuric acid should be avoided.

Thus, during plant shutdowns, when temperature is low enough, there is a danger of water absorption on the catalyst, deteriorating the mechanical strength.

Commercial catalysts used in abiabatic multibed reactors contain 4–9 wt% vanadium pentoxide, V_2O_5, as the active component, along with alkali-metal sulfate promoters. The reaction actually occurs within a molten salt consisting of potassium/cesium sulfates and vanadium sulfates, coated on the solid silica support. Vanadium is thus present as a complex sulfated salt mixture and not as vanadium pentoxide. Potassium sulfate is used most often in a K/V molar proportion of ca. 2.5–3.5. Cs compared to K does not directly influence catalyst activity; however, cesium sulfate, as a promoter, diminishes the melting point of the active components compared to potassium. This results in significantly lower inlet temperatures of the first bed, which is beneficial from the viewpoint of thermodynamics. Cs-promoted catalysts are also used in the final bed, minimizing SO_2 emissions.

The carrier material is silica in the form of diatomaceous earth or silica synthesized from water glass (sodium silicate). The catalyst manufacturers utilize predominantly kieselguhr. This diatomaceous earth (diatomite) is a form of silica composed of the siliceous shells of unicellular aquatic plants of microscopic size. Kieselguhr is heat resistant and besides its catalytic applications, it has been used as an insulator, a component in toothpaste, and as an abrasive in metal polishes. The specific features of the diatomaceous earth (surface area 20–40 m^2) are related to small amounts of alumina and iron as part of the skeletal structure and a broad range of pore sizes. Silica has a lower bulk density, which means that for the same active phase loading on the support, a higher reactor volume is needed. Silica is more liable to sintering above 900 K than alumina and is volatile in the presence of steam and elevated pressures. Different types of extrudated catalysts (Figure 9.7) in the form of cylindrical pellets, rings, or stars are used commercially, being different in pressure drop and reactor loading densities.

The final-step absorption of sulphur trioxide directly into water (eq. (9.7)) is not possible, as sulphur trioxide reacting with water vapor in equilibrium with the liquid near the surface forms sulphuric acid vapor. Condensation of the vapor makes a

Figure 9.7: Types of sulphuric acid catalysts. https://encrypted-tbn1.gstatic.com/images?q=tbn: ANd9GcRYOXMVyp-L6JtaQBiuKZWNdTd5R42LioLr_4D9hT0uy5AVTZ9M.

mist of submicron droplets, which are practically impossible to collect. On the contrary, sulphur trioxide reacts with sulphuric acid, forming disulphuric acid,

$$H_2SO_4 + SO_3 \rightarrow H_2S_2O_7 \tag{9.9}$$

which is converted back to sulphuric acid when reacted with water,

$$H_2S_2O_7 + H_2O \rightarrow 2H_2SO_4 \tag{9.10}$$

Sulphur trioxide is thus absorbed in sulphuric acid of 98–99% concentration. Due to low concentration of water in such acids, its partial pressure is very low and no mist is formed. The process gas withdrawn from the catalytic convertor is first cooled ca. 180–220 °C in a gas-gas heat exchanger or in a heat exchanger generating steam. If the gas is overcooled below the acid dewpoint (ca. 110–160 °C, depending on the gas composition), it can lead to unwanted mist formation and severe corrosion.

Figure 9.8 is a flow diagram of a sulphur-burning double-absorption plant with four beds and intermediate absorption after the third bed, giving ca. 98.5% sulphuric acid. Air after filtration, drying (pos. 3) with sulphuric acid, and compression to ca. 0.14 MPa is sent to the burner (pos. 1) to which sulphur is also fed. Combustion gases leaving the burner at ca. 1,100 °C are cooled in a waste-heat boiler (pos. 6). The inlet temperature in the reactor (pos. 1) should correspond to the type of catalyst and is ca. 420 °C for potassium-promoted vanadium catalyst. Due to the heat of the reaction, the temperature of the gas mixture at the outlet of the first bed increases to ca. 600–620 °C, giving ca. 60% conversion, a level essentially limited by equilibrium. The gases from the outlet of the first bed and the second bed are cooled in heat exchangers (pos. 6), while the gas after the third bed is routed through a heat exchanger to the intermediate absorber (pos. 4). Absorption is an exothermic process that is beneficial from a thermodynamic viewpoint at low temperature; thus, not only uniform acid and gas distribution in the absorber should be ensured, but the temperature and concentration of the absorber acid should also be carefully controlled. Temperature is kept in most plants at ca. 60–80 °C by indirect cooling, while process water addition through an acid tank (pos. 7) downstream the intermediate absorber is applied to control the acid concentration. The efficiency of SO_3 absorption is > 99.9%. The purified gas cooled in the absorber to about 70–80 °C is returned to the reactor through an intermediate heat exchanger where the cold gas is heated to the inlet temperature of the fourth bed (ca. 400 °C) by hot gas from the third bed. The outlet gas from the fourth bed is cooled in a heat exchanger to ca. 160 °C and is sent to the second absorber. The tail gas containing SO_2 in the amount of some hundred parts per million (ppm) is flared.

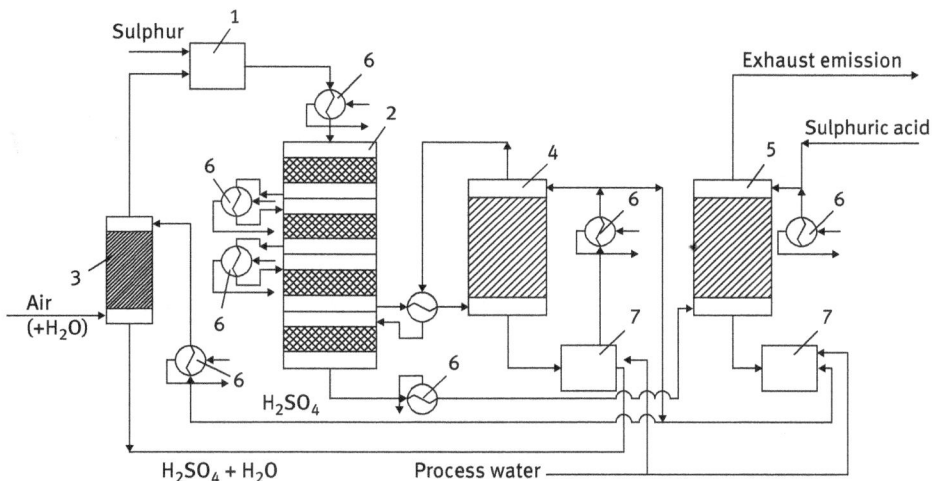

Figure 9.8: Sulphuric acid production: 1, combustion chamber of sulphur oxidation; 2, reactor with four beds; 3, air dryer; 4, intermediate absorber for SO_2; 5, final absorber for SO_2; 6, heat exchanger; 7, acid tanks.

Due to metal corrosion, which is a serious issue in sulphuric acid manufacturing, special alloy metals (e.g., Ni, Cr, Mo, Co) must be used to enhance corrosion resistance. The parameters influencing corrosion are, for example, acid concentration, temperature, and flow rates.

9.1.3 The Claus process

This process is applied industrially to recover sulphur from gases containing hydrogen sulfide exceeding 25% and is based on the following reaction

$$2H_2S + O_2 \rightarrow 2S + H_2O \qquad \Delta H = -186.6 \, kJ/mol \qquad (9.11)$$

In the first thermal part of the Claus process (Figure 9.9), hydrogen sulfide is partially combusted at temperatures above 850° C in the combustion chamber giving elemental sulfur which is precipitated downstream in gas cooler.

When extra amounts of oxygen are added to the feed hydrogen sulfide is oxidized further to sulfur dioxide

$$2H_2S + 3O_2 \rightarrow 2SO_2 + 2H_2O \qquad \Delta H = -518 \, kJ/mol \qquad (9.12)$$

The Claus process is organized in a way that ca. 1/3 of H_2S is converted to SO_2 ensuring that the stoichiometric ratio between the reagents in the Claus reaction per se

$$2H_2S + SO_2 \rightarrow 3S + 2H_2O \qquad \Delta H = -1,165.6 \, kJ/mol \qquad (9.13)$$

is maintained.

Typically, 60% to 70% of the total amount of elemental sulfur is generated in the first thermal step, while the rest is formed in the catalytic step with alumina or titania acting as catalysts.

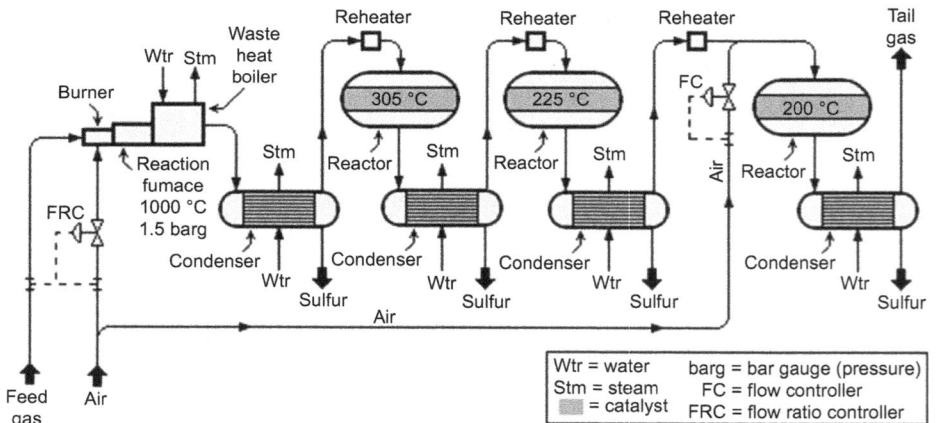

Figure 9.9: Process flow diagram of the Claus technology. By Mbeychok – Own work, CC BY-SA 3.0 (https://commons.wikimedia.org/w/index.php?curid=18564745).

The catalytic recovery of sulfur comprises heating, catalytic reaction, and condensation in two or three stages depending whether the tail-gas treatment unit is added downstream or not.

Reheating upstream a catalytic reactor is required to avoid condensation of sulfur which can be detrimental for catalytic performance. Several reheating options are available including fired heaters, heating with hot gas bypass or high pressure steam, and gas–gas heat exchanges.

The operating temperatures of the reactors are shown in Figure 9.9. The high temperature in the first stage helps to hydrolyze COS and CS_2 formed during combustion. Condensation is done to temperatures between 150 °C and 130 °C. Treatment of the tail gases still containing combustible components and H_2S includes incineration.

A typical Claus process with two catalytic stages gives the sulphur yields of 97% generating also 2.6 tons of steam per ton of sulphur.

9.1.4 Deacon reaction

In the Deacon process, hydrogen chloride gas is converted to chlorine gas at ca. 400–450 °C in the presence of different catalysts, including $CuCl_2$ and $Cr_2O_3 \cdot SiO_2$.

$$4\,HCl + O_2 \rightarrow 2\,Cl_2 + 2H_2O \tag{9.14}$$

As a technology for making chlorine, the Deacon process is outdated, being superseded by electrolytic processes for a number of reasons including a need to operate at high temperatures because of low catalyst activity.

The renewed interest in the Deacon reaction is related to formation of large amounts of hydrochloric acid as a by-product in manufacturing of various important industrial chemicals and consumer products, such as toluene diisocyanate or vinyl chloride. Subsequently, utilization of surplus hydrogen chloride by converting it into chlorine gas (Figure 9.10) can be a feasible industrial option.

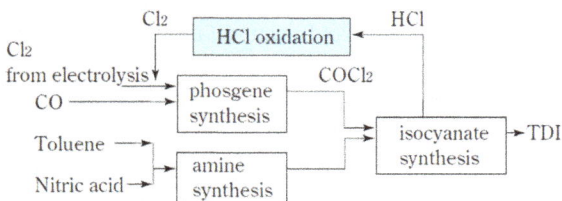

Figure 9.10: Oxidation of HCl as a part of toluene diisocyanate synthesis.

In response to a need for sufficiently active and, more importantly, stable catalytic materials, catalysts based on Ru(IV) oxide have been developed by Sumitomo. As support, TiO_2 and SnO_2 rutile-type carriers are used with silica or alumina as binders.

A catalyst with minimized Ru loading (0.5 wt %) can be utilized in fixed-bed reactors, exhibiting stable catalytic behavior.

Moreover, experience of Sumitomo Chemical, which started to operate a plant of 100,000 t/y capacity in 2003, indicates that chlorine is produced with higher purity than in alternative brine electrolysis technology. The process flow diagram is illustrated in Figure 9.11.

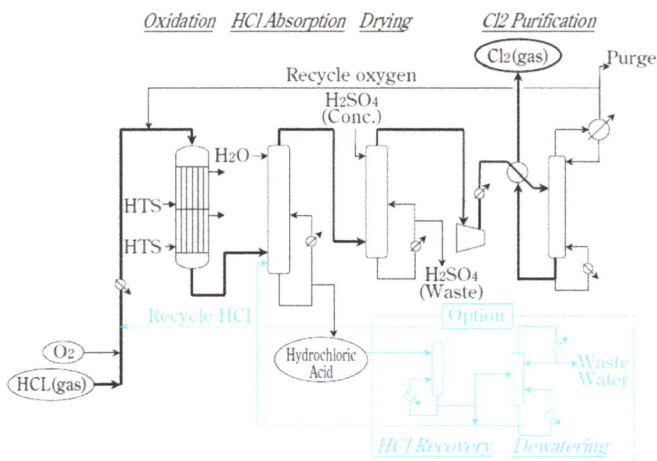

Figure 9.11: HCl oxidation process flowsheet.

Hydrogen chloride and oxygen are reacted in a fixed-bed tubular reactor in the gas phase generating chlorine and water. The reaction heat is removed using a heat transfer salt flowing through the tubes. After quenching the unreacted hydrochloric acid and produced water, there is a possibility of recovery and recycle of HCl. The gases taken from the top of the tower consist of oxygen and chlorine. In the subsequent drying process, water present in these gases is removed through exposure to concentrated sulfuric acid. The dried gas containing chlorine is compressed and cooled. Some of the uncondensed gases, primarily oxygen, is purged to avoid accumulation of impurities, with the remaining gases being recycled back into the reactor.

While Ru-based catalysts have good stability and catalytic activity at low temperatures, apparently problems with deactivation of the active component and high catalyst costs were pushing development of alternative materials. Ceria based catalysts having economic benefits operate in a wide temperature range, and are thus considered as a viable industrial alternative for Ru-based catalysts.

9.2 Oxidation of organic compounds

Such reactions could be classified either as complete or partial oxidation. The former results in the formation of CO_2 and water. In the current section, only selective or partial oxidation reactions will be discussed, as they are important for the synthesis of organic compounds.

Partial oxidation could happen without breaking carbon–carbon bonds, such as oxidation of toluene to benzaldehyde and benzoic acid:

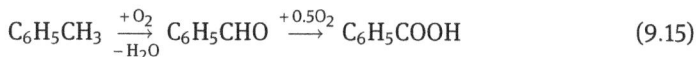

$$C_6H_5CH_3 \xrightarrow[-H_2O]{+O_2} C_6H_5CHO \xrightarrow{+0.5O_2} C_6H_5COOH \tag{9.15}$$

or synthesis of epoxides by oxidation of olefins

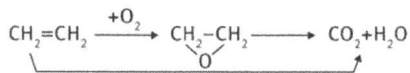

An example of oxidation with carbon–carbon bond breaking is oxidation of benzene to maleic anhydride or naphthalene to phthalic anhydride.

Finally, during oxidation, formation of carbon–carbon bonds can also occur. Such reactions are referred to as oxidation condensation:

$$2RH + 1.5O_2 + ROOR \rightarrow H_2O \tag{9.16}$$

The most important oxidant (oxidation agent) is obviously oxygen (as technical oxygen or air). Nitric acid could be also used for oxidation. In order to avoid nitration, diluted nitric acid (40–60%) is applied. Typically, nitric acid is not used to oxidize paraffins, while oxidation of cyclic compounds (cyclohexane) with carbon–carbon bond breaking leads to higher product yields with nitric acid than with oxygen. In addition to these oxidants, peroxy compounds (peracids or hydrogen peroxide) could also be used. Due to their higher costs and spontaneous decomposition, applications are limited to reactions that cannot be efficiently performed with either oxygen or nitric acid.

9.2.1 Heterogeneous catalytic oxidation

Several groups of catalytic materials are used for oxidation, such as metals (Cu, Ag, noble metals), oxides of transition metals, and mixed oxides of, e.g., V, Bi, Mo, W.

Typically, during oxidation reactions, complete oxidation can also take place. Selectivity control by variation of reactant concentration is difficult to achieve; thus, temperature remains as the main parameter to regulate selectivity. Typically, the activation energy of complete oxidation is much higher (by 20–40 kJ/mol) than for the selective oxidation; thus, with temperature increase, selectivity is diminished.

This points out on the need for efficient temperature control. Overheating in such exothermal processes should be avoided, as it worsens the overall performance. In addition, selectivity toward an intermediate product decreases with conversion; thus, an optimum conversion should be typically achieved, which is done by regulating the contact time (in continuous processes) and applying an excess of the organic compound, while oxygen is deficient.

Selectivity could be also controlled by a proper catalyst design and addition of inhibitors, which diminish the overall catalyst activity, improving selectivity. For example, some known catalyst poisons (halogens, selenium) are added to silver-based catalyst for oxidation of ethylene to ethylene oxide. In the same process, dichloroethane is introduced to the feed, diminishing many-fold activity and increasing selectivity toward the desired product of partial oxidation. Selectivity could be also regulated by selecting a proper type of the support, its porosity, or the size of the catalyst grains.

Owing to the high exothermal character of oxidation reactions, heat removal is essential. Moreover, mixtures of organic compounds with oxygen could be explosive; thus, either small amounts of oxygen are used, the mixture is diluted with steam, or on the contrary, the concentration of the organic substrate is kept very low (3–5 vol%).

For gas-phase processes with a stationary catalyst bed, typically, multitubular reactors are applied (Figure 9.12A), while either heat transfer fluids or water is used to remove heat. Typically, such reactors do not have a uniform temperature profile and hot spots exist, moving along the reactor length as the catalyst deactivates with time-on stream.

From the point of view of heat removal, fluidized-bed reactors (Figure 9.12B) can be considered as isothermal. On the other hand, such arrangement results in back-mixing diminishing selectivity. This could be partially avoided by having several segments (Figure 9.12C) or using a riser (Figure 9.12D), where the catalyst is separated in a cyclone and sent back to the reactor.

9.2.1.1 Ethylene and propylene oxide

One of the most versatile organic intermediates is ethylene oxide, with an epoxide ring, which can be readily opened by reacting with alcohols, fatty alcohols, and ammonia (amines) giving polyols, surfactants, and ethanolamines, respectively. Ethylene oxide is also an important monomer in the production of polyesters and polyurethanes. Oxidation of ethylene is done at elevated pressures (between 10 and 2.2 or 3 MPa, respectively, for oxygen- or air-based processes) using alumina-supported silver catalysts with metal loading 7–20 wt%. The surface area of the support is rather low (<2 m^2/g), and only high-temperature stable supports (such as α-alumina or SiC) are used.

Catalyst manufactures keep information about the composition of catalysts confidential, therefore, details of formulations are not readily available. Several promoters,

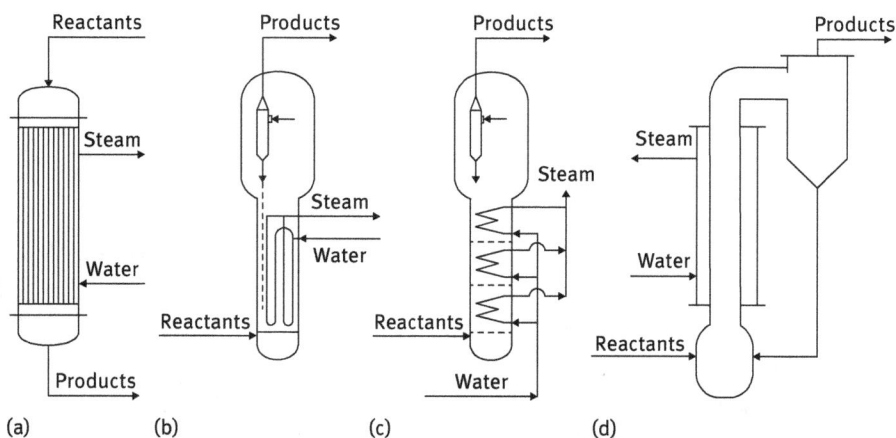

Figure 9.12: Reactors for heterogeneous catalytic oxidation: (a) multitubular reactor; (b) fluidized bed; (c) fluidized bed with different sections; (d) riser.

for example, salts or other compounds of alkali (Li, K, Na) and alkaline earth metals (Ba, Ca) are added to the catalyst significantly improving selectivity. Application of Rb or Cs oxide in combination with Re and its co-promoter (e.g., S, P, Mo, Mn, or Cr) was also reported. Light alkali (e.g., LiCl) can be used in combination with the heavy ones (e.g., CsCl). Moreover, chlorine compounds (e.g., 1,2-dichloroethane) are introduced to the reaction gases, having a negative impact on activity, improving, however, selectivity by suppressing total oxidation of ethylene to carbon dioxide and water.

There measures allow to reach 80–90% selectivity, which is declining with time on stream due to several reasons (poisoning, sintering, dust formation, etc.). The range of temperatures is 220–250 °C. A very high temperature obviously leads to overoxidation and formation of carbon dioxide, while catalysts might not be active enough at very low T. Oxidation of ethylene to epoxide is mildly exothermic ($\Delta H = -105$ kJ/mol) while complete oxidation of ethylene and ethylene oxide to carbon dioxide and H_2O is much more exothermic ($\Delta H = -1{,}324$ and $-1{,}220$ kJ/mol, respectively).

Besides CO_2, water, and ethylene oxide, traces of acetaldehyde and formaldehyde are formed and must be removed or separated from the recycled gas stream. Catalyst life is typically 2–5 years. Thereafter, silver is recovered from the spent catalyst.

Either air or oxygen is used for the oxidation of ethylene oxide, which is synthesized in a multitubular reactor (Figure 9.13) containing several (up to 20) thousands of 6- to 13-m-long tubes with internal diameter of 20–50 mm filled with the catalyst in the form of spheres or rings with a diameter of 3–10 mm. The contact time is ca. 1 s at 230–290 °C and 10–30 bar.

As in other selective oxidation processes a special care should be taken not only about conversion to avoid unwanted complete oxidation, but also about concentration

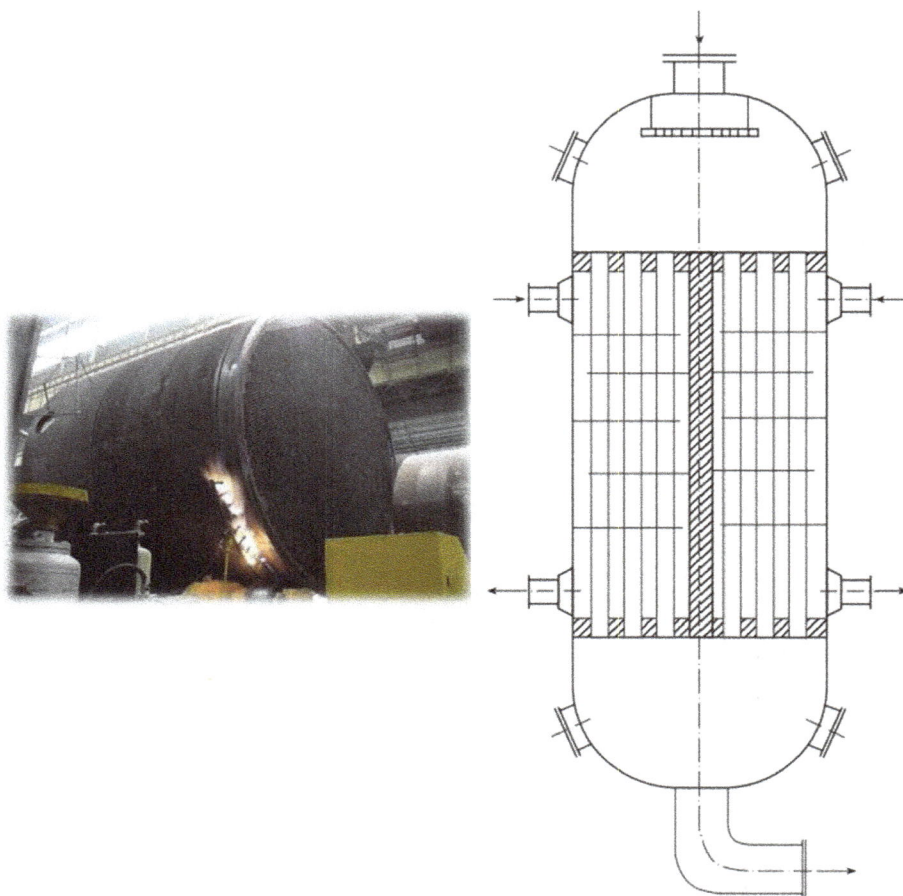

Figure 9.13: Reactor for ethylene oxide synthesis: (a) internals, (b) schematics.

range of ethylene in the mixtures with oxygen, since such mixtures are explosive. As a result, ethylene conversion per pass is limited being 20–65% in the air based compared with 7–15% for the oxygen-based process. The catalyst productivity is ca. two-fold higher in the oxygen-based process, which also displays better selectivity to ethylene oxide. The oxygen-based process is organized in the concentration domain above the explosion limit. In the oxygen- based process methane and argon are applied (Table 9.1) to improve the overall safety.

Incomplete conversion calls for utilization of gas recycling which in the case of air- based process means on one hand a recycle of a large amount of nitrogen in the recycle gas and on the other much larger amounts of purge gas to prevent build-up of nitrogen concentration in the recycle stream. Subsequently the off-gas leaving the primary reactor still contains substantial amount of ethylene, which should be processed before venting into the atmosphere.

Table 9.1: Process parameters in ethylene oxide production processes.

Parameter	Air-based	Oxygen-based
Ethylene concentration (vol%)	2–10	15–40
Oxygen concentration (vol%)	4–8	5–9
CO_2 concentration (vol%)	5–10	5–15
Ethane concentration (vol%)	0–1	0–2
Argon concentration (vol%)		5–15
Methane concentration (vol%)		1–60
Temperature, oC	220–277	220–235
Pressure, MPa	1–3	2–3
Ethylene conversion, %	20–65	8–12
Overall ethylene oxide yield	63–75	75–82

Figure 9.14 illustrates the flow scheme of an oxygen-based process.

When oxygen, being more expensive than air, is used, selectivity is higher, losses of ethylene are lower, and the equipment size is smaller. Pressure is typically 1–2 MPa, which does not influence selectivity in the oxidation reaction but improves downstream absorption. As follows from Table 9.1, ethylene is taken in excess compared to oxygen. Due to formation of CO_2, it should be efficiently removed. Recycled gas, along with fresh oxygen and ethylene (and inhibitor, not shown), is fed into the reactor (pos. 1). Steam is generated in the steam generator (pos. 3). Outlet gases are cooled in the heat exchanger (pos. 2), the cooler (pos. 4), and thereafter sent to the absorber (pos. 5), where ethylene oxide is dissolved in water due to its miscibility. In addition, a part of CO_2 is also adsorbed. A part of the gas (unreacted ethylene and CO_2, along with some impurities) is recycled while the rest is routed to the absorber (pos. 6), where CO_2 is removed using, for example, hot pot scrubbing or water solution of alkanol amines. Subsequently, in a stripper (pos. 9), higher temperatures than for absorption are applied and CO_2 is released. The lean solvent is sent back to the absorber. After a pressure release of ca. 0.5 MPa (expander 16), a water solution of ethylene oxide and CO_2 from the bottom of the absorber (pos. 5) passes through a heat exchanger (pos. 10) to the column 11 where ethylene oxide and CO_2 is distilled from water. The latter, after passing a heat exchanger (10) is sent back to the absorbers (5 and 13). After distillation (column 12), ethylene oxide (bottom) is sent to the

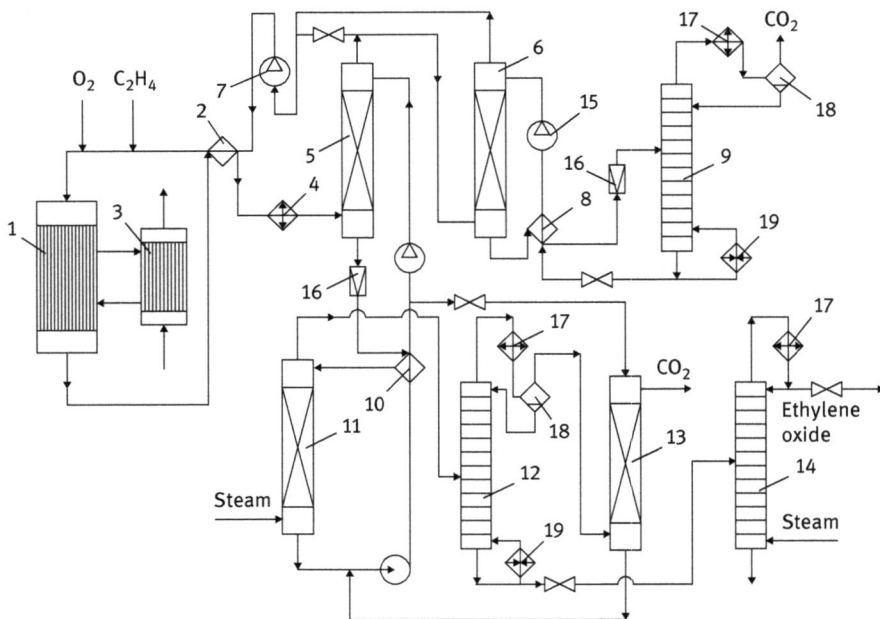

Figure 9.14: Oxygen-based synthesis of ethylene oxide: 1, reactor; 2, 8, and 10, heat exchanges, 3, steam generator; 4, cooler; 5, 6, and 13, absorbers; 7, compressor; 9, stripper (desorber); 11, column; 12 and 14, distillation columns; 15, pump; 16, valve; 17, condenser; 18, separator; 19, reboiler. After N. N. Lebedev, *Chemistry and Technology of Basic Organic and Petrochemical Synthesis*, Chimia, 1988.

final distillation (column 14), while a part of ethylene oxide (top of column 12) is captured in the absorber (13) with water and rerouted to column 11.

The first generation of high-selectivity catalysts introduced in the late 1980s allowed an increase of initial selectivity values by more than six percentage points (e.g., from 80%) to give start-of-cycle selectivity values of 86% or larger bringing enormous saving in ethylene feedstock costs. Activity and stability of these catalysts were lower compared with the traditional ones, thus selectivity declined more rapidly and the lifetime of high-selectivity catalysts was 1.5–2 years being approximately half of the lifetime of traditional ethylene oxide catalysts. More recently, developed so-called high-performance catalysts by CRI Catalyst Company (Figure 9.15) allowed approximately the same initial selectivity and a significantly slower performance decline. According to real plant data, the average selectivity over the lifetime of a catalyst is ca. 87%.

A similar process for propylene oxide has not been developed, since there are no catalysts available up to now that can afford selective oxidation of propylene. Thus, other routes have been practiced throughout the years, including the chlorohydrin routes, resulting in large amount of wastes (2.2 t of salt per ton of product):

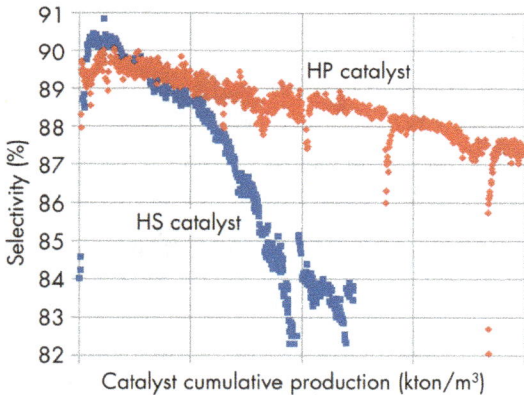

Figure 9.15: Selectivity curves comparing high selectivity (HS) and high performance (HP) catalysts over their lives at the same operating conditions.

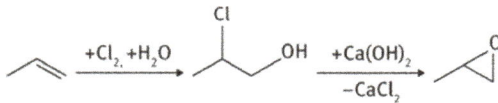

An alternative route is to use oxidation of ethylbenzene, giving a peroxy compound that subsequently reacts with propylene, resulting in propylene oxide and a secondary alcohol. The latter, after dehydration, gives styrene (2.5 t per ton of propylene oxide).

Recently, a more direct method for propylene oxide synthesis was developed based on hydrogen peroxide giving water as the only by-product.

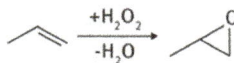

The process flow scheme is presented in Figure 9.16. Oxidation is done in methanol solution with titanium silicate TS-1 as the catalyst. Water is formed as a product from H_2O_2 along with glycols, which are generated in small amounts as by-products.

9.2.1.2 Acrylic acid
In the conventional vapor-phase air-based, two-stage propylene oxidation chemistry, propylene is oxidized in first stage to acrolein

$$CH_2 = CH - CH_3 + O_2 \rightarrow CH_2 = CH - CH = O + H_2O, \quad \Delta H = -340.8 \text{ kJ/mol} \quad (9.17)$$

followed by subsequent oxidation of the latter to acrylic acid in fixed-bed reactors

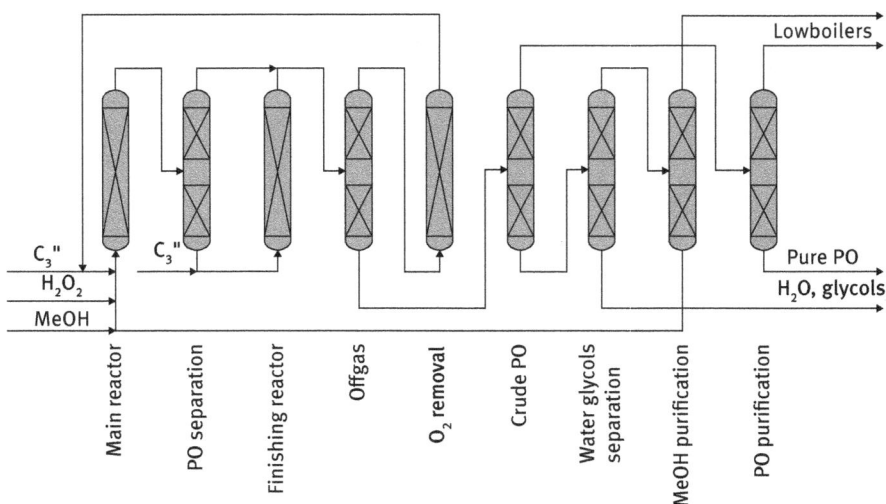

Figure 9.16: Oxidation of propylene to propylene epoxide with H_2O_2. http://www.metrohm-appli kon.com/Downloads/Process_Application_Note_AN-PAN-1007-HPPO-propylene-oxide-production. pdf.

$$CH_2 = CH - CHO + 0.5O_2 \rightarrow CH_2 = CH - COOH, \quad \Delta H = -254.1 \text{ kJ /mol} \quad (9.18)$$

The application areas of acrylic acid are related to production of superabsorbent polymers for diapers and other hygiene products as well as manufacturing of paint, coatings, and resin formulations.

The process flow diagram for the two-step air-based heterogeneous catalytic oxidation of propene in the gas phase is shown in Figure 9.17.

These two steps require different reaction conditions and catalysts to produce optimum conversion and selectivity in each step. For the first step, supported multicomponent systems containing Bi-Mo-Fe-Co mixed oxides are used. Acrolein yields depend not only on the chemical compositions of these catalysts, but also on their physical properties, such as shape, porosity, pore-size distribution, and specific surface area, as well as on the reaction conditions and construction of the reactor. Air, propylene, and steam are fed to reactors consisting of catalyst-filled tubes with molten salt circulating on the shell side for removal of the reaction heat. The salt passes through a steam boiler to produce steam for heat recovery. The gas composition is 5–10% propylene, 30–40% steam, and the rest is air. The oxygen-to-propylene molar ratio is generally 1.6 to 2.0 as an excess of air is used to keep the catalyst in the oxidized state. Steam can be replaced by inert gas. Other reaction parameters are temperature, 300–400 °C; contact time, 1.5–3.5 s; inlet pressure, 0.1–0.5 MPa. With time, the yield of the product and/or pressure drop is increasing to a point

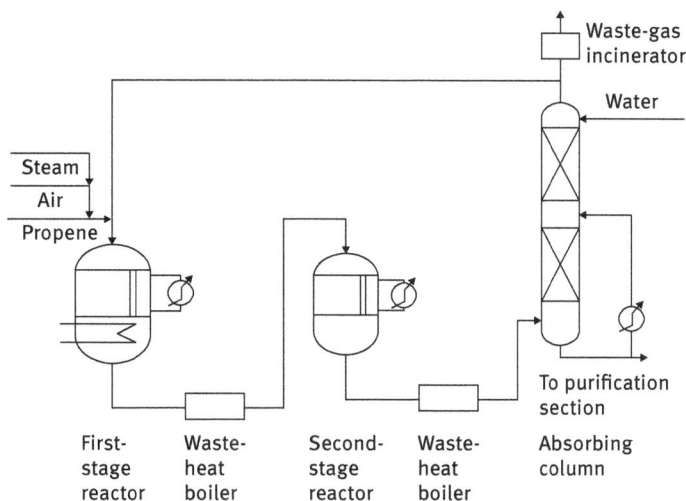

Figure 9.17: Schematic diagram of acrylic acid production (oxidation section).

when a catalyst charge is replaced. Such replacement can happen after a relatively long catalyst lifetime (up to 10 years).

At high one-pass propene conversion (up to 98%), acrolein is formed in the yields of 83 to 90%. The main by-product is acrylic acid (5–10%), with acetaldehyde and acetic acid being other by-products. The reactor effluent is rapidly quenched at the exit to prevent subsequent reactions of acrolein. The catalysts used in the second step (multi-component metal-oxide catalysts containing Mo, V, W, and some other elements) require reaction temperatures from 200 °C to 300 °C and contact times from 1 to 3 s. They give almost 100% conversion of acrolein and yields of acrylic acid larger than 90%. The effluent gas from the second-stage multitube reactor in Figure 9.17 is cooled to about 200 °C and then fed to the absorbing column to be scrubbed with water. Because the effluent gas contains a large amount of steam, acrylic acid usually is obtained as 20 to 70 wt% aqueous solution. The waste gas (nitrogen, excess oxygen, CO_2, and propylene), along with some residual organic compounds should be incinerated.

The next step is the separation of acrylic acid from acetic acid, maleic acid, and acetaldehyde by extraction (Figure 9.18).

After absorption in water, the acrylic acid is purified by extraction with an organic solvent and then by distillation. The solvents should have high affinity toward acrylic acid and low solubility in water. They must form an azeotrope with high water content. Solvents with the boiling points higher (*tert*-butyl phosphate, isophorone, aromatic hydrocarbons) or lower (ethyl acetate, butyl acetate, ethyl acrylate, and 2-butanone) than acrylic acid boiling point can be used. In the latter case, aqueous acrylic acid from the absorbing column is introduced into the extraction

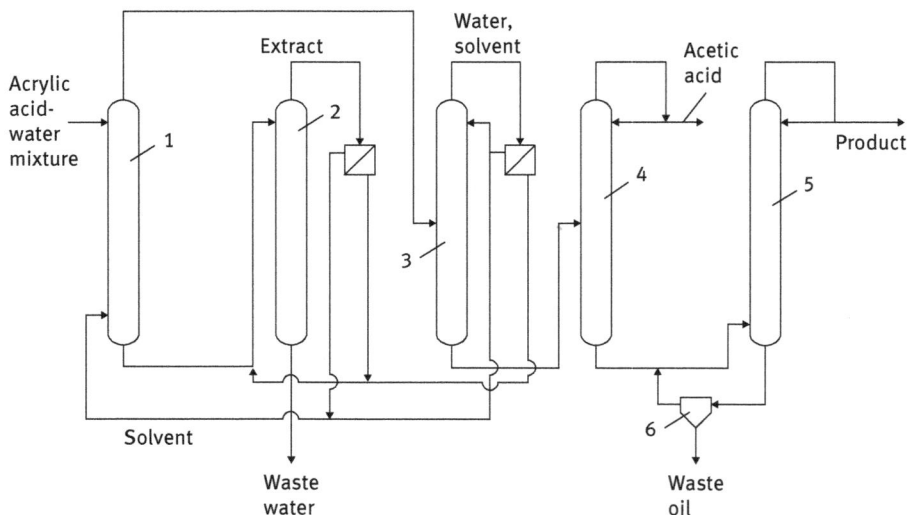

Figure 9.18: Schematic diagram of acid recovery and purification section: 1, extraction column; 2, raffinate-stripping column; 3, solvent separation column; 4, light-ends cut column; 5, product column; 6, decomposition evaporator.

column countercurrent to an organic solvent. The extract taken from the top of the column is separated from the solvent and water in the solvent separation column (pos. 3), where the solvent and water are distilled overhead. Subsequent organic phase-water phase separation gives the solvent, which is sent back to the extraction column, while the water phase is recycled to the raffinate-stripping column (pos. 2). The bottom stream from the extraction column is also sent to this column (pos. 2), which is needed to recover small amounts of the solvent. The top fraction of this column containing the solvent after a phase separation from the aqueous phase is recycled to the extraction column, while the aqueous phase is recycled. The wastewater from the raffinate-stripping column undergoes biological treatment or incineration. Distillation of the bottom fraction from the column (pos. 3) is done in the light-ends cut column (pos. 4), giving lower-boiling-point acetic acid at the top and crude acrylic acid at the column bottom. In the product column (pos. 5), the crude product is distilled, giving high-purity acrylic acid overhead. Distillation columns operate at reduced pressure, lowering the distillation temperature and in the presence of polymerization inhibitors, such as hydroquinone or hydroquinone monomethyl ether, to diminish acrylic acid polymerization. Evaporator 6 is installed for decomposition of acrylic acid dimer back into the monomer. Since some oligomerization and polymerization occur, the residue from the evaporator contains oligomers and polymers of acrylic acid as well as some inhibitors. This residue is burned as waste oil.

9.2.1.3 Formaldehyde

Formaldehyde is produced industrially from methanol by partial oxidation and dehydrogenation with air

$$CH_3OH \Rightarrow HCHO + H_2 \qquad\qquad \Delta H = +84\,kJ/mol \qquad (9.19)$$

$$CH_3OH + O_2 \Rightarrow HCHO + H_2O \qquad \Delta H = -159\,kJ/mol \qquad (9.20)$$

mainly in the presence of silver catalysts, steam, and excess methanol at either 680–720 °C with methanol conversion ca. 99% (so-called complete conversion) or at 550–650 °C giving incomplete conversion of methanol 70–80%. In the latter case, the product is distilled and the unreacted methanol is recycled. The advantage of lower temperature and incomplete conversion is in less prominent side reactions. The reaction temperature depends on the excess of methanol in the methanol – air mixture. The composition of the mixture must be outside the explosion limits. Addition of steam is needed to diminish coking on silver catalysts.

Oxidation can be done also without steam only with excess air in the presence of a modified iron – molybdenum – vanadium oxide catalyst at 250–400 °C with methanol conversion ca. 98–99%.

Among side reactions, oxidation of hydrogen to water, decomposition of formaldehyde to CO and hydrogen, and oxidation of methanol and formaldehyde to CO_2 and water should be mentioned.

Another important factor affecting the yield of formaldehyde and methanol conversion, besides the catalyst and temperature, is the addition of inert compounds to the reactants. Water is added to the spent methanol–water-evaporated feed mixtures, and nitrogen is added to air and air – off-gas mixtures, which are recycled to dilute the methanol – oxygen reaction mixture.

There are several options for the process based on silver catalysts. One variant used for many decades included incomplete conversion with a single-stage methanol separation and recycle (Figure 9.19).

First, a mixture of methanol and water is evaporated and sent to the adiabatic reactor containing a shallow bed (25–30 mm) of a silver catalyst. There is a separate flow to the vaporizer of the fresh process air. The content of the "inert" gases (nitrogen, water, and CO_2) should be selected to prevent formation of explosive methanol–oxygen mixtures. A typical methanol-rich feed (40–45% methanol, 20–25% air, and the rest steam) enables safe operation outside the flammability limits. The average lifetime of a catalyst bed depends on impurities such as inorganic materials in the air and methanol feed. The catalyst bed is located immediately above a cooler (water boiler). Temperature of the outlet gases diminishes to 150 °C and simultaneously superheated steam is produced. Such rapid cooling is needed to avoid oxidation of formaldehyde to CO_2. In the absorption tower, the process gas is flowing countercurrently with water. Methanol is recovered at the top of the distillation column and is recycled to the bottom of the evaporator. The product containing up to 55 wt%

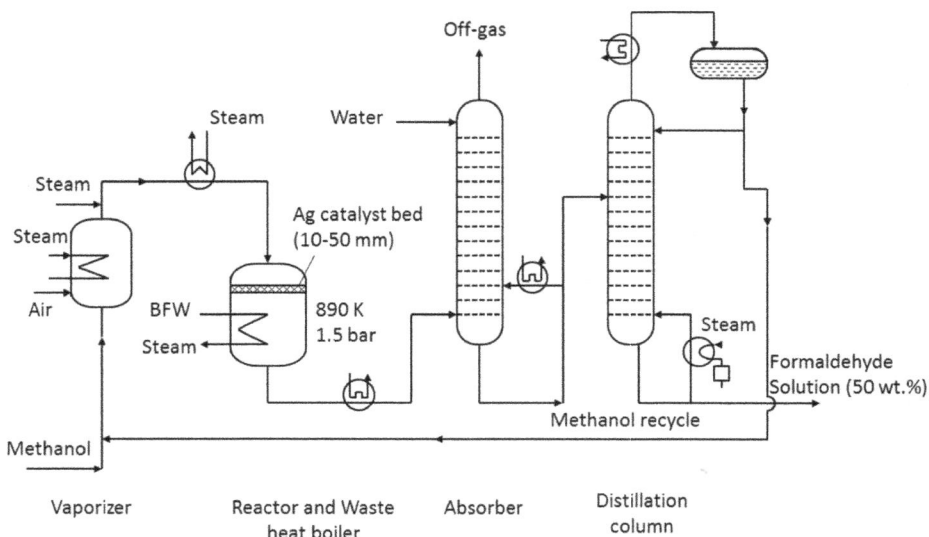

Figure 9.19: Silver-based methanol oxidation to formaldehyde. From J. A. Moulijn, M. Makkee, A E. van Diepen, *Chemical Process Technology*, 2013, 2nd Ed. Copyright © 2013, John Wiley and Sons. Reproduced with permission from Wiley.

formaldehyde and less than 1 wt% methanol along with ca. 0.01 wt% formic acid is taken from the bottom of the distillation column. Special care should be taken on preventing corrosion as formaldehyde solutions are corrosive.

Capital and operating costs for distillation and recycling equipment needed for recovery of unconverted methanol can be reduced by ca. 12% and 20% in the case of a complete conversion process (Figure 9.20), which also has several variants. The fixed-bed adiabatic reactor is, however, essentially the same as in incomplete conversion with a 15–30 mm deep, 2–3 m diameter bed of Ag catalyst (Figure 9.21). Several mesh fractions are typically used with the finest on the top and the largest granules on the bottom of the bed. The bed is arranged to prevent the finest silver grains from slipping through the structure.

The feed comprising methanol, air, steam, and tail-gas recycle is introduced to the reactor at 340–350 °C and ca. 1–1.5 bar. Safety considerations are important in methanol oxidation process. Addition of steam helps to increase selectivity and mitigates catalyst deactivation by lowering coke formation and sintering. There is, however, a limit on the amount of steam, because the target formaldehyde product strength of typically 55% should be achieved, and moreover, additional amounts of water increase the downstream treatment and recycling costs.

The residence time is ca. 0.01 s to minimize decomposition of formaldehyde. Methanol conversion is 98.5–99% with selectivity to formaldehyde of 90–92% with the rest of products being CO, CO_2, and small amounts of formic acid. Rapid cooling

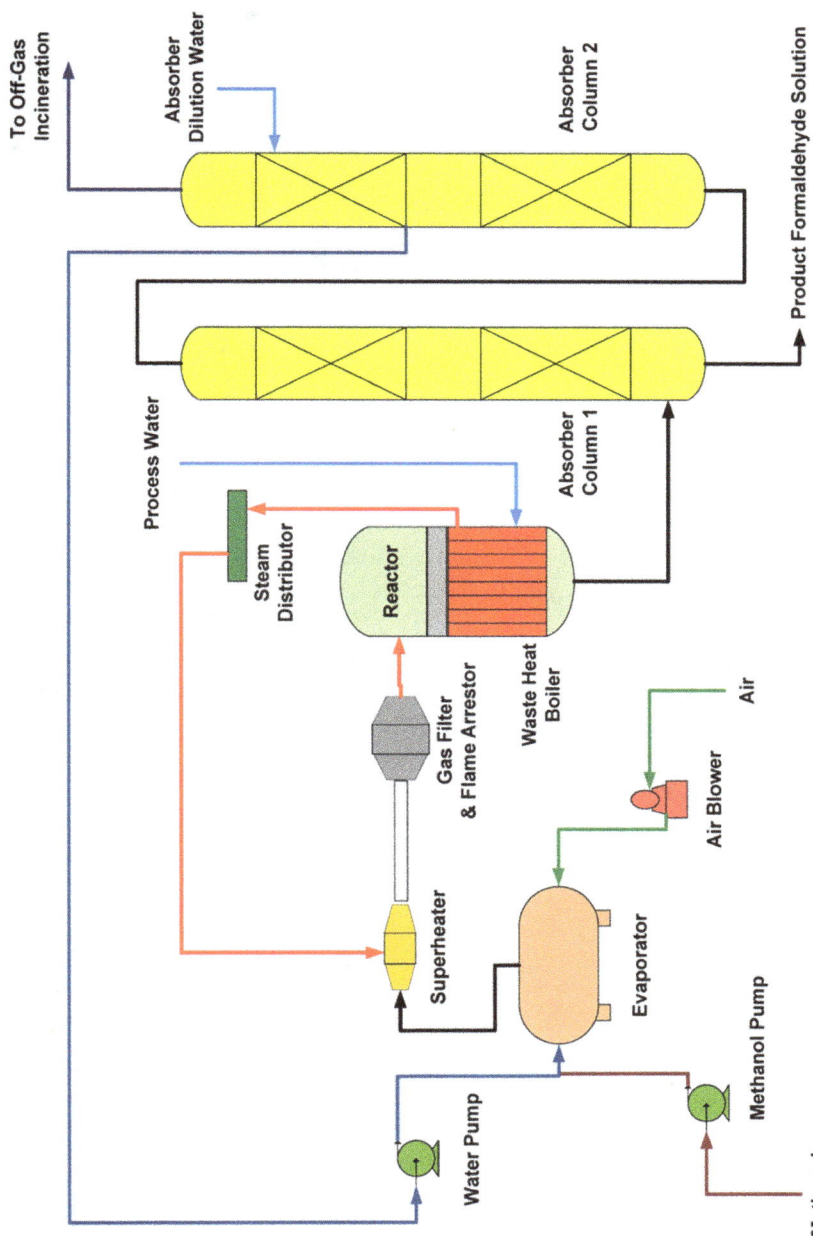

Figure 9.20: Silver-based complete methanol oxidation to formaldehyde process. Reprinted with permission from G.J. Miller, M. Collins, Industrial production of formaldehyde using polycrystalline silver catalyst, Ind. Eng. Chem. Res. 2017, 56, 9247–9265. Copyright American Chemical Society.

Figure 9.21: Left: Silver catalyst bed configuration; center: fresh bed; right: in operation. Reprinted with permission from G.J. Miller, M. Collins, Industrial production of formaldehyde using polycrystalline silver catalyst, Ind. Eng. Chem. Res. 2017, 56, 9247–9265. Copyright American Chemical Society.

of formaldehyde gas is done using a waste heat boiler system. The exit gas stream leaves at temperature between 120 °C and 200 °C, and then passes through the first absorption tower. Water containing recaptured methanol and formaldehyde from the second absorption column is sent to the reactor, enhancing the overall product yield.

The lifetime of the catalyst life is from few months to a year depending on the impurities in methanol and air, operation temperature, and the plant capacity. Sintering of silver at high reaction temperatures results in pressure drop elevation. As typical with many processes where catalyst deactivates at some point, a decision should be taken to replace the catalyst. Such decision is based on evaluation of the catalyst costs vs lower productivity, higher operations costs and safety considerations.

If higher capacities are required and a small number of reactors must be arranged in parallel, the process economics favors the silver-based technology compared to an alternative process where a metal oxide catalyst is applied.

The catalyst composition in the latter, so-called Formox process, is rather complex, for example, it can contain iron and molybdenum oxides $Fe_2(MoO_4)_3$ with an excess of MoO_3 giving Mo:Fe atomic ratio of 1.5–2.0. Iron molybdate structure is stabilized by additives such as V_2O_5, CuO, Cr_2O_3, CoO, and P_2O_5. The catalyst lifetime is typically 18–24 months because the catalyst is more tolerant to poisons.

An excess of air is used in this process with ca. 13:1 air-to-methanol molar ratio, thus the concentration of methanol is 6–9%. Very high amounts of methanol would lead to catalyst reduction and subsequent deactivation. Proper heat management by avoiding overheating is needed to prevent sintering. Methanol conversion is done at atmospheric pressure and much lower temperature (270–400 °C, a typical value 340 °C), compared to silver based process. Typical single-pass methanol conversion is 97–99% with formaldehyde selectivity of 92–95%. Lower temperature is beneficial from the viewpoint of higher selectivity. All these in addition to a simple method of steam generation contribute to the advantages of the Formox process. At the same time, the volume of gas is 3–3.5 higher than in the silver-based technology resulting in more expensive, larger capacity equipment.

Figure 9.22 displays the flow scheme for the Formox process. Methanol is fed through a steam-heated evaporator and after mixing with the fresh process air and the recycled off-gas passes through a heat exchanger heated by the product stream.

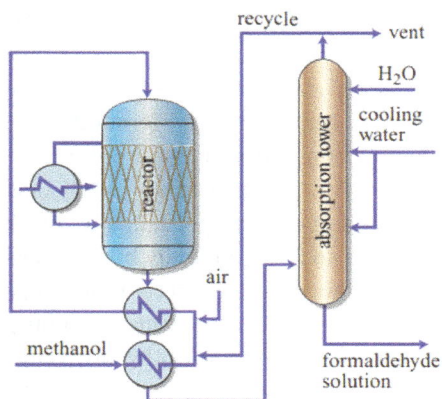

Figure 9.22: Synthesis of formaldehyde by oxidation on Fe–Mo catalyst (from http://www.treccani. it/portale/opencms/handle404?exporturi=/export/sites/default/Portale/sito/altre_aree/Tecnolo gia_e_Scienze_applicate/enciclopedia/inglese/inglese_vol_2/615-686_ING3.pdf&%5D).

The catalyst pellets or rings (Figure 9.23) of the size 3–4 mm are placed in multitubular reactors of a diameter ca. 2.5 m containing thousands of tubes (diameter 2–3 cm and 1.0–1.5 m length). The reaction heat is removed by a high-boiling heat-transfer fluid (oil, Dowtherm, molten salt) held at ca. 270–290 °C. This fluid is circulating outside the tubes generating steam in a boiler.

A typical feature of highly exothermal reactions in multitubular reactors are hot spots, which in the case of methanol oxidative dehydrogenation on iron molybdate reach 340–350 °C.

Figure 9.23: Fe–Mo catalysts of different shape (a) (https://www.clariant.com/en/Solutions/Prod ucts/2019/04/15/07/20/FAMAX-200-Series); (b) (http://www.glp.com.au/wp-content/uploads/ 2014/02/General-SC-Overview-BCT.pdf) for synthesis of formaldehyde by methanol oxidation.

The outlet gases are cooled to 110 °C to avoid formaldehyde and sent to the bottom of an absorber column to which the process water is added at the top. The product containing up to 55 wt% formaldehyde and 0.5–1.5 wt% methanol is further purified by ion exchange to diminish the formic acid content. The overall plant yield is 88–91 mol%. The absorber off gas (N_2, O_2, CO_2 with few percent of dimethyl ether, carbon monoxide, methanol, and formaldehyde) is non-combustible as such and requires catalytic incineration at 450–550 °C or alternatively combustion with additional fuel at 700–900 °C.

Both types of plans using either silver or iron catalysts despite operation at very different conditions give similar values of methanol conversion and formaldehyde yields. If cheap methanol is available, silver can be considered a better option, especially for smaller operations. For larger plants with capacity of up to 100,000 tons per year, the metal oxide catalysts can be more suitable compensating a higher capital expenditure related to fivefold higher air flow and subsequently higher compression costs. Plants using more productive silver catalysts may present lower operating costs and lower catalyst costs as silver can be regenerated contrary to only Mo from iron molybdate and lower steam production. The tail gas in the silver process contains about 20% hydrogen and can be burned directly eliminating carbon monoxide and other environmentally harmful organic compounds. In an alternative iron oxide process, burning of the tail gas with very low amounts of flammable compounds requires a catalytic incinerator or addition of extra fuel, which adds to the production costs. Metal oxide catalyst is more resistant to poisons and because of low temperature is less susceptible to thermal degradation giving a better quality formaldehyde product with less impurities such as formic acid, heavy metals, and unreacted methanol. A further advantage of the process using iron oxide with a substantially longer catalyst life are less downtime problems.

9.2.1.4 Maleic anhydride

Maleic anhydride has numerous industrial uses and is of significant commercial interest worldwide. The primary use of maleic anhydride is in the manufacture of polyester and alkyd resins. Benzene was used as the dominating feedstock

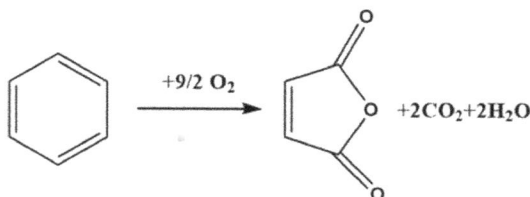

being in the recent years continuously replaced by oxidation of *n*-butane or *n*-butane-*n*-butene mixtures with a high paraffin content:

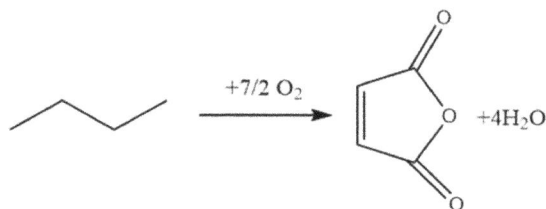

The traditional process begins by mixing benzene with an excess of air to give concentrations from 1 to 1.4 mol%. A low benzene concentration must be utilized in order not to exceed the flammability limit of the mixture. The reaction gas mixture then passes over the catalyst in a multitubular (diameter ca. 25 mm) fixed-bed reactor at an optimum pressure range of 0.15–0.25 MPa. The desired reaction is highly exothermic, causing hot spots of 340–500 °C. As a rule, commercial catalysts for benzene oxidation are supported by an alumina or silica carrier and have a surface area of 1–2 m^2/g. A typical catalyst is comprised of V_2O_5, MoO_3, and some other promoters, such as phosphorus, alkali and alkaline earth metals, tin, boron, and silver. Benzene is passed through the catalyst at the concentration of 60–130 g benzene per liter of catalyst per hour. Conversion 97–98% is achieved with an initial selectivity over 74%. The lifetime of the catalyst can be as long as 4 years depending mainly on the reactor operating temperature and the purity of the starting materials.

Recovery of the significant part of the product (40–60%) is done by cooling the reactor outlet stream to 55 °C. Condensed maleic anhydride must be removed as soon as possible to avoid prolonged contact with water in the reaction gas; otherwise, maleic acid formation will occur. The rest of maleic anhydride that cannot be recovered is washed out with water as maleic acid. Water scrubbing and subsequent dehydration of maleic acid at ca. 130 °C is required to purify and reform the remaining maleic anhydride. A side reaction is isomerization of maleic acid to fumaric acid.

The processes developed or in the developmental stage for the oxidation of C4 hydrocarbons are fixed-bed (Figure 9.24), fluidized-bed, and transport-bed reactors.

The reaction is the complex one, requiring extraction of eight hydrogen atoms and insertion of three oxygen atoms to butane along with a ring closure. A high concentration of butane, while advantageous, is dangerous, since butane and oxygen could be within explosion limits; thus, the concentration of butane in air is ca. 1.8 mol%. A major advantage of C4-based process compared to benzene oxidation is that no carbon is lost in the reaction to form maleic anhydride. Moreover, the feedstock is cheaper, not carcinogenic contrary to benzene, and the flammability limits for C4 hydrocarbons are lower. Oxidation of C4 hydrocarbons at 80% conversion and 70% selectivity gives comparable weight yields as for benzene oxidation. In the fixed-bed process (Figure 9.24), oxidation is done in tubular reactors similar to benzene oxidation with vanadium and phosphorus oxides (V-P-O) unsupported catalysts. In the catalyst formulations having phosphorus-vanadium ratio of 1.2, several promoters (Li, Zn, Mo, or Mg, Ca, and Ba) are applied, improving conversion and selectivity compared to unpromoted catalysts.

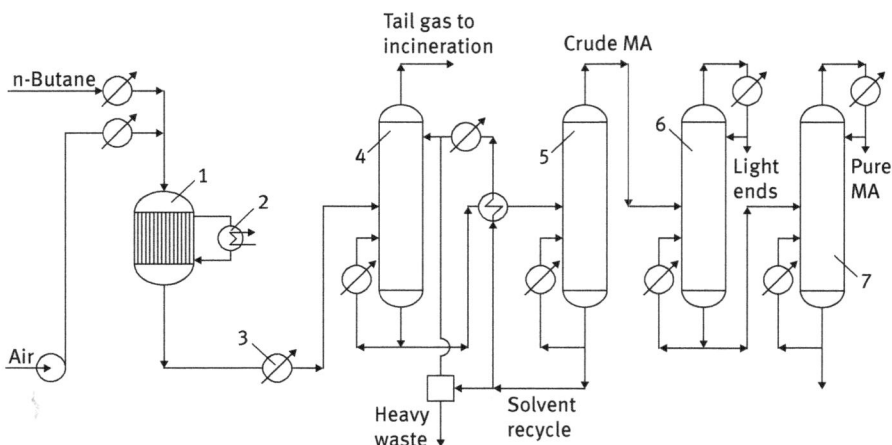

Figure 9.24: Synthesis of maleic anhydride by butane oxidation in a fixed-bed reactor.

Higher phosphorus/vanadium ratios increase conversion but lead to shorter catalyst lifetime. An oxidation reactor operates at 400–480°, and 0.3–0.4 MPa. Formation of water is higher in butane oxidation processes compared to benzene oxidation; therefore, only a small amount of maleic anhydride can be condensed from the reactor effluent. The remaining product is washed out as maleic acid in a scrubber and then dehydrated similar to the case of benzene oxidation. Tail gases after the absorption column are incinerated, as recovery of unreacted C4 hydrocarbons is substantially more difficult than in the case of benzene. An alternative recovery method utilizes absorption of the product from the reactor outlet by organic solvents and will be explained below for the case of the fluidized-bed reactor.

Application of such fluidized-bed reactors allows increasing the concentration of butane in the feedstock compared to fixed-bed operation by 50–100%. The fluidized-bed process has the advantage of a particularly uniform temperature profile without hot spots, thus improving process selectivity. A particular problem of this process is mechanical stress on the catalyst (V-P-O), its abrasion, and erosion at the heat-dissipating surfaces. On the other hand, the fluidized bed constitutes an extremely effective flame barrier, with the result that the process can operate at higher C4 concentrations (i.e., within the explosion range) than the fixed-bed process. The process flow scheme is given in Figure 9.25. In the recovery system, crude maleic anhydride is distilled from the high boiling solvent in a fractional distillation column (stripper in Figure 9.25) while the solvent is recycled. Further refining of the product is done by removing the light-ends and recycling heavier compounds (so-called heavies) from the final distillation column to the stripper. The tail gases are incinerated.

Butane oxidation can be also done utilizing a transport reactor (Figure 9.26) even if the industrial implementation of the process was not very successful with the

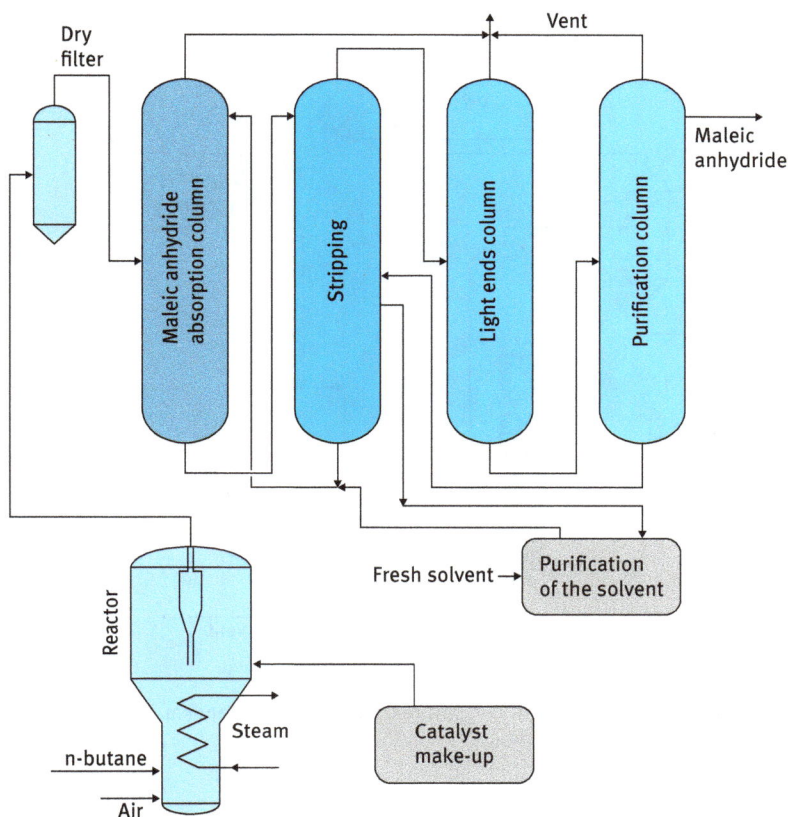

Figure 9.25: ALMA fluidized-bed process for oxidation of butane in maleic anhydride with solvent recovery. Modified after http://www.treccani.it/portale/opencms/handle404?exporturi=/export/sites/default/Portale/sito/altre_aree/Tecnologia_e_Scienze_applicate/enciclopedia/inglese/inglese_vol_2/615-686_ING3.pdf.

production being stopped after some time in operation. The advantage of such system, when a metal oxide oxidation state (in this case V_2O_5) is changing during the reaction (from V^{+5} to V^{+4}), is that donation of oxygen from the catalyst lattice to the substrate with subsequent reduction of V^{+5} to V^{+4} is separated in space from oxidation of V^{+4} to V^{+5}. The latter process is conducted in a separate reactor, which prevents butane from being in contact with air, thus making the process much safer.

9.2.1.5 Phthalic anhydride
Commercialization of phthalic anhydride (PA) production goes back to 1872 when BASF developed the naphthalene oxidation process

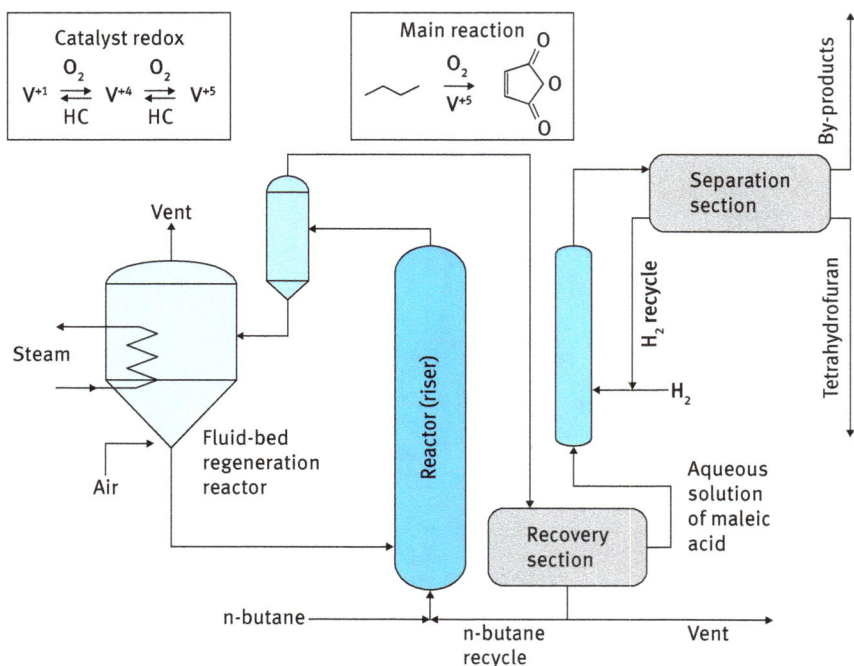

Figure 9.26: Maleic anhydride synthesis in a transport reactor-regenerator tandem.

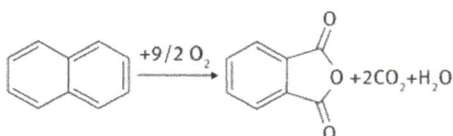

which is accompanied by total oxidation

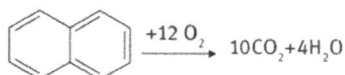

The most important derivatives of phthalic anhydride, which is produced globally with a capacity of ca. 4.5 million tons per year, are phthalic esters, which could be used as plasticizers, polyester resins, and dyes. In particular, the utilization of the diester dioctyl phthalate (DOP) is widespread, as it used to impart flexibility to polyvinylchloride, which otherwise is a rather rigid material. Until the early 1960s, coal-tar naphthalene was the dominant feedstock; later on, gas-phase oxidation of o-xylene over vanadium-based catalysts gained industrial importance and almost completely replaced oxidation of naphthalene.

In the industry, oxidation of o-xylene is done in fixed-bed tubular reactors (15, 000–30, 000 tubes) at 300–400 °C and 0.5–1.8 vol% of xylene in air. One of the reasons for the feedstock switch was o-xylene availability from petroleum, while quantities of naphthalene derived from coal tar were dependent on the production of coke and were not sufficient for the growing PA and plasticizer markets. The other reason is the so-called atom efficiency. As can be seen from eq. (9.21), naphthalene oxidation inevitably results in the consumption of two carbons and generation of 2 mol of carbon dioxide per mole of product formed. Oxidation of o-xylene does not lead to CO_2 release in the main reaction, but it could be certainly formed as a side product (Figure 9.27).

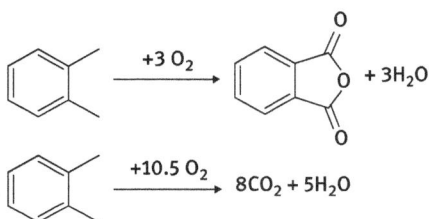

Figure 9.27: Oxidation of o-xylene to phthalic anhydride.

In addition to feedstock availability and unnecessary CO_2 emissions, in some areas, in particular those with cold climates, handling of naphthalene, which is solid at room temperature, was challenging. Although more than 90% of PA is nowadays produced from o-xylene, there are still few plants operating with naphthalene. These plants are located mainly close to major steel mills. Naphthalene is much cheaper than o-xylene, being a by-product of coke production. Few installations utilize mixed o-xylene-naphthalene feeds.

The oxidation reactions are not limited by equilibrium and are highly exothermic. High activation energy means that even with a small temperature increase, reaction rate, and thus heat production, increases substantially, leading to prominent hot spots in fixed-bed reactors. The reaction enthalpy of o-xylene oxidation is –1108.7 kJ/mol at 298 °C while it is –4,380 kJ/mol for complete oxidation. Since overall selectivity in o-xylene oxidation is ca. 80% at almost complete conversion, heat generation in this reaction is significant, reaching 1,300–1,800 kJ/mol o-xylene. Heat removal is thus essential in o-xylene oxidation. As already mentioned above, oxidation is performed in fixed-bed tubular reactors. Another option would be to utilize fluidized-bed reactors, which are very efficient in controlling temperature, maintaining a uniform temperature throughout the bed. Such option was in fact implemented in only few places in the case of naphthalene oxidation, where V_2O_5-K_2SO_4 on silica was applied, allowing good fluidization properties. Since such support does not allow high yields of phthalic anhydride in o-xylene oxidation, apparently, there are no industrial o-xylene-based fluidized-bed reactors currently in operation, although development work in companies and academia might still continue.

It should be noted that the feedstock as well the product can be explosive in mixtures with air. Thus, special care should be taken to ensure that the feedstock and product are not within explosion limits with air.

The lower explosion limit for *o*-xylene or naphthalene in air is 47 g/m³ (STP). Compression and preheating of air represent additional costs; thus, an increase in *o*-xylene loadings above this limit was addressed along the years. Many plants operate now at 60, 80, or even 100 g/m³, as it was shown that an explosion starting in the catalyst bed is quenched by the catalyst heat capacity. Extra thickness in the upper part of the reactor, combined with adequate rupture disks, is implemented to avoid explosions in the free space above the catalyst bed. The probability of phthalic anhydride auto-ignition is minimal since PA auto-ignition temperature (580 °C) is well above typical operation temperature (salt bath temperature 345–350 °C).

The reaction network showing formation of the main and side products is presented in Figure 9.28.

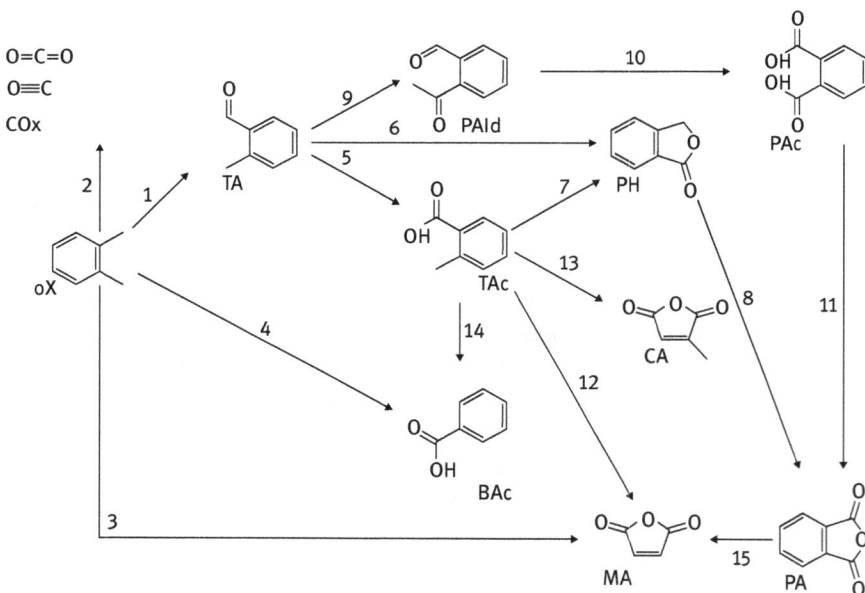

Figure 9.28: Oxidation of *o*-xylene. From R. Marx, H.-J. Wölk, G. Mestl, T. Turek, Reaction scheme of o-xylene oxidation on vanadia catalyst, *Applied Catalysis A: General*, 2011, 398, 37. Copyright Elsevier. Reproduced with permission.

The major intermediates in the selective oxidation path are tolualdehyde (TA, 0.005–0.02 wt%) and phthalide (PH, 0.02–0.07 wt%), while such intermediates as toluic acid (TAc), phthalic acid (PA), and phthalaldehyde (PAld) were also proposed. Non-selective route, i.e., a route that does not result in PA, includes formation of benzoic (BA) acid, 0.4–0.8 wt%; citraconic aldehyde, 0.35–0.5 wt%; maleic anhydride (MA, 3.3–4.0 wt%) and CO_x (CO and CO_2). The amounts of side products above are presented for the latest-generation catalysts with PA yield 94.5–95.6%. The byproduct formation is crucial not only in terms of product selectivity. In some instances, excessive presence of by-products deteriorates product quality, in particular, the so-called Hazen number, which reflects the PA color and should be in some applications rather low (5–10). Beside the main organic by-products, other carbon-containing by-products are carbon monoxide and carbon dioxide. In general, the reactor yields are not exceeding 112–115 kg PA/100 kg o-xylene, which corresponds to ca. 82% of stoichiometric yield. Taking into account losses during condensation and distillation, the yields of pure PA are about 112 kg PA/100 kg o-xylene.

For the sake of comparison, a few key numbers are also given for oxidation of naphthalene. The heat of reaction is 1,788 kJ/mol, while in complete combustion 5,050 kJ/mol is evolved. The main by-products are maleic anhydride and naphthoquinone (0.05–1.0 wt%) and the PA yield at the reactor outlet for the latest generation catalysts reaches 105–106 kg/kg naphthalene.

Oxidation reactions with either of the feedstock (o-xylene or naphthalene) are carried out in tubular reactors cooled by a molten salt. Ring types of catalysts are applied, which replaced spherical catalysts utilized in the old plants.

For naphthalene partial oxidation, V_2O_5 promoted by K_2SO_4 was applied on moderately porous silica carrier. For o-xylene, oxidation porous supports were found to be detrimental and largely non-porous carriers, such as silicon carbide, quartz, steatite ceramics, or porcelain with dimensions of 8×65 or $7 \times 7 \times 4$ mm and filling density of 0.88–0.93 kg/l are used. Low area catalyst supports are important for eliminating intraparticle concentration gradients and formation of side products. Finely divided titanium dioxide (anatase) and vanadium pentoxide oxide layer (0.02–2 mm), together with promoters (antimony oxide, rubidium oxide, cesium oxide, niobium, and phosphorous), are introduced on the support. In the monolayer formed on the surface, there are chemical bonds between V ion and Ti surface ions via oxo bridges. The layer is a two-dimensional sheet of the following formula $VO_{2.5}$, which is believed to contain both strongly bound tetravalent and pentavalent vanadium. Both of the valence states take part in the oxidation of o-xylene by a typical redox mechanism.

Titania is considered to inhibit desorption of the many intermediates involved in the overall reaction. Formation of bulk (crystalline or amorphous) vanadia, if it is introduced in excess, might have a negative impact on catalysts; thus, up to 8–10% of vanadium pentoxide in combination with titanium dioxide is used. In the past, some of the catalysts, such as silica gel-supported naphthalene catalyst containing V_2O_5 and K_2SO_4, required addition of SO_2 for activation and longer service life, an option that is not required with the modern catalysts.

In the industry, for low *o*-xylene loadings, two-zone catalysts (i.e., combination of catalysts with low and high activity) were used, while higher *o*-xylene loadings require three and even more catalysts (Figure 9.29).

(a)   (b)

Figure 9.29: BASF 04-68 catalyst for *o*-xylene load of 80 g/N m^3). From www.catalysis.basf.com/chemicals. (Copyright BASF SE).

The optimal amount of the active phase varies depending on which catalyst zone of the multitubular reactor the catalyst is located, as different reaction stages require different catalyst compositions. It could be easily anticipated that because of high reaction exothermicity, the top catalyst layer should be able to minimize the temperature increase in the hot spot and at the same time ensure fast start-up of cold reaction gas. In the second layer, most of *o*-xylene conversion takes place and the catalyst should have high selectivity with the lowest activity. Two final layers in the four-zone catalyst system should be more active and less selective, ensuring good conversion of remaining under-oxidation products to phthalic anhydride and complete combustion of over-oxidation products to CO_2. The amount of cesium in different layers is decreasing from top to bottom. Besides helping to avoid excessive hot spots, dopants such as Cs are also used to minimize formation of maleic anhydride and carbon oxides.

Such an operation with several layers allows for minimizing the negative influence of the hot spot, which is typically moving along the reactor length with time-on-stream (Figure 9.30).

BASF, one of the commercial catalyst suppliers, introduced in 2011 a five-layer catalyst 04-88 (Figure 9.31) with first three selective layers affording phthalic anhydride yield up to 116.5 wt%. The hot spot is located in the CL0 layer. Details on the composition are not available, while the patent literature disclosed, for example, the pre-catalyst layer of Ag/V(O) bronze (mixed oxides of silver and vanadium with atomic ratio Ag/V < 1) dispersed on Mg silicate rings. This formulation should not be, however, necessarily the one used commercially.

OxyMax phthalic anhydride catalysts from Clariant for selective oxidation of *o*-xylene, naphthalene, and mixed feedstocks, produced by fluid-bed coating process

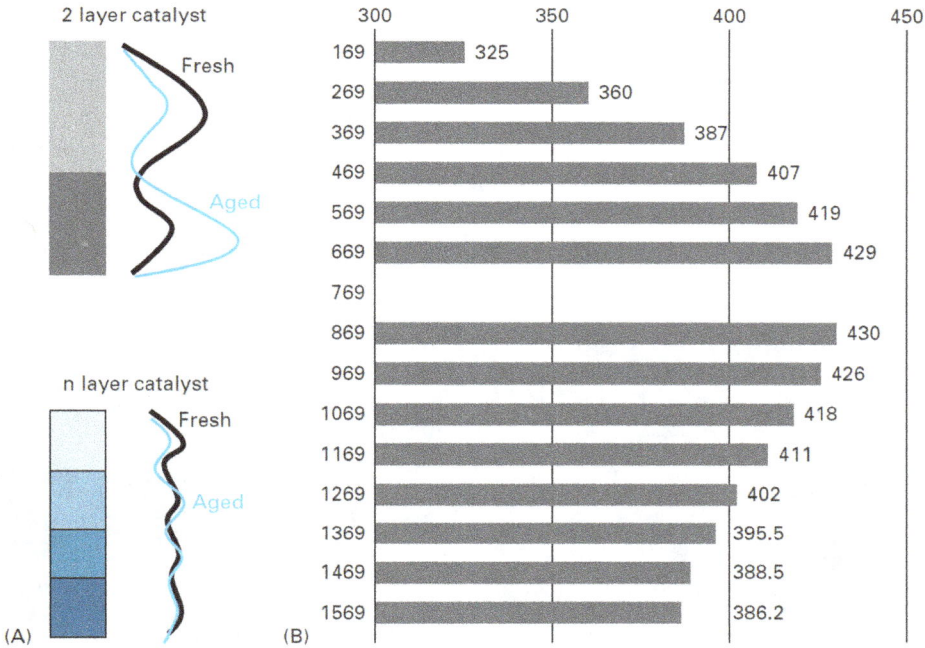

Figure 9.30: Temperature profile along the reactor length: (a) typical behavior, (b) experimentally observed in the first zone of a two-zone catalyst.

Figure 9.31: Five-zone BASF O4-88 catalyst for *o*-xylene load up to 100 g Nm^{-3} (From https://cata lysts.basf.com/public/files/literature-library/32012BF-9796-7152_BR_PUE_Catalysts_Ansicht.pdf).

also consist up to five catalyst layers. During catalyst preparation, the catalytically active materials titanium dioxide and vanadium pentoxide along with special promoters are coated in a thin shell on ceramic rings. The thin, porous shell is required to overcome mass and heat transport challenges during synthesis of phthalic anhydride. The catalyst reportedly allows 116 wt% reactor outlet phthalic anhydride yield as verified at Petrowidada plant (Figure 9.32).

Figure 9.32: Petrowidada phthalic anhydride plant in Gresik, Indonesia.

Oxidation of *o*-xylene is done industrially in multitubular reactors, containing up to 25, 000 tubes with ca. 2.5 cm internal diameter. In the initial design, reactors contained a salt bath circulation pump and a fitted internally cooler. This allowed temperature control by a molten salt, which is a eutectic of potassium and sodium nitrates, having a melting point of 141 °C. Later on, external cooling was introduced. The molten salts circulate through a waste heat boiler, which is used to generate saturated steam. Nowadays, in the case of utilized industrially high *o*-xylene loadings, reactors use radial salt flow (contrary to axial flow in previous designs) for improved heat transfer (Figure 9.33).

The oxidation air is preheated to about 180 °C and charged with feedstock in concentrations up to 100 g/N m^3. The air rate is typically 4 N m^3 per reactor tube and per hour.

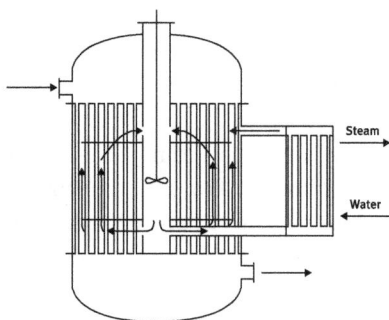

Figure 9.33: Reactor with radial salt flow (DWE type). From A. I. Anastasov, Deactivation of an industrial V2O5–TiO2 catalyst for oxidation of o-xylene into phthalic anhydride, *Chemical Engineering and Processing*, 2003, 42, 449. Copyright Elsevier. Reproduced with permission.

As could be seen from Figure 9.34, no recycling is needed, as high conversion levels are achieved. Operating salt temperatures are in the 325–425 °C range. Initial start-up temperature is ca. 375 °C, and it takes approximately 2 weeks to stabilize the catalyst activity and operate at 350–355 °C. When the reaction takes off, a hot spot is generated in the reactor, which exceeds the salt bath temperature by 50–60 °C, being in the 410–430 °C range. The temperature declines along the reactor length as the generated heat is related to the reaction rate, which declines with increased conversion. With operation time, catalyst deactivates and the hot spot is moving along the reactor downward. In order to compensate for the activity loss, the salt bath temperature is increased during operation to 360–365 °C at the middle of the catalyst lifetime

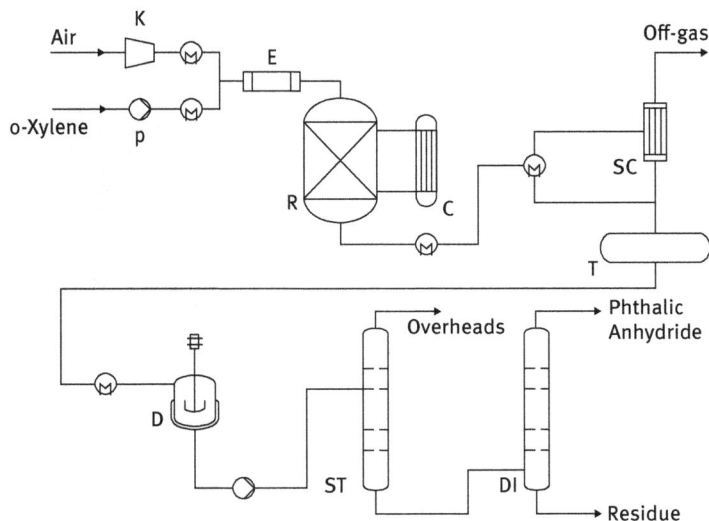

Figure 9.34: PA production technology. From http://d-nb.info/1030112800/34. K, Air compressor; P, *o*-xylene pump; E, evaporator; R, reactor; C, salt bath cooler; SC, switch condensers; T, crude phthalic anhydride tank; D, predecomposer; ST, stripper column; DI, distillation column.

and to 370–380 °C at the end. A higher operation temperature results in lower selectivity, higher phthalide yields, worsening of the product quality (which is seen as an increase in the color number), and finally in lower phthalic anhydride yields. Thus, after 4–5 years of operation, the catalysts have to be replaced.

Loading of catalysts in industrial multitubular reactors is a time-consuming procedure, as it should guarantee uniform flow distribution. For such multizone catalysts, loading is arranged by filling first one zone, measuring pressure drop in all tubes, leveling off, loading, and proceeding with another zone.

The reactor effluent gas leaving the reactor is cooled down first in the gas cooler (Figure 9.34). Downstream cooler switch condensers equipped with U-type finned tubes are applied. Such tubes are alternatively heated or cooled with a hot or cold heat-transfer medium. By cooling the gas, the crude PA desublimates on the fin tube surface.

Each switch condenser can be loaded with a certain quantity of crude phthalic anhydride, which is defined by the surface area of a switch condenser. After full loading, a switch condenser is isolated from the gas stream and heated up to melt the crude phthalic anhydride from the fin tubes and direct it to the purification section through a crude PA tank. The heat removed from the reactor process gas during the loading phase is absorbed by a heat transfer oil circulating through the fin tubes. The same oil at higher temperature is used for unloading. A scheme of a Rollechim switch condenser is presented in Figure 9.35.

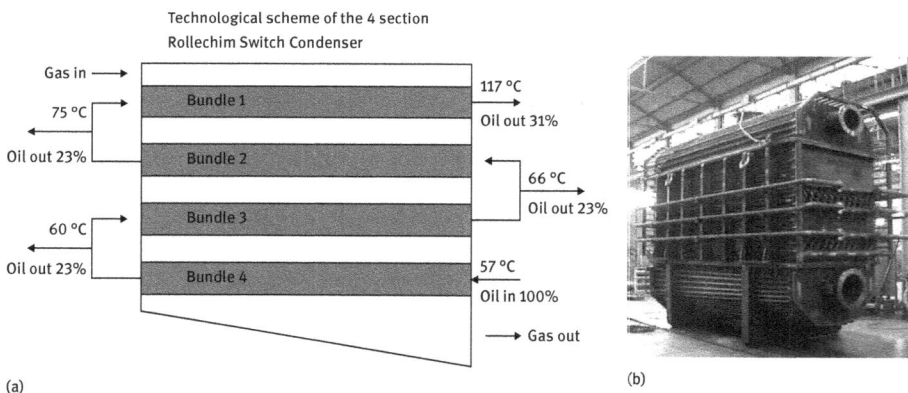

Figure 9.35: Switch condensers. Figures from http://www.rollechim.com/switch.htm.

The downflow mode of operation is applied, as the reaction gas enters at the top. Such operation allows deposition of the most of the solid product on the uppermost bundle. During the unloading stage, the lower tube bundle is washed by molten PA, this way removing highly corrosive maleic anhydride and phthalic acid from the tubes, which was challenging in up-flow operations. The high efficiency of the switch condenser affords more than 99.5% of phthalic anhydride recovery, and thus, less than 0.5% of the crude PA escapes with the off-gas. The off-gas leaving the switch condensers is passed through a scrubbing tower where the major part of the organic compounds, such as maleic anhydride, is removed by water scrubbing. Carbon monoxide cannot be removed by such scrubbers; thus, either thermal or catalytic combustion is utilized.

The product purification is performed first by thermal pretreatment and then by vacuum distillation. In the case of naphthalene oxidation, the pretreatment takes place in a cascade of vessels heated by outside coils with heat transfer oil at temperature 230–300 °C with 10–24 h of retention time. Such treatment can be applied for crude products with low contents of impurities. Chemical treatment was also used to destroy by-product naphthoquinone.

The purification in the case of o-xylene oxidation is intended to destroy phthalic acid by dehydration to PA and removal of water and low-boiling compounds such as maleic anhydride, o-tolualdehyde, and benzoic acid. After the predecomposer (where phthalic acid is dehydrated), phthalic anhydride is introduced into the first distillation column (Figure 9.34). It is separated in this column from the low boiling components, which are taken from the column top. In the subsequent column, purified phthalic anhydride is removed from the top and is delivered to flaking, liquid storage, or directly to downstream facilities, while the residue is discharged from the bottom. Both distillation columns operate under vacuum, which is generated by ejectors generally driven by steam.

9.2.1.6 Acrylonitrile

The average capacity of acrylonitrile plants is around 170,000 tons/year. The main application of acrylonitrile (CH_2=CH–CN) are in production of acrylic fibers, acrylonitrile–butadiene–styrene, adiponitrile, nitrile–butadiene copolymers as well as acrylamide for water-treatment polymers.

Prior to 1960s, the most popular process of acrylonitrile production was addition of hydrogen cyanide to acetylene. Currently, manufacture of acrylonitrile mainly follows the Standard Oil of Ohio (Sohio) process, based on ammoxidation of propylene in fluidized bed reactors using propylene, ammonia, and air as reactants and giving acetonitrile and hydrogen cyanide as by-products

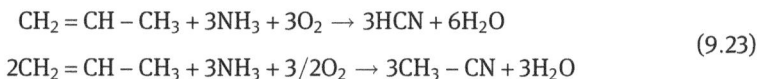

$$CH_2 = CH - CH_3 + 3NH_3 + 3O_2 \rightarrow 3HCN + 6H_2O$$
$$2CH_2 = CH - CH_3 + 3NH_3 + 3/2O_2 \rightarrow 3CH_3 - CN + 3H_2O$$

$$(9.23)$$

in commercially attractive quantities (ca. 25% total yield). The side reactions are favored at higher temperature and pressure, and longer residence time. The amounts of HCN are larger than that of acetonitrile. Besides these products, carbon dioxide is also formed. The complex reaction network in the synthesis of acrylonitrile is shown in Figure 9.36.

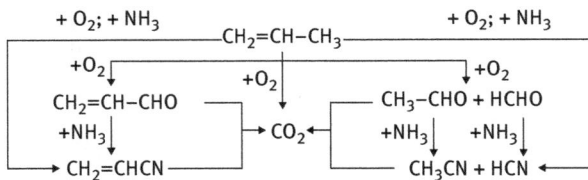

Figure 9.36: Reaction network in acrylonitrile synthesis.

The overall process has, nevertheless, high selectivity to acrylonitrile not requiring further recycling steps. The costs of acrylonitrile production are closely linked to the market price of reagents, in particular propylene, which contributed substantially (>70%) to the total cost.

A schematic view on the reaction mechanism is shown in Figure 9.37.

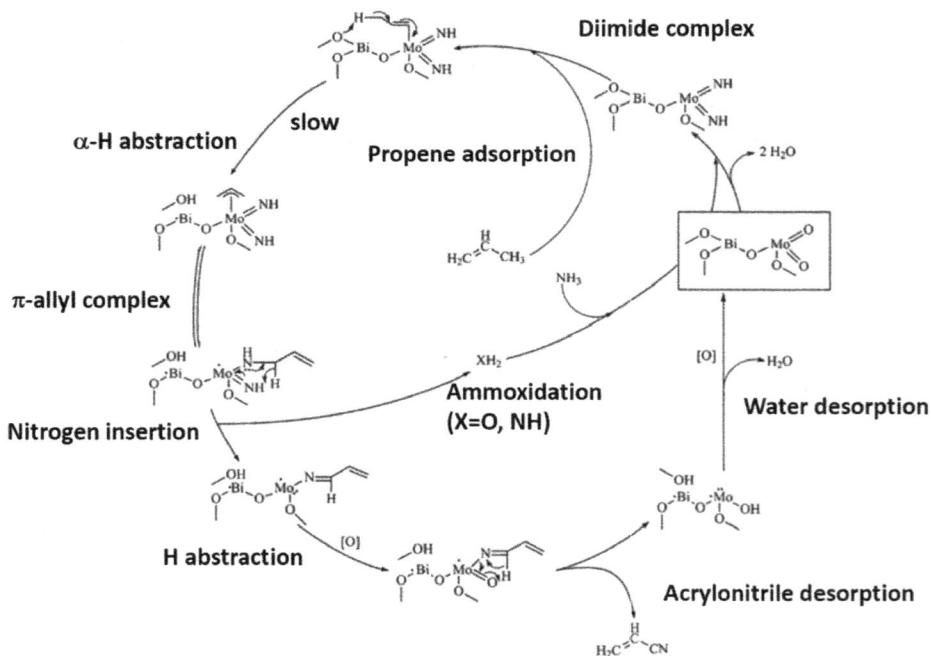

Figure 9.37: Ammoxidation mechanism. Modified from J.D. Burrington, C.T Kartisek, R.K. Grasselli, Surface intermediates in selective propylene oxidation and ammoxidation over heterogeneous molybdate and antimonate catalysts, J. Catal. 1984, 87, 363–380. Copyright Elsevier. Reproduced with permission.

Ammonia interacts with the bifunctional active centers, generating an extended ammoxidation site containing ammonia as =NH. On this site, propylene is inserted by α-hydrogen abstraction forming an allylic complex. Furthermore, nitrogen insertion and proton abstraction gives acrylonitrile, which then desorbs from the surface.

Regeneration of the reduced surface takes place by oxygen (O^{2-}) from the lattice oxygen, which is filled with oxygen from the gas phase. An ammoxidation catalyst thus possesses multifunctional and redox properties. As mentioned above, ammoxidation catalysts can also be used for synthesis of acrolein, even if in the presence of ammonia the yields of acrolein are small. An explanation for this is that surface species (e.g., π-allyl complex in Figure 9.37) which are precursors for acrolein react in a fast way to acrylonitrile. Thus, the rate of acrylonitrile formation is higher than that of acrolein.

The bismuth phosphomolybdate catalysts supported on silica are the dominant catalysts used commercially with such promoters as K, Cs, Mg, Mn, Co, Ni, Fe, W, and Re. Historically, to the mixed metal oxide Fe-Bi-Mo-O, first divalent cobalt and nickel molybdates were added improving activity and selectivity, followed by introduction of alkali metal promoters.

The catalyst should fulfil several functions, such as α-H abstraction (provided by Bi^{3+}, Sb^{3+}, or Te^{4+} in the composition), olefin chemisorption, and nitrogen insertion (Mo^{6+}, Sb^{5+}), while a redox couple (Fe^{2+}/Fe^{3+} or Ce^{3+}/Ce^{4+}) enhances the transfer of lattice oxygen between the bulk of the catalyst and its surface.

In the chemical composition of catalysts, multimetal molybdates are used comprising bismuth molybdate as α-$Bi_2Mo_3O_{12}$ phase, divalent metal (Ni, Co, Mg, Fe) molybdates $M^{2+}MoO_4$ of α and β structure, and trivalent $Fe_2Mo_3O_{12}$ dispersed in silica (50% w/w). The catalysts also contain antimonate with the rutile structure, made up of four metal antimonate cations and a redox couple of Fe, Ce, U, and Cr. As mentioned above, alkali metals (e.g., Rb, K, Cs) are added to the active phase along with other promoters (P, B, Mn). An example of a complex ammoxidation catalyst chemical composition of the active phase is given below: $(K, Cs)_{0.1}(Ni, Mg, Mn)_{7.5} (Fe, Cr)_{2.3}Bi_{0.5} Mo_{12}O_x$.

A coprecipitation procedure with aqueous solutions of bismuth nitrate and ammonium molybdate is used in catalyst preparation, followed by heat treatment of the dried particles at ca. 500 °C to crystallize the bismuth molybdate phase.

Because of high reaction exothermicity ($\Delta H_o = -515$ kJ/mol), ammoxidation reaction is conducted in a fluidized-bed reactor needed to remove excessive heat. Such operation imposes several restrictions on the catalyst size which should be in the range 40–100 μm and mechanical properties, which should account for the abrasive environment in a fluidized-bed reactor. The catalyst should be hard enough and resistant to attrition. To meet these requirements, bismuth molybdate catalyst is supported on silica.

The steady-state kinetics of propene oxidation is first order in propene and zero order in oxygen when oxygen concentration is above stoichiometry. The apparent activation energy for acrylonitrile formation is the same as for acrolein formation

(~83 kJ/mol), moreover, kinetic regularities are similar (first order in propene and zero order in oxygen and ammonia) pointing out on a similar mechanism.

The Sohio process (currently INEOS Technologies ammoxidation technology) operates with near stoichiometric amounts of reactants in temperature range of 400–510 ℃ and at 50–200 kPa (0.5 to 2 bar) giving propylene conversion above 95%, and the yield of acrylonitrile ca. 80%. In the Sohio process (Figure 9.38), propylene, ammonia after evaporation, and air are passed through a fluidized-bed reactor containing the catalyst.

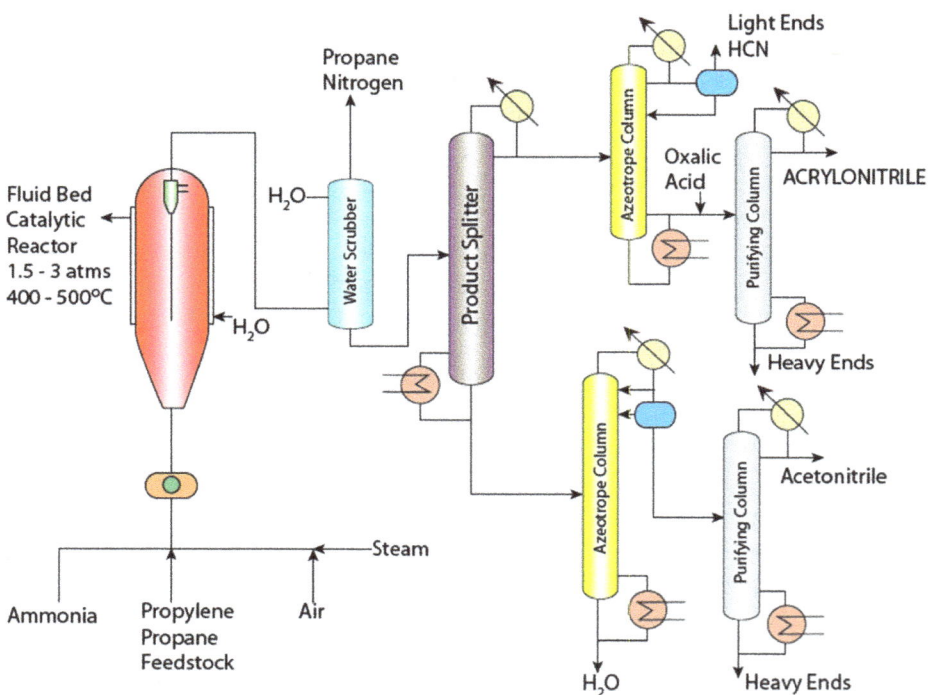

Figure 9.38: Flow scheme of acrylonitrile synthesis.

The ratio between reactants is 1:(0.9 ÷ 1.1):(1.8 ÷ 2.4) for propylene, ammonia, and oxygen, respectively. An excess of oxygen is required to keep the catalyst in a proper state needed for maintaining high activity and selectivity. The fluidized bed reactor is used to control temperature during this exothermal reaction (enthalpy −670 kJ/mol). As a general comment, it can be stated that typically a reaction heat above 500 kJ/mol should not be handled in fixed-bed multitubular reactors. Usually, the heat recovered is used to produce high-pressure steam.

Catalyst deactivation is a serious problem, however, by stabilization of the Mo phase, addition of excess of MoO_3 phase during preparation, and periodic addition of the catalyst during operation, it is possible to extend the catalyst lifetime to 10 years. Fine particles are required for proper fluidization and the optimal product yield. The contact time is 6–10 s.

The internal heat transfer coils in the reactor are used as baffles to improve fluidization, minimizing back-mixing. This finally leads to higher selectivity. An expansion chamber at the reactor top is used for separation of smaller and larger catalyst particles, the former ones are returned through the cyclones to the reactor inlet. The reactants pass through the reactor only once. A heat exchanger located downstream the reactor is used for generation of medium pressure steam. Thereafter, the reactants are quenched, and unreacted ammonia is neutralized in water to remove remaining ammonia. The product mixture containing water and organic compounds is split into two phases in the product splitter. Acrylonitrile, acetonitrile, and hydrogen cyanide are separated by multiple distillations. The column where HCN is removed, operates under slight vacuum to prevent release of toxic HCN to atmosphere.

The diameter of the reactor for acrylonitrile synthesis is ca. 9–10 m (Figure 9.39) and the height of the fluidized bed is 6,500 mm.

Figure 9.39: Fluidized-bed reactor for ammoxidation (https://corpwebstorage.blob.core.windows.net/media/30757/acn-reactor.jpg).

Direct ammoxidation of propane to acrylonitrile with a high active catalyst was developed by Asahi driven by a price differential between the conventional propylene feedstock and propane. The catalyst operating at 430 °C achieves a per-pass propane conversion of 87% at 60% selectivity to acrylonitrile giving subsequently 52 mol% yield. In 2013, Asahi started up a plant in Thailand using a propane based ammoxidation process, employing silica-based $MoVNbTeO_x$ as a catalyst and adding periodically so-called activators (telluric and molybdic acids) to the fluidized bed. These activators are needed to compensate for the eventual loss of the corresponding oxides. While addition of molybdenum compounds (e.g., simple MoO_3 or more complex Mo compositions) was done in all commercial reactors using SOHIO Bi–Mo oxide based catalysts, maintaining the Te content is much more challenging.

Figure 9.40: The propane-based plant of Asaki-Kaisei for manufacturing of acrylonitrile (https://www.asahi-kasei.co.jp/asahi/en/news/2012/images/e130213_1_02.jpg).

9.2.1.7 Synthesis of acetic acid by oxidation

The Japanese Showa Denko company developed a one-step direct gas-phase oxidation of ethylene into acetic acid at 160–210 °C using a palladium-based heteropoly acid Pd–Me–$H_4SiW_{12}O_{40}$ catalyst. A plant of annual capacity of 100,000 tons has been in operation since late 1990s. Without addition of a second metal, formation of carbon dioxide based on ethylene exceeds 10%, which implies loss of ethylene, a need to remove large amounts of heat by steam generation, and some constrains related to acetic acid production capacity because of safety. Addition of Se, Te, Sb, Bi, or Sn to Pd on a heteropolyacid was effective in minimizing complete oxidation. Besides carbon dioxide, small amounts of other by-products are formed (e.g., acetaldehyde).

The process flowsheet is illustrated in Figure 9.41.

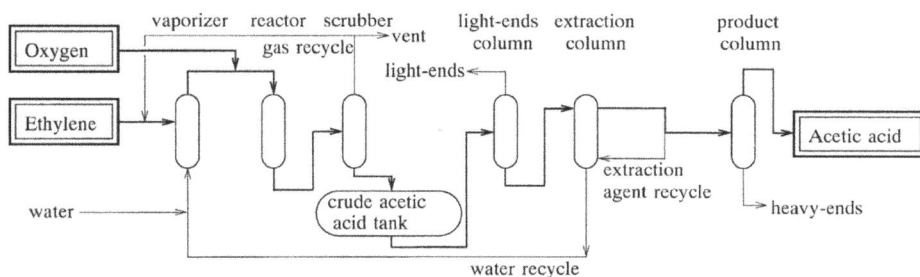

Figure 9.41: Process flowsheet of direct oxidation to ethylene into acetic acid. From K.-i.Sano, H. Uchida, S. Wakabayashi, Catalysis Surveys from Japan, 1999, 3, 5–60. Copyright 1999, Kluwer Academic Publishers. Reproduced with permission.

Acetic acid coming out of the reactor after cooling, condensation, and separation from incondensable gases undergoes further purification. The un-condensed gas is recycled to the reactor and a part is purged (vented). Purification of the crude acetic acid from the storage tank is done first by distillation removing the light-end by-products, such as acetaldehyde, ethyl acetate, and ethanol. This is followed extraction of water and the final purification also by distillation where highly pure acetic acid is obtained at the top and heavy ends up at the bottom of the distillation column. A specific feature of this technology is that stainless steel can be used as the corrosive compound is acetic acid per se.

Another emerging route for synthesis of acetic acid is the Sabic ethane-based process. It uses on one hand surplus feedstock that might otherwise be flared but on the other generates CO_2 as a by-product which can be an environmental burden. The process with annual capacity of 30,000 tons was commercialized by Sabic in 2005 with a calcined mixture of Mo, V, Nb, and Pd oxides as a catalyst affording selectivity to acetic acid of 80% and complete oxidation of carbon monoxide to carbon dioxide. Addition of Pd to MoVNb catalyst was instrumental in increasing selectivity to acetic acid from ca. 20% to ca. 80%. The process flow scheme is shown in Figure 9.42.

The flow scheme includes mixing of ethane and water with steam, compression and pre-heating (H2) to the reaction conditions (515 K, 16 bar). The molten salt is used to control temperature in the multitubular reactor (R3) with the tube diameter below 25 mm to avoid severe hot sports.

The downstream processing comprises cooling of the reaction mixture (H4 and H-4a), flashing of non-condensed components (S5) and distillation of the water–acetic acid mixture giving the final product with the content of water below < 0.1 mass%.

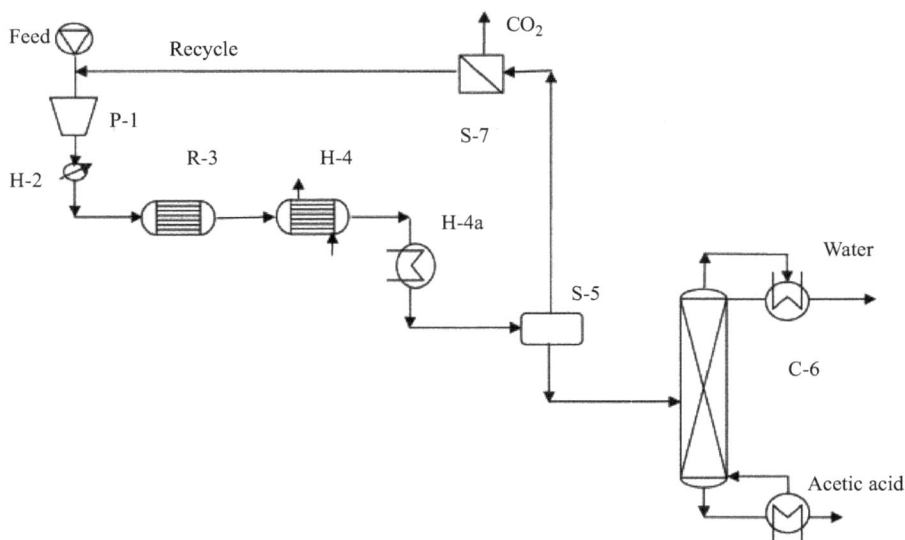

Figure 9.42: Acetic acid synthesis by ethane oxidation; 1: compressor, 2: pre-heater, 3: multi-tubular fixed-bed reactor cooled by molten salt, 4, 4a: cooler, 5: flash, 6: rectification of water–acetic acid (purification of acetic acid), 7: CO_2 separation, cooling 4, 4a: from reaction temperature to 303 K in two steps, $p = 16$ bar, flash: 303 K, $p = 16$ bar. C: rectification column, H: heat exchanger, P: compressor, R: reactor, S: absorber, separator, flash. From O. Smejkal, D. Linke, M. Baerns, Energetic and economic evaluation of the production of acetic acid via ethane oxidation, Chemical Engineering and Processing. Process Intensification, 2005, 44, 421–428. Copyright Elsevier. Reproduced with permission.

9.2.2 Liquid-phase oxidation

The following oxidation reactions occur in the liquid phase: oxidation of paraffins, cycloparaffins, and their derivatives and oxidation of the side chains of alkylaromatic compounds.

Such oxidation is mainly conducted in the liquid phase, either in autooxidation mode (chain reactions) or using catalysts – salts of metals capable of changing oxidation states (Co, Mn, etc.). The latter mode of operation affords a significantly lower activation energy (50–80 kJ/mol) compared to thermal oxidation with 100–150 kJ/mol.

Conversion is either close to 95–99% for processes when the product is stable to further oxidation (low-chain aliphatic and aromatic acids) or is kept rather low (5–30%) when the product can be further transformed (hydroperoxides, alcohols, ketones, or higher carboxylic acids). In the latter case, there are extra costs associated with the recycling.

Temperature elevates the activity and at the same time negatively influences selectivity (e.g., more side products at high temperature); thus, a certain optimum is needed. In addition, for the liquid-phase process, film diffusion could be prominent;

therefore, efficient mass transfer is typically required. Bubble columns are applied when the oxidant (air or technical-quality oxygen) is bubbled through the liquid, increasing the mass transfer area. Bubble columns of the height 10–15 m and 2–3 m diameter are used.

Heat removal is an important issue in the liquid-phase oxidation reactions. This is done either by applying internal heat exchanges complicating the reactor design or by having external heat exchangers (Figure 9.43a). A single bubble column (Figure 9.43b) could be used when a product is stable to subsequent oxidation (i.e., acetic acid). The reactant is fed to the bottom of the reactor and the product is taken from the top.

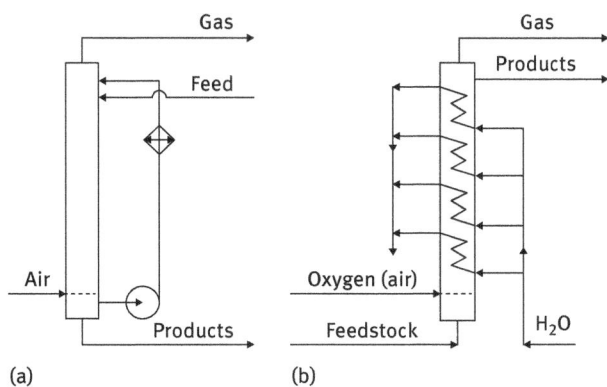

Figure 9.43: Liquid-phase oxidation reactors with (a) an external heat exchanger and (b) internal cooling.

When selectivity depends on conversion in an undesired way, i.e., decreasing with a conversion increase, back-mixing should be prevented, which can be achieved in a cascade reactor with, for example, cooling by evaporation of the solvent (Figure 9.44a) or reacting hydrocarbon.

A column with plates (Figure 9.44b) can be also used, with the liquid going from top to bottom and the gas flowing countercurrently upward. Coils with water can be used for cooling or, alternatively, external heat exchangers could be installed. For safety reasons, formation of explosive mixtures should be avoided in places where only gas phase is present (upper part of the bubble columns). This can be achieved by high oxygen conversion along with a proper choice of pressure and addition of nitrogen above the liquid phase on the plates.

9.2.2.1 Cyclohexane oxidation

The liquid-phase air oxidation of cyclohexane to cyclohexanol and cyclohexanone is usually carried out in a series of stirred reactors at 140–180 °C and 0.8–2 MPa

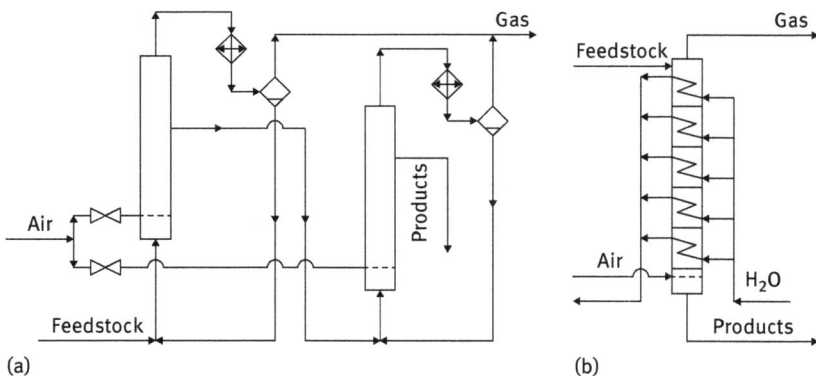

Figure 9.44: Reactors for oxidation (a) cascade with cooling by evaporation and (b) column-type reactor.

either in the presence of a homogeneous cobalt catalyst or without any catalyst. The most popular cationic catalysts are soluble Co^{2+} (cobalt stearate and cobalt naphthenate) and Cr^{3+} salts and their mixtures in the concentrations between ca. 10 ppm and sub-ppm levels. The total residence time in the oxidizers is 15–60 min.

A traditional catalyzed cyclohexane oxidation process consists of an oxidation and heat recovery section, a neutralization and decomposition section, a cyclohexane recovery section, a cyclohexanone separation and purification section, and finally cyclohexanol dehydrogenation (Figure 9.45).

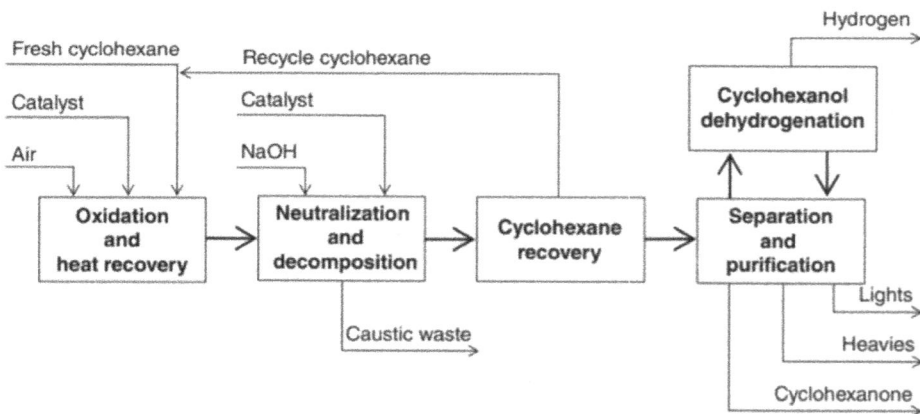

Figure 9.45: A simplified diagram of cyclohexane oxidation. From J.T. Tinge, *Cyclohexane Oxidation: History of Transition from Catalyzed to Noncatalyzed. In Liquid Phase Aerobic Oxidation Catalysis: Industrial Applications and Academic Perspectives,* 2016, 33–39. doi:10.1002/9783527690121.ch3. Copyright 2016 Wiley –VCH Verlag GmbH & Co. KGaA. Reproduced with permission.

A more detailed process flow diagram is presented in Figure 9.46.

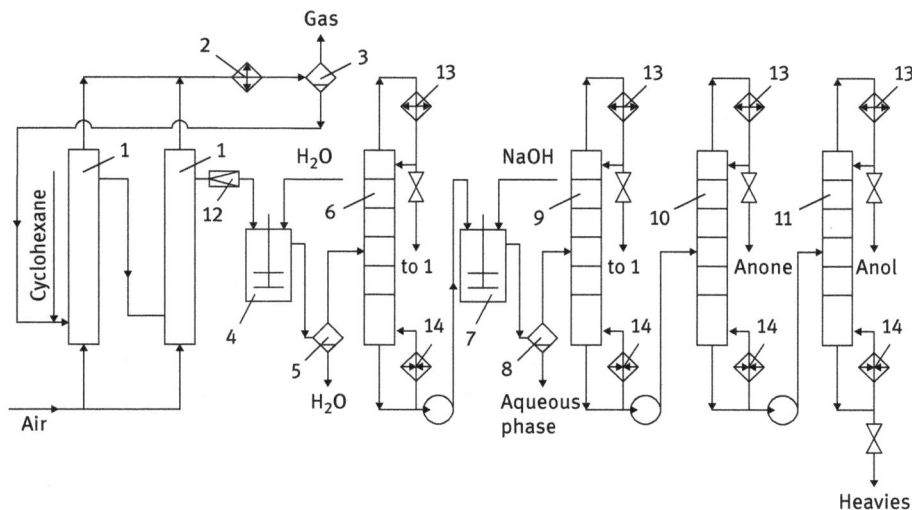

Figure 9.46: Oxidation of cyclohexane to a mixture of cyclohexanone and cyclohexanol: 1, reactor; 2, cooler; 3, 5, and 8, separators; 4 and 7, mixers; 6, 9, 10, and 11, distillation columns; 12, expander; 13, reflux cooler; 14, boiler. After N. N. Lebedev, *Chemistry and Technology of Basic Organic and Petrochemical Synthesis*, Chimia, 1988.

A cascade of bubble columns 1 is used with supply of air to each column. The selection of a catalyst or running a process without any catalyst is aimed at maximizing the amount of cyclohexyl hydroperoxide (CHHP), which is the reaction intermediate. This intermediate along with the products, cyclohexanol and cyclohexanone, is more reactive than cyclohexane; therefore, cyclohexane conversion is kept low (less than 6%).

Oxidation catalysts also catalyze decomposition of CHHP, reducing its residual concentration. For economic and safety reasons, this reaction intermediate is decomposed to mainly cyclohexanone and cyclohexanol in an after-reactor either in a monophasic system in cyclohexane or in a biphasic system with an aqueous caustic solution as the second phase.

During oxidation, a range of by-products, namely monocarboxylic and dicarboxylic acids, esters, aldehydes, and other oxygenated material, is produced. In particular, such by-products are obtained in various amounts: adipic, ε-hydroxycaproic, glutaric, succinic, valeric, caproic, propionic, acetic and formic acids, and noncondensable gases such as CO and CO_2. Esters are formed between mainly cyclohexanol and various carboxylic acids.

The approach of attaining the highest cyclohexyl hydroperoxide yield in the first reactor allows to decompose it downsteam into cyclohexanol and/or cyclohexanone under milder conditions in higher yields than in the case when only one reactor would be used with a target to maximize formation of cyclohexanol and cyclohexanone. Figure 9.46 displays only two reactors, however, a catalyzed cyclohexane oxidation section can consist of a series of three to eight reactors or compartments that are operated in plug flow in order to minimize by-product formation. An example of such system was discussed in Chapter 1 in connection with the accident at the Flixborough site for caprolactam production (Figure 1.21).

In order to remove the reaction heat, an excess of cyclohexane is evaporated and then condensed in a common cooler (2). It is then separated from the gas in the separator (3) and sent back to the reactor. The product from the last column is washed with water in a mixer (4) to remove lower acids and is separated from the aqueous layer in the separator (5).

From the top of the column (6), the major part of cyclohexane is removed. The bottom part containing small amounts of cyclohexane is washed with alkali in 7, which leads to saponification of esters and lactones as well as decomposition of hydroperoxide. The organic phase is separated from the aqueous phase in separator 8. The molar ratio between cyclohexanone and cyclohexanol, also called KA ratio, ranges from 0.3 to 0.8 in the cyclohexane stream containing cyclohexanol and cyclohexanone. This stream undergoes distillation in several columns, subsequently removing cyclohexane (pos. 9), cyclohexanone (pos. 10), and cyclohexanol (pos. 11). In this cyclohexane recovery section, cyclohexane is thus separated by overhead distillation from the resulting decomposed organic stream, giving a highly concentrated mixture of cyclohexanone and cyclohexanol (KA oil). The recovered cyclohexane is recycled back to the cyclohexane oxidation reaction section. The KA oil is fed to the separation and purification section. Distillation of cyclohexanone requires high vacuum conditions and a large number of separation stages because of cyclohexanone tendency to form oligomers at elevated temperatures. Subsequently, tall columns are needed to achieve high-purity end-product (purity > 99.9%). If the desired product is cyclohexanone, then a multitubular reactor for dehydrogenation of cyclohexanol is installed, which operates either with copper- or zinc-based catalysts.

The traditional catalyzed cyclohexane oxidation technology has apparently such disadvantages as low cyclohexane selectivity (76–80%), high caustic consumption, and by-product formation as well as energy consumption in the purification section. Moreover, a part of the homogeneous catalyst precipitates causing severe fouling, interruptions of the process for reactor cleaning. All these negatively influence variable and fixed production costs.

The noncatalyzed DSM oxanone cyclohexane oxidation process (Figure 9.47) was developed to overcome the disadvantages of high cyclohexane and NaOH consumption and significant downtime required for cleaning. This process after decomposition of CHHP gives a KA oil with a KA ratio exceeding 1.5, which in turn means a much smaller unit for cyclohexanol dehydrogenation.

Figure 9.47: Fibrant's oxanone technology of uncatalyst cyclohexane air oxidation process was implemented at Kuibyshev Azot in Togliatti (https://www.fibrant52.com/en/licenses).

In this technology, formation of the intermediate CHHP and decomposition of CHHP are separated in space to a large extent being an example of process extensification, which is opposite to more often desired process intensification. In the absence of any catalyst in the oxidation section, cyclohexane is mainly converted into CHHP at ca. 10 oC higher temperature compared to the catalyzed versions. After the last oxidation reactor, the hot oxidate is fed to a modified neutralization and decomposition section, where CHHP is decomposed in a biphasic system in the presence of a decomposition catalyst under lower temperature and high pH (>13) as illustrated in Figure 9.48.

9.2.2.2 Cyclohexanol oxidation

Adipic acid (hexanedioic acid) has several application areas including production of the polyamide nylon-6,6, polyester, and polyurethane resins. It is used as a plasticizer for polyvinyl chloride (PVC) and an approved additive in, e.g., cosmetics, lubricants, or adhesives.

Oxidation of cyclohexanol with 40–60% nitric acid gives adipic acid in a complex reaction pathway (Figure 9.49).

As side products, oxalic, glutaric (1, 3-propanedicarboxylic acid), and succinic (butanedioic acid) as well as valeric acids are formed. Utilization of nitric acid as an

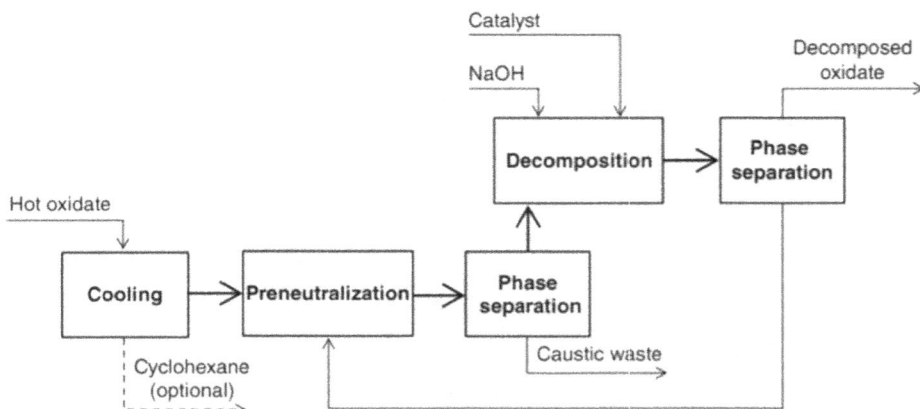

Figure 9.48: The modified neutralization and decomposition section as part of the noncatalyzed DSM oxanone® cyclohexane oxidation process. From J.T. Tinge, *Cyclohexane Oxidation: History of Transition from Catalyzed to Noncatalyzed. In Liquid Phase Aerobic Oxidation Catalysis: Industrial Applications and Academic Perspectives,* 2016, 33–39. doi:10.1002/9783527690121.ch3. Copyright 2016 Wiley –VCH Verlag GmbH & Co. KGaA. Reproduced with permission.

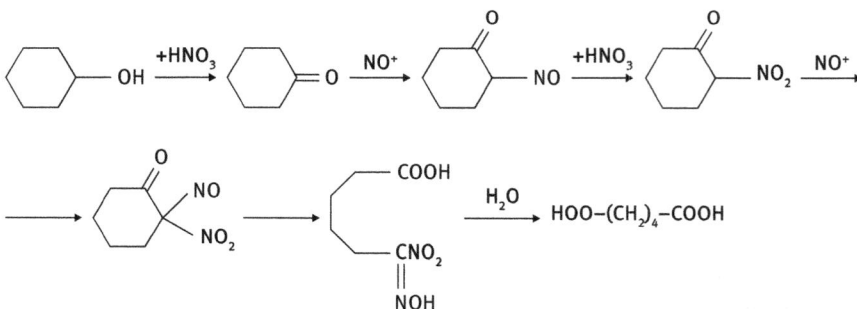

Figure 9.49: Reactions during oxidation of cyclohexanol.

oxidation agent also leads formation of substantial amounts of N_2O (in some plants 0.25 t of N_2O/ per tonne of adipic acid).

The yield of adipic acid is increased when a two-step approach is used; first, oxidation is done at 60–80 °C and thereafter at 100–120 °C, which is explained by the lower activation energy of primary products formation and a need for a higher temperature for subsequent hydrolysis. A catalyst containing copper oxide and ammonium metavanadate is typically added to bind NO_x and enhance the main reaction, respectively. The reaction is very exothermic (6,280 kJ/kg); thus, the reactor design is aimed at efficient removal of the reaction heat and minimization of the energy usage. In order to have more efficient recovery of NO_x, which are transferred downstream of the reactor into nitric acid by absorption, oxidation is carried out at elevated pressures (0.3–0.5 MPa). The process flow scheme is given in Figure 9.50.

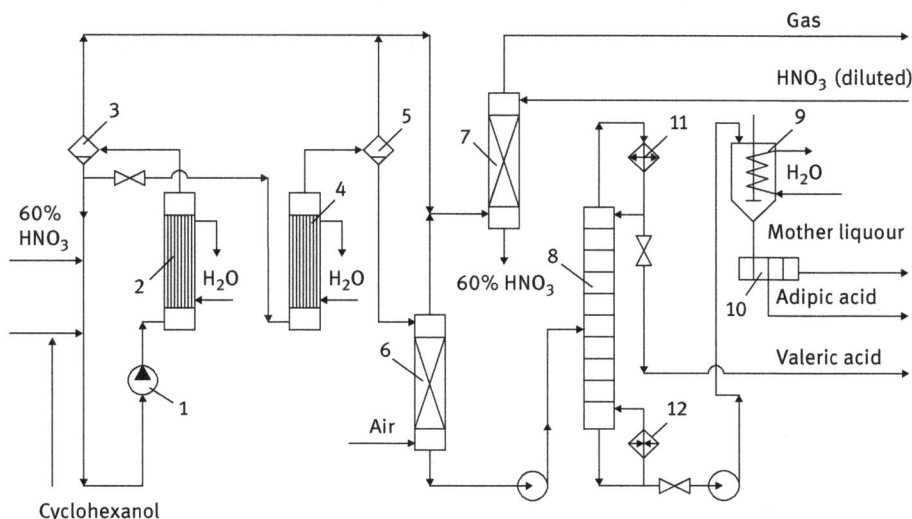

Figure 9.50: Oxidation of cyclohexanol to adipic acid: 1, pump; 2, 4, reactors; 3 and 5, separators; 6 and 7, scrubbers, 8, distillation columns, 9, crystallizer; 10, centrifuge, 11, reflux cooler. After N. N. Lebedev, *Chemistry and Technology of Basic Organic and Petrochemical Synthesis*, Chimia, 1988.

The product stream is passed through a scrubber-bleacher (6), in which the dissolved nitrogen oxides present in excess are removed with air and sent to the absorber (7), where they are reabsorbed and recovered as nitric acid. Crude adipic acid is removed from the bottom of a distillation column operated under vacuum (8) and sent to a crystallizer (9) followed by subsequent filtration or centrifugation (10).

Further purification of adipic acid is done by recrystallization. The mother liqueur contains a mixture of oxalic, succinic, and glutaric acids as well as some amounts of adipic acid dissolved in low concentrated nitric acid. Several methods of mother liqueur processing have been developed.

9.2.2.3 Xylene oxidation to terephthalic acid

Terephthalic acid (TFA) production is driven by a growing demand for textile fibers and polyethylene terephthalate bottles. Crude acid is produced by homogenous catalytic oxidation of *p*-xylene followed by purification by hydrogenation. An alternative option is production of medium-quality TFA with a post-oxidation system instead of the entire purification section of the conventional purified terephthalic acid (PTA) production process. Although the impurity levels are higher in this case when compared to PTA, the product is marketed as compatible with PTA for most polyester applications.

Catalytic oxidation of *p*-xylene, being a complex process (Figure 9.51), gives a nearly quantitative oxidation of the methyl groups, leaving the benzene ring virtually untouched.

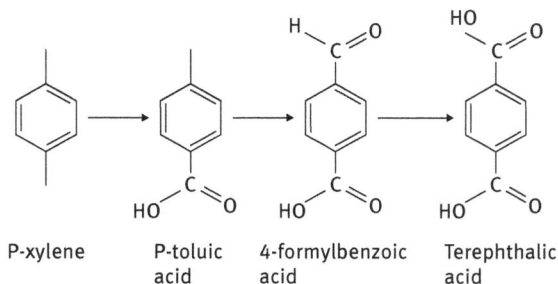

P-xylene   P-toluic   4-formylbenzoic   Terephthalic
            acid   acid   acid

Figure 9.51: Reactions during oxidation of *p*-xylene.

As catalysts, combinations of cobalt, manganese, and bromine (ca. 70% of production units) or cobalt with a co-oxidant, e.g., acetaldehyde are used with oxygen as the oxidant. Acetic acid is applied as the reaction solvent. In a soluble cobalt-manganese-bromine catalytic system, various salts of cobalt and manganese can be utilized, and the bromine source can be HBr, NaBr, or tetrabromoethane. The process flow diagram is illustrated in Figure 9.52.

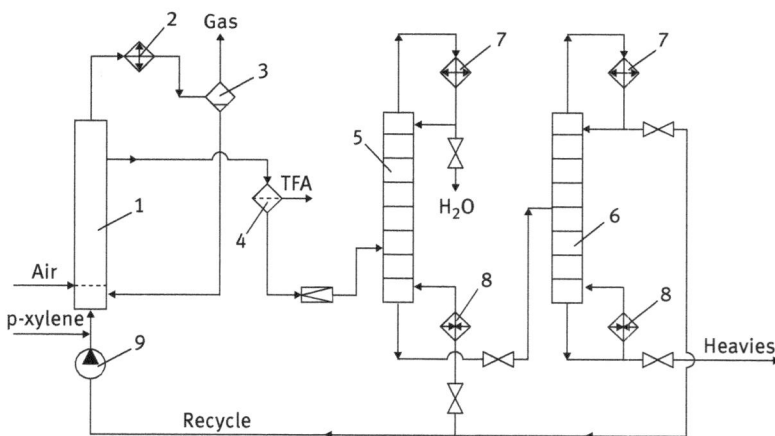

Figure 9.52: One-step synthesis of terephthalic acid: 1, reactor; 2, cooler; 3, separator; 4, centrifuge; 5, distillation column; 6, column for acetic acid recycling; 7, condenser; 8, boiler; 9, pump. After N. N. Lebedev, *Chemistry and Technology of Basic Organic and Petrochemical Synthesis*, Chimia, 1988.

A feed mixture of *p*-xylene, acetic acid, and the catalyst is continuously fed to the oxidation reactor 1 (a bubble column or a slurry reactor with a residence time up to 2 h) where the heat of the highly exothermic reaction is removed by evaporation of hydrocarbons, acetic acid, and water. After condensation in 2 and phase separation in 3, they are sent back to the reactor. Compressed air is added to the reactor in excess from stoichiometry requirements to afford higher oxygen partial pressure resulting in high *p*-xylene conversion (98%) with the yield of terephthalic acid above 95 mol%. Since terephthalic acid is poorly soluble in the solvent, it precipitates, giving a three-phase system: crystals of solid terephthalic acid, acetic acid with some dissolved terephthalic acid, and the vapor phase consisting of nitrogen, acetic acid, water, and a small amount of oxygen.

Suspension of TFA is separated by either filtration or centrifugation (pos. 4). Water should be removed from the filtrate (column 5) since it influences the reaction in a negative way. The catalyst and acetic acid should undergo separation from the heavies, which are generated in the process (column 6), and are recycled. The oxidation of the second methyl group starts when the first oxidation step to *p*-toluic acid is completed. In the second oxidation step, 4-formylbenzoic acid is formed, which, due to structural similarity with terephthalic acid, co-crystallizes with the latter and is not oxidized further. Crude terephthalic acid, containing up to 5,000 ppm 4-formylbenzoic acid and residual amounts of catalyst metals and bromine, is unsuitable as a feedstock for polyester.

The Amoco purification process (Figure 9.53), which removes 4-formylbenzoic acid to a level below 25 ppm, starts with complete dissolution of TFA in water at elevated temperature, ≥ 260 °C, giving a solution of at least 15 wt%.

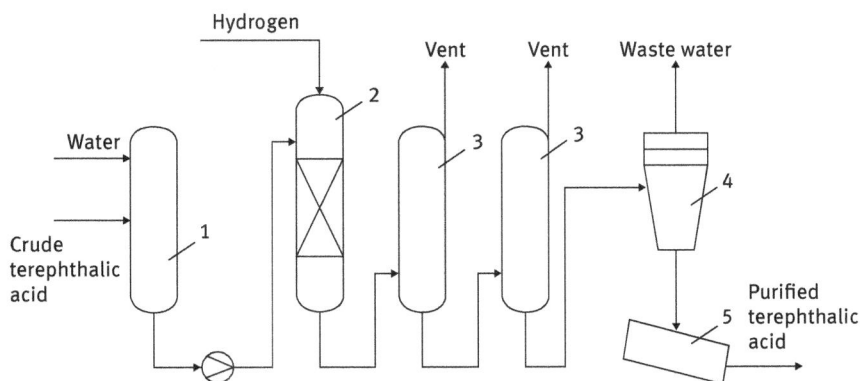

Figure 9.53: Purification of terephthalic acid by the Amoco process: 1, slurry drum; 2, hydrogenation reactor; 3, crystallizers; 4, centrifuge; 5, dryer.

In the subsequent liquid-phase hydrogenation step carried out over a Pd/C catalyst, 4-formylbenzoic acid is selectively hydrogenated to *p*-toluic acid. In a series of crystallizers where the pressure is sequentially decreased, terephthalic acid is crystallized, while a more soluble *p*-toluic acid and other impurities stay in the mother liquor. Crystallization is followed by centrifugation and/or filtration, giving a wet cake, which is subsequently dried affording terephthalic acid in the powder form in > 98 wt% yield from the crude terephthalic acid.

Some recent developments in xylene oxidation to TFA are related to attempts of Samsung General Chemicals to carry out in the presence of carbon dioxide. It is claimed that the oxidation rate is 26% higher and that a higher quality crude PTA with less partially oxidized components is produced. The industrial implementation of this technology has not been reported.

Some PTA plants have been modified to manufacture isophthalic acid using *m*-xylene instead of *p*-xylene.

9.2.2.4 Synthesis of acetaldehyde by oxidation: the Wacker process

Acetaldehyde is mainly produced by oxidation of ethylene

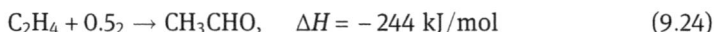

$$C_2H_4 + 0.5_2 \rightarrow CH_3CHO, \quad \Delta H = -244 \text{ kJ/mol} \tag{9.24}$$

in the so-called homogeneous catalytic Wacker process, where palladium and copper chloride complexes are present in an aqueous solution of HCl. Acetaldehyde formation is a result of the reaction between ethylene and aqueous palladium chloride, which determines the overall rate of the direct oxidation of ethylene in the Wacker–Hoechst process:

$$C_2H_4 + PdCl_2 + H_2O \rightarrow CH_3CHO + 2HCl + Pd \tag{9.25}$$

Metallic palladium is then reoxidized by $CuCl_2$, which is further regenerated from CuCl with oxygen:

$$Pd + 2CuCl_2 \rightarrow PdCl_2 + 2CuCl \tag{9.26}$$

$$2Cu^+ + 1/2O_2 + 2H^+ \rightarrow 2Cu^{2+} + H_2O \tag{9.27}$$

In the one-stage method (Figure 9.54), an ethylene–oxygen mixture reacts with the catalyst solution. During the reaction, a stationary state is established in which the "reaction" (or formation of acetaldehyde and reduction of $CuCl_2$) and the so-called oxidation (reoxidation of CuCl) proceed at the same rate. This stationary state is determined by the degree of oxidation of the catalyst. The reactants (ethylene and oxygen) are introduced into the lower part of the reactor (1). The catalyst is circulated via the separating vessel (2). Reaction conditions are about 100–130 °C and 0.3–1 MPa. Heat removal is done by evaporation of water and acetaldehyde as the reaction occurs at the boiling point of the reaction mixture. The product mixture (acetaldehyde–water and unconverted reagents) after passing through the separating vessel is cooled (cooler 3) and washed with water (scrubber 4). Washing is done to dissolve acetaldehyde, while unreacted ethene is recycled. Ethene conversion per pass is only 30%, as ethane is used in excess, while, otherwise, ethylene and O_2 can form explosive mixtures. Purge is needed to remove N_2 and/or CO_2; thus, a small portion is removed from the recycled gas. In the catalyst regenerated unit (11), by-products that have accumulated in the catalyst are decomposed at 160 °C. Purification of crude acetaldehyde is done in two stages. Extractive distillation (or washing) with H_2O is done to remove light products as solubility of acetaldehyde in H_2O is higher than for lights ends (chloromethane and chloroethane), but boiling points are similar. Water and higher-boiling by-products, such as acetic acid (from oxidation of acetaldehyde), crotonaldehyde (a result of acetaldehyde condensation), and chlorinated acetaldehydes (from reaction of acetaldehyde with $CuCl_2$), are withdrawn together with acetaldehyde at the bottom of column 7. In the second column (9), acetaldehyde is purified by distillation.

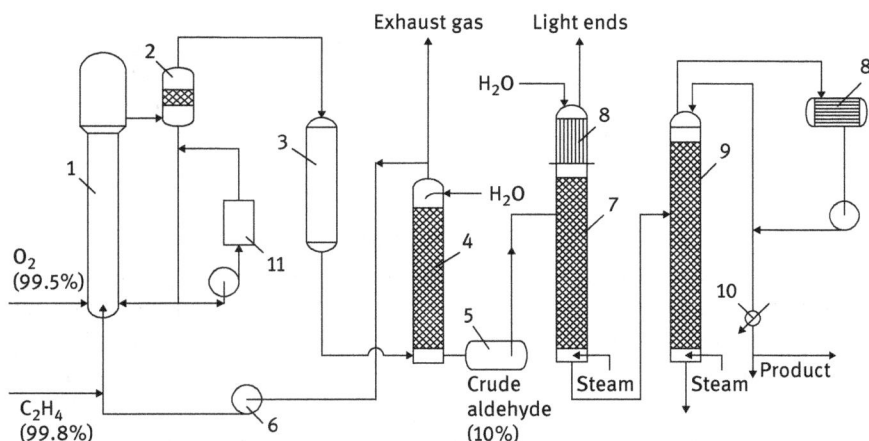

Figure 9.54: One-stage process of ethylene oxidation to acetaldehyde: 1, reactor; 2, separating vessel; 3, cooler; 4, scrubber; 5, crude aldehyde tank; 6, cycle-gas compressor; 7, light-ends distillation; 8, condensers; 9, purification column; 10, product cooler; 11, regeneration.

In the two-stage process (Figure 9.55), the reaction is carried out with ethylene and then with oxygen in two separate tubular reactors (1 and 4). The catalyst solution is alternately reduced in reactor 1 and oxidized in reactor 2. Air is used instead of pure oxygen for the catalyst oxidation.

Ethylene oxidation takes place at 105–110 °C and ca. 1.1 MPa, giving ca. 98% conversion. Higher temperatures could lead to chlorination of acetic acid and higher consumption of HCl, which is in fact added to compensate for chloride losses. Moreover, high temperature will lead to an increase in CO_2 formation.

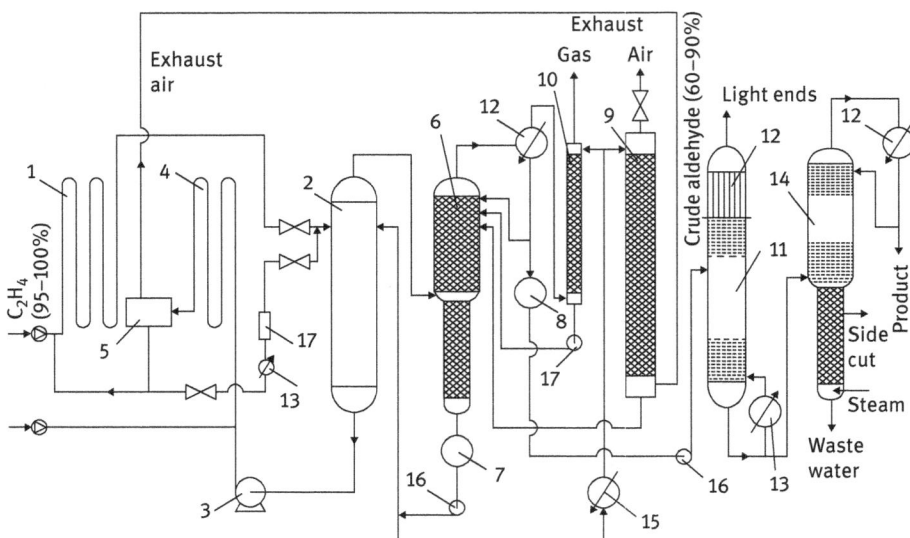

Figure 9.55: Two-stage process of ethylene oxidation to acetaldehyde: 1, reactor; 2, flash tower; 3, catalyst pump; 4, oxidation reactor; 5, exhaust-air separator; 6, crude-aldehyde column; 7, process-water tank; 8, crude-aldehyde container; 9, exhaust-air scrubber; 10, exhaust-gas scrubber; 11, light-ends distillation; 12, condensers; 13, heater; 14, purification column; 15, cooler; 16, pumps; 17, regeneration.

The catalyst solution containing acetaldehyde and a significant excess of Cu salts compared to Pd chloride (as oxidation of Cu^+ is a slow process) is expanded in a flash tower (pos. 2) where the pressure is reduced to atmospheric level. A delicate balance between $Cl^-/Cu^+/Cu^{2+}$ is needed since free hydrochloric acid retards oxidation, a high ratio between Cl^- and Cu^{2+} decreases ethylene conversion, while the opposite, along with high concentration of Cu^+, leads to extensive foaming. The latter can lead to catalyst carry-over. Moreover, a high concentration of $CuCl_2$ in the catalyst results in a lower yield of acetic acid and an increase in chloro-organic impurities.

The vapor mixture of acetaldehyde and water is distilled overhead while the catalyst is sent through the pump (3) to the oxidation reactor (4), where Cu^+ is oxidized to Cu^{2+} with air at ca. 1–1.3 MPa at 113 °C. The tubular reactor is in fact a ca. 150-m-long tube with a diameter of 500 mm. The contact time is ca. 60 s. Oxygen conversion is almost complete. In the separator (5), the oxidized catalyst solution is separated from exhaust air and sent for the reaction with ethylene. After flashing in (2), the vapor mixture of acetaldehyde and water is preconcentrated in column (6) to give 60–90% acetaldehyde. From the bottom of crude aldehyde column (6), the process water is taken and is partially returned to the flash tower to keep a constant catalyst concentration and is also used for scrubbing exhaust air in (9) and exhaust gas in (10). Scrubber water is sent to the crude aldehyde column (6). The flow scheme is completed by a two-stage distillation of the crude acetaldehyde, separating first low-boiling compounds (chloromethane, chloroethane, and carbon dioxide) in column 11, and subsequently, in column 14, water and higher-boiling by-products (chlorinated acetaldehydes and acetic acid) from acetaldehyde, which is taken overhead. The purity of acetaldehyde is around 99.6%, with such impurities present as acetic acid (0.1%), crotonaldehyde (0.1%), and water (0.2%). Chlorinated acetaldehydes are concentrated within the column as medium-boiling substances. The thermal treatment of a partial stream of the catalyst is done using HCl at about 160–165 °C and 0.8–1.0 MPa (regeneration unit 17) in order to decompose insoluble copper oxalate formed in reactor 1.

From the viewpoint of the acetaldehyde yield, both one- and two-stage processes afford a 95% yield. In the first alternative, only one reactor is required for processing a very corrosive reaction mixture, diminishing the investment costs, while the two-stage alternative is safer, as ethylene and air (cheaper oxidant) are used in separate reactors and no recycling of ethylene is needed.

Both methods require expensive construction material. Those parts, which are in contact with the extremely corrosive aqueous $CuCl_2$-$PdCl_2$ solution, are made entirely from titanium or the reactor is lined with acid-proof ceramic material, while the tubing is made of titanium.

The environmental burden of the Wacker oxidation process is high since chlorinated hydrocarbons, chlorinated acetaldehydes, and acetic acid are formed as by-products. Products in the waste air (ethane, chloromethane, chloroethane, and methane) can be oxidized over a heterogeneous catalyst, giving water, CO_2, and HCl. The latter is removed by washing. Chlorinated aldehydes, which end up in the wastewater, must be treated, for example, by alkaline hydrolysis before entering the wastewater plant, as they are not biodegradable.

9.2.2.5 Synthesis of phenol and acetone by isopropylbenzene oxidation

The main part of phenol (90%) is produced by the so-called cumene method, with a global annual capacity of ca. 7 million ton. Cumene is oxidized by air to cumene hydroperoxide (CHP) and cleaved in the presence of an acid catalyst. The process was invented in the 1940s independently in the former USSR and Germany. Oxidation of cumene is a free-radical chain reaction that starts by the formation of a cumene radical (Figure 9.56).

Chain initiation requires that an initiator be present in the reaction system. In the particular case of cumene oxidation, such initiator is the main product of the reaction – cumene hydroperoxide. In a classical way of process implementation starting with pure cumene, it would result in a long induction period. Industrial oxidations therefore are conducted in a series of continuous reactors with at least 8 wt% of cumene hydroperoxide in the first reactor. The cumene radical, formed by abstraction of the tertiary benzylic

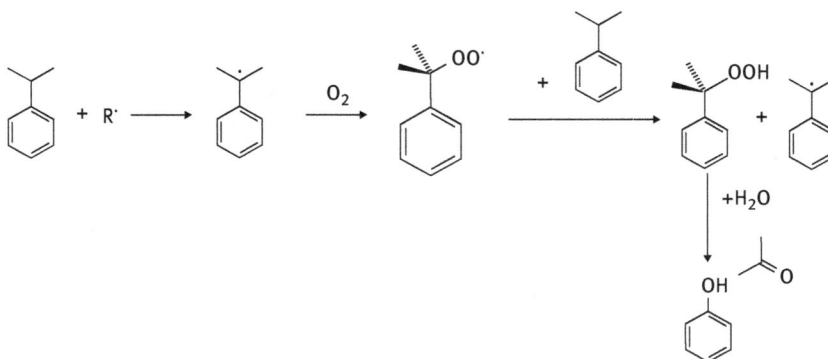

Figure 9.56: Mechanism of isopropylbenzene oxidation to phenol and acetone.

hydrogen, is subsequently oxidized, forming cumene hydroperoxide radical. The latter reacts with another molecule of cumene, giving cumene hydroperoxide and recovering cumene radical. In this way, a radical straight-chain reaction of cumene oxidation can be sustained. The final step in the formation of phenol and acetone is hydrolysis of cumene hydroperoxide, which can be done using homogeneous or heterogeneous acid catalysts. Besides the main chain reaction, there is a secondary one resulting in formation of dimethylphenylmethanol and acetophenone. This reaction is accelerated with an increase in the chain initiator (cumene hydroperoxide) concentration. Other minor by-products are dicumyl, formaldehyde, and formic acid. Formation of formic acid negatively influences oxidation, since it is an oxidation inhibitor; therefore, formic acid is neutralized.

As mentioned above, oxidation is carried out in a train of oxidation reactors with the product concentration being 8–12 wt% already in the first reactor. The concentration of cumene hydroperoxide is gradually increasing, reaching 25–40 wt% in the last one. Air is supplied to each reactor and vented at the top after removal of organic vapors.

Several versions of the cumene oxidation process have been implemented. In the Hercules process, in order to keep unstable peroxide in the liquid state, oxidation is done under pressure (0.6 MPa) at 90–120 °C. This process is carried out in the presence of sodium carbonate buffer, which was supposed to suppress the inhibiting role of phenol. Due to sodium carbonate recycling, there is formation of $NaHCO_3$, which is also an inhibitor. Significant reactor volume leads to an increase in capital costs. The residence time in the cascade of reactors operating in the Hercules process is 4–8 h. A hydroperoxide molar selectivity of 90–94% is reached. The spent air after removal of organic organics is vented.

Atmospheric pressure with the absence of buffers or promoters is applied in an alternative Allied/UOP process (Figure 9.57), which operates at 80–100 °C.

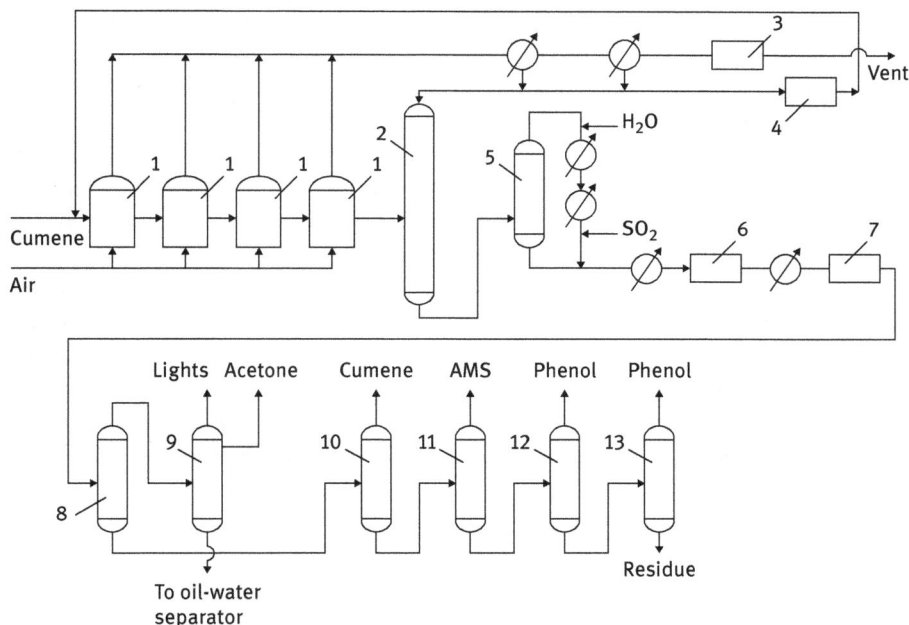

Figure 9.57: Transformations of cumene to phenol and acetone.

Decrease in temperature was aimed to improve selectivity. Lower reaction rates, on the other hand, resulted in larger reactors, which are less flexible in terms of safety keeping in mind a large amount of CHP present in reactors. Residence time in the cascade (pos. 1) is 10–20 h with CHP molar selectivity of 92–96%. Although the low-T process is typically used to improve selectivity and decrease formation of tarry compounds, generation of the latter (so-called phenol tar) is in fact very high, remaining at the level of 150–200 kg/t of phenol. Similar to cumene oxidation under pressure, in the atmospheric oxidation process the spent air is vented. Prior to venting, organic compounds are recovered by condensation of organic vapors with subsequent air cleaning by activated carbon (pos. 3). Condensed organic compounds are recycled to the oxidizer after washing with aqueous sodium hydroxide and water (pos. 4). Such alkali supply to cumene oxidation reactors partly solved the problem of the neutralization of organic acids negatively influencing selectivity in cumene oxidation and at the CHP concentration steps.

Cumene hydroperoxide is concentrated to over 80 wt% by evaporation of excess cumene (pos. 2) followed by hydrolysis (pos. 5) in the presence of sulphuric acid, generated *in situ* from sulphur dioxide. In order to keep a strict control on the reaction rate and prevent unwanted runaways during decomposition of CHP, which accelerates rapidly with increasing temperature, the concentration of cumene hydroperoxide should be maintained at a low level. Therefore, the reaction is conducted in a continuous stirred reactor where the reaction rate is defined by the outlet reactant concentration, but not the inlet. Conversion is thus not complete, and after cleavage, a certain amount (up to 5 wt%) of CHP remains unreacted. When decomposition temperature is below 70 °C, very high selectivity to phenol and acetone can be achieved (>99.5%). Therefore, as illustrated in Figure 9.57, decomposition temperature is controlled by heat exchangers in the loop at the level of 60–80 °C. Under these conditions, several reactions besides cleavage of CHP to phenol and acetone are possible. Acetone is oxidized by cumene hydroperoxide, resulting in hydroxyacetone (0.2–0.5% of acetone present).

Dimethylphenylmethanol and acetophenone, besides being already present in inlet feed of reactor 5, are also formed. The former compound gives undesirable condensates but at the same time is dehydrated to a desirable product α-methylstyrene. Condensation of dimethylphenylmethanol with cumene hydroperoxide results in formation of dicumyl peroxide, which is decomposed into phenol, acetone, and α-methylstyrene in a plug-flow reactor (pos. 6) with a short residence time at 110–140 °C. In the subsequent treatment, the acid catalyst is removed by ion exchange (pos. 7). Crude acetone distillation column 8 removes acetone from the product mixture, which is further distilled from the light-end in the acetone-refining column (pos. 9).

Cumene, α-methylstyrene, and phenol from the bottoms of the crude acetone column (pos. 8) are separated by sequential distillation in columns 10, 11, and 12. Removal of residues from phenol in column 13 is required, as only top-grade phenol having not more than 0.01% impurities and meeting the stringent color standards can be used for the manufacture of carbonate plastics.

Hydrogenation of α-methylstyrene to cumene is done using a very robust Pd catalyst at moderate pressure (Figure 9.58). Nearly complete conversion is achieved with high selectivity to cumene. In the first reactor, there is an excess of hydrogen, typical for exothermal hydrogenation reactions as will be discussed in the following chapter on hydrogenation. In the second reactor, complete conversion of α-methylstyrene is achieved with a once through hydrogen flow having only a slight stoichiometric excess. After cooling and flashing away the dissolved gases in the flash drum, cumene as a liquid product is recycled to oxidation.

Alternative routes to phenol production without generating acetone as a by-product are direct oxidation of benzene with nitrous oxide or hydrogen peroxide or oxidation of toluene. The former technology afford phenol selectivity of 97–100% at 100% N_2O conversion over an iron modified zeolite catalyst, requiring, however, frequent regeneration. Such technology is strongly relying on availability of N_2O, which preferably should not be produced on-purpose, but utilized as a cheap by-product of adipic acid manufacture.

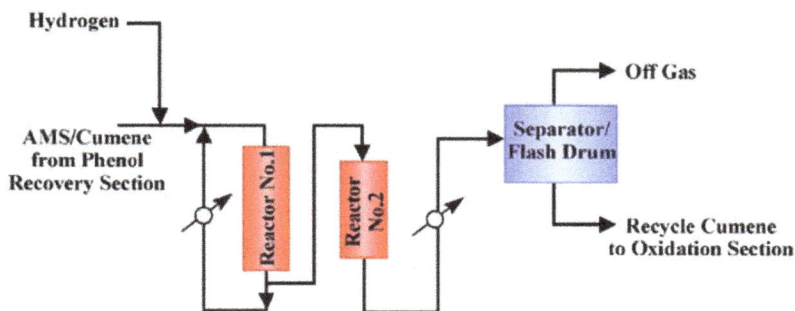

Figure 9.58: Cumene phenol-acetone process (Allied process): 1, oxidizers; 2, flash column; 3, carbon adsorber; 4, alkaline extraction and wash; 5, cumene hydroperoxide decomposer; 6, dicumyl peroxide decomposer; 7, ion exchange; 8, crude acetone column; 9, acetone-refining column; 10, cumene column; 11, α-methylstyrene column; 12, phenol column; 13, phenol residue topping column; AMS, α-methylstyrene.

9.2.2.6 Hydrogen peroxide

Hydrogen peroxide is manufactured using the anthraquinone process comprising a cyclic operation where the alkyl anthraquinone is reused. The cycle consists of sequential hydrogenation, oxidation, and extraction apart from some other steps shown in Figure 9.59.

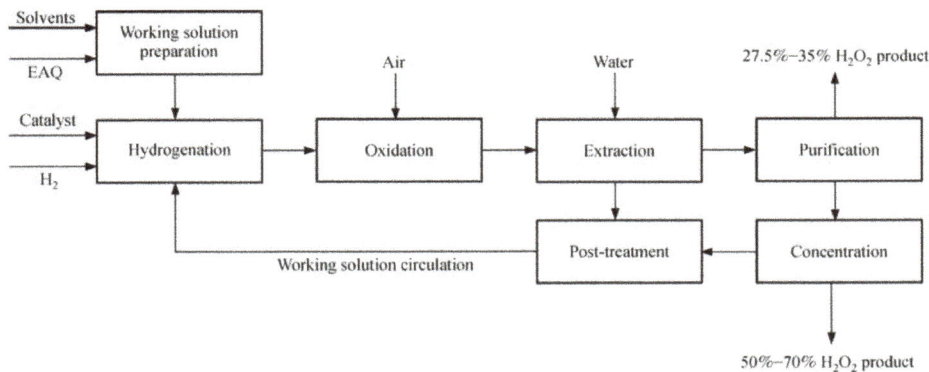

Figure 9.59: Huels process for α-methylstyrene (AMS) hydrogenation. From R.J. Schmidt, Industrial catalytic processes-phemol production, Applied Catalysis A: General, 2005, 280, 89–103. Copyright Elsevier. Reproduced with permission.

For hydrogenation (Figure 9.60), an alkyl anthraquinone is dissolved in two mutually soluble solvents of different polarity (i.e., nonpolar and polar). This working solution, comprising anthraquinone and solvents, is recycled. Aromatic, aliphatic, naphthenic hydrocarbons, their mixtures (a mixture of C9–C10 aromatics – AR), as well as fatty alcohols, organic acids, and organic acid esters are used as solvents by different manufacturers ensuring high hydrogenation efficiency and stability under both hydrogenation and oxidation conditions, low water solubility, and low concentration of organics in the final product – hydrogen peroxide. Moreover, they should afford good solubility of both quinone and hydroquinone, sufficiently lower density than water for the phase separation during extraction, low volatility and toxicity. Examples of solvents include trioctyl phosphate, 2-methyl cyclohexyl acetate, diisobutylcarbinol, or tetrabutylurea (TBU). The content of anthraquinone in the working solution depends on the solvent and can be in the range of 120 to 180 g/L.

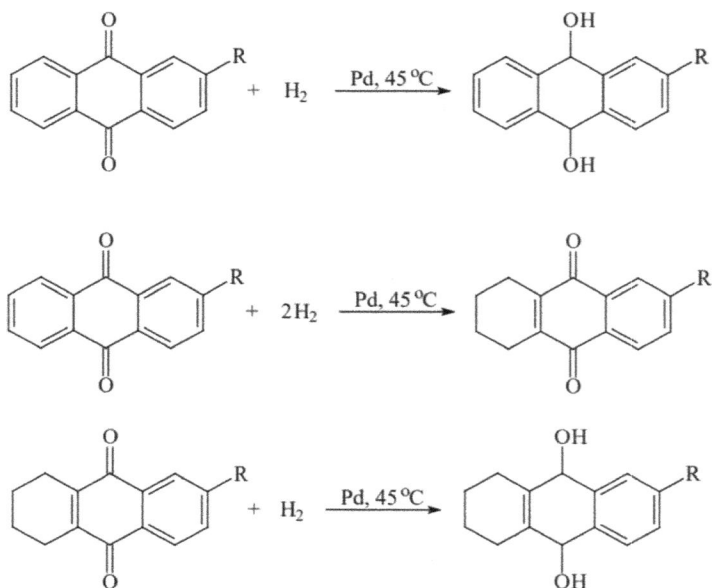

Figure 9.60: The anthraquinone process for hydrogen peroxide manufacture. From H. Li, B. Zheng, Z. Pan, B. Zong, M. Qiao, Advances in the slurry reactor technology of the anthraquinone process for H_2O_2 production. *Frontiers of Chemical Science and Engineering, 2018, 12,* 124–131, https://doi.org/10.1007/s11705-017-1676-5. Copyright © 2017, Higher Education Press and Springer-Verlag GmbH Germany. Reproduced with permission.

Hydrogenation of the working solution containing dissolved anthraquinone can be done in fixed-bed and a slurry-bed reactors. In fixed-bed reactors, an egg shell 0.1–0.5 wt% Pd/Al_2O_3 catalyst with the grain size in the range of 3–5 mm is used. The catalyst requires regeneration every 4–6 months with steam. The operation conditions were reported to be a liquid/gas ratio not exceeding 1.5/10, temperatures of 45–75 °C and pressures of 0.18–0.5MPa. A fixed-bed reactor with honeycomb monoliths having thin layers of porous silica gel, coated with palladium was also practiced industrially.

Typical problems with fixed-bed operations are difficulties in heat management, hot-spots in the catalyst bed, formation of degradation products, lowering catalyst stability, limited hydrogenation efficiency, high pressure drop, and influence of mass transfer. In a fixed-bed reactor with an anthraquinone solubility of 120 g/L, the hydrogenation efficiency is at the level of 5–8 g/L, while the theoretical value is 17 g/L.

Slurry bubble columns are thus applied in hydrogenation operating with the catalyst powders of the size 20 to 300 μm and thus minimizing the impact of internal mass transfer resistance. Moreover, the fluidized-bed reactor are much more efficient in terms of heat transfer. Such reactors offer higher hydrogenation efficiency, lower catalyst consumption per ton of hydrogen peroxide. Online removal and addition of supported palladium catalysts is much easier compared to the fixed bed.

A comparison of the technological parameters in fluidized- or fixed-bed reactors is summarized in Table 9.2.

Table 9.2: Comparison of hydrogenation steps in hydrogen peroxide manufacture.

		Chinese companies	FMC	MGC	Solvay	Degussa	Arkema
Working solution	Solvent 1	AR	AR	AR	AR	AR	AR
	Solvent 2	TOP	TOP	TBU	TBU	TOP	2-MCHA
	Solvent 3						
	AQ	EAQ	EAQ	AAQ	EAQ	EAQ	EAQ
	AQ solubility g/L	125–140	160–180	250–300	160–180	160–180	160–180
	Pd mass fraction, %	0.3	0.3	1–2	1–2	1–2	1–2
Hydrogenation	Reactor	Fixed bed	Fixed bed	Fluidized bed	Fluidized bed	Fluidized bed	Fluidized bed
	Efficiency g/L	7–8	10–12	15–18	12–15	11–15	11–14
Extraction	H_2O_2, %	27.5–35	27.5–35	45–48	43–46	40–45	40–45

After hydrogenation, the working solution containing hydrogenated anthraquinone is filtered from the catalyst and is oxidized in a non-catalytic step with slightly pressurized air up to 0.5MPa forming hydrogen peroxide in an organic phase

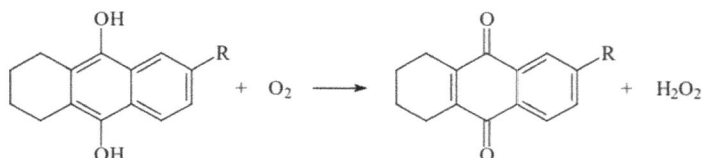

An example of the oxidation technology is presented in Figure 9.61.

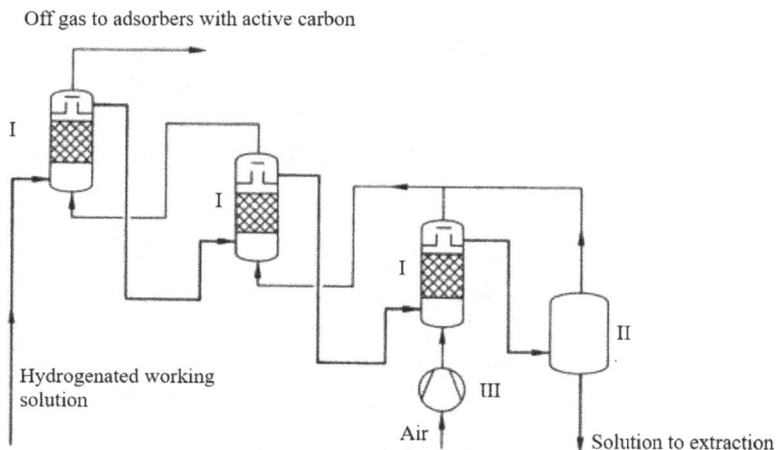

Figure 9.61: Degussa technology for oxidation (I) Concurrent-flow oxidation reactors; (II) separator; (III) air compressor.

The hydrogenated working solution flows concurrently upward with the oxidizing air through the first reactor. After separating air and the working solution in the upper part of the reactor, the offgas is treated by adsorption with activated carbon, while the liquid phases passes through two downstream reactors working in series. The air after pressuring in a compressor is fed to the third reactor and in this sense, the working solution and air flow are arranged countercurrently from the process flow viewpoint. At the reactor level both streams operate concurrently.

The subsequent step is extraction of hydrogen peroxide from the organic phase which is done in various liquid–liquid extractors such as sieve-tray extraction, packed columns, or pulsed packed columns. Efficiency of extraction, which is done by adding demineralized water to the top of a liquid–liquid extraction column, depends on working solution composition. At the bottom of the extractor, an aqueous solution containing 25–35% w/w crude hydrogen peroxide is obtained, while the working solution free from hydrogen peroxide is taken from the top of the extractor and is routed back to the hydrogenation step. In the BASF technology, after extraction, the working solution is passed through an aqueous potassium carbonate solution for drying.

Vacuum distillation of the crude hydrogen peroxide gives the concentrated product with up to 70% w/w concentration. To prevent decomposition of hydrogen peroxide, various proprietary stabilizers are added.

Chapter 10
Hydrogenation and dehydrogenation

10.1 General

Hydrogenation reactions involving typically a catalytic reaction between molecular hydrogen (H_2) and another compound or element. A large number of homogeneous and heterogeneous hydrogenation catalysts are available, and they are optimized according to a specific application. Homogeneous hydrogenation catalysts are seldom used in the industry when there is a need to process large quantities. They are difficult to separate from the reaction media and are not employed if a suitable heterogeneous catalyst is available. Nonetheless, they are useful in a variety of hydrogenation processes that are not possible with heterogeneous counterparts. Heterogeneous catalysis based on platinum group metals – platinum, palladium, rhodium, and ruthenium – are particularly active as catalytic materials. Non-precious metal systems, especially those based on nickel, copper, transition metals like molybdenum, and cobalt have also been developed as economical alternatives. The choice of the catalyst takes into account the tradeoff between the cost of the catalyst and cost of the required equipment (size, high operating pressure, etc.).

Hydrogenations can be subdivided into reactions with elements such as nitrogen (ammonia synthesis), hydrogenolysis

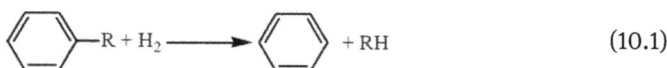

$$\qquad (10.1)$$

addition of hydrogen to unsaturated organic compounds (CO, alkenes, alkynes, aldehydes, ketones, imines, amides, nitriles, etc.); hydrogenation with addition of hydrogen, and release of water in reduction of organic compounds, such as acids, alcohols, or nitrocompounds

$$RCOOH + 2H_2 \rightarrow RCH_2OH + H_2O \qquad (10.2)$$

$$RCONH_2 + 2H_2 \rightarrow RCH_2NH_2 + H_2O \qquad (10.3)$$

$$ROH + H_2 \rightarrow RH + H_2O \qquad (10.4)$$

$$RNO_2 + 3H_2 \rightarrow RNH_2 + 2H_2O \qquad (10.5)$$

or hydrogenation with release of other molecules (ammonia, HCl, or H_2S)

$$RCOCl + H_2 \rightarrow RCHO + HCl, \quad RSH + H_2 \rightarrow RH + H_2S \qquad (10.6)$$

Hydrogenation reactions are exothermal, while dehydrogenation is endothermal. Values of $- \Delta H^{\circ}_{298}$, kJ/mol are presented in Tab. 10.1.

https://doi.org/10.1515/9783110712551-010

For hydrogenation of triple bonds, the heat release is much higher than for double-bond hydrogenation, while reduction of a carbonyl group releases less heat, being at the same time slightly more exothermal than reduction of a keto group.

An important aspect in hydrogenation reactions is its reversibility. Hydrogenation reactions are favored at low T, while an increase in temperature will drive the reaction backward. The most thermodynamically favored case is dehydrogenation of hydrocarbons with formation of conjugated and triple bonds as well as dehydrogenation of six-member-ring cycloalkanes to aromatic compounds.

In hydrogenation processes with release of water, equilibrium is shifted more to the right side, for example, in hydrogenation of alcohols to hydrocarbons or nitrocompounds to amines. An exception is hydrogenation of acids to alcohols with a small change in Gibbs energy (ca. 40 kJ/mol).

Hydrogenation is thus preferred at low temperatures, as is typically done in the range of 50–300 °C depending on the substrate, although lower and higher temperatures are possible.

Table 10.1: Heats of some hydrogenation reactions, $-\Delta H°_{298}$, kJ/mol.

$RCH = CH_2 \xrightarrow{+H_2} RCH_2 - CH_3$	113–134
$CH \equiv CH \xrightarrow{+2H_2} CH_3 - CH_3$	311
$C_6H_6 \xrightarrow{+3H_2} C_6H_{12}$	206
$RCHO \xrightarrow{+H_2} RCH_2OH$	67–83
$R_2CO \xrightarrow{+H_2} R_2CHOH$	≈58
$RCN \xrightarrow{+2H_2} RCH_2NH_2$	134–159
$RCOOH \xrightarrow{+2H_2; -H_2O} RCH_2OH$	38–42
$RNO_2 \xrightarrow{+3H_2; -H_2O} RNH_2$	439–472
$-CH_2 - CH_2 - \xrightarrow{+H_2} CH_3 + -CH_3$	42–63

Since, in hydrogenation, the number of moles in the product is less than in the reactants, high pressures (from 1 to 5 or even to 30 MPa) are used.

On the contrary, high temperature and low pressure (by dilution, for example, with steam) are preferred for various dehydrogenation reactions: dehydrogenation of paraffins (butane to butane)

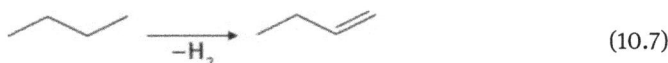

$$\text{/\textbackslash/} \xrightarrow{-H_2} \text{/\textbackslash} = \qquad (10.7)$$

alkylaromatics (ethylbenzene to styrene)

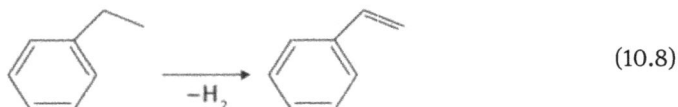

$$(10.8)$$

dehydrocyclization

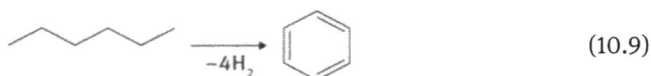

$$(10.9)$$

and synthesis of ketones and aldehydes from alcohols

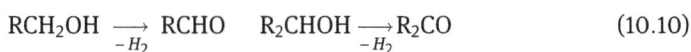

$$RCH_2OH \xrightarrow[-H_2]{} RCHO \quad R_2CHOH \xrightarrow[-H_2]{} R_2CO \tag{10.10}$$

In hydrogenation reactions, in many cases, an important issue is selectivity, since, for example, hydrogenation of alkynes can result not only in olefins, but also in alkanes, hydrogenation of acids gives aldehydes, alcohols, and hydrocarbons, etc., and even hydrogenolysis of organic compounds with undesired products can occur. A careful choice of catalysts is thus needed to ensure a proper selectivity. Independent on catalyst type, temperature increase typically leads to deterioration of selectivity due to side reactions (hydrogenolysis, cracking, etc. with higher activation energy than the main hydrogenation reaction). For consecutive reactions, selectivity obviously depends on conversion, thus the reaction (or residence) time should be carefully controlled.

In the gas-phase hydrogenation, multitubular reactors (Figure 10.1a) where cooling is done by condensate in between the tubes and multibed reactors with cold gas quenching (Figure 10.1b) are applied.

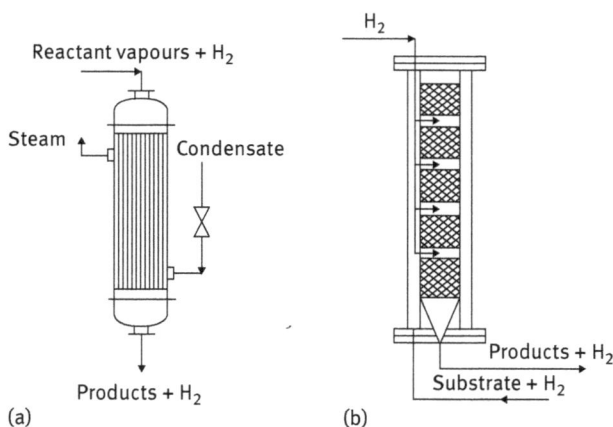

Figure 10.1: Multitubular reactor (a) and reactor with cold gas injection-quenching (b).

An alternative for less exothermal reactions is to utilize a multibed reactor with interned heat exchangers.

An important parameter in regulating the heat removal is the application of an excess of hydrogen, which ranges from 5:1 to 30:1.

10.2 Ammonia synthesis

Large-scale production of ammonia is one of the most important catalytic processes. The process was discovered by F. Haber and his coworkers, who demonstrated in 1909 the possibility of producing ammonia at a scale of 2 kg/day using osmium as a catalyst. The technology was further developed by BASF under the leadership of C. Bosch. Such development required construction of new high-pressure equipment and application of a less exotic and expensive catalyst. Efforts of A. Mittash and co-workers, who tested over 2,500 formulations, resulted in Fe catalyst, promoted with alumina, calcium oxide, and potassium. The very same catalyst is essentially used in hundreds of industrial installations with capacity that grew from several tons a day in the original column erected in Oppau (now part of BASF site in Ludwigshafen) to single trains with 1,500–1,800 tons/day. Haber and Bosch were later awarded Nobel prizes, in 1918 and 1931, respectively. Development of much more active Ru/C catalysts led to commercial application of Ru catalysts in few plants in the 1990s.

Although the chemistry of ammonia synthesis is superficially very simple, this reaction, which operates at high pressures and temperatures gave rise to development of very many fundamental concepts, such as surface heterogeneity, structure sensitivity, virtual pressure, rate determining steps, microkinetics, etc. Reactor design is also very rich and sophisticated with different types of reactors employed in the industry.

The ammonia synthesis reaction

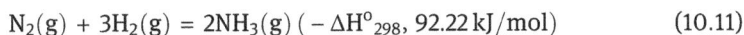

$$N_2(g) + 3H_2(g) = 2NH_3(g) \, (-\Delta H^{\circ}_{298}, 92.22 \, kJ/mol) \qquad (10.11)$$

is exothermic with a volume decrease ($\Delta V < 0$) and is limited by thermodynamics. Therefore, high P and low T should be preferred (Tab. 10.2).

Equilibrium concentrations of ammonia as a function of temperature and pressure are given in Figure 10.2.

A catalyst of high activity is thus needed, which should be able to operate at low temperature. Unfortunately, high activity could only be achieved for rather inexpensive iron catalysts at 400–450 °C, leading to severe equilibrium limitations. The process is thus conducted industrially at 15–30 MPa between 300 °C and 550 °C. High temperature and pressure are beneficial from the point of view of kinetics, although at the expense of conversion limitations because of thermodynamic restrictions and costs, associated with high pressure technology.

Table 10.2: Variation in equilibrium constant K_{eq} for the ammonia synthesis as a function of temperature.

Temperature (°C)	$Keq = \dfrac{P^2_{NH_3}}{P_{N_2}\, P^3_{H_2}}$
300	4.34×10^{-3}
400	1.64×10^{-4}
450	4.51×10^{-5}
500	1.45×10^{-5}
550	5.38×10^{-6}
600	2.25×10^{-6}

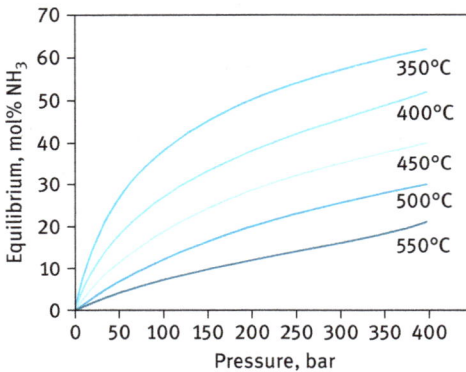

Figure 10.2: Effect of pressure and temperature on equilibrium ammonia concentration at an inlet hydrogen/nitrogen ratio of 3:1. C. H. Bartolomew, R. J. Farrauto, *Fundamentals of Industrial Catalytic Processes*, 2nd Edition, 2006, Copyright 2010, John Wiley and Sons. Reproduced with permission.

Ammonia synthesis reaction is exothermic; thus, in order to operate efficiently, equilibrium should be continuously shifted and heat should be removed. This is done by organizing the process in several stages. After one bed of catalyst, concentration of ammonia is becoming rather substantial (Figure 10.3), coming closer to the equilibrium line. Thereafter, either intermediate cooling is done by heat exchangers or ammonia is removed through quenching (adding cold reactants). In this case, the concentration line (dotted) is not going parallel to the x-axes in Figure 10.3. In should be noted that at the inlet of the catalyst bed reaction is far from equilibrium, while at the outlet, due to high ammonia concentration, the reverse reaction is becoming more prominent, influencing in a negative way the overall reaction.

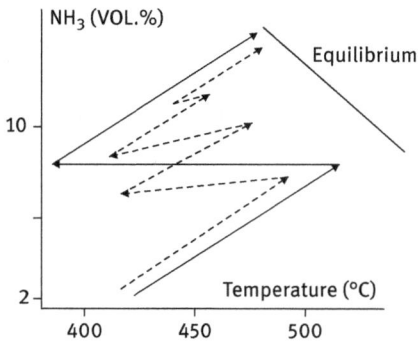

Figure 10.3: Concentration-temperature diagram in ammonia synthesis.

The single pass through a reactor yield is then ca. 15%, calling for the recycling of the reactants. This is done by cooling gases in order to liquefy ammonia, removing it as a liquid, and returning hydrogen and nitrogen into the reactor.

In a simple form, the kinetics of ammonia synthesism based on experimental data in the pressure range from below 1 to 500 atm, can be expressed near equilibrium by the Temkin-Pyzhev equation,

$$r = k_+ P_{N_2} \left(\frac{P_{H_2}^3}{P_{NH_3}^2} \right)^m - k_- \left(\frac{P_{NH_3}^2}{P_{H_2}^3} \right)^{1-m}, \tag{10.12}$$

where m is a constant $(0 < m < 1)$, $k_+/k_- = K$, where K is the equilibrium constant. Only one of the constants, either k_+ or k_-, together with m should be determined from the experimental data as they are related through an equilibrium constant. At high pressures, eq. (10.12) should be modified to include the deviations from the laws of ideal gases and to incorporate the effect of pressure on the reaction rate depending on the volume change at activation. A more recent approach including microkinetic analysis using rate and equilibrium constants from surface science studies demonstrated the applicability of such analysis to rather accurately describe experimental data.

Typical commercial unreduced iron-based catalysts contain ca. 0.5–1% K_2O, 2–4% Al_2O_3, 2–4% CaO, while the rest is magnetite Fe_3O_4 (and some wustite FeO). Alumina functions as a structural promoter, enabling the production and preparation of an open and porous structure. During the initial catalyst activation, reduction starts on the outside of the granule, and dissolved alumina separates out of solid solution, and in the pores, forming between the iron crystallites. This minimizes further growth during completion of reduction and later operation giving small 20- to 49-nm Fe crystallites. Calcium oxide, and other basic promoters (for example, MgO, < 0.5%), react with silica impurities in the raw materials to form glassy silicates, which themselves can enhance the thermal stability of the reduced iron. The main benefit is to minimize any neutralization of the K_2O promoter, which would diminish its effectiveness. The role of K is more complicated. This component is considered to be an electronic promoter, functioning by considerably increasing the intrinsic activity of the high-surface-area iron particles produced on reduction. Such enhancement of activity due to formation of potassium ferrites is seen in up to about 0.8% potash, while above this level, activity falls. The chemical action of potassium is associated with a decrease of the adsorption energy on iron, lowering the concentration of adsorbed ammonia, thus increasing the number of sites available for nitrogen adsorption.

Catalysts are prepared by the fusion of a mixture of magnetite and the promoters (added as carbonates, oxides, or hydroxides, or as nitrate in case of potassium). Natural magnetite with low impurity levels is used nowadays. The resulting catalyst is a solid solution of wustite and alumina in the magnetite crystal lattice. The surface area of the final catalysts is rather small, being in the range of 1–2 m^2/g. The concentration of alumina should be less than the solubility of alumina in magnetite, which corresponds to a maximum content of about 3% alumina.

As-prepared catalyst in the form of magnetite must be activated by reduction to metallic iron before use. The reduction process in done when the unreduced catalyst is loaded in ammonia converter prior to operation. On reduction, the catalyst becomes porous and develops a surface area around 20 m^2/g. In order to diminish reduction time, ammonia synthesis catalysts are manufactured in the pre-reduced form. After such reduction is done by a catalyst manufacturer, the catalyst is carefully stabilized by a controlled flow of a mixture of nitrogen and air, resulting in less than 10% of iron reoxidation to thin oxide film around iron particles. This procedure allows starting reduction at lower temperature (330 °C *versus* 400 °C for non-reduced catalyst), giving a shorter reduction time at a plant site. Such shorter procedure brings substantial savings, as during catalyst reduction, the ammonia synthesis train including steam reforming of natural gas is in full operation consuming several dozens of thousands cubic meters per hour of natural gas. It is thus desired to reach the full production capacity as soon as possible, without compromising, however, long-term catalyst activity. The pre-reduced catalysts are more expensive than the unreduced ones; thus, operating costs are balanced with the higher catalyst price. Typically, pre-reduced catalysts are only used in the top reactor sections, simplifying the beginning of the reduction process.

Ammonia synthesis is one of several catalytic processes where catalyst manufacturers guarantee catalyst performance. A comparison between guaranteed parameters, mean values during warrant tests, and performance at nominal capacity are given in Tab. 10.1.

Part of the industrial production now takes place with a more expensive ruthenium-based catalyst (the KAAP process) rather than with cheaper iron because this more active catalyst allows reduced operating pressures.

Ruthenium, promoted by alkali metals and barium and then supported on a cheap carbon support, has provided a significant increase in activity compared

Table 10.3: Warranty testing data for iron-based catalyst.

Parameter	Units	Guarantee	Mean values	Values at nominal capacity
Inlet gas flow	10^3 m^3/h	790	932	790
Inlet reactor T, not more	°C	175	172	172
Outlet reactor T, not more	°C	329	327	329
ΔP	MPa	0.56	0.601	0.42
Outlet pressure	MPa	21.0	22.1	20.0
Ammonia output	t/d	1,375	1412.7	1,375
Inert gases	%	12–14	15	14

Modified after D. B. Shepetovsky, I. G. Brodskaya, and P. A. Ozolin, *Catalysis in Industry*, 2007, 42–47.

with iron catalysts. K and Cs apparently facilitate nitrogen adsorption, allowing electron charge transfer to Ru, similar to iron-based catalysts. A magnesium oxide support has been claimed to give a longer life. Ru/C demonstrated much higher activity than iron at moderate hydrogen to ammonia ratios, and ammonia content, while the rate of ammonia production is strongly inhibited by ammonia. Operation at lower temperature and pressures is thus favorable over Ru; thus, this catalyst should not be used at higher ammonia concentrations.

In 1998, two grassroots KAAP plants producing ammonia at 1,850 t day^{-1} began operating in Trinidad and Tobago using one bed of iron catalyst and three beds of ruthenium catalyst. A third KAAP plant in Trinidad started up in 2002 and a fourth began operating in 2004.

Startup of other KAAP plants in Trinidad, Egypt, and Venezuela was in 2008–2010. In new designs, lower pressure (90 bar) and temperature, less gas compression, and less high-pressure equipment lead to lower operating and capital costs, less downtime while still being capable of maintaining required the ammonia yield per pass.

The ammonia synthesis reaction occurs in a loop where hydrogen and nitrogen synthesis gas is circulated continuously through a converter containing the catalyst (Figure 10.4).

Ammonia is condensed (condensation temperature in industry is typically −6 °C) and removed from the loop. In order to avoid accumulation of inert gases in the loop, such as methane and argon, they are continuously removed in a purge gas stream.

In the past, axial flow reactors were used, operating at high pressure (300 bar). Larger catalyst particles (6–10 mm) were required in order to limit the pressure drop through the catalyst. As the catalyst effectiveness factor (at large values of Thiele modulus) is inversely proportional to particle size, strong diffusional limitations were present. By designing converters in such a way that gas flows radially through the catalyst bed, it is possible to decrease the overall pressure drop, using smaller catalyst particles (1.5–3 mm) exhibiting higher activity per unit volume. Revamp of axial flow plants to radial flow operation mode does not lead, however, to higher ammonia production. Instead, pressure in the converter is decreased to 20–22 MPa, lowering production costs.

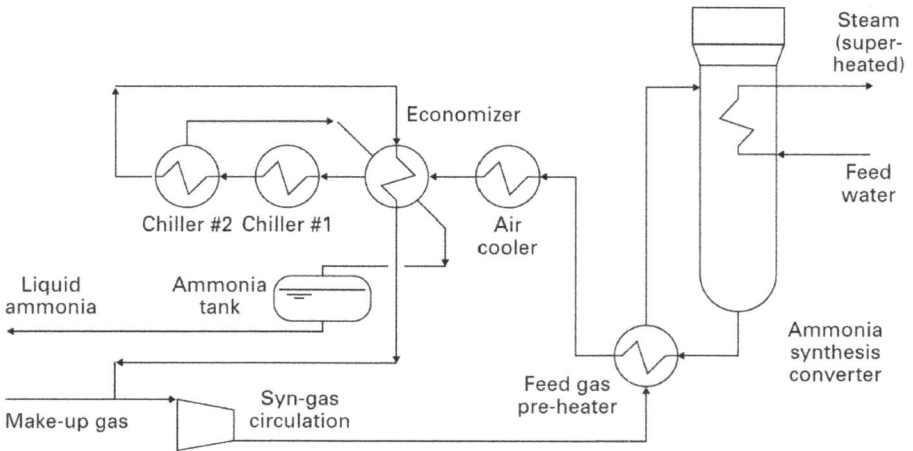

Figure 10.4: Ammonia synthesis loop (redrawn from A. Luzzi, K. Lovegrove, E. Filippi, H. Fricker, M. Schmitz-Goeb, M. Chandapillai, S. Kaneff, Technoeconomic analysis of a 10 MWe solar thermal power plant using ammonia-based thermochemical energy transfer. *Solar Energy*, 1999, 66, 91–101. Copyright 1999 Elsevier Science Ltd. Reproduced with permission).

As explained above, because of equilibrium limitations, reactors typically include either interstage cooling or the addition of cold synthesis gas, an operation known as quenching.

Several reactor options are installed in commercial ammonia plants. In the Haldor Topsøe S-200 radial converter, two catalysts beds are applied with an intermediate cooling (Figure 10.5A). The S-300 reactor from the same company has three beds and two interbed exchangers. This converter enables a higher conversion for the same catalyst volume or smaller catalysts volumes and lower investment costs for new plants. Similar designs (with three beds and radial flow from outside to inside) are available from other companies (for example, Uhde GmbH, Figure 10.5B). Kellogg axial converters (Figure 10.5C) have several beds with quenching.

In Figure 10.6, a horizontal ammonia synthesis reactor that allows removal of the catalyst basket from the converter without aid of heavy cranes usually needed to lift the catalyst basket out of vertical converters is shown.

In the reactors discussed above, the catalysts were distributed in several beds operated adiabatically in an attempt to follow the optimal equilibrium curve (Figure 10.7).

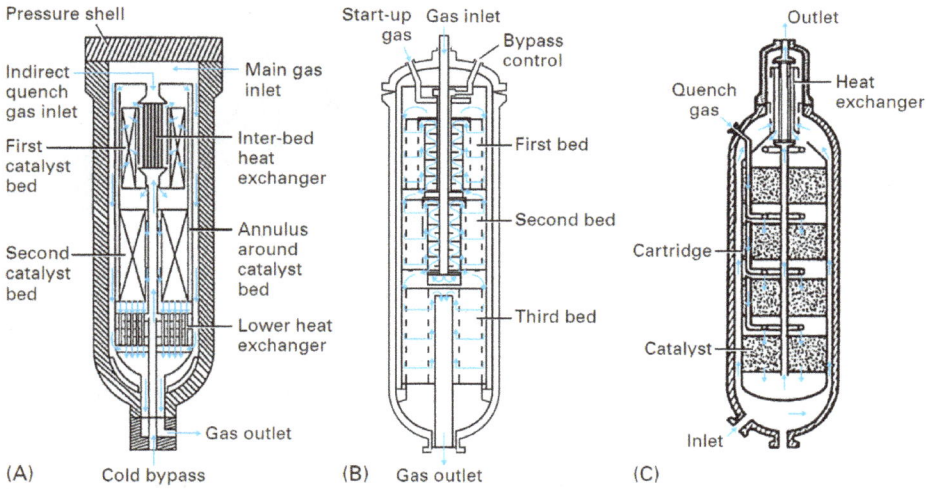

Figure 10.5: Ammonia synthesis converters with: (A) reactor with two beds, from Fertilizer Manual. Kluwer, 1998, copyright UNIDO and IFDC, (B) reactor with three beds (from http://www.thyssenk rupp-uhde.de/de/kompetenzen/technologien/fertiliser/ammoniakharnstoff/75/94/ammonia-conversion.html.), (C) reactor with three beds and quenching.

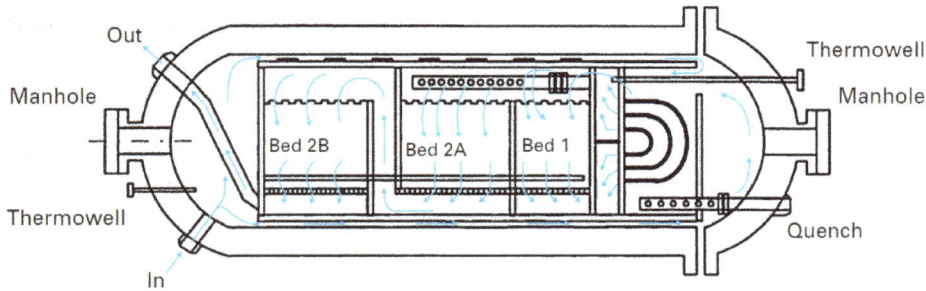

Figure 10.6: KBR horizontal ammonia synthesis reactor (from Fertilizer Manual. Kluwer, 1998).

In a departure from this concept, Ammonia Casale developed a pseudo-isothermal ammonia synthesis converter. The new design is based on the use of cooling plates immersed in the axial–radial catalytic bed to remove the reaction heat while it is formed. This is a very efficient design and can help to build mega capacity ammonia plants. So far, Ammonia Casale revamped a plant in Thailand running at capacity of 753 mtpd to boost it to 953 mtpd with energy savings of 0.96 Gcal MT^{-1}. Significant reductions in the circulation rates and specific duties of heat exchangers were obtained. Lately, reactor internals were installed in another plant in Canada.

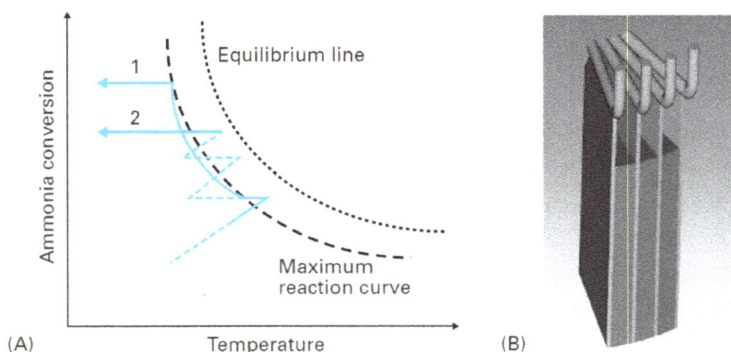

Figure 10.7: Isothermal ammonia converter with (A) temperature profiles with (1) multitubular reactor, (2) isothermal design; (B) cooling plates.

In ammonia dual pressure process by Uhde (Figure 10.8), the synthesis section comprises a once-through section followed by a synthesis loop. In this way, large capacities can be achieved (4,000–5,000 metric tons per day).

The ammonia synthesis loop consists of two stages. An indirectly cooled once-through converter operates at ca. 110 bar. The major part of ammonia produced in this converted after cooling is separated from the gas products. Overall, the once-through converted generates one third of the total ammonia. The ammonia synthesis loop per se operates at ca. 21 MPa.

Although many ammonia plants were revamped to improve performance efficiency, there are still some barriers to their implementation of new technologies. Among these barriers, limited knowledge of process technologies outside the plant should be mentioned, as small operators are not aware of all the best technologies. Plant operators also want more data on new technologies to prove its reliability and are not eager to be the first to make risky investments. In addition, key barriers such as low energy costs, lack of corporate R&D, capital availability, and corporate priorities could hinder more widespread implementation of new technologies.

10.3 Gas-phase hydrogenation

As an example gas-phase hydrogenation of phenol will be considered. One of the options for hydrogenation of phenol to cyclohexanol is to carry it out in the gas phase over a supported nickel catalyst at 140–150 °C and 1–2 MPa. Cyclohexanone, cyclohexene and cyclohexane are formed as side products. The flow scheme is presented in Figure 10.9.

Hydrogen is compressed (pos. 2) and heated (pos. 3) by the outlet gases coming from the reactor (pos. 8). Thereafter, it is sent to evaporator-saturator 6. Phenol (solid at room temperature), after melting (pos. 4), is pumped (pos. 5) to the saturator

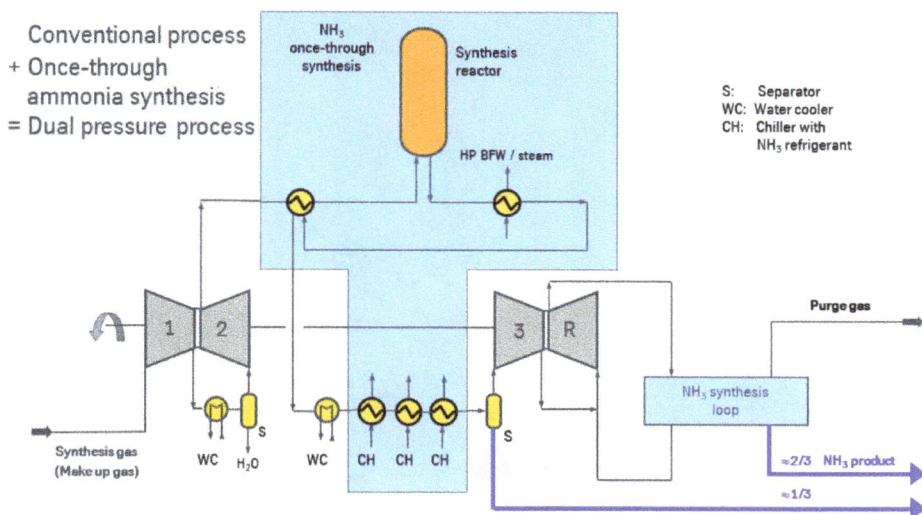

Figure 10.8: Ammonia dual pressure process by Uhde (http://www.iffcokandla.in/data/polopoly_fs/1.3253588.1472738687!/fileserver/file/681710/filename/4F.pdf).

(pos. 6), where the temperature (120–125 °C) and the level of phenol are regulated to ensure an excess of hydrogen (ca. 10). After the heater (pos. 7), phenol and excess of hydrogen pass through a reactor with a catalyst, affording conversion of 85–99%. After cooling in 3 and further in 10, hydrogen is separated in a high-pressure separator 11 from the condensate. Hydrogen is recycled through a compressor (pos. 2), while the condensate is expanded to atmospheric pressure (pos. 15). In a low-pressure separator (pos. 12), the product is separated from the gas, containing also methane. Cyclohexanol with unreacted phenol is distilled in column 13, where from the column top, an azeotropic mixture of cyclohexane, cyclohexene, and water is removed. In the subsequent distillation, column 14 cyclohexanol is separated from unreacted phenol, which could be recycled. Cyclohexanone is not separated from cyclohexanol since, for example, when the latter is used for production of caprolactam, the subsequent step in processing of cyclohexanol is its dehydrogenation to cyclohexanone.

10.4 Liquid-phase hydrogenation

Liquid-phase hydrogenation can be done in the substrate *per se*, in an inert solvent, or in emulsions. The first approach is the most commonly applied. Hydrogenation in a solvent can be done, for example, in cases when the substrate is solid at the reaction temperature or prone to formation of undesired products at high concentrations. Hydrogenation of aldehydes in the solutions of corresponding alcohols helps to avoid

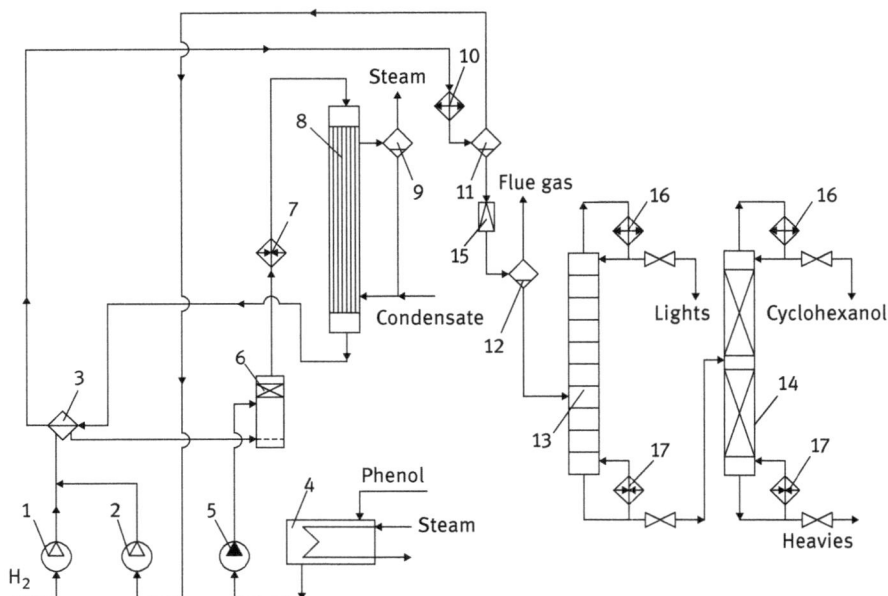

Figure 10.9: Gas-phase phenol hydrogenation: 1 and 2, compressors; 3, heat exchanger; 4, vessel; 5, pump; 6, evaporator-saturator; 7, heater; 8, reactor; 9, separator; 10, heat exchanger; 11 and 12, separators; 13 and 14, distillation columns; 15, expansion valve; 16, reflux vessel; 17, boiler. After N. N. Lebedev, *Chemistry and Technology of Basic Organic and Petrochemical Synthesis*, Chimia, 1988.

aldol condensation. Hydrogenation of nitro-compounds is done in water emulsions, which helps in efficiently removing heat and separating diamine soluble in water.

In liquid-phase hydrogenation, the catalyst is either suspended in the liquid as powder or applied in the form of extrudates or tablets in stationary beds.

Heat is removed either by evaporation of the solvent, using coils or heat exchanger, or by applying quenching with cold hydrogen. Several of these and other possible arrangements are illustrated in Figure 10.10.

For example, conventional slurry reactors operated in the batchwise mode could be applied. Alternatively, a batchwise column with a suspended catalyst (Figure 10.10a) can be applied with the upper part of larger size operating as a demister (with trays or packings). Unreacted hydrogen applied in excess is separated from the condensate and recycled back to the reactor. The operation of such reactor in a continuous mode is inefficient, as the concentration of the substrate in the CSTR in fact corresponds to the outlet concentration, meaning that the reaction rate would be rather low for non-zero-order reactions. Therefore, a cascade of reactors (3–5) is typically used (Figure 10.10b). Adiabatic fixed-bed reactors for not extremely exothermal reactions can be utilized either with intermediate cooling (Figure 10.10c) or quenching by cold hydrogen (Figure 10.10d).

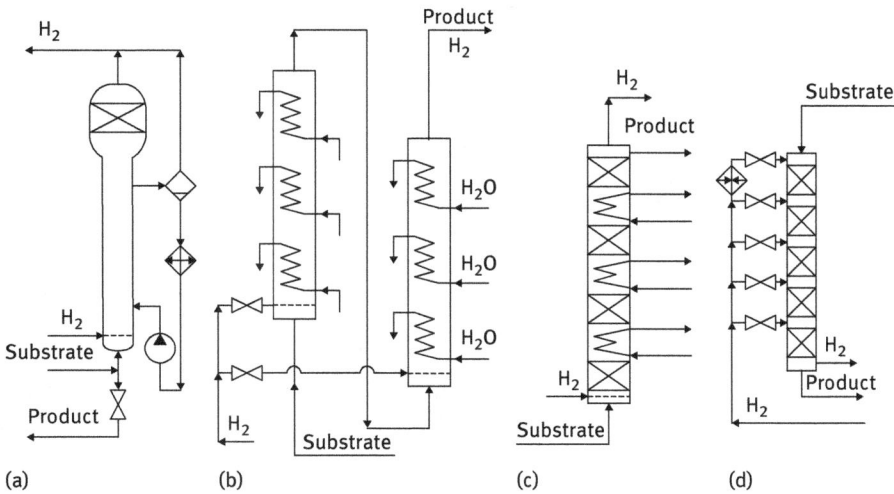

Figure 10.10: Liquid-phase hydrogenation: (a) column with a suspended catalyst and external heat exchanger; (b) cascade of reactors operating with suspended catalysts and internal coils for cooling; (c) a multibed reactor with stationary catalyst beds and interstage heat exchangers; (d) reactor with stationary catalyst beds and quenching by cold hydrogen.

It is also possible to combine hydrogenation in a reactor with a suspended catalyst with a stationary bed catalyst (Figure 10.11). The main conversion occurs in the first reactor with heat removal done by one of the methods described above. Since the concentration of the substrate after the first reactor is much smaller than at the inlet, it is possible to furnish hydrogenation in an adiabatic bed reactor, which can handle the temperature rise without deteriorating the catalyst performance.

As a typical example, liquid-phase hydrogenation of C10–C18 fatty acid (methyl) esters can be considered (Figure 10.12).

Reaction is conducted at high pressure (30 MPa) and 300 °C over a copper-chromite (Adkins) catalyst. Since the reaction heat is moderate, it is possible to use adiabatic reactors with a suspended catalyst. Fresh and recycled hydrogen is compressed (pos. 1, 2) and heated in a heat exchanger (pos. 3) with the hot outlet gases and further in a furnace (pos. 4). The substrate is delivered through a heater operating with steam (pos. 6), while the catalyst is added as a suspension from the mixer (pos. 15) by a pump (pos. 16). After the reactor (pos. 7), the reaction mixture (unreacted ester, the product – corresponding alcohol, methanol, and side products – hydrocarbons and water) is separated (pos. 8) in the liquid and gas phases. The gas phase is cooled (pos. 3 and 9), and after further separation (pos. 10) hydrogen is recycled, while the condensate is expanded (pos. 20) with further separation of water (pos. 18 and 19) from flue gases. The liquid phase after separator (pos. 8) is expanded to ca. atmospheric pressure (pos. 20), cooled (pos. 11), and further separated from flue gases (pos. 12). The catalyst

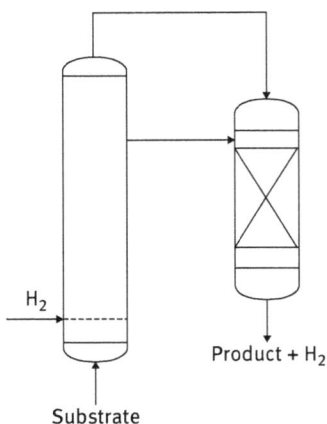

Figure 10.11: Hydrogenation in two reactors (with stationary and suspended catalysts).

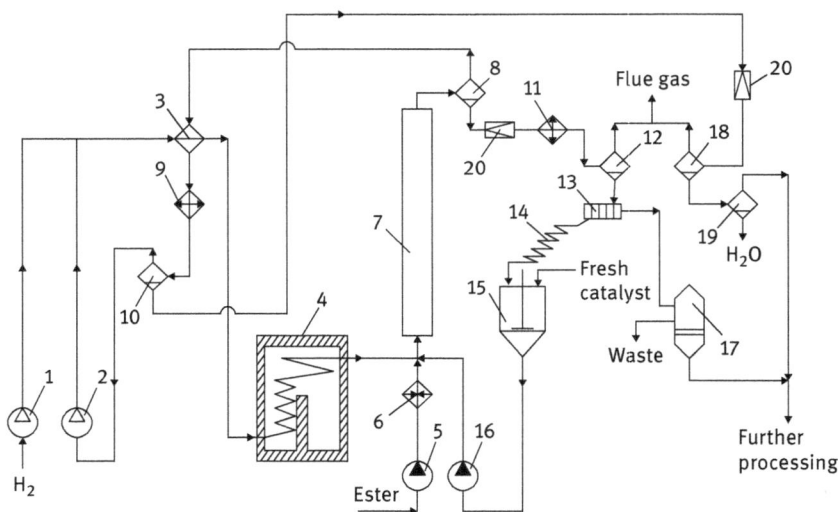

Figure 10.12: Hydrogenation of C10–C18 fatty acids esters: 1 and 2, compressors; 3, heat exchanger; 4, furnace; 5 and 16, pumps; 6, steam heater; 7, reactor; 8, 10, 12, 18, and 19, separators; 9 and 11, heat exchanger; 13, centrifuge; 14, screw conveyor; 15, mixer; 17, filter; 20, pressure expansion. After N. N. Lebedev, *Chemistry and Technology of Basic Organic and Petrochemical Synthesis*, Chimia, 1988.

is centrifuged (pos. 13) and delivered (pos. 14) to the catalyst mixer (pos. 15), where the fresh catalyst (15% of the required amount) is added. Small catalyst particles that, due to attrition, cannot be recycled are filtered (pos. 17) and disposed. The liquid after filtration is further processed. Methanol is removed and routed to the ester synthesis step, unreacted ester is saponified with sodium hydroxide, and the desired product – fatty acid alcohol – is distilled from the heavies.

10.4.1 Nitrobenzene hydrogenation

This reaction producing aniline is predominantly needed as a step in manufacturing corresponding isocyanates, intermediates for polyurethane synthesis. Both liquid- and gas-phase hydrogenation is practiced. The liquid-phase hydrogenation illustrated in Figure 10.13 comprises hydrogenation per se in a tubular reactor containing, e.g., a noble metal (Pt–Pd) catalyst supported on carbon (DuPont technology). Nitrobenzene conversion to aniline of nearly 100% in a single pass is followed by separation of the hydrogen excess from the reactor effluent and removal of water from the liquid product in a dehydration column. The bottoms stream flows to the purification column, where heavy impurities (tars) are separated from the crude aniline stream. The final product – high-quality aniline has purity exceeding 99.95 wt% and contains less than 0.1 wppm of nitrobenzene.

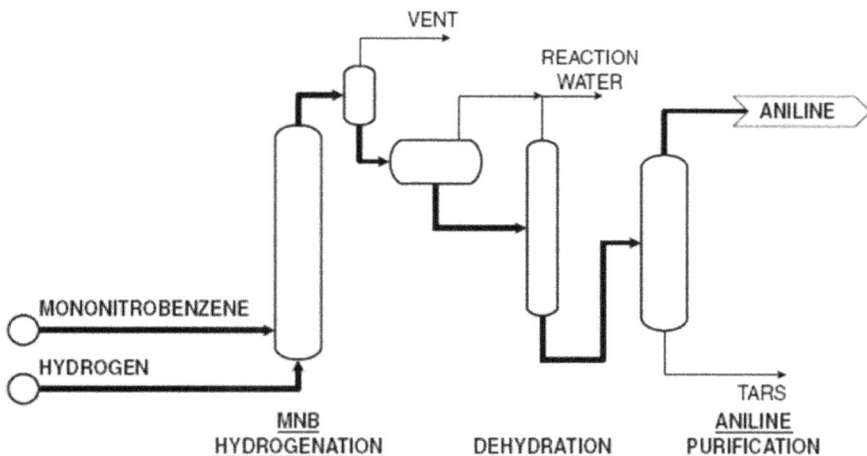

Figure 10.13: Aniline production by liquid-phase nitrobenzene hydrogenation (https://www.acces sengineeringlibrary.com/binary/mheaeworks/d49066840171883e/0de7ed764b2f478adb59 f5119277a23d1b615d00bbffc753f0a3f2dd0113b3a4/02_1x03.png).

Aniline produced by liquid-phase hydrogenation gives constant high product purity without the variations originating from changes in catalyst activity typical of vapor-phase fixed-bed hydrogenation. In the latter case (Lonza process), a fixed bed of copper on pumice is used at ca. 215 °C and hydrogen to nitrobenzene molar ratio of ca. 100:1 at the reactor inlet. Because of activity decline by coking regeneration after ca. 6 months is required. The total catalyst lifetime can be up to 5 years. In another technology developed by Bayer, Pd on alumina with vanadium and lead as activity modifiers is used in fixed-bed reactors at 100–700 kPa and an inlet temperature of 250–350 °C. Because of the adiabatic temperature rise, the maximum outlet temperature can be 460 °C. A vapor-phase, fluidized-bed hydrogenation was developed by BASF utilizing copper (ca. 15 wt%) on a silica support, promoted with chromium, zinc, and barium, as a catalyst. The operation conditions are 250–300 °C and 400–1,000 kPa under hydrogen excess.

The liquid-phase process with a highly selective catalyst affording complete conversion has a very simple product purification system and does not require large hydrogenation gas recycle systems, typical of vapor-phase technologies. Moreover, contrary to vapor-phase fluidized-bed or fixed-bed technologies, parallel reactor trains or multiple reactor stages are not required. Finally, since shutdown for catalyst regeneration is not required, the on-stream time of the liquid-phase nitrobenzene hydrogenation plant is very high.

10.4.2 Liquid-phase C5+ olefins hydrogenation

Gas-phase and liquid-phase hydrogenation is often used as a means to treat various streams from steam cracking. A simplified process diagram of such hydrogenation units shown in Figure 10.14 illustrates that an adiabatic fixed-bed reactor is used. Because of exothermicity, the adiabatic temperature rise should be moderate, thus recycling of the effluent and utilization of cooling water is required. Typical inlet temperatures are in the range 40 °C to 100 °C depending on the feed and presence of impurities, which deteriorate catalyst activity.

Such activity decline over time due to coking is at least partially compensated by the temperature or hydrogen feeding increase. At some point, this approach is no longer feasible as, for example, temperature increase impairs selectivity requiring removal of coke by in situ or ex situ regeneration. Hot hydrogen stripping or oxidative regeneration are available for restoration of catalyst activity. Regenerations are performed after several months of service and thus to ensure uninterrupted operation a spare reactor is available.

Downflow and upflow modes of operation are practiced industrially. The hydrogen supply and the process pressure are carefully controlled for selective hydrogenation of dienes and alkynes to avoid over-hydrogenation to alkanes, while for complete hydrogenation, the limitations are less stringent. Hydrogenation of hydrocarbons is an exothermal reaction thus hot spots and potential thermal runaways are monitored with a possibility to automatically shut down the feed and the hydrogen supply in case of excessive temperatures.

Selective hydrogenation of C5+ streams is used to hydrogenate polyunsaturated C5+ hydrocarbons. Conversion of dienes and selectivity to the different olefins depends on the type of olefins and are different for linear, branched, and cyclic dienes, etc. Branched C5 olefins undergo hydroisomerization of, e.g., 3-methyl-1-butene to 2-methyl-butenes.

One technological option (Axens) is presented in Figure 10.15.

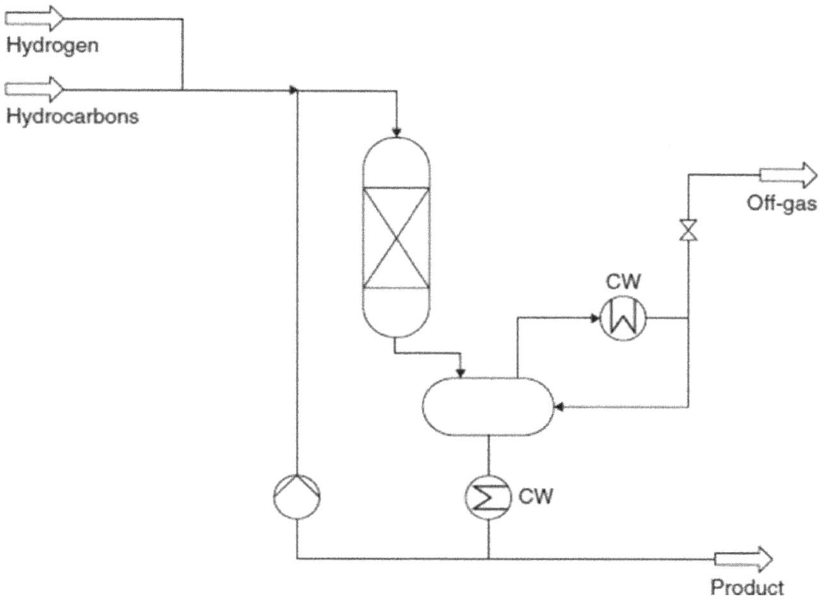

Figure 10.14: A typical flow scheme for selective hydrogenation (from https://www.accessengineer inglibrary.com/highwire/markup/item_fulltext/4922686).

In the first stage, diolefins and styrenics are selectively hydrogenated in a fixed-bed reactor over Pd or Ni based catalysts (e.g., Pd/Al$_2$O$_3$) while minimizing formation of heavy products by polymerization. In the second stage of pygas hydrogenation (Figure 10.16), the C$_6$–C$_8$ cut is further processed by selectively hydrogenating the olefins and performing hydrodesulphurization, at the same time minimizing hydrogenation of aromatics, which are recovered downstream.

10.5 Hydrotreating

Hydrotreating is a catalytic process mainly aimed at the removal of unwanted components such as sulphur (hydrodesulphurization, HDS) to form hydrogen sulphide. Typically, Co-Mo, Ni-Mo, and Ni-W sulphide catalysts supported on refractory oxides such as γ-alumina are used with hydrogen pressure and temperature varying between 0.5 and 4 and even 8 MPa and 300–400 °C depending on the feedstock. For example, heavy residues require the highest temperature and hydrogen pressure. Typically, catalysts in the form of trilobe or pentalobe extrudates or hollow cylinders with the diameters of 1.5–2 mm and lengths of 3–5 mm length are used in oil refining (Figure 10.17) with extended external surface area. Catalysts are typically loaded

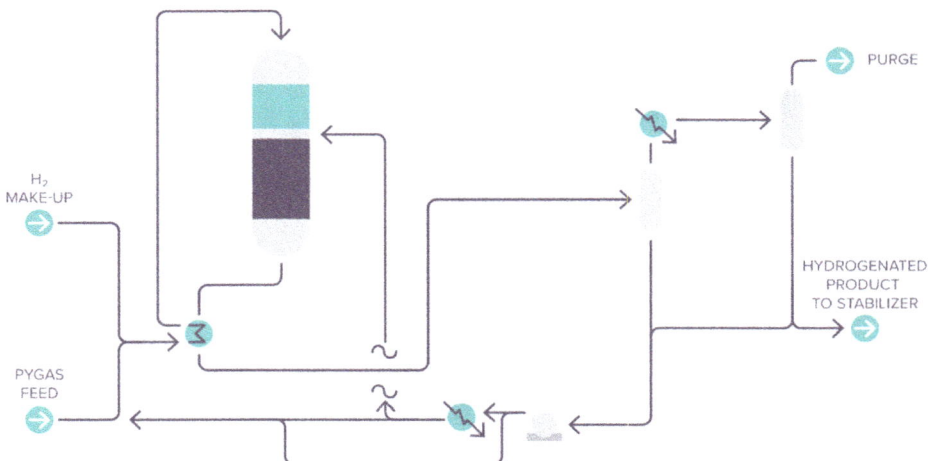

Figure 10.15: First stage of pyrolysis gasoline hydrogenation (from https://www.axens.net/mar kets/petrochemicals/olefins-selective-hydrogenation).

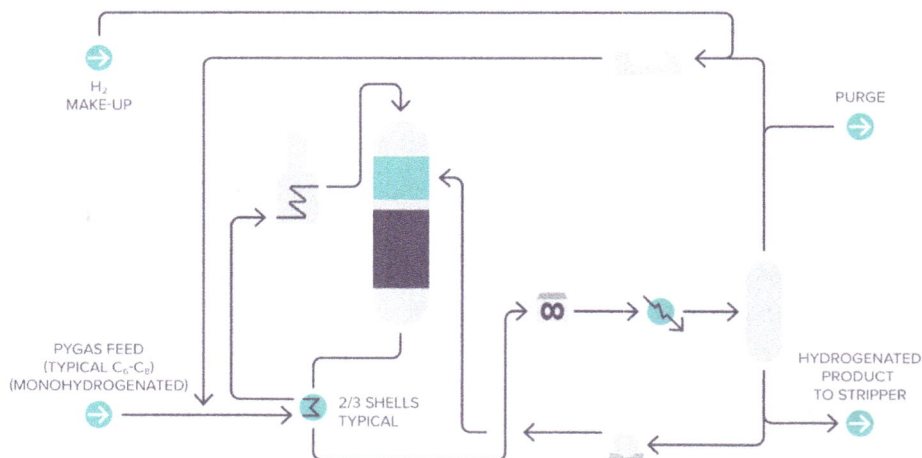

Figure 10.16: Second stage of pyrolysis gasoline hydrogenation (from https://www.axens.net/ sites/default/files/2021-05/owls_typicalghu-2_processscheme_.svg).

into the reactor in the form of oxides, followed by sulfidation, transforming oxides into sulphides.

Catalyst deactivation due to coking requires periodic regeneration of the catalyst, which is carried out ex situ taking into account the hazardous nature of sulfur in the coke and the catalyst. Subsequently, refiners often have two loads of catalyst; one in a hydrotreating reactor while another one is in the regeneration factory.

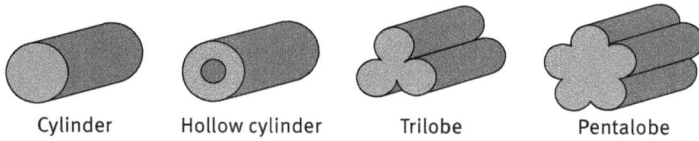

Cylinder   Hollow cylinder   Trilobe   Pentalobe

Figure 10.17: Shapes of hydrotreating catalysts.

The vapor-phase process for hydrotreating of gasoline and the liquid-phase process in trickle-bed reactors for hydrotreating and desulphurizing of straight-run middle distillates or vacuum gas oils (Figure 10.18). The major concern in trickle-bed reactors is uneven distribution (maldistribution) of the feed over the cross section and only partial wetting of the catalyst, leading to dry regions, which might lead to undesired coking.

The process schemes of both process types are similar, starting from mixing the feed with hydrogen-rich gas (fresh gas from catalytic reforming as well recycled gas from hydrotreating process) and heating up prior to introduction to a multibed adiabatic reactor.

Reaction exothermicity calls for the heat removal, which is done by quenching with cold hydrogen. Multibed reactors for hydrotreating of middle distillates can also contain not only HDS but also isomerization catalysts.

The main operating variables in hydrotreating process are the feed quality, temperature, hydrogen pressure, and residence time or space velocity. Removal of sulfur and nitrogen compounds is favored at higher hydrogen partial pressures diminishing also coke formation. Temperature increase is beneficial for kinetics leading at the same time to thermal cracking and coke formation. Finally, high space velocity results in low conversion, low hydrogen consumption, and low coke formation.

The average temperature of the reactor at the start of a run depends on feedstock as mentioned above and is generally between 330 °C and 370 °C. With time-on-stream catalysts deactivate due to coking. Decline of activity is compensated by temperature increase, which obviously should be controlled to avoid excessive feedstock cracking with even more severe deactivation. At the end of the run, temperature can be 380–400 °C and even 410 °C. At the end of a run, the catalyst is regenerated, with the start-of-the-run temperature gradually increasing in successive cycles. This also means that end-of-the-run temperatures are increasing, and at some point, when a maximum level in terms of the catalyst performance and/or process economics is reached, the catalyst is replaced.

Hydrogen partial pressure depends on the feedstock, with lighter feeds requiring lower hydrogen pressure. Higher hydrogen pressure, in fact, is beneficial from the viewpoint of catalyst stability by diminishing deactivation. The hydrogen/hydrocarbon ratio should be thus selected to be large enough to prevent deactivation not only by suppressing coking but also by diminishing partial pressure of H_2S,

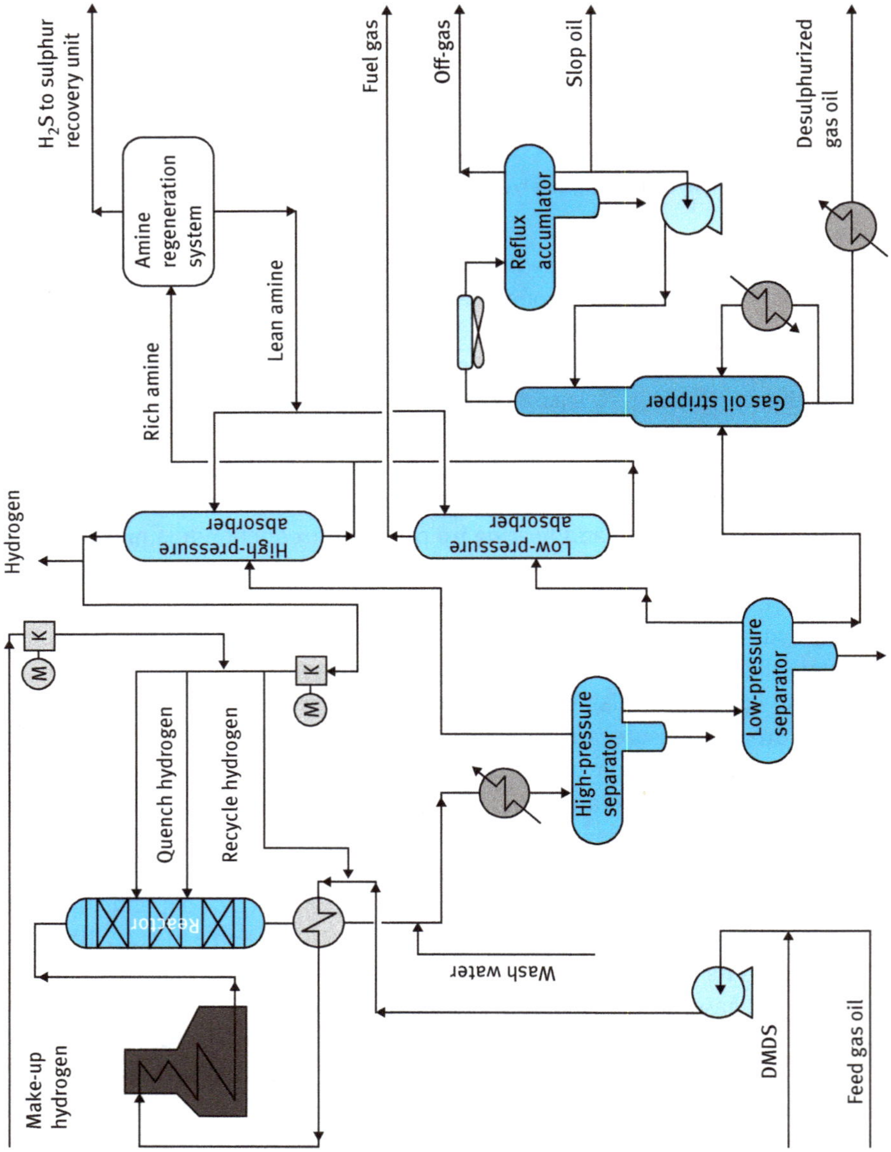

Figure 10.18: Flow scheme of HDS of middle distillates and gas oil. Modified after http://www.trec cani.it/portale/opencms/handle404?exporturi=/export/sites/default/Portale/sito/altre_aree/Tec nologia_e_Scienze_applicate/enciclopedia/inglese/inglese_vol_2/113-136ING.pdf.

which has a considerable inhibiting effect on desulphurization. Too high hydrogen/hydrocarbon ratio would result in very large gas quantities to be recycled. Deep desulphurization requires purification of the recycled gas from H_2S by, for example, amine washing. A small quantity of hydrogen sulphide is needed to maintain the catalyst in the sulfidic form.

Downstream the reactor, the product is cooled and further separated in two consecutive high- and low-pressure separators into a gas and a liquid phase. Hydrogen-containing gas from the high-pressure separator is purified first from H_2S by amine washing and then recycled. Hydrogen sulphide should be also removed by the same method from the low-pressure separator off-gas and then converted into elemental sulphur by a Claus process:

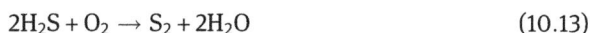

$$2H_2S + O_2 \rightarrow S_2 + 2H_2O \tag{10.13}$$

The liquid stream from low-pressure separator is stripped with steam giving desulphurized product at the column bottom.

Hydrotreating in oil refineries is an efficient process allowing, for example, to achieve 10 ppm of sulphur in the treated feed starting, for example, with ca. 7,500 ppm of S and 18% polyaromatics in a combined feed containing straight-run middle distillates and gas oil.

10.6 Dehydrogenation

Dehydrogenation, being opposite to hydrogenation discussed above, is highly endothermic, meaning that if run in adiabatic mode, a strong temperature decrease will be observed. For example, an adiabatic temperature decrease of 200 °C just for 25% conversion in dehydrogenation of propane is too high for a single adiabatic reactor.

From the thermodynamic point of view, it is necessary to operate dehydrogenation processes at low pressure, which can be achieved by either using lower pressure or a diluent. For example, steam can be used as a diluent, at the same time, preventing too strong catalyst deactivation. If it is not sufficient and the catalyst gradually deactivates, a periodic regeneration of the catalyst might be necessary, normally by burning coke in air.

Typically, group VIII metals (alumina-supported platinum promoted with tin and supported on alumina-chromia) are used for dehydrogenation of light alkanes (C2–C5) to olefins. The former catalyst is applied for dehydrogenation of long-chain alkanes. Supported iron oxides with promoters are utilized for synthesis of styrene by ethylbenzene hydrogenation, while copper is applied in dehydrogenation of alcohols to aldehydes.

10.6.1 Dehydrogenation of light alkanes

Conversion in dehydrogenation of light alkanes is limited by thermodynamics, increasing with T increase (Figure 10.19).

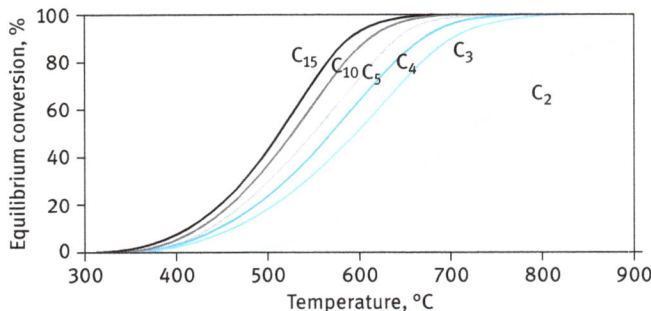

Figure 10.19: Equilibrium conversion in light alkane dehydrogenation depending on temperature.

Side reactions (cracking, oligomerization, alkylation, aromatization) have higher activation energy (more prominent with T increase) and are largely consecutive. The number of moles in dehydrogenation increases; thus, lower pressure is needed to shift equilibrium. The heat to the process streams is supplied either directly for tubular reactors in fired furnaces or adiabatic reactors with interstage heating or indirectly by using catalyst as a heat storage, for example, circulating the catalyst.

An industrially important process among dehydrogenation of alkanes is dehydrogenation of isobutane, which constitutes the basis for similar processes such as propane dehydrogenation.

Two groups of catalysts for alkane dehydrogenation based, respectively, on Pt and Cr differ more in the formation of by-products and approach to regeneration than in activity and selectivity. In PtSn/alumina catalysts, tin is added as a promoter to improve activity, selectivity, and stability by neutralizing support acidity and diminishing coke formation. Chromium-based catalysts contain thermodynamically stable chromia (Cr_2O_3) on transition aluminas ($\delta - \theta$). The active phase is promoted with alkali metals (K) decreasing the acidity. Irreversible deactivation of the catalyst is related to the formation of a-alumina- chromia (a -Al_2O_3-Cr_2O_3), which is catalytically inactive. Besides phase transformations, sintering and volatilization of the catalyst components can also happen, restricting the catalyst lifetime for both catalyst types to few years.

A periodical regeneration with air is mandatory for both catalysts to burn off the coke, which is done either by alternating the gas atmosphere (switching from hydrocarbon stream to air and/or steam) or by circulating the catalyst between a reactor and a regenerator.

In Catofin technology (Figure 10.20) operated in adiabatic fixed-bed reactors with chromium oxide/alumina catalyst, the heat is stored in the catalyst during the

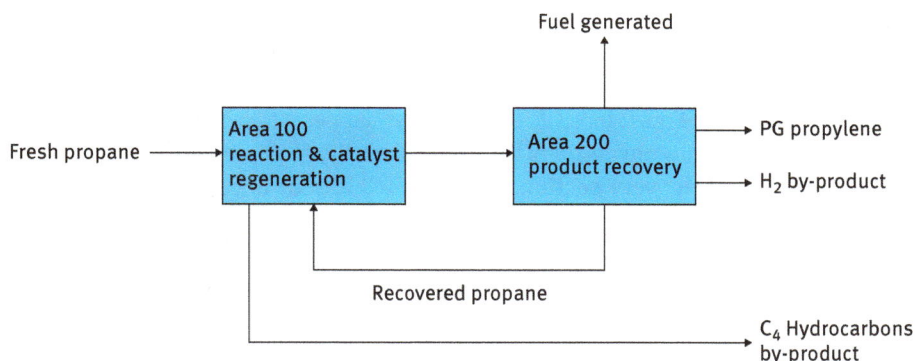

Figure 10.20: Catofin propane dehydrogenation process. http://base.intratec.us/home/chemical-processes/propylene/propylene-from-propane-via-dehydrogenation-2.

regeneration step and released during the reaction period. Such technology with re-actors operating under vacuum (0.01–0.07 MPa) switching from reaction, purging, and regeneration (heating) can be organized in a continuous operation using several (3–8) parallel reactors, which are alternatively on stream for reaction and off-stream for regeneration. The reaction temperature is 860–920 K, giving conversion of propane of 48–65% and selectivity to propene of 82–87%.

In the Oleflex process (Figure 10.21), commercialized by UOP, Pt/Sn on the alumina catalyst is used in adiabatic, mobile beds in series with interstage reheating in fired furnaces. Hydrogen is recycled, helping in the reduction of coke formation. The catalyst flows slowly, pulled by gravity through the reaction system, while the reagents

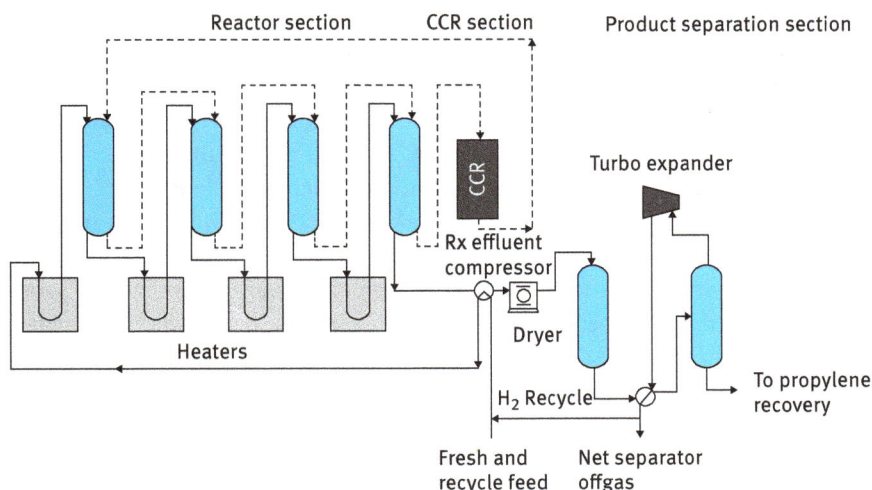

Figure 10.21: UOP Oleflex Process for propane dehydrogenation. © 2004 UOP LLC.

flow radially within the various reactors. The catalyst is collected on the bottom of the last reactor, transported pneumatically to the regenerator, and then sent back into the first reaction zone. The process is conceptually similar to the continuous regenerative reforming process developed by UOP. The reaction temperature is 820–890 K, and the process affords propane conversion of 25% and selectivity to propene of ca. 90%.

In the product recovery section, the reactor effluent is cooled, compressed, dried, and sent to a cryogenic system to separate hydrogen (85–93 mol% purity) from hydrocarbon. After separation, the liquid is first selectively hydrogenated to remove first diolefins and acetylenes and thereafter separated from ethane and further from propane, which is recycled to the reactor section.

In the steam-active reforming (STAR) process by Krupp Uhde (developed by Phillips), Pt/Sn on Mg/Zn aluminates catalyst is used in tubular, fixed-bed reactors (Figure 10.22).

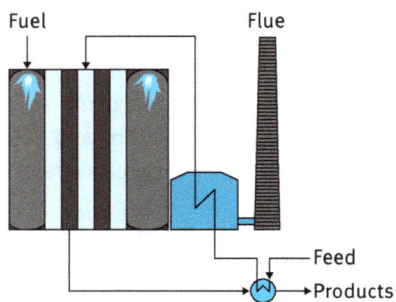

Figure 10.22: Steam-active reforming for dehydrogenation of isobutane. Modified after http://www.treccani.it/export/sites/default/Portale/sito/altre_aree/Tecnologia_e_Scienze_applicate/enciclopedia/inglese/inglese_vol_2/687-700_ING3.pdf.

The heat supply is arranged in the same way as in methane steam-reforming by burning fuel in a fired furnace. The feed is diluted with steam, and the total operating pressure is usually below 0.5 MPa. Catalyst regeneration is done by operating several reactors in parallel. The supply of steam (used in excess with a molar ratio between 3.5 and 4.2 with respect to the hydrocarbon) prevents an excessive decrease in temperature, which is detrimental for the reaction. Moreover, it inhibits the side reactions, which otherwise lead to coke deposition. The reaction cycle is ca. 7 h, while regeneration requires 1 h. The reaction temperature is 750–890 K, giving a propane conversion 30–40% with selectivity to propene 80–90%.

A new development in this technology resulting in increasing conversion of paraffin was the introduction of a secondary adiabatic reactor for selectively burning hydrogen from the first reactor with O_2 (90% purity) supplied in a molar ratio of between 0.08 and 0.16 with respect to hydrocarbon supply. The starting temperature to the main reactor is about 880 K, with an exit temperature ranging between 920 and 940 K, while the exit temperature for the oxy-dehydrogenation reactor is about 950 K.

Propane dehydrogenation (PDH) technology by Linde-BASF–Statoil (Figure 10.23) originally applied chromia on alumina and lately Pt on magnesium aluminate.

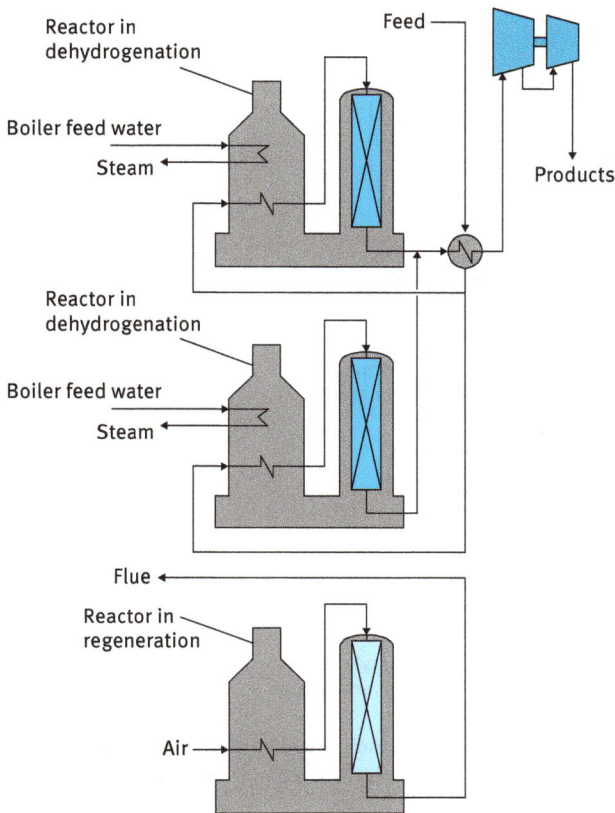

Figure 10.23: PDH technology for paraffin dehydrogenation. Modified after http://www.treccani.it/ export/sites/default/Portale/sito/altre_aree/Tecnologia_e_Scienze_applicate/enciclopedia/in glese/inglese_vol_2/687-700_ING3.pdf.

The reaction section includes three identical dehydrogenation reactors of the steam – reformer-type being similar in this sense to STAR process. Two reactors operate under dehydrogenation conditions, while the third reactor is regenerated by combusting coke with a steam/air mixture. One of the features of PDH technology different from the STAR process is the absence of reagent dilution with steam, leading to a reduction of the reactor dimensions and simplification of the product purification.

Fluidized-bed dehydrogenation (FBD) by Snamprogetti-Yarsintez (Figure 10.24) is similar to old FCC units with continuous catalyst circulation between a fluidized-bed reactor and a regenerator. As a catalyst, chromium oxide on alumina in a microspherical form with an average diameter of particles less than 0.1 mm is used.

The reaction heat is supplied by the heat capacity of the hot (>650 °C) catalyst. The spent catalyst at < 560 °C flows downward and is sent from the bottom of the reactor to

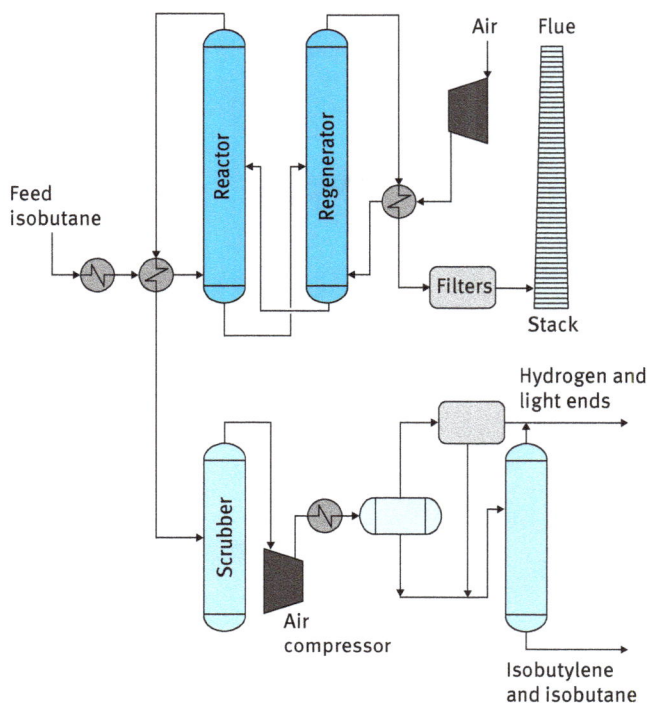

Figure 10.24: FBD process scheme for butane dehydrogenation. Modified after http://www.trec cani.it/export/sites/default/Portale/sito/altre_aree/Tecnologia_e_Scienze_applicate/enciclope dia/inglese/inglese_vol_2/687-700_ING3.pdf.

the top of the regenerator through a transfer line system (Figure 10.25). Both reactors operate in a countercurrent mode with respect to the flows of the gas and the solid.

The overall pressure in the reactor is 0.11–0.15 MPa. The conversion of propane is 40% with selectivity to propene 89%, while even higher conversion of butane (50%) is achieved, giving selectivity to isobutene of 91%.

Qualitative conversion-temperature profiles in isobutane dehydrogenation are given in Figure 10.26. Curve E represents the equilibrium conversion curve as a function of temperature. Curve A corresponds to a fired tubular reactor displaying an almost isothermal situation, while curve B illustrated a multibed reactor with the reaction heat supplied to gas in interstage heat exchangers. The thermal profile for curve C represents a situation when the catalyst quantity varies in the reactor and the solid is periodically heated. Due to such cyclic operation, several parallel reactors are needed for continuous operation. Introduction of baffles (Figure 10.27) limiting the internal mixing into a well-mixed and isothermal fluidized-bed reactor shifts the thermal profile to curve D. This makes a fluidized-bed operation similar to a tray distillation column. Comparison of the reactors shows that they have different average temperature. Higher

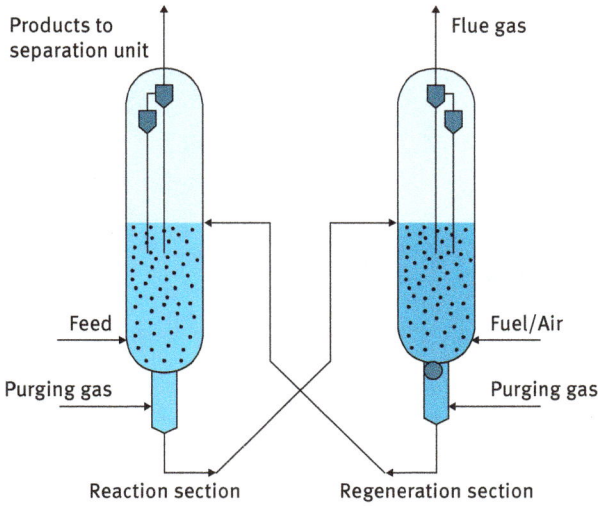

Figure 10.25: Reactor-regenerator system in butane dehydrogenation. http://www.pa.ismn.cnr.it/scuolagic2010/presentazioni_docenti/Sanfilippo.pdf.

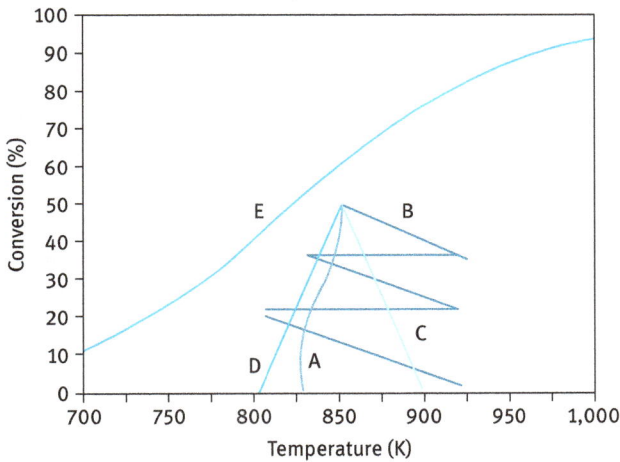

Figure 10.26: Qualitative profiles of reactor temperature conversion in isobutane dehydrogenation: (a) fired tubular reactor, almost isothermal – heat to gas through wall; (b) adiabatic multibed reactor – reaction heat supplied to gas in interstage heat exchangers; (c) adiabatic fixed-bed – reaction heat from solid heat capacity; (d) staged fluidized bed – reaction heat from countercurrently circulating catalyst; (e) equilibrium. Modified after http://www.treccani.it/export/sites/default/Portale/sito/altre_aree/Tecnologia_e_Scienze_applicate/enciclopedia/inglese/inglese_vol_2/687-700_ING3.pdf.

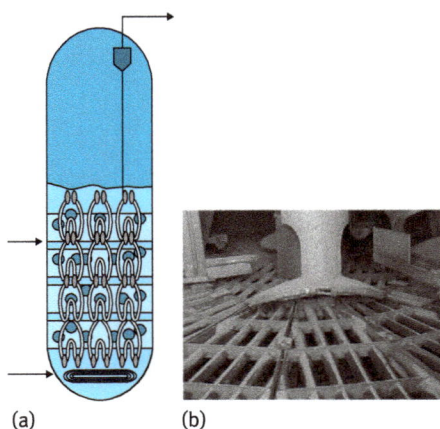

(a)   (b)

Figure 10.27: Operation of (a) fluidized-bed reactor with (b) baffles. http://www.pa.ismn.cnr.it/scuolagic2010/presentazioni_docenti/Sanfilippo.pdf.

values imply higher activity and low catalyst quantity, as well as the reactor volume, at the expense of lower selectivity toward the desired product, because a high temperature would promote secondary reactions, such as cracking.

Several advances in dehydrogenation of light alkanes have been implemented in industry in the recent years. The circulating fluidized-bed reactor (Figure 10.28) is in the heart of propane/butane dehydrogenation (ADHO) technology of the China University of Petroleum, industrialized in 2016 at Shandong Hengyuan Petrochemical Company Limited.

The catalyst is a non-noble metal oxide, which can operate with propane and butane separately or with their mixtures.

A circulating fluid bed is also applied in the fluidized catalytic dehydrogenation (FCDh) technology of Dow Chemical Company to produce propylene from the shale gas (Figure 10.29). This technology allows to achieve at the reactor outlet a pressure of 1.3–1.75 bar ca. 93 mol% selectivity to propylene at 45% propane conversion over

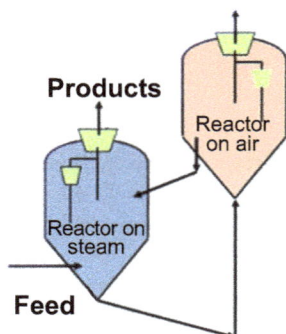

Figure 10.28: The circulating fluidized-bed reactor as a part of ADHO technology for propane/butane dehydrogenation. Reproduced with permission of Royal Society of Chemistry from S. Chen, X. Chang, G. Sun, T. Zhang, Y. Xu, Y. Wang, C. Pei, J. Gong, Propane dehydrogenation: catalyst development, new chemistry, and emerging technologies, *Chemical Society Reviews*, 2021, 50, 3315–3354.

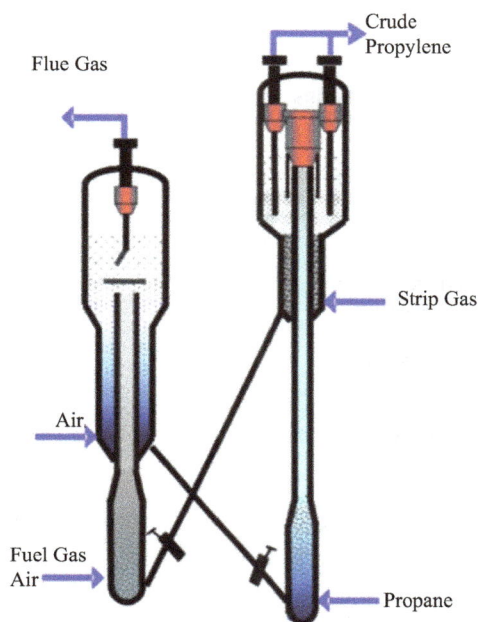

Figure 10.29: The reactor regeneration part of FCDh technology for propane dehydrogenation.

GaO$_x$ on alumina catalyst promoted with Pt and K. The energy demand per kilogram of propylene is diminished because of the higher propane conversion at moderate operating pressure. One of the specific features of the FCDh process technology is a short path of the deactivated catalysts between the dehydrogenation and regeneration units. The same technology can be applied not only for butane but also for ethylbenzene dehydrogenation to styrene which will be covered in the subsequent section.

The propane dehydrogenation process developed by KBR in 2019 relies on the Orthoflow reactor (Figure 10.30) with continuous catalyst regeneration developed originally by M.W. Kellogg Company for fluid catalytic cracking. As can be seen in Figure 10.30, the disengager, stripper, and regenerator vessels are combined in a single vessel. In the reaction section, fresh and recycled propane are fed into the reactor, operating at 1.5 bar, while the effluent gas flows into the compressing unit, low temperature section, and subsequent pressure swing adsorption and purification sections. The catalyst which is devoid of Pt and chromium is claimed to afford 87–90% selectivity at 45% conversion.

ORTHOFLOW™ FCC

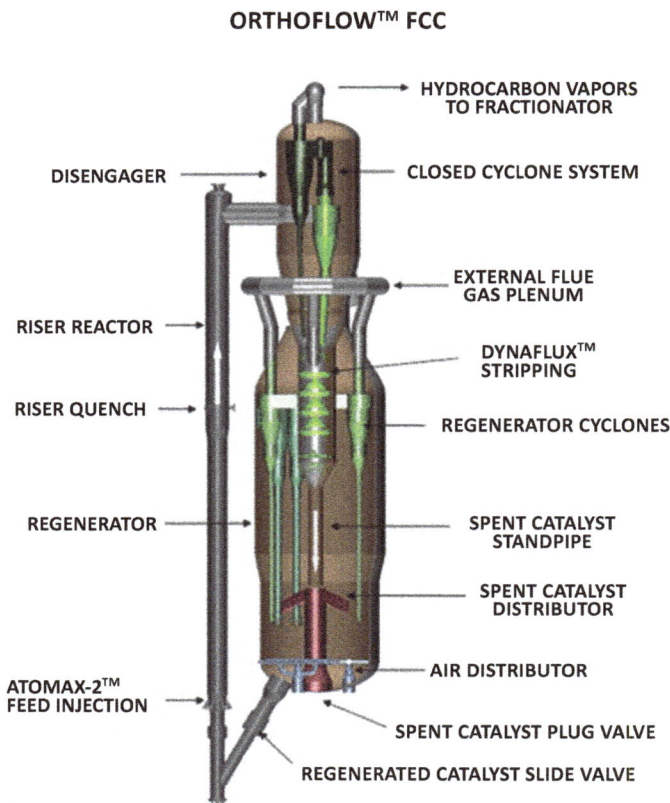

Figure 10.30: Orthoflow FCC reactor (https://www.gulfoilandgas.com/main/images/catalog/5107_P04.gif).

10.6.2 Dehydrogenation of ethylbenzene to styrene

Styrene is a monomer for the production of polystyrene and co-polymers, such as acrylonitrile-butadiene-styrene (ABS) and styrene-butadiene-rubber (SBR). The dominant process technology is synthesis through reversible, endothermic dehydrogenation of ethylbenzene:

$$C_6H_5CH_2CH_3 \leftrightarrow C_6H_5CH = CH_2 + H_2, \quad \Delta H(600°C) = 124.9 \text{ kJ/mol} \tag{10.14}$$

Among the side reactions are cracking of ethylbenzene to benzene

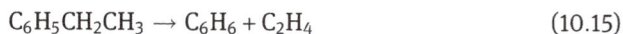

$$C_6H_5CH_2CH_3 \rightarrow C_6H_6 + C_2H_4 \tag{10.15}$$

and catalytic hydrogenolysis to toluene

$$C_6H_5CH = CH_2 + 2H_2 \rightarrow C_6H_5CH_3 + CH_4 \qquad (10.16)$$

Iron catalysts used for dehydrogenation of ethylbenzene typically contain ca. 84% of Fe_2O_3 and 2.5% Cr_2O_3. Promotion with potassium (ca. 13%) introduced as potassium carbonate aids in the gasification of carbon/coke with steam giving carbon dioxide.

Equilibrium conversion under typical conditions of ethylbenzene dehydrogenation (550–630 °C) is 80%, with most commercial units operating at 50–70 wt% conversion. The reaction temperature interval is selected to ensure meaningful reaction rate, while application of higher temperature results in cracking.

High endothermicity requires in the case of adiabatic processes reheating, therefore, ethylbenzene dehydrogenation is mainly done in several adiabatic reactors. Steam is also added (steam/ethylbenzene ratio, 12–17 mol/mol) to the feed in order to lower partial pressure of ethylbenzene shifting equilibrium and at the same time to diminish deactivation. Direct injection of superheated steam is used to supply the heat, which alternatively can be done also by indirect heat exchange. Different reactor arrangements are presented in Figure 10.31.

In Figure 10.31a, only superheated steam is added, resulting in conversion of ca. 40%. A concept with two reactors and reheating of the outlet of the first reactor is illustrated in Figure 10.31b. Conversion of ethylbenzene is typically 35% in the first reactor and 65% overall. Either a low positive pressure or even vacuum is applied to ensure more favorable thermodynamic conditions. In addition to adiabatic reactors, other options had been explored commercially. BASF has operated multitubular reactors with much lower steam to ethylbenzene mass ratio (1:1) than in adiabatic reactors. More expensive reactor design restricts the annual production capacity to ca. 150, 000 tons.

The overall process flow diagram of styrene production by ethylbenzene hydrogenation is presented in Figure 10.32.

After heating fresh and recycled ethylbenzene in heat exchangers 3 and 4 up to 520–570 °C, the feed enters the reactor (pos. 5) (or several reactors) along with generated in boiler (pos. 2) steam, which is heated to 700 °C. After the reactor, the product is cooled in heat exchangers 3 and 4 and the same boiler generating low-pressure steam. After cooling in heat exchanger 6 by water and separation from the gas (pos. 7), the liquid fraction is further separated into aqueous and organic phases (pos. 8). The organic fraction-crude styrene containing unreacted ethylbenzene, styrene, and by-products (benzene, toluene) is separated in a series of distillation columns. The steam condensate is recycled, and the gas containing hydrogen and carbon dioxide is first treated to recover aromatics and can be further used as a fuel or as a stream in a hydrogen plant.

A typical crude styrene from the dehydrogenation process consists mainly of 64% of styrene with boiling point of 145 °C and 32% of ethylbenzene with a somewhat lower boiling point, 136 °C. There are some minor impurities such as benzene, toluene, and some heavies. The separation of these components by distillation is straightforward, although the residence time at elevated temperature needs to be

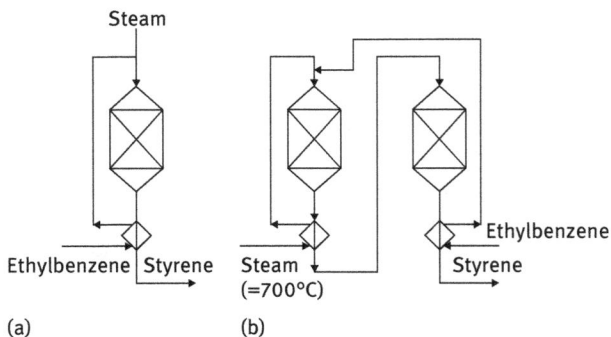

Figure 10.31: Different reactor arrangements for ethylbenzene dehydrogenation: (a) one- and (b) two-reactor systems.

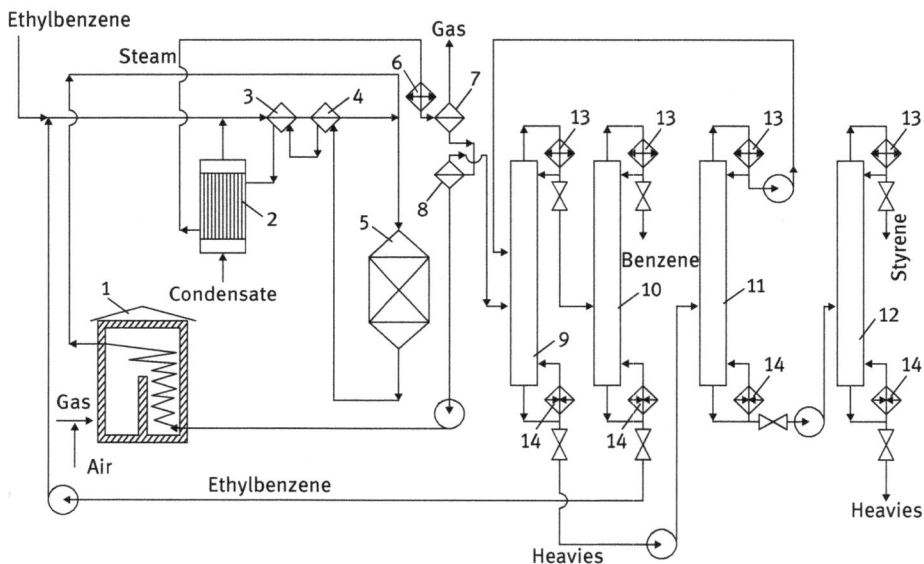

Figure 10.32: Ethylbenzene dehydrogenation to styrene production. 1, furnace; 2, boiler; 3 and 4, heat exchangers; 5, reactor; 6, cooler; 7 and 8, separators; 9–12 distillation columns; 13, reflux; 14, boiler. After N. N. Lebedev, *Chemistry and Technology of Basic Organic and Petrochemical Synthesis*, Chimia, 1988.

minimized to minimize styrene polymerization. A polymerization inhibitor (distillation inhibitor) is needed therefore throughout the distillation train. Aromatic compounds with amino, nitro, or hydroxy groups can be used. Moreover, temperature should be controlled; thus, distillation under vacuum (pos. 9) is done first.

The main part of ethylbenzene along with benzene and toluene is separated from the heavier fraction containing styrene, ethylbenzene, and some heavier compounds. In column 10, ethylbenzene at the bottom is separated from benzene and toluene and recycled. The higher boiling fraction from 9 is sent to distillation in column 11, operating under vacuum, where the remaining part of ethylbenzene is distilled away and routed back to the column 9. The bottom fraction is further distilled (pos. 12) into styrene, with 99% + purity as overhead. The heavies are styrene polymers.

The separation section can be arranged in a different way as realized in the Badger/ATOFINA process operating using potassium promoted iron catalyst with a cumulative annual production of 9 million tons. As can be seen in Figure 10.33, benzene and toluene are separated from styrene in the first distillation column downstream of the settling drum followed by separation of ethylbenzene and styrene. The former is recycled back to the reactor, while the latter is separated from the residues in the final column, which operates as all other columns below atmospheric pressure to prevent formation of polymers.

In the so-called SMART technology developed by Lummus/UOP and implemented in several plants worldwide with an annual overall capacity exceeding 1 million tons, the dehydrogenation section contains an extra reactor between the existing dehydrogenation reactors. This additional reactor is operating under oxidative dehydrogenation conditions in the presence of air (Figure 10.34).

The advantages of this technology are apparent as the equilibrium can be shifted more to ethylbenzene, resulting in a higher conversion per pass of up to 75%. This gain, however, comes together with the safety risks explaining a limited number of plants using this technology, which involves a high temperature mixture of oxygen and hydrogen.

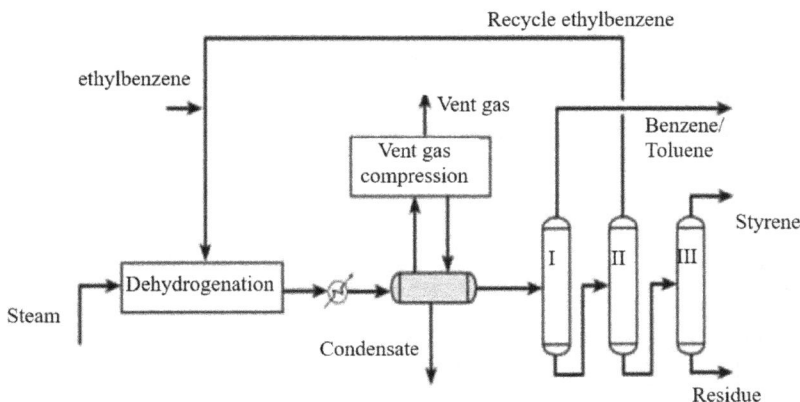

Figure 10.33: Process flow diagram of the Badger/ATOFINA styrene process. (I) benzene-toluene column, (II) ethylbenzene recycle column, (III) styrene distillation column.

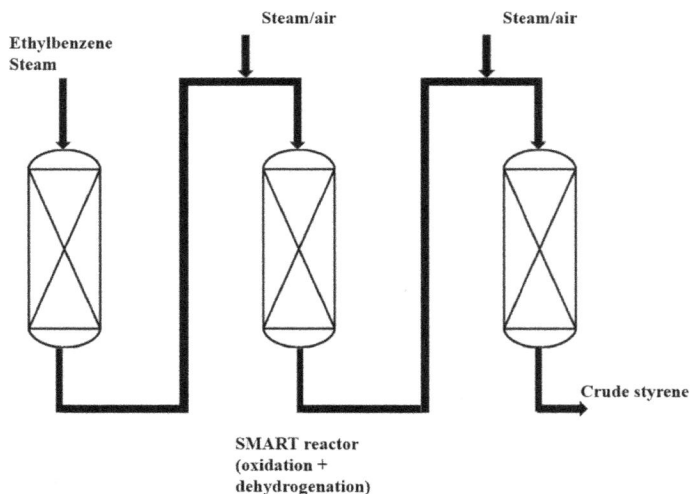

Figure 10.34: Dehydrogenation section of Lummus/UOP SMART process.

A novel development in styrene synthesis is the process developed by Snamprogetti and Dow, coined as the SNOW process (Figure 10.35), when ethane and ethylbenzene are used as a feed for dehydrogenation section. The SNOW process has been commercialized under the name Advanced Styrene Monomer technology and has been proven at a semicommercial or large-process demonstration unit scales.

Ethane and ethylbenzene are dehydrogenated in an FCC-type riser reactor with regenerator at 590–700 °C forming ethylene and styrene. The reactor section comprises of a typical riser as in FCC where the heat is supplied by the regenerated catalyst particles (Figure 10.36).

Higher temperatures are favorable from the thermodynamic and kinetic viewpoint, resulting in more severe deactivation. This is overcome in the process by a short residence time in the riser similar to FCC (1–5 s) combined with continuous regeneration. A special feature of the regenerator arrangement is that in addition to

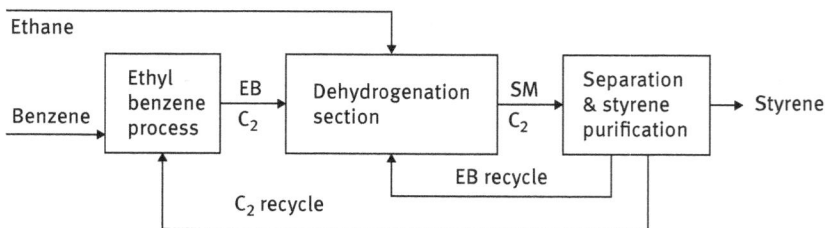

Figure 10.35: Block diagram for SNOW process. EB, ethylbenzene; SM, styrene monomer. From http://www.pa.ismn.cnr.it/scuolagic2010/presentazioni_docenti/Sanfilippo.pdf.

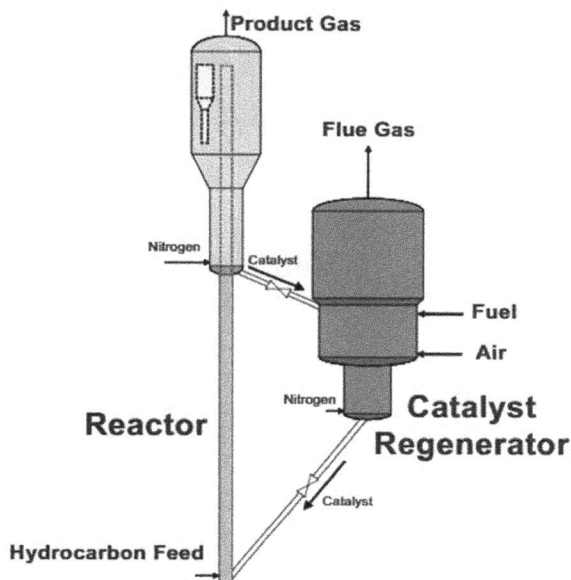

Figure 10.36: The reactor section of the SNOW process. From D. Sanfilippo, G. capone, A. Cipelli, R. Pierce, H. Clark, M.Pretz, SNOW: Styrene from ethane and benzene, Studies in Surface Science and Catalysis, 2007, 167, 505–510. Copyright © 2007 Elsevier B.V. Reproduced with permission.

heat generated by combustion of coke, extra fuel has to be burned in the regenerator to close the heat balance. For this purpose, hydrogen, generated in hydrogenation, can be used with an advantage of lowering overall carbon dioxide emissions.

Another advantage of the process is that there is no need to lower partial pressure of hydrocarbons for the purpose of improving selectivity or minimizing deactivation. Steam dilution is thus not applied. A platinum-promoted gallium oxide catalyst being active and selective in dehydrogenation of ethylbenzene is also active in dehydrogenation of ethane. Dehydrogenation of ethane and ethylbenzene is still limited by thermodynamics. Ethane is more difficult to dehydrogenate; thus, the ratio between the components in the reactor should be adjusted because ethane is then sent to a benzene alkylation section. Moreover, different residence times for both compounds are possible with separate feeding points. Similar to conventional processes, ethylbenzene (as well as unreacted ethane) is recycled.

Chapter 11
Reactions involving water: hydration, dehydration, etherification, hydrolysis, and esterification

11.1 Hydration and dehydration

Addition of water to olefins (hydration) follows the Markovnikov rule, leading to, for example, isopropanol and isobutanol from propene and butene-1, respectively.

$$CH_3CH = CH_2 + H_2O \rightarrow CH_3CH(OH)CH_3 \tag{11.1}$$

$$CH_3CH_2CH = CH_2 + H_2O \rightarrow CH_3CH_2CH(OH)CH_3 \tag{11.2}$$

Hydration of triple bonds in acetylene

$$CH \equiv CH \xrightarrow{+H_2O} CH_3 - CHO \tag{11.3}$$

or nitriles

$$RC \equiv N \xrightarrow{+H_2O} RCONH_2 \tag{11.4}$$

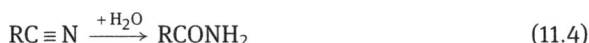

results in aldehydes and amides. All these reactions are reversible. Dehydration of alcohols, however, can be either intermolecular or intramolecular:

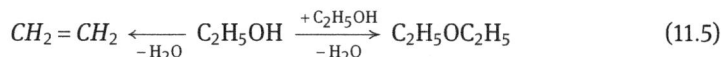

$$CH_2 = CH_2 \xleftarrow{-H_2O} C_2H_5OH \xrightarrow[-H_2O]{+C_2H_5OH} C_2H_5OC_2H_5 \tag{11.5}$$

Hydration of olefins is an exothermal reaction (ca. 50 kJ/mol); thus, it is favored at low temperature from the point of view of thermodynamics. Gibbs energy shows a minor dependence on the nature of olefin. The equilibrium can be also shifted by increasing pressure in accordance with Le Chatelier principle.

The first step in the reaction mechanism can be viewed as an attack of proton to the olefins. Hydration and dehydration reactions are examples of acid catalysis, requiring typically Brønsted acids (phosphoric, sulphuric, or heteropolyacids), applied as such or on supports. The rate of these reactions is determined by the stability of the intermediate carbenium ion (tertiary > secondary > primary). Therefore, hydration of isobutene proceeds at room temperature in the presence of low H^+ ion concentrations owing to the relative stability of the intermediate tertiary carbenium ion.

Pressure of ca. 7–8 MPa is required for ethylene hydration at 250–300 °C, giving conversion of 7–22%, which is still practical from the industrial implementation viewpoint. At high temperatures, there could be, however, prominent side reactions, such as oligomerization to higher olefins. This reaction has a higher activation energy compared to hydration. Higher olefins can also in turn undergo hydration.

https://doi.org/10.1515/9783110712551-011

As an example of a hydration reaction, direct gas-phase hydration of ethylene using acidic heterogeneous catalyst is considered in Figure 11.1. Typically, phosphoric acid (35 wt%) on a wide-pore silica gel is used. The catalyst is prepared by impregnating a support with phosphoric acid with a subsequent drying at 100 °C. Ethylene and deionized water (molar ratio range 1:0.3–1:0.8) are compressed (pos. 1, 2) and heated to 250–300 °C at 6–8 MPa by passage through a heat exchanger (pos. 4) and a superheater (pos. 3). Heat integration is important, since recovery of ethanol in this process with a limited conversion is done by condensation of ethanol-steam mixtures, where water is present in huge excess. Thus, ca. 95% of overall heat demand is used to generate steam, which is then condensed.

Conditions of the reaction result in 8–10% equilibrium conversion of ethylene. In practice, conversion level is ca. 4%. Since conversion is rather low and the reaction is not very exothermal, there are no special measures with respect to heat removal in the reactor *per se* and a simple adiabatic reactor can be used. Heat integration is done by heating the feed with the reaction product. Olefin purity is typically 97–99%, with such impurities as ethane, methane, and hydrogen. In order to avoid buildup of impurities, the recycling loop also contains purge.

The reactors could have a diameter of 1 m and height of 10 m. The internals of the reactor should be lined with copper to prevent corrosion by phosphoric acid. Heat exchangers and connected pipes are either made of copper or also lined with it for the same reason.

Neutralization of phosphoric acid entrained by the gas stream is done by injecting a dilute solution of NaOH downstream of the reactor. Because of such partial entrainment of phosphoric acid, the catalyst loses its activity after 400–500 h. The catalyst activity could be extended to 1,500 h using such preventive measures as replacing the acid continuously or periodically by spraying it on the catalyst bed. It addition to entrainment, catalyst activity declines due to formation of coke.

After cooling (pos. 4) and condensation (pos. 7), the gas is separated from the liquid (pos. 8). Since some amount of ethanol is still present in the gas, absorption with water (pos. 9) is used. The recycled gas is recompressed and sent back to the reactor. The liquid, after passing through the high-pressure separator (pos. 8), is decompressed (pos. 15) and further separated from the remaining gas in a low-pressure separator (pos. 10). After condensation, the concentration of ethanol is ca. 15 wt%. Selectivity to ethanol is ca. 95%; thus, this solution also contains diethylether (selectivity 2–3%), acetaldehyde (1–2%), and oligomers (1–2%). Presence of acetaldehyde is undesirable in the final product; thus, it can be hydrogenated even prior to distillation. Purification can be done in two distillation columns (pos. 11, 12), where in the first one the lights, e.g., diethylether, and acetaldehyde (if there is no upstream hydrogenation) are taken as a top fraction. Diethylether could be also recycled back to the reactor. The bottom fraction is distilled in column 12 with a direct steam injection, giving ethanol-water azeotrope (95% ethanol). The bottom fraction

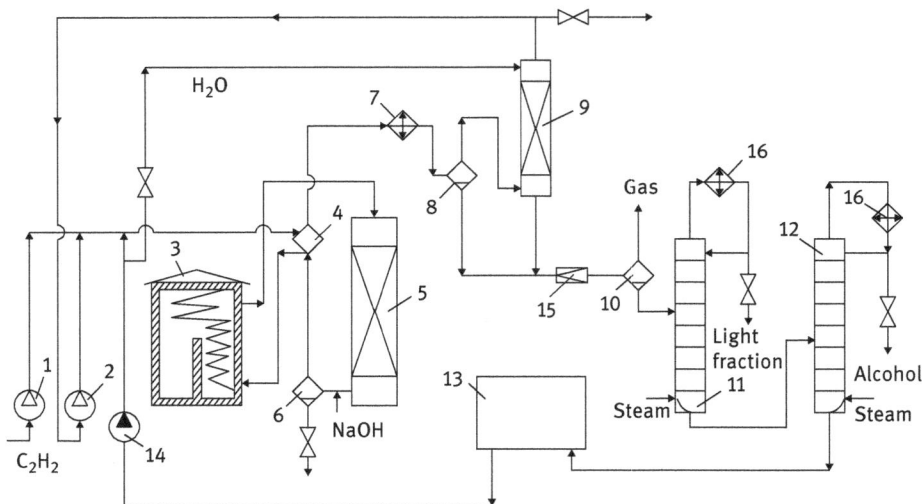

Figure 11.1: Ethanol production by gas-phase ethylene hydration. 1 and 2, compressors; 3, furnace; 4, heat exchanger; 5, reactors; 6, salt removal; 7, heat exchanger; 8 and 10, separators; 9, absorber; 11 and 12, distillation columns; 13, ion exchange unit for recycled water; 14, pump; 15, decompressor; 16, reflux condenser. After N. N. Lebedev, *Chemistry and Technology of Basic Organic and Petrochemical Synthesis*, Chimia, 1988.

containing water is removed from salts using ion exchange (pos. 13) and recycled back to the reactor.

Another example of a hydration reaction is related to synthesis of acrylamide which is a water-soluble monomer primarily consumed in the production of polyacrylamides and in a wide range of other application areas. The main route historically was hydration of acrylonitrile in the presence of sulphuric acid followed by separation of the product from its sulfate salt using, for example, base neutralization or ion exclusion. The reaction temperature was typically 90–100 °C. Such relatively high temperature along with a long residence time (1 h) resulted in formation of impurities, especially polymers and acrylic acid, influencing the properties of subsequent polymer products. Recovery of the acrylamide product was challenging and expensive. Application of, for example, ion exclusion (a sulfonic acid ion-exchange resin) for purification produced a dilute solution of acrylamide in water. Treatment of a dilute sulphuric acid waste stream was problematic increasing production costs.

In the 1970s a catalytic route was implemented using a fixed bed of copper catalyst at 85 °C, giving a solution of acrylamide in water at high conversion and selectivity to acrylamide. Several options for the hydration including utilization of Raney copper catalyst in both slurry and fixed-bed reactors have been implemented industrially. For example 50 wt% solution of acrylonitrile in water undergoes hydration in a slurry reactor with Raney copper at 120 °C with selectivity to the amide of nearly 100%.

The other alternative developed in 1985, by Nitto Chemical Industry is to perform hydrolysis with a nitrile hydratase at low temperatures, which is much more cost-effective and is a predominant option for new installations (Figure 11.2).

Figure 11.2: Enzyme-based production plant for acrylamide in Nanjing, China. From https://renewable-carbon.eu/news/media/2017/10/BioACMPlantOpeningNanjingOct23_Plant-300x200.jpg.

The most recent reports describe Rhodococcus rhodochrous bacteria as the catalyst for the hydration reaction

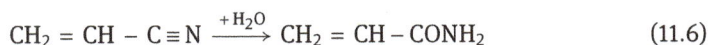

$$CH_2 = CH - C \equiv N \xrightarrow{+H_2O} CH_2 = CH - CONH_2 \qquad (11.6)$$

which is run at 0–20 °C and pH 7–9 in a fed-batch reactor for several hours giving almost complete conversion with very small amounts of by-products such as acrylic acid. The space time yield was reported to be 1,920 g/L/d. The heat of the reaction is 69–79 kJ/mole, which is similar to the heat of the side undesired reaction, i.e. formation of acrylic acid:

$$CH_2 = CH - C \equiv N \xrightarrow{+H_2O} CH_2 = CH - C(O)OH + NH_3 \qquad (11.7)$$

The flow scheme of a fed-batch process is shown in Figure 11.3. The blocks with the catalyst (i.e. *Rhodococcus rhodochrous*) are introduced in vessel 2 along with water up to 20–25% concentration. After homogenization for 45–60 min, the suspension is diluted to concentration 10–12% and fed to the main reactor 3 to which acrylamide is introduced at an average temperature of 20 °C. The same temperature is maintained in the reactor. Feeding of acrylonitrile is done typically in several portions. The downstream treatment includes averaging of the output (pos.5) and purification (pos.6) only for concentrated acrylamide solutions and final storage (pos.7).

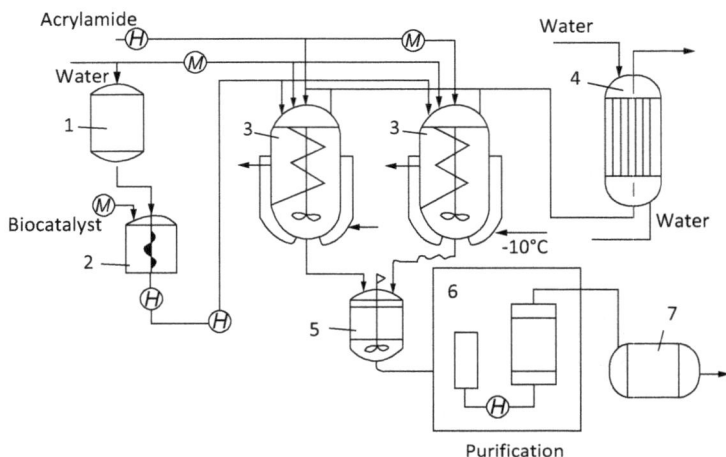

Figure 11.3: The fed-batch hydration of acrylonitrile (from Russian patent 2,112,804). 1-vessel for water, 2-vessel for preparation of biocatalyst, 3-reactors, 4-heat exchanger, 5-averaging reactor, 6-purification, 7-storage vessel.

The enzyme-based biocatalytic production method results in lower amounts of waste compared to a heterogeneous catalytic alternative, requiring also higher temperature and being thus more energy-intensive.

Dehydration reactions can be done either in the liquid or in the gas phases. The former approach can be used when the reactants are not stable enough at elevated temperatures needed for gas-phase dehydration. Examples of the liquid-phase processes are given in Figure 11.4.

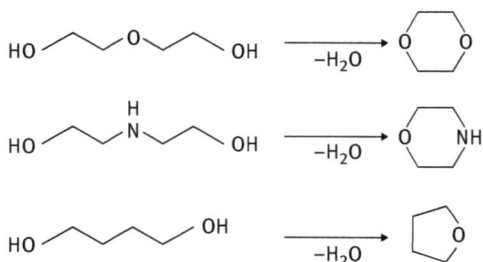

Figure 11.4: Examples of the liquid-phase dehydration reactions.

In the case of di-alcohol cyclization sulphuric acid, phosphoric acid, or K and Mg phosphates, etc. can be used as catalysts at temperatures between 100 and 160–200 °C.

The process can be organized by continuously distilling lower boiling point components (the target product, ether as such or an azeotrope with water). The reflux (Figure 11.5) is typically done to regulate the catalyst concentration. Liquid-phase dehydration can be also done in a multitubular reactor.

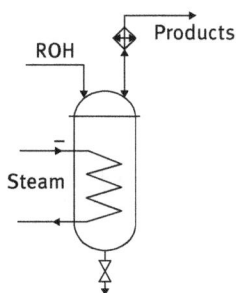

Figure 11.5: Liquid-phase dehydration with product reflux.

Gas-phase intramolecular dehydration can be used, for example, in the synthesis of isobutene from *tert*-butanol. Temperature can vary from 225 to 250 °C for diethylether synthesis to 700–750 °C in the case of ketene synthesis; dehydrating acetic acid with triethylphosphate as a catalyst under reduced pressure allows isolation of ketene prior to its reaction with acetic acid at 45–55 °C and low pressure (0.005–0.02 MPa):

$$H_2C = C = O + CH_3COOH \rightarrow (CH_3CO)_2O \qquad (11.8)$$

An example of intramolecular dehydration is production of ethylene from ethanol, which is generated from different type of biomass, including readily available non-food raw materials such as agricultural waste. There are several companies globally producing ethylene from bioethanol which capacities of single units ranging between ca. 60 and 200 kt/a.

The flowsheet of a typical bioethanol-to-ethylene process includes feedstock preparation, synthesis per se, and product purification (Figure 11.6).

After pre-evaporation the feed is dehydrated and the product is washed with water and subsequently with sodium hydroxide. Drying is followed by separation into light and heavy fractions by distillation giving polymer-grade ethylene. The operation conditions and the catalysts are summarized in Table 11.1. As can be seen from Table 11.1, the process is conducted in the gas phase at 200–450 °C giving high bioethanol conversion and product selectivity. Various types of catalysts can be applied requiring regeneration with a steam–air mixture after 1–12 months. Most of current ethanol-to-ethylene dehydration processes employ alumina catalysts and

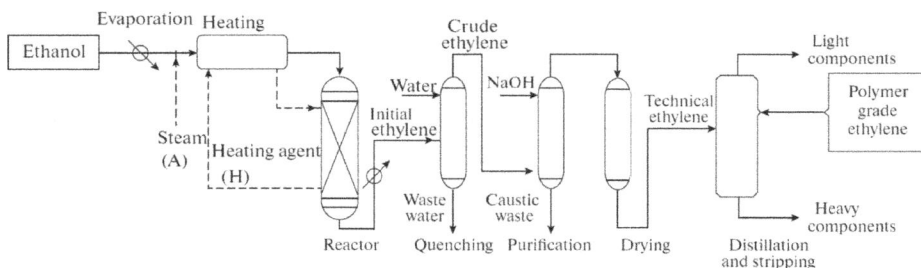

Figure 11.6: Flowsheet for ethylene production from bioethanol. From A. Morschbaker, Bio-ethanol based ethylene, *J. Macromol. Sci., Polym. Rev.*, 2009, 49, 79–84. Copyright Taylor & Francis Group. Reproduced with permission.

are carried out in tubular or adiabatic fixed-bed reactors. Among other catalysts a heteropoly acid is used by BP in the reactor operating at 160–270 °C and 1–45 bar. The unreacted ethanol in recirculated to the reactor. The zeolite catalyst NKC-03A (ZSM-5) was reported to be used in a tubular reactor.

Table 11.1: Ethanol dehydration processes. From I. S.Yakovleva, S. P. Banzaraktsaeva, E. V. Ovchinnikova, V. A. Chumachenko, L. A. Isupova, Catalytic dehydration of bioethanol to ethylene. *Catalysis in Industry*, 2016, *8*, 152–167. Reporduced with permission.

Company	Reactor	Catalyst	T, °C	P, atm	WHSV,h^{-1}	Ethanol, vol.%	Conversion/ Selectivity, %
Braskem	Adiabatic (4 laters)	Al_2O_3-MgO/SiO_2	450	–	0.56	95	99/97
Solvay Indupa	Adiabatic	Syndol	200–400	5.9–6.4	0.33–0.43	95	99/97
Lummus	Adiabatic (4 laters)	γ-Al_2O_3	400	0.66	–	50	99/95
Petrobras	Parallel adiabatic reactors	Al-Si (aluminosilicate)	300–440	0.84–7	0.03–0.7	–	99/98
China	Tubular	HZSM-5	260	–	2.4		98/99

Despite substantial efforts and apparent commercial success bioethylene production from bioethanol cannot replace the large-scale petrochemical route for ethylene manufacturing.

The last examples of dehydration are the methanol-to-gasoline (MTG) and methanol-to-olefins (MTO) processes (Figure 11.7).

$$2\,CH_3OH \xrightleftharpoons[+\,H_2O]{-\,H_2O} CH_3OCH_3 \longrightarrow C_{2\text{-}5}\ Olefins \longrightarrow \begin{array}{l} Higher\ Olefins \\ Naphthenes \\ Aromatics \\ Paraffins \end{array}$$

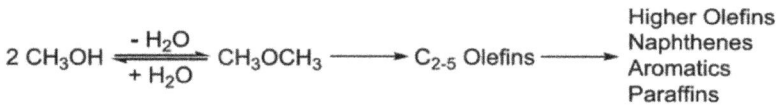

Figure 11.7: Dehydration of methanol with subsequent formation of olefins (MTO) and gasoline (MTG).

In MTG process an equilibrium mixture of methanol, dimethyl ether and water formed by dehydration of methanol over the acidic HZSM-5 is further converted to a mixture of olefins, aliphatics, and aromatics (<C10) at 350–400 °C and atmospheric pressure. The narrow range of hydrocarbons, high selectivity for isoparaffins, and aromatics of higher octane value along with slow catalyst deactivation are clear advantages of this technology.

The methanol-to-olefins technology with SAPO-34 as a catalyst gained a lot of attention since 2000 as a means to produce ethylene/propylene (Table 11.2).

As can be seen from Table 11.2, an acid catalyzed reaction with 100% methanol conversion occurs at ca. 400–550 °C and 0.1–0.5 MPa. The process flow diagram of the so-called D-MTO/D-MTO-II process operating with a fluidized catalytic reactor, a catalyst regenerator, and a separation unit apart from peripheral equipment (e.g. utilities, air compression units) is shown in Figure 11.8. The properties of SAPO-34 catalyst can be adjusted to change the ratio of ethylene and propylene from 0.6 to 1.3.
The fluidized-bed reactor needed for efficient removal of heat generated in a highly exothermal MTO reaction is combined with a regenerator as SAPO-34 catalyst experiences rapid catalyst deactivation.

In the second generation of DMTO process (DMTO-II), the by-products (C_{4+}) are separated from the products stream and further converted to ethene and propene in the fluidized-bed cracking reactor (Figure 11.9).

The current view on the reaction mechanism implies generation of the hydrocarbon pool (Figure 11.10) where $(CH_2)_n$ reflect the species adsorbed on the catalyst, which undergo successive methylation reactions and subsequent elimination of the side chain to produce light olefins. Methylbenzene, methylcyclopentadiene, and their corresponding carbenium ions are considered the active hydrocarbon pool species.

Such olefins as ethene, propene, and butene are further transformed by condensation, alkylation, cyclization, and hydrogen transfer reactions in secondary transformations resulting in higher olefins, alkanes, and aromatic hydrocarbons.

Another approach for the reaction section is adopted in the Lurgi MTP (methanol to propylene) process (Figure 11.11) where methanol is first pre-heated to 260 °C before entering the DME reactor. In the latter, 75% of the methanol feed is converted to DME and water over an acidic heterogeneous catalyst with unreacted methanol being the remaining part. After heating to 470 °C, the reaction mixture is fed into the MTP reactor with steam (0.75–2 mol steam/mol reaction mixture). Already in the first MTP

Table 11.2: Some industrial installations of methanol to olefins technology. From M. R. Gogate, Methanol-to-olefins process technology: current status and future perspectives, *Petroleum Science and Technology*, 2019, 37, 559–565. Copyright Taylor & Francis. Reproduced with permission.

No.	MTO technology	Licensor/operator	Scale	Catalyst	Process conditions (T, P, WHSV)	Catalyst performance (C, S, Y)
1	UOP/hydro advanced MTO (with olefin cracking)	UOP (Feluy, Belgium)	0.2 MM MTPA	SAPO-34 (attrition-resistant formulation)	400–550 °C, 1–4 atm	~100%, 80%+ C-selectivity to C2, C3, olefins
2	D-MTO (Dalian Institute of Chemical Physics)	Shenhua China Energy (Mongolia, China)	0.6 TPA	SAPO-34	400–550 °C, 4–5 atm	N/A
3	S-MTO (Sinopec MTO)	Sinopec	0.2 MM TPA	SAPO-34	400–550 °C, 1–5 atm	N/A
4	MTP (Lurgi)	Shenhua Group with Ningxia Provincial Govt. (Ningxia, China)	0.5 MM TPA	SAPO-34	N/A	N/A
5	MTP (Lurgi)	Datang Int'l Power with China Datang (Mongolia, China)	0.5 MM TPA	SAPO-34	N/A	N/A
6	Honeywell UOP	Jiangsu Sailboat Petrochemical company (Jiangsu Province, China)	0.8 MM TPA	SAPO-34	400–550 °C, 1–5 atm	~100%, ~85% to C2 + C3
7	Honeywell UOP	Wison China Energy (Nangxin Province, China)	0.3 MM TPA	SAPO-34	400–550 °C, 1–5 atm	~100%, 85% to C2 + C3

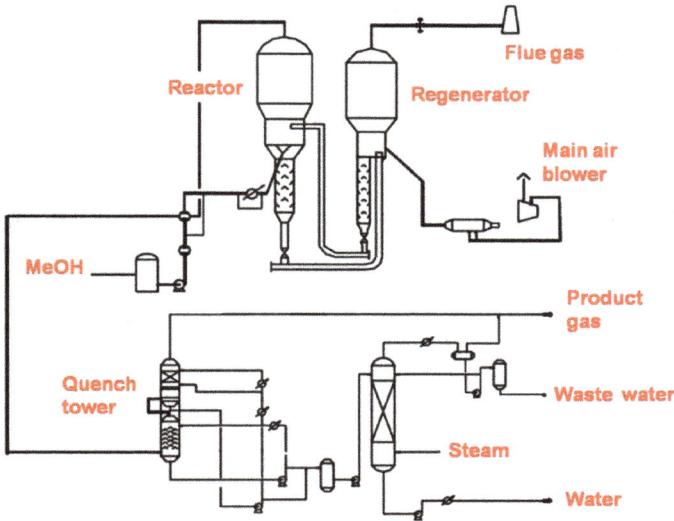

Figure 11.8: Flow diagram of DMTO process developed by DICP.

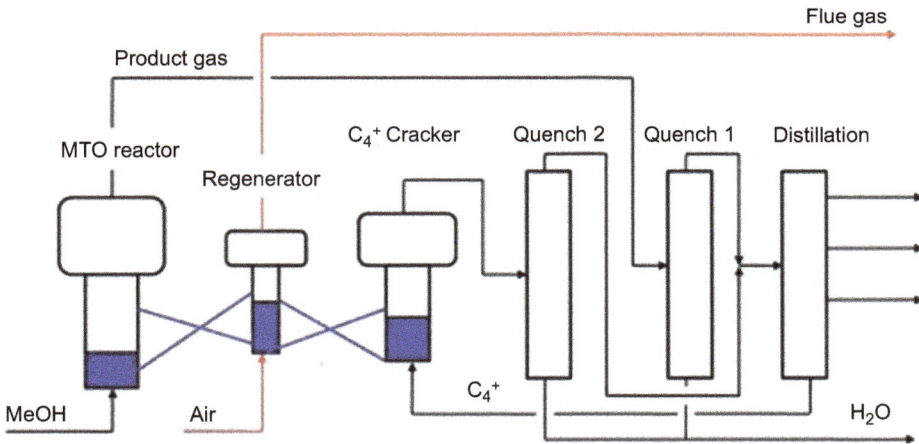

Figure 11.9: Flow scheme of DMTO-II process for conversion of methanol to olefins. https://ars.els-cdn.com/content/image/1-s2.0-S0065237715000095-f05-03-9780128038451.jpg.

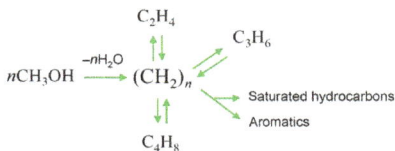

Figure 11.10: Hydrocarbon pool mechanism.

reactor more than 99% of the methanol/DME mixture is converted into propylene followed by reactions in downstream reactors increasing the propylene yield.

11.2 Hydrolysis

Acid-base-catalyzed hydrolysis (addition of water) includes such reactions as hydrolysis of amides

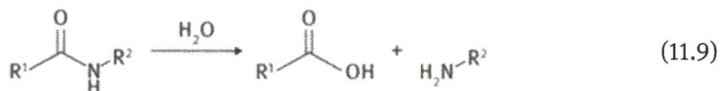

$$ \tag{11.9} $$

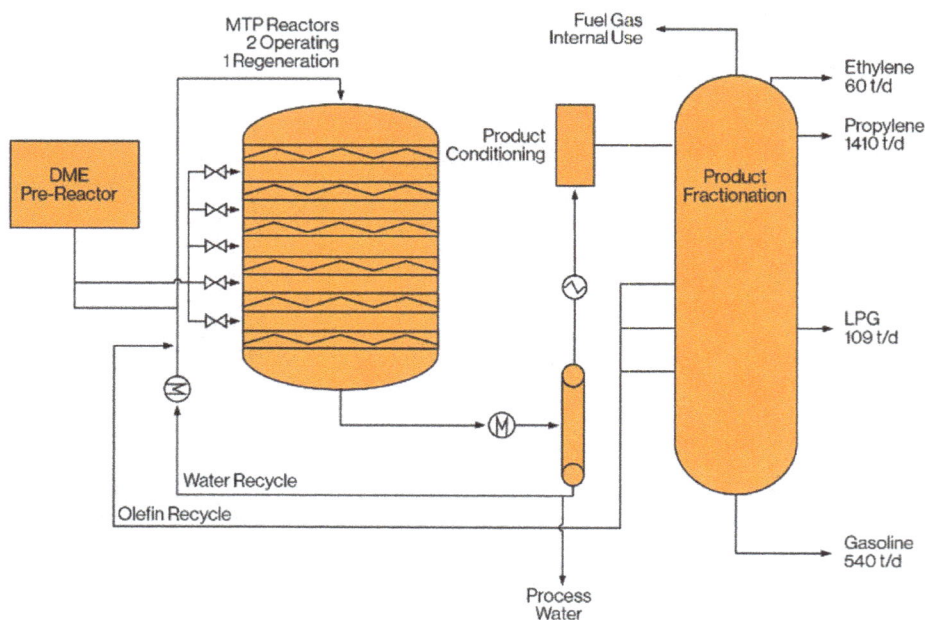

Figure 11.11: An overview of the Lurgi MTP process. https://www.engineering-airliquide.com/sites/activity_eandc/files/styles/938w/public/2018/07/11/lurgi-mtp-air-liquide-938_0.jpg?itok=icUHEYj5.

esters, triglycerides, or hydrolysis of glycosidic (C–O–C) bonds in oligosaccharides and polysaccharides.

$$(11.10)$$

11.2.1 Acid-catalyzed hydrolysis of wood

Such oligosaccharide and polysaccharide hydrolysis is the first step in synthesis of ethanol, where the second step is fermentation of hydrolytic sugars. Hydrolysis is typically done using diluted sulphuric acid at 120–190 °C and 0.6–1.2 MPa. As a result of the process, the product contains, besides glucose, also furfural, organic acids, and some other impurities. The reason of by-product formation is that, besides the main reaction, side reactions also occur, which are more prominent at elevated temperatures. As an example, a plant with a capacity 13.5 million liters of ethanol per year that used woody biomass as a feedstock (Figure 11.12) will be considered.

Of 18 hydrolysis reactors of 40-m^3 volume, each operating parallel in a batchwise mode, only 7 are in use at any time. Hydrolysis stats by loading 4.5 to 5 tonnes of wood chips and sawdust into the reactor, which corresponds to 11–12 m^3 of solid wood in its original solid state. First, 11.5 m^3 of diluted sulphuric acid (0.75 wt%) is added into the reactor (pos. 1) over a period of 20 min. The reactor is heated with steam for 40 min with purging and venting, finally reaching the pressure of 0.6 MPa. At this point, vertical percolation is commenced. The principle of the process is the addition of diluted sulphuric acid to the top of the reactor with a corresponding amount of liquid being withdrawn from the bottom of the reactor. The following is a typical sequence: add 13 m^3 of 0.5 wt% of sulphuric acid for 20 min at 0.7 MPa, 12 m^3 of 0.46 wt% of sulphuric acid for 20 min at 0.9 MPa, two 10 m^3 of 0.48 wt% of sulphuric acid for 20 min first at 1.2 MPa and then at 1.25 MPa; finally, at 1.25 MPa, add 10 m^3 of 0.32 wt% of sulphuric acid for another 20 min.

The reactor operating conditions such as pressure may vary, depending on the feedstock, namely which type of mix of coniferous and non-coniferous wood is applied. Higher content of non-coniferous wood requires slightly lower pressure. The percolation stops after 100 min of operation followed by removing the solids from the reactor to a cyclone tank (pos. 3). The amount of solid is ca. 8 tonnes, containing lignin at 60–80% moisture. The lignin residue was either dried to lower moisture content (ca. 40%) and burned or dumped. For other commercial purposes, further drying to ca. 5–10% moisture is needed, which is an expensive operation. Since there is accumulation of tarry caramel deposits in the bottom of the reactor and in the piping from the reactor, special cleaning (manual and with water) is done after

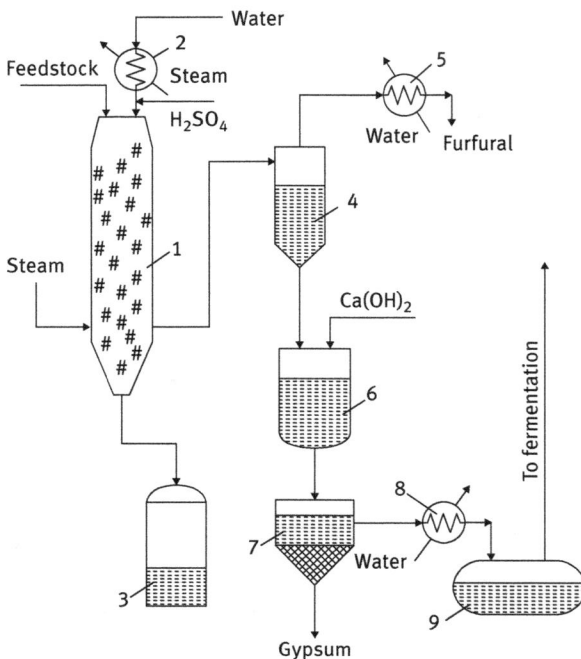

Figure 11.12: Flow scheme for hydrolysis of wood: 1, reactors; 2, heater; 3, cyclone tank (lignin collector); 4, evaporator; 5, condenser; 6, neutralization tank; 7, precipitator tanks; 8, vacuum cooler; 9, holding tank.

every 300 usage cycles. The acidic liquid drawn off from the reactor is sent to two flash tanks (each 16 m³), which operate in series, serving as evaporators (4). The liquid contains pentoses from the hydrolysis of the wood hemicellulose, some furfural, and dissolved, partially hydrolyzed cellulose. Due to the drop in pressure between the reactor and the tanks, some 10% of the liquid is flashed off as steam. The main part of furfural is removed by evaporation and in principle could be recovered. The extent of furfural removal is important since it is negatively influences subsequent fermentation if hydrolysate is used for the synthesis of ethanol.

The concentrated liquid from the second flash tank is pumped to a 1,000-m³-inverter tank (not shown), where it is held for 4.5 h at 105 °C, to complete the hydrolysis of the cellulose to glucose and other hexoses. The hydrolysate emerges from the inverter at about pH 1.6 and is sent to the neutralization tanks. There are seven neutralization tanks (pos. 6) (five with 34-m³ capacity and two with 100-m³ capacity) fitted with vertically mounted agitators. Typically, two smaller tanks are kept on standby. The hydrolysate enters the first neutralization tank, where it is mixed with milk of lime (a suspension of calcium hydroxide in water) to raise the pH to 3.2. This lime is made in a lime kiln, where calcium carbonate limestone is processed by coke firing. It is then transferred to a holding tank, where it is agitated, before going to a second neutralization tank. There, it is mixed with ammonium hydroxide to raise the pH to 3.8–4.0. Ammonium phosphate and a mixture of ammonium and potassium salts are added to provide nutrients for the subsequent fermentation. The liquid is then transferred to one of five shallow precipitator tanks (pos. 7), which have a capacity of 285 m^3. The solids from the bottom of the precipitator are composed of about 70% calcium sulfate, with the remainder being mostly tars, lignin, and some sand. After passing through a vacuum belt filter, the solids are separated and sent to lignin-waste dumpsite. This clarification is carried out to remove gypsum and lignohumates and improve the quality of the hydrolysate, reducing its toxicity for yeast. The supernatant liquid that is drawn off from the top of the precipitator at 80–100 °C goes first into a 40-m^3 holding tank (not shown), from where it is pumped through a four-stage vacuum cooler (pos. 8), giving 32–34 °C to a holding tank of 160-m^3 capacity (pos. 9). The main flow of hydrolysate (called also wort) is sent to fermentation.

An alternative to hydrolysis of diluted mineral acids, the enzymatic hydrolysis process has higher yields of glucose under moderate temperatures (40–50 °C) and atmospheric pressure. Long processing time, inhibition of enzymatic activity as well as high cost of enzymes prevent economically viable enzymatic hydrolysis of lignocellulosic biomasses.

11.2.2 Enzymatic hydrolysis of acyl-l-amino acids

As an example of application of enzymes, hydrolysis of acyl-l-amino acid (Figure 11.13) by immobilized aminoacylase is presented in Figure 11.14.

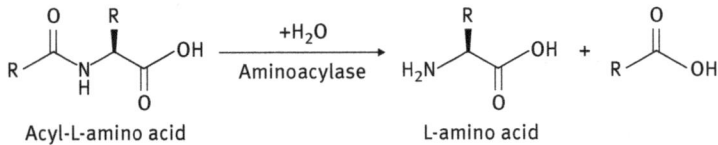

Figure 11.13: Hydrolysis of amino acid.

Only one of the amino acids undergoes hydrolysis reaction in the presence of the immobilized enzyme aminoacylase. After the reaction is carried out at a relatively low temperature, water is separated by evaporation, while separation of the desired product from the unconverted acyl-D-amino acid is done by crystallization. Racemization of acyl-D-amino acid gives a racemate that is recycled back through a storage tank to the reactor along with the make-up racemic mixture of acylamino acids.

Mixture of
acyl-L and
D-amino acid

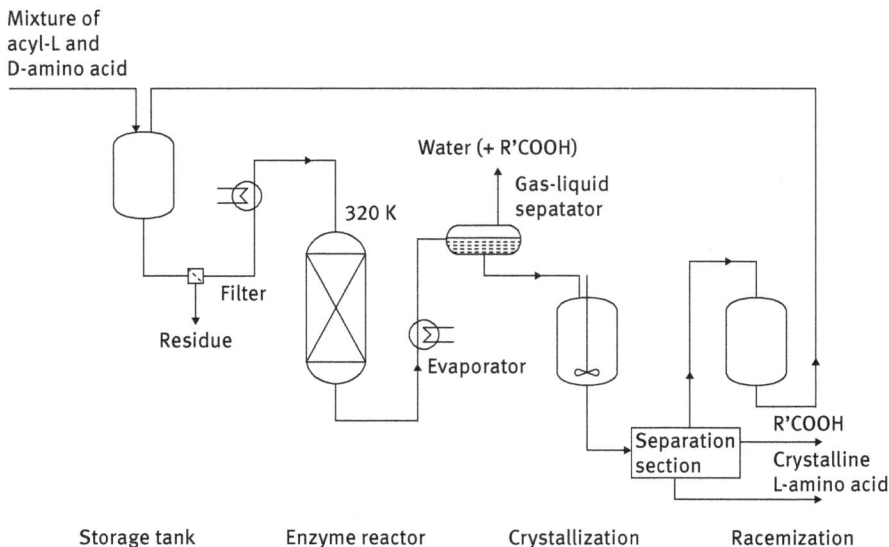

Figure 11.14: Flow scheme for amino acid hydrolysis with immobilized enzymes.
From J. A. Moulijn, M. Makkee, A E. van Diepen, *Chemical Process Technology*, 2013, 2nd Ed.
Copyright © 2013, John Wiley and Sons. Reproduced with permission from Wiley.

11.2.3 Hydrolysis of fatty acid triglycerides

Enzymatic hydrolysis of esters can be done by lipases at temperatures below 50 °C
and is applied to special fats, which can be normally hydrolyzed only by saponification and neutralization. Enzyme-catalyzed hydrolysis is carried out under conditions milder than non-catalytic or acid-catalyzed reactions, which substantially
diminishes the formation of secondary products. This technology is still mostly in
the development stages.

Mineral acids can be also used in hydrolysis, which, at the same time, promote
corrosion of the equipment.

Hydrolysis can also be done at 210–260 °C and 1.9–6.0 MPa in a battery of autoclaves operating batchwise or continuously with concurrent or countercurrent
streams cascade and in the absence of any catalyst. An alternative approach is to
apply a spray column with a countercurrent flow of liquids (Figure 11.15) when the
fat is introduced at the bottom and water is fed on the top with the oil/water feeding
ratio of 2:1. High-pressure steam (ca. 0.2 t per ton of oil) is admitted directly to the
reaction zone. The residence time in the reactor is ca. 4 h.

The bottom fraction contains glycerol and water and is separated from water by
evaporation from 15–20% glycerol concentration to 88% crude glycerol.

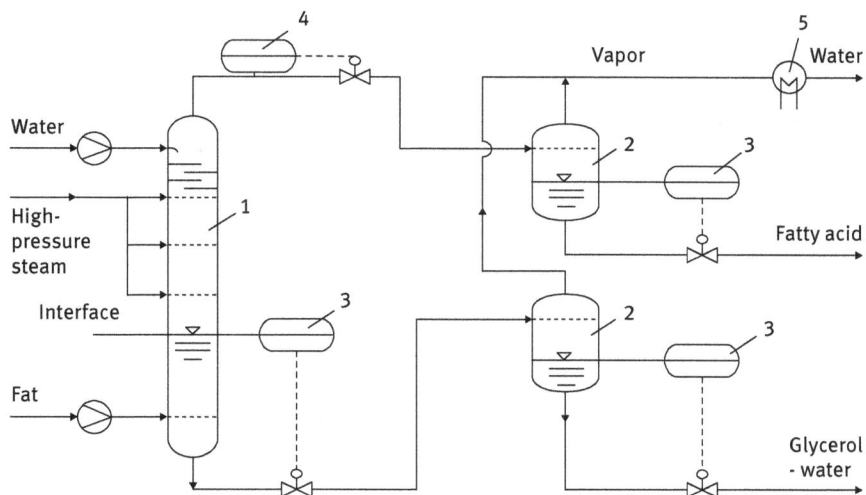

Figure 11.15: Hydrolysis of fatty acids triglycerides in a spray column: 1, splitting column (250–260 °C, 5–6 MPa); 2, flash tank; 3, level control; 4, pressure control; 5, condenser.

Crude fatty acid mixtures taken from the column top contain a number of low- and high boiling compounds. Since fatty acids can easily undergo various transformation reactions (oxidation, thermal decomposition, dehydration, oligomerization, etc.), distillation should be performed under high vacuum and therefore low temperature using predominantly falling-film evaporators.

11.3 Esterification

Esterification can be considered as a joint dehydration of acids and alcohols:

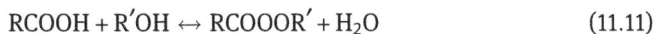

$$RCOOH + R'OH \leftrightarrow RCOOOR' + H_2O \tag{11.11}$$

In the presence of acid catalysts (sulphuric acid, HCl, ion exchange resins, etc.), the reaction can be carried out in the liquid at 70–150 °C depending on the type of reactants. The reaction heat is rather small; thus, equilibrium constants only mildly, if at all, depend on temperature, being at the same time dependent on the nature of reactants. This implies that with an increase in chain length and branching of the alcohol, the reaction is less thermodynamically favored. The reaction rate is influenced also by the structure of the alcohol in the same way as the esterification thermodynamics. Tertiary alcohols and phenols are the most difficult to esterify, and instead of alcohols in these cases, either anhydrides or chloroanhydrides are used.

While the reaction rate is also decreasing with an increase in the chain length in the acid similar to changes in the structure of an alcohol, the equilibrium constant has an opposite behavior, increasing with branching and chain length increase.

The shift in equilibrium for esterification reactions can be done by removing the lower boiling point product (water or ester) from the mixture. In some cases, an azeotrope of alcohol and water is removed (i.e., butanol and water), which, upon condensation, could be separated into two layers. The upper (organic) layer is recycled. When a homogeneous condensate is formed, components forming low-boiling-point azeotropes with water (benzene or dichloroethane) can be added. When methanol and ethanol are used as alcohols, water is removed along with the alcohol and separated by subsequent distillation.

In the case of esters forming azeotropes with water when water is present in excess, the reaction mixture would be enriched with the ester. After separation of the aqueous layer from the organic one (ester and alcohol), the latter is further separated to recycle alcohol. An interesting case is formation of low-boiling-point esters such as ethyl acetate, which forms an azeotrope (boiling-point, ca. 70.3 °C) containing 83.2 wt % ethyl acetate and 7.8% water. This corresponds to 2.4:1 molar excess of the ester and implies that during removal of the ester, the reaction mixture is enriched with water. The final product (ester) contains some amount of water and alcohol.

Another option is to use an excess of one of the reactants, which is typically the cheapest one.

Esterification combined with distillation can be arranged either continuously or discontinuously. Some arrangements are presented in Figure 11.16.

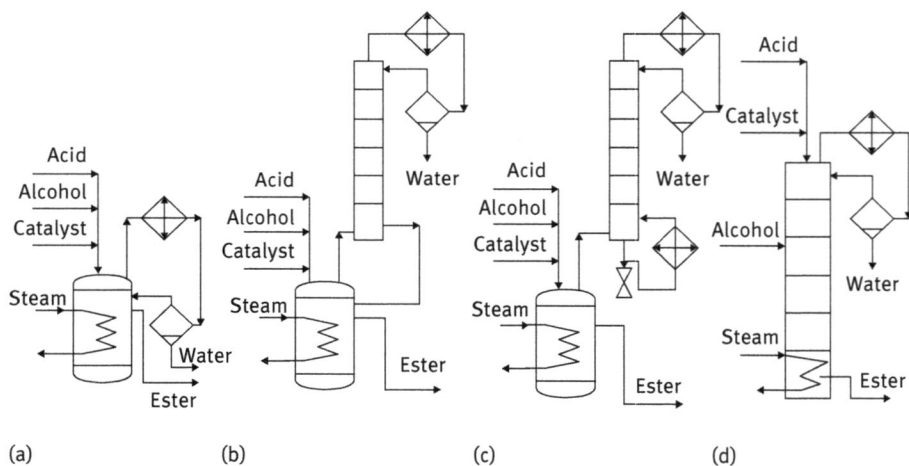

Figure 11.16: Reactors for liquid-phase esterification with azeotrope removal: (a) steam boiler reactor with condenser; (b) steam boiler reactor with a reflux column; (c) steam boiler with a distillation column; (d) tray column reactor. After N. N. Lebedev, *Chemistry and Technology of Basic Organic and Petrochemical Synthesis,* Chimia, 1988.

The first three arrangements (Figure 11.16a–c) have reactors of large volume heated with steam. In the case of Figure 11.16a, only a condenser is used, while for the case in Figure 11.16b, a distillation column is also installed. The latter can have its own boiler (Figure 11.16c), which improves separation. Thus, an arrangement as in Figure 11.16c is used for the case of the lowest difference in the boiling points of the reaction liquid and the liquid that is distilled away. For continuous esterification, a significant reaction rate is required; thus, a cascade of reactors is typically applied. A different arrangement is show in Figure 11.16d, when the reaction is conducted in a distillation column. The catalyst and a higher boiling point compound are typically introduced at the top of the column, while the alcohol is introduced at a certain place in the column.

As an example, one of the flow schemes for synthesis of ethyl acetate, which is in general made from ethanol and acetic acid in batch or continuous processes, is presented in Figure 11.17. The reaction mixture and the catalyst (sulphuric acid) from vessel 1, through a heat exchanger (pos. 2), is sent to reactor 4. Steam is introduced directly to the bottom of the reactor. The distillation process separates the ester as the alcohol-ester-water azeotrope (70% of ethanol and 20% of ester). Typically, the reaction time is selected to ensure that the reactor bottom contains only acetic acid with the catalyst.

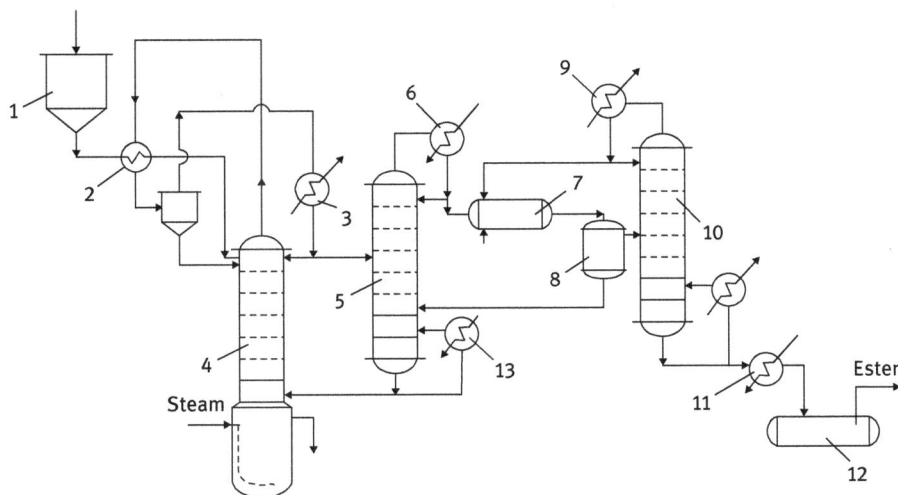

Figure 11.17: Synthesis of ethyl acetate from ethanol and acetic acid: 1, vessel for reactants; 2 and 11, heat exchanger; 3 and 11, cooler; 4, reactor; 5 and 10, distillation columns; 6 and 9, reflux condensers; 7, mixer; 8, separator; 12, collecting vessel; 13, boiler. After N. N. Lebedev, *Chemistry and Technology of Basic Organic and Petrochemical Synthesis*, Chimia, 1988.

The vapors from the reactor are first cooled in the heat exchanger (pos. 2) and then in the condenser (pos. 3) and sent back to the reactor. A part of the stream after condenser 3 is distilled in column 5, where the azeotrope is separated from the mixture containing mainly alcohol and water. This mixture from the bottom of the distillation column is sent back to the reactor (pos. 4) to ensure sufficient amount of alcohol at the bottom part of the reactor.

The low-boiling-point fraction from column 5, after cooling in 6, is sent to a mixer 7 to which additional amount of water is added (1:1). Such addition is needed; otherwise, it is difficult to separate water from a mixture of ester and alcohol. The formed emulsion is separated in a separator (8) into two layers. The upper layer contains the desired ester with some amounts of dissolved alcohol and water, while the bottom layer is a water solution of ester and alcohol.

The purification of ester is done in a distillation column (pos. 10) by removing a lower boiling azeotrope of ester, alcohol, and water. A part of the azeotrope is refluxed, while the other part is sent to the mixer (pos. 7). Ethyl acetate is taken from the bottom of the column (pos. 10) and after cooling (pos. 11) is collected in the vessel (pos. 12). After decanting and purification, the final yield of the ester can be as high as 95%.

Chapter 12
Alkylation

Alkylation is defined as the introduction of alkyl groups into organic molecules and is applied in synthesis or alkylaromatics, alkylation of isoparaffins, or transformations of epoxides.

Alkylation reactions can be classified based on the type of the formed bond: alkylation by substituting hydrogen located at a carbon atom (C-alkylation) with an alkyl group, substituting oxygen or sulphur (O- and S-alkylation), or nitrogen (N-alkylation). Alkyl groups can also have substituents, such as hydroxyl or carboxy groups. In the current chapter, C-, N-, and O-alkylation as well as β-oxyalkylation will be described.

Alkylation can be done using unsaturated compounds such as olefins and acetylene; chloro-containing compounds with active Cl, which can be replaced; aldehydes or ketones, for example, in N-alkylation; epoxides when alkylation proceeds with a carbon-oxygen bond rupture.

Olefins are mainly used for C-alkylation, while they are typically not used for N-alkylation and not effective for O- and S-alkylation. Alkylation activity of olefins depends on their ability to form stable carbocations; thus, chain increase and branching is favorable for reactivity improvement.

12.1 Alkylation of aromatics

For alkylation of aromatics with olefins, either $AlCl_3$ or other acid catalysts (HF, sulphuric acid, supported phosphoric acid, zeolites) can be used. Either liquid- or gas-phase alkylations are applied over a broad range of temperature and pressure as discussed in detail below.

Side reactions include formation of polyalkylaromatics, as well as generation of coke, cracking, and polymerization of olefins. The reaction network is given in Figure 12.1.

The traditional process for the production of EB was developed around 1930. The catalyst used was $AlCl_3$-HCl, and all operations were carried out in agitated reactors, under moderate conditions: 130–170 °C (depending on the transalkylation arrangement) and ca. 0.7 MPa. Transalkylation can be done either in the same reactor, or alternatively, after separation, the polyethylaromatics are transalkylated with benzene. A substantial amount of ethylbenzene is still produced by this technology utilizing $AlCl_3$, even if disposal of waste streams and corrosion are serious shortcomings.

Aluminum chloride is very poorly soluble in hydrocarbons and is a bad catalyst *per se*. However, upon exposure to HCl the complex of $AlCl_3$ and HCl with one to six molecules of aromatic hydrocarbon is formed. The catalyst, dark in color, has the

https://doi.org/10.1515/9783110712551-012

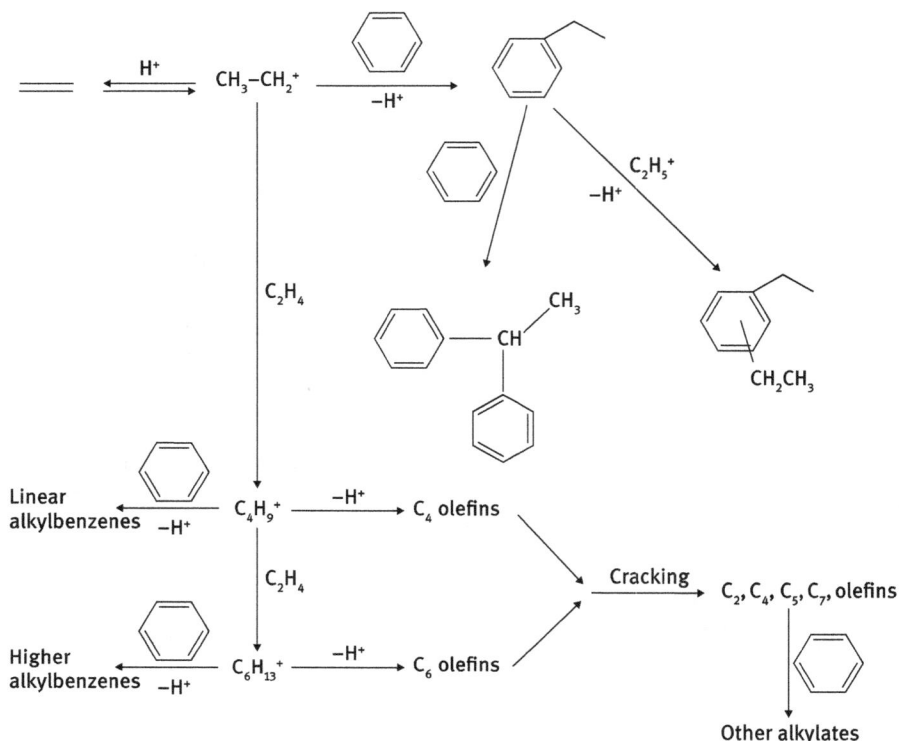

Figure 12.1: Main reactions during alkylation of benzene with ethylene.

composition presented in Figure 12.2. In practice, ethylchloride is added to the reaction mixture.

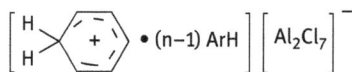

Figure 12.2: Composition of an alkylation catalyst.

Because the alkylation reaction with a catalyst complex is very fast, the overall process is rather limited by mass transfer (diffusion of the olefin through a film around the catalyst complex) and is enhanced with extensive stirring or vigorous bubbling of olefins through the reaction mixture. A low value of activation energy also reflects the significance of mass transfer. On the contrary, the transalkylation reaction, i.e., reaction of dialkylbenzene with benzene (Figure 12.3), has an activation energy of ca. 63 kJ/mol and becomes more prominent with temperature increase.

From the point of view of thermodynamics, transalkylation is independent on temperature and pressure, while alkylation is exothermic and is favored at low T and elevated P.

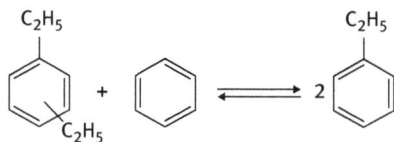

Figure 12.3: Transalkylation of diethylbenzene.

Both reactions (alkylation and transalkylation) are influenced by catalyst deactivation, with the latter having a stronger dependence. This implies that along with deactivation, formation of polyalkylaromatics is becoming more prominent. Since at higher temperature, formation of polycyclic compounds and coke is detrimental to the catalyst, a very high temperature (favorable for transalkylation) cannot be applied and a certain optimum gives a possibility to sustain a reasonable transalkylation rate without extensive deactivation.

The process flow with alkylation and transalkylation in the same reactor is given in Figure 12.4. The catalyst complex, prepared in reactor 6 is fed to the alkylation reactor (pos. 9), is more clearly shown in Figure 12.5, along with dry benzene and ethylene. Drying of benzene, even if expensive, is essential since benzene should be free of water; otherwise, the catalyst complex will be destructed forming aluminum hydroxide. Drying is done by azeotrope distillation (column 3). A low-boiling azeotrope of benzene and water is condensed (pos. 4) and separated (pos. 5) into two layers. The organic (benzene) layer is refluxed. Dried benzene (water content 0.002–0.005%) from the bottom of column 3 is heated in a heat exchanger (pos. 2) and through a collecting vessel (8) sent to the alkylation reactor (9). The reactor itself is an empty vessel with an internal lining made from corrosion-resistant bricks. Olefin (ethylene or propylene for ethylbenzene or cumene production, respectively) is sparged to the lower part of the reactor into the liquid, which consists of the catalyst complex (20–40 vol%) and a mixture of aromatics basically insoluble in this complex. Olefins should be purified by monoethanol amine scrubbing from sulphur-containing compounds capable of forming complexes with the catalyst, thus deactivating it.

Intensive mixing is done by bubbling ethylene through the liquid phase. The reaction temperature is typically limited (130–140 °C), and the reaction heat is removed by evaporation of benzene. The latter is condensed (pos. 10) and after separation (pos. 11) is recycled to alkylation while the gases are further treated. Since the gas contains substantial amounts of benzene especially when operating with diluted olefin fractions, it is scrubbed in absorber 13 with polyalkylaromatics and routed back to reactor 9. The gases after absorber 13 are scrubbed with water (pos. 14) to remove HCl.

The liquid from the alkylation reactor goes to a separator (12), where a heavier catalyst complex, after settling, goes to the reactor, while the alkylate (containing benzene, monoaromatics, diaromatics, and even polyaromatics, and tars), after passing through a cooler (15), is further separated (pos. 16) from the catalyst complex. The latter is sent back to the reactor.

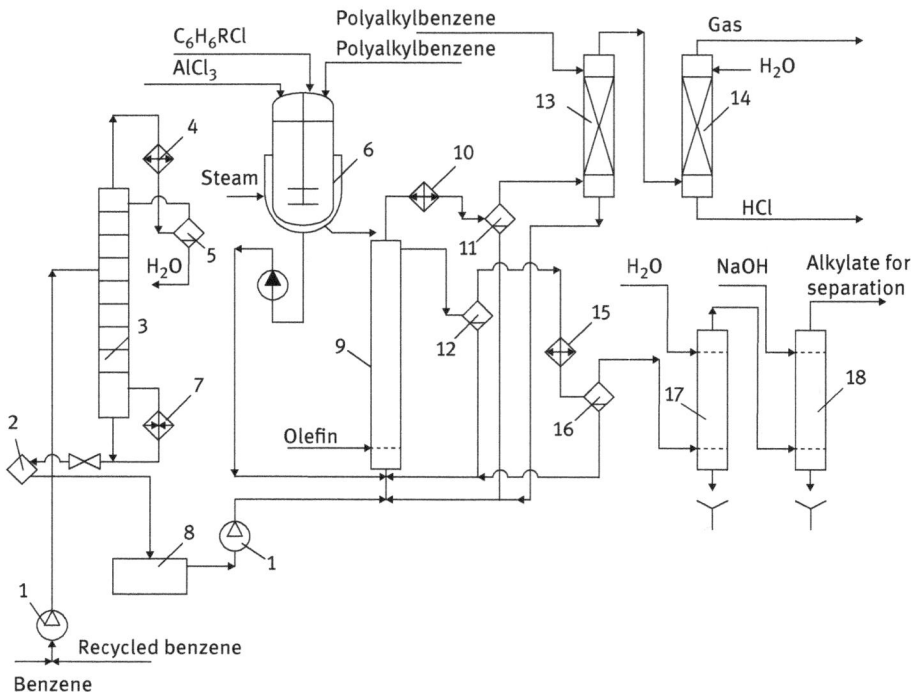

Figure 12.4: Alkylation of benzene: 1, pump; 2, heat exchanger; 3, benzene drying column; 4 and 10, reflux condensers; 5, separator; 6, reactor for catalyst preparation; 7, boiler; 8, collector; 9, alkylation reactor; 11, phase separation; 12 and 16, separator; 13, absorber; 14, water scrubber; 15, cooler; 17 and 18, washing columns. After N. N. Lebedev, *Chemistry and Technology of Basic Organic and Petrochemical Synthesis*, Chimia, 1988.

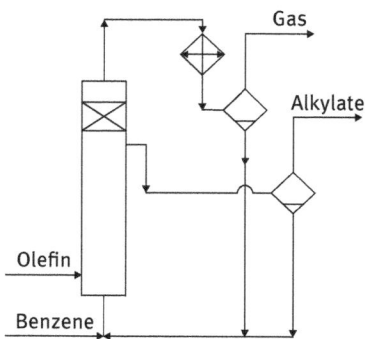

Figure 12.5: A bubble column for alkylation with aluminum chloride.

The alkylate should be cleaned from dissolved HCl and AlCl₃ by first washing with water (pos. 17) and thereafter with sodium hydroxide. The alkylate is then distilled in a series of distillation columns, recovering unconverted benzene at the top of the first column and separating heavies from ethylbenzene in the second column. In the final column, these heavy recyclable polyalkylbenzenes are taken at the top, while non-recyclable high-molecular-mass residue stream containing polycyclic aromatics is burned as fuel.

The fresh catalyst complex is made in a stirred tank (6) from technical-grade aluminum chloride, diethylbenzene, or equal amounts of benzene and dialkylbenzene. This is needed because the complex is not formed when only benzene is applied. In addition, small amounts of ethylchloride or even dry HCl if available at a plant site are added.

In the alkylation *per se*, an excess of benzene (2–2.5 mol/mol ethylene is used) to diminish oligomerization of ethylene and avoid extensive formation of polyalkylaromatics.

The conventional process was improved by Monsanto by careful control of ethylene addition, increase in temperature (160–180 °C), and minimization of the concentration of aluminum chloride, thus eliminating a presence of two phases in the reactor.

In this homogeneous catalytic process (Figure 12.6), only dry benzene, ethylene, and the catalyst are introduced to the alkylation reactor because presence of substantial amounts of polyalkylbenzenes is detrimental for alkylation at low catalyst concentrations. The outlet of the alkylation reactor is mixed with recycled polyethylbenzene and processed in a separate transalkylation reactor, which operates at much lower temperature than the primary alkylation reactor. Transalkylation products are washed and neutralized in order to remove residual $AlCl_3$. Purification of the product mixture free of the catalyst is done in the same way as in the conventional process.

Globally, ca. 25% are still using the aluminum chloride-based process.

Heterogeneous multibed fixed-bed catalytic alkylation process for production of ethylbenzene in the gas phase operating under high temperature (390–450 °C) and pressure (1.5–2 MPa) using a zeolitic catalyst ZSM-5 was developed by Mobil-Badger, and introduced into commercial practice in 1976. Coke deposition is an issue, and after 40–60 days, *in situ* regeneration by coke combustion with air had to be performed. This required two reactors, one for the regeneration and another for the reaction. Currently, much longer time-on stream prior to regeneration has been achieved (18–24 months). An advantage of ZSM-5 operating in the gas phase compared to liquid-phase alkylation is less sensitivity to water and some other poisons such as sulphur, therefore not requiring drying of benzene. Temperature control of the exothermal alkylation is done by cooling through quenching with cold ethylene after each bed. Significant overall excess of benzene relative to ethylene (8–15 mol/mol) ensures complete conversion of ethylene, increasing, however, the costs of benzene recovery and recycling.

The polyalkylates after separation are transalkylated in a transalkylation reactor, which can operate at a lower pressure. Application of non-corrosive substances eliminated the need for high-alloy materials and brick lining or equipment for catalyst recovery.

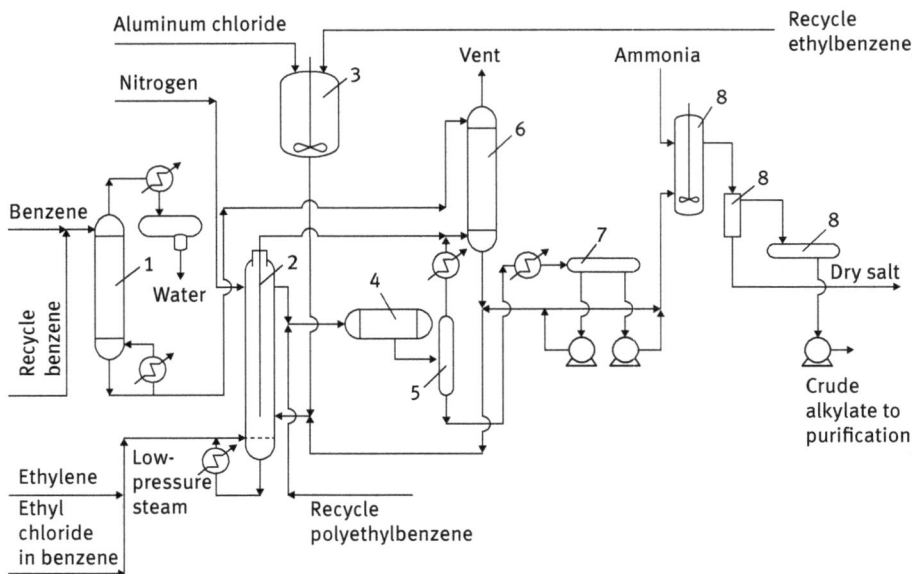

Figure 12.6: Homogeneous liquid-phase alkylation process for ethylbenzene production:
1, benzene drying column; 2, alkylation reactor; 3, catalyst preparation tank; 4, transalkylator;
5, flash drum; 6, vent gas scrubbing system; 7, decanter; 8, neutralization system.

The separation from polyethylbenzene is done basically in the same way as in a liquid-phase alkylation using three distillation columns. Benzene and ethylbenzene are distilled overhead in the first and second distillation columns, respectively. Benzene is recycled to the reactor. In the last column, the recyclable alkylbenzenes and polyalkylbenzenes are separated from heavy non-recyclable residue.

Further development of alkylation processes was done by UOP/Lummus/Unocal performing the reaction in the liquid phase and allowing a better thermal control and prolonged catalyst lifetime, and subsequently, regeneration *ex situ* rather than *in situ*. Instead of medium-pore zeolites (ZSM-5), a large-pore zeolite Y was applied, being replaced in the 1990s by modified zeolite β.

A zeolite of medium pore size, MCM-22, characterized by two systems of channels independent of each other, was applied in a liquid-phase process called EBMax, which has been commercialized by Exxon/Mobil since 1995. This catalyst, being active in alkylation, displays negligible oligomerization activity, thus allowing operation at low benzene/ethylene ratio and resulting in the highest yield and product purity compared to other processes.

The flowsheets of the zeolite-based liquid processes (Figure 12.7) are rather similar. A multibed catalytic reactor with multiple-point ethylene injection operates with the excess of dried benzene at ca. 4 MPa and a temperature below the critical temperature of benzene (289 °C). Separation is done with three distillation columns in the same manner as already discussed above. Higher alkylbenzenes are reacting with benzene in the liquid-phase transalkylation reactor. The transalkylation product is sent to the distillation train.

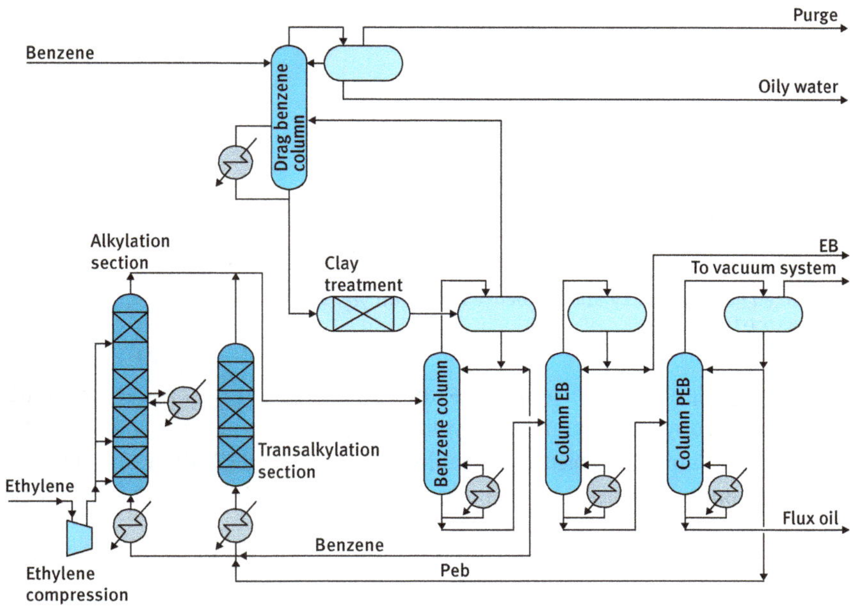

Figure 12.7: Flow scheme for the liquid-phase process for the production of EB. Polimeri Europa zeolite-based liquid-phase process. Modified after http://www.treccani.it/export/sites/default/Portale/sito/altre_aree/Tecnologia_e_Scienze_applicate/enciclopedia/inglese/inglese_vol_2/591-614_ING3.pdf.

A mixed-phase (gas/liquid) ethylbenzene process (Figure 12.8) combines an alkylator and a benzene stripper in one piece of equipment, which is a reactive distillation column. A zeolitic catalyst is positioned on the plates of the distillation column. The overhead benzene is used for a liquid-phase transalkylation, whereas bottoms from the benzene stripper are fractionated into ethylbenzene product as the overhead product column and polyethylbenzenes, which are further separated from the heavies and returned to transalkylation reactor. The transalkylation effluent is routed to the catalytic distillation column.

Alkylation of benzene with propylene is conceptually very similar to that with ethylene. A critical point in alkylation of propylene is the need to obtain lower quantities of *n*-propylbenzene, which cannot be separated from isopropylbenzene (cumene) by simple distillation.

In the supported phosphoric acid process developed by UOP in the 1940s, a multibed reactor operated in the liquid phase (180–240 °C, 3–4 MPa) with quenching by

Catalytic   Transalkylator   Ethylbenzene   PEB
distillation   column   column
column

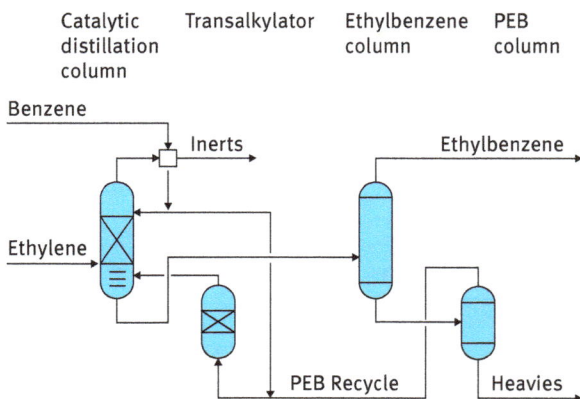

Figure 12.8: Mixed gas-liquid alkylation of benzene with ethylene. http://www.cdtech.com/ techProfilesPDF/CDTECHEB.pdf.

cold propylene. A high benzene/propylene ratio (from 5 to 10) was needed to minimize formation of propylene oligomers as well as polyalkylates because the SPA catalyst, contrary to aluminum chloride, is unable to make transalkylation. In addition, such a catalyst generates corrosion problems because of leaching and could not be regenerated at the end of the life cycle.

In the 1970s, a technology based on $AlCl_3$-HCl was introduced, which allows the transalkylation of the polyalkylates. Further development resulted in several processes applying different zeolites (MCM-22, β, Y, or mordenite).

The process flow scheme developed by UOP using β-zeolite as an alkylation catalyst is shown in Figure 12.9.

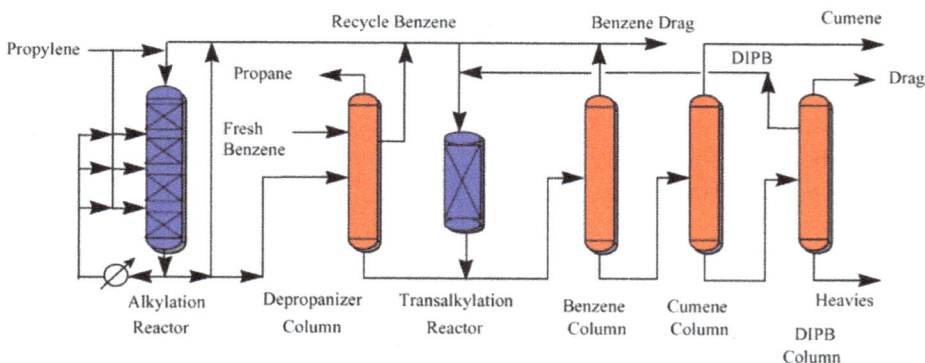

Figure 12.9: Production of cumene by alkylation of benzene.From R. J. Schmidt, Industrial catalytic processes-phemol production, *Applied Catalysis A: General*, 2005, 280, 89–103. Copyright Elsevier. Reproduced with permission.

Feeding of benzene is done first to the upper-mid section of the depropanizer column to remove excess water and then to the alkylation reactor having four catalyst beds. Recycle benzene together with the fresh benzene is fed to the alkylation reactor in the donwflow mode along with propylene. The latter is split between the beds ensuring complete consumption in each catalyst bed. Minimization of olefin oligomerization and as well as polyalkylation is done by using an excess of benzene. Heat control of the exothermic alkylation is done by recycling a part of the reactor effluent and quenching by some cooled reactor effluent between the beds. In the depropanizer column, propane which entered with the propylene feed is removed along with excess water. The bottoms stream of this column is sent to the benzene column where cumene is separated from benzene. The latter is recycled while the former goes to the cumene column where the heavies (including diisopropylbenzene DIPB) are separated from cumene. DIPB after separation from heavy aromatic components is routed to the transalkylation reactor. The bottoms of the DIPB column are blended into fuel oil.

In the transalkylation reactor, DIPB and benzene react to form additional cumene. Zeolitic catalysts are used in both alkylation and transalkylation reactors. After 2–4 years in service, the catalyst is regenerated by burning coke either *ex situ* or even *in situ*. The catalysts can withstand several regeneration cycles. The requirements for the catalysts are low Lewis acidity promoting oligomerization and coking while keeping the Brønsted acidity high. This is achieved by avoiding formation extra-framework aluminum in zeolites during the catalyst preparation.

The process presented in Figure 12.9 gives the levels of cumene between 85 and 95 mol% and DIPB between 5 and 15 mol%. Transalkylation over a beta-zeolite is done at a sufficiently low temperature giving high conversion and minimizing formation of side products.

12.2 Alkylation of olefins

Among paraffins, only those having a tertiary carbon atom can undergo alkylation. Generally, different olefins could be applied, but in practice, mainly butenes (or butane-butene fraction after cracking and separation from butadiene) are used, reacting with isobutane and giving C8 hydrocarbons (Figure 12.10).

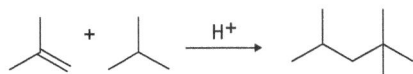

Figure 12.10: Alkylation of isobutene with isobutane with formation of 2, 2, 4-trimethylpentane.

An excess of isoparaffin is beneficial from the viewpoint of selectivity because it suppresses side reactions and improves octane number. Otherwise, polymerization of olefins gives undesirable low-octane number, high-boiling-point components. The heavier polymerization products (with more than 10 carbon atoms) known as acid-soluble oils tend to deactivate the catalyst. At the same time, if the excess is too high, it would lead to unnecessary high costs for recycling. In industrial conditions, typically, the mole excess of isoparaffin is varied from 6 to 20 depending on the process. The usual I/O ratio ranges from 5 to 8 in sulphuric acid plants and from 10 to 15 in HF plants.

As in other alkylation processes, the reaction starts with generation of a carbocation by protonation of the olefin. After that, it can abstract a hydride from isobutane, forming a *tert*-butyl carbenium ion:

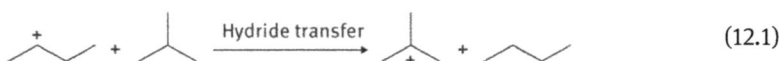

(12.1)

This is followed by an exothermal step of electrophilic addition to another olefin, giving a larger carbenium ion:

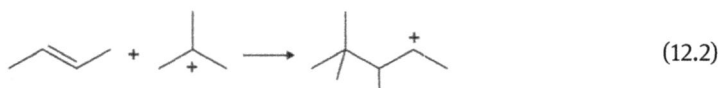

(12.2)

Subsequent hydride transfer sustains the catalytic cycle, resulting primarily in 2, 2, 3 trimethylpentane:

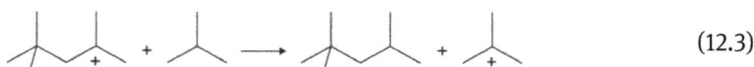

(12.3)

Alkylation reactions are highly exothermic (on average, 75–96 kJ/mol); therefore, heat removal is essential. Low temperatures such as 0–10 °C with sulphuric acid and ca. 30 °C for HF is preferred, because they minimize the formation of polymerization and cracking by-products. A very low temperature in case of sulphuric acid would result in unwanted increase in viscosity, while an increase in temperature would lead to formation of tars and sulphur dioxide. Since HF is not an oxidant, a somewhat higher temperature can be used, simplifying obviously the cooling arrangements, while for the process with sulphuric acid, cryogenic cooling is needed.

The influence of pressure on the reaction equilibrium is less straightforward. From the viewpoint of thermodynamics, high pressure should be applied, influencing in a negative way side reactions; thus, in practice, mildly elevated pressures (0.2–2 MPa) are used.

A requirement of a low temperature implies that strong acids such as sulphuric and hydrofluoric acids are used as catalysts even if such processes have inherent corrosion, pollution, and safety problems.

The traditional alkylation reaction takes place in a medium in which the hydrocarbon drops are dispersed in a continuous acid phase. Thus, alkylation is often treated as a process where the reaction rate is proportional to the interfacial area.

Application of HF results in a higher octane number product due to the hydrogen transfer reactions having lower catalyst consumption and higher isobutane consumption. Total acidity is decreased during operation by contamination with water and organics. Contamination with organics is more pronounced in HF process due to the higher (order of magnitude compared to sulphuric acid) solubility of isobutane.

Catalyst activity decreases with time due to dilution, formation of so-called red oil, and impurities. Although HF can be purified by fractionation to remove water and red oil at the plant site, some losses are inevitable, as HF forms an azeotrope with water. Contrary to HF, complete removal of H_2SO_4 is needed, requiring regeneration by complete decomposition of the acid, which is done outside of refineries, leading to high overall H_2SO_4 consumption (100 kg/t of product) compared to HF losses.

In the sulphuric acid process developed by Stratco, the reactor is a horizontal pressure vessel with an inner tube bundle acting as a heat exchanger (Figure 12.11). Temperature is also controlled by a high recycling ratio of the emulsion, which is formed when the acid and the hydrocarbon feed are vigorously stirred. Pressure is kept at about 0.35 to 0.5 MPa to ensure the liquid-phase operation. Short contact time is applied, preventing side reactions.

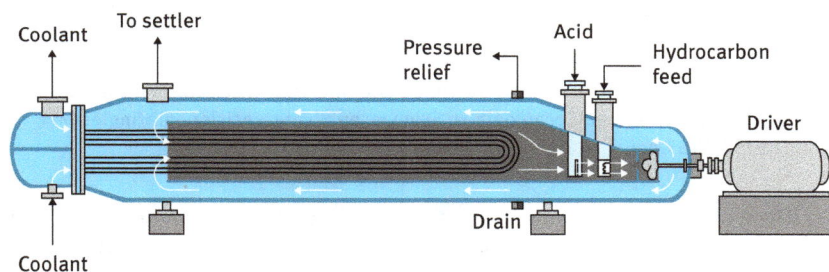

Figure 12.11: Horizontal alkylation reactor. Modified after http://www.treccani.it/export/sites/de fault/Portale/sito/altre_aree/Tecnologia_e_Scienze_applicate/enciclopedia/inglese/inglese_ vol_2/181-192_ING3.pdf.

The process of ExxonMobil (Figure 12.12) applies a horizontal reactor with a cascade of compartments. Mixers are present in each compartment to emulsify the hydrocarbon-acid mixture. Reaction temperature is kept at 4–5 °C, while low pressures (ca. 0.15 MPa in the stage rich in isobutane and 0.05 MPa) are applied. Heat removal is done by evaporation of isobutane, which, along with the acid, passes concurrently from one compartment to another. Olefin, premixed with recycled isobutane, is added.

After cooling first the vapors and condensation of water with its subsequent removal in the coalesce, the vapors are compressed and condensed. Economizer (an intermediate pressure flash drum) reduces the power requirements of the refrigeration compressor. Caustic and water washing of isobutane is done prior to its separation from propane and recycling isobutane back to the process.

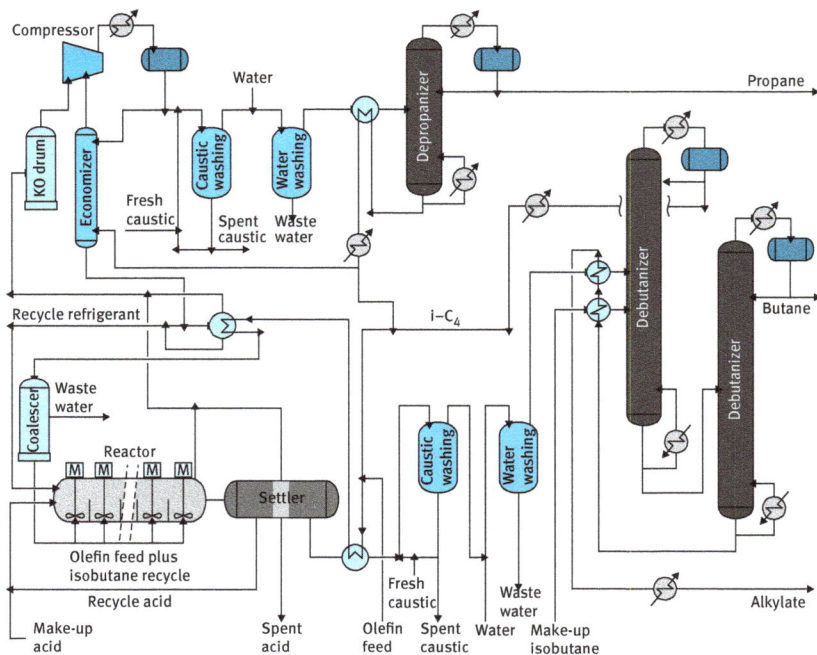

Figure 12.12: ExxonMobil alkylation technology. Modified after http://www.treccani.it/export/
sites/default/Portale/sito/altre_aree/Tecnologia_e_Scienze_applicate/enciclopedia/inglese/
inglese_vol_2/181-192_ING3.pdf.

After settling of the liquid product, the acid is returned to the reactor. A certain
amount of acid is purged in order to remove feed impurities and polymerized olefins ("red oil"), while fresh acid is added to keep the required acid strength. Further
removal of acid components after settling is needed; thus, caustic and water wash
is applied. In the downstream deisobutanizer, the overhead isobutane is recycled,
while the bottom (alkylate and *n*-butane) is fractionated in the debutanizer for the
separation of the alkylate product from butane.

Despite higher volatility of HF and a higher isobutane/olefin ratio, there are more
processes operating with HF compared to sulphuric acid, even if a larger excess of
isobutane implies a larger fractionation section. At the same time, a clear advantage
of applying HF is its low viscosity and as well as better solubility of isobutane, which
is translated to much easier emulsification and therefore absence of stirring devices.

A process flow diagram is given in Figure 12.13, highlighting a reactor system
with a riser reactor, acid cooler, and a settler. Acid circulation is done by gravity
only, without using an acid circulation pump.

The feed after dehydration, which is needed to minimize corrosion, is mixed with
HF at a pressure level high enough to ensure that all components are in the liquid
phase and admitted to a tubular reactor. The residence time is ca. 0.5 min. Two

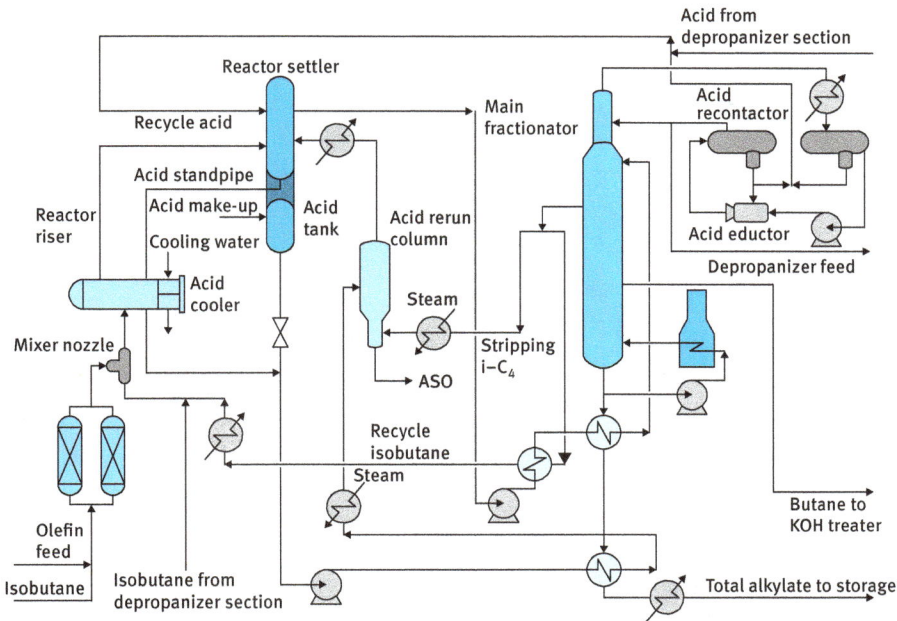

Figure 12.13: Phillips process for alkylation of isobutene using HF. Modified after http://www.trec cani.it/export/sites/default/Portale/sito/altre_aree/Tecnologia_e_Scienze_applicate/enciclope dia/inglese/inglese_vol_2/181-192_ING3.pdf.

phases are formed from the reaction mixture after settling. The acid is taken from the bottom, cooled, and recycled. Removal of water and polymerized hydrocarbons is done in the acid rerun column after withdrawing a slipstream of acid from the settler. Such removal is needed to maintain activity of the acid, which is returned to the system after condensation. The bottom product is a mixture of red oil and HF-water azeotrope, which are further separated by an additional settling. The organic component can be used as fuel.

An alkylate, along with propane, isobutane, and normal butane, is taken from the top of the reactor settler and separated by fractionation. Isobutane is recycled, while propane and *n*-butane are washed with caustic soda to remove traces of HF. Alkylate is taken from the bottom of the fractionator.

There could be different reactor arrangements in HF alkylation process.

Thus, in UOP technology, the hydrocarbon phase is dispersed in the acid continuous phase. The inlet pressure is provided by a pump. Feeding of the dried olefin and isobutane is done at different reactor heights, while the acid is fed at the reactor bottom (Figure 12.14). The acid is returned to the reactor after settling. The hydrocarbon phase after settling is processed in the isostripper. The alkylate is taken from the bottom. Isobutane and *n*-butane are taken as a side streams, with isobutane being recycled.

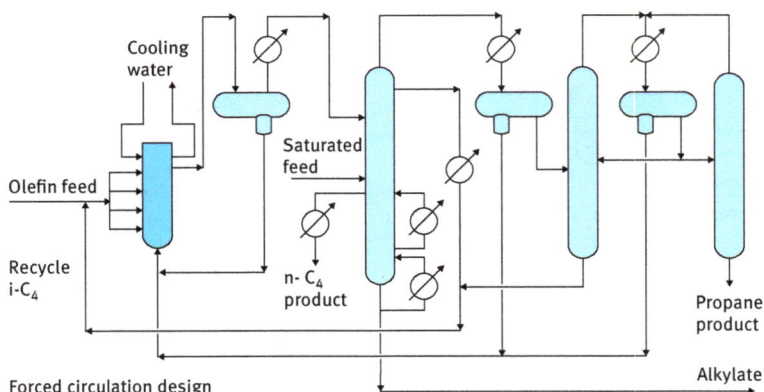

Figure 12.14: UOP alkylation process.

HF is separated from the overhead of the isostripper. The acid is recycled, while the organic phase (isobutane and propane) is further treated in a depropanizer in the case of C3–C4 olefins feed and even for C4 olefins with high propane content. From the bottom of the depropanizer, isobutane is recycled to the reactor, while the top, after a cooler, passes through a depropanizer receiver drum and further to an HF stripper, from where propane is taken as a bottom product. Treatment of the effluent streams and process vents is done with, for example, KOH, which is periodically regenerated with lime.

Since the 1990s, a number of companies have proposed alkylation processes based on solid catalysts, both using fixed-bed or riser reactors. The AlkyClean process developed by Lummus Technology, Albemarle Catalysts, and Neste Oil employs a zeolite catalyst. In the patent describing the process, a Pt/USY zeolite with alumina binder is employed. A simplified block diagram is given in Figure 12.15.

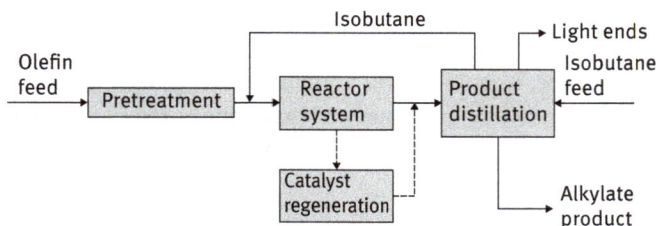

Figure 12.15: A simplified diagram for alkylation with a solid acid catalyst (AlkyClean process). http://www05.abb.com/global/scot/scot271.nsf/veritydisplay/436e0e59f93e253b c1256ef500466fd0/$File/71-76%20-%20M610.pdf.

The demonstration unit incorporated three reactors, with one used for alkylation, another one in mild regeneration, and third one undergoing high-temperature regeneration with hydrogen at 250 °C completely restoring catalyst activity. Regeneration was required after 4–30 days depending on the operation severity. The arrangement of regeneration is presented in Figure 12.16.

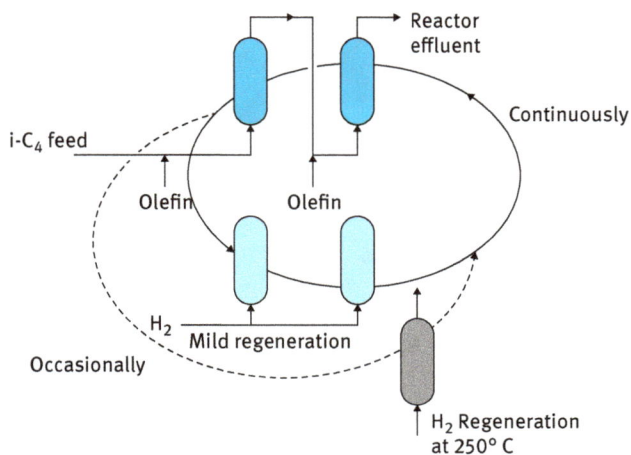

Figure 12.16: Catalyst regeneration in alkylation with a solid acid catalyst (AlkyClean process). https://afdelingen.kiviniria.net/media-afdelingen/DOM100000170/Archief-2005/AlkyClean.pdf.

The total installed cost of the commercial unit should be significantly lower than in the current liquid-acid processes, such as those based on HF and sulphuric acid. With no liquid acids or chlorides in the system, no product treatment or disposal of acids or chlorides is required. Absence of corrosive acids in the system and mild operating conditions allow for carbon steel construction, while limited pretreatment and no post-treatment result in fewer equipment outside of the boundary limits.

The world's first commercial-scale, solid acid catalyst alkylation unit based on AlkyClean technology with the annual capacity 100,000 metric tons was started up in 2015 in China, giving an octane value (RON) between 96 and 98.

In addition to the efforts in replacing the liquid-phase alkylation with solid catalysts, there are also several commercial ionic liquid-based alkylation units. Ionic liquid salts made by pairing organic cations with organic or inorganic anions exhibit low (below – 100 °C) melting points and low vapor pressure. The overall scheme of the ISOALKY process developed by Chevron- Honeywell UOP and commercially implemented is presented in Figure 12.17. The alkylate yield per reacted olefin in this technology is basically the same as for other liquid-phase alkylations.

As Figure 12.17 illustrates, the chloroaluminate ionic liquid which has strong acid properties acts as a catalyst replacing in this capacity hydrofluoric and sulphuric acids used in alkylation. The reaction occurs at the interface of the ionic liquid catalyst droplets and the hydrocarbon phase.

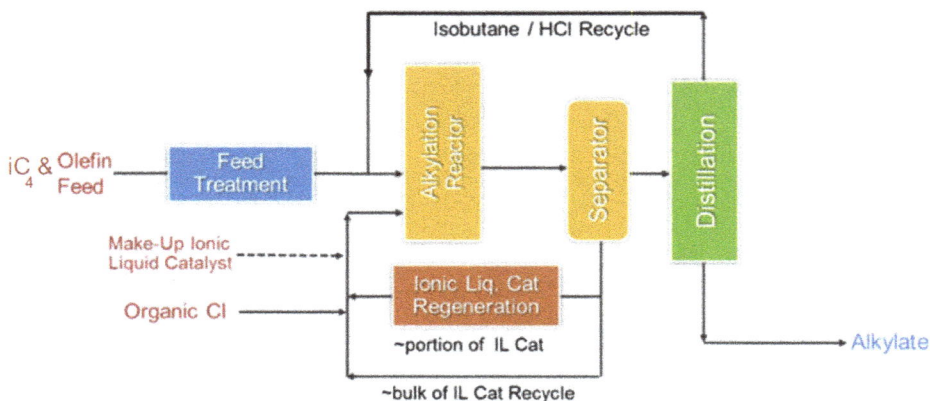

Figure 12.17: Overall process scheme of ISOALKY Technology. From H. K. Timken, H. Luo, B.-K. Chang, E. Carter, M. Cole, ISOALKY™ technology: Next-generation alkylate gasoline manufacturing process technology using ionic liquid catalyst, in M. B. Shiflett (ed.), *Commercial Applications of Ionic Liquids, Green Chemistry and Sustainable Technology*, 2020, 33–47, https://doi.org/10.1007/978-3-030-35245-5_2. Copyright Springer. Reproduced with permission.

The advantages of the ionic liquid as a strong acid compared to other alternatives are its lower concentration, improving safety, diminishing handling and the reactor volume requirements in the reactor, lower consumption in alkylation, and a possibility of on-site inexpensive regeneration.

In addition to ionic liquid alkylation catalyst, which has the general structure

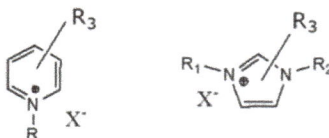

with $X^- = Al_2Cl_7^-$, trace amounts of anhydrous HCl co-catalyst are added *in situ* through continuous introduction of an organic chloride. Thus the combined superacid catalyst system comprises an anhydrous $AlCl_3$-based Lewis acid catalyst promoted with anhydrous HCl Brønsted acid.

Low solubility of the ionic liquid in the hydrocarbon ensures an easy separation and recovery of the ionic liquid catalyst. A specific feature of the catalyst is its hydrophilicity; therefore, to avoid hydrolysis in the presence of water the feed is dried with high-performance dryers.

The reaction temperature and pressure are respectively – 1 to 50°C and 2.7 to 17 bar, while the external isobutane to olefin mole ratio is 8–10. At these conditions, the olefin conversion exceeds 99.9%. The volume of the ionic liquid catalyst is 3–6 vol% contrary to, for example, 50% using sulphuric acid as a catalyst.

The technology requires, however, feed moisture specification of less than 1 ppm, which is more stringent compared to 10 ppm with HF. The moisture content is not critical for sulphuric acid-based units. The level of the conjunct polymer formation ("red oil") being in the range of 0.3–0.5 wt% of olefins is lower than for sulphuric acid units (1–1.5) and similar for HF-based units (ca. 0.5).

Efficient separation of the ionic liquid from hydrocarbons is needed to minimize the losses and avoid contamination and/or degradation of the products. Separation is done using a coalescer element pad material with a stronger affinity for the ionic liquid than for the hydrocarbons. After capture and coalescence, the ionic liquid droplets fall by gravity from the material. Distillation of the hydrocarbons is done as in the conventional technology.

A part of the ionic liquid catalyst is sent to the regeneration unit. Recycling of HCl co-catalyst is incorporated in the technology to minimize the makeup of organic chloride promoter.

The ISOALKY process does not employ washing with caustic solutions. The residual chlorides in the alkylate, n-butane, and propane are removed with an aid of solid adsorbents.

A similar Ionikylation process (Figure 12.18) has been used in several Chinese state-owned refiners starting from a retrofit of an existing 65,000 tones per year sulphuric acid alkylation unit in 2005, followed by a start-up of the grass route plant and subsequent commissioning of other units with annual capacity of 50,000 to 300,000 tones. There is also an example of the first commercial revamp of an HF alkylation process.

The process flow diagram of the unit in Figure 12.18, presented in Figure 12.19, comprises the feed pretreatment, the reactor section, catalyst regeneration, and product separation and purification sections.

In the commercial operation RON of 97.6 was achieved at 100% olefins conversion, 77% of the C_8 alkylate yield and absence of $C_{12}+$ compounds. During operation, the majority of the catalyst is regenerated onsite. The catalyst make-up is 2.5 kg per 1 ton of alkylate. The catalyst activity is maintained by addition of an organic chloride activator (1.1. kg/t alkylate).

12.3 O-Alkylation

This reaction (addition of alcohols to a tertiary olefin) is very important in refineries for production of oxygenates such as methyl-*tert*-butyl ether (MTBE) or methyl-*tert*-amyl ether with excellent octane characteristics. Presence of MTBE in few drinking water supplies in the USA mainly due to improper infrastructure eventually led to its complete elimination in the USA in 2005, while it continues to be used in other regions of the world.

Figure 12.18: PetroChina Harbin Petrochemical Co. Ltd.'s Heilongjiang plant of 150,000 tones per year capacity which uses Ionikylation process. From https://www.hydrocarbonprocessing. com/magazine/2020/april-2020/special-focus-clean-fuels/safe-and-sustainable-alkylation-performance-and-update-on-composite-ionic-liquid-alkylation-technology.

Addition of methanol to an isoolefin such as isobutene with a double bond on a tertiary carbon atom giving MTBE

$$\ce{}\qquad +CH_3OH \rightleftharpoons \qquad\qquad OCH_3 \qquad\qquad (12.4)$$

is driven by equilibrium; thus, due to exothermicity, relatively low temperatures (40–70 °C) are applied. Higher pressures also favor this reaction, which is performed in the industry at 0.8–2 MPa, keeping the reactants in the liquid phase. Selectivity for MTBE production exceeds 99% with the following by-products: dimers of isobutene, *tert*-butyl alcohol, and methyl-*sec*-butyl ether (Figure 12.20).

Although formation of dimethylether and diisobutenes is thermodynamically favored in the reaction conditions, it is possible to limit their formation to a few hundred ppm. Moreover, due to a limited amount of water in the feed, a much faster formation of *tert*-butyl alcohol (TBA) by addition of water to isobutene is not that profound, giving concentration of TBA in MTBE below 1 wt%. Reactivity of linear olefins, present in the FCC and steam cracking streams, is much lower than that of isobutene, resulting in only a few hundred ppm of a more linear ether (MSBE).

In order to determine the equilibrium constant for a liquid-phase reaction, there is a need to accurately calculate activity coefficients for such non-ideal system, the topic that is addressed in detail in specialized literature.

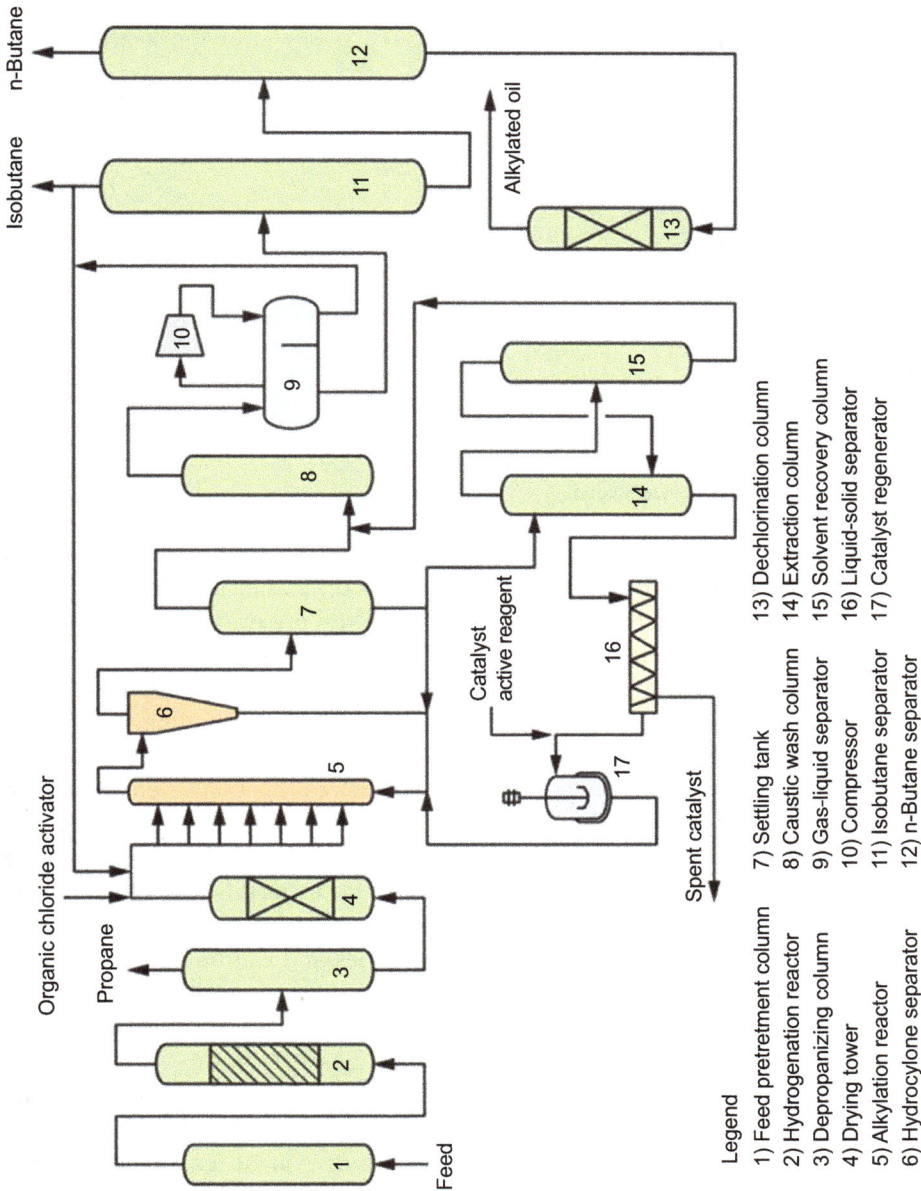

Figure 12.19: Process flow diagram for the Ionikylation liquid-phase alkylation technology based on an ionic liquid. From https://www.hydrocarbonprocessing.com/magazine/2020/april-2020/special-focus-clean-fuels/safe-and-sustainable-alkylation-performance-and-update-on-composite-ionic-liquid-alkylation-technology.

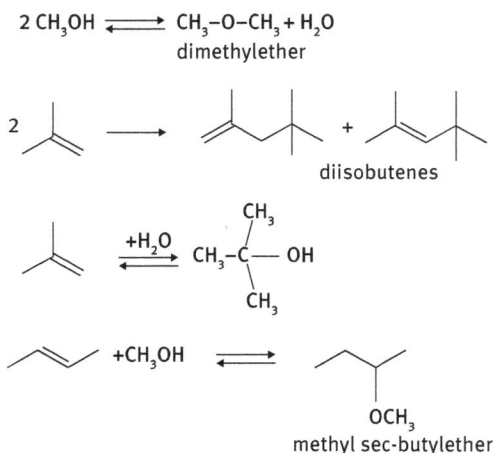

Figure 12.20: Side reactions in alkylation of isobutene with methanol.

The acidic catalysts used in industrial etherification processes consist of cation exchange macroporous resins (sphere size between 0.3 and 1.2 mm), which are prepared by a suspension styrene polymerization in the presence of a cross-linking agent (e.g., divinylbenzene). The resin, insoluble in water and organic solvents, is further functionalized by sulfonation with sulphuric or chloro-sulfonic acid. The resins can either swell or shrink depending of the degree of cross-linking, the type of ion bound to the functional group, and solvent polarity. In the case of industrial etherification catalysts, there is ca. 20–30% volume contraction when a hydrated form of the resin is introduced in a much less polar reacting environment consisting of hydrocarbons and methanol. The lifetime of industrial catalysts depends on the process parameters and the amount of impurities in the feed being approximately 6–12 months in refineries and up to 2 years in steam cracking plants. The difference in lifetime is mainly associated with the impurities and can be extended to 4–5 years for cleaner feeds. Due to a limited thermal stability of catalysts with degradation starting at 130 °C, this temperature should not be exceeded. Pressure is typically above 0.8 MPa to keep the olefin in the liquid phase.

Adiabatic reactors can be used either as a front-end reactor or for "finishing" being the simplest and the most economical configuration option. The temperature increase is controlled by using only low concentrations of isobutene in the feed and limiting the conversion. Since the thermal stability of the resins is limited, adiabatic reactors are suitable for treating feeds from, for example, FCC with low concentration of isobutene. Otherwise, the feed must be diluted with C4 compounds, or alternatively, external cooling should be done with a part of the outflowing stream being cooled and recycled.

An alternative option is to use a boiling point reactor operating at a lower pressure than conventional adiabatic reactors, which results in heat removal by partial evaporation of the reactants. Moreover, an expanded bed reactor could be applied where the catalyst is kept in motion by the reactants, avoiding local overheating or improper liquid distribution.

Finally, water-cooled tubular reactors (front end arrangement) or a reactor comprising also distillation are applied (finishing step). In the latter case of the reactor column, the products are removed, thus shifting thermodynamic equilibrium and allowing higher conversion of isobutene. Removal of products from the bottom implies higher reflux ratio than a traditional fractionation column and thus a larger column size.

When MTBE production is integrated in a refinery and uses FCC cuts, an olefin alkylation unit can be installed downstream of the MTBE unit; thus, a high conversion level of isobutene is not needed and two adiabatic reactors in series with intercooling can be used. One potential arrangement is shown in Figure 12.21, with an expanded bed front-end reactor 1 and the second adiabatic fixed-bed reactor 2.

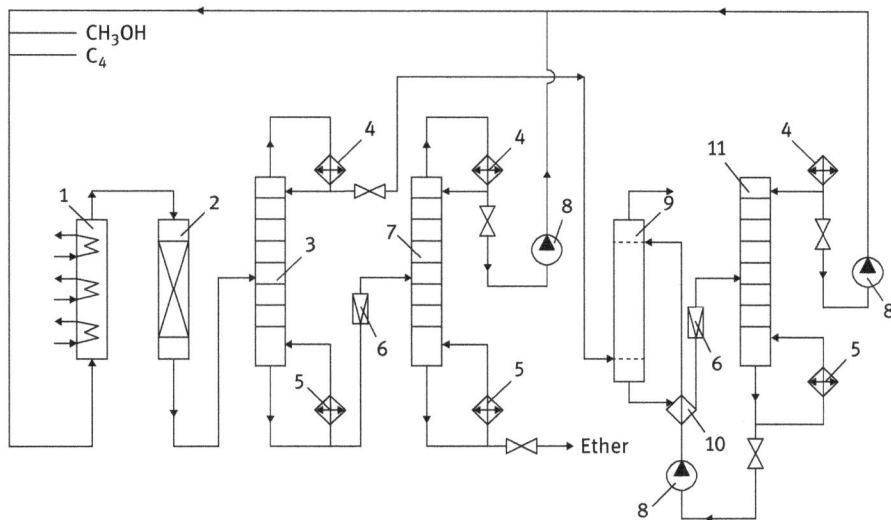

Figure 12.21: MTBE production scheme by O-alkylation: 1 and 2, reactors; 3, column for primary separation; 4, reflux condenser; 5, boiler; 6, expander; 7 and 11, distillation columns; 8, pump; 9, extraction column; 10, heat exchanger. After N. N. Lebedev, *Chemistry and Technology of Basic Organic and Petrochemical Synthesis*, Chimia, 1988.

In column 3, the light fraction (azeotrope of C4 hydrocarbons with a small amount of methanol) is separated from the heavy one (MTBE with the major part of methanol). In subsequent distillation column 7, MTBE is taken as a heavier fraction, while methanol from the top is recycled. The light fraction after column 3 is washed with water in extraction column 9, removing methanol from C4 hydrocarbons. These gases (from the top of column 9) are further purified (not shown) by distillation to remove minor amounts of water and dimethylether. Water and methanol fraction from the bottom of column 9 passes through a heat exchanger (pos. 10) and is distilled in column 11, from where the low-boiling-point fraction (methanol) is recycled and water is taken from the bottom and sent to extraction.

The reaction is typically done in a slight excess of methanol compared to isobutene in order to obtain C4 fraction with only limited content of isobutene.

In the case of feeds from steam cracking or isobutane dehydrogenation plants, a much higher level of conversion of isobutene is required (>99%). The distillation column is then installed after the first reactor (Figure 12.22). This fractionation step is needed to remove the ethers by separating the C4/methanol azeotrope (from the top) from MTBE taken from the bottom. The second distillation column also recovers C4/methanol azeotrope from the top and MTBE from the bottom, which is routed to the first columns. Similar to Figure 12.21, the scheme consists of a washing tower with water to remove methanol from C4 and a distillation column for separate water from methanol.

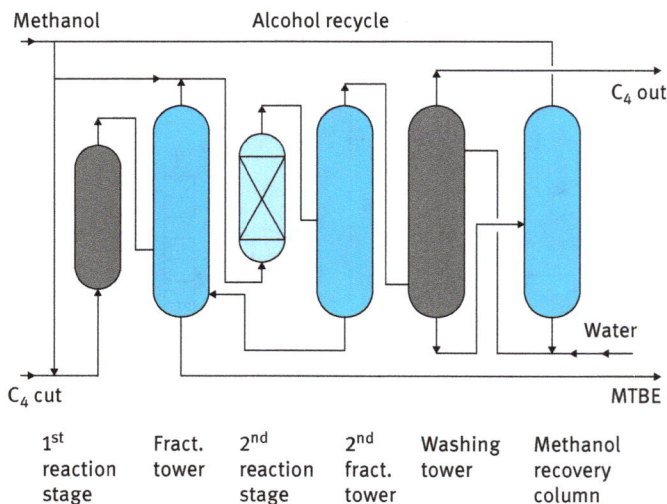

Figure 12.22: MTBE production scheme with a distillation column after the first reactor. Modified after http://www.treccani.it/portale/opencms/handle404?exporturi=/export/sites/default/Portale/sito/altre_aree/Tecnologia_e_Scienze_applicate/enciclopedia/inglese/inglese_vol_2/193–210_ING3.pdf.

The technology of Neste Jacobs (NexETHERS) differs from the other processes by the absence of a traditional recovery section with washing and distillation. This technology (Figure 12.23) is aimed for the combined production of MTBE, TAME, and heavier ethers or their ethanol-based counterparts ETBE, TAEE, and heavier ethers in one unit and combines alcohol recovery and circulation with oxygenate removal. After the reactor operating with a commercial cation exchange resin the effluent is separated in a distillation column into ether products and heavy hydrocarbons (C5s and heavier) taken from the bottom and unreacted C4 hydrocarbons and lighter components (the column top). Most of the alcohol and unreacted isoolefins are re-cycled to the reactors *via* a side draw-off from this column. The remaining alcohol and light oxygenates (dimethylether and water) are taken as the column overhead prod-uct. After the second fractionation unreacted alcohol is returned to the reactor section, thus removing it from the C4 stream (the bottom product). The latter is sent to alkyl-ation without an additional purification. The overhead stream contains only an azeo-tropic amount of alcohol, allowing an almost complete conversion of the feed alcohol.

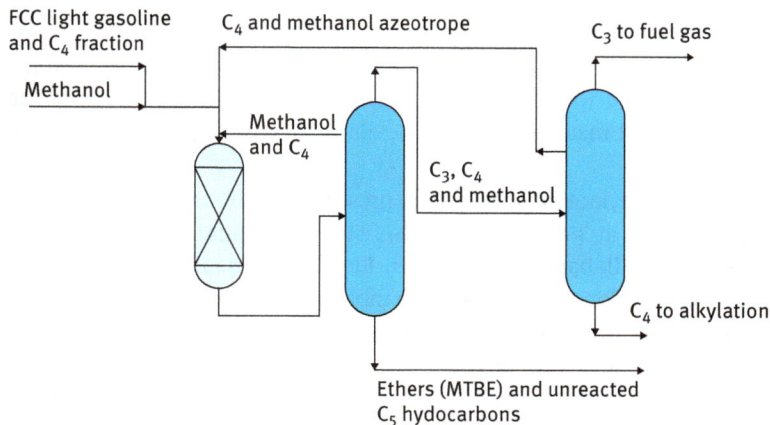

Figure 12.23: Neste Jacobs (NexETHERS) technology for MTBE synthesis. Modified after http://www.treccani.it/portale/opencms/handle404?exporturi=/export/sites/default/Portale/sito/altre_aree/Tecnologia_e_Scienze_applicate/enciclopedia/inglese/inglese_vol_2/193-210_ING3.pdf.

12.4 N-Alkylation

For alkylation of ammonia or amines, typically either alcohols

$$ROH + NH_3 \rightarrow RNH_2 + H_2O \tag{12.5}$$

or chloro-containing compounds are used. Contrary to other alkylation reactions, when olefins are widely utilized, application of the olefins will lead only to minor formation of amines, with nitriles being the dominant products.

Reaction of ammonia or amines with alcohols is an exothermal and thermodynam-ically favored. As catalysts, mineral acid (sulphuric acid for synthesis of methylaniline) or solid acids (alumina, aluminosilicates, aluminophosphates) can be used. In the lat-ter case, the process can be arranged in a gas phase at 350–450 °C. As side reactions, formation of ethers from alcohols or dehydration of an alcohol to the corresponding olefin can occur. While an ether can alkylate ammonia or amines, formation of ole-fins as mentioned before should be avoided since they are inactive in alkylation and could moreover lead to catalyst deactivation. To counterbalance unwanted dehydration, N-alkylation is typically done in the excess of amines.

Another side reaction, when the target is a primary amine, is consecutive alkylation, which is kinetically more favored

$$NH_3 \xrightarrow[-H_2O]{+ROH} RNH_2 \xrightarrow[-H_2O]{+ROH} R_2NH \xrightarrow[-H_2O]{+ROH} R_3N \tag{12.6}$$

In addition to catalysts by solid acids, it is possible to arrange N-alkylation through the so-called hydrogen borrowing concept, when an alcohol is first dehydrogenated to an aldehyde (for primary alcohols) or a ketone (for secondary alcohols), which react with ammonia or amine, resulting in the formation of an imine. The latter is hydrogenated using hydrogen, which was "borrowed" in the first step. For such type of the reaction pathway, supported metals (for example, alumina supported nickel, copper) are used as catalysts.

A flow scheme for methylamine synthesis is presented in Figure 12.24. The reaction occurs in the gas phase at 380–450 °C and 2–5 MPa with alumina or alumophosphate as a catalyst in an adiabatic fixed reactor. Elevated pressures are needed to suppress unwanted dehydration of methanol. The mole excess of ammonia is typically 4:1 in relation to the alcohol (methanol for methylamine or ethanol for ethylamine). As mentioned above, the reaction is a consecutive one, giving not only monomethylamines, but also dimethylamines and trimethylamines (TMAs). The overall process can be tuned to yield a particular amine by recycling other "unwanted" products.

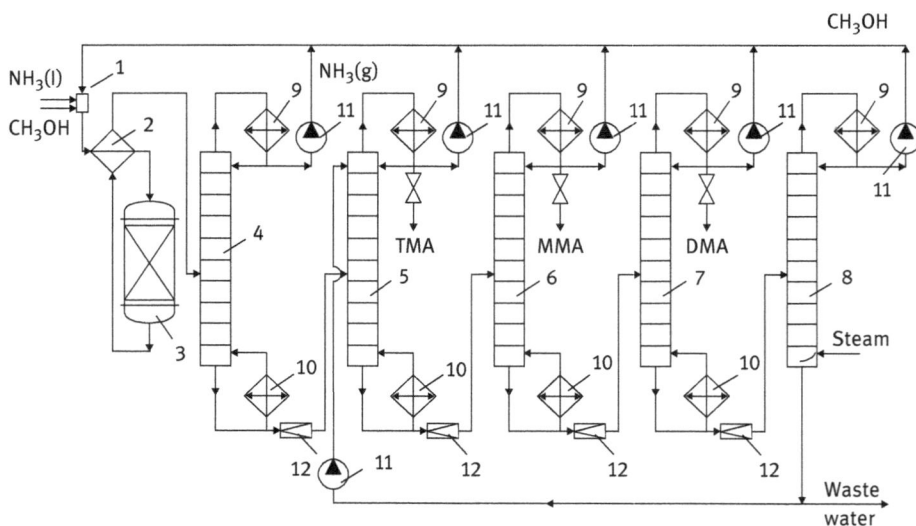

Figure 12.24: Methylamine synthesis: 1, mixer; 2, heat exchanger; 3, reactor; 4 and 8, distillation columns; 9, reflux; 10, boiler; 11, pump; 12, expansion. After N. N. Lebedev, *Chemistry and Technology of Basic Organic and Petrochemical Synthesis*, Chimia, 1988.

After mixing the reactants in mixer 1, they are evaporated in a heat exchanger (2) and routed to the reactor (3). The effluent is separated in a series of distillation columns. In the first one (pos. 4), ammonia is separated and recycled while the heavier fraction is sent to the extractive distillation column (pos. 5) to which water is added. In the presence of water, TMA is becoming the most volatile and taken as the overhead product and is recylced. Two other amines (methylamines and diethylamines) are separated in columns 6 and 7, respectively. These amines are either used as final products (the same is valid for TMA) or (partially) recycled. In column 8, unreacted methanol is separated from water and recycled back to the reactor, while the bottom fraction is sent to extractive distillation column 5.

Recent process improvements for synthesis of methylamines include utilization of shape-selective synthetic or natural zeolites such as a mordenite catalyst modified with alkali metals.

In Mitsui Chemical process, methanol amination is done over a silylated mordenite catalyst prepared by liquid-phase silylation of H-mordenite with tetraethoxysilane limiting selectivity to TMA. Methylamines selectivities (in wt%) are reported to be 33.3% MMA, 63% DMA, and 3.7% TMA at 90% methanol conversion, nitrogen/carbon ratio of 1.9, 310 °C, 1.86 MPa, and gas hour space velocity of 590 h^{-1}. Recovery of TMA is done not by distillation, saving on capital costs, but as an azeotropic mixture with ammonia. This mixture is processed in a disproportionation reactor with a non-selective proton form of mordenite. The effluent from this reactor after combining with fresh methanol is fed to the reactor with the shape-selective catalyst.

12.5 Oxyalkylation

In this process, epoxides (for example, ethylene oxide) react with water or alcohols. Such reactions are either non-catalytic and can be performed at 180–220 °C or require the presence of a nucleophile and proceed at somewhat lower temperature (100–150 °C).

In a non-catalytic reaction of ethylene oxide with water at ca. 200 °C not only the desired ethylene glycol, but diethylene, triethylene, tetraethylene, and polyethylene glycols are also formed, however, with decreasing yields.

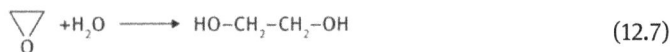

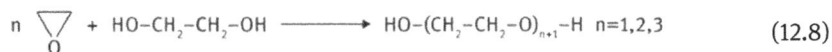

$$\text{▽O} + H_2O \longrightarrow HO-CH_2-CH_2-OH \tag{12.7}$$

$$n\, \text{▽O} + HO-CH_2-CH_2-OH \longrightarrow HO-(CH_2-CH_2-O)_{n+1}-H \quad n=1,2,3 \tag{12.8}$$

In order to keep the reactants in the liquid phase, a pressure of ca. 2 MPa is required. A large excess of water (15–20 m) is needed to minimize the formation of higher homologues because ethylene oxide reacts with ethylene glycols much faster than with water. Nevertheless, selectivity to diethylene and triethylene glycol is ca. 10%. Even if a high excess of water leads to substantial energy consumption during distillation, at the same time, it helps to remove the reaction heat, leading to a temperature rise of just 40–50 °C.

In some cases, the consecutive reactions could be, on the contrary, desirable as in the synthesis of non-ionic surfactants:

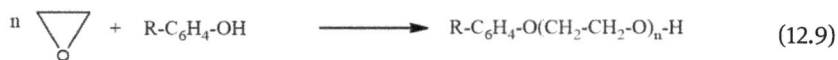

$$n \; \triangledown\!\!\!/_O \; + \; R\text{-}C_6H_4\text{-}OH \longrightarrow R\text{-}C_6H_4\text{-}O(CH_2\text{-}CH_2\text{-}O)_n\text{-}H \qquad (12.9)$$

For oxyalkylation, different reactors can be used (Figure 12.25). In a simple adiabatic reactor, the reaction mixture is introduced in the reactor through a central tube being heated by the product. The contact time in such reactors is relatively long (20–30 min), implying some deterioration of selectivity due to axial back mixing. Such mixing is less prominent in multitubular reactors, which are used for the production of ethanolamines or synthesis of glycols using phosphoric acid.

When the excess of the other reactant is not very high (3–5 to 1), as in the production of thioglycols or oxyalkylated amines, the heat release is significant. In order to cope with such heat release, either multitubular reactors (Figure 12.25b) or reactors with an external heat exchanger are used (Figure 12.25c). The former option requires operation under pressure to keep the reactants in the liquid phase.

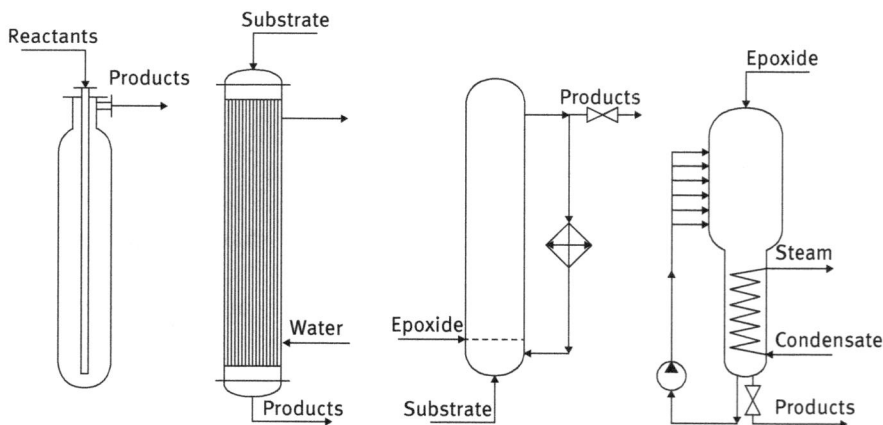

Figure 12.25: Oxyalkylation reactors: (a) adiabatic reactor; (b) multitubular reactor; (c) reactor with an external heat exchanger; (d) batch (discontinuous) reactor with liquid spraying for synthesis of non-ionic surfactants.

The option with an external heat exchanger is used for the synthesis of thioglycols or alkylenecarbonates:

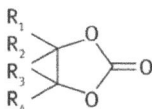

$$R_1 \\ R_2 \\ R_3 \\ R_4 \qquad \begin{array}{c} O \\ \diagdown \\ \diagup \\ O \end{array} = O \qquad\qquad (12.10)$$

The reaction milieu is typically the product itself through which a reactant (CO_2 or H_2S along with ethylene oxide) is bubbled. The efficiency of such bubbling is substantially decreased in a batch reactor mode when viscous products are formed, as in the case of non-ionic surfactant production. An alternative (Figure 12.25d) is spraying of the reaction mixture into gaseous ethylene oxide, which increases the contact area and allows to substantially reduce the batch time.

The scheme in Figure 12.26 shows ethylene glycol synthesis, ethylene oxide, and condensate being sent through a mixer (1) and a boiler (2) (heating the mixture to ca. 130–150 °C) to the adiabatic reactor (3). Besides the target products monoethylene, diethylene, and triethylene glycols, acetaldehyde (from isomerization of ethylene oxide) and its condensation products are also formed. After leaving the reactor, the product mixture having a temperature of 200 °C undergoes expansion to atmospheric pressure. A part of water is evaporated and the liquid is cooled to 105–110 °C. The mixture passes through a series of boilers (only two are shown in the scheme) with decreasing pressures. The heavier fraction after evaporator 5 undergoes distillation in the column (pos. 7) operating under vacuum. All water condensate fractions are combined and recycled to mixer 1, while monoethylene glycol is separated from diethylene and triethylene glycols by vacuum distillation in column 8. These polyglycols are also separated by vacuum distillation (not shown). The yield of tetraethylene glycol is too low for separate isolation.

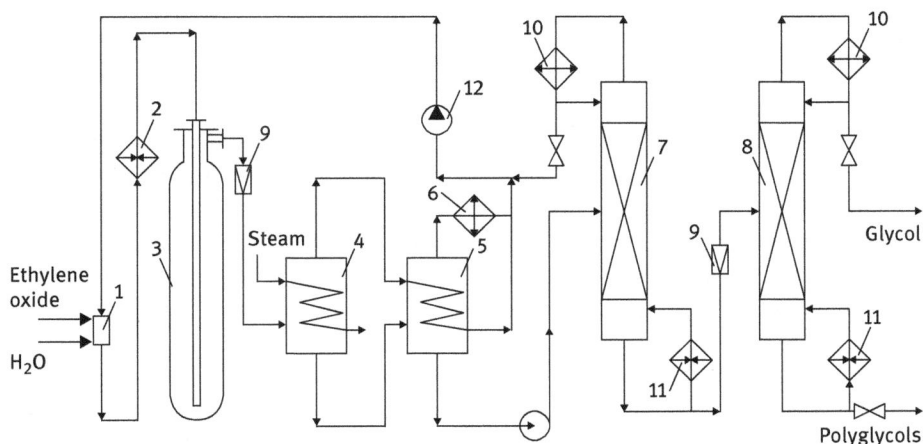

Figure 12.26: Flow scheme for ethylene glycol synthesis: 1, mixer; 2, team boiler; 3, reactor; 4 and 5, boilers; 6, condenser; 7 and 8, distillation columns; 9, expansion, 10, reflux; 11, boiler; 12, pump. After N. N. Lebedev, *Chemistry and Technology of Basic Organic and Petrochemical Synthesis*, Chimia, 1988.

In synthesis of glycol ethers (cellosolve), an alcohol is taken in excess and re-cycled. Since the difference in boiling points of alcohols and cellosolve are not large, their separation is done not by evaporation, as in Figure 12.26, but by distillation.

Somewhat similar to oxyalkylation of alcohols with ethylene oxide is synthesis of ethanol amine by reacting ethylene oxide with ammonia (Figure 12.27).

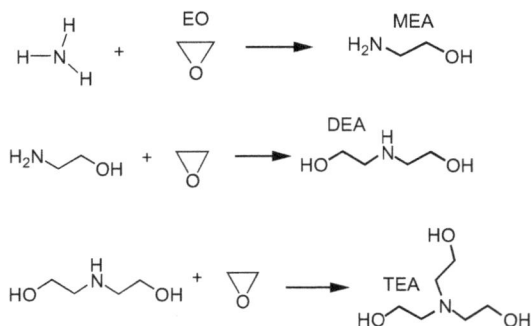

Figure 12.27: Synthesis of ethanol amines from ethylene oxide and ammonia.

Small quantities of water are needed to promote reaction, at the same time significant amounts of water favor a side reaction of H_2O with ethylene oxide giving ethylene glycol.

In the past monoethanolamine (MEA) was used for CO_2 removal in substantial amounts, while the current market favors only diethanolamine (DEA) production. From the operational viewpoint selectivity can be regulated by residence time, temperature, pressure, ammonia concentration, and the reactor type. Obviously, the product composition is especially sensitive to the ratio of ammonia to ethylene oxide and with the higher ratios leading to a higher monoethanolamine yield.

The process flow diagram is presented in Figure 12.18.

Ethylene oxide reacts with aqueous ammonia (molar ratio 1:40 with ammonia in excess) at 60 to 150 °C and 30 to 150 bar in a tubular or an adiabatic reactor to form three possible ethanolamines. The operation pressure must be sufficiently large to prevent evaporation of ammonia. The product stream is cooled prior the first distillation column where any excess ammonia is removed overhead and recycled. In the subsequent columns, ammonia and water are removed followed by separation of ethanolamines in a series of vacuum distillation columns.

Nippon Shokubai developed a process industrialized in 2003 with a capacity of 50 000 t/y using a highly selective pentasil-type zeolite binder-free catalyst modified with rare earth elements. The process is aimed at production of DEA in an adiabatic reactor without increasing the yield of triethanolamine (TEA) as the shape selective properties of the microporous zeolites restrict formation of TEA. Because of catalyst deactivation regeneration is required, which is by ammonia done after several days of operation. An illustration of the reaction/regeneration is provided in Figure 12.29.

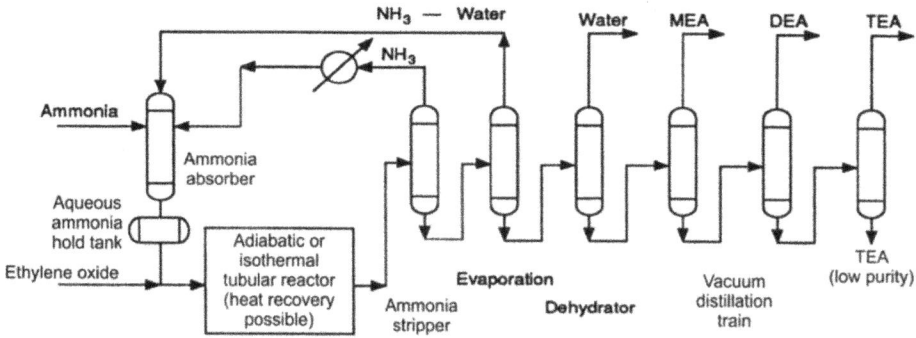

Figure 12.28: A process flow diagram of ethanol amine production plant. From G. Zahdei, S. Amraei, M. Biglari, Simulation and optimization of ethanol amine production plant, *Korean Journal of Chemical Engineering*, 2009, 26, 1504–1511. Copyright Springer. Reproduced with permission.

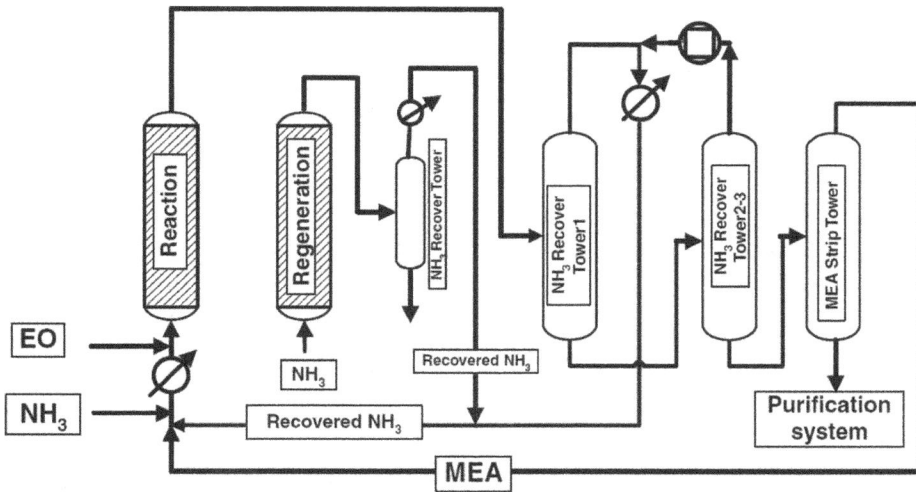

Figure 12.29: A process flow diagram of the reaction/regeneration section of ethanolamine production plant using a shape selective zeolite. From T. Hideaki, K. Masaru, O. Tomoharu, Development of 2,2′-iminodiethanol selective production process using shape-selective pentasil-type zeolite catalyst, *Bulletin of the Chemical Society of Japan,* 2007, 80, 1075–1090. doi:10.1246/bcsj.80.1075. Copyright 2007 The Chemical Society of Japan.

Chapter 13
Reactions with CO, CO_2, and synthesis gas

In this chapter, a range of industrially important reactions with CO, CO_2, and synthesis gas (mixture of CO and hydrogen) will be discussed.

13.1 Carbonylation

Methanol carbonylation to acetic acid

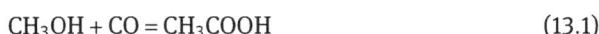

$$CH_3OH + CO = CH_3COOH \tag{13.1}$$

is an exothermal reaction favored at low temperature, affording almost complete conversion. Since there is a volume decrease in the reaction, it is favored at high pressures. The original BASF process commercialized in the 1960s operated at 70 MPa and 200–240 °C with cobalt (II) iodide used for *in situ* generation of $[Co_2(CO)_8]$ and hydrogen iodide. The rate of the reaction depended strongly on both the partial pressure of carbon monoxide and methanol concentration.

The low-pressure Monsanto process developed in the late 1960s operated at much milder conditions (3–6 MPa and 150–200 °C) with much more active Rh-based catalysts. Even if the process can operate at much a lower pressure, elevated pressure is applied to keep the reaction mixture in the liquid phase. Moreover, the active catalyst complex is not stable at low carbon monoxide pressure.

The reaction is first order with respect to methyl iodide (much less corrosive than HI but still requiring expensive corrosion-resistant construction materials) and the catalyst being zero order with respect to reactants. From the reaction engineering viewpoint, it implies that even a relatively small volume of CSTR can afford high conversion. Since polar solvents enhance the reaction rates, an acetic acid/water solvent medium is preferentially used.

The yields in the high- and low-pressure processes are 90% and 99.5%, respectively, based on methanol. On the contrary, the yields based on CO are much lower (ca. 70%) due to the water-gas shift reaction consuming carbon monoxide. For a high-pressure process, the by-products, besides CO_2, are methane, acetaldehyde, ethanol, propionic acid, alkyl acetates, and 2-ethyl-1-butanol.

The main by-products in the low-pressure Monsanto process are CO_2 and hydrogen, while methane, acetaldehyde, and propionic acid are formed in minor amounts even with significant presence of hydrogen in CO.

The reaction mechanism is given in Figure 13.1. The catalytically active species is the anion *cis*-$[Rh(CO)_2I_2]^-$ reacting with methyl iodide in the oxidative addition step to form the hexacoordinate species $[(CH_3)Rh(CO)_2I_3]^-$, which is rapidly transformed through methyl migration to the pentacoordinate acetyl complex $[(CH_3CO)$

https://doi.org/10.1515/9783110712551-013

$Rh(CO)I_3]^-$. Further reaction of this complex with CO gives the six-coordinated dicarbonyl complex, which undergoes reductive elimination, releasing acetyl iodide (CH_3COI) and regenerating the catalytically active anion cis-$[Rh(CO)_2I_2]^-$. Hydrolysis of acetyl iodide results in acetic acid.

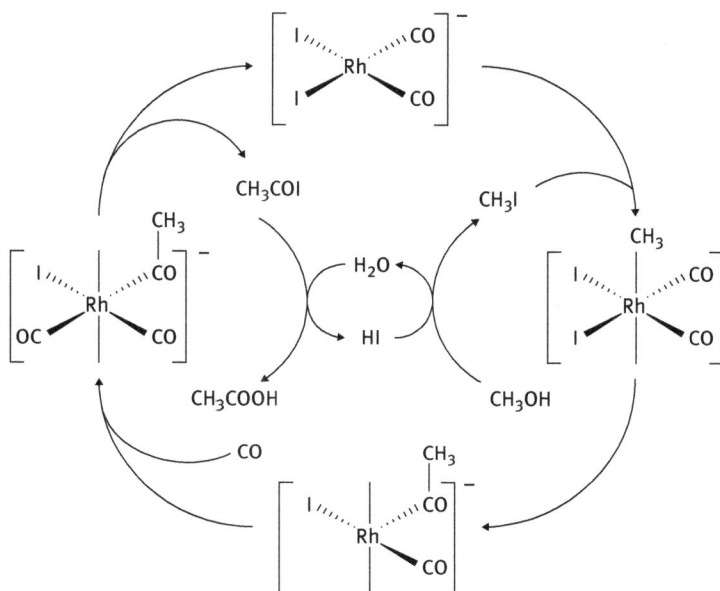

Figure 13.1: Reaction mechanism of methanol carbonylation with Rh catalysts. http://patenti mages.storage.googleapis.com/WO2006122563A1/imgf000011_0001.png.

Since the reaction rate does not depend on the concentration of reactants, being the first order in catalyst, and methyl iodide, the rate-determining step is proposed to be the oxidative addition of methyl iodide to cis-$[Rh(CO)_2I_2]^-$.

In the Monsanto process (Figure 13.2), the reactants, CO, and methanol, in equimolar amounts, are introduced in a sparged CSTR (2) continuously at ca. 150–200 °C and 3–6 MPa.

The non-condensable by-products such as CO_2, hydrogen, and methane are vented from the reactor. Unreacted CO along with the vapors is cooled in (pos. 3). The condensate from (4) is recycled, while the off-gas from the reactor and the off-gas from the purification section (light-ends column 8) are combined and sent to a vent recovery system (nor shown), where the light-ends, including volatile and toxic methyl iodide, are scrubbed by methanol and recycled back to the reactor, while the non-condensable gases are flared. The reactor solution is sent through a pressure reduction valve (5) to a flash vessel (6) where the liquid phase containing the dissolved catalyst is separated from the gas phase and recycled to the reactor

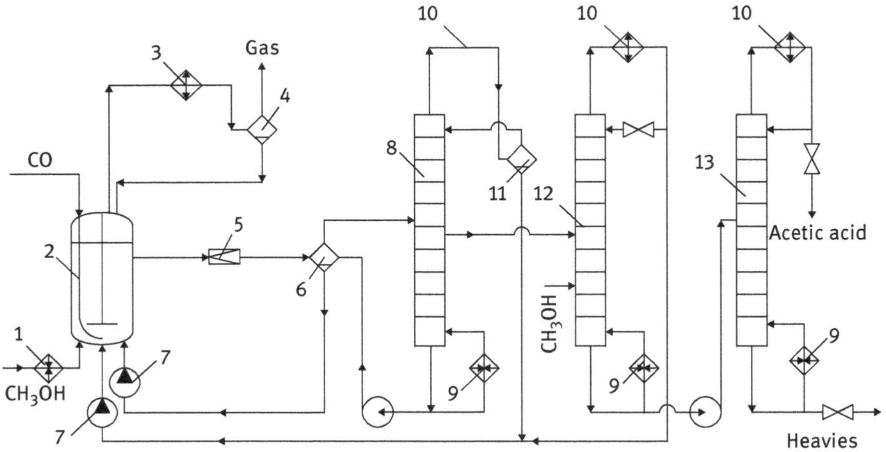

Figure 13.2: Monsanto methanol carbonylation process: 1, heater; 2, reactor; 3, cooler; 4 and 11, separator; 5, pressure release valve; 6, flash vessel, 7, pump; 8, 12, and 13, distillation columns; 9, boiler; 10, reflux. After N. N. Lebedev, *Chemistry and Technology of Basic Organic and Petrochemical Synthesis*, Chimia, 1988.

with the pump (7). The reaction heat is controlled by evaporation and can be regulated by the temperature of incoming methanol. The crude acetic acid containing methyl iodide, methyl acetate, and water is taken overhead from the flash vessel and sent to the light-ends column (8) where the light components (methyl iodide, methyl acetate, and water) are recycled back to the reactor as a two-phase overhead stream. The side stream from the light-end column (wet acetic acid) is sent to the dehydration (or drying column) column (12). As an overhead stream from this column, an aqueous acetic acid (35% solution) is recycled to the reactor, meaning that a fixed amount of acetic acid-water mixture is continuously circulating.

More efficient removal of HI is achieved by adding methanol as a side stream to this column (12) resulting in the formation of methyl iodide. Dry acetic acid from the column bottom is routed to a heavy-ends column (pos. 13). In the latter column, the major liquid by-product propionic acid along with other higher-boiling carboxylic acids are removed as the bottom stream and eventually incinerated. Acetic acid is removed overhead in the product (heavy end) column and further purified in the finishing column (not shown) being taken as a side-stream, while the overhead stream is recycled to the purification section of the process.

Superior catalyst stability and three to four times higher productivity compared to Rh catalyst was achieved in an iridium-based process developed by BP and known as the Cativa process. The reaction cycle for the process is given in Figure 13.3. The slowest step is insertion of CO rather than oxidative addition of methyl iodide, as in the Rh-based process. Enhancement of ionic iodide removal is thus beneficial for the overall process. To this end, simple iodide complexes of, i.e., rhodium are applied as promoters.

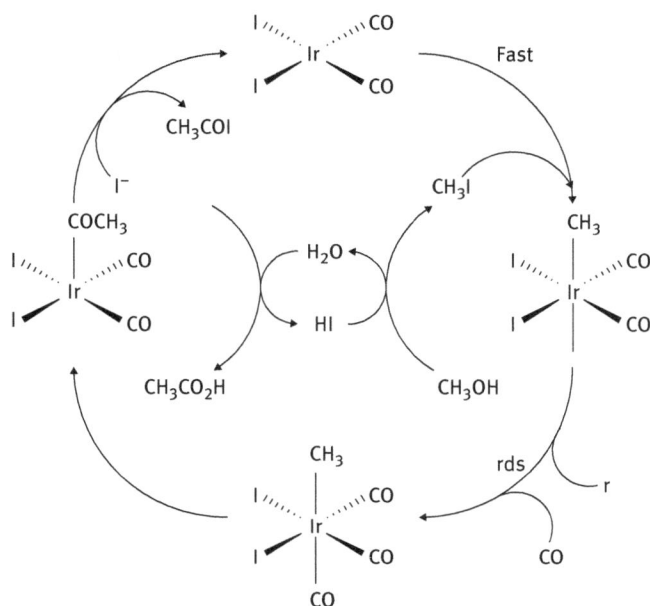

Figure 13.3: Reaction cycle for the iridium-catalyzed methanol carbonylation reaction (Cativa process).

This process operates at low water concentration (0.5 wt%), while much larger water content was required in the original Monsanto process (ca. 10–15 wt%). Such high concentration was needed to keep Rh in the solution because the carbon monoxide pressure is low in the flash vessel required for Rh complex recycling. This may lead to loss of CO ligands and precipitation of insoluble rhodium species (e.g., RhI_3).

Dilution with water also suppresses the formation of methyl acetate (enthalpy − 15.3 kJ/mol at 298 K)

$$CH_3COOH + CH_3OH \leftrightarrow CH_3COOCH + H_2O \tag{13.2}$$

and dimethylether (enthalpy − 23.6 kJ/mol at 298 K)

$$2CH_3OH \leftrightarrow CH_3OCH_3 + H_2O \tag{13.3}$$

Although there are apparent advantages with using extra amounts of water, such water excess results in elevation of water-gas shift reaction and a need to separate water and acetic acid, bringing extra costs to the process.

In the Cativa process (Figure 13.4), the iridium catalyst is stable under the low water conditions, resulting in lower energy requirements. Moreover, higher selectivity to the target product and subsequently much lower formation of higher acids allows having a plant with a simpler flow scheme than the Monsanto process. The flow scheme of the Cativa process contains the jet loop reactor operating at 3 MPa and ca. 180 °C, a finishing plug-flow reactor for increased CO conversion, a flash vessel, a drying column (performing also the functions of the light-end column in the Monsanto process) and the product column. The latter has a smaller size, owing to better catalyst selectivity and thus smaller amounts of formed heavier carboxylic acids.

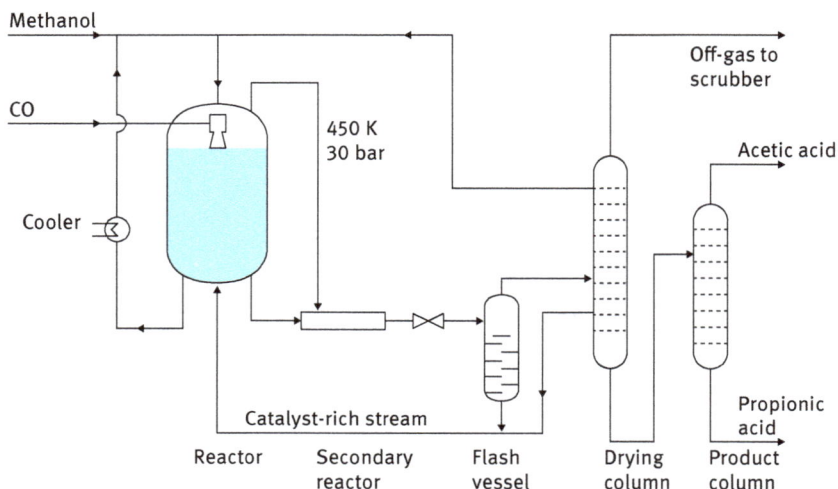

Figure 13.4: The Cativa process for the manufacture of acetic acid. From J. A. Moulijn, M. Makkee, A E. van Diepen, *Chemical Process Technology*, 2013, 2 nd Ed. Copyright © 2013, John Wiley and Sons. Reproduced with permission from Wiley.

13.2 Carboxylation

The use of CO_2 in the synthesis of chemicals is challenging since CO_2 is well known to be a stable molecule ($\Delta G_f^\circ = -396$ kJ/mol). Carboxylation reactions resulting in production of inorganic and organic carbonates $[RO]_2CO$ as well as carbamates and acids can in principle be performed. This chapter addresses thermal processes of carboxylation, namely the Kolbe-Schmidt synthesis and production of urea from ammonia and CO_2, as well as synthesis of ethylene glycol by carboxylation/hydrolysis.

13.2.1 Kolbe-Schmidt synthesis

The synthesis of salicylic acid (Kolbe-Schmidt synthesis) is based on the reaction of sodium (or potassium) phenoxide in a stream of CO_2 (Figure 13.5) under 0.5 MPa at ca. 150–190 °C (for Na) and ca. 220 °C for K salt, giving a yield of ca. 90%.

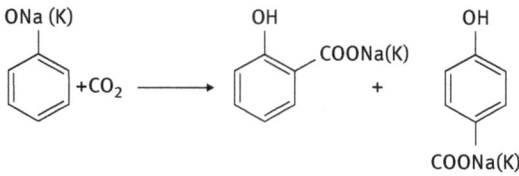

Figure 13.5: Synthesis of salicylic acid.

The *o*- or *p*-isomers of the salt have quite different applications, with the *o*-isomer being used for the synthesis of aspirin (ca. 20 kt/year) and the *p*-isomer is applied as a monomer for specialty optical polymers. Powerful mill autoclaves operating in a batchwise mode are used for carboxylation (Figure 13.6), which is an exothermal reaction ($\Delta H = -90.1$ kJ/mol).

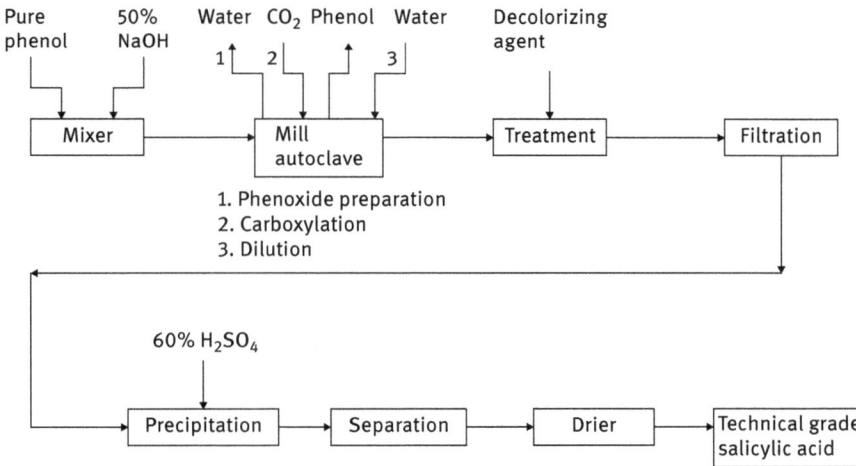

Figure 13.6: Simplified representation of salicylic acid production by the Kolbe-Schmitt method.

Sodium phenoxide is prepared first with a 1–2% molar excess of caustic soda while larger amounts of alkali would lead to water formation, which should be avoided. Anhydrous sodium phenoxide is prepared either in the autoclave mixer itself by evaporation of an aqueous solution of phenoxide, starting at normal pressure and then gradually introducing vacuum, or in a special drying equipment. Presence of water results in formation of alkali-metal hydroxide, which converts CO_2 into carbonate with the regeneration of water. Carbon dioxide contains less than 0.1% of oxygen to prevent discoloration and tar formation. Phenol formed by a side reaction along with disodium salicylate is separated by distillation. Formed phenol is negatively influencing carboxylation by wetting the solid anhydrous phenoxide and diminishing the interfacial area. Therefore, typically at the end of the carboxylation cycle, the CO$_2$ pressure can be increased to enhance the reaction. If conversion is not sufficient (as, for example, in carboxylation of 2-naphthol salts), the steps of carboxylation and 2-naphthol removal under vacuum are repeated.

To the crude sodium salicylate, a mixture of water and a decolorizing agent (e.g., activated carbon) is added, and after filtration, salicylic acid is precipitated with sulphuric acid.

13.2.2 Synthesis of ethylene glycol

Synthesis of ethylene glycol by oxyalkylation was discussed in Chapter 12. In an alternative method for the production of monoethylene glycol (MEG) developed by Mitsubishi Chemical Corporation, selectivity to the desired product exceeds 99%, while the MEG selectivity in the conventional non-catalytic process is ca. 89% because of di-ethyleneglycol and tri-ethyleneglycol formation. High selectivity is a result of the two-step synthesis through ethylene carbonate, where first ethylene oxide reacts with CO$_2$ in the presence of a phosphonium salt catalyst of $R_1R_2R_3R_4P^+X^-$ type (Ri; alkyl or allyl, X$^-$ halide, e.g., iodides and bromides) type. This is followed by subsequent hydrolysis of the carbonate to ethylene glycol releasing simultaneously carbon dioxide.

The simplified process flow diagram is shown in Figure 13.7.

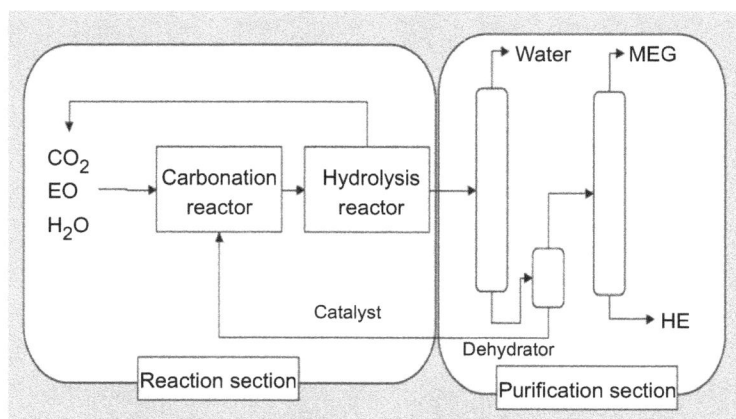

Figure 13.7: Synthesis of ethylene glycol from ethylene oxide. From K. Kawabe, Development of highly selective process for mono-ethylene glycol production from ethylene oxide via ethylene carbonate using phosphonium salt catalyst. *Catalysis Surveys from Asia, 2010,* 14, 111–115. Copyright Springer Science Business Media. Reproduced with permission.

The feed comprises 60 wt% aqueous solution of ethylene oxide and gaseous CO$_2$, a by-product of ethylene oxidation to epoxide, as well as the catalyst (a phosphonium salt) solution. A sufficient supply of carbon dioxide to the liquid phase is needed otherwise resulting in lower selectivity and deterioration of the product quality.

In addition to formation of ethylene carbonate through an exothermal addition of carbon dioxide to ethylene oxide, hydrolysis also happens to a certain extent resulting in small amount of ethylene glycols. The reaction temperature is reported to be ca. 50 °C lower that in the oxyalkylation process. Unreacted carbon dioxide is separated from the liquid, compressed, and recycled, while the liquid phase is routed to the hydrolysis reactor where ethylene carbonate is hydrolyzed to ethylene glycol generating CO_2, which is recycled apart from purge. The latter is needed to prevent impurity accumulation in the recycle loop. Even if the water content in ethylene oxide feed is sufficient for hydrolysis, additional water feed is preferred accelerating the reaction.

The catalyst separation downstream water removal is done by evaporation, allowing catalyst recycle back to the carbonation reactor. Monoethylene glycol is recovered at the top of the distillation column, while small amounts of diethyleneglycol and heavier glycols are obtained from the bottom and disposed. The catalysts for carboxylation are selected because of their high solubility in the reaction, low melting and high boiling points, good stability under the reaction conditions, and a non-corrosive behavior.

High activity of the phosphonium catalyst ensures minimal unselective ethylene oxide hydration with dimerization, the latter generating also undesired ethyleneglycol. Such differences in the rates allow utilization of an aqueous ethylene oxide solution directly obtained from ethylene oxide manufacturing.

The technology discussed above is aimed at manufacturing of monoethylene glycol. However, standalone production of ethylene carbonate, an important ingredient of Li battery cells, is also possible and has been commercially implemented.

13.2.3 Urea from CO_2 and ammonia

Urea can be used to produce urea-ammonium-nitrate, which is one of the most common forms of liquid fertilizers. The existing urea capacity is 190 Mt/year, with the largest plant having a capacity of 1 Mt/year.

Synthesis of urea involves a fast non-catalytic reaction of ammonia and carbon dioxide (no catalysts are used due to the corrosiveness of the reaction mixture) at high pressure (preventing the backward reaction) to form ammonium carbamate NH_2COONH_4 ($\Delta H = -159$ kJ/mol at 25 °C):

$$2NH_3 + CO_2 \leftrightarrow NH_2COONH_4 \qquad (13.4)$$

For example, at 147 °C pressures of 13 MPa are required to prevent ammonium carbamate dissociation into ammonia and carbon dioxide. Elevated pressure is needed also to ensure proper dissolution of otherwise gaseous ammonia and CO_2, while elevated temperature is required to perform the reaction in the liquid phase because the melting point of urea is 153 °C. Subsequent slow endothermal decomposition of ammonium carbamate (the so-called Basarov reaction, with $\Delta H = 31.4$ kJ/mol at 25 °C) results in the target product-urea:

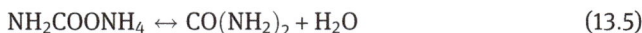

$$NH_2COONH_4 \leftrightarrow CO(NH_2)_2 + H_2O \qquad (13.5)$$

Thermodynamically, the urea yield goes through a maximum at ca. 177–207 °C. The location of the maxima in Figure 13.8 is composition dependent.

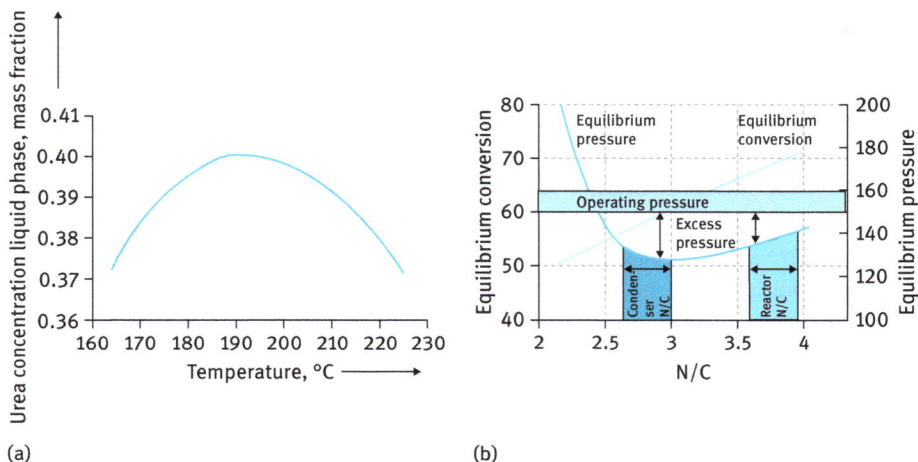

(a)   (b)

Figure 13.8: Thermodynamic data for urea synthesis: (a) urea yield at chemical equilibrium as a function of temperature at NH_3/CO_2 ratio = 3.5 mol/mol (initial mixture) and H_2O/CO_2 ratio = 0.25 mol/mol (initial mixture) and (b) equilibrium conversion and pressure as a function of nitrogen/carbon ratio. Adapted from http://www.toyo-eng.com/jp/ja/products/petrochmical/urea/technical_paper/pdf/2000_Development_of_the_ACES%2021_Process.pdf.

Ammonia-carbon dioxide system is characterized by a strong positive azeotrope, whose mole ratio is approximately 3 (Figure 13.8B), being far from the stoichiometric ratio of 2 (eq. (13.4)). Pressure rise at the carbon dioxide-rich side of the azeotrope is much steeper than for ammonia-rich side, and it is thus unpractical to perform the process at such high pressures. All commercial processes operate therefore in the synthesis step at a higher than the stoichiometric ratio between ammonia and carbon dioxide. Conversion also increases with increasing NH_3/CO_2 ratio and decreases with increasing H_2O/CO_2 ratio. Since there is a detrimental effect of excess water on urea yield, one of the targets in the process design is to minimize water recycling. Very high NH_3/CO_2 ratio reduces urea yield; thus, typically a molar ratio of 3–5 mol of ammonia per 1 mol of carbon dioxide is used (between 3 and 3.7 in stripping processes and between 4 and 5 in conventional recycling ones).

Conversion of CO_2 to urea reaches 50–80%; thus, carbamate is separated from urea by thermal decomposition of the former back to ammonia and carbon dioxide.

Recycling of ammonia/carbon dioxide mixtures is not straightforward (Figure 13.9), requiring recompression to reaction pressure with subsequent formation of either droplets or small crystals of carbamate, damaging compressor components.

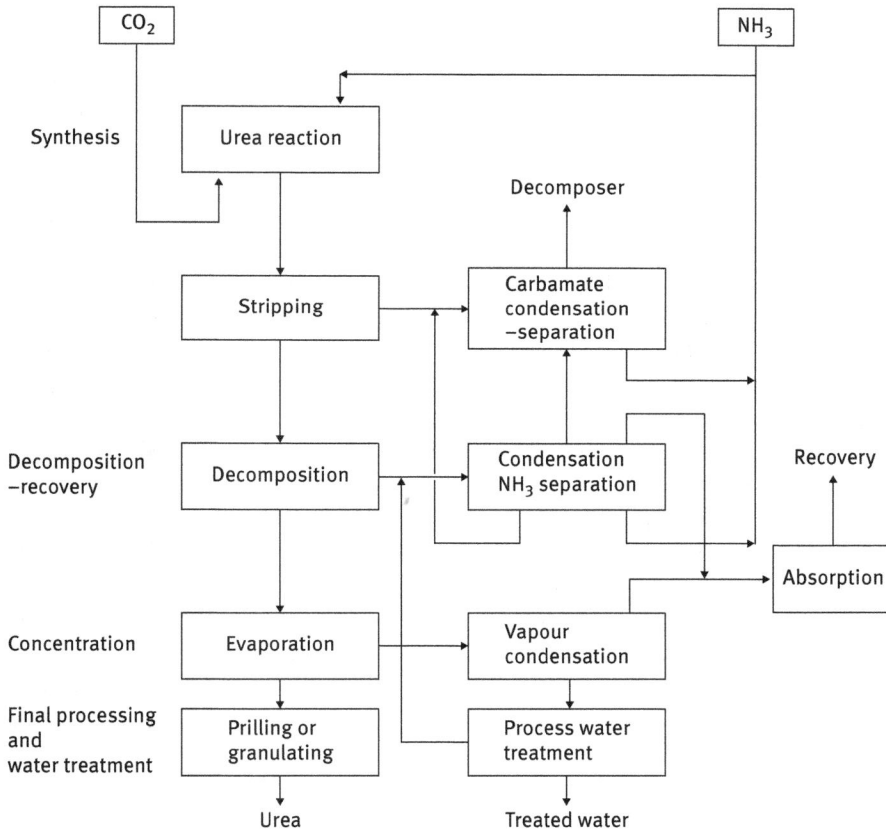

Figure 13.9: Overall flow diagram for urea synthesis. http://3.bp.blogspot.com/_jwzEb3tcs7U/TR1P8nkyQDI/AAAAAAAAALg/o3hgB_V1ar0/s1600/block+diagram+of+total+recycle+ammonia+stripping + urea + process.gif.

In the production of urea, side reactions, which should be minimized, are urea hydrolysis

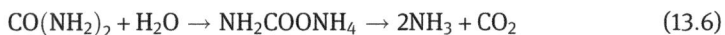

$$CO(NH_2)_2 + H_2O \rightarrow NH_2COONH_4 \rightarrow 2NH_3 + CO_2 \tag{13.6}$$

formation of biuret

$$2CO(NH_2)_2 \rightarrow NH_2CONHCONH_2 + NH_3 \qquad (13.7)$$

and isocyanic acid

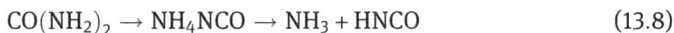

$$CO(NH_2)_2 \rightarrow NH_4NCO \rightarrow NH_3 + HNCO \qquad (13.8)$$

The extent of hydrolysis depends on temperature, implying that the urea-containing solutions should not be kept at high temperature for a long time. The biuret reaction, which is endothermic, should be minimized, as biuret is detrimental for crops. Even if equilibrium is reached for this reaction in the reactor, only small amounts of biuret are formed due to high ammonia concentration. However, downstream processing requires removal of ammonia from the urea solutions, calling for minimizing of urea solution exposure to high temperature for long time. Generation of isocyanic acid is relevant for conditions when pressure of ammonia is low, shifting equilibrium to the right, e.g., in evaporation section. Isocyanic acid is collected in the process condensate from the vacuum condensers. Low temperature promotes backward reaction of HNCO with ammonia forming urea.

As mentioned above, solutions containing ammonium carbamate are corrosive. Stainless steel, having a protective oxide layer, is much more corrosion resistant than austenitic stainless steel; therefore, a small amount of air is added to the carbon dioxide feed to keep such layer intact. Duplex (austenitic/ferritic) stainless steel was also introduced in urea plants, having an advantage of considerably less requirements in oxygen or even no necessity to use air for preventing corrosion.

In conventional recycling processes, unconverted ammonia and carbon dioxide were recycled to the urea reactor. One of the options was to operate the first recirculation stage at medium pressure (1.8–2.5 MPa) where carbamate was decomposed into gaseous ammonia and CO$_2$, with simultaneous evaporation of excess NH$_3$ in a decomposition heater.

Rectification of the off-gas done at lower pressure resulted in formation of relatively pure NH$_3$ taken from the top of the rectification column and an aqueous ammonium carbamate solution at the bottom. Both streams were recycled separately to the reactor. Such approach resulted in recycling of the main fraction of non-converted NH$_3$ without an associated water recycling. High NH$_3$/CO$_2$ ratios (4–5 mol/mol) were used to maximize CO$_2$ conversion per pass, as the target was to achieve a minimum CO$_2$ recycling. Otherwise, if all non-converted CO$_2$ is recycled as an aqueous solution, the detrimental effect of water on conversion would be profound.

Stripping processes requiring external heat supply were developed starting from the 1960s and have replaced the conventional processes owing to its numerous advantages, such as recycling the major part of both non-converted NH$_3$ and CO$_2$ through the gas phase and not requiring large water recycling to the synthesis zone. Another difference between the processes is related to the way the heat is supplied.

Stripping the product mixture is done either with carbon dioxide (Stamicarbon and Toyo Engineering) or with ammonia (Snamprogetti/Saipem). In stripping plants, recycling of unconverted ammonium carbamate, excess ammonia, and CO_2 takes place at a pressure that is close to the urea synthesis pressure, which in fact is supercritical for both reactants. Downstream the urea synthesis reactor, the solution is heated first at virtually reactor pressure, resulting in endothermal decomposition of ammonium carbamate into ammonia and carbon dioxide in the liquid phase and partial evaporation of ammonia and carbon dioxide into the gaseous phase. The urea-containing liquid phase is then separated from the gaseous phase. The latter is cooled, partially condensing carbon dioxide and ammonia and resulting in exothermal formation of ammonium carbamate, which is recycled into the reaction zone. Such recycling option does not require any water addition to the recycling, avoiding the negative effect of water on overall conversion.

Condensation and carbamate formation in the stripping processes take place at elevated pressures and temperatures, allowing recovery of the heat of condensation and ammonium carbamate formation as low-pressure steam, which can be used either in the process itself or outside of the urea plant. Stripping processes differ in terms of the stripping agent, ratio of recycled amounts of ammonium carbamate and excess of ammonia in the high-pressure recycling loop to the amounts recycled in subsequent low(er)-pressure stage(s), reaction parameters (temperature, pressure, and composition), driving forces for the recycling (gravity or power-driven devices for keeping the flow, such as liquid-liquid ejectors with pressurized ammonia as driving medium), and construction materials.

In the late 2000s, in the Snamprogetti/Saipem stripper design, the full zirconium stripper was applied where both lining and tubes were made of resistant to erosion and corrosion zirconium. Alternatively, Omegabond tubes obtained by extrusion of titanium (external) and zirconium (internal) billets can be used where a metallurgical bond is formed of the two materials. In both options, strippers can withstand more severe bottom temperatures, thus prolonging equipment lifetime and minimizing maintenance.

Stamicarbon CO_2-stripping processes (the original one with a vertical film condenser; The Urea 2000plus concept, applying pool condensation in the high-pressure carbamate condensation step and, more lately, the Avancore process) use CO_2 as stripping agent in a high-pressure stripper, gravity flow to maintain the main recycling flow in the high-pressure loop, and azeotropic N/C ratio (3:1) in the reactor. High conversion of both reactants allows to have only one small low-pressure carbamate recycling loop.

In the original Stamicarbon CO_2 stripping process, a falling-film evaporator was used for stripping with the urea synthesis solution flowing as a falling film along the inside of the vertical heat-exchanging tubes countercurrent to carbon dioxide entering the stripper from the bottom. External heat removal was done by steam at the shell side of such shell-and-tube heat exchanger. Unconverted ammonium carbamate was decomposed to ammonia and CO_2, while unconverted ammonia was stripped from the solution to the gas phase.

The location of the reactor above the stripper allows to capitalize on the difference in the density of the liquid flowing down from the reactor and the gaseous components flowing upward from the stripper, generating a gravity-based driving force in the high-pressure synthesis loop.

Catalytic removal of hydrogen from CO_2 streams by oxidation with air is needed since carbon dioxide generated from located nearby ammonia plants contains hydrogen, whose presence can lead to explosions of the urea plant tail gas. In addition to the air required for hydrogen removal, extra air is supplied as already mentioned with CO_2 to the synthesis section to preserve a corrosion-resistant layer on the stainless steel.

In Stamicarbon 2000plus technology, first, a pool condenser and, later, a horizontal pool reactor (Figure 13.10) were introduced. In pool condensation, the liquid phase is continuous, while the gases to be condensed are present as bubbles, rising through the liquid phase. Such arrangement enhances both heat and mass transfer. An increased residence time of the liquid phase in the condenser allows slow dehydration of ammonium carbamate to take place already in the pool condenser, thereby decreasing the required volume in the urea reactor. Since the urea and water that formed in the pool condenser have a higher boiling temperature than ammonia and ammonium carbamate, this gives a higher net boiling temperature of the liquid mixture in the condensation step. Subsequently, there is a higher temperature difference between the process side and the cooling side, which results in a smaller size heat exchanger and reduction in capital investment costs.

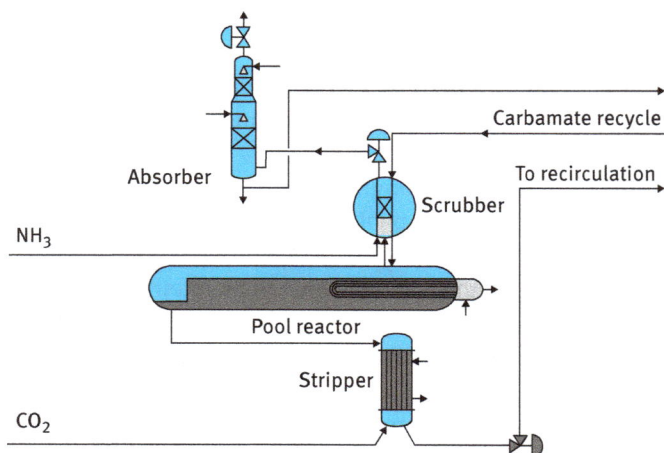

Figure 13.10: Pool reactor in Stamicarbon 2000plus technology. From J. Meessen, Urea synthesis, *Chemie Ingenieur Technik*, 2014, 86, 2180–2189, Copyright © 2014 WILEY-VCH Verlag GmbH & Co. KGaA, Weinheim. Reproduced with permission.

By combining the functions of the urea reactor and the carbamate condenser in one vessel, a further development of urea technology was made and the number of high-pressure items in the synthesis unit has been reduced from 4 to 2 (stripper and pool reactor), making a considerable amount of interconnecting high-pressure piping obsolete. The condensing section of the reactor contains the U-tube bundle, while the rest of the pool reactor forms the reaction zone for the dehydration of the carbamate.

The Avancore urea process (Figure 13.11) is a further development of Stamicarbon. It uses a corrosion-resistant Safurex material (special duplex-austentic/ferritic-stainless steel) in an oxygen-free carbamate environment, which eliminates the need for air passivation. Moreover, hydrogen or any other combustible gases present in the feed do not posses any risk of explosion. A low-elevation layout of the synthesis section in Avancore process with the reactor located at a ground level still relies on a gravity flow in the synthesis recycling loop but allows less investment and easier maintenance. The pool condenser off-gas cannot flow into this low-level reactor anymore; thus, such low-level reactor arrangement, diminishing the overall plant height to just 25 m, requires another heat source for the endothermic dehydration reaction. At the same time, most of the urea formation takes place already in the pool condenser; thus, a minor amount of CO_2 supplied to the reactor is sufficient to close the heat balance.

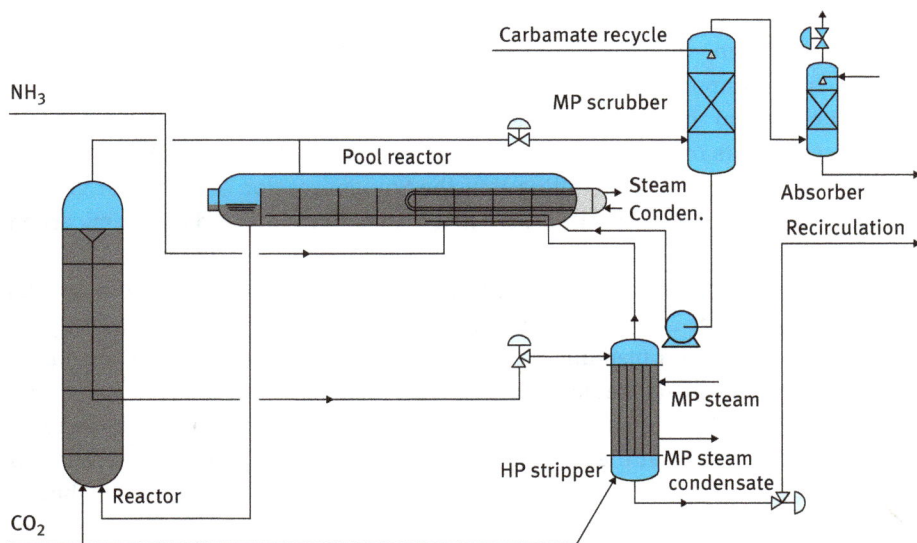

Figure 13.11: Avancore technology with pool condenser variant. From J. Meessen, Urea synthesis, *Chemie Ingenieur Technik*, 2014, 86, 2180–2189, Copyright © 2014 WILEY-VCH Verlag GmbH & Co. KGaA, Weinheim. Reproduced with permission.

Another feature of the Avancore process is that the vapor from the urea synthesis section is treated in a scrubber, which operates at a reduced pressure. A carbamate solution coming from the downstream low-pressure recirculation stage is used to absorb most of ammonia and carbon dioxide left after scrubbing. No additional water needs to be recycled to the synthesis section.

Ammonia was used as stripping agent in the first generation of Snamprogetti/ Saipem urea process, which led to large amounts of dissolved ammonia in the stripper effluents. The self-stripping variant of the same process was introduced later, relying only on thermal ("self") stripping but still requiring an ammonia-carbamate separation section due to relatively high ammonia/CO_2 ratio (Figure 13.12). A significant number of plants (more than 100) have been designed using either ammonia- or self-stripping processes. Thermal stripping is done at relatively high temperatures (200–210 °C); therefore, instead of stainless steel titanium and other materials are used. Medium-pressure purification and recovery section typically operate at 1.8 MPa, allowing decomposition of carbamate, evaporation of ammonia, and separation of gaseous ammonia from liquid ammonium carbamate, which is recycled as the liquid ammonium carbamate-water mixture. In a second low-pressure decomposition step (Figure 13.12), ammonium carbamate from urea solution is further decomposed, resulting in an almost carbamate-free aqueous urea solution, while the off-gas from this low-pressure decomposer after condensation is recycled through the medium-pressure recovery section to the synthesis section in the form of an aqueous ammonium carbamate solution. Evaporation of water from urea is done in a single evaporation step when fluidized-bed granulation is used for finishing or in a two-stage evaporator when finishing is done by prilling. Urea and ammonia are recovered from the process condensate.

Typical layouts of a Stamicarbon PoolCondenser plant and a Saipem plant are presented in Figure 13.13 illustrating that the Saipem reactor is located at ground level. The Stamicarbon reactor is located at a higher elevation, and the Stamicarbon PoolCondenser is located at the third floor.

Technological approaches in Stamicarbon and Saipem are different as CO_2 stripper in the former is considered to be more efficient than the ammonia stripper in the latter, requiring a medium-pressure recirculation section to further separate carbamate from the urea/water mixture. Moreover, better efficiency implies lower temperatures and thus less expensive construction materials. The medium-pressure recirculation section allows, at the same time, condensation of the medium-pressure off gases saving low-pressure steam consumption. Moreover, the inerts can be washed in the medium-pressure section, while a high-pressure scrubber is applied in Stamicarbon technology leading to an extra high-pressure heat exchanger. There are also few other minor differences in the technologies and operation conditions as can be seen from Table 13.1.

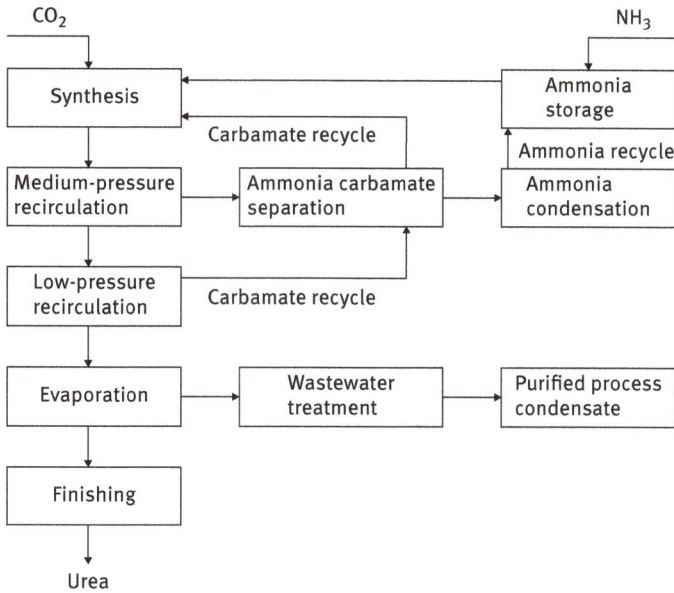

Figure 13.12: Flow diagram of Snamprogetti/Saipem urea technology.

Figure 13.13: Typical layouts of a Stamicarbon PoolCondenser plant and a Saipem plant. From https://www.researchgate.net/profile/Prem-Baboo/publication/309385422_The_Comparison_of_Stamicarbon_and_Saipem_Urea_Technology/links/580ce51b08ae2cb3a5e3c195/The-Comparison-of-Stamicarbon-and-Saipem-Urea-Technology.pdf.

Table 13.1: Comparison between Stamicarbon and Saipem urea manufacturing technologies. From https://www.researchgate.net/profile/Prem-Baboo/publication/309385422_The_Comparison_of_ Stamicarbon_and_Saipem_Urea_Technology/links/580ce51b08ae2cb3a5e3c195/The-Comparison- of-Stamicarbon-and-Saipem-Urea-Technology.pdf.

Parameter	Units	Stamicarbon	Saipem
Synthesis layout		Vertical	Horizontal
Synthesis loop driver		Gravity	High pressure ammonia ejector
High pressure equipment items		4	4
Medium-pressure section		No	Yes
Pure ammonia recycle		No	Yes
Reactor pressure	bara	140	158
Reactor outlet temperature	°C	184	188
Reactor outlet N/C ratio		3.0–3.1	3.3–3.6
Reactor CO$_2$ conversion	%	60	64–67
Reactor NH$_3$ conversion	%	40–41	36–37
Stripper pressure	bara	140	148–150
Stripper temperature range (top-bottom)	°C	187–167	190–208
Synthesis CO$_2$ conversion	%	79–81	84–86
Synthesis NH$_3$ conversion	%	79–81	47–50
CO$_2$ consumption	kg/mt	567	566
NH$_3$ consumption	kg/mt	733	735

ACES21 process (Figure 13.14) was introduced by Toyo Engineering. Feeding of liquid ammonia to the reactor is done by a high-pressure carbamate ejector providing the driving force for circulation in the synthesis loop.

A minor part of carbon dioxide is also introduced to the reactor, while most of the carbon dioxide is admitted to the stripper along with passivation air. The duplex stainless steel material DP28W used in this process is claimed to possess excellent corrosion resistance and passivation properties requiring less passivation air. A drastic decrease of the inert gas feed to the reactor, being just 20% of the conventional CO$_2$ stripping process, leads to a substantial decrease of the vapor phase in the reactor. In the Toyo process, the nitrogen/carbon ratio is 3.7, which, along with other operation parameters (182–184 °C and 15.2 MPa), gives CO$_2$ conversion of 63–64%. This in turn results in less decomposition heat in the HP stripper and less energy for compression of CO$_2$ and pumping of liquid ammonia and carbamate solution. After the reaction, unconverted ammonium carbamate from the urea synthesis solution is

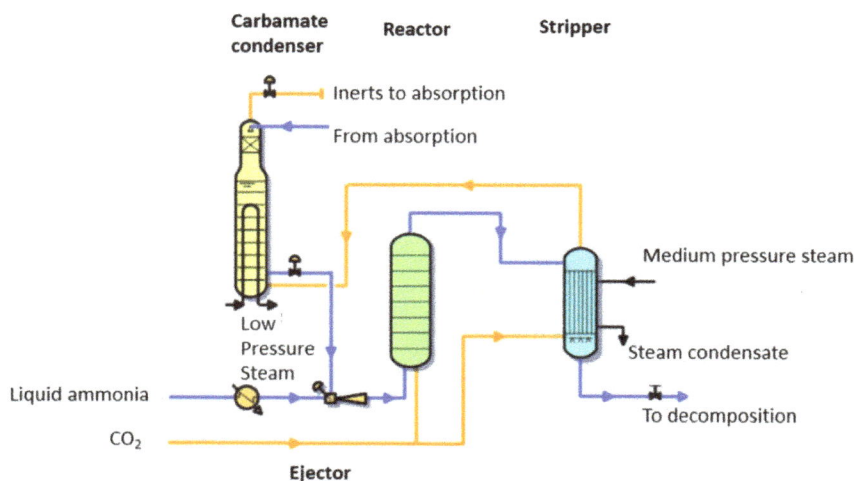

Figure 13.14: ACES21 process synthesis section. From http://www.toyo-eng.com/jp/en/products/ petrochmical/urea/technical_paper/pdf/ACES21_Brochure.pdf.

decomposed in a stripper. Excess ammonia and carbon dioxide are separated by CO_2 stripping. A vertical submerged condenser (bubble column reactor with boiler tubes) used in the Toyo process (Figure 13.15) operates at N/C ratio of 2.8–3.0, 180–182 °C, and 15.2 MPa, resulting in efficient ammonium carbamate dehydration to form urea.

A vertical submerged condenser is designed to condense ammonia and carbon dioxide gas mixture, forming ammonium carbamate with subsequent dehydration to urea on the shell side as well to remove the reaction heat by generating steam in boiler tubes. It allows high gas velocity, appropriate gas hold-up, and sufficient liquid depth in the bubble column promoting mass and heat transfer. Moreover, efficient distribution of bubbles owing to baffle plates is achieved without pressure losses.

ACES process uses medium-pressure and low-pressure decomposition stages for the treatment of the urea solution from the stripper and evaporation, giving a concentrated urea melt.

Either prilling or granulation (Figure 13.16) is used for the final shaping technology for urea.

Prilling, which is a low-investment and variable-costs option, has been in operation for a long time. It uses the distribution of the urea melt in the form of droplets in a prilling tower. Cooling is done with upflowing air, resulting in the solidification of urea droplets that fall down the tower through showerheads or a rotating prilling bucket with holes. Prilling technology gives a limited maximum average size (ca. 2.1 mm) of the product. Larger and less stable sizes would require uneconomically high prilling towers. Formation of fine dust (0.5–2 µm) is a clear disadvantage of the process. Such dust is technically difficult and expensive to remove because dry

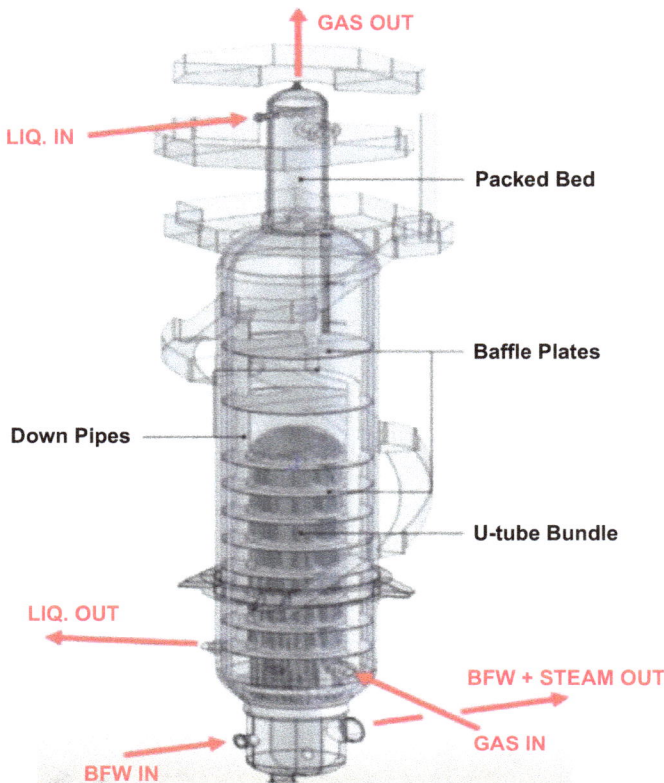

Figure 13.15: A vertical submerged condenser used in the Toyo process. From https://www.toyo-eng.com/jp/ja/products/petrochmical/urea/technical_paper/pdf/ACES21_Brochure.pdf.

cyclones cannot be used and wet impregnation should be implemented. Moreover, a limited crushing strength resistance of prills results in problems with their long-distance transportation; therefore, many new urea plants use granulation instead of prilling. Fluidized-bed or drum granulation improves the size and strength of the product and, due to lower contact time, results in much coarser dust, allowing much simpler dust emission control compared to prilling. In granulation, the urea melt is sprayed on granules, which gradually increase in size. Removal of the solidi-fication heat is done by cooling with air or alternatively by water evaporation. As an example of granulation, the Toyo spouted-bed granulation technology is pre-sented in Figure 13.16.

After addition of formaldehyde or formaldehyde-containing components re-quired for granulation, the solution of urea is sent into a granulator operating at 110–115 °C to which fluidization air is supplied to ensure fluidization of granules. An aftercooling section inside the granulator is used to cool the enlarged urea

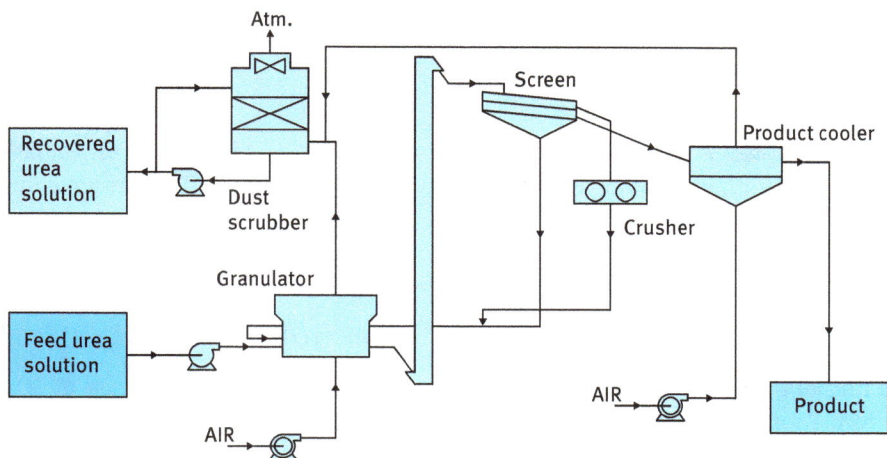

Figure 13.16: Schematic of the TEC spouted-bed granulation technology. http://www.toyo-eng.com/jp/en/products/petrochmical/urea/technical_paper/pdf/ACES21_Brochure.pdf.

granules to ca. 90 °C prior to transport to the screening section. The larger granules are crushed and recycled along with the undersized granules. Urea granules of the desired size after screening are cooled to 60 °C and sent to storage. Dust from the granulator and the product cooler is scrubbed with water in a scrubber. This allows recovering ca. 3–4% of total urea produced, which is recycled in the form of 45 wt% urea solution back to the urea plant.

Several existing urea plants revamp strategies have emerged recently. In the high-efficiency combined urea process, two urea reactors are placed in parallel, with one of them operating without carbamate recycle. This once-through reaction line is installed in parallel to an existing plant. An alternative approach is to add a new decomposition section to treat carbamate from the low-pressure and/or medium-pressure recirculation stages. This diminishes the amount of recycled water and allows higher conversion. The vapor phase after decomposition contains ammonia and carbon dioxide and is sent to the synthesis section, while the purified solution is returned to the back end of the plant.

13.2.4 Synthesis of melamine

Besides being used as a fertilizer, urea is a feedstock for synthesis of melamine (Figure 13.17a), which is often integrated with production of urea starting from ammonia as feedstock.

Melamine resins are produced by the reaction of melamine with formaldehyde, while melamine foams can be used for insulation and some other applications, such as fire-retardant additives in paints or plastics and paper. Melamine is produced from urea in an overall endothermic process ($\Delta H = 629$ kJ/mol)

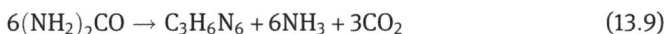

$$6(NH_2)_2CO \rightarrow C_3H_6N_6 + 6NH_3 + 3CO_2 \tag{13.9}$$

using either a gas phase catalytic technology operating at ca. 1.0 MPa or a high-pressure liquid-phase option requiring pressure above 8.0 MPa. Dry and aqueous recovery can be used in both technologies.

In the catalytic processes, the first reaction step is decomposition of urea to isocyanic acid and ammonia

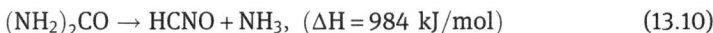

$$(NH_2)_2CO \rightarrow HCNO + NH_3, \quad (\Delta H = 984 \text{ kJ/mol}) \tag{13.10}$$

with further transformation of the acid into the final product through release of CO_2 and formation of cyanamide H_2NCN or carbodiimide $HNCNH$, which trimerizes to melamine according to the following overall reaction:

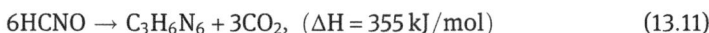

$$6HCNO \rightarrow C_3H_6N_6 + 3CO_2, \quad (\Delta H = 355 \text{ kJ/mol}) \tag{13.11}$$

In the catalytic fluidized-bed process, which uses alumina or aluminosilicates catalysts and operates at 390–410 °C, the fluidizing gas is either ammonia or a mixture of NH_3 and CO_2 formed in the reaction. Separation of gaseous melamine from ammonia and carbon dioxide is done with water quenching followed by crystallization or desublimation when the cold reaction gas is applied for quenching.

The yield of melamine based on urea conversion is ca. 90–95%. Some of the by-products formed either during synthesis or at melamine recovery due to ammonia release or hydrolysis are illustrated in Figure 13.17 and include melam, melem, melon (poly(tri-s-triazine)) as well as oxotriazines (ammeline, ammelide, and cyanuric acid). Ureidotriazine is a product of a reaction between melamine and isocyanic acid.

Figure 13.18 illustrates a one-stage catalytic vapor phase process developed by BASF. The advantage of using a single stage is in the transformations of the corrosive intermediate isocyanic acid to melamine in the same reactor and better heat integration. The heat of this exothermic reaction is in fact utilized for the endothermic decomposition of urea occurring in the same reactor as the first step in melamine synthesis.

In this process, molten urea is fed to the reactor (1) operating with alumina at 395–400 °C and atmospheric pressure. Make-up ammonia is added to the reactor besides the fluidizing gas (process off-gas mixture of ammonia and CO_2), which is preheated to 400 °C in a preheater (3). Internal heating coils (2) with a molten salt are used to sustain the reaction temperature. The outlet gases contain, besides melamine and unreacted urea (as isocyanic acid and ammonia), ammonia and CO_2 (formed and introduced as the fluidization gas) and some by-products as well as entrained catalyst fines. Coarser catalyst particles are retained by cyclone separators

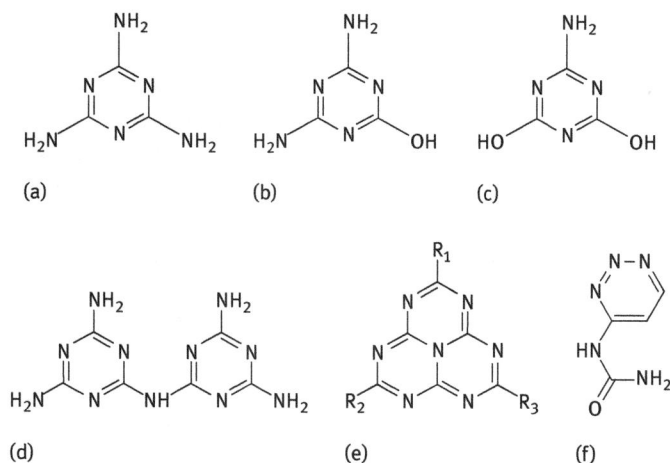

Figure 13.17: (a) Melamine (1, 3, 5-triazine-2, 4, 6-triamine), (b) ammeline (4, 6-diamino-2-hydroxy-1, 3, 5-triazine), (c) ammelide (6-amino-2, 4-dihydroxy-1, 3, 5-triazine), (d) melam (N^2-(4, 6-diamino-1, 3, 5-triazin-2-yl)-1, 3, 5-triazine-2, 4, 6-triamine), (e) melem, and (f) ureidotriazine.

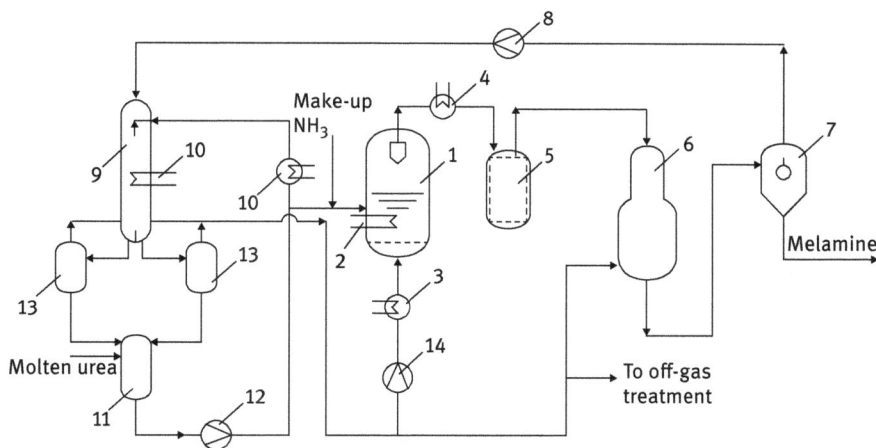

Figure 13.18: Low-pressure vapor-phase process for melamine synthesis developed by BASF: 1, reactor; 2, heating coils; 3, fluidizing gas preheater; 4, gas cooler; 5, gas filter; 6, crystallizer; 7, cyclone; 8, blower; 9, urea washing tower; 10, heat exchanger; 11, urea tank; 12, pump; 13, droplet separator; 14, compressor.

inside the reactor. Cooling in a gas cooler (4) is done to a by-product melem crystallization temperature. This by-product, as a fine powder, is removed together with the entrained catalyst fines in gas filters (5).

Crystallization of melamine with 98% efficiency is organized in the crystallizer (6) to which counter-currently the recycled off-gas is added at 140 °C, decreasing the temperature in the crystallizer to 190–200 °C. Fine crystals of melamine are recovered in a cyclone (7), giving at the end a minimum product purity of 99.9%. In the urea washing tower (9), scrubbing of almost melamine-free gas stream coming from (7) is done with molten urea (135 °C). This is followed by separation of the droplets in (13) and recycling of the clean gas to the reactor as the fluidizing gas and to the crystallizer as quenching gas. An off-gas treatment unit is used for the cleaning of the surplus.

DSM Stamicarbon process is similar to BASF technology involving also a single catalytic stage. The differences are in pressure (0.7 MPa), fluidizing gas (pure ammonia), catalyst type (silica-alumina type), and melamine recovery from the reactor outlet gas (water quench and recrystallization).

An alternative to BASF and DSM Stamicarbon processes but still a low-pressure process is the two-stage Chemie Linz process, where molten urea is decomposed in a fluidized sand-bed reactor to ammonia and isocyanic acid at ca. 350 °C and 0.35 MPa. Ammonia is used as the fluidizing gas, while molten salt circulating through the internal heating coils is applied for heat supply. The gas stream is routed to a fixed-bed catalytic reactor for conversion of isocyanic acid to melamine at near atmospheric pressure and ca. 450 °C. Fast quenching of melamine by water is needed to prevent significant hydrolysis of melamine to ammelide and ammeline (see Figure 13.17). Further cooling of the melamine suspension completes melamine crystallization, which is followed by centrifugation, drying, milling, and finally storage. Exhaust gas from the quencher contains CO_2 and ammonia. After washing with a lean carbamate solution, ammonia containing gas, cleaned from CO_2, is dried with make-up ammonia and partly recycled to the urea decomposition reactor after compression or used for other purposes such as urea production.

In high-pressure (>7 MPa) non-catalytic melamine synthesis technology, melamine is produced in the liquid phase at temperatures above 370 °C, generating high-pressure off-gas, which is more suitable for use in urea production. The overall purity of melamine in such high-pressure processes is above 94%. Technically, the process is organized by injecting molten urea at high pressure into a reactor with a molten melamine-urea mixture. Although, as typical with liquid-phase processes, smaller reactor volumes can be used, expensive corrosion-resistant construction materials such as titanium are required because of a highly corrosive nature of the system.

In a high-pressure process, cyanic acid HOCN is formed first

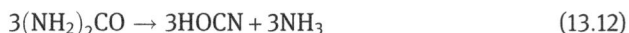

$$3(NH_2)_2CO \rightarrow 3HOCN + 3NH_3 \qquad (13.12)$$

followed by exothermal transformation to cyanuric acid

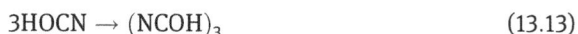

$$3HOCN \rightarrow (NCOH)_3 \qquad (13.13)$$

which condenses with ammonia, forming melamine and water

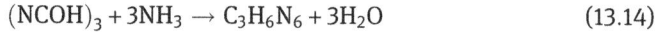

$$(NCOH)_3 + 3NH_3 \rightarrow C_3H_6N_6 + 3H_2O \tag{13.14}$$

Hydrolysis reactions subsequently generate carbon dioxide and ammonia.

Several technologies have been applied for production of melamine at high pressure, differing in pressure, temperature as well as separation. For example, a single-stage process of Melamine Chemicals operates at 11–15 MPa and 370–425 °C, giving the product yield of ca. 96–99.5%. In the cooling unit, liquid ammonia is used to solidify crystals from the liquid melamine. A somewhat lower pressure (10 MPa) is applied in the Nissan melamine process operating 400 °C, where ammonia is also fed to the reactor. Melamine and unreacted urea removal from the reactor off-gas in the Nissan process is done by washing with urea pressurized to 10 MPa. The process of Montedison (Figure 13.19) operates at 370 °C and 7 MPa.

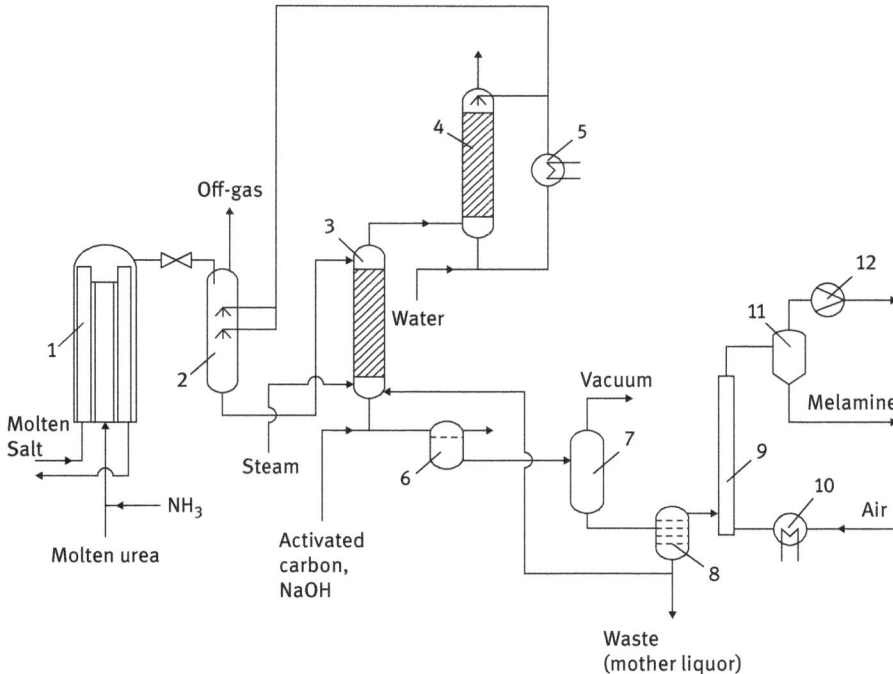

Figure 13.19: Montedison process for high pressure melamine production: 1, reactor; 2, quencher; 3, stripper; 4, absorption column; 5, heat exchanger; 6, filter; 7, vacuum crystallizer; 8, filter; 9, pneumatic dryer; 10, heat exchanger; 11, cyclone; 12, blower.

Preheated ammonia is fed along with molten urea (at 150 °C) to the reactor (1) heated by a molten salt. After the reactor, which operates with a residence time of ca. 20 min, the reaction mixture is expanded to a pressure of 2.5 MPa and quenched at 160 °C in (2) with an aqueous solution of ammonia and carbon dioxide, resulting in melamine precipitation. Unconverted urea as well as biuret and triuret are decomposed in the quencher to ammonia and carbon dioxide. Removal of the remaining NH_3 and CO_2 is done in the steam stripper (3). The quencher off-gas is recycled to urea or fertilizer production, while the stripper off-gas is first dissolved in water in an absorption column (4) and then recycled to the quencher as a solution. Dissolution of melamine from its slurry after ammonia and carbon dioxide removal is done by adding water to the stripper bottom followed by treatment with active carbon and sodium hydroxide in (6). After this clarification and subsequent crystallization in a vacuum crystallizer (7) operating adiabatically under vacuum, the mother liquor is separated from the crystals of melamine in a rotary filter (8). Downstream treatment of melamine includes drying with air in a pneumatic conveyor-dryer (9) and its separation in a cyclone (11) prior to storage. Crystallization and washing of melamine generates a considerable amount of wastewater, which is concentrated prior to disposal into a solid (1.5–5% of the weight) containing, besides melamine (ca. 70%), oxytriazines and some minor amounts of polycondensate.

In the process of Eurotecnica, which is also a single-stage liquid-phase noncatalytic process, the contaminants in the wastewater are decomposed to NH_3 and CO_2 and recycled to the urea synthesis; therefore, the wastewater can be recycled to the melamine plant itself or used as clean cooling water make-up.

13.3 Methanol from synthesis gas

Methanol is synthesized from CO and hydrogen according to the following reversible exothermal reaction:

$$CO + 2H_2 \leftrightarrow CH_3OH, \quad \Delta H = -90.8\,kJ/mol \tag{13.15}$$

Since the reaction is exothermal, equilibrium constant is decreasing with the temperature increase. Elevation of pressure results in shifting equilibrium toward the product side.

Another reaction leading to methanol is related to hydrogenation of carbon dioxide:

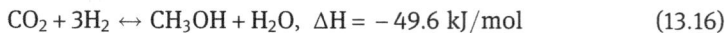

$$CO_2 + 3H_2 \leftrightarrow CH_3OH + H_2O, \quad \Delta H = -49.6\,kJ/mol \tag{13.16}$$

These two reactions are coupled by the water-gas shift reaction (eq. (5.5)), discussed in Chapter 5:

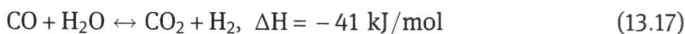

$$CO + H_2O \leftrightarrow CO_2 + H_2, \quad \Delta H = -41\,kJ/mol \tag{13.17}$$

By-products in this process are higher alcohols and hydrocarbons. Formation of dimethylether is also possible due to methanol dehydration. Application of active catalysts based on copper (CuZn/Al$_2$O$_3$) allowed to decrease the operation pressure (25–35 MPa) used in the classical gas phase processes with ZnO-Cr$_2$O$_3$ catalysts to ca. 5–10 MPa. Selectivity toward the desired product in low-pressure plants is above 99%. It should be kept in mind that modern catalysts allow to obtain such high selectivity toward the product, which is not the most thermodynamically stable. In fact, methane by methanation of CO is a more thermodynamically favored product than methanol. The conversion of CO and CO$_2$ to methanol is limited by chemical equilibrium (Table 13.2); thus, a temperature rise, being, in principle, beneficial from the viewpoint of kinetics, negatively influences thermodynamic equilibrium. In addition, high-activity catalysts are sensitive to temperature rise because they promote irreversible sintering and thus catalyst deactivation (Figure 13.20). Although initial activity declines substantially during operation as illustrated for different commercial catalysts in Figure 13.20, with a careful catalyst design, the lifetime can range from 4 to 6 years and could be even extended to 8 years.

Table 13.2: Conversion of CO and CO$_2$ at equilibrium conditions (syngas: 3 vol% CO$_2$, 27 vol% CO, 64 vol% H$_2$, 6 vol% CH$_4$ + N$_2$).

Temperature (°C)	Pressure (MPa)									
	2.5		5		7		15		30	
	CO	CO$_2$	CO	CO$_2$	CO	CO$_2$	CO	CO$_2$	CO	CO$_2$
275	14.4	3.91	39.1	4.18	53.9	4.47	81.9	6.23	95.8	13.5
300	5.94	5.55	21.7	5.68	35.1	5.88	69.2	7.13	90.1	11.6
325			10.4	7.63	19.5	7.76	53.2	8.64	81.8	11.7
350					9.41	10.1	36.4	10.7	70.1	12.8

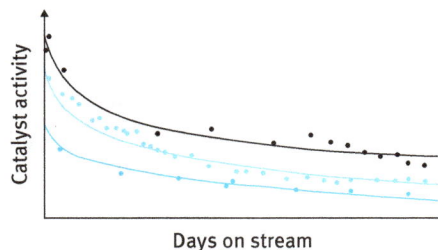

Figure 13.20: Dependence of catalytic activity in methanol synthesis with time on stream for different catalysts.

A typical measure for counterbalancing deactivation in various catalytic processes is to increase temperature, restoring activity but often compromising selectivity. In the case of methanol synthesis, such approach cannot be easily applied, as the temperature should not exceed ca. 270 °C. Because copper is marginally active below 230 °C, the temperature window for the process is rather low.

Increase in pressure is an alternative way of compensating for activity loss due to sintering. At the same time, too high pressures of CO and CO$_2$ (favoring conversion from the thermodynamic viewpoint) increase equipment costs in the synthesis loop and syngas compressor. The synthesis loop is thus required, as pressure in the modern plants of 5–10 MPa gives only moderate conversion levels (15–30% in adiabatic reactors). Unreacted gas is recycled back, acting as a syngas quench cooler. The ratio between the recycle gas and the fresh feed ranges from 3:1 to 7:1, which, along with the purge, allows to prevent buildup of impurities (methane and argon) in the loop.

An important theoretical and practical issue is related to a question of which reactant leads to methanol. A long controversy surrounded this topic, and either CO or CO$_2$ or both were proposed as the true reactants. Methanol can be produced from both H$_2$-CO and H$_2$-CO$_2$ mixtures, while a mixture containing H$_2$, CO, and CO$_2$ gives much higher yields of methanol. Isotopic labeling studies suggest that the source of carbon in methanol is CO$_2$, while CO is mainly converted to CO$_2$ via a water-gas shift reaction. CO$_2$ also influences the properties of the catalyst, keeping an intermediate oxidation state of copper (Cu°/Cu$^+$) and preventing reduction of ZnO. High concentrations of CO$_2$, however, inhibit methanol synthesis, whose rate drops slightly up to 12 vol% of CO$_2$ and thereafter more steeply.

A typical gas composition (can be different depending on the syngas generation procedure) could be thus 67.5% H$_2$, 21.5% CO, 8% CO$_2$, and 3% CH$_4$ for high-capacity plants and 69% H$_2$, 18% CO, 10% CO$_2$, and 3% CH$_4$ for lower- and medium-capacity plants.

Modern commercial catalysts for methanol synthesis from various suppliers applied in the form of tablets contain above 55 wt% CuO, 20–25% ZnO with 8–10% Al$_2$O$_3$, and also catalyst promoters, as well as catalyst binders (for example, graphite). For such a structure-insensitive reaction as methanol synthesis, the activity is dependent only on the total exposed copper area and is not affected by the structure of the crystallites. This means that large loading of copper (reaching 64% in some commercial formulations) and small cluster sizes are in fact needed for efficient catalysts. High metal dispersion as such is not sufficient for successful industrial operation, as the catalyst should be stabilized against sintering. It was mentioned above that thermal sintering is a key mechanism for deactivation with temperature approaching 315 °C depending on the reactor type. Moreover, sulphur and, in some cases, iron and nickel carbonyls introduced into the loop with fresh syngas contribute to catalyst deactivation. ZnO, used in the commercial formulations for many decades, is a textural and chemical promoter being introduced as small crystallites (2–10 nm). It helps to stabilize copper against sintering, facilitating the formation of small copper clusters and also scavenging sulphur. The alumina needed in the catalyst to stabilize both ZnO and copper oxide might, in principle, lead to formation of dimethylether; however, the presence of ZnO neutralizes acidic sites of alumina. Other promoters (such as MgO) were also introduced in commercial formulations.

Utilization of commercial catalysts in the form of cylindrical pellets of 5–12 mm implies that diffusional limitations can be significant.

A number of reactor designs and synthesis flowsheet arrangements for methanol production can be utilized. Reactor choice depends on plant size and the syngas generation method. In reaction selection, conversion temperature profiles should be optimized, being close to the equilibrium and affording lower peak temperature. Moreover, in addition to optimized temperature profile, proper mixing and reactant distribution allow higher selectivity, thus diminishing amounts of by-products, with substantial savings in product purification as well as lower deactivation and longer catalyst life.

Mainly multibed (3–4) adiabatic fixed-bed reactors are applied in low-pressure methanol processes with heat removal either by quenching with the cold feed (quenching) or using heat exchangers. The temperature profiles are far from the maximum rate curve as illustrated in Figure 13.21 for a four-bed adiabatic reactor with heat exchangers.

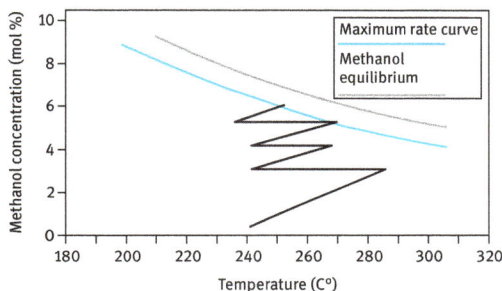

Figure 13.21: Temperature profile for multibed adiabatic methanol synthesis reactor with heat exchangers. Adapted from http://www.gbhenterprises.com.

Such multibed reactors represent, however, an attractive low-cost reactor concept when there is no need for steam generation. Not only cylindrical but also spherical adiabatic reactors (Figure 13.22c) are applied when the catalyst is located between two perforated spherical shells. Such a reactor type allows a decrease in vessel wall thickness at a given pressure and thus affords lower reactor costs. The flow in such reactors is organized from the outside of the catalyst layer to the center of the spherical core. Pressure drop is minimized as a relatively thin catalyst layer is used.

In a tube-cooled converter (Figure 13.22a), the feed enters the reactor at the bottom and flows upward through the tubes with minimum thickness, becoming preheated by the product gas flowing downward through the catalyst bed. This way of

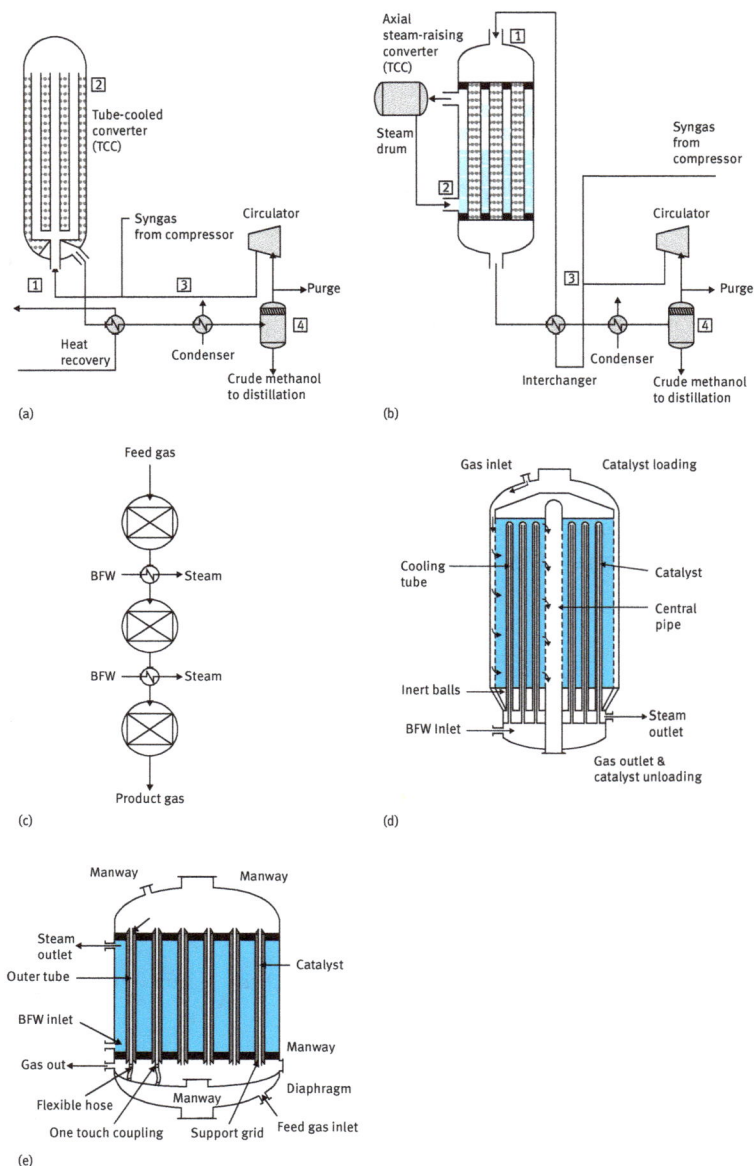

(a)

(b)

(c)

(d)

(e)

Figure 13.22: Special reactor types used for methanol synthesis: (a) tube-cooled converter and (b) radial-flow converter with axial steam rising (Johnson Matthey Davy design; from http://www.davy protech.com/wp-content/themes/davy/images/flowsheet-rollover/web-A-SRC-flowsheetBASE. png), (c) spherical adiabatic reactors, (d) Toyo's MRF-Z reactor (adapted from http://www.gbhenter prises.com/methanol%20converter%20types%20wsv.pdf), (e) Mitsubishi superconverter (adapted from http://www.slideshare.net/GerardBHawkins/methanol-flowsheets-a-competitive-review).

arranging the heat exchange gives a temperature profile (Figure 13.23) much closer to the maximum rate curve than the case of a multibed reactor with heat exchangers. The catalyst amount in such a tube-cooled reactor with the axial flow is limited by pressure drop considerations. For large-capacity plants, several reactors might be needed.

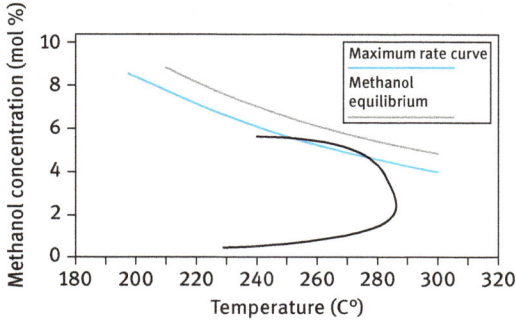

Figure 13.23: Methanol concentration profile in tube cooled converter. Adapted from http://www.gbhenterprises.com.

Near-isothermal operation is provided in a tubular boiling water reactors with axial flow where the catalyst is located on the tube side. Temperature is controlled by the pressure of water, which is circulated on the shell side, generating steam at the maximum possible pressure without overheating the catalyst. The temperature profile shown in Figure 13.24 is close to the maximum rate curve and allows somewhat low temperatures than in tube cooler converters, still requiring a significant recycle ratio. High investment costs for this reactor concept limit the maximum plant size to ca. 1,500 tpd and require several reactors in series for larger capacity.

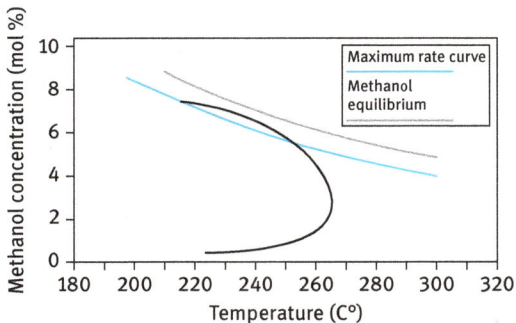

Figure 13.24: Concentration profile in a steam generating multitubular reactor. Adapted from http://www.gbhenterprises.com.

Not only an axial- but also radial-flow steam-raising converter can be used for methanol synthesis with the catalyst outside and steam inside the tubes. In the Johnson Matthey Davy design (Figure 13.23b), the fresh feed gas enters the reactor at the bottom through a central perforated-wall distributor pipe afterward flowing in the radial direction. Removal of heat is done by partial evaporation of water, which is fed upward through the tubes. Similar to a tubular boiling water reactor, control of temperature is done by varying the steam pressure.

A specific feature of Toyo's MRF-Z reactor (Figure 13.22d) is multistage indirect cooling and a radial flow facilitating the capacity increase in methanol plants. This reactor type generates steam of ca. 3 MPa and has a good approach to equilibrium (Figure 13.25), a small number of tubes, and a low pressure drop (0.05–0.075 MPa), while it can be in the range of 0.3–1 MPa for fixed-bed adiabatic reactors.

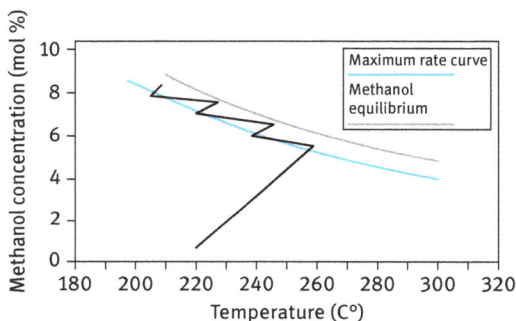

Figure 13.25: Concentration profile in Toyo's MRF-Z reactor (Figure 13.22d) reactor. Adapted from http://www.gbhenterprises.com.

Mitsubishi reactor (Figure 13.22e) for methanol synthesis can be viewed as an integration of interchange and steam rising. The design is rather complex, consisting of a large number of tubes, a manifold, and two tube sheets. It generates ca. 4 MPa of steam, closely following the maximum rate line (Figure 13.26) and thus allowing high conversion per pass and a lower recycling rate.

As follows from the description of reactors presented above, only gas-phase processes have been implemented. An alternative liquid-phase process for methanol production was developed by Air Products and Chemicals (Figure 13.27).

The technology relies on a bubble slurry reactor, in which an inert hydrocarbon acts a reaction medium and a heat sink. As the feed gas bubbles through the catalyst slurry forming MeOH, the mineral oil transfers the reaction heat to an internal tubular boiler where the heat is removed by generating steam. The reactor operates at isothermal conditions being able to handle CO-rich (in excess of 50%) syngas with wide compositional variations. Such operation mode allows to reach much higher concentration of methanol (ca. 15%) than in the gas-phase process increasing

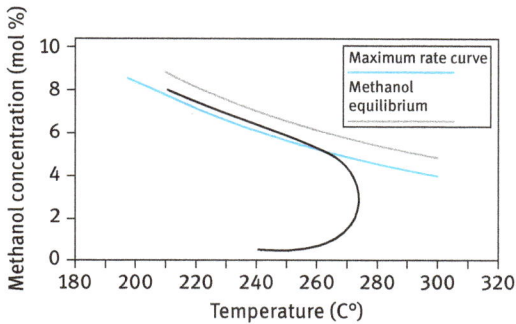

Figure 13.26: Concentration profile in Mitsubishi superconverter (Figure 13.22e). Adapted from http://www.gbhenterprises.com.

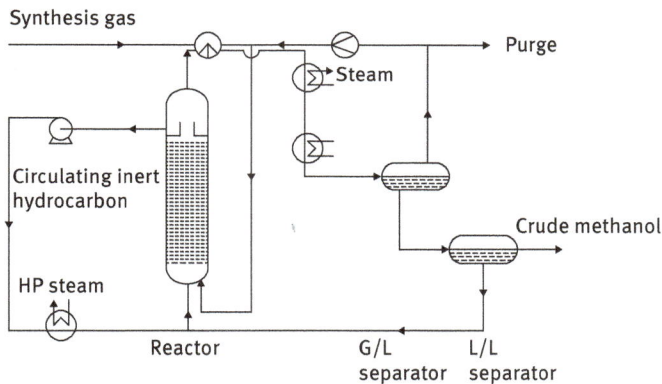

Figure 13.27: Air Products and Chemicals' liquid-phase process for methanol production. From J. A. Moulijn, M. Makkee, A E. van Diepen, Chemical Process Technology, 2013, 2nd Ed. Copyright © 2013, John Wiley and Sons. Reproduced with permission from Wiley.

conversion from 15% to ca. 35%. This technology was proven at the demonstration plant level but has not yet been commercialized.

Due to limited per pass conversion (8–15%) and moderate methanol concentration at the reactor outlet (5–7% in most processes), a recycle is necessary and conventional methanol synthesis processes employ a synthesis loop shown in Figure 13.28. Converters can be of different types as described above. The inlet temperature is ca. 220 °C and the pressure of the syngas is ca. 5 MPa. Variations in temperature and pressure are possible, depending on the process technology.

Most often, syngas is generated directly from steam reforming of natural gas with subsequent adjustment of hydrogen rich composition by addition of carbon dioxide. The feed and recycle rate depends on the process and its capacity. Typical values of flow rates are 8,000–12, 000 h^{-1}.

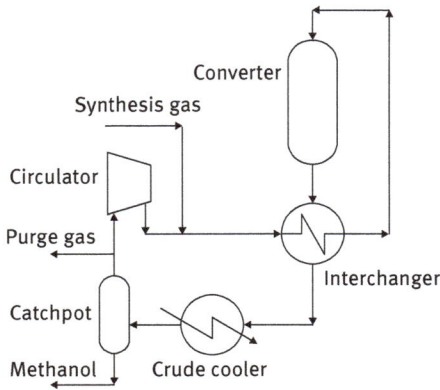

Figure 13.28: Recycle loop in methanol synthesis, From http://www.slideshare.net/GerardBHaw kins/methanol-synthesis-theory-and-operation.

High overall methanol yields are realized through recycling of unreacted CO and hydrogen and removal of methanol and water. Raw methanol containing water and impurities is condensed and sent to the distillation unit, whose design depends on the desired product purity. Typically, one to three distillation columns are used, with the first one (the so-called topping column) acting as stabilizer for removal of dissolved gases (CO, CO$_2$, H$_2$, N$_2$, and CH$_4$) and some of the light by-products (aldehydes, ketones, and dimethylether). In the downstream columns, raw methanol containing, besides water, minor amounts of higher alcohols is fractionated (Figure 13.29). The heat input is in fact optimized in the three-column system.

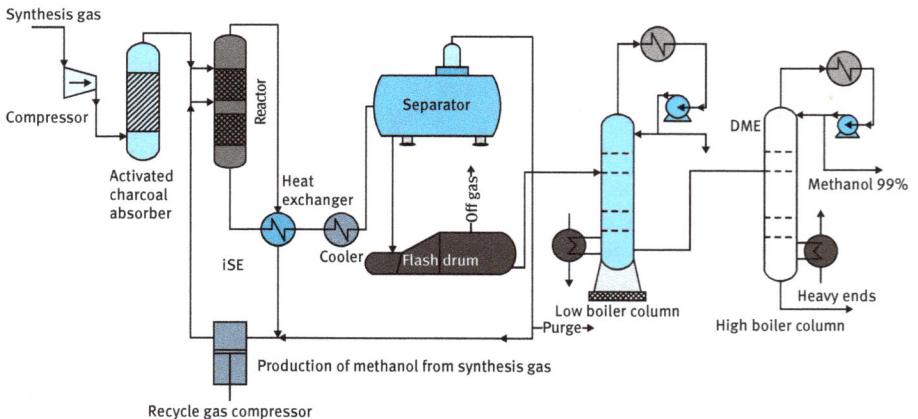

Figure 13.29: Methanol production flow scheme with purification section. http://www.inclusive-science-engineering.com/wp-content/uploads/2012/01/Production-of-Methanol-from-Synthesis-Gas.png.

13.4 Hydrocarbons from synthesis gas: Fischer-Tropsch synthesis

The Fischer-Tropsch (FT) synthesis coverts synthesis gas (mixture of hydrogen and CO with a stoichiometric ratio of 2:1) to a wide range of hydrocarbons. This process originally developed for production of synthetic fuels from coal (coal-to-liquid, CTL) had limited application outside of Sasol Company in South Africa for many decades due to a number of political and technical reasons. In the recent years, there is, however, a renewed interest in using FT process for synthesis of gasoline and diesel from primarily natural gas (gas-to-liquid, GTL) but also from biomass (BTL). These processes (GTL, CTL, BTL) consist of principally four steps (Figure 13.30), namely (A) syngas generation from coal, natural gas, or biomass, (B) cleaning of the gas, (C) FT synthesis reaction generating hydrocarbons, and (D) separation and upgrading of the products.

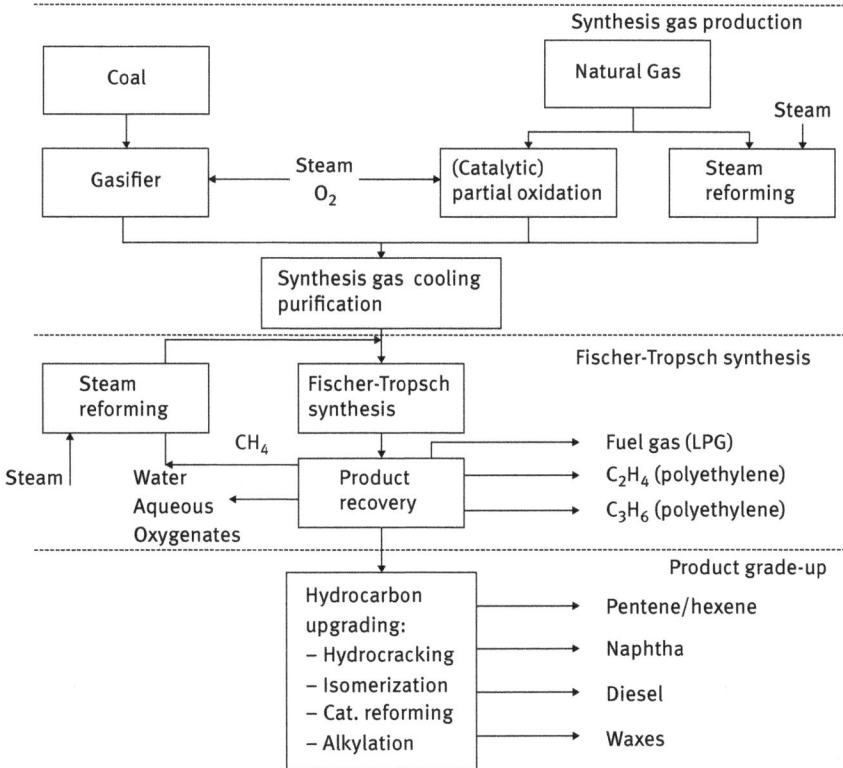

Figure 13.30: The main steps in conversion of various feedstock to alkanes by FT synthesis. From G. P. van der Laan, Kinetics, Selectivity and Scale Up of the Fischer-Tropsch Synthesis, PhD thesis, University of Groningen, 1999.

When coal is used as a feedstock, it is gasified with oxygen and steam. The syngas is purified from sulphur and nitrogen compounds, which deactivate the FTS catalyst. The synthesis per se can be performed in various reactors (fixed-bed, fluidized-bed, slurry bubble column) using either Fe- or Co-based catalysts. The latter option is utilized for highly pre-purified gas, when the hydrogen/CO ratio is adjusted to 2.0–2.1 by performing additional water-gas shift reaction. Among the products, hydrocarbons ranging from methane to liquid hydrocarbons and waxes are formed and separated. With natural gas as the feedstock, the synthesis gas has a favorable H_2/CO ratio of 2 and undergoes an FTS in a slurry bubble column over Co or Fe, resulting in heavy liquid and waxes. Subsequent hydrocracking and hydro-isomerization generate high-quality middle distillates.

The overall expression for FTS is

$$nCO + 2nH_2 \rightarrow n(-CH_0-) + nH_2O \, (\Delta H = -167 \text{ kJ/mol}) \tag{13.18}$$

with the stoichiometric ratio between H_2 and CO being equal to 2.

Other reactions that also occur at the same time are methanation, water-gas shift, Boudouard reaction, and generation of coke:

$$H_2 + CO \rightarrow C + H_2O \, (\Delta H = -133 \text{ kJ/mol}) \tag{13.19}$$

Among the side reactions, the most detrimental is methanation, which is, however, favorable from the thermodynamics viewpoint. It reduces the overall selectivity to oligomers. Selectivity to C_{2+} hydrocarbons depends on the catalyst and reaction conditions, decreasing with an increase in hydrogen/CO ratio, increase in temperature, and a decrease of pressure. Co, Fe, and Ru favor the formation of higher hydrocarbons, while nickel promotes mainly methanation. The products in FTS are predominantly normal paraffins, while significant quantities of α-olefins

$$nCO + 2nH_2 \rightarrow C_nH_{2n} + nH_2O \tag{13.20}$$

and/or alcohols can be also formed

$$nCO + 2nH_2 \rightarrow C_nH_{2n+1}OH + nH_2O \tag{13.21}$$

Even if the exact mechanism is very complex and still under debate, the main reaction in FTS follows a polymerization-like mechanism when a monomer CH_x species (x = 1–2) is added stepwise to a growing aliphatic chain. Chain termination by desorption of unsaturated surface species and hydrogenation with subsequent desorption of saturated species relative to chain propagation determines process selectivity. The weight fraction of a product with a carbon number n is defined through Anderson-Schulz-Flory distribution $W_n = na^{n-1}(1-a)^{2n}$, where parameter a is the chain growth probability and is the ratio of the chain propagation to the sum of chain propagation and chain termination (Figure 13.31a). This parameter is supposed to be independent on the carbon number.

Product composition is strongly influenced by the catalyst type (with cobalt giving more paraffins and iron resulting in the product higher in olefins and oxygenates) and operating conditions (temperature, pressure, and CO/hydrogen ratio). Under typical operation conditions, with a typical catalyst, the degree of polymerization a ranges from 0.7 to 0.95. The analysis of product distribution (Figure 13.31a, b) clearly shows that even with $a = 0.95$, a range of different products is generated with predominant formation of high-molecular-weight linear waxes. Because it is impossible to produce directly a well-defined range of products (i.e., middle distillates), the concept for newer and more economical FT processes relies on hydroprocessing of waxes to optimize the overall liquid production, and thus, the strategy in catalyst and process optimization is to increase the value of a.

Low temperature (200–240 °C) and medium pressure (2–3 MPa) are selected for the FT process along with active catalysts based on iron and cobalt to get high selectivity to heavier products. Alternatively, utilization of nickel as a catalyst results in mainly methanation.

It is important to note that in FT synthesis, essentially no aromatic compounds are formed except for high-temperature processes. The product is also free from sulphur and nitrogen compounds.

The so-called carbine mechanism, which is supported by the vast majority of studies, assumes CO adsorption with dissociation, hydrogenation of C to CH_x species, and insertion of CH_x monomers (CH_2 in Figure 13.32) into the metal-carbon bond of an adsorbed alkyl chain.

Co-based catalysts are generally preferred for natural gas-based syngas giving FT stoichiometric H_2/CO ratio or close to it (Figure 13.33).

Metallic cobalt, which is considered to be the active phase in FT catalysts, has low water-gas shift activity. Earlier Co catalysts were prepared by co-precipitation, while novel generation is mainly synthesized by impregnation of oxides with aqueous or organic solutions of cobalt nitrates and other additives. Calcination of Co nitrates results in the formation of Co oxide, which is reduced in a hydrogen-containing gas. During catalysis Co metal crystallites are largely covered by active and inactive carbonaceous species. Co catalysts are more expensive than iron-based ones, at the same time possessing 10–20 times higher activity (calculated per weight for promoted Co versus promoted Fe catalysts), high selectivity to long-chain paraffins (C5 +), and low selectivity to olefins and oxygenates, being also resistant to deactivation. The metal loading is typically 35 wt% with metal dispersion ca. 8–10%. A range of metal promoters (0.1–0.3 wt% Pr, Re, or Ru) is used to increase reducibility and dispersion of Co, improve stability against carbon buildup, and increase C5 + selectivity. Oxide promoters (i.e., 1–3% BaO or La_2O_3 or other additives) are used to stabilize cobalt crystallites and support and promote hydrocarbon chain growth. As a support, δ-alumina (ca. 150 m^2/g) stabilized with lantana is used. The support should be chemically and physically stable during catalyst preparation, activation,

(a)

(b)

(c)

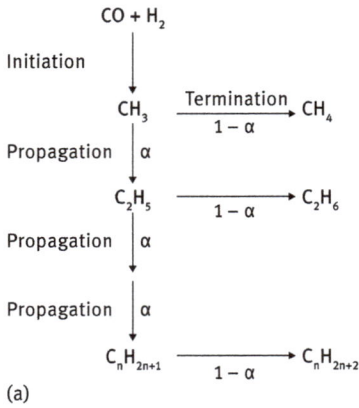

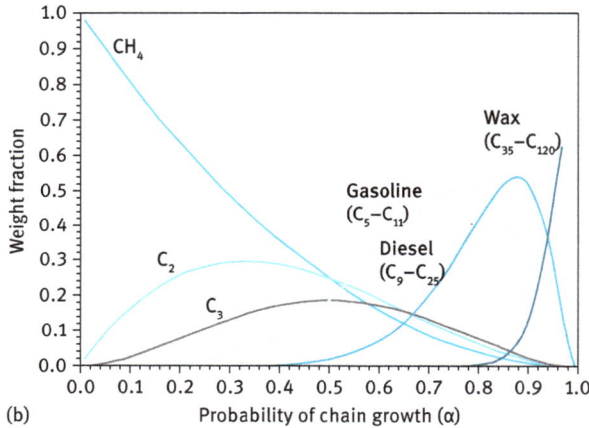

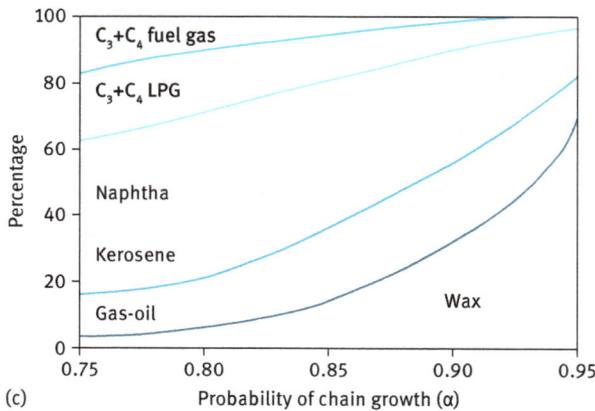

Figure 13.31: Illustration of chain growth, weight fraction of hydrocarbons, and percentage of different hydrocarbons products as a function of chain growth.

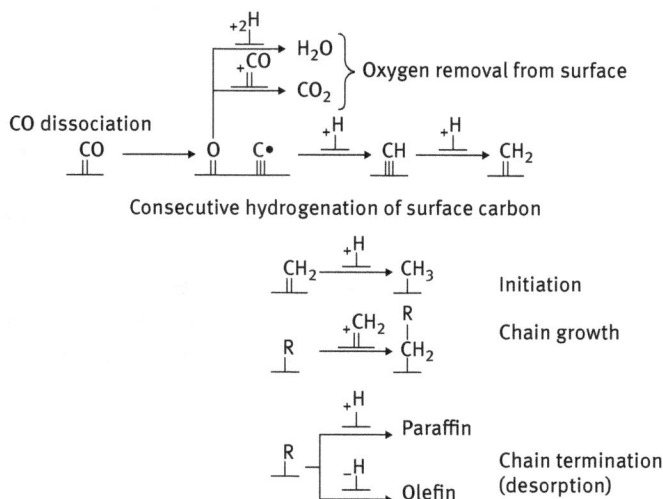

Figure 13.32: The carbene mechanism. From S. B. Ndlovu, N. S. Phala, M. Hearshaw-Timme, P. Beagly, J. R. Moss, M. Claeys, E. van Steen, Some evidence refuting the alkenyl mechanism for chain growth in iron-based Fischer–Tropsch synthesis, Catalysis Today, 2002, 71, 343–349. Copyright Elsevier. Reproduced with permission.

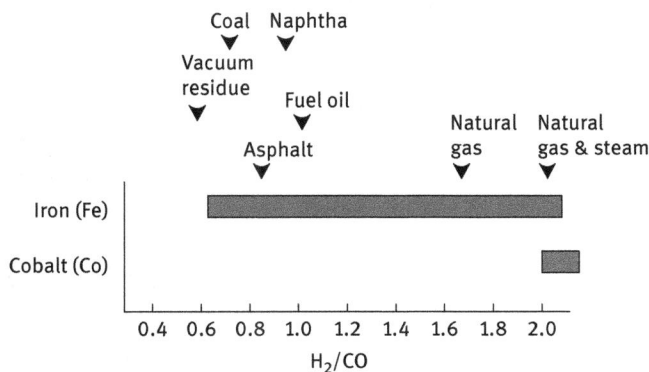

Figure 13.33: Ratios of H_2/CO ratio for different feedstock and catalysts. From G. P. van der Laan, Kinetics, Selectivity and Scale Up of the Fischer-Tropsch Synthesis, PhD thesis, University of Groningen, 1999.

regeneration, and reaction. The support can be stabilized with some other oxides. Besides alumina, such supports as silica and titania can be applied.

Fe-based catalysts are less expensive than cobalt and were used commercially in Sasol plants. They are generally preferred for coal-based plants with lower hydrogen/CO ratio. The active phase is Fe carbides (Fe_xC, $x < 2.5$), covered by various carbonaceous species. Fused iron oxide catalysts, when promoters were added in the fusion step, were an economic option and were applied in the so-called high-temperature FT synthesis using fluid bed (Sasol Advanced Synthol) reactors. Precipitated iron oxide with addition of promoters (additives and modifiers) during precipitation is more expensive to prepare and has less structural strength. Such catalysts can be used in low-temperature FT synthesis organized in trickle-bed (ARGE) and slurry reactors. The catalysts are prone to deactivation and a gradual loss of activity with a possibility for regeneration. Copper is added to the iron catalyst in order to increase the rate of iron reduction and catalyst activity. Selectivity toward longer hydrocarbons (waxes) is improved through the addition of alkali (K$_2$O). In general, iron catalysts are more economical than cobalt-based catalysts, possess low selectivity to long-chain paraffins and high selectivity to olefins and oxygenates, promote water-gas shift reaction, and are prone to fast deactivation by generation of coke. Silica is mainly used as a support for precipitated iron catalyst affording the highest activity and selectivity to waxes.

Catalyst deactivation in FT synthesis occurs through poisoning of the active metal by sulphur or nitrogen compounds. This can be prevented by desulphurization of the syngas feed with formation of H$_2$S, which is captured by ZnO installed upstream a FT reactor. Prevention of fouling due to blockage of pores with hard waxes can be achieved by operating at lower value of parameter a, control of hydrogen/CO ratio, lowering temperature, and in situ treatment of the deactivated catalyst in hydrogen at temperature 10–20 °C higher than the reaction temperature. Hydrothermal sintering of Fe happens at high steam pressure. Keeping the latter below 0.5–0.6 MPa or operating below 50–60% of CO conversion prevents sintering as well as formation of iron oxides. Application of multiple reactors with intermediate removal of H$_2$O and stabilization of the supports with Ba, Zr, or La oxides serve as a preventive measure against deactivation. The same approach is used to prevent formation of cobalt oxides. Generation of inactive cobalt carbides is minimized by keeping hydrogen to CO ratio above 2.1 in all reactor parts.

Loss of catalytic material due to abrasion and erosion in the case of fluidized or slurry reactors can be prevented by adequate preparation methods, including sol-gel granulation, application of binders, etc.

High exothermicity is a typical feature of the process with a heat release of 165 kJ per mole of –CH$_2$– formed. The choice of the catalyst and the process conditions (pressure, temperature, hydrogen/CO ratio) influence the product's molecular weight distribution. The products in FT reaction are mainly n-paraffins, although terminal olefins and alcohols could also be formed. More expensive cobalt-based catalysts operating at lower temperatures (200–250 °C) favor long-chain paraffins. They are more robust and have low water-gas shift (WGS) activity, contrary to a cheaper alternative, iron. The latter as already mentioned needs higher temperature (220–350 °C or higher for fluidized beds), possesses high selectivity to olefins and oxygenates, is a WGS catalyst, and readily deactivates.

Due to the exothermicity of FT reaction, heat removal is of major concern in the reactor design. In addition, selectivity to the unwanted product, methane, is increased with temperature; thus, efficient temperature control is needed to achieve the desired selectivity.

Several types of reactor systems presented in Chapter 3 (tubular fixed-bed, circulated fluidized bed, and slurry bubble columns) are used in industrial practice. A general comparison was given in Table 13.2. Tubular reactors or trickle fixed beds with downward flow through the catalyst bed (Figure 13.34A) were first to be used commercially.

Although this is a simple easy-to-scale-up design, construction was rather expensive due to a large number of tubes needed for the industrial reactor. As already mentioned above, catalyst replacement is an issue, since iron catalysts should be periodically replaced due to deactivation. Contrary to iron, cobalt-based catalysts are more robust with a lifetime of several years and could be regenerated. The catalyst size is above 1 mm to avoid extensive pressure drop; thus, the effectiveness factor is certainly below unity and mass transfer limitations are present. Possible temperature gradients in the tubes can lead to sintering and deactivation.

Sasol Arge trickle fixed-bed reactor with a 3-m-diameter shell contains 2050 tubes with diameter of 5.5 cm in and length of 12 m. Lower temperature of operation (230–235 °C) favors formation of heavier hydrocarbons. Typical conversion levels are ca. 50%. While no catalyst losses because of attrition and higher conversion due to close to plug-flow regime are clear advantages of trickle fixed-bed reactors, there are also disadvantages related to low heat transfer coefficients and subsequent limited productivity along with a complicated construction.

Alternatives to tubular reactor are circulating fluidized-bed (CFB, Figure 13.35a) reactor (Sasol Synthol reactor) and fixed fluidized-bed reactor (Sasol Advanced Synthol reactor, Figure 13.35b). CFB reactor provides better heat removal and temperature control with near-isothermal operation at higher temperature (exit T of 320 °C for CFB and 340 °C for fixed fluidized-bed reactor), experiencing fewer pressure drop problems than a tubular reactor even if the pressure drop was relatively high due to a large catalyst inventory. Catalyst removal and addition online are possible, being clear advantages of fluidized-bed reactors (and slurry phase reactors described below) compared to fixed beds.

The major disadvantage of fluidized beds for FT applications is that a low- molecular-weight product (gasoline) is obtained while the concentration of diesel range products and waxes cannot be high, since products must be volatile at the reaction conditions. If non-volatile hydrocarbons accumulate on the catalyst particles, fluidization behavior is worsened, as the particles stick to each other. Very high temperatures cannot be used in order to avoid excessive carbon formation.

Scaling up of such reactors is more difficult in comparison to tubular reactors as mentioned in Table 13.2. The Sasol Advanced Synthol reactor with elimination of circulation is a fixed fluidized-bed reactor operating at similar operating conditions

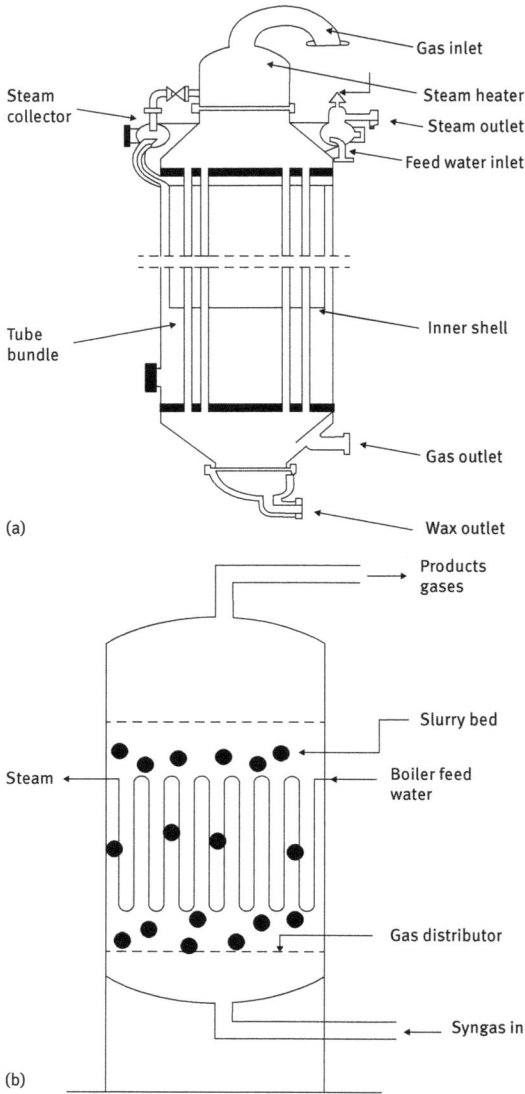

Figure 13.34: Reactor for FT synthesis: (a) a multitubular trickle fixed-bed reactor (from M. E. Dry, The Fischer-Tropsch process:1950–2000, Catalysis Today, 2002, 71, 227–241, copyright Elsevier, reproduced with permission), (b) slurry bubble column (from S. Saeidi, M. T. Amiri, N. A. S. Amin, M. R. Rahimpour, Progress in reactors for high-temperature Fischer–Tropsch process: determination place of intensifier reactor perspective, International Journal of Chemical Reactor Engineering, 2014, 12, 639, copyright De Gruyter).

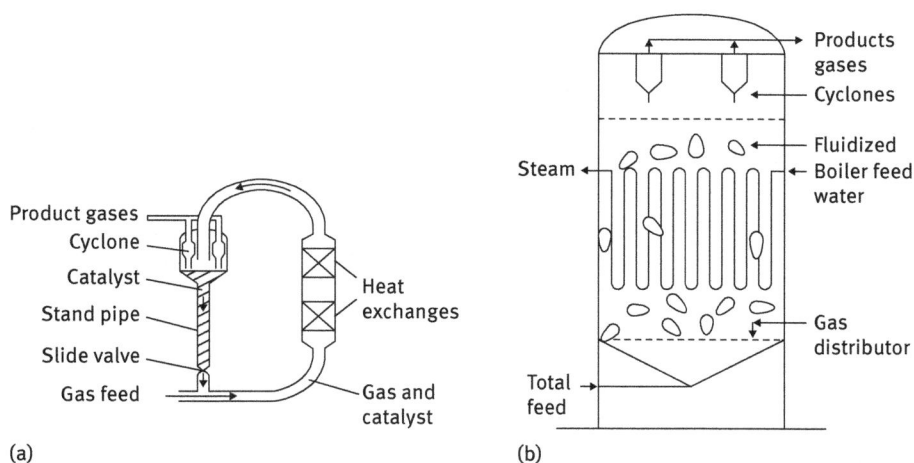

Figure 13.35: (a) Circulating and (b) fixed fluidized-bed reactor for FT synthesis. From S. Saeidi, M. T. Amiri, N. A. S. Amin, M. R. Rahimpour, Progress in reactors for high-temperature Fischer–Tropsch process: determination place of intensifier reactor perspective, International Journal of Chemical Reactor Engineering, 2014, 12, 639. Copyright De Gruyter.

as the circulating reactor, allowing, however, a significant size reduction and thus the capital costs for the same capacity. Fused and reduced iron catalyst is applied in SAS reactor. The feed is distributed through a gas distributor. The products and unconverted gases along with the catalyst pass through internal cyclones, where the catalyst is separated and returned to the process. An advantage of these reactors is that due to efficient catalyst separation, scrubber towers used in the CFF reactor are not needed for removal of traces of the catalyst. Moreover, this reactor is simpler and more cost-effective, as catalyst recycling is absent, has lower operating costs, and better maintenance. Higher conversions at higher gas loads along with more efficient heat removal through cooling coils give either a capacity increase in SAS compared to CFB or lower operating costs at the same capacity.

Slurry phase bubble columns or SPBC (Figure 13.34b) are considered as the choice for newer FT reactors (low-temperature FT synthesis) with more active cobalt catalyst, which is suspended in a slurry. The synthesis gas is bubbled through this slurry containing hydrocarbon waxes, liquid at reaction conditions, and the catalyst particles of the size 50–80 µm diminishing substantially mass transfer limitations. The height of such reactor, weighing ca. 2,200 tons, could be up to 30–40 and even 60 m with an outer diameter of 6–10 m. Operation conditions are 2–4 MPa of pressure and temperature 230–250 °C. The temperature cannot be too low; otherwise, the reaction mixture (the liquid wax) becomes very viscous. High T, on the contrary, leads to hydrocracking. In SPBC, heat is removed through internal cooling coils. Such reactors provide good heat transfer and temperature control, low pressure drop, and are suited for synthesis of higher boiling products. This could be an advantage, since it gives more overall flexibility if there is a downstream cracking unit. The design is rather simple, allowing easy addition and removal of catalysts. The gaseous products are removed from the top, while there is a need to separate (Figure 13.34b) waxes from the catalyst, which is sent back to the reactor. The cooling coils and a gas distributor are cheaper than the arrangement with tubes in a trickle fixed-bed reactor and easier to scale up.

Due to much better mixing compared to TFB reactor and near isothermal conditions without axial and radial temperature gradients, a higher average temperate can be used. As a consequence, an order of magnitude higher production capacity can be achieved compared to TFB; moreover, the pressure drop in a slurry reactor is less than 0.2 MPa, being 0.3–0.7 MPa in TFB. Thus, clear advantages of slurry bubble columns are simpler, cheaper construction with lower capital costs and easier maintenance, also allowing online catalyst replacement; lower pressure drop; and higher production rates for the same reactor dimensions.

Low-temperature TBR and slurry bubble columns can be used in both CTL and GTL processes, aiming at production of waxes, diesel fuel, and lubricants. Details of operational conditions and the product composition for most important FT plants are given in Table 13.3.

Table 13.3: The most important FT plants. From Advanced liquid biofuels synthesis. Adding value to biomass gasification. ECN-E–17-057 – February 2018, www.ecn.nl.

Company	Plant, location, date	Feed	Technology/ reactor	Catalyst	Conditions	Products and capacity
Sasol	Sasol I, Sasolburg, South Africa, 1955	Natural gas	LT/slurry phase distillate + multitubular fixed bed	Precipitated Fe/K	220–250 °C	5,000 bbl/d Paraffin, waxes, oxygenates, and fuels gas
Shell	Bintulu, Malaysia, 1993	Natural gas	LT/multi-tubular fixed bed	Co/SiO$_2$	220 °C 25 bar	SMDS, 14,700 bbl/d LPG (0–5%), naphtha (30–40%), distillate (40–70%), amd oils (0–30%)
Sasol	Oryx GTL Ras Laffan Industrial City, Qatar, 2007	Natural gas	LT/slurry phase distillate	Co/Pt/Al$_2$O$_3$	230 °C 25 bar	34, 000 bbl/d LPG, naphtha, and distillate (diesel blend)

Table 13.3 (continued)

Company	Plant, location, date	Feed	Technology/ reactor	Catalyst	Conditions	Products and capacity
Shell	Pearl GTL Qatar, 2009	Natural gas	LT/multi-tubular fixed bed	Co/SiO_2	220 °C 25 bar	SMDS, 140,000 bbl/d LPG (0–5%), naphtha (30–40%), distillate (40–70%), amd oils (0–30%)
Chevron	Escravos GTL, Nigeria, 2014	Natural gas	LT/similar to Oryx	$Co/Pt/Al_2O_3$	230 °C 25 bar	34,000 bbl/d LPG, naphtha, and diesel blend
Sasol	Sasol 2 and 3 (Synfuels), South Africa, 1980	Coal	HT/fixed fluidized bed	Fused Fe/K	350 °C 24 bar	160,000 bbl/d. Fuel gas, oils, alpha-olefins, ammonia, gasoline, jet fuel, diesel
PetroSA	Mossgas, Mossel Bay, South Africa	Natural gas	HT/ circulating fluidized bed	Fused Fe/K	330–360 °C 25 bar	30,000 bbl/d. LPG, gasoline, diesel, fuel oil, kerosene, aromatics, alcohols

Table 13.4 illustrates the operation conditions and product composition for different reactor types, clearly showing that product composition and operation parameters are reactor dependent. Thus, in the fluidized-bed reactor, operating at a much higher temperature than slurry bubble column and the fixed-bed reactor, the major product is gasoline, and large amounts of light products, such as methane and lower alkanes, are produced.

Table 13.4: Comparison between different three-phase reactors for FT synthesis.

Application criteria	Slurry bubble column	Riser	Fixed bed
Conditions			
Inlet/outlet T (°C)	260/265	320/325	223/236
Pressure (MPa)	1.5	2.3	2.5
H_2/CO ratio	0.68	2.5	1.7
Conversion (%)	87	85	60–66

Table 13.4 (continued)

Application criteria	Slurry bubble column	Riser	Fixed bed
Products (wt.%)			
CH₄	6.8	10.0	2.0
C2	4.4	8.0	1.9
C3	9.3	13.7	4.4
C4	8.0	11.3	4.5
C5–11 (gasoline)	18.6	40.0	18.0
C12–18 (diesel)	14.3	7.0	14.0
C19 + waxes	37.6	4.0	52.0
Oxygenates	1.0	6.0	3.2

After J. A. Moulijn, M. Makkee, A. van Diepen, *Chemical Process Technology*, Wiley, 2001.

A more detailed composition of gasoline and diesel products is presented in Table 13.5.

Table 13.5: Typical product composition in various reactors for iron catalysts.

	LT fixed bed		LT slurry		HTFT Synthol	
Compounds	C5–12	C13–18	C5–12	C13–18	C5–12	C13–18
Paraffins	53	65	29	44	13	15
Olefins	40	28	64	50	70	60
Aromatic	0	0	0	0	5	15
Oxygenates	7	7	7	6	12	10
n-Paraffins	95	93	96	95	55	60

A. De Klerk. Fischer-Tropsch Refining, PhD thesis, University of Pretoria, 2008.

As also shown in Tables 13.4 and 13.5, in general, two temperature regimes are used in FT synthesis. High-temperature operations require 300–350 °C, giving mainly short-chain alkanes and gasoline. Reforming and isomerization are upgrading technologies to improve the low octane number. Low-temperature operation (200–250 °C) resulting in diesel oil and wax needs hydrocracking of wax to generate additional amount of diesel oil, which has in fact excellent properties due to high cetane number and absence of sulphur or aromatics.

The LT slurry synthesis product (Table 13.5) is more olefinic than the fixed-bed product. The amount of olefins can be diminished as mentioned above by changing the catalyst to cobalt-based ones.

The first FT plants began their operation in Germany in 1938. There were nine low-temperature cobalt-based plants, which eventually closed after WWII with a total annual product capacity of 660, 000 tonnes. The first Sasol plant in Sasolburg, South Africa, started operation in 1953 and had annual production of million of tonnes of FT products using coal as a feedstock and operating five tubular fixed-bed (ARGE) reactors for wax production and three circulating fluid-bed reactors. A slurry reactor of the same production capacity replaced five ARGE reactors in 1993. In 2004, natural gas reforming was introduced, instead of coal gasification, by transforming the plant technology from CTL to GTL producing waxes and chemicals.

Further expansions by Sasol in Secunda, South Africa, were done in 1980-s utilizing high temperature Synthol reactors with improved heat exchange, thereby boosting threefold the capacity compared to the first generation CFB reactors. The main focus of the production site of Sasol in Secunda is motor gasoline and diesel as well as some chemicals. A high-pressure distillate hydrogenation section was also added to the tar refinery of Sasol III, processing gasification-derived coal pyrolysis liquids from Sasol II and Sasol III. Sasol Advanced Synthol reactors eventually replaced sixteen second generation CFB reactors with eight fixed fluid-bed (FFB) reactors, decreasing the operation costs at the same capacity.

In the early 1990-s, Mossgas started up a natural gas, 1-million-ton-per-year FT plant in South Africa using a high-temperature process with an iron catalyst for making motor gasoline, distillates, kerosene, alcohols, and LPG, while Shell put on stream 500, 000 tons/year natural gas-based FT plant using the Shell middle distillate synthesis (SMDS) process for automotive fuels, specialty chemicals, and waxes (Figure 13.36). The strategy of the low-temperature FT synthesis illustrated in Figure 13.36 is to produce heavier products with a cobalt catalyst, when formation of long-chain waxes is favored (*a* value of 0.9 and higher). The heavy alkanes are converted through mild hydrocracking to the desired carbon number range with subsequent product distillation.

The Sasol Onyx GTL plant operating in Qatar using a cobalt catalyst at low temperature was commissioned in 2006, producing 34, 000 barrel per day (bpd) of mainly diesel fuel and naphta as by-product. The FT syncrude is similar to the SMDS process and is processed in a similar way. Shell Pearl GTL plant put on stream in 2011 in the same location in Qatar with a capacity of 120, 000 bpd of petroleum liquids relies on the same low-temperature FT technology to produce distillate and base oils. Other projects based on coal, shale gas, and biomass have been announced, with some of them already postponed or delayed.

A general overview of the process flow with coal as a feedstock is given in Figure 13.37.

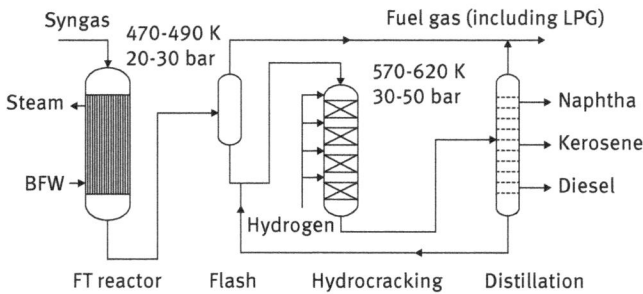

Figure 13.36: Simplified flow scheme of the Shell Middle Distillate synthesis plant. From J. A. Moulijn, M. Makkee, A E. van Diepen, *Chemical Process Technology*, 2013, 2nd Ed. Copyright © 2013, John Wiley and Sons. Reproduced with permission from Wiley.

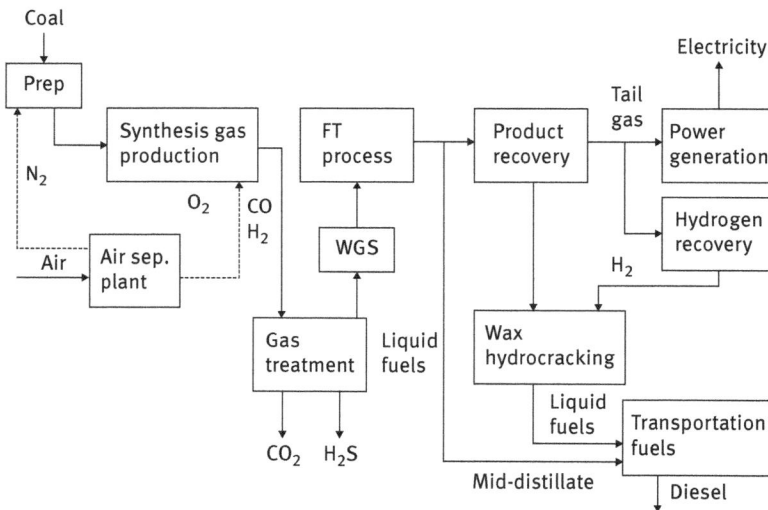

Figure 13.37: A general overview of the process flow. http://what-when-how.com/energy-engineering/coal-to-liquid-fuels-energy-engineering/.

A high-temperature FT (HTFT) syncrude conducted in CFB or FFB reactors with Co catalyst obviously has a higher naphtha yield, while application of lower temperatures (LTFT) using slurry (Co catalyst) or tubular (Fe catalyst) results in higher boiling point hydrocarbons. LT FT refineries are less complex and typically have hydroprocessing (wax hydrocracking) steps and fractionation to produce naphtha and middle distillates. At the same time, LTFT refineries are making fuel-blending stocks rather than final fuels.

On the contrary, diesel production can be achieved in HTFT process, as the syncrude contains aromatics and naphthenes, giving a diesel density closer to the required specification. The process workup system used in the Sasol II Secunda plant in South Africa is shown in Figure 13.38.

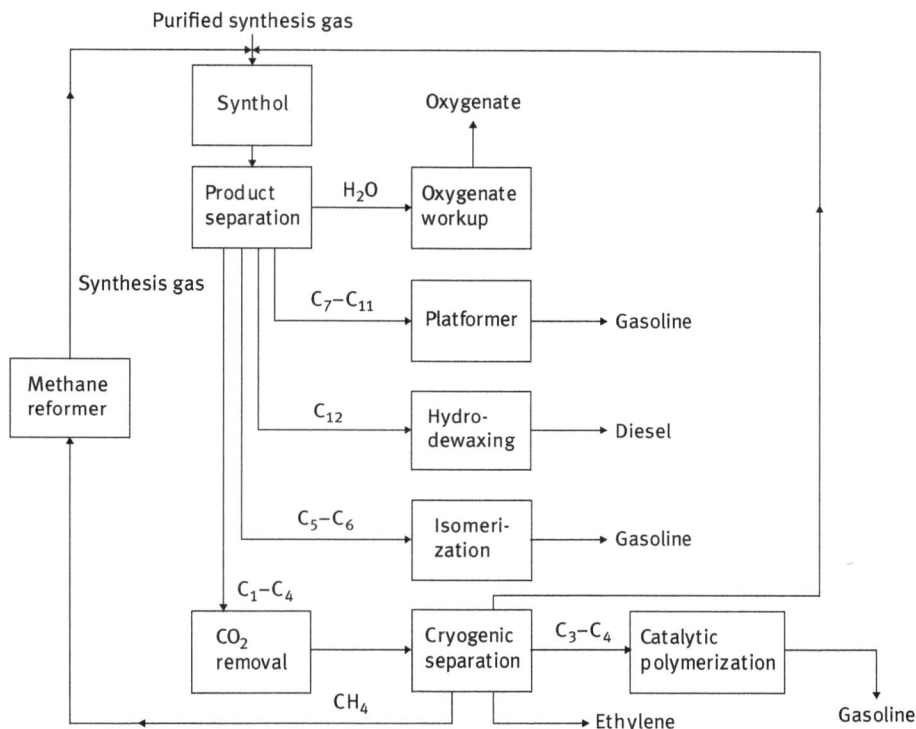

Figure 13.38: Flowsheet for Sasol II products workup.

After product separation, the gas is scrubbed to remove the CO_2 and fractionated in a cryogenic unit. Methane is autothermally reformed to synthesis gas and recycled and ethylene and propene can be used as a (petro)chemical feedstock, while butene is alkylated to gasoline. Fractions heavier than C3 and C4, such as pentane and hexane, are isomerized or undergo reforming over Pt increasing gasoline octane number (C7–C11 fraction). Even higher carbon number fractions are catalytically hydrodewaxed, generating a "zero-sulphur diesel fuel". Such α-olefins as 1-hexene and 1-octene are used as petrochemical feedstock.

13.5 Reactions of olefins with synthesis gas: hydroformylation

The reaction of olefins with the synthesis gas (oxo synthesis) in the presence of homogeneous catalysts discovered 1938 by Otto Roelen leads to aldehydes containing

one additional carbon atom (Figure 13.39), as exemplified below for hydroformylation of propene, giving normal and *iso*-products.

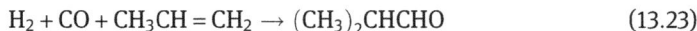

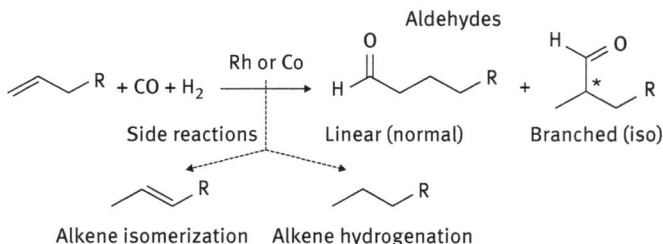

$$H_2 + CO + CH_3CH = CH_2 \rightarrow CH_3CH_2CH_2CHO \qquad (13.22)$$

$$H_2 + CO + CH_3CH = CH_2 \rightarrow (CH_3)_2CHCHO \qquad (13.23)$$

Figure 13.39: Hydroformylation of olefins.

Side reactions are olefin (propene) hydrogenation and double-bond migration to
form less reactive in hydroformylation internal olefins in the case of higher carbon
chain olefins. The oxo-synthesis reactions are exothermal, with the heat released
ranging between ca. 115 and 145 kJ/mol. The most important product is *n*-butyralde-
hyde formed by hydroformylation of propene with capacity, exceeding 4 million t/
year. Normal butyraldehyde has a higher market value, and it is used for the production of 2-ethylhexanol (2-EH) through aldol condensation in the presence of alkali with subsequent catalytic hydrogenation (Figure 13.40).

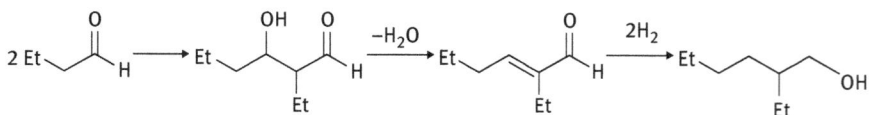

Figure 13.40: Synthesis of 2-ethylhexanol from *n*-butyraldehyde.

2-EH is applied in the synthesis of plasticizers, such as dioctyl phthalate (DOP) from
phthalic anhydride and 2-EH (Figure 13.41).

Initially, the cobalt-based catalysts were used, giving a mixture of normal and
iso-aldehyde. In the mid-1970s, the quest for higher selectivity toward the desired
normal aldehyde led to the introduction in the industrial practice of rather expensive
rhodium-based catalysts, which outperform cobalt in terms of activity and selectivity.
Efficient catalyst recovery and extension of its lifetime were serious issues that had to
be solved for successful implementation of homogeneous catalysis with such expensive catalysts. It is fair to say that cobalt catalysts requiring much higher pressures are
still used industrially since there is a commercial interest in *iso*-butyraldehyde.

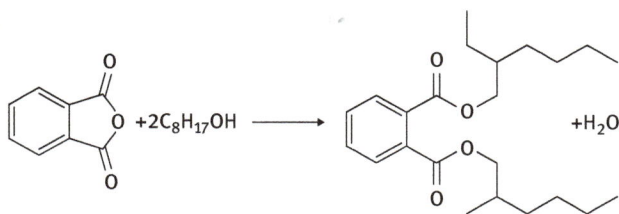

Figure 13.41: Synthesis of dioctyl phthalate from phthalic anhydride and 2-EH.

Figure 13.42: Neopentyl glycol.

Namely, it is used for production of neopentyl glycol (Figure 13.42), which is needed in the synthesis of polyesters, paints, lubricants, and plasticizers.

In aldol reaction, *iso*-butyraldehyde reacts with formaldehyde leading first to hydroxypivaldehyde, which can be converted to neopentyl glycol with either excess formaldehyde or by catalytic hydrogenation of the aldehyde group.

From the thermodynamic viewpoint, hydroformylation requires low T and elevated pressures, which shift conversion to the product side. An *iso*-product is more thermodynamically favored (Figure 13.43).

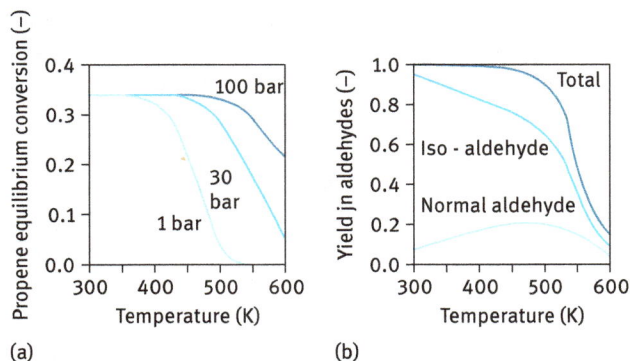

(a)   (b)

Figure 13.43: Thermodynamic data for hydroformylation of propene. From J. A. Moulijn, M. Makkee, A E. van Diepen, *Chemical Process Technology*, 2013, 2nd Ed. Copyright © 2013, John Wiley and Sons. Reproduced with permission from Wiley.

The most important catalysts are Rh and Co, which are introduced as carbonyls. Cobalt hydridocarbonyl, $HCo(CO)_4$, was the catalyst introduced in the 1940s requiring high pressures of several tens of megapascals or hundred bars to afford the required catalyst activity and stability. The mechanism for hydroformylation using Co catalysts is illustrated in Figure 13.44.

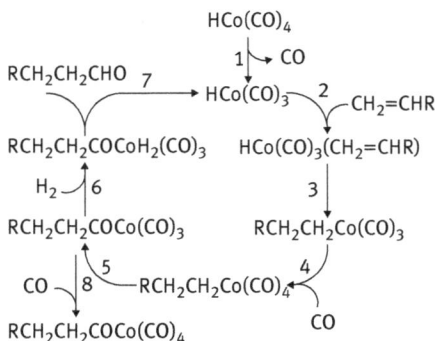

Figure 13.44: The mechanism of Co-catalyzed hydroformylation. http://en.wikipedia.org/wiki/Metal_carbonyl#mediaviewer/File:Hydroformylation_Mechanism_V.1.svg.

While for lower alkenes, Co has been mainly substituted by Rh catalysts, for higher alkenes, cobalt is still the preferred catalyst. The reason for utilization of cobalt is that the higher alkene feed (C10–14) for the production of detergent alcohols contains internal alkenes being either a product from the wax-cracker (terminal and internal alkenes) or the by-product of the ethene oligomerization process (internal alkenes).

Similar to hydroformylation of lower alkenes, the desired product has a linear structure; thus, the catalyst, besides hydroformylating only the terminal bond to get an acceptable concentration of linear products, should also isomerize the internal alkenes to the terminal ones. These features can be achieved with HCo(CO)$_4$, while the activity of transition metal complexes for isomerization of alkene in the presence of carbon monoxide is low. In the Kuhlmann process, now Exxon, one organic phase consisting of higher alkene and aldehyde is present in a reactor, with an external loop operating with a cobalt catalyst (Figure 13.45). After the reaction and a gas/liquid separator, the liquid phase is treated with aqueous Na$_2$CO$_3$, thus transforming the acidic HCo(CO)$_4$ into the water-soluble conjugate base NaCo(CO)$_4$. In this way, a liquid/liquid separation of the product and the catalyst can be done. Further treatment of the basic solution containing NaCo(CO)$_4$ with sulphuric acid and extraction in the presence of the fresh olefin allows to regenerate HCo(CO)$_4$ without its decomposition. The aqueous phase after extraction of the catalyst contains Na$_2$SO$_4$ in stoichiometric amounts to the Co catalyst and is sent to the wastewater treatment. The organic phase after the phase separation is distillated, giving a crude aldehyde.

A trialkylphospine (Figure 13.46a)-substituted cobalt carbonyl catalyst, giving higher regioselectivity than the classical catalyst but possessing lower activity and forming side products, was developed in the 1960s.

High regioselectivity was achieved in the 1970s with rhodium catalysts, which also displayed low hydrogenation and double-bond migration activity. Therefore,

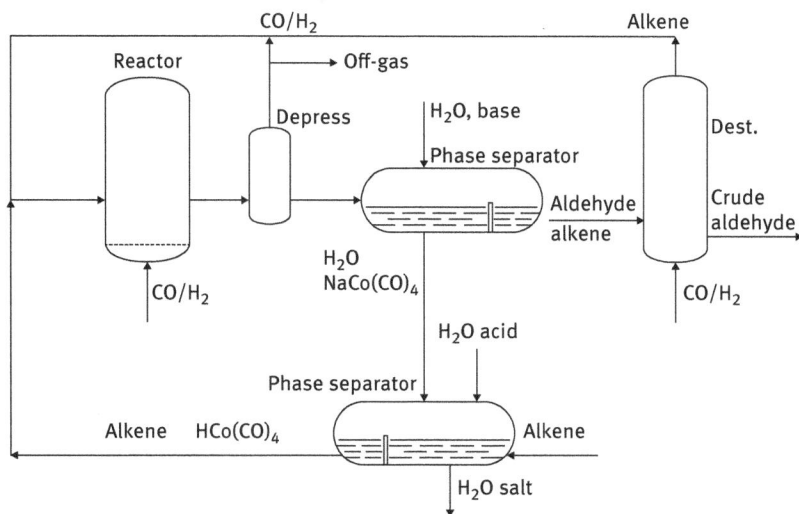

Figure 13.45: Kuhlmann hydroformylation process.

despite their high price (ca. 1,000 times more expensive than Co), such homogeneous Rh catalysts operating at lower pressures with lower energy consumption in compression units and in smaller reactors started to be employed industrially. An overview of different catalysts and process conditions is given in Table 13.6. As can be seen from this table, the catalyst in the case of Rh is modified with different ligands, such as triphenylphosphine (TPP) or water-soluble trisodium salt (Figure 13.46b and 13.46 c, respectively), affording very high regioselectivity but lower activity and thus higher T and pressure than with TPP. Utilization of ligands used in excess (in the Union Carbide process, the ratio of PPh_3/Rh is 400:1; in the Ruhrchemie/Rhône-Poulenc process, TPPTS/Rh ≥ 100:1) allowed to avoid losses of the expensive metal even if pure Rh carbonyls without any ligands are the active hydroformylation catalysts.

A catalytic cycle for Rh catalysts is shown in Figure 13.47, while the overall mechanism that is able to describe propene hydroformylation kinetics and regioselectiviy is given in Figure 13.48. The mechanism is based on the concept of one cycle selective to normal aldehyde (III) and two cycles leading to mixed aldehydes (cycles I and II). This approach was required to explain regioselectivity dependence on the ligand concentration. A selective cycle (Figure 13.43) consists of alkene addition to $HRh(CO)(L)2$ (1–0); forming a π-alkene complex 1–1 with 18 electrons, isomerization to a 16 electrons σ-complex 1–2, reaction with CO forming an alkyl complex 1–3, isomerization to a σ-acyl complex 1–4, addition of hydrogen, and final release of normal aldehyde 1–5, returning to the initial complex 1–0. In order to explain the dependence of regioselectivity on only the ligand concentration but not on the partial pressure of CO, a

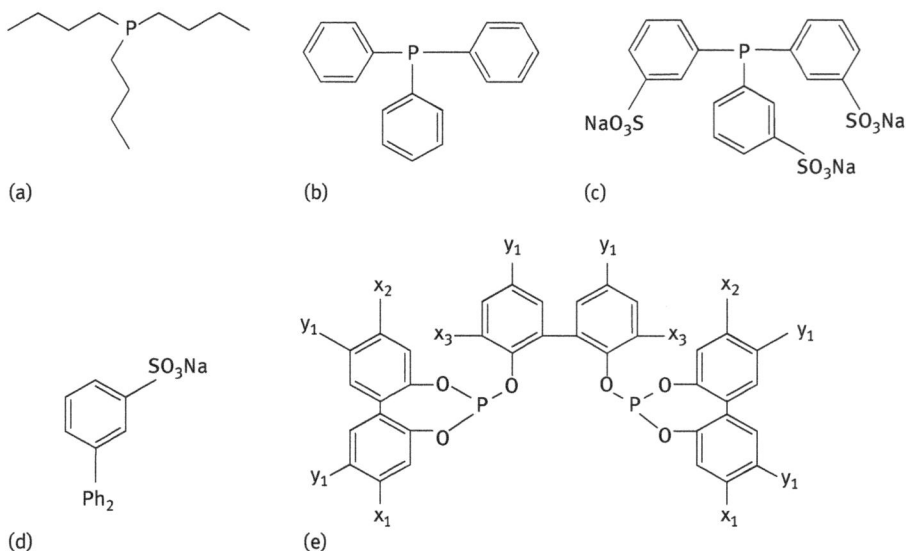

Figure 13.46: Ligands used with cobalt (a) tributylphospine and rhodium catalysts, (b) triphenylphosphine, (c) sulfonated triphenylphosphine, and (d) monosulfonated triphenylphosphine ligand (e) biphosphate.

Table 13.6: Different hydroformylation technologies.

	Ruhrchemie, Kuhlmann	Shell	Union Carbide, Davy Powergas, and Johnson Matthey	Ruhrchemie/ Rhône-Poulenc
Feed	Internal C4–C17	Internal C4–C17	Propene	Propene
Catalyst	HCo(CO)$_4$	HCo(CO)$_3$L	HRh(CO)L$_3$	HRh(CO)L$_3$
Ligand	None	Tributylphospine	Tributylphospine	Sulfonated triphenylphosphine
Temperature (°C)	110–180	160–200	85–115	110–140
Pressure (MPa)	20–30	5–10	1.5–2	4–5
n/*iso* ratio	4.0	7.3	11.15	19
Alkane yield (%)	2	10–15	0	0

From A. Jess, P. Wassersheid, *Chemical Technology: An Integrated Textbook*, 2013. Copyright Wiley. Reproduced with permission.

sequence of equilibrium steps was assumed between Rh species, leading to an overall mechanism (Figure 13.48).

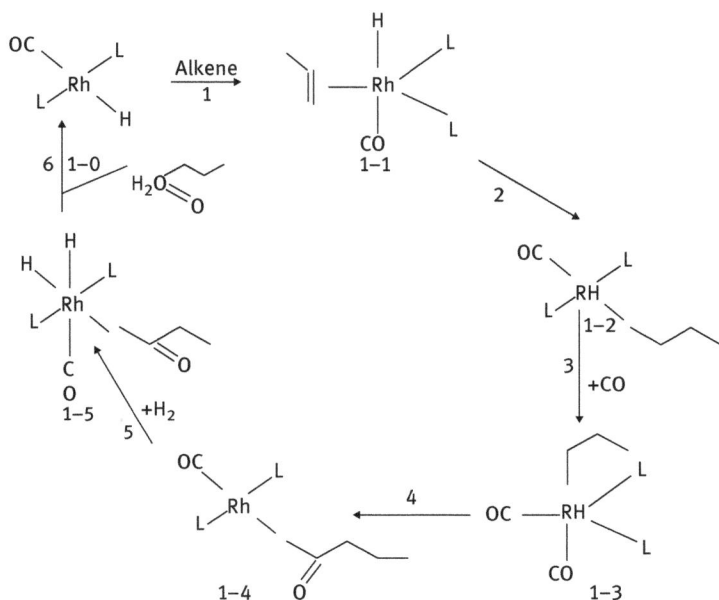

Figure 13.47: Hydroformylation cycle. From D. Yu. Murzin, A. Bernas, T. Salmi, Mechanistic model for kinetics of propene hydroformylation with Rh catalyst, *AIChE Journal*, 2012, 58, 2192–2201. Copyright © 2011 American Institute of Chemical Engineers (AIChE). Reproduced with permission from Wiley.

Parameters for the low-pressure oxo process in the case of propene hydroformylation are shown in Table 13.6. Due to very high sensitivity to the impurities of the ligand and rhodium used at a low concentration (300 ppm), the feed should be very thoroughly purified. The linearity (expressed through normal/*iso* ratio) depends on the ligand concentration and type. The catalyst in a continuous stirred tank reactor is dissolved in the solvent, which is an oligomer (trimer and tetramer) of the product butyraldehyde and by-products. The heat is removed by cooling through the reactor jacket and by-product evaporation. An important feature is the almost non-existent selectivity for hydrogenation of olefins or aldehydes in the presence of CO. The product aldehyde is removed with the gases by evaporation (Figure 13.49). The alkene conversion per pass is ca. 30%, and therefore, a significant amount of propene is recycled after cooling and condensing the reaction products. Such operation mode obviously leads to substantial energy consumption in compression and cooling units. Moreover, a large reactor volume is required due to a large gas flow. Operating conditions, in particular, the gas-recycling rate, are set in a way that all liquid

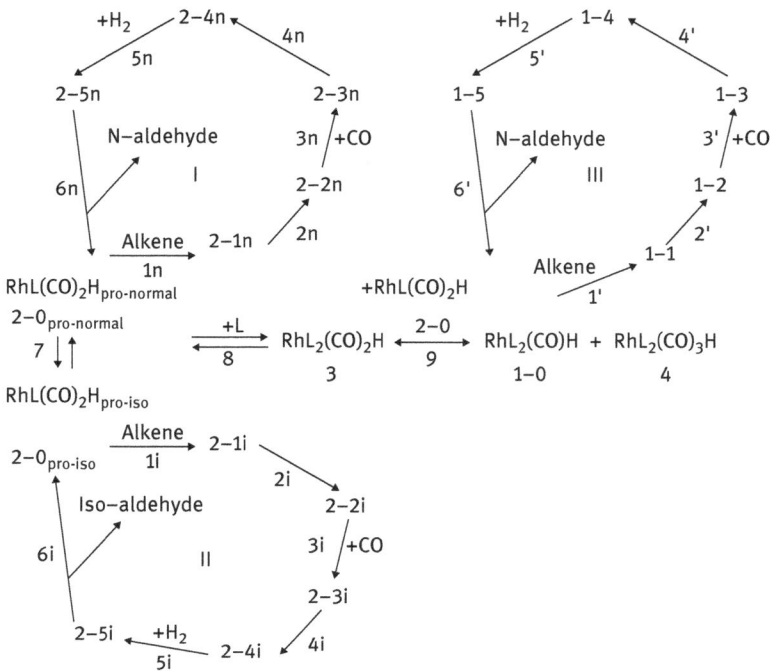

Figure 13.48: Mechanism of hydroformylation. From D. Yu. Murzin, A. Bernas, T. Salmi, Mechanistic model for kinetics of propene hydroformylation with Rh catalyst, *AIChE Journal*, 2012, 58, 2192–2201. Copyright © 2011 American Institute of Chemical Engineers (AIChE). Reproduced with permission from Wiley.

products leave the system at the same rate at which they had been formed; thus, the reactor inventory remains constant.

Unconverted propene after recompression is recycled with a small amount purged. Product separation is done by first removing the residual propene in a stabilizer column with subsequent distillation. In low-pressure oxo synthesis, gas-phase recycling process, the droplets of the catalyst are removed in the demister and sent back to the reactor. A slow decomposition of the ligand triphenylphosphine in the rhodium-catalyzed process to very stable but inactive phenyl and diphenylphosphido fragments calls for a small catalyst recycling. Formation of inert rhodium complexes is also influenced by feed impurities. Almost all the catalysts stay in the reactor, operating at identical conditions and improving overall catalyst efficiency; however, there are no options to remove heavier by-products formed by condensation reactions.

In an alternative to the gas recycling process, namely the liquid recycle process (Figure 13.50), the product is taken out of the reactor as a liquid containing the catalyst. Separation of the product from the catalyst happens independently on reaction conditions. This leads to a situation when vaporization parameters can be

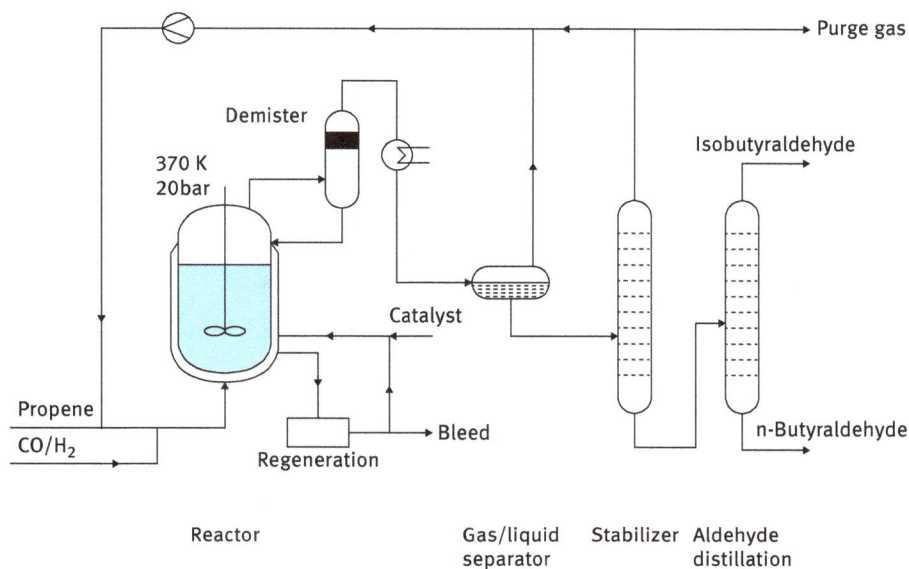

Figure 13.49: Low-pressure hydroformylation with gas recycling. From J. A. Moulijn, M. Makkee, A E. van Diepen, *Chemical Process Technology*, 2013, 2nd Ed. Copyright © 2013, John Wiley and Sons. Reproduced with permission from Wiley.

chosen independent on the reaction parameters (temperature and concentration), which could be optimized in their own way without consideration of catalyst/product separation. Care should be taken, however, regarding the optimal conditions for this separation. More efficient separation with more severe distillation conditions leads to higher catalyst concentration in the reactor and higher concentration of the product at the reactor outlet. At the same time, faster catalyst deactivation would be a negative consequence of more severe distillation; thus, there should be a trade-off between process productivity and the catalyst stability. This task was successfully solved and almost all low-pressure plants designed since the mid-1980s utilize the liquid recycle; moreover, gas-recycling designs have been revamped to the liquid recycle. After the gas-liquid separation and recycling of unreacted olefin and synthesis gas, the products are taken at the product/catalyst separation column top, while the catalyst remains at the bottom. In a subsequent crude aldehyde distillation column, further purification of aldehydes from heavier by-products takes place. Finally, a mixture of aldehydes is sent to a splitter column where the *n*-isomer is separated from the *iso*-product. The liquid recycle approach allowed to reduce the size of the reactor compared with gas recycling, when a significant excess reaction volume was required because of the entrainment of bubbles from a large gas flow, leading to the liquid-phase expansion. In the case of revamping of gas-recycling plants to the liquid recycle, an almost twofold production capacity could be reached with the same reactor size.

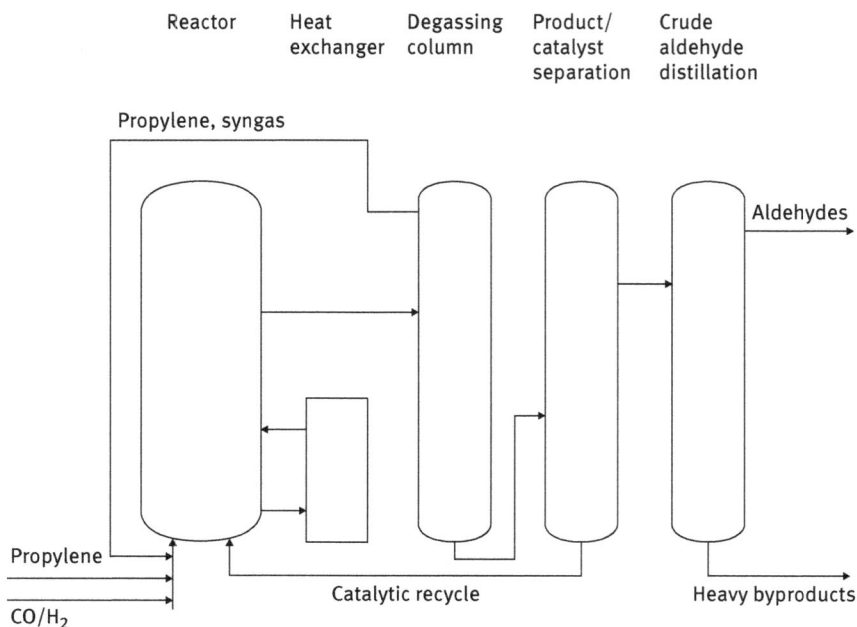

Figure 13.50: LPO process scheme with removal of product in the liquid phase. From A. Jess, P. Wassersheid, *Chemical Technology:An Integrated Textbook*, 2013. Copyright Wiley. Reproduced with permission.

In general, two classes of ligands can be considered in hydroformylation: phosphines with P–C bonds and phosphites with P–O bonds. Initially, in the 1960s, the latter have also been considered as potential ligands for rhodium hydroformylation; however, triphenylphosphine (Figure 13.46b) was preferred partly because of the instabilities of conventional phosphites in the presence of aldehydes. Bisphosphite ligands (Figure 13.46e) do not suffer from these limitations. A high normal to *iso*-selectivity is determined by a choice of substituents X3, X4, Y3, and Y4. Nowadays, several plants operate using a bisphosphite-modified rhodium catalyst, giving a possibility to achieve the ratio of normal/*iso*-butyraldehyde of 30:1. This catalyst is more active than the triphenylphosphine-based one, which gives the possibility to apply lower concentrations of rhodium and thus diminish the rhodium inventory of a plant to less than one third. This obviously saves on costs to establish certain rhodium inventories for running the process.

The Ruhrchemie/Rhône-Poulenc process uses an alternative approach to catalyst/product separation at much milder conditions operating in the water-organic, liquid-liquid reaction medium with rhodium present in the water phase and the substrate and the product in an organic phase. The catalyst used is the rhodium complex with a trisulfonated triarylphosphine (Figure 13.46c), which is highly soluble in water (about 1 kg/l) but not in the product. Sulfonation provides hydrophilic properties to the organometallic complex. Such catalytic system can, in fact, be considered as a heterogeneous one, as the catalyst is quantitatively immobilized in an aqueous phase. This ligand is used in ca. 50-fold excess, suppressing catalyst leaching. In Ruhrchemie/Rhône-Poulenc process, the reactants are used in stoichiometric ratios. Besides a mixture of butyraldehyde and *iso*-butyraldehyde in the ratio 96:4, few by-products such as alcohols, esters, and higher-boiling fractions are also formed in this TPPTS butanal-from-propene process, first commercialized by Ruhrchemie with the initial work done at Rhône-Poulenc. From the mid-1990s, there is also industrial experience with hydroformylation of 1-butene, which, however, is not recycled, but, being partially isomerized, is sent to a reactor operating with a cobalt catalyst.

In general, the two-phase process is not suited for higher alkenes because of the low solubility of olefins in water with increasing C-atom number (Figure 13.51).

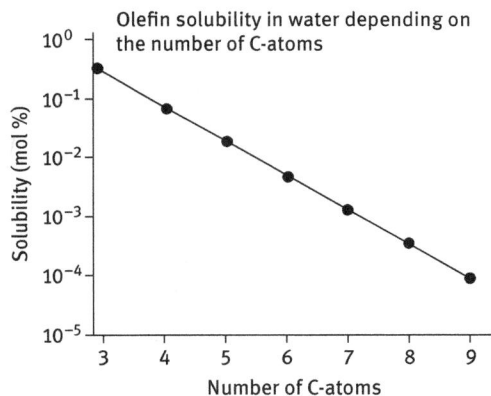

Figure 13.51: Solubility of olefins in water depending in the number of carbon atoms.

The reaction rate has a first-order dependence on alkene concentration; thus, with an increase in carbon number, the reaction productivity diminishes substantially. This limits the applicability of the process to hydroformylation of propene and butene. Even for these reactants, a relatively large reactor and high Rh inventory are required due to somewhat lower reactivity.

Separation of butanal from the aqueous/catalyst phase in the Ruhrchemie/Rhône-Poulenc process (Figure 13.52) is done by phase separation, with the aqueous catalyst phase remaining in the reactor. The process requires intensive stirring (with subsequent energy costs) in a tank reactor to which the olefin and the syngas are bubbled from the bottom. This is needed for efficient mass transfer of the syngas to the aqueous phase and the olefin from the organic phase to the water phase. Typical reaction conditions are 120 °C and 5 MPa of total pressure, with the water/organic phase ratio equal to 6. Some typical process data are presented in Table 13.7.

Table 13.7: Typical process data for RCH/RP process.

	Unit	Typical value	Variance
n-Butyraldehyde	(%)	94.5	92–97
iso-Butyraldehyde	(%)	4.5	3–8
n-Butyralcohol	(%)	0.5	0.5
iso-Butyralcohol	(%)	<0.1	<0.1
Butyl formats	(%)	Traces	Traces
Heavy ends	(%)	0.4	0.2–0.8
n/iso ratio	–	21	11–32
Selectivity toward C_4 products	(%)	>99.5	>99
Selectivity toward C_4 aldehydes	(%)	99	99
Temperature	(°C)	120	110–130
Total pressure	(MPa)	50	30–60
CO/H_2 ratio	–	1.01	0.98–1.03
Aqueous/organic phase ratio	–	6	4–9
Heat recovery without radiation losses	(%)	>99	>99
Conversion	(%)	95	85–99
Propylene quality	(% propene)	95	85–99.9

From C. W. Kohlpaintner, R.W. Fischer, B. Cornils, Aqueous biphasic catalysis: Ruhrchemie/Rhone-Poulenc oxo process, *Applied Catalysis A*, 2001, 221, 219–225. Reproduced with permission from Elsevier.

The product mixture is depressurized. The off-gas coming from the gas-liquid separator contains olefin. The off-gas is recycled and a part of it is purged. The two-phase liquid mixture containing crude aldehyde is separated at the top from the aqueous phase in a settler tank. The aqueous catalyst-containing solution after re-heating in a heat exchanger (not shown) is pumped back into the reactor. The aldehyde phase is separated in the absence of a catalyst from the excess olefin and syngas in a stripper. As shown in Figure 13.52, syngas is introduced through a stripping column countercurrent to the crude aldehyde to aid in recovering unreacted propene, which, together with syngas, is fed to the reactor. Potential catalyst poisons coming with the syngas are removed during stripping with the crude aldehyde. Distillation of the latter allows separation of the organic phase into butyraldehyde (92–95%) and *iso*-butyraldehyde (5–8%). Efficient heat integration is thus an essential advantage of this process technology. There is no accumulation of catalyst poisons because of their removal during stripping and the inability of high-boiling-point compounds to be dissolved in the aqueous catalyst phase. Part of the water remains in the crude aldehyde being compensated by extra water supply to the reactor. Losses of rhodium are at parts per billion level, according to the industrial experience.

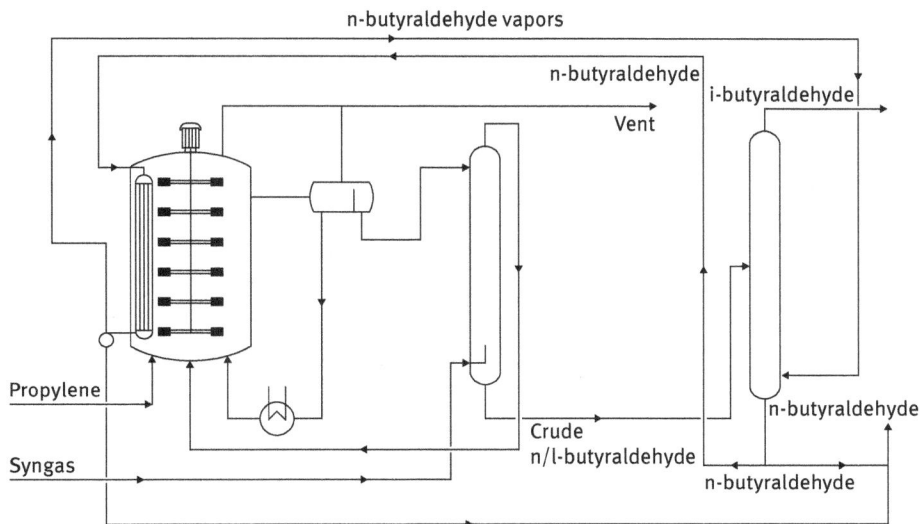

Figure 13.52: Ruhrchemie/Rhône-Poulenc process. From C. W. Kohlpaintner, R.W. Fischer, B. Cornils, Aqueous biphasic catalysis: Ruhrchemie/Rhone-Poulenc oxo process, *Applied Catalysis A*, 2001, 221, 219–225. Reproduced with permission from Elsevier.

An overall conclusion when comparing the liquid recycle and the Ruhrchemie/Rhône-Poulenc processes is that they have their own advantages and disadvantages. A detailed techno-economical analysis taking into account licensing conditions, site infrastructure, etc. is therefore needed when selecting a technology for grass-roots plants.

The description of hydroformylation technologies above was limited to low alkenes (C3 and C4) even if higher olefins represent a significant part of the global oxo business. Low reactivity of higher olefins (C5 +) using two-phase catalyst and reaction systems and Rh-based catalysts can, in principle, be overcome by several methods, including utilization of amphiphilic water-soluble ligands or applying co-solvents. The preferred option seems to be hydroformylation in one phase allowing high olefins concentration and thus high rates. In the second step, the product aldehyde should be separated from the unreacted olefin and the rhodium catalyst.

For sensitive-to-temperature high-boiling products and Rh complexes, instead of distillation, another option, namely extraction of the catalyst, can be applied. This approach (Figure 13.53) developed in the mid-1990s by Union Carbide (now Dow Chemical) introduced hydroformylation using a monosulfonated triphenylphosphine ligand (Figure 13.46d) dissolved in N-methyl-pyrrolidone (NMP), which is miscible both with water and with apolar feedstock and products. After completion of the reaction and removal of syngas, water is added to the product mixture containing, besides the Rh catalyst, unreacted olefins, aldehydes, and higher-boiling products dissolved in the solvent. This results in liquid-liquid separation with the rhodium-ligand complex being extracted to the aqueous phase. The organic reaction products and unconverted olefin remain in the organic phase, which is immiscible with the water phase. The solvent is partitioned between two phases. After decanting, the organic phase, still containing some water with dissolved Rh, undergoes another extraction step with freshly distilled water, which is added from the water-removal section of the catalyst phase. The water phase, after such second water extraction step, is sent to the induced phase separator while the crude aldehyde undergoes subsequent distillation. The catalyst phase containing NMP and water should be dried to avoid liquid-liquid separation in the reactor. Such drying, which is inevitably energy intensive due to high heat of water evaporation, is done in two steps. This technology can be used, for example, in Rh-catalyzed hydroformylation of 1-alkenes (C11–C14) to C12–C15 detergent alcohols.

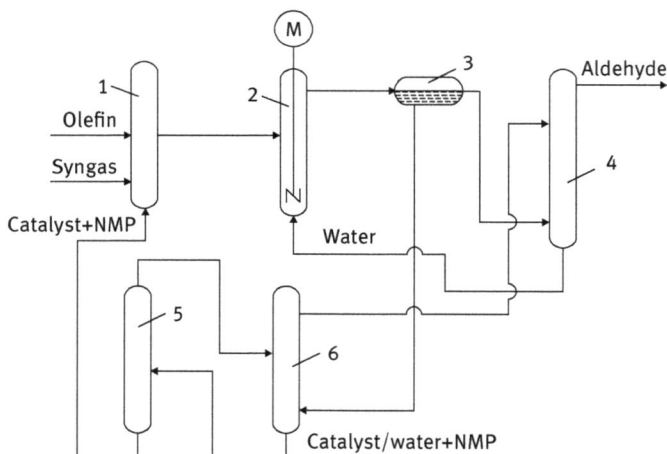

Figure 13.53: Hydroformylation of higher olefins (Union Carbide process): 1, reactor; 2, induced phase separator; 3, decanter; 4, water extractor; 5, catalyst drying; 6, primary water-catalyst separator. From A. Jess, P. Wassersheid, *Chemical Technology: An Integrated Textbook*, 2013. Copyright Wiley. Reproduced with permission.

Chapter 14
Key reactions in the synthesis of intermediates: nitration, sulfation, sulfonation, alkali fusion, ketone, and aldehyde condensation

14.1 Nitration

Nitration is defined as a substitution of one or several hydrogen atoms by a nitro group. As a nitration agent, typically, nitric acid or products of its transformations can be used. This reaction contrary to sulfation discussed also in this chapter is irreversible. As a nitration agent, nitric acid is often combined with concentrated sulphuric acid and water due to a fact that this mixture is not that aggressive from the corrosion viewpoint. One of the typically used compositions contains 20% of nitric acid, 60% of sulphuric acid with water being the rest. Stronger acid concentration would lead to oxidative side reactions. The active agent is thus nitronium cation formed according to

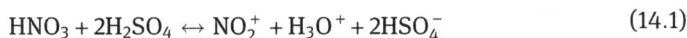

$$HNO_3 + 2H_2SO_4 \leftrightarrow NO_2^+ + H_3O^+ + 2HSO_4^- \tag{14.1}$$

which in fact is a combination of many equilibria. The amount of nitration agent is often close to the stoichiometric ratio required for nitration. In nitration, a careful temperature control should be done, since the reaction is exothermal. Thus, mono-nitration of benzene

$$ArH + HNO_3 \rightarrow ArNO_2 + H_2O \tag{14.2}$$

has the reaction enthalpy is – 117 kJ/mol, while for naphthalene, it is even higher (–209 kJ/mol). Heat also evolves when water formed in the nitration reaction dilutes sulphuric acid.

Temperature increase would be typically seen as an increase in the formation of NO_x. Moreover, higher temperature results in decreased formation of the nitronium ion.

Strict requirements in terms of heat management call for a careful temperature control, achieved by intensive mixing and a slow addition of reactants. A necessity of mixing stems also from the fact that the nitrating agent in an aqueous phase, while the reactant and the product form a separate organic phase. It has been argued therefore that the reaction in industrial conditions is controlled by mass transfer rather than kinetics. In particular, for nitration of aromatic compounds, it was shown that the reaction occurs mainly in the aqueous layer, while the rate is much slower in the organic phase. It should not be forgotten either that some polynitrocompounds (e.g., trinitrotoluene, TNT) are highly explosive.

Either batch or continuous reactors are used for nitrations. The former can be applied in the case of low tonnage and a need to produce a variety of products. In

https://doi.org/10.1515/9783110712551-014

order to achieve high productivity in the nitration processes, it is necessary to have high heat transfer coefficient (influenced by mixing efficiency), high heat transfer area (influenced by the reactor design, installation of internal coils, etc.), and significant temperature difference between the reaction mixture and the cooling agent. The last parameter is more difficult to modify. Continuous operation can be achieved, for example, in a cascade of small agitated reactors, allowing much lower material inventory than in batch reactors and thus easier temperature control.

Due to high heat involved and explosive nature of nitration, the size of vessels is typically limited to 6 m^3. In addition to this safety aspect, the operation policy in batch type reactors also takes into account safety by adding the nitrating agent in a semi-batch way with the rate of addition determined by the efficiency of cooling of the reaction mixture. The apparent danger of equipment malfunctioning is associated with a sudden stop of mixing, since it would result in the accumulation of the nitrating agents. In order to circumvent such unwanted situations, the addition of the nitrating agent is stopped when a certain temperature is reached or stirring ceased to operate.

A traditional batch-wise nitration of benzene, when the latter was added at 50–55 °C in a slight excess compared to the nitrating agent and the reaction completion was achieved by increasing the temperature to 80–90 °C, was replaced by a continuous process under similar conditions. A reactor for nitration can be a cylindrical vessel 1 (Figure 14.1) with tubes 2 and an impeller. The nitrating agent is introduced through valves 4 and 5. The substrate through a pipe (8) is introduced in the reactor and flows to the bottom. In the lower part of the reactor, the nitrating agent is mixed with the substrate and the product flows through tube 2. In order to remove heat, there is a cooling agent between the tubes. The product in a form of emulsion is taken from 7.

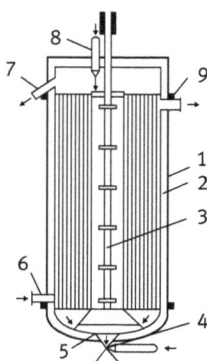

Figure 14.1: Reactor for nitration: 1, vessel; 2, tubes; 3, impeller; 4, valve; 5, inlet of nitrating agent; 6 and 9, inlet and outlet of cooling agent; 7, outlet of the product and spent acid; 8, inlet for substrate.

The reactor of a cylindrical type can be also separated into several zones to avoid slippage of the substrate from the reactor. An alternative to a cylindrical reactor could be a reactor with impeller, jacket, and coils operating alone as a CSTR or with a second nitrator of a similar type. Utilization of a second reactor allows to finish nitration at a temperature different from the one in the first reactor. An example of a flow scheme for benzene nitration is presented in Figure 14.2.

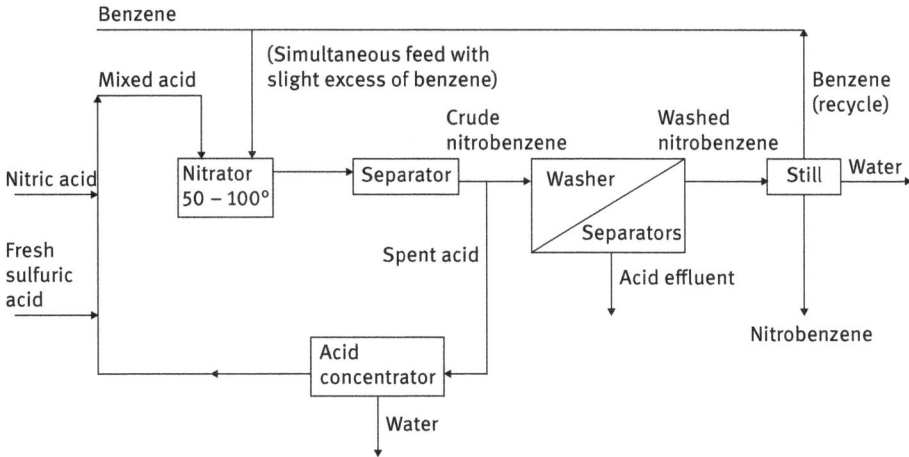

Figure 14.2: Nitration of benzene.

If the nitrator is a CSTR operating at temperature level 65–68 °C, the product concentration in the reactor is ca. 5%. The reaction mixture, containing crude nitrobenzene, prior to separation can pass through heat exchangers, where the reaction is completed. The consumption of nitric acid is ca. 90% in the reactor and 9–9.5% in the first heat exchanger. In the second heat exchanger, the reaction mixture is cooled to ca. 30 °C.

Separation of crude nitrobenzene from the spent acid, continuous concentration and addition of the fresh acid as well as neutralization and washing of the product are essential stages of the workup process. Crude nitrobenzene is first separated from the spent acid taken as the top layer from the separator. Separation is based on the differences in the density of nitrobenzene and spent acid and occurs relatively fast (residence time in a separator is 5–10 min). An example of the continuous separator is shown in Figure 14.3, where internal elements are installed to increase the path of the reaction mixture, and thus residence time.

The aqueous spent acid is recycled to the reactor after concentration. Residual acid is washed from the crude nitrobenzene, first with dilute alkali (1–2 wt% sodium carbonate or 5% ammonium hydroxide) and then with water. Removal of water and benzene is done in a still. Benzene is recycled, while nitrobenzene is distilled under vacuum, leading to a pure product with ca. 96% overall yield. Wastewater treatment is necessary to remove nitrobenzene.

In Meissner units, there is nitrogen blanketing for additional safety. The spent acid is extracted with benzene to remove both residual nitrobenzene and nitric acid.

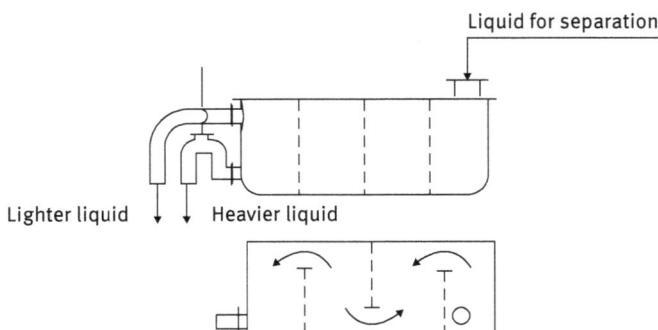

Figure 14.3: Continuous separator with internal elements.

Scrubbing of the residual waste gases is done by a mixed acid loop. Nobel Chematur developed the pump nitration circuit when nitration takes place in the pump itself, resulting in very short reaction times (<1 s) due to intensive mixing. Large residence times (ca. 4 min) are used in a vigorously agitated tubular reactor of the American Cyanamid adiabatic process when benzene (1.1 mol per mole of HNO_3) is fed concurrently with the mixed acid stream entering the reactor at 60–80 °C, which is below the benzene boiling point. The outlet temperature is ca. 120 °C. Recycling in this process, operating with much weaker sulphuric acid, is much more economical.

Nitration at higher temperature (120–160 °C) and distilling water excess from the nitrator as an azeotrope with benzene eliminate a need for sulphuric acid reconcentration in a separate step. In duplex azeotropic nitration, the benzene azeotrope from the first high-temperature nitration is fed to a lower-temperature second stage completing the reaction. Despite considerable energy savings, this technology seems not to be competitive with adiabatic nitration. The work-up of different processes is very similar (Figure 14.2).

Since introduction of the second nitro group is two orders of magnitude slower than the nitration of benzene, it is possible to make nitration with a spent acid from the second nitration step mixed with a conventional nitration mixture. The process starts at 25–40 °C and continues at 60 °C. Thereafter, the product mixture is separated from the spent acid; the latter is used for the preparation of the nitrating agent used in the first step. Crude nitrobenzene is nitrated with a fresh nitrating agent (33% nitric acid and 67% sulphuric acid), first at 25–40 °C and subsequently at 90 °C. Such process results in excessive formation of NO_x, which should be first separated from the spent acid and thereafter efficiently treated.

Large-scale nitration of another aromatic compound – toluene – is done with a mixed acid at 25–40 °C and a nitric acid/toluene molar ratio close to 1. The nitration rate of toluene is ca. 17 times faster than that of benzene; thus, continuous reactors with a low residence time are applied. Generation of readily formed by-products along with a faster rate requires the application of low temperatures. An isolated yield of 96% for all isomers is obtained, with the ratio between *ortho* and *para* isomers being 1.6. Product separation after nitration is done in a continuous, usually centrifugal, separator with subsequent washing with dilute alkali and water similar to benzene nitration. Fractional distillation of nitrotoluene after the still giving fairly pure *o*-nitrotoluene is done under vacuum.

Another example is production of nitrocellulose starting from preferably cotton-base cellulose by nitration with a mixture of nitric and sulphuric acids. Chemically, nitration of hydroxyl groups in a macromolecule (cellulose) is similar to the nitration of hydroxyl groups in the low-chain molecules (Figure 14.4) giving also water as a product and releasing heat. As can be seen from Figure 14.4, up to three hydroxyl groups can be substituted. When all three groups are substituted the nitrogen content in nitrocellulose is 14.14%.

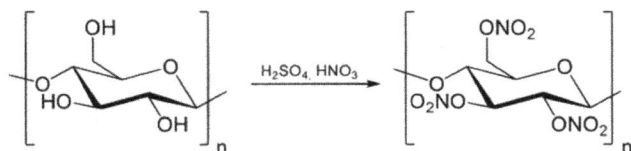

Figure 14.4: Nitration of cellulose.

Nitration reaction of cellulose is a reversible reaction, where equilibrium can be shifted by increasing the concentration of nitric acid and removing the reaction product – water. Therefore, manufacturing of nitrocellulose is done under the excess of the nitration agent.

The heterogeneous nature of the process is defined by presence of reactants in different phases (i.e., solid cellulose and liquid acid mixture). Subsequently, nitration occurs at the interface requiring efficient diffusion of the acids to the external surface of the cellulose fibers and then to the internal structure of the biopolymer. The structural characteristics of the feedstock cellulose have thus a large impact on the overall rate, which is defined by the diffusion rate of the acids and the reaction rate per se. If diffusion is the rate-limiting process, then nitrocellulose becomes inhomogeneous in terms of the degree of substitution, otherwise when the rate is determined by the rate of nitration a uniform product in terms of substitution degree is obtained.

In addition to nitration, the acidic media can also promote cleavage of the glycosidic bonds leading to shortening of the macromolecular chain as well as oxidation of some hydroxyl groups to aldehydes, ketones, and acids. The products of these side reactions can in turn be nitrated. In addition to nitration, also sulfation of the cellulose can take place forming cellulose sulfate. Moreover, oxidative and hydrolytic products can also be sulfated. The side reactions in the cellulose nitration process become more prominent with the increase of the moisture and sulfuric acid content in the mixed acid. Nevertheless, presence of sulfuric acid is needed to form the nitronium cation as the active agent, promote swelling of cellulose and capture of released water.

Poor water absorption capacity along with an elevation of nitration temperature also results in increased side reactions.

A higher content of sulfuric acid, which is less expensive than nitric acid, in the mixed acid results in a slower nitration rate, larger amounts of sulfates, and more severe degradation of the substrate, namely its fiber structure and subsequently the yield of nitrocellulose. The ratio of sulfuric acid to nitric acid is controlled in the range of 2–3. The water content is typically kept in the range of 4–20%, as lower amounts of water in the acid mixture are less reactive because of poorer wettability. On the contrary, very high water content will lead to the product with a lower nitrogen content and inhomogeneous character because of enhanced hydrolysis and oxidation reactions.

The overall process rate is substantially dependent on diffusion of the acid mixture, thus the temperature effect is less prominent than in diffusion limitations free cases. Moreover, higher temperature leads to more prominent side reactions of oxidation and hydrolysis and thus cellulose degradation. The equilibrium character of cellulose nitration implies that elevation of either temperature or process time do not significantly influence the nitrogen content in the product, thus nitration is typically done at 20–40°C.

As mentioned above, cotton based-cellulose is preferred compared to the wood based as it has less impurities (e.g., fats, lignin, tars), which can otherwise influence wettability and retard diffusion inside the fibers.

U-type continuous nitration process is shown in Figure 14.5.

First, the mixed acid is passed through a thermostat 1 and pumped into a tank 2 by a pump. The dried cellulose from a cyclone 3 falls into a U-type nitrator 6. The mixed acid is sprayed from the sprayer 5 impregnating cellulose. The feed containing the acid and cellulose flows along the U-tube reactor 6 being discharged in the last U-tube.

Acid removal and washing are essential in the manufacture of nitrocellulose, which is related to equipment and involves production costs. The waste acid recovery rate is not sufficient if only filtration through the layer of nitrocellulose is used, thus the acid removal is done by the horizontal piston discharge centrifuge 7, which is followed by several steps of acid replacement and washing. The washer 8 is a stainless steel cylindrical container with, e.g., an agitated propeller. The washed nitrocellulose is conveyed to the digestion process by the pump 9 for stability treatment. The process flow scheme (Figure 14.5) also features several collection tanks (pos. 10 to 14, 16, 17) which are used for collecting acids after different process steps and storing them. In the continuous nitration process, the number and length of the U-tubes depend on product specification. It is sufficient to have 4–5 U-shaped tubes for nitrocellulose with low nitrogen content, while a high nitrogen content requires more U-tubes (i.e., 5–7).

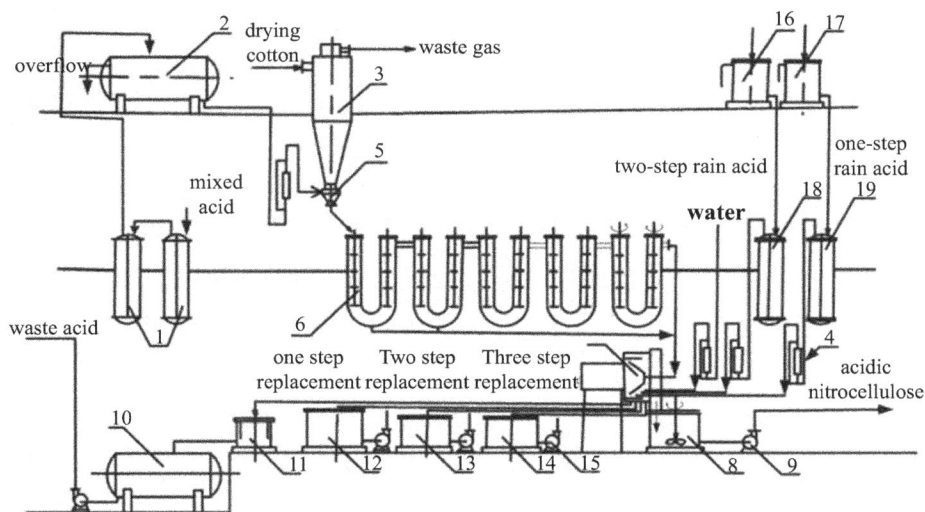

Figure 14.5: U-type continuous nitration process. From J. Liu, Nitrocellulose. In Nitrate Esters Chemistry and Technology, 2019, Springer. Singapore. https://doi.org/10.1007/978-981-13-6647-5_10. Copyright Springer. Reproduced with permission.

14.2 Sulfation and sulfonation

14.2.1 Sulfation

Sulfation involves forming a carbon-oxygen-sulphur bond resulting in an alcohol sulphuric acid, which is not hydrolytically stable, being prone to decomposition unless neutralized.

The sulfation reaction

$$ROH + H_2SO_4 \leftrightarrow ROSO_2OH + H_2O \tag{14.3}$$

is reversible and, at equimolar ratio of alcohol and acid, gives a conversion level of 40% to 65% for secondary and primary alcohols, respectively. Primary alcohols are much more reactive than secondary ones. The reaction is exothermal mainly due to the heat released when alcohol and formed water dilute concentrated sulphuric acid. In order to shift the equilibrium, concentrated sulphuric acid is used in excess to the alcohol (ca. 2:1 molar ratio). Moreover, formed water can be removed to shift equilibrium. Sulfation of long-chain alcohols not soluble in sulphuric acid is influenced by mass transfer requiring, thus efficient stirring. Since sulphuric acid can promote dehydration of alcohols and their oxidation to aldehydes and acids, a careful temperature control (20–40 °C) is needed to avoid these side reactions. For the same reason of diminishing side reactions, oleum is not applied as a sulfation agent.

In sulfation of alcohols with sulphur trioxide, alcohol sulphuric acid is produced

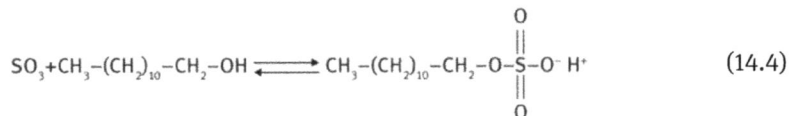

$$SO_3 + CH_3-(CH_2)_{10}-CH_2-OH \rightleftharpoons CH_3-(CH_2)_{10}-CH_2-O-\overset{\displaystyle O}{\underset{\displaystyle O}{\overset{\displaystyle \|}{\underset{\displaystyle \|}{S}}}}-O^- \ H^+ \qquad (14.4)$$

It is important to control the ratio between sulphur trioxide and alcohol, since excess SO_3 promotes side reaction. Formation of by-product can be avoided using diluted SO_3 or complexing it with hydrochloric acid, giving chlorosulphuric acid ($ClSO_3H$). When the latter acid is applied, hydrochloric acid, besides the target compound, is released as a reaction product, which should be scrubbed with water or a diluted basic solution. In the case of chlorosulphuric acid (which can be used in batch or continuous processes), complete conversion of the substrate is achieved. A flow scheme of a continuous chlorosulphuric acid sulfation process applied by several companies for making detergents is shown in Figure 14.6. This scheme illustrates that the reactants (the alcohol and chlorosulphuric acid) are first mixed, which is followed by degassing under a slight vacuum to enhance removal of HCl from the reaction products. Two degassers are used to achieve complete separation of HCl, which is then absorbed in water. The product is continuously neutralized.

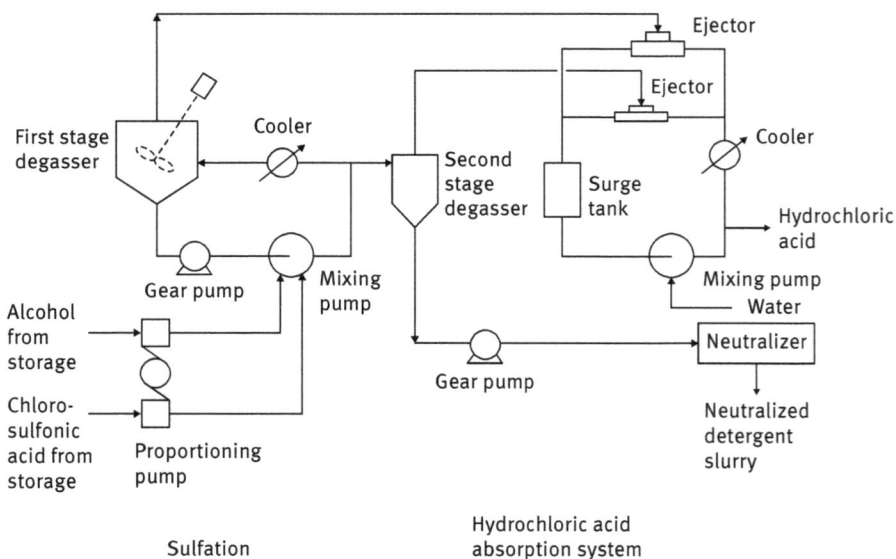

Figure 14.6: Continuous alcohol sulfation with chlorosulfonic acid. http://www.chemithon.com/Resources/pdfs/Technical_papers/Sulfo%20and%20Sulfa%201.pdf.

Sulfation with sulphur trioxide (obtained either from liquid SO_3 or from burning sulphur) has an advantage of irreversibility, lower costs, and absence of HCl utilization. The main difficulty in this technology is high exothermicity and high reaction rates. This leads to overheating and generation of side products with subsequent browning of the reaction mixture. To overcome these problems, sulphur trioxide is diluted with dry air up to 4–7 vol%, making heat removal much easier. Multitubular sulfonation reactors are typically applied in a way that the organic substrate and SO_3 flow concurrently down the tubes of ca. 25 mm diameter and 7 m length (Figure 14.7), with a residence time 2–3 min. The heat of the reaction is taken by cooling water.

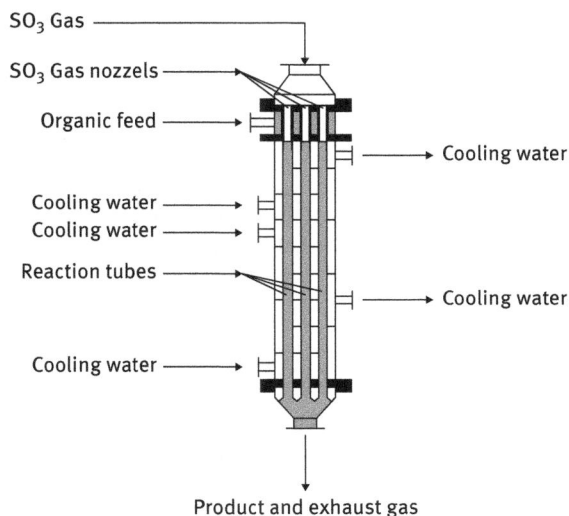

Figure 14.7: Reactors for sulfonation with SO_3. http://www.chemithon.com/Resources/pdfs/Technical_papers/Sulfo%20and%20Sulfa%201.pdf.

A flow scheme for the synthesis of alkyl sulfates, which are applied as surfactants and active ingredients in laundry detergents and shampoos, is given in Figure 14.8.

Alcohol, air, and SO_3 vapors diluted with air are sent to a tubular reactor. The process stream after passing the reactor is separated in separator 2 and the gas phase is washed with water in absorber 3. The formed alcohol sulphuric acid is neutralized with alkali in vessel 4, which has an external cooler (pos. 5) to keep the reaction mixture below 60 °C. Final neutralization is done in reactor 6. In the mixer (pos. 7), additives (phosphates, carboxymethylcellulose, etc.) are added to the solution of neutralized alkylsulfate in water. The obtained mixer is dried in a spray dryer 8 with the aid of hot flue gases. The solid particles are separated in a cyclone 9 and transported by a conveyer 10 to packaging.

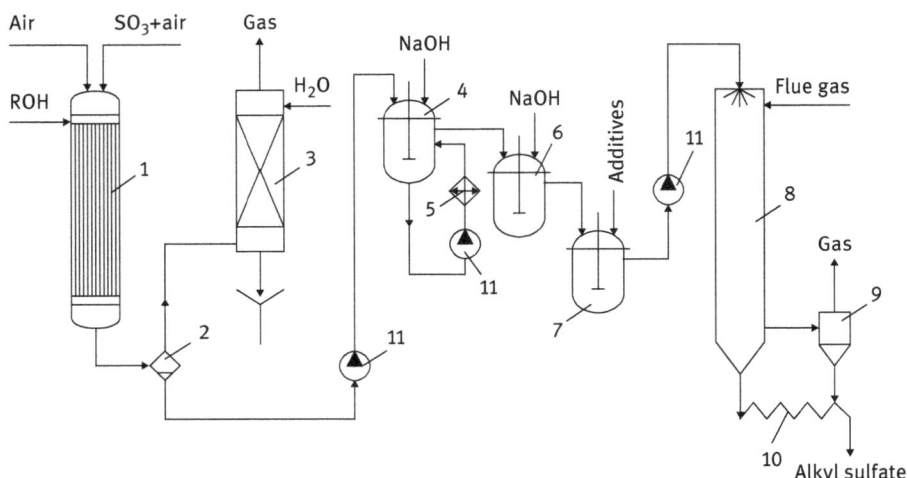

Figure 14.8: Flow scheme for synthesis of alkylsulfate: 1, reactor; 2, separator; 3, absorber; 4 and 6, neutralizers; 5, cooler; 7, mixer; 8, spray dryer; 9, cyclone; 10, screw conveyer; 11, pump. After N. N. Lebedev, *Chemistry and Technology of Basic Organic and Petrochemical Synthesis*, Chimia, 1988.

14.2.2 Sulfonation

Sulfonation is a reversible exothermal reaction of replacing a hydrogen atom typically in an aromatic compound with a sulfonic acid functional group. The products are intermediates in the preparation of dyes and pharmaceuticals, such as *p*-aminobenzenesulfonic acid, which is the basis for sulfa drugs made by aniline sulfonation (Figure 14.9).

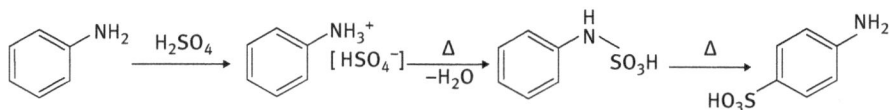

Figure 14.9: Sulfonation of aniline.

Typically, concentrated sulphuric acid or even oleum are used as sulfonation agents. The equilibrium of the reaction

$$ArH + H_2SO_4 \leftrightarrow ArSO_2OH + H_2O \tag{14.5}$$

is shifted to the right; however, at higher temperature and simultaneous removal of hydrocarbons, for example, in the separation of alkylbenzene isomers, the reverse process of sulfonic acid hydrolysis occurs. The real sulfonation agents are sulphur trioxide and the cation HSO_3^+, which is formed when sulphuric acid dissociates

$$2H_2SO_4 \Leftrightarrow SO_3 + H_3O^+ + HSO_4^- \tag{14.6}$$

$$2H_2SO_4 \Leftrightarrow H_3SO_4^+ + HSO_4^- \tag{14.7}$$

$$H3SO_4^+ \Leftrightarrow + HSO_3^+ + H_2O \tag{14.8}$$

Diluted sulphuric acid cannot be used, since in the presence of water, it dissociates according to

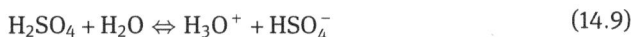

$$H_2SO_4 + H_2O \Leftrightarrow H_3O^+ + HSO_4^- \tag{14.9}$$

Formation of water during sulfonation diminishes the generation of sulfonation agents (SO_3 and HSO_3^+); thus, for sulfonation, typically an excess of sulphuric acid is used. A mixture of sulfa acids is produced during sulfonation, and in the sulfonation of benzene, one acid is formed $RH_x + xSO_3 \rightarrow R(SO_3)H_x$. The introduction of the second is slower than the first one while diacids react even slower than the mono acid. This allows finding conditions when derivatives of benzene and naphthalene can be synthesized with a desired number of functional groups.

Sulfonation is typically done without any catalyst in a broad temperature range, from – 20 °C to 200 °C. Temperature influences not only the rate but also regioselectivity. Thus, sulfonation of phenol at low temperature drives the substitution in *ortho*-position, while an increase in temperature to 100 °C results in *para*-substitution. Naphthalene-1-sulfonic acid is produced by adding naphthalene to 96% sulphuric acid at 20 °C and slowly increasing the temperature to 70 °C (Figure 14.10). The amount of the β – isomer is ca. 15%. Naphthalene-1-sulfonic acid is not usually isolated on a large scale because 1-naphthol, an important derivative of this acid, is produced by other routes. The same sulfonation reaction at ca. 160 °C gives predominantly naphthalene-2-sulfonic acid (β acid) with a minor yield of the α-acid. Contrary to naphthalene-1-sulfonic acid, naphthalene-2-sulfonic acid is isolated for conversion to 2-naphthol. The sulfonation mass can be also further sulfonated *in situ* for the production of disulfonic acid and trisulfonic acid.

Sulphuric acid (96%) from vessel 1 and melted naphthalene from vessel 2 are introduced in the reactor 3, where sulfonation occurs at 160–165 °C. The reaction mixture is then transferred to reactor 4, where, after addition of water, hydrolysis of naphthalene-1-sulfonic acid (ca. 15%) happens at 140–150 °C. Thereafter, sulfonation mass containing merely 2% of naphthalene-1-sulfonic acid, ca. 9% naphthalene (unreacted and formed by hydrolysis of naphthalene-1-sulfonic acid), 65% of naphthalene-2-sulfonic acid, and sulphuric acid is transferred to reactor 6 to which sodium sulfate is added along with steam. After final hydrolysis, naphthalene is removed with steam, condensed in 7 (temperature not higher than 40 °C), and sent through the bunkers 8 and 9 for further collection. The sulfonation mass free from naphthalene and naphthalene-1-sulfonic acid goes to 10, where it is neutralized with 15% solution of sodium sulfite,

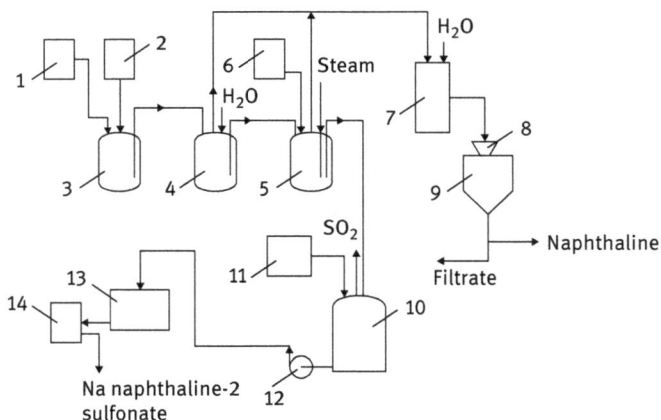

Figure 14.10: Naphthalene sulfonation: 1 and 2, vessels for sulphuric acid and naphthalene; 3, sulfonation reactor; 4, hydrolysis reactor; 5, removal of unreacted naphthalene, 6, vessel for sodium sulfate; 7, condenser; 8, bunker; 9, melting; 10, neutralization; 11, vessel for sodium sulfite; 12, pump; 13, crystallization; 14, filter. Adapted from V. N. Lisytsin, *Chemistry and Technology of Intermediates*, Chimia, 1987.

which is heated to 90 °C in vessel 11. Released SO_2 is used in the alkali fusion process. After neutralization, the sulfonation mass is crystallized in 13 to give a sodium salt, which is separated by filtration, or with a centrifuge and used in alkali fusion.

Further elevation of reaction temperature during sulfonation gives substitution of naphthalene in several positions (Figure 14.11). Formation of six diacids (1, 3-, 1, 5-, 1, 6-, 1, 7-, 2, 6-, and 2, 7-) (4)–(7), three trisubstituted acids (1, 3, 5-, 1, 3, 6-, and 1, 3, 7-) (9)–(11) and 1, 3, 5, 7-naphthalenetertrasulfonic acid (12) is possible.

Sulfonation reactions are exothermal; thus, for reactions at low temperature, the heat should be removed, while for sulfonation at higher temperature (>100 °C), the heat supply should be arranged. Sulfonation with oleum results in the dilution of sulfating agent with water to ca. 90–95%, which does not lead to extensive corrosion. At the same time, application of chlorosulfonic acids gives HCl, which, in presence of water, promotes very strong corrosion of iron-based alloys.

Several options for sulfonation reaction technology can be considered. For liquid-liquid systems, when sulfonation of liquid aromatic compounds that are non-soluble in the sulfation liquid agent is done, efficient stirring is required, as the density of the organic phase is lower. For such heterophase systems, reactions occur in the phase of the sulfation agent; therefore, the solubility of the substrate as well as the mass transfer and the interphase area are of importance.

In the case of the solid feedstock and liquid sulfation agent, the reaction mixture is very viscous and anchor-type impellers are used, while installations of coils should be avoided. Grinding of the solid to a required size is needed to ensure a constant reaction rate with each addition of the substrate.

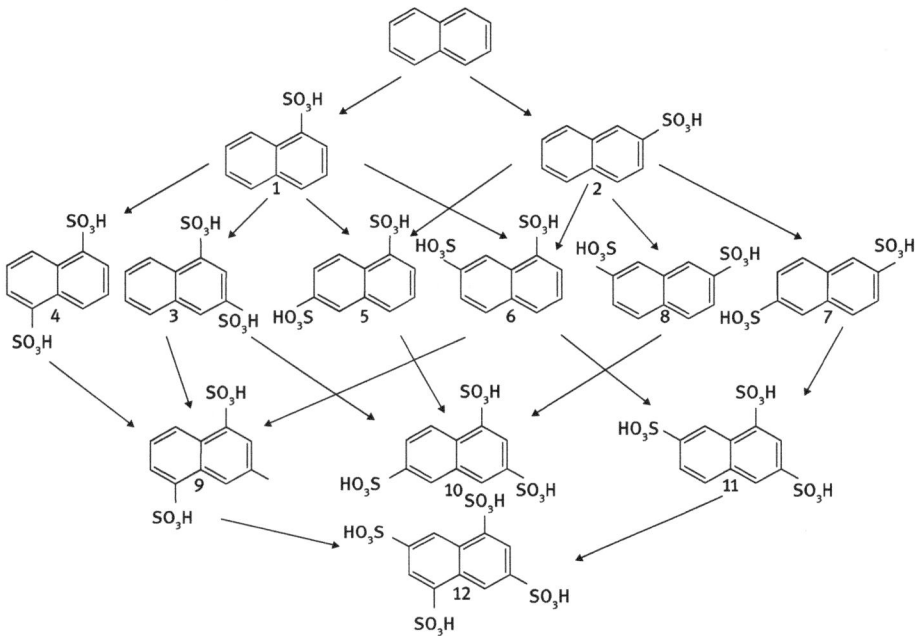

Figure 14.11: Reaction network in sulfonation of naphthalene. Adapted from V. N. Lisytsin, *Chemistry and Technology of Intermediates*, Chimia, 1987.

Sulfonation in the vapor phase is done for some aromatic compounds (benzene, toluene) by the vapors of sulphuric acid in the excess of the substrate. Water formed in the reaction

$$RH + H_2SO_4 \rightarrow RSO_3H + H_2O \tag{14.10}$$

is detrimental for sulfonation, as explained above, and should be thus removed. This is done by azeotropic removal of water with benzene or toluene. Stirring is arranged by bubbling vapors of the organic compound.

A scheme illustrating the sulfonation of benzene in the vapor phase is given in Figure 14.12.

Benzene is heated with the excess of sulphuric acid, and then vapors or benzene are bubbled through the reaction mixture containing benzenesulfonic acid. Continuous sulfonation (Figure 14.13) of benzene allows diminishing costs related to evaporation of benzene at the expense of worsening selectivity.

Fresh and recycled benzene is overheated in 5 and sent to reactor 1, which does not have any impeller. 90–93% sulphuric acid is continuously added. Benzene vapors are introduced in fourfold to sixfold excess. Liquid from 1 goes to reactor 2, and simultaneously, benzene vapors are also introduced in the same tray column. The composition of sulfonation mass travelling from top to bottom is changing along the column. It

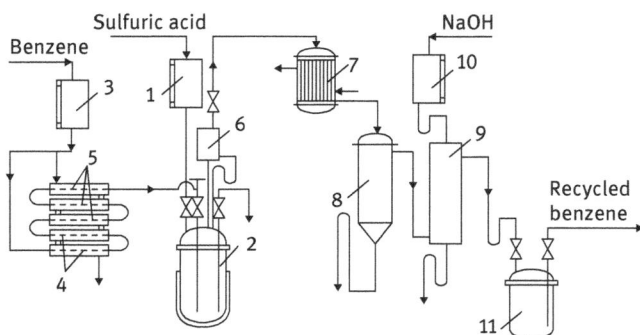

Figure 14.12: Sulfonation of benzene in vapors: 1, vessel for sulphuric acid; 2, reactor; 3, vessel for benzene; 4, evaporator; 5, benzene vapors generation (160–170°C); 6, demister; 7, tubular condenser; 8, continuous separator operated at 45–55°C; 9, columns for neutralization of recycled benzene, 10, vessel for alkali; 11, collection of neutralized recycled benzene.

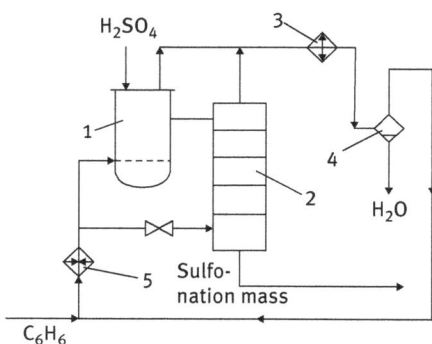

Figure 14.13: Continuous sulfonation: 1 and 2, reactors; 3, condenser; 4, separator; 5, evaporator.

is more and more enriched with the benzenesulfonic acid. Benzene vapors from reactor 2 are condensed in 3. Downstream of the condenser, the benzene layer is separated from water and after neutralization recycled.

Finally, sulfonation by gaseous sulphur trioxide can be efficiently performed when the substrate is in the liquid phase and a propeller type of stirrer can be used. Sulfonation of benzene

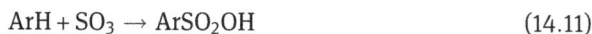

$$ArH + SO_3 \rightarrow ArSO_2OH \tag{14.11}$$

is an exothermal reaction that generates the reaction heat at 25 °C of – 217 kJ/mol.

Note that sulfonation with sulphur trioxide is an addition reaction and does not result in formation of water and subsequently spent sulphuric acid. Sulphur trioxide can be used as such or as its derivative with a base

$$\text{(pyridine)} + SO_3 \longrightarrow \text{(pyridinium)} \quad N^{\oplus} - SO_3^- \tag{14.12}$$

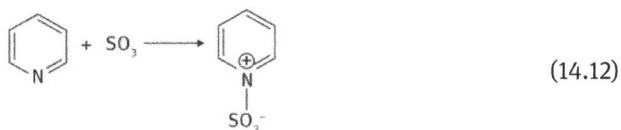

Sulfonation with sulphur trioxide dissolved (10–15%) in the liquid SO_2 can be also done at the temperature of liquid dioxide (–10 °C) when the heat is removed by evaporation of sulphur dioxide.

Separation of the sulfonation reaction products depends on their properties. When sulfonic acids are poorly soluble in aqueous sulphuric acid, dilution with water or ice can be used. However, in many cases, separation can be done by formation of salts. For example, a common method for isolating naphthalenesulfonic acids and substituted derivatives is the process when the sulfonation mass, after addition to excess water, is neutralized with lime (calcium carbonate). Precipitated calcium sulfate is filtered off when still hot. Dissolved calcium salt of the product reacts with sodium carbonate to form the sodium salt of the sulfonic acid, while precipitated calcium carbonate is removed by filtration. The solution can be evaporated to give the solid sodium salt.

Another option is to use sodium sulfite

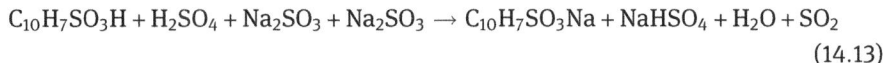

$$C_{10}H_7SO_3H + H_2SO_4 + Na_2SO_3 + Na_2SO_3 \rightarrow C_{10}H_7SO_3Na + NaHSO_4 + H_2O + SO_2 \tag{14.13}$$

which results in formation of sulphur dioxide. The latter can be used in the alkali fusion process, as will be explained in Chapter 14.3.

14.3 Alkali fusion

Alkali fusion is typically applied to replace a sulfonic acid group with a hydroxyl group in a substituted aromatic ring. For this reaction, as alkali, potassium, sodium, and calcium hydroxides can be used. For example, 2-naphthol is synthesized by alkali fusion of naphthalene-2-sulfonic acid salt (whose production is explained in Chapter 14.2).

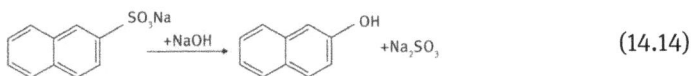

$$\text{(naphthalene-}SO_3Na) \xrightarrow{+NaOH} \text{(naphthalene-}OH) + Na_2SO_3 \tag{14.14}$$

An important issue in the process in feedstock purity. The substrate should be free from mineral salts, which would not be melted during the process, influencing homogenization of the mixture and resulting in local overheating. Alkali fusion can be done either in open vessels with direct contact to air or in autoclaves. The former option allows higher productivity operating with higher concentrations of reactants but is more energy-consuming and can result in oxidation products.

The reaction rate of alkali fusion depends on the concentration of the acid (first order) and the alkali (between the first and the second reaction orders, depending on the acid); thus, an excess of the latter is applied (2.1–3 mol per 1 mol of acid).

The production scheme of 2-naphthol is given in Figure 14.14.

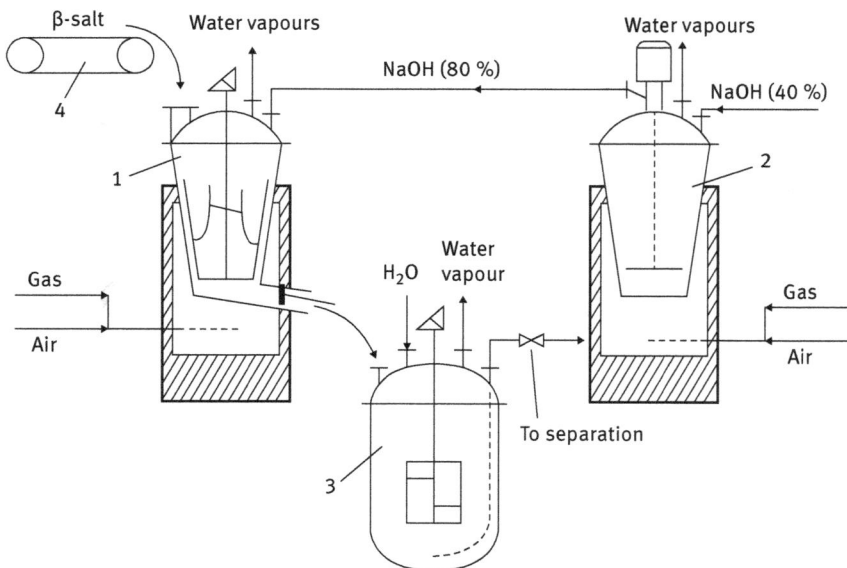

Figure 14.14: Alkali fusion: 1, reactor; 2, vessel for concentrating alkali; 3, dilution with water. Adapted from V. N. Lisytsin, *Chemistry and Technology of Intermediates*, Chimia, 1987.

Gradual addition of the sulfonic acid sodium salt solid powder or 85% paste to 80–85% sodium hydroxide is done at 300 °C in the case of 2-naphthol followed by subsequent heating at 320 °C in a gas-fired iron vessel with efficient stirring (pos. 1). The alkali is produced by the evaporation of 40% solution of NaOH in vessel 2 in Figure 14.14. The reaction is controlled by monitoring the concentration of OH^-. When the reaction is completed, giving the concentration of sodium hydroxide and unreacted salt of ca. 3% and ca. 1–2%, respectively, in the melt, the latter, containing ca. 40% of 2-naphthol and 35% of sodium sulfite, is exposed to an excess of water in reactor 2. Thereafter, the naphtholate solution is neutralized (not shown) to pH 8 with dilute sulphuric acid or SO_2. The latter is formed during sulfonation of naphthol, as discussed in Chapter 14.2. The crude product comes as an oil if kept at > 100 °C during neutralization and is separated, washed with hot water, and finally distilled under vacuum. This is followed by flaking of the pure molten 2-naphthol, resulting in the final product ready for packaging. The overall yield in the process based on naphthalene is 70%, with the yield of the alkali fusion *per se* being 80% of the theoretical. The high temperature of the process in the case of 2-naphthol is related to the low reactivity of the leaving group in position 2. For naphthalene-1-sulfonic acid salt alkali fusion to 1-naphthol, the reaction temperature would have been much lower (160–250 °C). This process is, however, not used commercially since there are more economically attractive routes for synthesis of 1-naphthol by, for example, nitration of naphthalene to 1-nitronaphthalene

$$C_{10}H_8 + HNO_3 \rightarrow C_{10}H_7NO_2 + H_2O \qquad (14.15)$$

followed by hydrogenation

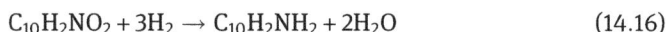

$$C_{10}H_2NO_2 + 3H_2 \rightarrow C_{10}H_2NH_2 + 2H_2O \qquad (14.16)$$

and hydrolysis:

$$C_{10}H_7NH_2 + H_2O \rightarrow C_{10}H_7OH + NH_3 \qquad (14.17)$$

14.4 Carbonyl condensation reactions

Addition and condensation reactions involving carbonyl groups are important in industrial organic synthesis. Two types of condensation reactions can be considered. Aldehydes and ketones can react with acids according to the so-called aldol condensation. Weak acids (HCN) or pseudo-acids can be applied. The latter include carbonyl compounds with activated hydrogen since protons attached to carbons adjacent to a carbonyl group are weakly acidic. In fact, in solutions, aldehydes and ketones exist not only in the keto but also in the enol form. A reaction scheme is given in Figure 14.15. Aldol condensation reactions are reversible with not very high reaction enthalpy (ca. 20–60 kJ/mol).

As follows from Figure 14.15, aldol condensation *per se* is followed by dehydration, with the overall thermodynamics depending substantially on the second step.

In the case of nitrogen-containing compounds, condensation is accompanied with intramolecular dehydration:

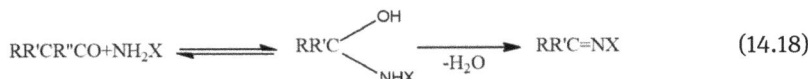

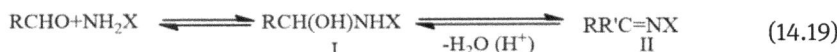

$$RR'CR''CO + NH_2X \rightleftharpoons RR'C\begin{smallmatrix} OH \\ \\ NHX \end{smallmatrix} \xrightarrow{-H_2O} RR'C=NX \qquad (14.18)$$

$$RCHO + NH_2X \rightleftharpoons \underset{I}{RCH(OH)NHX} \underset{-H_2O\,(H^+)}{\rightleftharpoons} \underset{II}{RR'C=NX} \qquad (14.19)$$

With very reactive molecules such as formaldehyde, the first stage resulting in substrate I does not need a catalyst, while the second step is accelerated in the presence of acids, which also catalyze the first step. Formaldehyde can be thus

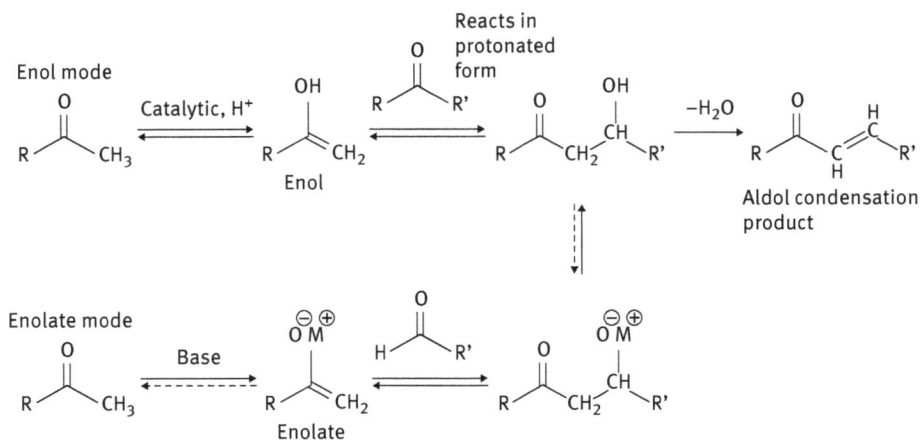

Figure 14.15: A reaction scheme of aldol condensation.

condensed with ammonia without any additional catalyst in the liquid phase, giving hexamethylenetetramine (urotropine)

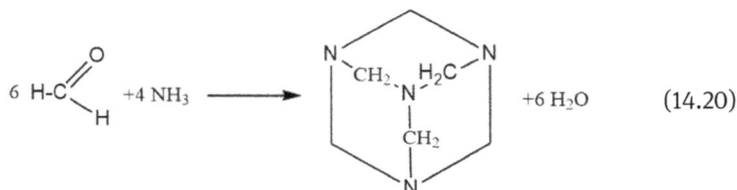

$$6\ H\text{-}C + 4\ NH_3 \longrightarrow \qquad + 6\ H_2O \qquad (14.20)$$

which is used in the synthesis of plastics, pharmaceuticals, and rubber additives. A particular case of cyclic ketones condensation with hydroxylamine for oxime synthesis will be considered in Chapter 14.5. Condensation of aldehydes and ketones with aromatic compounds and olefins is much more exothermal (ca. 100 kJ/mol). Activation of the carbonyl group in this case is done with protic acids, such as sulphuric or hydrochloric acid, or by solid acids (ion exchange resins).

14.4.1 Condensation with olefins (Prins reaction)

The Prins reaction, or condensation of an aldehyde or ketone with an olefin or alkyne, can give diols or dioxane (Figure 14.16).

Condensation of isobutylene with formaldehyde in the presence of an acidic catalyst such as diluted sulfuric gives 4, 4-dimethyldioxane-1,3 (DMD) which is decomposed into isoprene on a solid phosphate catalyst such as calcium phosphate (Figure 14.17). The second reaction is performed at 350–370 °C using dilution with water vapor at the mass fraction between steam and DMD of 1.5–2.0.

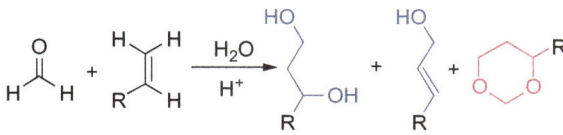

Figure 14.16: Prins condensation reaction.

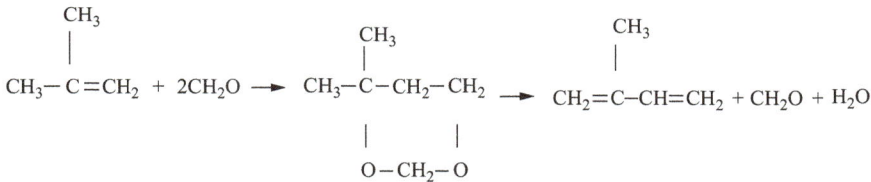

Figure 14.17: Condensation of isobutylene with formaldehyde with subsequent decomposition to isoprene.

$$Ca_3(PO_4)_2 + H_3PO_4 \rightarrow 3CaHPO_4$$

Beside the main product of the first step DMD, a range of by-products, such as trimethyl carbinol, methylal, dioxane alcohols, diols, and ethers, are also formed. Moreover, some side products are formed because of impurities in isobutylene, originating from steam cracking of naphtha. Side products are also generated during the second step of DMD decomposition to iso-butylenes and formaldehyde, giving dihydromethylpyran, hexadiene, piperylene, terpene compounds, green oil, etc. in addition to the desired product.

These numerous by-products can amount to 0.5 ton per 1 ton of isoprene.

Selectivity in the second stage can be improved by the introduction of small amounts of phosphoric acid vapors directly into the catalysis zone leading to formation of acidic phosphates on the surface of the calcium phosphate.

Catalyst regeneration after 2–3 h is done by burning off coke with air/steam at temperatures above 500°C.

While the first stage in the overall process is rather efficient, the second stage is much more energy intensive with DMD decomposition selectivity in terms of isoprene of 70–80%. This is in part related to selection of the gas phase decomposition technology on a heterogeneous catalyst.

The process flow diagram of the second step of isoprene production, i.e., DMD decomposition, is illustrated in Figure 4.18.

In the reactor (2), dimethyldioxane vapor is mixed with steam supplied from the steam heater (1). The product mixture passes through a cascade of heat exchangers (3) to the settler (4) resulting in a biphasic (organic/aqueous) condensate. Dissolved formaldehyde is extracted from the oily phase in (5) with water. In the distillation column (6), the lighter products, mainly isobutylene and isoprene, are separated from the

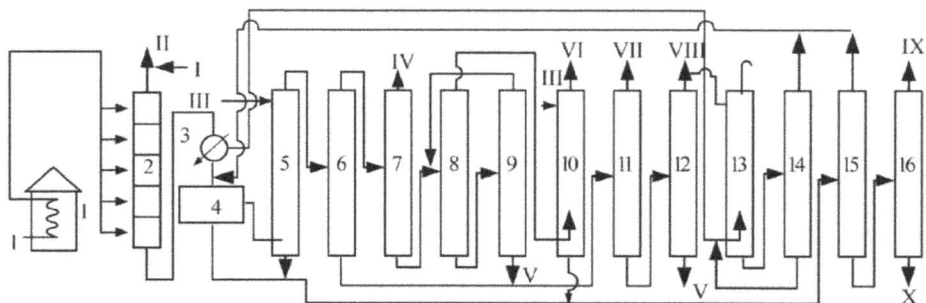

Figure 14.18: The process flow diagram of 4, 4-dimethyldioxane-1,3 decomposition. From N.A. Plate, E.V. Slivinskii, Fundamentals of Chemistry and Technology of Monomers, Nauka/ Interperiodika Publishing, 2002, pp. 131–162. 1: Steam heater, 2: reactor, 3: condenser, 4: settler, 5, 10: columns for washing, 6: crude isoprene distillation column, 7: distillation column for recycled iso-butylenes, 8, 9: column for isoprene rectificate, 11: MDGP (methyldihydromorphine) fraction column, 12: recycled DMD (dimethyldioxane) column, 13: absorber, 14: desorber, 15: distillation column for light organic compounds, 16: recovery column for formaldehyde. Streams: I steam, II- DMD vapour, III: washing water, IV: recycled isobutylene, V: boiling impurities, VI: isoprene rectificate, VII: MDGP (methyldihydromorphine) fraction; VIII: recycled DMD; IX: recovered formaldehyde, X: waste water.

decomposed dimethyl dioxane and other less volatile substances. To prevent isoprene polymerization, inhibitors are added to the distillation columns. The bottom fraction from (6) goes to column (11), while the light fraction from (6) enters the distillation column (7) giving highly concentrated isobutylene at the top, which is recycled for synthesis of dimethyl dioxane. The bottom from (7) containing isoprene is purified in columns 8, 9 by removing impurities with high boiling points, mainly cyclopentadiene and carbonyl compounds, which are washed out in column (10).

In column (11), methyldihydromorphine is taken from the top, while the bottom fraction from column (11) enters the vacuum column (12) where high boiling points byproducts (e.g., isoprene oligomers or green oil) are removed from the recycled DMD.

Apparent deficiencies in the industrial method of isoprene manufacturing prompted development of several alternatives. In the Kuraray's one-stage process not open to the general public *t*-butanol (TBA) is used along with formaldehyde. In another alternative technology implemented industrially by the Russian company EuroChim, isobutylene, formaldehyde, and water form a substituted dioxane compound that subsequently reacts with TBA to two molecules of isoprene and water. All reactions are carried out in the liquid phase. According to the developer, there are several advantages of this process compared to the traditional two-step DMD-process (Figure 14.18) including lower energy consumption and emissions as well as better atom economy

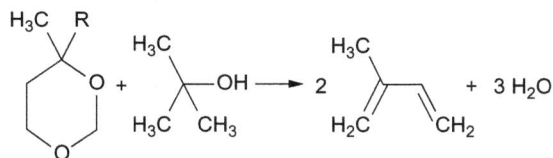

Figure 14.19: The second step in the EuroChim isoprene process.

14.4.2 Condensation with aromatic compounds

This reaction is an example of electrophilic substitution reaction:

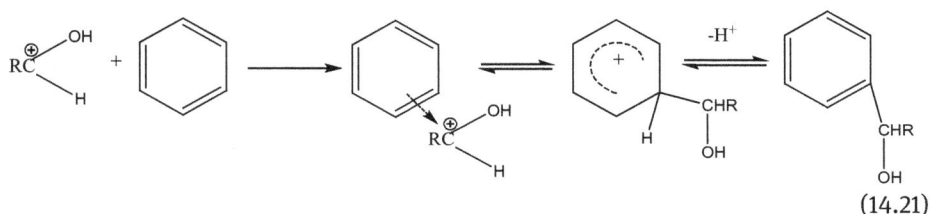

$$(14.21)$$

The formed alcohol reacts further, forming a carbocation, which in turn undergoes alkylation:

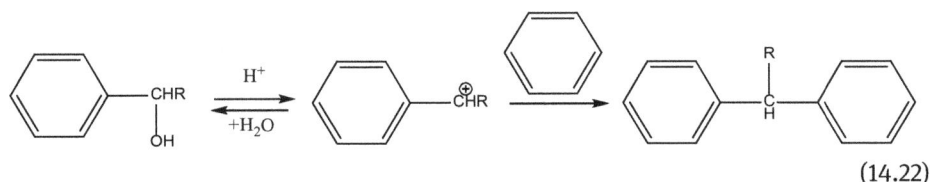

$$(14.22)$$

An example of such condensation is the production of bisphenol A (BPA), giving in fact a range of

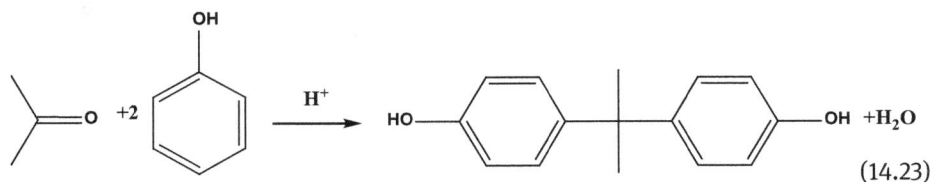

$$(14.23)$$

BPA is mainly used in advanced plastics, such as polycarbonates and epoxy resins. Reaction of phenol and acetone, obtained as a mixture in the cumene hydroperoxide process described in Chapter 9, is catalyzed by a strong mineral acid (HCl or H_2SO_4) or solid acids giving first a carbonium ion, which subsequently reacts with phenol in a stepwise fashion, first by the formation of another carbonium ion through addition of phenol, followed by the elimination of water and addition of a second phenol molecule. *Ortho/para*-isomer, along with some other compounds, is formed as by-products (Figure 14.20). The former product can be partially isomerized to the desired *para/para* isomer.

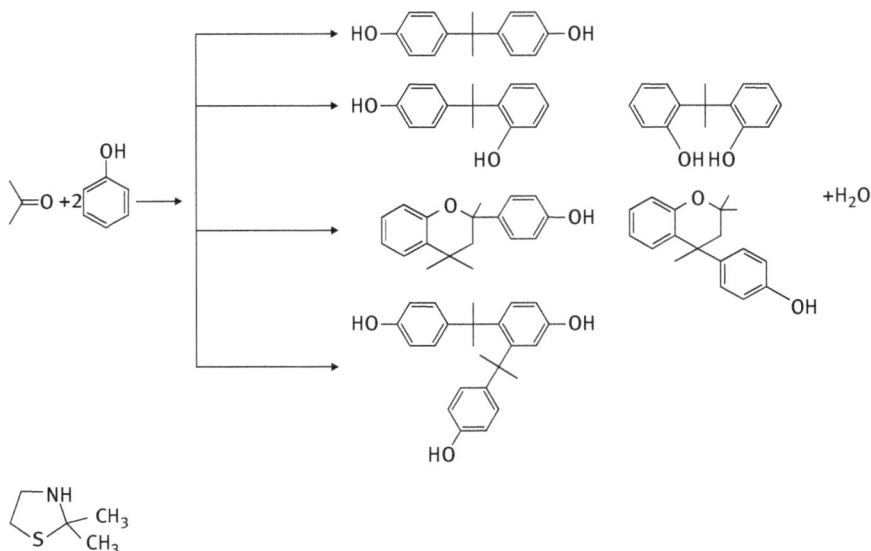

Figure 14.20: Products of the condensation of acetone with phenol.

Hydrochloric acid is preferred as catalyst, compared to sulphuric acid, because of easier separation. Although application of this acid requires lower T (ca. 50 °C) than an alternative process with solid acids (strong acidic cation exchange resins with or without activity enhancing modifiers), the yield of BPA is also lower not because of reactivity but rather because of BPA decomposition during distillation in the presence of acids. Application of solid acids compared to mineral acids has the advantages of no catalyst recycling and mitigation of equipment corrosion and problems with wastewater treatment. Elevation of temperature to 70–80 °C is needed to counterbalance lower activity.

The molar ratio between phenol and acetone ranges from 3:1 to 10:1. One of the reasons for such molar ratio is to suppress the formation of side products generated from mesityl oxide. The latter is formed by the self-condensation of acetone in the presence of ion exchange resins.

$$2CH_3COCH_3 \rightarrow (CH_3)_2C = CH(CO)CH_3 + H_2O \qquad (14.24)$$

For acidic ion-exchange resins, two options are applied. In the free co-catalyst method, this co-catalyst is an organic methyl- or ethyl-mercaptan. It is applied to enhance the selectivity and/or activity by freely circulating in the reactor. Recycling of the co-catalyst (also called promoter or modifier) can be done. Another option is to immobilize the mercaptan promoter groups to the backbone sulfonate ion of the resin by covalent or ionic nitrogen linkages. One technology implemented at Blachownia Chemical Works in Poland utilized two reactors: one with a sulfonated styrene-divinylbenzene copolymer catalyst for the recycled process streams and an ion-exchange resin catalyst with chemically bound 2, 2-dimethyl-1, 3-tiazolidyne promoter for reaction with acetone. A further development of the process is based on the application of only the promoted catalyst. One of the generic options of BPA production technology is given in Figure 14.21.

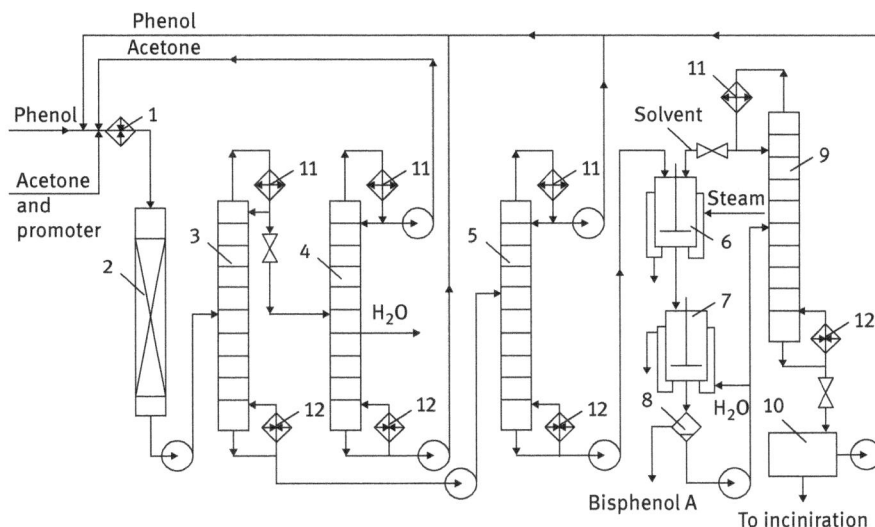

Figure 14.21: Production of BPA: 1, heater; 2, reactor; 3, 5, and 9, distillation columns; 6, vessel for dissolution, 7, crystallizer; 8, centrifuge; 10, treatment of the bottoms from column 9; 11, reflux condenser; 12, boiler. After N. N. Lebedev, *Chemistry and Technology of Basic Organic and Petrochemical Synthesis*, Chimia, 1988.

Fresh and recycled streams of acetone and excess phenol, along with the modifier after heating (pos. 1) to the desired temperature, are sent to the reactor system (2) with an ion exchange resin catalyst. In the new design options, two consecutive reactors are applied and acetone is introduced separately into both stages to ensure the optimum acetone/phenol ratio. In the crude distillation column (3), water, acetone, and unreacted phenol are removed from the reactor effluent and further separated from water into acetone and phenol fractions, which are recycled. Water is withdrawn for purification. The bottoms of the crude distillation are sent to the distillation column (5) operating under vacuum, where phenol is distilled away and BPA is concentrated to a level suitable for crystallization. Crystallization is done with an organic solvent. First, in vessel 6, BPA is dissolved in the solvent at elevated temperature and recrystallized in 7. The crystals of BPA are separated by centrifugation (8). The BPA finishing system removes phenol from

the product and solidifies the resulting molten BPA, making BPA prills. The mother liqueur from the purification system is distilled in the solvent recovery column 9. The solvent is sent back to 6, while the solvent-free mother liqueur containing BPA and some
side products (isopropenylphenol), which can be transformed to diphenylol propane, is
recycled. Some heavier products are incinerated.

14.4.3 Aldol condensation

From the viewpoint of reaction technology, aldol condensation reactions can be
conducted either separately or simultaneously with the subsequent reactions. In
the former case, typically low temperature (0–30 °C) and long residence times are
applied, giving a moderate yield of a condensation product (10–40% conversion)
due to thermodynamic limitations. Plug-flow reactors are applied. The product of
aldolization reaction after addition of organic acid to decrease pH is sent to a separate dehydration reactor where the latter reaction is conducted at 100–130 °C. An
alternative approach is to combine condensation with dehydration. In the synthesis
of 2-ethylhexanol (Figure 13.37), aqueous sodium hydroxide is used as a catalyst for
the condensation of butyraldehyde giving 2-ethyl-2-hexenal (Figure 14.22).

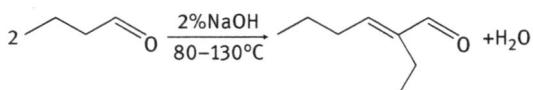

Figure 14.22: Butyraldehyde aldolization.

Special care should be taken to ensure efficient mixing of the two-phase system and
also to avoid local overheating, leading otherwise to side reactions and a decrease of
the yield. Reaction heat of aldolization is used for steam generation. Conversion
higher than 99% is achieved with the ratio of aldehyde to aqueous sodium hydroxide
solution of 1:10–1:20. In general, different reactors (mixing pump, packed columns,
stirred vessels) could be applied. The process flow scheme is given in Figure 14.23.
The aldolization and dehydration reactions are done at 100–130 °C in reactor 1 in the
presence of 40% solution of sodium hydroxide with an external cooling (pos. 2). The
mixture is separated in a phase separator (pos. 3) into an upper organic phase and a
lower aqueous phase containing the aldolization solution. The organic layer is distilled
in columns 5 and 6. In the first column, the product is separated from the lights (unreacted butyraldehyde, some water), while the second column operating under vacuum
is needed to remove heavy products. The resulting 2-ethylhexenal is hydrogenated in a
single stage or in two stages into 2-ethylhexanol (Figure 13.40), which reacts with
phthalic anhydride, giving bis(2-ethylhexyl) phthalate plasticizer (Figure 13.41). The aldolization solution contains valuable products that can be partially recycled.

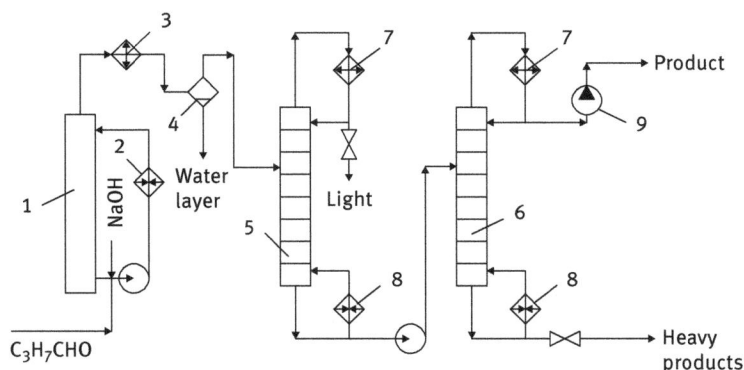

Figure 14.23: The flow scheme for butyraldehyde aldolization: 1, reactor; 2 and 3, coolers; 4, separator; 5 and 6, distillation columns; 7, reflux; 8, boiler; 9, pump. After N. N. Lebedev, *Chemistry and Technology of Basic Organic and Petrochemical Synthesis*, Chimia, 1988.

14.5 Caprolactam production

14.5.1 Condensation of cyclohexanone to cyclohexanone oxime and subsequent Beckmann rearrangement

A very industrially important reaction of ketones condensation with nitrogen-containing compounds is the synthesis of oximes from cycloalkanones and hydroxylamine in a reversible oximation reaction with subsequent acid catalyzed (Beckmann) rearrangement into lactams (Figure 14.24).

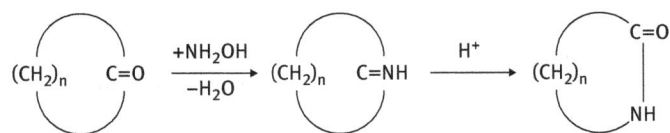

Figure 14.24: Synthesis of lactams from cyclic ketones.

For the synthesis of oximes, typically aqueous solutions of hydroxylamine sulfate are used.

A particular important reaction is the synthesis of caprolactam (Figure 14.25).

Beckmann rearrangement in the presence of mineral acids gives in fact bisulfate salt of caprolactam, which requires a subsequent step of neutralization with ammonia, resulting in lactam and at the same time generating undesired ammonium sulfate. One of the main driving forces in the development of alternatives

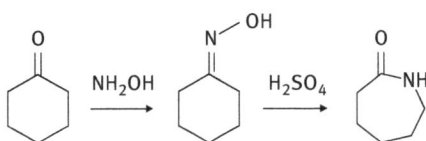

Figure 14.25: Synthesis of caprolactam from cyclohexanone.

routes for caprolactam manufacturing was a need for minimizing or completely avoiding generation of this unwanted ammonium salt.

One of the main reaction by-products in the first oximation steps are the products of self-condensation of cyclohexanone (Figure 14.26).

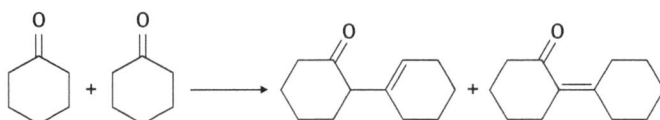

Figure 14.26: Self-condensation of cyclohexanone.

The yield of such by-products increases with increase in temperature, acidity, and cyclohexanone concentration. In order to avoid formation of these by-products, oximation can be done under excess of ketone at low temperature (ca. 40 °C) with subsequent oximation under excess hydroxylamine (hydroxylammonium sulfate in the BASF process) with temperature increase to 75–80 °C, thereby avoiding crystallization of caprolactam and reaching 99% of yield.

The Beckmann reaction is strongly exothermal (−235 kJ/mol) with the rate increasing with an increase in acidity and temperature. When oleum is used as a catalyst, the reaction temperature is ca. 125 °C. Heat removal should be properly addressed, and thus, intensive stirring is used with careful cooling of the reaction mixture done through an external heat exchanger. Because of exothermicity, molten cyclohexanone oxime and concentrated oleum (27%) having a molar ratio of 1–1.05 are introduced simultaneously in a relatively large amount of the already formed product. The generic scheme is given in Figure 14.27. Cyclohexanone is continuously fed in reactor 1, where it reacts at 40 °C with hydroxylammonium sulfate in a water solution of ammonium sulfate generated in the second oximation stage. Manufacturing of hydroxylammonium sulfate is done by hydrogenation of NO in the presence of sulphuric acid over a carbon-supported platinum catalyst. In separator 2, cyclohexanone-containing oxime is separated from ammonium sulfate. An excess of poorly soluble in water cyclohexanone helps to extract partly water-soluble oxime from a water-sulfate layer. The second oximation step is conducted at 75–80 °C in a cascade of several reactors (pos. 3 and 4). Ammonium hydroxide or ammonia is added into these reactors in order to regulate the pH and avoid decomposition of hydroxylamine sulfate with a subsequent decrease of acidity.

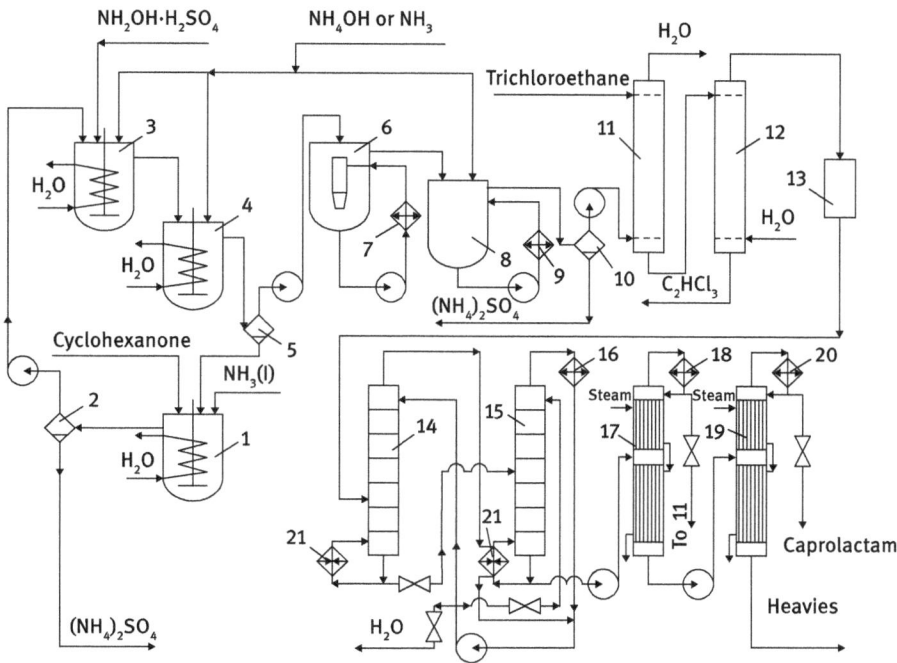

Figure 14.27: Scheme of caprolactam production: 1, oximation reactor, 1 stage; 2, 5, and 10, separators; 3 and 4, reactors of second oximation stage; 6, reactor for rearrangement; 7 and 9, external heat exchangers; 8, neutralizer; 11 and 12, extractors; 13, purification section; 14 and 15, evaporators; 16, 18, and 20, condensers; 17 and 19, rotary film evaporators; 21, boiler. After N. N. Lebedev, *Chemistry and Technology of Basic Organic and Petrochemical Synthesis*, Chimia, 1988.

The reaction mixture after reactor 4 practically does not contain cyclohexanone and is separated into an aqueous layer (unreacted hydroxylamine, which is then sent to reactor 1) and crude oxime containing ca. 5% of water, minor amounts of ammonium sulfate, cyclohexanone, and side products. Crude oxime is directly sent to Beckmann rearrangement reactor (pos. 6) containing a circulation pump and external heat exchanger 7. Oleum is introduced upstream the pump. The reaction mixture after reactor 6 is neutralized in 8 with ammonium hydroxide solution. The temperature during neutralization is kept at 40–50 °C by circulation of the reaction mixture with pump 9. Neutralization is followed by separation in 10 of the lactam oil from the water solution of ammonium sulfate. An additional extraction step of lactam from the latter is done with an organic solvent (not shown). The crude lactam contains, besides 60–65% lactam, water (30–35%), up to 2% of ammonium sulfate and some minor amounts of side products. Further processing of lactam is done by first extracting it with an organic solvent (benzene, toluene, or trichloroethane), removing in 11 impurities nonsoluble in the solvent. This is followed by re-extraction with water in 12 to further remove impurities soluble in organic solvent. This technology with two extraction steps was developed by DSM. Purification section 13 includes, e.g., treatment with ion-exchange resins and hydrogenation.

Hydrogenation is typically done using Raney nickel catalyst thus requiring catalyst separation from caprolactam. SINOPEC Research Institute of Petroleum Processing developed a technology for hydrogenation where the Raney Ni was replaced by amorphous Ni catalyst. The latter having magnetic properties operates in a magnetically stabilized bed reactor (Figure 14.28) with a uniform magnetic field. Such operation combining the advantages of the fixed-bed and the fluidized-bed reactors effectively prevents fine particles from leaving the reactor.

Figure 14.28: Magnetically stabilized bed reactor unit in caprolactam production with 100 kt/a capacity. From B. Zong, B. Sun, S. Cheng, X. Mu, K. Yang, J. Z. Zhao, X. Zhang, W. Wu, Green production technology of the monomer of Nylon-6: caprolactam, Engineering, 2017, 3, 379–384. http://dx.doi.org/10.1016/J.ENG.2017.03.003. Open access.

An amorphous Ni alloy with initially poor thermal stability and low specific surface area was modified with several additives including large-radius rare-earth atoms to retard migration of nickel. The catalyst preparation method was further developed to substantially increase the surface area. All these measures allowed improvement of caprolactam quality by more efficient elimination of unsaturated impurities. According to industrial experience of Sinopec group companies, catalyst consumption can be diminished by 50% in industrial magnetically stabilized bed reactors operating at 80°C and 0.7 MPa with magnetic field intensity of 20 kA/m.

Removal of water from lactam is done in columns 14 and 15, resulting in 95–97% of lactam. Final distillation is done under vacuum in rotary film evaporators. Initially, in 17, water is removed with some amounts of lactam. This fraction is sent to either extraction (pos. 11) or neutralization (8). Subsequently, in evaporator 19, lactam is removed from the heavies, which still contain some quantities of caprolactam, recovered in either 13 or 11.

If a caprolactam polymerization plant is located nearby, then the molten monomer can be directly transported to that plant; otherwise, caprolactam should undergo crystallization in a flaker. As mentioned above, oxime formation by reacting cyclohexanone with hydroxylamine sulfate inevitably results in the formation of sulphuric acid, which is removed to maintain the desired reaction pH by continuous addition of ammonia and ammonium hydroxide.

Several technologies were developed to diminish the formation of ammonium sulfate by conducting, for example, acidic oximation with ammonium hydroxylammonium sulfate

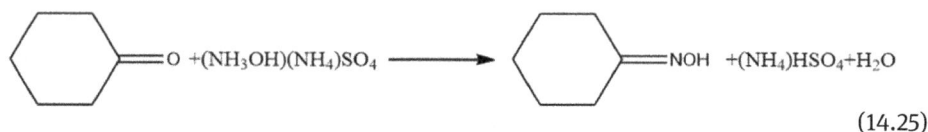

$$\text{cyclohexanone} =O +(NH_3OH)(NH_4)SO_4 \longrightarrow \text{cyclohexanone oxime} =NOH +(NH_4)HSO_4+H_2O \tag{14.25}$$

which is formed by the hydrogenation of nitric oxide in an ammonium hydrogen sulfate solution over a graphite-supported platinum catalyst:

$$NO + 3/2H_2 + (NH_4)HSO_4 \rightarrow (NH_3OH)(NH_4)SO_4 \tag{14.26}$$

Because cyclohexanone oxime recovery does not require neutralization of ammonium hydrogen sulfate, the latter is directly recycled into hydroxylamine production.

Another option to diminish formation of ammonium sulfate is to replace sulphuric acid in the process. For example, oximation can be done with hydroxyl amine phosphate:

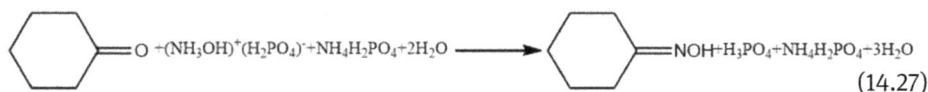

$$\text{cyclohexanone} =O +(NH_3OH)^+(H_2PO_4)^-+NH_4H_2PO_4+2H_2O \longrightarrow \text{cyclohexanone oxime} =NOH+H_3PO_4+NH_4H_2PO_4+3H_2O \tag{14.27}$$

In this process, pH is maintained by using the regenerable phosphate buffer, which can be recycled to the stage of hydroxylamine phosphate synthesis. This cannot be done if a conventional method is applied. Oxime generated in the oximation step can be recovered from a weakly acidic phosphate buffer without neutralization using only toluene extraction. Such extraction is followed with the replacement of the consumed nitrate ions by addition of 60% nitric acid:

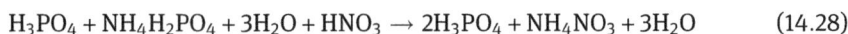

$$H_3PO_4 + NH_4H_2PO_4 + 3H_2O + HNO_3 \rightarrow 2H_3PO_4 + NH_4NO_3 + 3H_2O \tag{14.28}$$

Reduction of the phosphoric acid/ammonium nitrate buffer solution is done at pH 1.8 with hydrogen in the presence of a carbon or alumina-supported palladium catalyst with formation of hydroxylammonium phosphate:

$$NH_4NO_3 + 3H_2 + 2H_3PO_4 \rightarrow (NH_3OH)(H_2PO_4) + NH_4H_2PO_4 + 2H_2O \qquad (14.29)$$

Overhydrogenation results in the formation of excess ammonium ions, which should be removed, as such excess deteriorates the pH of the phosphate buffer. This is achieved by treating the solution with nitrous gases from the ammonia combustion step, which is an integral part of caprolactam production:

$$2NH_4H_2PO_4 + NO + NO_2 \rightarrow 2N_2 + 2H_3PO_4 + 3H_2O \qquad (14.30)$$

Handling of the excess of nitrogen oxides is done by adsorbing them in a downstream column and subsequent recycling for hydroxylamine synthesis.

The flow scheme is presented in Figure 14.29. An excess of hydrogen used in the reaction after separation (pos. 3) and compression (pos. 1) is recycled to reactor 2. The catalyst is filtered in 4 and recycled, while the hydroxylamine buffer solution is sent to the reactor cascade (5) operating at pH 2 for oximation with cyclohexanone, which is supplied countercurrently. This reaction occurring in toluene as a solvent results in cyclohexanone oxime and release of phosphoric acid. In the cascade, the overall conversion is 98%; the remaining part of cyclohexanone reacts in 6 with hydroxyl amine (ca. 3% of the hydroxyl amine flow) at pH = 4.5 with ammonia addition. After separating 30% cyclohexanone oxime solution in toluene from the aqueous buffer solution in 6, the organic phase is distilled in 7. Cyclohexanone oxime is used for Beckmann rearrangement process, while toluene is recycled. Since the solvent contains residual organics (cyclohexanone and oxime), they are extracted in extraction column 8 with the spent buffer solution. The residual toluene still present in the exhausted buffer solution is stripped with steam in 9. The process results in formation of ca. 1.8 t of ammonium sulfate per ton of caprolactam.

In the 1980s, EniChem Company developed a highly selective ammoximation reaction where cyclohexanone oxime was produced in a one-pot reaction of cyclohexanone, ammonia, and hydrogen peroxide over TS-1 (titanium silicate with MFI framework) zeolite at ca. 90 °C:

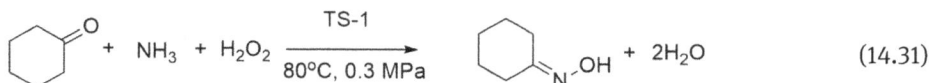

$$(14.31)$$

The reaction proceeds by first oxidation of ammonia to hydroxylamine with a subsequent reaction of the latter to cyclohexanone oxime. In the original concept, spherical catalysts with a diameter of 20 μm and tanks-in-series slurry-bed reactors were applied affording cyclohexanone conversion of at least 99.9% and selectivity to cyclohexanone oxime of at least 99.3%. In the further development, Sinopec

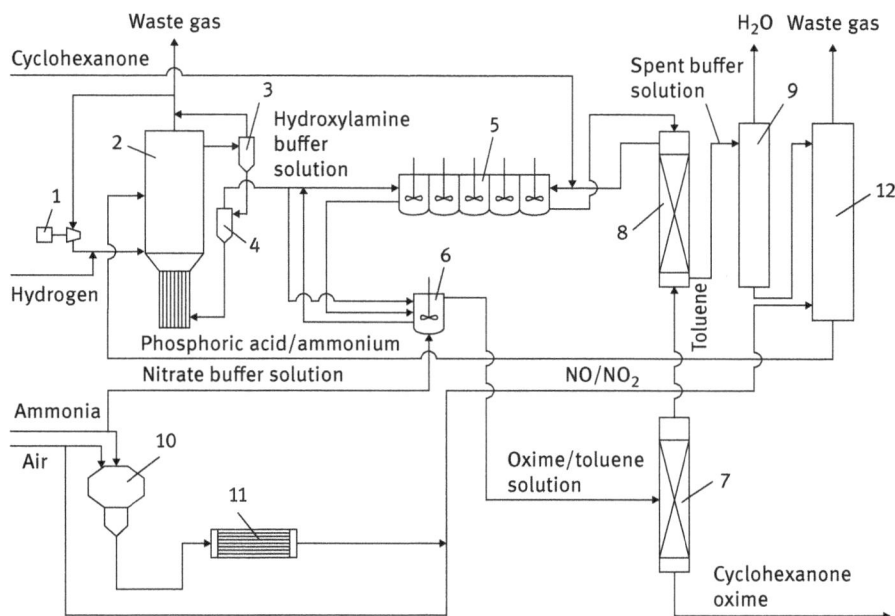

Figure 14.29: DSM HPO hydroxylamine and cyclohexanone oxime production: 1, compressor; 2, hydroxylamine generator; 3, separation; 4, filtration; 5, oximation; 6, neutralization; 7, solvent distillation; 8, extraction; 9, toluene stripping; 10, ammonia combustion; 11, condenser; 12, decomposition and absorption column.

RIPP introduced micro-sized hollow TS-1 zeolite operating in a slurry-bed reactor combined membrane separation (Figure 14.30). The same levels of cyclohexanone conversion and cyclohexanone oxime selectivity were achieved in a commercial process (Figure 14.31) as in the original more complicated concept of cyclohexanone ammoximation.

An interest to gas-phase heterogeneous catalytic Beckmann rearrangement was, for many years, linked to a need of diminishing ammonium sulfate formation. Sumitomo Company developed a vapor-phase Beckmann rearrangement process shown in Figure 14.32.

High-silica MFI zeolite is used as a catalyst in the vapor-phase Beckmann rearrangement in the presence of methanol vapors at 350–400 °C and ambient pressure. After cooling, the product methanol is recovered and recycled. A fluidized-bed reactor with a regenerator (Figure 14.33) is used for oximation since it is necessary to regenerate the deactivated catalyst continuously. The reactor-regenerator system basically operates in the same way as other fluidized-bed reactors such as FCC.

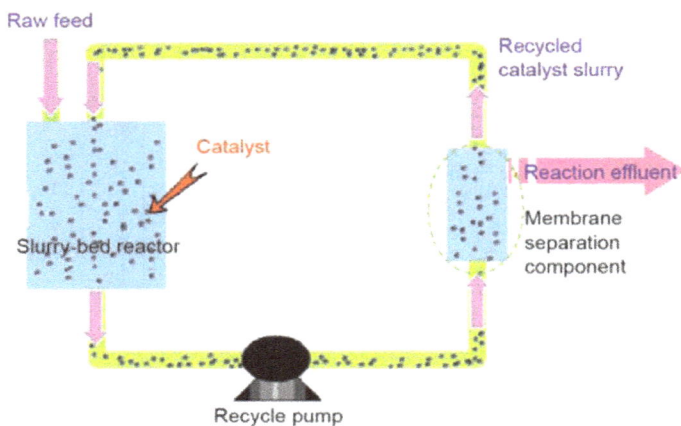

Figure 14.30: Schematic diagram of the ammoximation of cyclohexanone technology developed by RIPP. From B. Zong, B. Sun, S. Cheng, X. Mu, K. Yang, J. Z. Zhao, X. Zhang, W. Wu, Green production technology of the monomer of Nylon-6: caprolactam, Engineering, 2017, 3, 379–384. http://dx.doi. org/10.1016/J.ENG.2017.03.003. Open access.

Figure 14.31: The 200 kt/y cyclohexanone oxime industrial production unit. From B. Zong, B. Sun, S. Cheng, X. Mu, K. Yang, J. Z. Zhao, X. Zhang, W. Wu, Green production technology of the monomer of Nylon-6: caprolactam, Engineering, 2017, 3, 379–384. http://dx.doi.org/10.1016/J. ENG.2017.03.003. Open access.

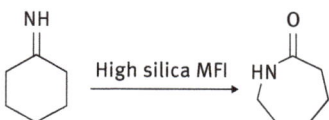

Figure 14.32: Vapor-phase Beckmann rearrangement.

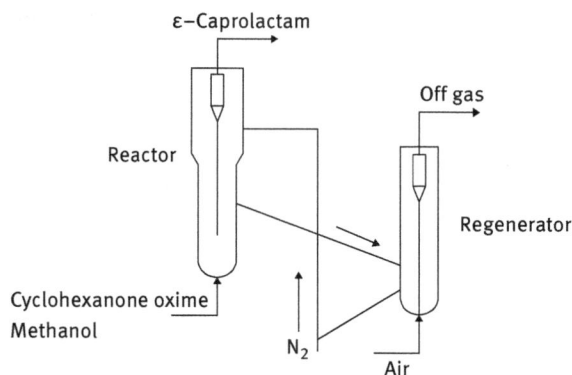

Figure 14.33: A fluidized-bed reactor with regenerator for oximation. From
H. Ichihashi, M. Kitamura. Some aspects of the vapor phase Beckmann rearrangement for the
production of ε-caprolactam over high silica MFI zeolites, *Catalysis Today*, 2002, 73, 23–28.
Copyright Elsevire. Reproduced with permission.

14.5.2 Methods for caprolactam production

ε-Caprolactam production methods can be divided into several main groups. In the
methods presented in Section 14.5.1, cyclohexanone oxime was made from a corresponding ketone, i.e., by the reaction between cyclohexanone and hydroxylamine.
Cyclohexanone is mostly produced by the oxidation of cyclohexane with air at
125–165 °C and 0.8–1.5 MPa, giving a ketone/alcohol (cyclohexanone/cyclohexanol)
mixture applying Mn or Co salts as homogeneous catalysts. Conversion of cyclohexane is restricted (6%) to afford reasonable selectivity toward the desired products
and avoid overoxidation. This is done, as not only cyclohexanol and cyclohexanone,
but also the intermediate cyclohexyl hydroperoxide are more readily oxidized than
cyclohexane (Figure 14.34). Oxidation by-products include a wide range of monocarboxylic acid and dicarboxylic acid, esters, aldehydes, and other oxygenates.

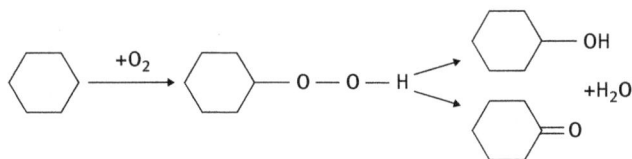

Figure 14.34: Oxidation of cyclohexane.

The ratio between cyclohexanol and cyclohexanone is ca. 3.5; therefore, after initial
separation of the unreacted cyclohexane from the products by distillation, subsequent distillation of cyclohexanol and cyclohexanone is done. Cyclohexanol is then

dehydrogenated to cyclohexanone (Figure 14.35) in the vapor phase on either copper- or zinc-based catalysts.

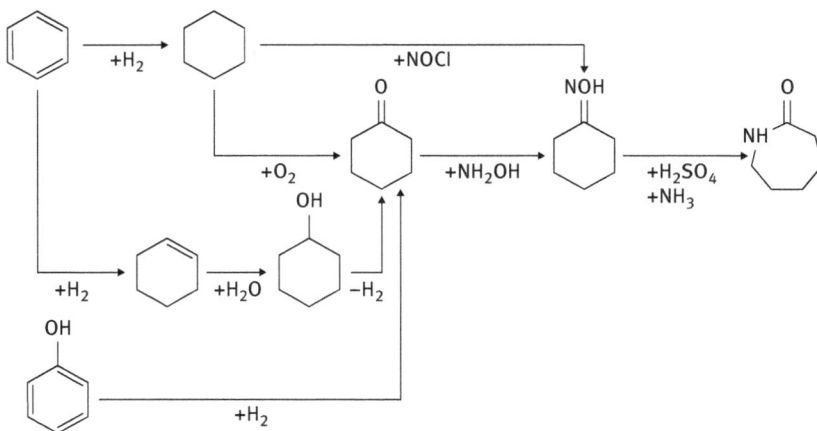

Figure 14.35: Reactions in synthesis of caprolactam starting from benzene or phenol.

Cyclohexane in turn is made by hydrogenation of benzene (Figure 14.35) either in the liquid or in the vapor phase. A part of cyclohexanone is produced from phenol by selective gas-phase hydrogenation over palladium catalysts at 140–170 °C and 0.1–0.2 MPa. Alternatively, a two-step process is used, when phenol is first hydrogenated over nickel catalysts at 140–150 °C and 1.5 MPa, followed by dehydrogenation of formed cyclohexanol to cyclohexanone over either copper- or zinc-based catalysts.

A new process of Asahi Chemical relies on partial hydrogenation of benzene to cyclohexene on a ruthenium catalyst with further hydration of cyclohexene to cyclohexanol with an acid catalyst.

Another conceptually different method of caprolactam production completely avoids the formation of cyclohexanone. In Toray's photonitrosation of cyclohexane (PNC), cyclohexane is reacted with nitrosyl chloride with the aid of UV radiation to give cyclohexanone oxime hydrochloride:

$$(14.32)$$

A gas mixture containing HCl and nitrosyl chloride is introduced into cyclohexane at temperature below 20 °C and the reaction is initiated by UV light. The lamp cooler is washed periodically with concentrated sulphuric acid to prevent deposition of the oxime salt and resinous coating. Unreacted cyclohexane and nitrosyl chloride are recycled. NOCl is formed by reacting HCl with nitrosylsulphuric acid:

$$NOHSO_4 + HCl \rightarrow NOCl + H_2SO_4 \qquad (14.33)$$

The latter in turn is prepared from sulphuric acid and nitrous gases (obtained in ammonia combustion):

$$2H_2SO_4 + NO + NO_2 \rightarrow 2NOHSO_4 + H_2O \qquad (14.34)$$

The product of the photochemical reaction is in fact oxime dihydrochloride

$$(14.35)$$

which separates at the bottom of the reactor as a lower, heavy oily phase in cyclohexane. This phase is rearranged to caprolactam in the excess of sulphuric acid or oleum and is subsequently neutralized with water solution of ammonia, giving crude lactam and ammonium sulfate. The latter formed in the amounts of ca. 1.6 tons per ton of caprolactam is crystallized by evaporation. The process flow scheme is given in Figure 14.36.

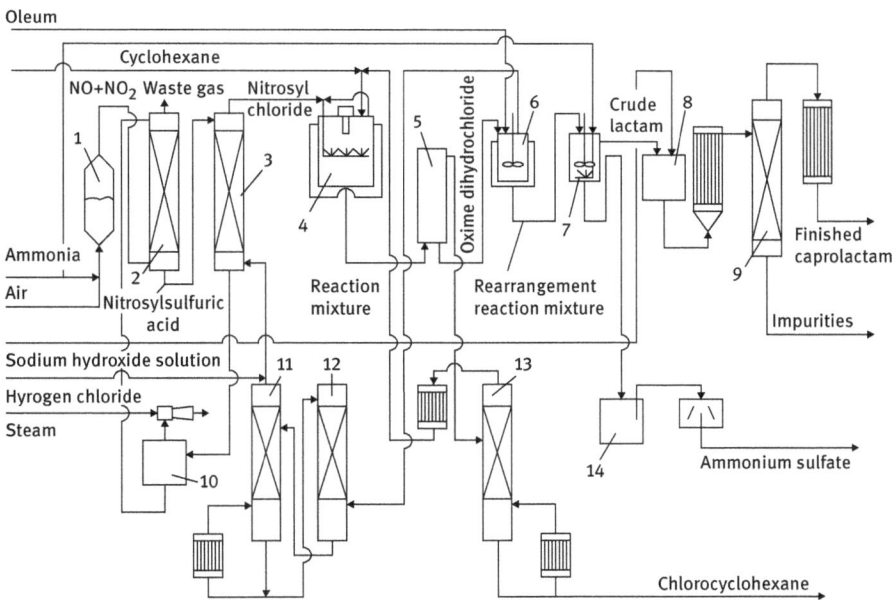

Figure 14.36: Toray PNC caprolactam production: 1, ammonia combustion; 2, nitrosylsulphuric acid generator; 3, nitrosyl chloride generator; 4, photonitrosation; 5, cyclohexane/cyclohexanone oxime separation; 6, rearrangement; 7, neutralization; 8, chemical treatment; 9, drying and lactam distillation; 10, dewatering of sulphuric acid; 11, hydrogen chloride regenerator; 12, hydrogen chloride recovery; 13, cyclohexane recovery; 14, ammonium sulfate recovery.

Overall, in the processes described above, large amounts of ammonium sulfate are produced as a by-product through oximation and Beckmann rearrangement reactions ranging from 1.6. to 4.4 tons per ton of caprolactam.

An alternative way of caprolactam production is to avoid completely the formation of cyclohexanone oxime and subsequent Beckman rearrangement. Among technologies implemented industrially, the Snia Viscosa cyclohexane carboxylic acid process (Figure 14.37) will be described below.

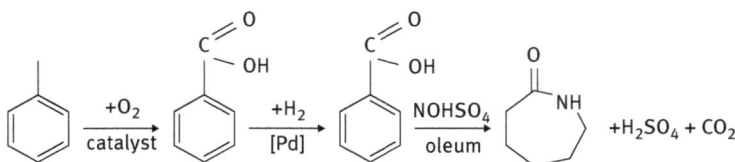

Figure 14.37: Snia Viscosa process for cyclohexanone oxime synthesis.

Oxidation of toluene is done with the air in the liquid phase using a cobalt catalyst at 160–170 °C and 0.8–1 MPa pressure with > 90% overall yield. The gases containing mainly nitrogen with small amounts of oxygen, carbon dioxide, and carbon monoxide are cooled to 7–8 °C in order to recover unreacted toluene. The flow scheme is presented in Figure 14.38.

Water and toluene are removed as overhead from the reactor (pos. 2). After separation in a separator drum (pos. 3), toluene is recycled back to the oxidation reactor. The liquid-phase product stream contains, besides ca. 30% benzoic acid, various side products as well as toluene and the cobalt catalyst. From the top of the distillation column (pos. 4), the light compounds and toluene are recycled in the reactor, while the vapor phase benzoic acid is removed as a side stream and high-boiling by-products as the residue. Liquid-phase hydrogenation of benzoic acid to cyclohexanecarboxylic acid is done over Pd/C catalyst in a cascade of stirred reactors (pos. 7) at ca. 170 °C and 1–1.7 MPa, giving almost complete conversion (99.9%). The separation of the catalyst for further reuse is done by centrifugation (pos. 8). The product – cyclohexanecarboxylic acid – is distilled (pos. 9) under reduced pressure. Nitrosation of cyclohexanecarboxylic acid is performed in a multistage reactor, giving complete conversion of 73% nitrosylsulphuric acid solution in sulphuric acid and ca. 50% conversion of cyclohexanecarboxylic acid. For efficient heat removal, the reaction in (pos. 16) is done in boiling cyclohexane at atmospheric pressure. Subsequently, the products are hydrolyzed with water at low temperatures (pos. 17). Unreacted cyclohexanecarboxylic acid is extracted with cyclohexane and recycled into the process. In neutralization stage (pos. 19), the acidic caprolactam solution containing excess sulphuric acid is neutralized with ammonia directly in a crystallizer under reduced pressure, giving two liquid layers. The first is a saturated ammonium sulfate solution

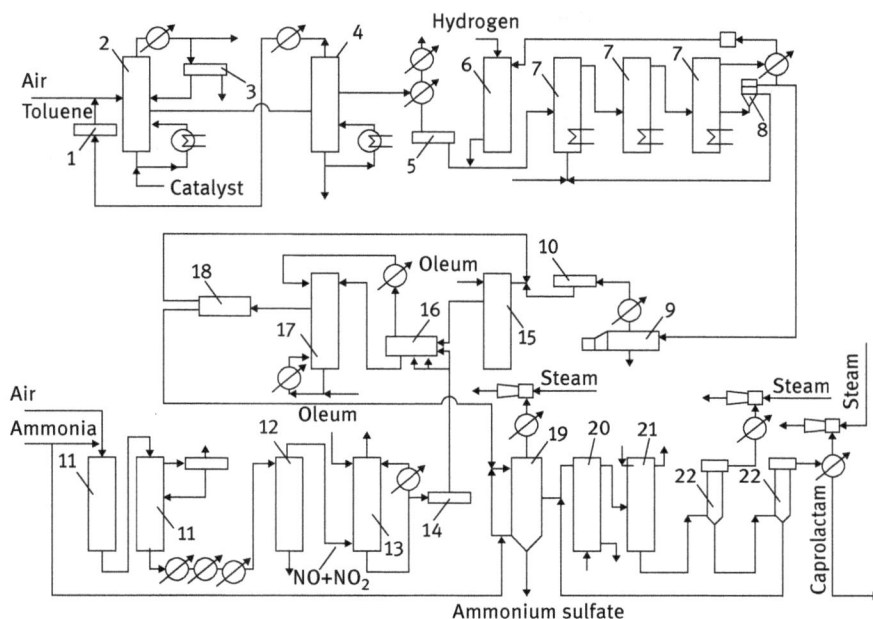

Figure 14.38: SNIA caprolactam production: 1, toluene tank; 2, oxidation; 3, separation; 4, rectification; 5, benzoic acid tank; 6, benzoic acid/hydrogen mixture; 7, benzoic acid hydrogenation; 8, removal of catalyst; 9, cyclohexanecarboxylic acid distillation; 10, cyclohexanecarboxylic acid tank; 11, ammonia combustion; 12, separation; 13, nitrosylsulphuric acid generator; 14, nitrosylsulphuric acid tank; 15, cyclohexanecarboxylic acid/oleum mixture; 16, rearrangement; 17, hydrolysis; 18, solution of cyclohexanecarboxylic acid in cyclohexane; 19, neutralization, and ammonium sulfate crystallization; 20, solvent extraction; 21, water extraction; 22, lactam distillation.

that is crystallized. Even if there is no oxidation and Beckmann rearrangement of oxime, ca. 4.1 tons of ammonium sulfate per ton of caprolactam are produced in the original Snia Viscosa process. A concentrated aqueous caprolactam solution is first purified by extraction with toluene (pos. 20), thereby removing water-soluble by-products. Subsequent counterextraction of the caprolactam-toluene solution with water (pos. 21) results in an aqueous caprolactam solution, leaving toluene-soluble by-products in the organic layer. Pure caprolactam is produced by distilling the aqueous caprolactam solution (pos. 22).

It is possible to eliminate the formation of ammonium sulfate in this technology by modification in the separation procedure. Extraction of caprolactam dissolved in sulphuric acid can be done by diluting this solution with small amounts of water, which is thereafter extracted with an alkylphenol. Thermal cracking of the remaining sulphuric acid destroys the impurities and recovers SO_2, which is recycled. This option does not lead to formation of ammonium sulfate, thus avoiding waste disposal problems with impurities.

Ammonium sulfate free process was developed in 1990s by several companies starting from then cheap butadiene by first hydrocyanation followed by partial hydrogenation of adipidinitrile to 6-aminocapronitrile and subsequent cyclization (Figure 14.39). Such technology, not based on aromatics derived from crude oil, was not, however, commercially implemented for a number of feedstock price related issues.

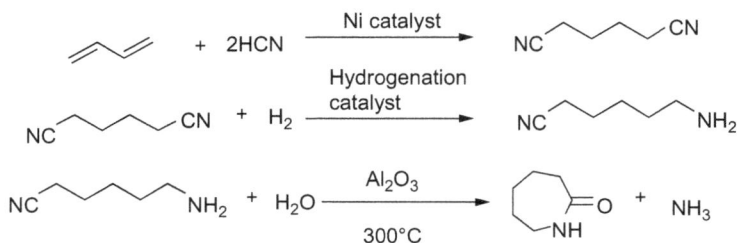

Figure 14.39: Technology for caprolactam production from butadiene.

Among the research efforts aimed at improving the process sustainability by using biomass derived feedstock, synthesis of caprolactam from renewable resources such as HMF (5-hydroxymethylfurfural) could be mentioned (Figure 14.40).

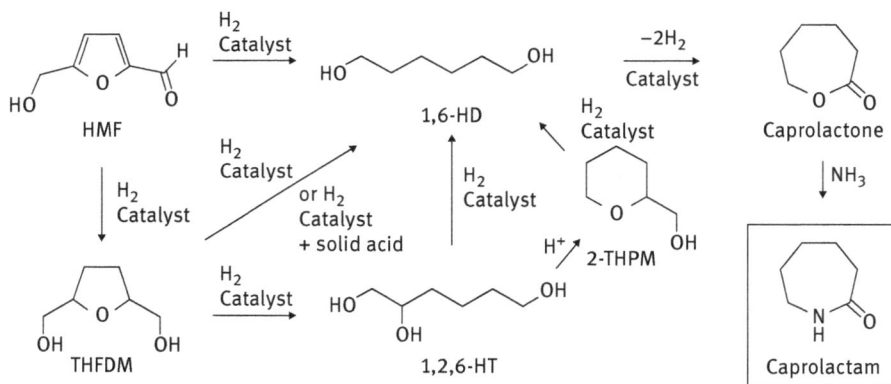

Figure 14.40: Synthesis of caprolactam from HMF. From T. Buntara, S. Noel, P. H. Phua, I. Melián-Cabrera, J. G. de Vries, H. J. Heeres, Caprolactam from renewable resources: catalytic conversion of 5-hydroxymethylfurfural into caprolactone, Angewandte Chemie International Edition, 2011, 50, 7083–7087. Copyright 2011 WILEY-VCH Verlag GmbH & Co. KGaA, Weinheim. Reporduced with permission.

The biobased caprolactam manufacturing technology comprising fermentation of sugars to a caprolactam precursor was developed by Genomatica which previously scaled up to a commercial scale production of 1,4-butanediol and 1,3-butylene glycol using microorganisms.

Chapter 15
Oligomerization and polymerization

15.1 Combining double bond isomerization, oligomerization, and metathesis: production of linear alkenes (SHOP)

In this section, production of linear 1-alkenes will be discussed covering Shell Higher Olefins Process (SHOP), which comprises ethene oligomerization, double bond isomerization, and metathesis giving C10-C14 internal olefins starting from ethylene. The process of the annual capacity exceeding a million ton of capacity was developed to manufacture olefins, a feedstock for making linear primary detergent alcohols through hydroformylation.

Oligomerization of ethylene to α-olefins (i.e. 1-alkenes) occurs at 80 to 90 °C and 10 to 11 MPa in polar solvents (1,4 butanediol) using a homogeneous catalyst – nickel phosphine complex. Figure 15.1 illustrates the reaction with the catalyst a [Ni(P,O)Ph (PPh$_3$)] type, i.e., a complex that may typically be employed in the SHOP process

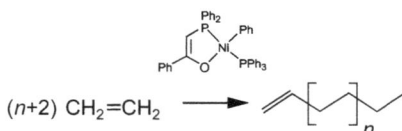

Figure 15.1: Oligomerization of ethylene over a homogeneous catalyst – nickel phosphine complex.

Oligomerization occurs through a catalytic chain-growth reaction, which will be discussed below in connection to polymerization, giving, however, much shorter chains because the catalyst prevents chain growth to the polymer stage. The catalyst is prepared *in situ* from nickel chloride NiCl$_2$, a chelating phosphorous–oxygen ligand (e.g. Ph$_2$PCH$_2$COOH), and sodium boron hydrate (NaBH$_4$) as a reducing agent. A broad distribution of immiscible in the solvent even-numbered C4–C40 linear α (terminal) olefins with a Flory-Schultz distribution is obtained, while only a certain fraction (ca. C10–C14 or C10–C18) is required for synthesis of detergents. A typical general distribution is as follows: C4–C8 40%, C10–C18 40%, and C20 + 20%. Higher olefins (C20 +) have almost no commercial applications, while the C4–C8 fraction has a limited commercial value. Subsequently, the product mixture is fractionated collecting the target olefins, the light and the heavy fractions.

Immiscibility of the formed olefins in the solvent allows facile separation of the product and the catalyst phases, and therefore recycling of nickel. Valorization of light and heavier fractions is done first by double bond migration over a solid potassium metal catalyst to give an equilibrium mixture of internal alkenes (Figure 15.2).

https://doi.org/10.1515/9783110712551-015

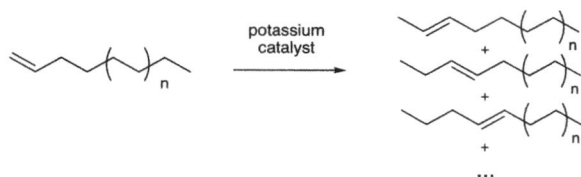

Figure 15.2: Double bond migration in olefins.

The double bond migration (often called isomerization) is followed by metathesis over an alumina-supported molibdate catalyst giving a broad mixture of linear internal olefins (C11–C14) with both odd and even numbers of carbon atoms, of which 10–15% are in the desired range. Both isomerization and metathesis catalysts operate at 100–125 °C and 10 bar. One example of the metathesis reaction is given below:

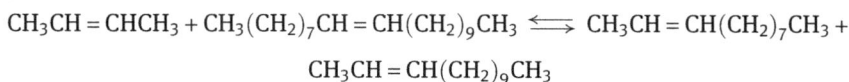

$$CH_3CH = CHCH_3 + CH_3(CH_2)_7CH = CH(CH_2)_9CH_3 \rightleftharpoons CH_3CH = CH(CH_2)_7CH_3 + $$
$$CH_3CH = CH(CH_2)_9CH_3$$

The desired product consisting of > 96% of linear internal C_{11}–C_{14} alkenes is separated by distillation and converted either to detergent alcohols by hydroformylation or into detergent alkylates. All other linear olefins, which are formed as by-products, can be recycled by isomerization and metathesis.

The process flow diagram is shown in Figure 15.3 featuring only one oligomerization reactor, while a series of reactors are used with water-cooled heat exchanges in between to remove the heat.

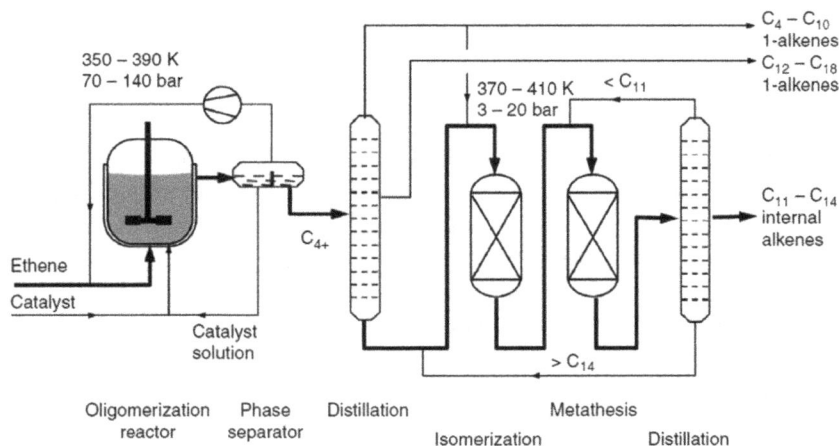

Figure 15.3: Simplified flow scheme of the Shell Higher Olefins Process (SHOP). From J. A. Moulijn, M. Makkee, A. E. van Diepen, *Chemical Process Technology*, 2013, 2nd Ed. Copyright © 2013, John Wiley and Sons. Reproduced with permission from Wiley.

As can be seen from Figure 15.3 in the high-pressure phase-separator, the unconverted ethene is separated from the two liquid phases and recycled back to the reactor, along with the catalyst solution. The second liquid phase containing 1-alkene product is sent to the distillation columns, where C4–C10 1-alkenes are removed as light products. A part of this C4–C10 stream is taken for isomerization. The C20 + fraction taken as the bottom of the distillation column reacts in the isomerization reactor with the C4–C10 fraction, which is followed by metathesis. The latter reaction gives a mixture with the broad carbon number distribution, thus requiring downstream distillation. The heavier components are isomerized, while the lighter ones are sent back to the metathesis reactor. A combination of three different reactions makes the SHOP process very flexible allowing control on the carbon number distribution and the quantity of the desired product.

15.2 Polymers

Different types of polymers or macromolecules, composed of many repeated subunits (monomers), are produced industrially, such as polyolefins, polyamides, polyurethanes. In polyamides, the repeating units are linked by amide bonds, which are generated by polycondensation of dicarbonic acids (adipic acid) with diamines (hexamethylene diamine), making nylon 6–6 (Figure 15.4), by polycondensation of amino acids, and ring-opening polymerization of lactams.

$$NH_2(CH_2)_6NH_2 + HO\overset{\overset{\displaystyle O}{\|}}{C}-(CH_2)_6\overset{\overset{\displaystyle O}{\|}}{C}OH \longrightarrow \left[-NH(CH_2)_6NH\overset{\overset{\displaystyle O}{\|}}{C}-(CH_2)_6\overset{\overset{\displaystyle O}{\|}}{C}-\right]_n +H_2O$$

Figure 15.4: Synthesis of Nylon 6–6.

Synthesis of polyamide 6 (Nylon 6 or Perlon) by ring-opening polymerization of caprolactam and production of the latter will be addressed in detail in this chapter. Polyurethane consists of units with carbamate (urethane) links formed in reactions of diisocyanate or polyisocyanate with a polyol (Figure 15.5).

Mechanistically, polymerization can be divided into two categories, step-growth and chain-growth polymerization. Many polymers can be synthesized by both methods; thus, this classification is based not on the structure of repeating units but on the synthesis mechanism.

In the case of chain-growth polymers (Figure 15.6), such as polyethylene, polystyrene, polypropylene, poly(vinyl chloride), poly(methyl methacrylate), poly(tetrafluoroethylene), or poly(acrylonitrile), monomers are added to the chain one at a time only. Polymer molecules can grow to the full size in a few seconds, as in the case of

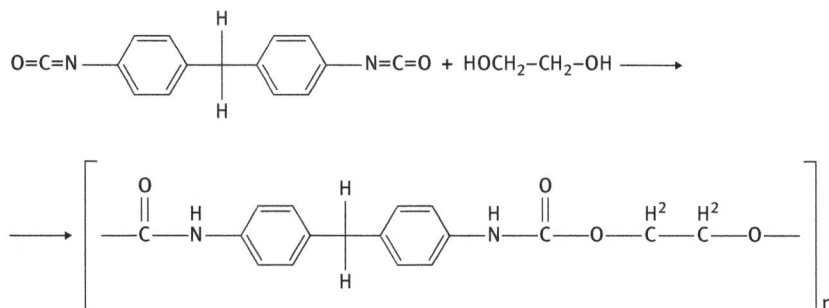

Figure 15.5: Synthesis of polyurethane.

free-radical polymerization. Active centra, such as free radicals, cations, or anions, are required for chain-growth polymerization.

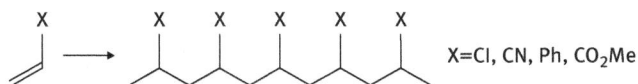

Figure 15.6: Chain-growth polymers.

15.3 Step-growth polymerization

Examples of step-growth polymerization are syntheses of polyamides, polyurethanes, and polyesters (Figure 15.7)

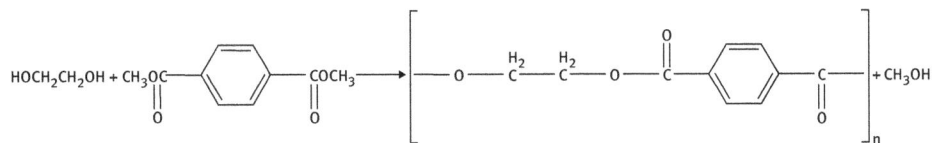

Figure 15.7: Synthesis of polyester.

An interesting case is the single-monomer polyamide of AB-type Nylon 6,

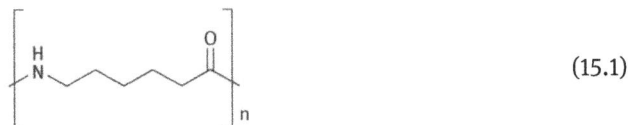

$$(15.1)$$

is not a condensation polymer being synthesized by ring-opening polymerization of caprolactam in the presence of water.

In step-growth polymerization, chains of monomers can combine with one another directly with generally only one type of chemical reaction linking molecules of all sizes m-mer + n-mer→$(m + n)$-mer. The process of polymer growth is relatively slow, being in the order of hours. The rate constant is effectively independent of the chain length; thus, for the determination of the molecular mass distribution, it can be considered that a randomly selected functional group is reacting and statistical methods can be used. Slow step-growth polymerization also implies that high-molecular-mass polymers are usually not produced until the final stage of reactions, where high viscosity can be an issue and special reactors should be used, which are capable of handling high-viscosity products.

Batch and continuous polymerization processes are used for synthesis of Nylon 6. In a batch process used only for the production of specialty polymers (e.g., very high molecular weight), the monomer caprolactam and water (2–4%) are initially heated to 250 °C for 10–12 h in an inert atmosphere to produce 6-aminohexanoic acid (Figure 15.8), which further reacts with caprolactam (Figure 15.9).

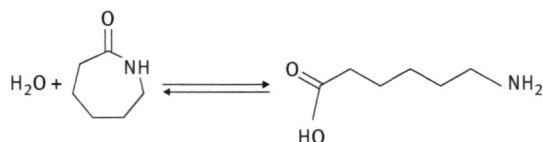

Figure 15.8: Hydrolysis of caprolactam.

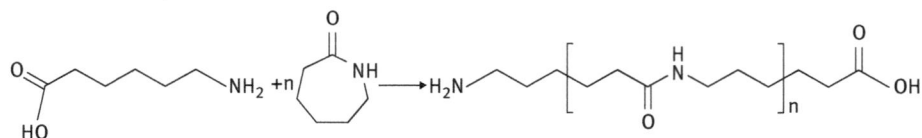

Figure 15.9: Synthesis of Nylon 6 by reaction of caprolactam with 6-aminohexanoic acid.

Reversibility of caprolactam hydrolysis results in incomplete conversion; therefore, the crude polymer containing some 10% of caprolactam and cyclic low-molecular-weight oligomers is heated at 180–200 °C in a partial vacuum to complete polymerization and increase the polymer molecular weight if desired.

Continuous processes mainly used for production of polyamide 6 can be done in a vertical tube (VK, or Vereinfacht Kontinuierliches) reactor (Figure 15.10), which operates at atmospheric pressure. Heating to ca. 220–270 °C and prepolymerization take place in the upper part while the polymer is formed in the lower section. Initially, water is needed to initiate hydrolysis. Thereafter, a low water environment is required to complete polymerization, approaching equilibrium.

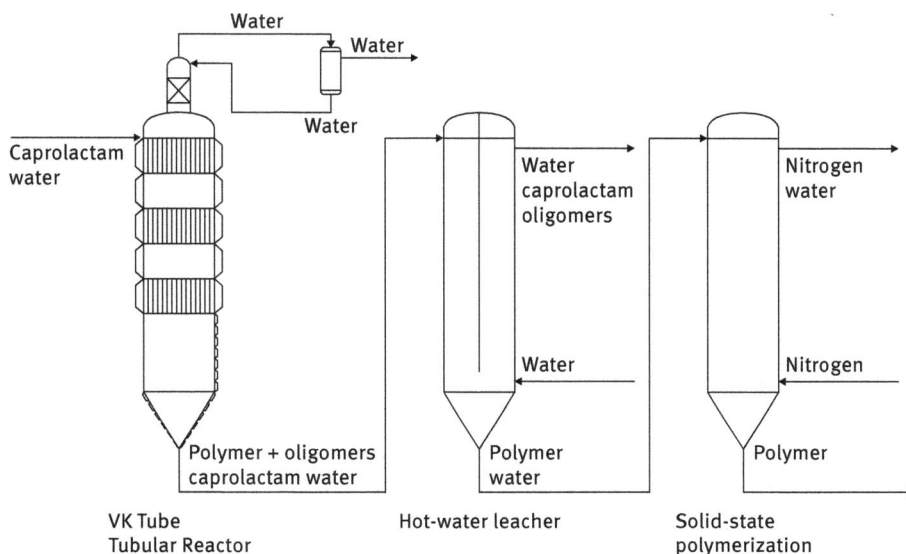

Figure 15.10: Continuous production of polyamide 6 in VK reactor. From Cakir, S. (2012). Polyamide 6 based block copolymers synthesized in solution and in the solid state. Technische Universiteit Eindhoven. https://doi.org/10.6100/IR730916.

The VK tube is followed by a hot-water leacher, where water flows in a countercurrent fashion to remove unreacted monomer and oligomers. At the end of the process, polymer pellets laden with water enter the top of a solid-state polymerization reactor where dry gas enters the bottom of the reactor and flows counter-currently with respect to the polymer phase. As the polymer travels down the reactor, it is dried and increases in temperature. Drying the polymer at high temperature drives the reaction equilibrium toward a higher polymer molecular weight.

Separation of the product from the monomer and oligomers can be also done by a continuous vacuum stripping process when nylon 6 is still in the melt. The polymer with the oligomer content of ca. 2–3% can be spun directly into fiber. This technology avoids water quench, extraction, drying, and remelting.

In the two-step polymerization process (Figure 15.11), the solid and/or liquid lactam feedstock is fed to the pre-polymerizer for the ring opening reaction of lactam. This is followed by the second polymerization stage when the chains grow to the desired length until reaching the specified viscosity.

In the AB-type monomer (caprolactam), both functionalities are combined in the same molecule, while AABB monomers, to which nylon 6–6 resin belongs, require

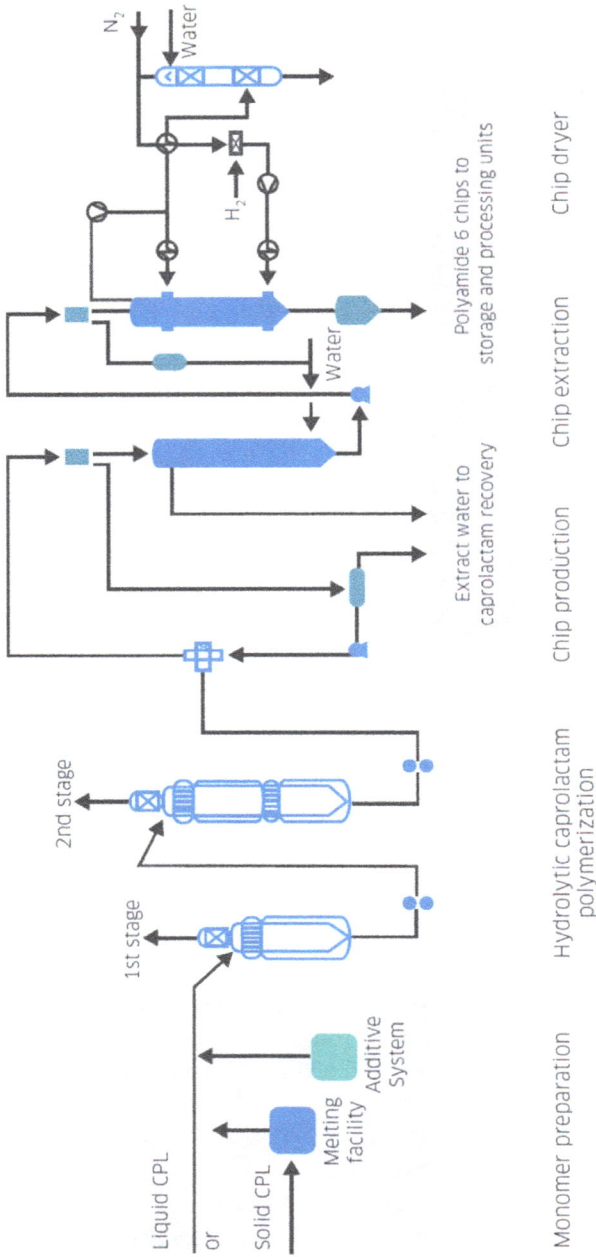

Figure 15.11: Two-stage polyamide process of Zimmer (now Technip). https://www.technipfmc.com/media/ffjb5iaa/zimmer-polymer-technologies_210x270-web.pdf.

two monomers for polymerization, with one monomer containing two amine functionalities (A) and the second one containing two carboxylic moieties (B). Nylon 6–6 is produced by the condensation reaction of hexamethylenediamine with adipic acid (Figure 15.4). The first step in the process is generation of a balanced (1:1) salt in an aqueous solution with pH serving to control stoichiometry. Differences in volatility of the acid and diamine compromise the exact ratio between the components; thus, some excess of diamine is used. Both polymerization processes are non-catalytic, even if some amounts of, e.g., aminocaproic acid can be added to caprolactam water mix to diminish the induction time in lactam hydrolysis.

Hexamethylenediamine (HMD) is synthesized by the hydrogenation of adiponitrile (ADN) under high pressure of ca. 60 MPa and 100–130 °C over Co-Cr catalysts or at somewhat lower pressures of 30 MPa and 100–180 °C over Fe-based catalysts:

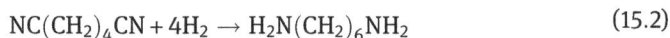

$$NC(CH_2)_4CN + 4H_2 \rightarrow H_2N(CH_2)_6NH_2 \qquad (15.2)$$

Hydrogenation is done in molten adiponitrile diluted with ammonia; the latter is needed to suppress formation of polyamines and partially hydrogenated intermediates: hexa-methyleneimine and triamine bis(hexamethylenetriamine).

An alternative process operates in diluted ADN conditions using HMD itself as a solvent and Raney Ni as a catalyst. This process does not need ammonia and works at lower pressure and temperature.

ADN, in turn, is formed by dehydrative amination of adipic acid (the acid produced by oxidation of a cyclohexanol and cyclohexanone mixture with nitric acid)

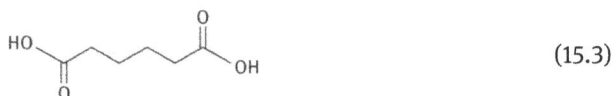

$$(15.3)$$

with ammonia in, for example, the gas phase in fixed or fluidized bed reactors with supported phosphoric acid or by direct hydrocyanation of butadiene with HCN. The latter liquid-phase process operating at 30–150 °C at atmospheric pressure in a solvent such as THF consists of two steps. In the first step of HCN addition to butadiene, a mixture of pentene nitriles and methylbutene nitriles isomers is formed, which is further isomerized into mainly 3- and 4-pentene nitriles. Anti-Markovnikov addition of HCN in the second step results in the formation of the product with a high overall selectivity. As catalysts complexes of Ni^0 with phosphine and phosphite ligands and metal salt promoters (aluminum or zinc chlorides) are suitable.

Polyamidation to produce nylon 6–6 (PA 6.6) is done by first performing polymerization of 60–80% water slurry of 1:1 nylon salt at 200 °C and > 1.7 MPa to conversion of 80–90%. An elevated pressure is required to keep water, which is needed for better heat transfer and mixing, in the liquid phase, as well as to minimize excessive loss of diamine. Finishing of polymerization is continued at 270–300 °C with a release of steam and simultaneous decrease of pressure avoiding cooling. Holding the batch at atmospheric or reduced pressure finalizes the formation of the polymer with the target molecular mass. The last step of polymerization is done above the melting point of polymer (250 °C); thus, the overall process is referred to as melt polymerization. The polymer is extruded under inert gas pressure.

An illustration of Zimmer multi-autoclave process for production of PA6.6 chips is given in Figure 15.12, which illustrates that the feedstock (i.e. the AH-salt solution) is formed either from adipic acid and hexamethylenediamine or by dissolving solid AH-Salt with water. Additives can be blended to the concentration unit, where water is evaporated, and to the polycondensation step per se. Continuous operation is ensured by operating with multiple autoclaves in parallel.

Continuous polymerization can be also done in a reactor system with initial evaporation of water to form a prepolymer and minimize loss of diamine. Polymerization proceeds in a long tube under controlled evaporation.

Another example of step-growth polymerization is the synthesis of polyesters, when the equilibrium is much less favorable than for the synthesis of polyamides. In polyester production, equilibrium therefore should be shifted by continuous removal of the condensation product usually by the application of high vacuum and high temperature. Poly(ethylene terephthalate) (PETP) is synthesized from dimethyl terephthalate (DMT) and ethylene glycol (Figure 15.7) as well as by direct esterification of terephthalic acid.

In the former process (Figure 15.13), dimethyl terephthalate reacts with excess of ethylene glycol at 150–200 °C in the melt containing a basic catalyst at atmospheric pressure. In order to shift equilibrium, methanol is distilled from the reactor while ethylene glycol is recycled (pos. 1). In column 2, operating under vacuum [(13–133) × 10^2 Pa], the excess of ethylene glycol is distilled off at higher temperatures, 265–285 °C. Continuous distillation of ethylene glycol is also done in the second transesterification step (pos. 3), which proceeds at the same temperature but under higher vacuum (≪6 × 10^2 Pa). In this final polycondensation stage, a very high vacuum is required because of high viscosity.

When dimethyl terephthalate reacts with excess of ethylene glycol the first reaction is transesterification:

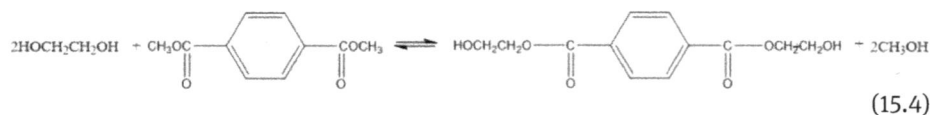

$$2HOCH_2CH_2OH + CH_3OC-\!\!\!\bigcirc\!\!\!-COCH_3 \rightleftharpoons HOCH_2CH_2O-C-\!\!\!\bigcirc\!\!\!-C-OCH_2CH_2OH + 2CH_3OH$$

(15.4)

followed by polymerization:

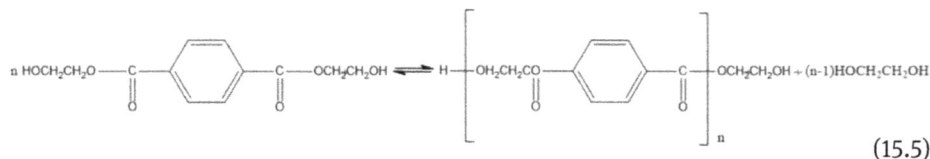

$$n\ HOCH_2CH_2O-C-\!\!\!\bigcirc\!\!\!-C-OCH_2CH_2OH \rightleftharpoons H\!-\!\!\left[OH_2CH_2CO-\!\!\!\bigcirc\!\!\!-C\right]_n\!\!OCH_2CH_2OH + (n-1)HOCH_2CH_2OH$$

(15.5)

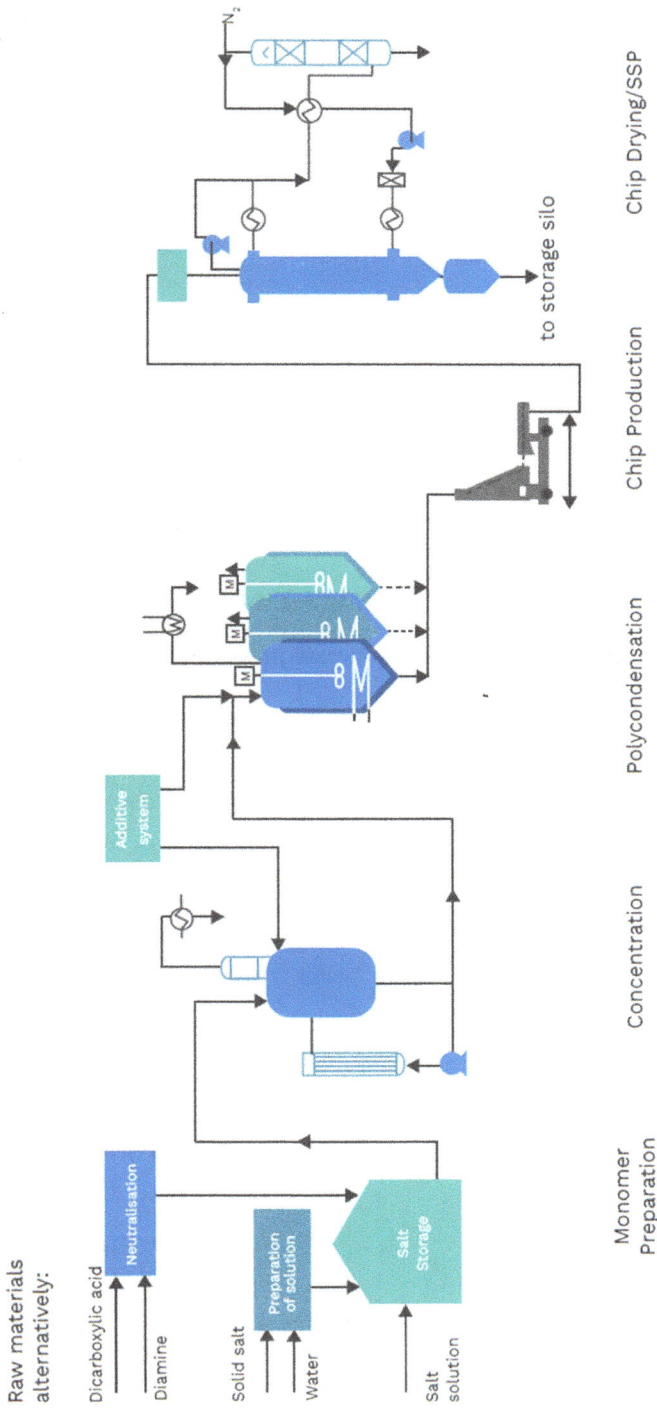

Figure 15.12: Three-reactor PA6.6 polycondensation process for chip production. https://www.technipfmc.com/media/ffjb5iaa/zimmer-polymer-technologies_210x270-web.pdf.

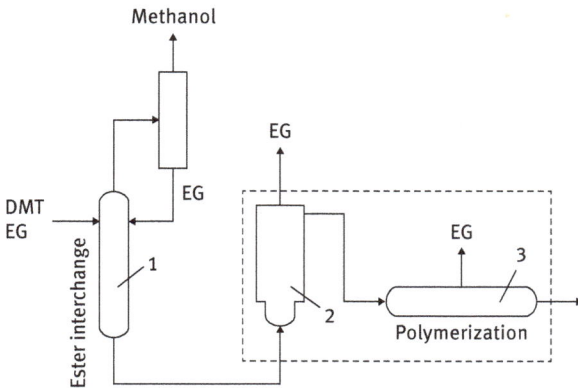

Figure 15.13: Continuous polymerization process of PETP *via* ester interchange route: 1, reactor; 2, distillation; 3, second transesterification.

In an alternative terephthalic acid process, esterification of ethylene glycol and terephthalic acid

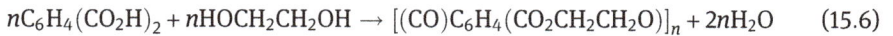

$$nC_6H_4(CO_2H)_2 + nHOCH_2CH_2OH \rightarrow [(CO)C_6H_4(CO_2CH_2CH_2O)]_n + 2nH_2O \quad (15.6)$$

is done at high temperature (220–260 °C) and moderate pressure (0.27–0.55 MPa) with continuous water removal by distillation.

Esterfication of terephthalic acid can be done also with other glycols. An example of polytrimethylene terephthalate (PTT) technology is shown in Figure 15.14.

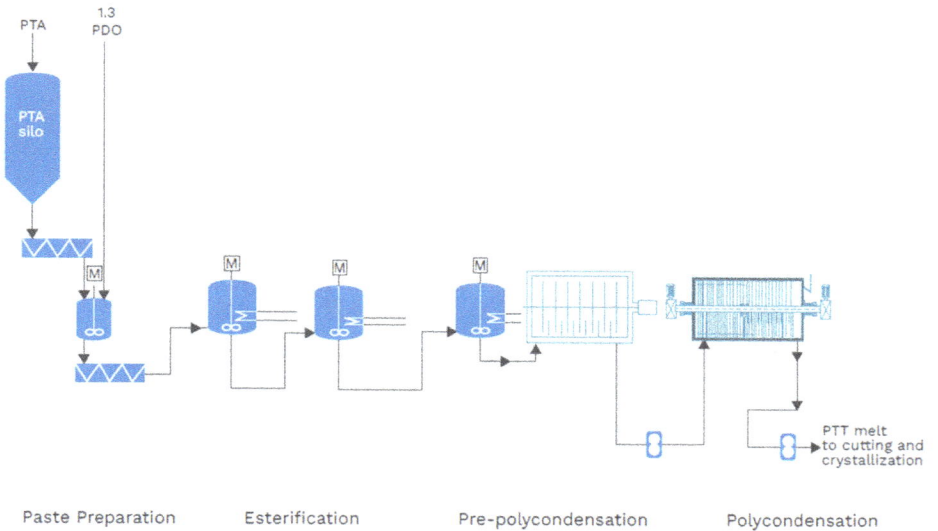

Figure 15.14: Five-reactor PTT polycondensation process for chip production by Zimmer.
https://www.technipfmc.com/media/ffjb5iaa/zimmer-polymer-technologies_210x270-web.pdf.

An interesting feature of this technology is the Double Drive Reactor where vapors comprising precondensate components are distributed as aerosols and the precondensate components condense on the reactor walls and on a separator in an exit chamber of the reactor. The condensates are guided to the unstirred discharge. Sump and the upper layers of the discharge sump are continuously recirculated into the stirred reactor area, promoting reconversion and additional polycondensation.

15.4 Polymerization process options

Polymerization processes are highly exothermal, and it is also important for the process to continue at the same rate after consumption of the monomer since the reactivity does not practically depend on the chain length. Moreover, polymers are much more viscous than corresponding monomers, making efficient heat transfer challenging. Generation of hot spots negatively influences molecular weight distribution (MWD), promotes decomposition reactions, and has an impact on product color, in general, lowering product quality. From the process viewpoint, polymerization can be either homogeneous or heterogeneous. In the first option, polymerization occurs either in the substance (the monomer or the formed polymer) or in the solvent. Precipitation, slurry-phase, suspension, emulsion, and gas-phase polymerization belong to heterogeneous polymerization.

15.4.1 Homogeneous polymerization in substance

Polymerization in substance has the clear advantage of utilizing the whole reactor volume and the absence of any substance other than the monomer. Due to a constant increase in viscosity and subsequently problems with efficient heat removal, conversion should be limited to ca. 50%. As examples, synthesis of polystyrene, polyesters, and polyamides and high pressure polymerization of ethylene to low-density polyethylene (LDPE) can be mentioned. This latter high-pressure process leads to the formation of LDPE-grade polymer. Other grades, such high-density and linear low-density polyethylene (Table 15.1), are produced with catalysts and will be discussed later.

Two types of high-pressure polymerization reactors are used operating between 150 and 200 MPa for autoclaves and between 200 and 350 MPa for tubular reactors. Radical polymerization requires an initiation process

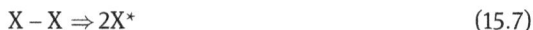

$$X - X \Rightarrow 2X^* \tag{15.7}$$

Table 15.1: Conditions for production of various polyethylene grades.

	LDPE	HDPE	LLDPE
Initiator or catalyst	Oxygen or organic peroxide	Ziegler or Phillips catalyst	Ziegler or Phillips catalyst
Reaction temperature (°C)	200–300	As low as 60	As low as 60
Pressure (bar)	1,300–2,600	1–300	1–300
Structure	Branched	Linear	Linear with short branches
Approximate crystallinity (%)	55	85–95	55

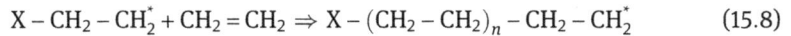

$$X - CH_2 - CH_2^* + CH_2 = CH_2 \Rightarrow X - (CH_2 - CH_2)_n - CH_2 - CH_2^* \qquad (15.8)$$

followed by propagation

$$X{-}CH_2 - CH_2{^*} + CH_2 = CH_2 \Rightarrow X - (CH_2 - CH_2)n - CH_2 - CH_2{^*} \qquad (15.9)$$

The chain process is also terminated when the radical recombines on the, e.g., reactor walls:

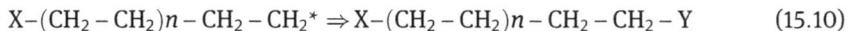

$$X{-}(CH_2 - CH_2)n - CH_2 - CH_2{^*} \Rightarrow X{-}(CH_2 - CH_2)n - CH_2 - CH_2 - Y \qquad (15.10)$$

Typically, adiabatic autoclaves of CSTR type (Figure 15.15) with limited conversion (20%) have two zones, operating at 180 °C and 290 °C, respectively. The reaction temperature is regulated by the amount of the radical initiator, such as benzoyl peroxide, which is injected at several points initiating the radical process through self-decomposition:

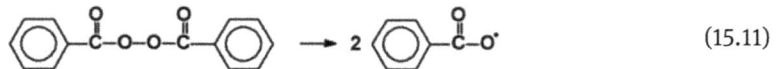

$$(15.11)$$

Volume of this reactor with wall thickness of 10 cm and a high length/diameter ratio (4:1 to 18:1) is ca. 1 m³. Removal of heat is done by quenching with the fresh monomer.

An alternative to an autoclave is a tubular reactor (Figure 15.15) with jacketed tubes (200–1,000 m in length) of internal diameter between 25 and 75 mm. The flow scheme for polymerization in tubular reactors is presented in Figure 5.16. Similar to autoclaves, the temperature is kept constant by regulating feeding of the radical imitator, which is either oxygen or peroxydicarbonate. The starting polymerization temperature is 190 °C and 140 °C, respectively. Reactors have cooling jackets to remove the heat. With multiple injection points, conversion levels of 35% are achieved in tubular

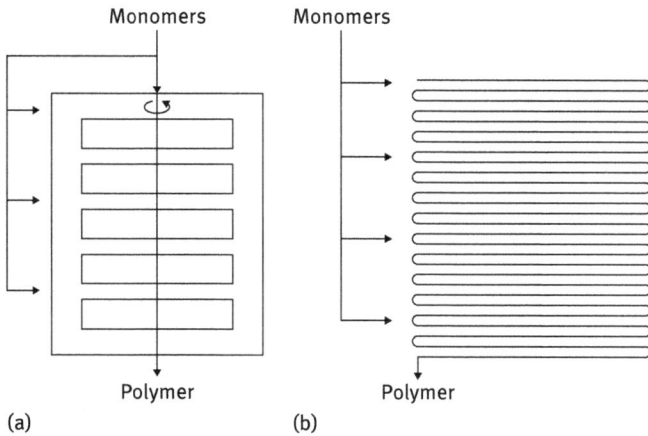

Figure 15.15: High-pressure polymerization reactors: (a) autoclave and (b) tubular reactor.

reactors. The reactor pressure is periodically reduced from ca. 300 to 200 MPa with the aid of a cycle valve. This is needed to assure the high velocity required for efficient heat transfer and removal of contaminants on the reactor walls. Obviously, such cyclic operation with large pressure and temperature changes requires that the tube materials are stable enough to withstand such gradients. As conversion in an autoclave or a tubular reactor is rather low, ethylene recycling is required, which calls for the installation of a high-pressure separator. Ethylene with low-molecular-weight polymers (waxes) is taken as the overhead stream, cooled, and further separated from the waxes. A polymer stream after a low-pressure separator is separated from the remaining monomer, which is recycled back to the reactor. As typical for recycling processes, some ethylene is purged to avoid accumulation of feed impurities. The polymer, in the molten form, undergoes shaping (e.g., extrusion) and subsequent drying.

15.4.2 Homogeneous polymerization in solution

Polymerization in solution is applied to decrease the viscosity in the reactor and arrange heat removal by evaporation of the solvent. Obviously, extra costs are involved for the separation of the solvent from the polymer. Moreover, residual monomer should be also recovered. Due to dilution of the monomer, the reaction rate is decreasing, as the concentrations of monomer and polymer are lower, lowering the space-time yield. Acceptable mixing in the manipulation of highly viscous solutions and melts is achieved when the length/diameter ratio is < 2. To maintain isothermal polymerization, jacket cooling can be insufficient and additional methods of heat removal should be used such as external or reflux cooling.

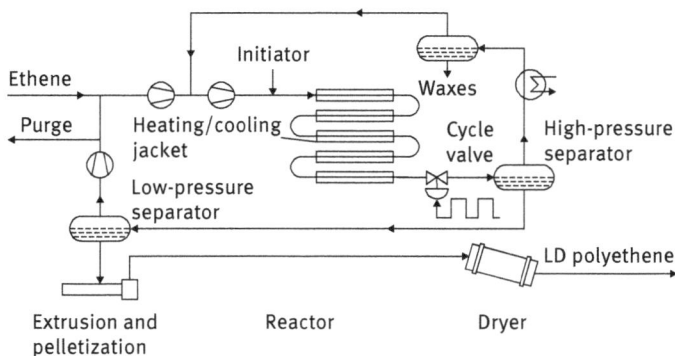

Figure 15.16: Polymerization of ethylene in a tubular reactor. From J. A. Moulijn, M. Makkee, A E. van Diepen, *Chemical Process Technology*, 2013, 2nd Ed. Copyright © 2013, John Wiley and Sons. Reproduced with permission from Wiley.

Internal cooling coils interfere with stirring and highly viscous solutions can be difficult to pump in the case of external cooling; thus, reflux cooling, which removes the polymerization heat by solvent and monomer evaporation, can be the most efficient option.

15.5 Heterogeneous polymerization

This type of polymerization includes precipitation, suspension, dispersion, emulsion, and slurry polymerization distinguished by the initial state of the polymerization mixture, kinetics, and mechanism of particle formation as well as the size and shape of the final polymer particles (Figure 15.17).

15.5.1 Precipitation polymerization

A special variation of solution polymerization is precipitation polymerization. During the process, the polymer becomes increasingly insoluble and can be isolated from the solution by filtration of the polymer precipitate. Due to precipitation, the viscosity of the solution is almost constant. The method can be used for solvent-free polymerization of vinyl chloride and synthesis of poly(acrylonitrile) in water. PVC can be also produced in a liquid-liquid biphasic system, where the other liquid phase is water. This process will be discussed further along with other suspension polymerization processes.

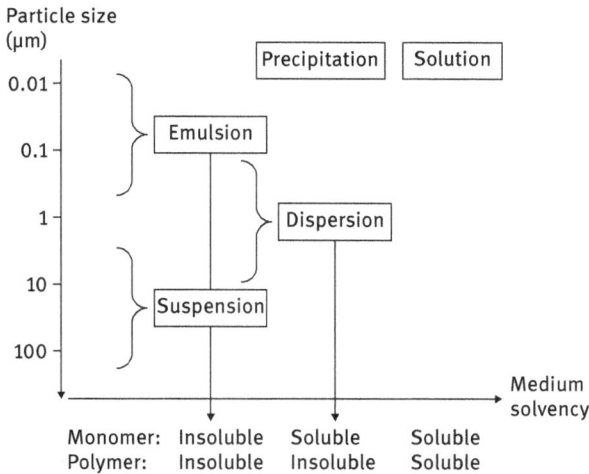

Figure 15.17: Particle size for different types of heterogeneous polymerization.

Dispersion polymerization is a particular type of precipitation polymerization with addition of stabilizers, thus making smaller and more regular particles compared to classical precipitation polymerization.

Another special type of precipitation polymerization is gas-phase polymerization, when polymerization occurs within the polymer particle to which the monomer is supplied from the gas phase. A flow scheme of the low-pressure Unipol process for making polyethylene powder in a fluidized-bed reactor with a modified chromium catalyst and ethylene as a fluidizing gas is shown in Figure 15.18.

High-density polyethylene (HDPE) has a very broad molecular mass distribution and linear chains. Short-chain branching and lower polymer density between HDPE and LDPE made in the high-pressure free-radical processes can be achieved using copolymerization of ethylene with propene, 1-butene, and 1-hexene. In the gas phase Spherilene process of Basell Polyolefins (Figure 15.19) a fluidized-bed reactor, operating at 80–100 °C and 0.7–2 MPa, is used with microspheroidal catalysts of Ziegler type or chromia. By applying these catalysts, monomodal products such as LLDPE (linear low density) for film, HDPE for injection molding, and MDPE (medium density) for rotomolding and textiles are produced. Control of temperature is important since the reaction temperature is not far from the polymer melting temperature. Such temperature control is done by reversible poisoning of the catalyst with CO_2.

Another example of precipitation polymerization is a polymer-monomer-precipitant system where a non-solvent for the polymer is applied, while the precipitant is miscible with the monomer. In the synthesis of high-density polyethylene (HDPE), soluble Ziegler-Natta (transition metal catalysts, e.g., $TiCl_4/MgCl_2/Et_3Al$ catalyst) can be used. Slurry polymerization can be considered as a special version of HDPEproduction

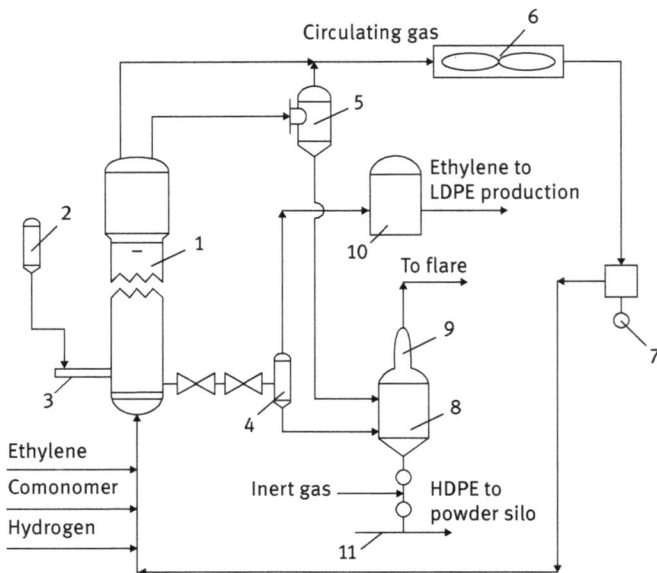

Figure 15.18: Gas-phase Unipol polymerization process: 1, fluidized-bed reactor; 2, catalyst transfer tanks; 3, catalyst feeders; 4, product discharge tanks; 5, multicyclone dust separator; 6, air coolers; 7, compressor; 8, product degassing tank; 9, filter; 10, ethylene tank; 11, pneumatic transport system.

when immobilized solid Ziegler-Natta catalysts are applied and the solid is introduced from the beginning of the process.

First-generation catalysts for polypropylene polymerization were applied in a cascade of CSTR at ca. 50–100 °C and 0.5–3 MPa using n-hexane or n-heptane as hydrocarbon diluent. They were characterized by low productivity and insufficient isotacticity (e.g., not the same configuration at successive positions along the polymer chain). Therefore, there was a need to remove the catalytic residues and the atactic amorphous polymer, which could not crystallize.

Such processes have been replaced by other types of processes for the synthesis of polypropylene. Spheripol process (Figure 15.20), developed by Himont, is a hybrid process consisting of polymerization in the liquid monomer and subsequent copolymerization in the gas phase.

The Spheripol process includes polymerization in the liquid monomer with short residence times, separation of unreacted monomer with subsequent gas-phase polymerization-making heterophase copolymers, the polymer-finishing section comprising stripping monomer with steam and finally drying. Fourth-generation catalytic systems developed in the 1980s, such as $MgCl_2/TiCl_4$/phthalate + AlR_3/silane systems, afforded high production rates (40–70 kg polypropylene/g catalyst) with high crystallinity (isotacticity of 99%). These catalysts had medium hydrogen response and

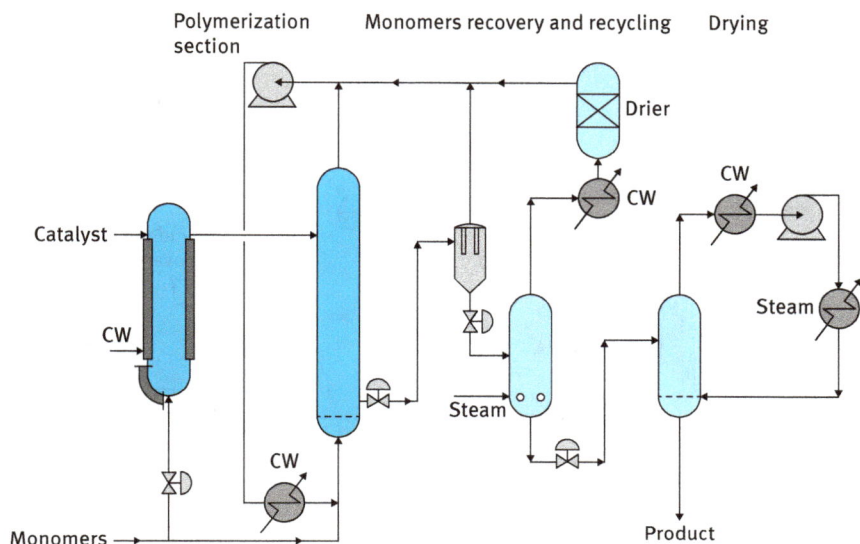

Figure 15.19: Spherilene process for production of HDPE, LLDPE, and very low density polyethylene (VLDPE). CW, cooling water. Modified after http://www.treccani.it/portale/opencms/handle404?ex porturi=/export/sites/default/Portale/sito/altre_aree/Tecnologia_e_Scienze_applicate/enciclope dia/inglese/inglese_vol_2/759-788_ING3.pdf.

MWD. The polymer is made by growing it on suspended catalyst particles, whose removal from the polymer is not required due to high catalyst productivity. Further catalyst development was aimed either at increased activity to 130 kg PP/g cat with lower MWD and excellent hydrogen response or at achieving broader MWD. The latter case belongs to so-called sixth-generation catalysts when phthalate is replaced by succinate. Typical process parameters for Spheripol process are given in Table 15.2.

A wide range of PP-based polymers can be made when polymerization in the liquid monomer is combined with gas-phase polymerization. The modular technology with up of three mutually independent gas-phase polymerization reactors (Figure 15.21) allows polymerization of different monomers either separately or in series on the same growing spherical particles, which are transported between the reactors. The flexibility of the process is in the independence of the gas-phase composition in these three polymerization reactors and a possibility to add co-monomers.

Tailor-made PP can be also produced using Borstar technology (Figure 15.22), wherein the catalyst is introduced only in the loop reactor with supercritical propane as the polymerization medium.

Such medium allows diminished polymer solubility above the critical point and also decreased fouling. The first polymerization step is followed by polymerization in gas-phase reactors, to which addition of catalysts is not needed. With addition of hydrogen, a chain is detached from the active catalytic site. In the loop reactor of

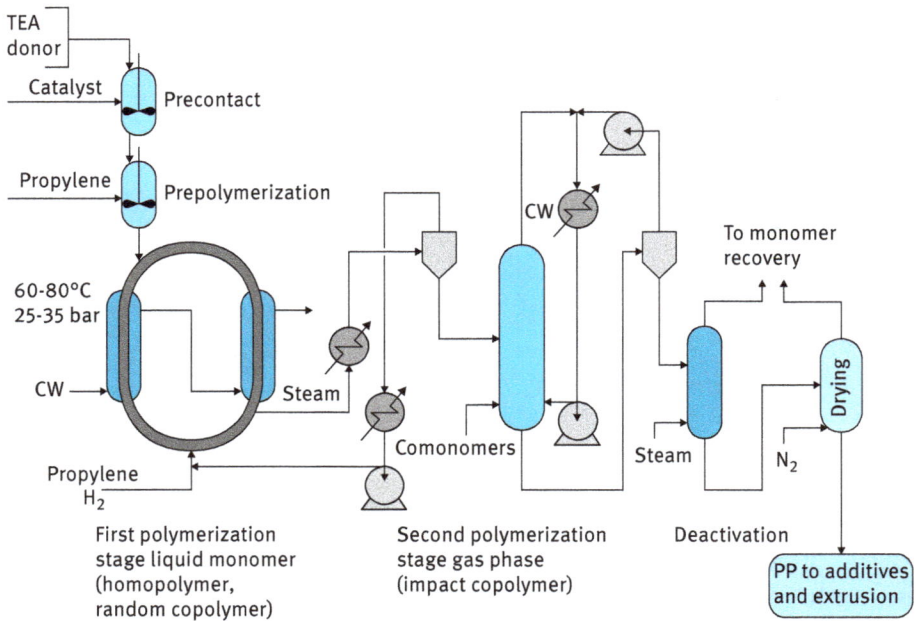

Figure 15.20: Spheripol process for polypropylene production. Modified after www.treccani.it/por tale/opencms/handle404?exporturi=/export/sites/default/Portale/sito/altre_aree/Tecnologia_e_ Scienze_applicate/enciclopedia/inglese/inglese_vol_2/759-788_ING3.pdf.

Table 15.2: Process parameters for the Spheripol process.

Process step	Temperature (°C)	Pressure (MPa)
Catalyst activation	10	4.0
Prepolymerization	20	3.5
Polymerization-loop reactor	70	3.4
High-pressure separation	90	1.8
Polymerization-gas-phase reactor	75–80	1–1.4
Steaming	105	0.02
Drying	90	0.01

the Borstar process, metallocene catalysts (Figure 15.23) are applied. These catalysts are made from well-defined organometallic compounds with their active sites having group IV metals (Ti, Zr, Hf) with a least one π ligand (such as cyclopentadienyl). The high regularity of polymers made with metallocene catalysts is often their key shortcoming. This is due to their exceptionally narrow MWDs, which leads to poor

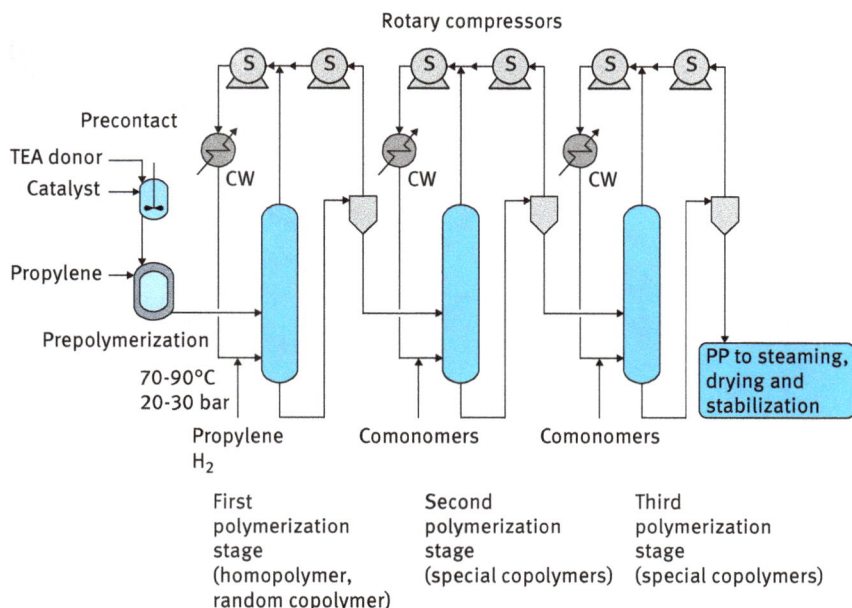

Figure 15.21: Modular technology for production of various PP polymers. Modified after http://www.treccani.it/portale/opencms/handle404?exporturi=/export/sites/default/Portale/sito/altre_aree/Tecnologia_e_Scienze_applicate/enciclopedia/inglese/inglese_vol_2/759-788_ING3.pdf.

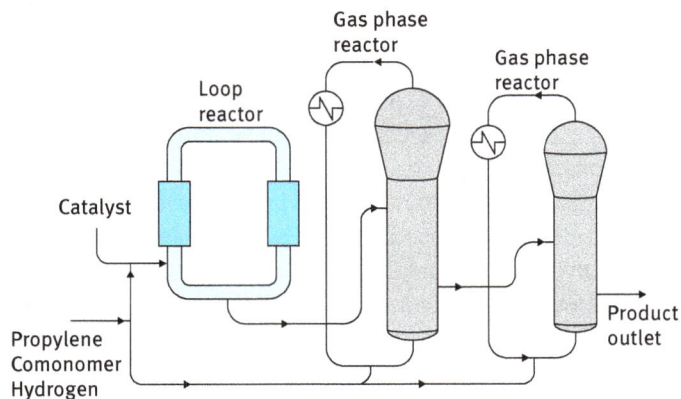

Figure 15.22: Technology of Borealis for PP production. http://guichon-valves.com/wp-content/up loads/Polypropylene-PP-liquid-phase-process-example.jpg.

extrusion characteristics in many applications. Thus, metallocene catalysts are currently applicable when only a narrow MWD is needed, or when the products are used in blends to reduce the effects of their poor processability. The Borstar process gives a possibility to make bimodal polymers, when lower-molecular-weight polymers are formed in the loop, while high-molecular-weight polymers are generated in the gas-phase reactor. Bimodal capability ensures good processability of the polymer.

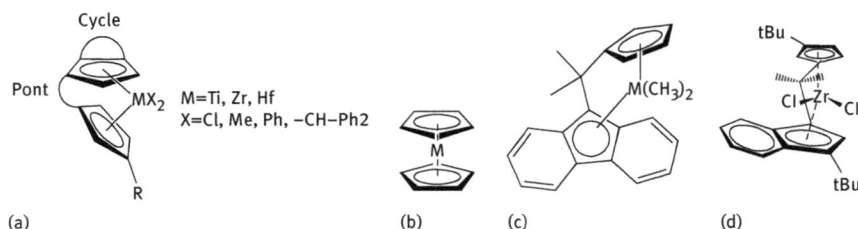

Figure 15.23: Metallocene catalysts: (a) general formula and (b–d) examples.

15.5.2 Suspension and emulsion polymerization

In suspension polymerization typically done in stirred tanks, there is a liquid-liquid biphasic system where small droplets of monomer are first formed in the aqueous phase. Upon addition of an initiator, polymerization in these droplets starts. Surrounding water acts as a heat-transfer fluid. In the absence of any coagulation, the size of polymer beads is the same as the size of the initial monomer droplets. Dispersants (protective colloids) are specifically added to prevent coagulation. These additives are either insoluble macromolecules or inorganic powders, the so-called Pickering emulsifiers, such as barium sulfate, talc, aluminum hydroxide, hydroxyapatite, tricalcium phosphate, calcium oxalate, magnesium carbonate, and calcium carbonate. Modified cellulose (carboxymethylcellulose, hydroxyethylcellulose, and methylcellulose) and natural products (alginates, agar, starch) having amphipathic character concentrate at the monomer-water interface, lowering the interfacial tension. The suspending medium for suspension polymerization is predominantly water. Hydrophobic organic suspending media in the presence of water in the disperse phase can be also used in inverse-suspension polymerizations for manufacturing very-high-molecular-weight polymers and copolymers with acrylamide as a co-monomer. Typical concentrations of dispersants relative to the aqueous phase are 0.1–5 wt% for protective colloids and 0.1–2 wt% for Pickering emulsifiers, while a typical initiator concentration relative to the monomer is 0.1–1 wt%.

During polymerization, the initially non-viscous liquid monomer (such as styrene) is converted into a polymer solution in its monomer with increasing solution viscosity.

Similar to suspension polymerization is polymerization in emulsions, where water-organic phase systems are also used. Contrary to suspension polymerization, the radical initiator added to the reaction mixture is soluble in water, but not in the monomer droplets. This requires addition of an emulsifier (natural or synthesis detergents), forming micelles where polymerization occurs, leading to a polymer particle suspended in water having a smaller size than the initial monomer droplet. By choosing the emulsifier, its amount, and the mode of addition, the particle size of the dispersion can be adjusted in the range of 50–1,000 nm. Anionic emulsifiers used in concentrations of 0.2–5 wt% related to monomers include alkali salts of fatty and sulfonic acids, C12 – C16-alkyl sulfates, ethoxylated and sulfated or sulfonated fatty alcohols, alkyl phenols, and sulfodicarboxylate esters. Ethoxylated fatty alcohols and alkyl phenols with 2–150 ethylene oxide units per molecule are examples of non-ionic emulsifiers. Less often used cationic emulsifiers comprise ammonium, phosphonium, and sulfonium compounds containing at least one hydrophobic long aliphatic hydrocarbon chain.

Viscosity changes are very small in suspension and emulsion polymerization; therefore, heat removal is efficient even at high monomer concentration.

Batch, semi-continuous (continuous emulsion or monomer feed) and continuous processes are applied for the synthesis of PVC, styrene copolymers, polyacrylates, etc. These processes are performed under nitrogen, as oxygen inhibits polymerization. Batch processes can be conducted below 50 °C with redox initiators (for example, hydrogen peroxide and alkali persulfates as oxidizing agents and Fe(II) sulfate, sodium bisulfite orodium thiosulfate as reducing agents) or between 50 °C and 85 °C with water-soluble peroxo compounds as initiators (e.g., alkali persulfates, ammonium persulfate, or hydrogen peroxide) in amounts of 0.2–0.5 wt% related to monomer. The heat of polymerization should be effectively removed. Despite a larger space-time yield, there are clear disadvantages of batch processes such as poor reproducibility and far from efficient utilization of the cooling capacity.

In the semi-batch production mode, a part (5–10%) of the monomer emulsion and the initiator solution can be introduced initially in a batch mode, determining the number of particles and their size in the polymer dispersion, while the rest of the emulsion is fed in the subsequent 15–30 min in which polymerization is completed. Alternatively, water and emulsifiers are first added to the reactor followed by a continuous feed of the monomer mixture, initiator solution, and the solution of auxiliaries.

An example of suspension polymerization giving typically a broad particle size distribution is the synthesis of expandable polystyrene (Figure 15.24), when blowing agents are used in a second step after polymerization.

The volume of the beads can be increased by a factor of ca. 30–50 after elevation of temperature to 80–110 °C. The monomer styrene, benzoyl peroxide as the primary initiator, and a finishing initiator (di-*tert*-butyl peroxide or *tert*-butyl peroxybenzoate) are introduced into the reactor containing water along with either an inorganic or a polymeric steric stabilizer. Polymerization starts at 75–95 °C, depending on the initiator, proceeding through a sticky stage (32–35% conversion), when the beads grow from

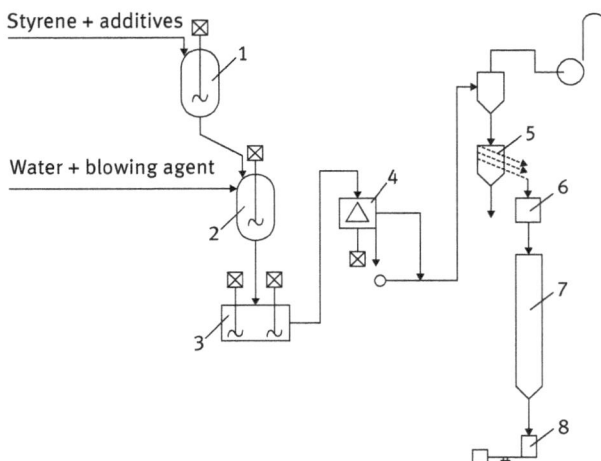

Figure 15.24: Schematic representation of the manufacture of Styropor by batch suspension polymerization: 1, mixing tank; 2, stirred reactor; 3, tank; 4, centrifuge; 5, sieving; 6, drying; 7, silo; 8, packaging.

a size of ca. 0.2 mm to the desired size, to 65–68% conversion level when, at sufficiently large particle viscosity, the particle size growth stops. The next step is heating the reaction mixture above the polystyrene glass transition temperature (>100 °C) with simultaneous pressurizing in the presence of a blowing agent, such as pentane or other C4–C7 hydrocarbons taken in 5–8% with respect to the polymer. During the final impregnation stage lasting for 3–8 h and conducted under nitrogen pressure of 700–950 kPa, the blowing agent diffuses into the beads, the volume is expanded, radicals are rapidly generated, and the final monomer conversion reaches ca. 99.9%. Subsequent cooling to 20–30 °C, depending on the blowing agent, is needed to prevent bead expansion during handling.

Suspension polymerization is also used for production of polyvinylchloride (PVC) and is the most important technology for making of this polymer (95% of PVC is made by suspension and the rest by emulsion polymerization). Important issues in VCM (vinyl chloride monomer) polymerization are steady heat removal, morphology control of the polymer to avoid formation of non-homogeneous PVC agglomerations with different density. VCM is virtually insoluble in water, and therefore, its polymerization in an aqueous suspension is carried out with small drops of VCM dispersed in a continuous medium made up of water, which enables removal of the reaction heat and generates regular polymerization in isothermal and controlled conditions. The operation parameters regulating properties of the final polymer granules (in current processes 95% of them are in the range 100–200 μm) are heat removal, mechanical energy input, additives having an impact on size and its distribution as well as morphology, and finally, the reaction temperature influencing

PVC molecular weight. The flow scheme of VCM suspension polymerization is presented in Figure 15.25.

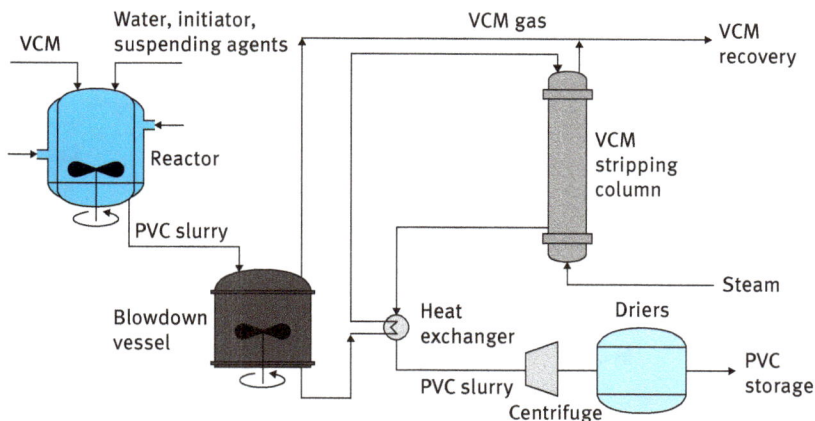

Figure 15.25: Diagram of the process for the polymerization in suspension of VCM (Ineos Vinyls, Italy). Modified after http://www.treccani.it/export/sites/default/Portale/sito/altre_aree/Tecnolo gia_e_Scienze_applicate/enciclopedia/inglese/inglese_vol_2/863_884_ING3.pdf.

Industrial polymerization of VCM in aqueous suspensions, performed between 50 °C and 70 °C in pressurized vessels because of relatively high vapor pressure of the monomer (0.7–1 MPa), is affected, in addition to volume contraction during polymerization, also by the insolubility of the polymer in the monomer, which in itself has a limited solubility (ca. 30%) in PVC. This allows to describe VCM polymerization in terms of two phases in each drop of the aqueous suspension: monomer with minor amounts of polymer and monomer dissolved in the polymer.

Final words

It is the author's opinion that the content of textbooks for graduate students should be accurate and precise rather than original. Therefore, during the writing of this textbook, the author had consulted different sources including various encyclopedia, textbooks, review papers, original articles, companies websites, information portals, etc., with the aim of giving as precise and correct information as possible.

During preparation of the second edition, the author was updating the text by introducing description of processes/technologies that have appeared recently. It was amazing and somewhat unexpected to witness how dynamic is the field and how substantial was the progress in developing new technologies for seemingly mature processes!

Despite all efforts to update the textbook, some relevant information on novel technologies might be missing or obsolete technologies are described; therefore, any comments, corrections, and updates would be highly appreciated.

In very many cases, while compiling the text, difficult decisions had to be taken regarding the level of details that should be included for particular processes. There are many textbooks and reference books that provide much more extensive descriptions. The author has, for example, books of several hundreds of pages devoted just to ammonia or methanol synthesis, hydroformylation, oxychlorination, oil refining, etc. A helicopter view showing how general principles of chemical reaction technology are utilized for particular cases was thus adapted in the textbook. More specialized literature should be consulted in a quest for more precise descriptions of various chemical processes.

The author is grateful to the editorial team at De Gruyter for efficient collaboration.

Finally, the author hopes that the knowledge on chemical reaction technology can be transferred from one process to another and that the book could be interesting not only for students, but also for professionals working in chemical process industries.

https://doi.org/10.1515/9783110712551-016

Index